Electrodynamics of Solids and Microwave Superconductivity

OPTOELECTRONIC PACKAGING • *A. R. Mickelson, N. R. Basavanhally, and Y. C. Lee (eds.)*

OPTICAL CHARACTER RECOGNITION • *Shunji Mori, Hirobumi Nishida, and Hiromitsu Yamada*

ANTENNAS FOR RADAR AND COMMUNICATIONS: A POLARIMETRIC APPROACH • *Harold Mott*

INTEGRATED ACTIVE ANTENNAS AND SPATIAL POWER COMBINING • *Julio A. Navarro and Kai Chang*

FREQUENCY CONTROL OF SEMICONDUCTOR LASERS • *Motoichi Ohtsu (ed.)*

SOLAR CELLS AND THEIR APPLICATIONS • *Larry D. Partain (ed.)*

ANALYSIS OF MULTICONDUCTOR TRANSMISSION LINES • *Clayton R. Paul*

INTRODUCTION TO ELECTROMAGNETIC COMPATIBILITY • *Clayton R. Paul*

INTRODUCTION TO HIGH-SPEED ELECTRONICS AND OPTOELECTRONICS • *Leonard M. Riaziat*

NEW FRONTIERS IN MEDICAL DEVICE TECHNOLOGY • *Arye Rosen and Harel Rosen (eds.)*

ELECTROMAGNETIC PROPAGATION IN MULTI-MODE RANDOM MEDIA • *Harrison E. Rowe*

ELECTROMAGNETIC SYSTEM DESIGN USING EVOLUTIONARY OPTIMIZATION • *Yahya Rahmat-Samii and Eric Michielssen*

NONLINEAR OPTICS • *E. G. Sauter*

InP-BASED MATERIALS AND DEVICES: PHYSICS AND TECHNOLOGY • *Osamu Wada and Hideki Hasegawa (eds.)*

DESIGN OF NONPLANAR MICROSTRIP ANTENNAS AND TRANSMISSION LINES • *Kin-Lu Wong*

FREQUENCY SELECTIVE SURFACE AND GRID ARRAY • *T. K. Wu (ed.)*

ACTIVE AND QUASI-OPTICAL ARRAYS FOR SOLID-STATE POWER COMBINING • *Robert A. York and Zoya B. Popović (eds.)*

OPTICAL SIGNAL PROCESSING, COMPUTING AND NEURAL NETWORKS • *Francis T. S. Yu and Suganda Jutamulia*

SiGe, GaAs, AND InP HETEROJUNCTION BIPOLAR TRANSISTORS • *Jiann Yuan*

ELECTRODYNAMICS OF SOLIDS AND MICROWAVE SUPERCONDUCTIVITY • *Shu-Ang Zhou*

Electrodynamics of Solids and Microwave Superconductivity

SHU-ANG ZHOU
Ericsson Components AB

A Wiley-Interscience Publication
JOHN WILEY & SONS, INC.
New York • Chichester • Weinheim • Brisbane • Singapore • Toronto

This text is printed on acid-free paper. ♾

Published simultaneously in Canada.

For ordering and customer service, call 1-800-CALL-WILEY.

Library of Congress Cataloging in Publication Data:

Zhou, Shu-Ang.
Electrodynamics of solids and microwave superconductivity / Shu-Ang Zhou.
p. cm. — (Wiley series in microwave and optical engineering)
"A Wiley-Interscience publication."
Includes bibliographical references and index.
ISBN 0-471-35440-6 (alk. paper)
1. Electrodynamics. 2. Superconductivity. I. Title.
II. Series.
QC631.Z565 1999
537.6—dc21 99-25308

Printed in the United States of America

10 9 8 7 6 5 4 3 2 1

Contents

Foreword

Beginning in 1960, research in the field of electrodynamics of continua was reopened by a group of scientists in the field of mechanics who had seldom engaged in research in the field of electrodynamics of continua since the time of Maxwell. Industrial demands in micro-chip and computer technology and the fundamental ideas regarding the unified nature of science for a better understanding of physical phenomena at their root system, encouraged researchers to return to the unity of science. Indeed, the unified approach bore fruit in the frontier fields of engineering and applied science, and it has opened new fields.

The cross-fertilization of scientific disciplines brings life to new coupled physical phenomena that escape our attention when they are studied individually. Through this coupling, it is possible to make measurements in one field to assess the properties of the second field. Coupling leads to new fields of research and industries as demonstrated by high-temperature superconductivity, plasma physics, etc..

Continuum theory of electrodynamics requires consideration of deformation, superconductivity, and thermal interactions of electromagnetic (EM) fields. Ultra high sensitivity of magnetic micro-chips to heat and EM fields demands these unifications. *Electrodynamics of Solids and Microwave Superconductivity* by Dr. Shu-Ang Zhou is a welcome addition to the concept of unification.

This book covers several interesting and important fields in engineering and applied science. In particular, there are extensive accounts on the subject of superconductivity and application to various fields (Josephson Junctions and deformable superconductors) which bring to focus the intricate interrelations not discussed in other books in a unified fashion.

A.C. ERINGEN

Preface

Currently, rapid development of high-technologies and advanced material science is requiring interdisciplinary research in many areas. In the areas of classical electromagnetics and mechanics, the situation has developed for many years that researchers and engineers in mechanical engineering seldom consider electromagnetic phenomena, while physicists and electrical engineers often ignore the mechanical effects. However, such traditionally separated subjects of electromagnetics and mechanics are now becoming fused by the needs of modern science and technology, requiring novel utilization of the effects of electromagnetic–mechanical interaction for electronic instruments and optimal balance between basic mechanical structures and overall electromagnetic behaviors of engineering devices and systems. For instance, we may already observe the fact that the rapid development of semiconductor processing technologies has led to the emergence of the promising micromachining technology, which has, in turn, stimulated the rapid development of micro-electromechanical systems and micro-transducer technologies, bridging the electronic world with its environments, including the classical mechanical world. Today, with advanced low-cost electronics and well-developed manufacturing technologies, it has become possible to explore and make use of various electromagneto–thermomechanical effects, that previously were commercially impossible.

The emerging interdisciplinary subject of electrodynamics of solids, which deals with electromagnetic phenomena, mechanical motion and deformation, and heat conduction in material solids, has recently been studied extensively. However, little work has so far been carried out systematically on electromagnetics and mechanics of superconductors, except for the preliminary work of Zhou (1991a). At present, there already exist a number of excellent textbooks on superconductivity for students and engineers. However, none of them has given a comprehensive description of electromagnetics and thermomechanics of superconductors, which is often required in the design of practical systems. Although there are some professional books on applied superconductivity, they deal usually with some discrete and specialized subjects at a level somewhat difficult for non-specialists to appreciate. The situation is somewhat similar in the field of continuum electromagnetomechanics.

Published books on continuum electrodynamics are usually of such a high academic nature that they are relatively difficult to understand for university students and especially electrical engineers, since most discussions on the subject come from the communities of continuum mechanics and/or continuum physics. Besides, high-frequency behaviors of superconductors have recently been studied extensively in the world, owing to possible applications of superconductivity in microwave (radio) communications, high-capacity internet switches, ultra-high speed digital processors, and supercomputers. The subject of microwave superconductivity is therefore of great interest in both theory and application.

This book is intended to provide an introductory and comprehensive theoretical foundation of solid electrodynamics and microwave superconductivity for applied physicists, electrical and mechanical engineers, university students, and post-graduate students, who are likely to be involved in the subjects of theoretical and applied electromagnetics, continuum mechanics, applied superconductivity, engineering physics, high-speed electronic circuit design, microwave engineering, and transducer technology.

In view of the interdisciplinary nature of the subject, I have tried using language that could be understood by both mechanical and electrical engineers. In particular, I have provided sufficiently broad background materials on classical electromagnetics, superconductivity, and continuum electromagneto–mechanics for students and engineers who just start the subject, and for both non-specialists and specialists, who might be specialized only in a certain sub-field, so that they might have a profound grasp of the basic concepts, theoretical models and, more importantly, quickly get a global view of this interdisciplinary subject, and its natural harmony and intrinsic relationship among classical electromagnetics, mechanics, and material physics.

The summary of knowledge in the text of this book is based on original research results of the author and his collaborators and on a literature review of several hundreds of relevant references. The book is unique in the sense that it gives a unified and comprehensive treatment of the interdisciplinary subject, covering not only the classical electromagnetics, continuum electrodynamics, solid mechanics, and superconductivity, but also relevant materials which may be involved in engineering design of electromagneto–mechanical systems, microwave components, and superconducting devices. Different from usual textbooks of physics, which mainly introduce various individual physical effects in simple cases and with simple geometries, this book emphasizes the synergy of various physical effects and their theoretical formulations applicable to engineering problems, often involving coupling of various physical effects and complex geometries in three dimension. Greater emphasis will also be given to the introduction of basic concepts and general methodology, which are of particular utility due to the rapid broadening of engineering fields and their interdisciplinary nature in recent years.

In writing this book, I have chosen the way of studying, in great detail, electromagnetic and mechanical behaviors of superconductors to illustrate the interdisciplinary subject of solid electrodynamics, since the superconductor not only has its unique electromagnetic properties, but also its very own special thermal and mechanical properties. In other words, I have tried to bridge the subjects of electromagnetics and mechanics with the aid of superconductors, since a complete appreciation and practical utilizations of superconductive materials require indeed such an interdisciplinary study. Besides, there is another motivation for studying superconductors, that is, if one knows how to solve electrodynamic problems for superconductors, one should also be able to solve corresponding problems involving non-superconducting (normal) conductors, since mathematically they can be treated as the special case of superconductors at the normal state, with perhaps different conductivities numerically, but not vice versa. Furthermore, one will have the advantage of being able to compare results from different technologies used in one´s own work. The book is, therefore, written not only for superconductivity specialists but also for readers who are not specialized in superconductivity, but working on electrodynamic problems with materials other than superconductors. Hopefully, readers will find that it is an efficient way of learning that it is also stimulating and interesting, especially for those having open minds toward different technologies.

The text of the book will start from basic concepts and principles, and give a rather detailed derivation of important theoretical results. This includes examples of applications that are easy to understand, which may be used for self-study by interested readers who have reasonably good mathematical backgrounds (usual level of university mathematics). The concepts and theories of fields are emphasized in the book. Attempts are made at correlating traditional circuit and advanced field analyses, which is becoming an important subject in future high-speed electronic circuit design and microwave component design. Recently, rapid increasing computational power, available almost to everyone, has made it possible to perform advanced field analyses, which has opened a new and cost-effective way of design in both electrical and mechanical industries.

Since this interdisciplinary subject involves very many basic concepts, theories, and methods, I have chosen to not present technical details in mathematics (commonly known at the university level) for calculating explicitly, for instance, line and surface integrals, linear algebra, vector analysis problems, as most standard textbooks do. Instead, I shall present as clearly as possible the physical concepts, theories, analytical methods, and necessary hints and steps to get the right results. The comprehensive picture of this interdisciplinary subject is presented as complete as possible in a limited number of pages. SI units are used throughout the text, which should be particularly convenient for applied physicists and for electrical and mechanical engineers studying either small-scale or large-scale phenomena.

The text is divided into seven chapters. Chapter 1 presents a review of classical electrodynamics, by introducing its basic concepts and theoretical principles, for the study of electromagnetic phenomena in material media. The basic concepts, definitions, and principles introduced in this chapter are of importance for later discussion. Emphasis is given here to the general methodology used in formulating macroscopic electromagnetic phenomena in material media, including moving media. Readers may find that some of the derivations for the classical subject are in a new form, which may be interesting and stimulating. Some examples of practical interest are also provided for illustrating applications of classical electromagnetics.

Chapter 2 introduces continuum electrodynamics of deformable solids, which is a subject covering electromagnetic phenomena as well as their interaction with mechanical phenomena in material solids which may be deformed or in motion under electromagnetic and/or mechanical loadings. The purpose of this chapter is to give a brief introduction to this subject, its basic concepts, and theoretical principles, for the study of phenomena of electromagneto–thermoelastic interaction in material solids. Particular emphasis will be put on electro-quasistatic and magneto-quasistatic problems, which are of practical interest for many engineering applications of electromagnetic–mechanical devices involving small velocities, compared to the speed of light. Illustratively, continuum models for thermoelastic conductors and dielectrics are formulated in some detail, to demonstrate the methodology used in continuum electrodynamics. Continuum models for photothermoelastic solids and elastic dielectric composites are also introduced, which are useful for a variety of engineering problems of practical interest. Some boundary value problems are provided and solved as examples.

The discovery of superconductivity and its recent advances in high-temperature superconducting materials have led to the rapid development and advancement of superconductivity technology. Although the development of quantum mechanics in the 1920s led to the understanding of the normal process of electrical conduction in metals and the BCS microscopic mechanism in 1957 for conventional superconductors, our understanding of the microscopic mechanism of recently discovered high-temperature oxide superconductors remains uncompleted. In order to develop the new technology utilizing superconducting materials, a complete appreciation of the macroscopic electromagnetic behaviors of superconductors is needed before one can undertake the most basic engineering calculations for any concrete problem in application. This situation encouraged the development of phenomenological theories, which had established that most superconducting phenomena can be derived from a small number of empirical postulates. The purpose of Chapter 3 is therefore to introduce some basic phenomena of superconductivity and to formulate macroscopic electrodynamic theories for superconductors in weak fields. The well-known BCS microscopic mechanism on superconductivity will also be introduced briefly. Some examples are given to illustrate the use of electrodynamic models, especially for superconductors operating at microwave frequencies.

In Chapter 4, some theoretical models will be presented for the study of electromagnetic properties as well as thermomagnetoelectric effects in some superconductors, called type II superconductors, which may maintain their superconducting state in very strong magnetic fields. Recently discovered high-T_c oxide superconductors are examples of such superconductors. Both the classical models, such as the Ginzburg-Landau theory, the London-Bean model, and some recently developed electrodynamic models for type II superconductors in the mixed state at microwave frequencies are introduced. Illustrative examples in applying these theories are also provided.

Chapter 5 introduces the phenomenon of tunneling of electron pairs across a superconductor–insulator–superconductor (SIS) junction, call the Josephson effect, and formulates electrodynamics of the Josephson junctions. Several circuits of practical interest are outlined. Some electrodynamic analyses are given to superconducting quantum interference devices (SQUIDs). Some possible applications of SQUIDs are discussed briefly. To compare the Josephson technology with the existing and forthcoming semiconductor technologies from a system point of view, some physical limits of switching technologies are also discussed in this chapter.

Chapter 6 introduces, at first, the theory of transmission lines based on a distributed-circuit model. A relationship between the circuit theory and the more general electromagnetic field theory is then established for the transmission lines. By the field theory based on Maxwell's equations, I shall demonstrate how the distributed circuit elements, such as the line resistance, line inductance, line capacitance, and line conductance of the transmission lines of complex geometry and material composites can be analyzed numerically with the use of the finite element method. To illustrate the limit of the classical quasi-TEM approximation often used in studying lossy transmission lines, a full-wave analysis is carried out to investigate properties of anisotropic superconducting transmission lines which may operate at extremely high frequencies. Electromagnetic behaviors of coupled transmission lines and directional couplers are also studied. Full-wave analysis for waveguide with conducting boundary is introduced. Electromagnetic properties of microwave resonators, such as discrete element resonant circuits, transmission line resonators, and cavity resonators are also discussed.

In many engineering applications of superconductivity, superconductors may not only conduct electric currents but they may also be subject to thermal and mechanical loadings. Since the discovery of hard superconductors, the mechanical behaviors of superconducting materials have been of interest and concern to researches for many years because of large mechanical forces expected in foreseen applications of superconducting devices, such as magnets for plasma confinement and energy storage, superconducting generator, electrical transmission lines, magnetic levitating trains, electromagnetic-propulsion ship, or still unimagined new devices. In particular, the latest generation of superconductors uses ceramic materials which are brittle and much harder to work with than metals, where a considerable amount of work is needed to improve the mechanical behaviors of the new superconductors. Furthermore, dynamic effects in non-equilibrium superconductors may introduce interesting acoustic phenomena in superconductors, which are of importance for potential technological applications. Obviously, the possibility for wider applications of superconductivity hinges on success in improving our understanding of superconductivity and its related electrodynamic and mechanical problems. Chapter 7 is, therefore, devoted to the introduction of some theoretical models developed recently for the study of the interaction between electromagnetics and the mechanics of superconductors.

This text may be used either as reference material or as course material. For instance, according to the contents of the book, one may use Chapter 1 and Chapter 6 for a course on classical electromagnetics, and Chapter 2 for a course on continuum electrodynamics of solids. Chapters 3, 4, and 5 may be used for an introductory course on the electrodynamics of superconductors. Chapter 7 may be used for a more advanced course on the electrodynamics of deformable superconductors.

Because of the enormous volume of literature on the subjects of electromagnetics, continuum electrodynamics, solid mechanics, superconductivity, material physics, and microwave engineering, it is difficult to cite all the relevant articles and books, so the author hopes to have his readers' understanding if they discover that some of their own works have gone unmentioned.

Here, I would like to thank Dr. A. Cemal Eringen, emeritus professor of Princeton University, who not only has kindly written the Foreword of this book, but also has constantly encouraged me to work in this field during the past several years. I must also thank Dr. Guo-Chun Liang at Conductus Inc., for his interest, recommendation, and useful suggestions on this book, which stimulated me to complete the writing of this book, which has lasted several years. Also, I would like to express my gratitude to several individuals who have provided valuable advice, encouragement, and support. They are Prof. Kenzo Miya at University of Tokyo, Prof. Göran Grimvall at Royal Institute of Technology in Stockholm, Mr. Bernt Ericson, Director of Ericsson Research and Innovation, and especially Ms. Sigrun Hjelmquist, President of Ericsson Components AB. Thanks also to Prof. Kai Chang at Texas A&M University for his encouragement and recommendation, and to Mr. George J. Telecki, Executive Editor and Mr. Andrew Prince, Managing Editor at John Wiley & Sons, Inc., for their careful examination and efficient management of this book. I thank also the IEE Publishing Department for permitting me to use some of its copyrighted text and illustrative materials.

Finally, I must give my hearty thanks to my wife, Yatong, who has given me so much of her precious time and energy, which she could otherwise have contributed to her own artistic creation during my many years of writing, in the evenings and on holidays.

SHU-ANG ZHOU

Stockholm 1999

1

Introduction to Classical Electrodynamics

The development of electromagnetic theory has a long history, beginning perhaps with the ancients´ experimentation with the electrical properties of amber and the magnetic properties of lodestone. There exists an overwhelming amount material associated with the subject gathered over several hundreds of years. Today, new advances are still being made in electromagnetic theory due, in part, to new applications of the theory to many practical applications involving novel materials and devices. The purpose of this chapter is to present a review of this classical subject by introducing its basic concepts and theoretical principles for the study of electromagnetic phenomena in material media. The basic concepts, definitions, and principles introduced in this chapter are of importance for later discussion. Some emphasis is given here to the general methodology used in formulating macroscopic electromagnetic phenomena in material media. Readers may find that some of the derivations for the classical subject are in a new form, which may be interesting and stimulating. Some examples of practical interest are also provided for illustrating applications of classical electromagnetics.

1.1 CHARGES AND CURRENT

1.1.1 Charges and Charge Density

Charge is the source and the object of action of an electromagnetic field. In nature charge is found in two forms, called positive charge and negative charge. The numerical value of a charge can only be an integral multiple of the elementary charge which has the numerical value $|e| = 1.6022\times10^{-19}$ C (Coulomb). Examples of charged particles are, for instance, electrons and protons. The electron is the material carrier of an elementary negative charge (having a mass of 9.1×10^{-31} kg), which is usually assumed as a structureless point particle. The proton is the material carrier of an elementary positive charge (having a mass of 1.7×10^{-27} kg), which is not considered as a point particle. The entire proton charge is practically concentrated in a sphere of radius of about 10^{-15} m.

Many charged particles besides electron and proton, discovered by nuclear physicists, share the property of carrying charges $\pm e$. It follows that the total charge carried by a piece of matter must be an integral multiple of the electronic charge e. A situation like this one, in which a physical quantity is not allowed to have a continuous range of values, but is restricted to a set of definite discrete values, is referred to as a quantum phenomenon. So far, no one knows why electric charge should obey this quantum rule; it is an experimental fact. Nevertheless, since an elementary charge is very small and most of macroscopic phenomena in electromagnetics involve a huge number of electric charges, such a discrete nature is not manifested in any way in the domain of macroscopic phenomena. Hence, we may assume the charge to be continuously distributed in space and disregard its discreteness. In the continuum model, we introduce a volume density of charges by

$$\rho_e = \lim_{\Delta V \to 0} \frac{1}{\Delta V} \sum_{\alpha} e_{\alpha} \qquad (\mathrm{C/m^3}) \tag{1.1.1}$$

where e_α are the elementary charges within the volume ΔV. Here we speak of the volume ΔV as an infinitely small volume in the physical sense, which means that it is very small and hence its position in space is characterized with a sufficiently high accuracy by the coordinate of a point lying inside this volume. In addition, to be valid the limit $\Delta V \to 0$ must represent a volume large enough to contain a large number of charges e_α, yet small enough to appear infinitesimal when compared with the characteristic dimensions of the system considered. For example, a very small cube with each side of 1 micron has a volume of 10^{-18} m^3, which can still contain about 100 billion atoms. Thus it is expected that the smoothed-out function ρ_e defined by Eq.(1.1.1) may yield accurate macroscopic results for nearly all practical purposes.

1.1.2 Current Density

If charges contained in a volume ΔV are moving with velocities which may differ in magnitude and in direction, the motion of a charge will result in a transport of the charge in the direction of its velocity. Consequently, various movements of charges contained in the volume ΔV result in a certain average transport of the charges contained in this volume. The intensity of this transport of charges may be characterized by introducing a vector quantity $\boldsymbol{J}$ called the current density, defined by

$$\boldsymbol{J} = \lim_{\Delta V \to 0} \frac{1}{\Delta V} \sum_{\alpha} e_{\alpha} \mathbf{v}_{\alpha} \qquad (\mathrm{A/m^3}) \tag{1.1.2}$$

where $\mathbf{v}_\alpha$ denotes the velocity of the charge e_α.

The unit of current (ampere) and the unit of charge (coulomb) can be found to be related by 1 coulomb = 1 ampere·second. The direction of current density of positive charges coincides with the direction of their average velocity, while for negative charges, the current density has a direction opposite to that of the average velocity. We may, therefore, write

$$J = \lim_{\Delta V \to 0} \frac{1}{\Delta V} \left\{ \sum_{\alpha} e_{\alpha}^{+} \mathbf{v}_{\alpha} + \sum_{\beta} e_{\beta}^{-} \mathbf{v}_{\beta} \right\} \tag{1.1.3}$$

where we have separated the contributions from the positive charges $e_{\alpha}{}^{+}$ with velocity $\mathbf{v}_{\alpha}$ and from the negative charges $e_{\beta}{}^{-}$ with velocity $\mathbf{v}_{\beta}$. If we define a charge-average velocity $\mathbf{v}^{+}$ and $\mathbf{v}^{-}$ by

$$\mathbf{v}^{+} = \frac{\sum_{\alpha} e_{\alpha}^{+} \mathbf{v}_{\alpha}}{\sum_{\alpha} e_{\alpha}{}^{+}} \quad \text{and} \quad \mathbf{v}^{-} = \frac{\sum_{\beta} e_{\beta}^{-} \mathbf{v}_{\beta}}{\sum_{\beta} e_{\beta}^{-}} \tag{1.1.4}$$

the total current density may be expressed as

$$J = \rho^{+} \mathbf{v}^{+} + \rho^{-} \mathbf{v}^{-} \tag{1.1.5}$$

where ρ^{+} and ρ^{-} are respectively the density of positive charges and the density of negative charges. The net charge density is given by $\rho_e = \rho^{+} + \rho^{-}$. In many cases, the net charge may be zero. However, a current can still be present due to the difference between charge-average velocities.

The electric current describing the flow of, for instance, positive charges with an average velocity $\mathbf{v}$ across a planar surface S with its unit normal vector $\boldsymbol{n}$ parallel to $\mathbf{v}$ can be found by

$$I = \int_S J \cdot \boldsymbol{n} dS = \frac{\rho \mathrm{v} \Delta t S}{\Delta t} = \frac{\Delta q}{\Delta t} \tag{1.1.6}$$

where Δq is the total charge flowing across the surface S during the time interval Δt. Eq.(1.1.6) shows that the electric current defines the rate of flow of charge.

Quantitatively, we may consider an example of finding the average velocity of the free electrons in a copper wire whose cross-sectional area is $S = 1\ \text{mm}^2$ when the wire carries a current of $I = 1$ A. Here, assuming that each copper atom contributes one electron to the free electron gas, and that the number of free electrons per unit volume in copper is about $n = 8.5 \times 10^{28}$ electrons/m^3, we may find by Eq.(1.1.6)

$$\mathrm{v} = \frac{I}{neS} \approx 7.4 \times 10^{-4} \quad (\text{m/s})$$

which shows that the average velocity (also called the drift velocity) v of the free electrons in the copper wire due to an applied electric field is very small. One may wonder why an electric appliance of large size may go on as soon as its switch is closed and not minutes or hours later. The answer is that applying a potential difference across a circuit very rapidly creates an electric field in the circuit, and as a result, all the free electrons begin their drift almost simultaneously.

1.1.3 Conservation Law of Charge

Charges are conservative, that is, charges can neither be created nor destroyed. Such a postulate has been confirmed by experimental evidence. The mathematical statement of this postulate on the conservation of charge is expressed by the equation of continuity that may be obtained by the following consideration. Consider a volume V enclosed by a stationary surface S. The conservation of charge means that the net outflow of charge current through the closed surface S must equal the rate at which the total charge in the volume V decreases, that is,

$$\int_S \boldsymbol{J} \cdot \boldsymbol{n} dS = -\frac{\partial}{\partial t} \int_V \rho_e dV \tag{1.1.7}$$

where $\boldsymbol{n}$ is the unit outward normal vector of the surface S.

For a smooth continuous distribution of charges, since charges are conserved for any volume element, including an arbitrarily small one located anywhere, we obtain the local equation of conservation of charge as

$$\nabla \cdot \boldsymbol{J} + \frac{\partial \rho_e}{\partial t} = 0 \tag{1.1.8}$$

which is the equation of continuity. This conservation law of charge must be satisfied at all times and under any circumstances.

At a discontinuous surface where surface charges accumulate, we have the following interface condition:

$$\boldsymbol{n} \cdot [J] = -\frac{\partial \alpha_f}{\partial t} \tag{1.1.9}$$

where α_f is the surface charge density. Here, $[F] = F^+ - F^-$ denotes the jump of the quantity F across the interface S with $\boldsymbol{n}$ being the unit normal vector always drawn from S^- to S^+, as shown in Fig. 1.1.

For steady currents, charge density does not vary with time, and therefore, Eq.(1.1.8) is reduced to

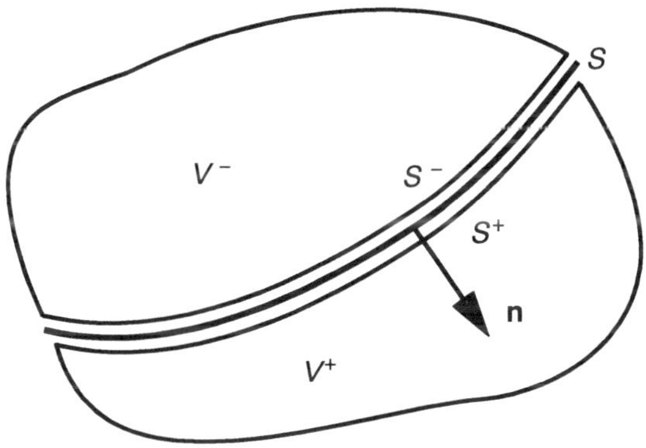

Figure 1.1 A discontinuous surface.

$$\nabla \cdot \boldsymbol{J} = 0 \tag{1.1.10}$$

which implies that the stream lines of steady currents are closed loops. The integral form of Eq.(1.1.10) may be written

$$\int_S \boldsymbol{J} \cdot \boldsymbol{n} dS = 0 \tag{1.1.11}$$

a particular form of which reads

$$\sum_k I_k = 0 \tag{1.1.12}$$

This equation expresses a well-known law in electric circuit theory, that is, the *Kirchhoff current law,* which states that the algebraic sum of all the currents flowing out of a junction in an electric circuit is zero. As an illustration, surface S in Fig. 1.2 encloses a volume that contains four conductors meeting at a junction point. Taking the current flowing away from the junction as positive and the current flowing toward the junction as negative, we have, from Eq.(1.1.12),

$$I_1 + I_2 - I_3 + I_4 = 0 \tag{1.1.13}$$

1.1.4 Coulomb´s Law

Charges in space interact with each other. The interaction between charges in free space (vacuum) was first studied quantitatively by Coulomb in 1785. He found an experimental law for describing the force (electrostatic force) between

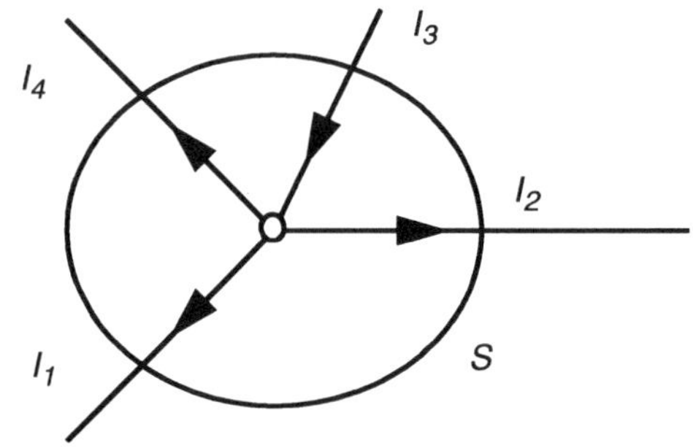

Figure 1.2 A junction point of several conductors.

two stationary (point) charges q_1 and q_2 in free space, which is now called the *Coulomb law*. Coulomb's law was originally formulated as an action at a distance with the magnitude

$$F = \frac{1}{4\pi\varepsilon_o} \times \frac{q_1 q_2}{r^2} \quad \text{(N)} \tag{1.1.14}$$

where $\varepsilon_o = 8.854\times10^{-12}$ Farad/meter (1 F = 1 C^2/N·m) is called the permittivity of free space (vacuum) and r is the distance between the two point charges. The direction of the force is along the line between the two charges. For both charges of the same sign, the force is repulsive, and for charges of opposite sign, the force is attractive. In addition, the Coulomb forces acting on the two charges are equal and opposite even though the individual charges may differ greatly in magnitude. Equation (1.1.14) indicates that the force is inversely proportional to the square of the distance between the charge particles. Coulomb's law is known experimentally to be valid at a distance down to the order of 10^{-17} m [Jackson (1975) and Matveev (1986)].

Similar to the electrostatic forces between different charges, there exists also gravitational forces between different masses at rest, which obeys also the inverse square law of force, discovered long ago by Newton in 1665. A comparison of the Coulomb force F_e and the Newton gravitational force F_g between two charged particles, for instance, an electron and a proton, may be made by

$$\frac{F_e}{F_g} = \frac{1}{4\pi\varepsilon_o} \times \frac{e^2}{G m_e m_p} \tag{1.1.15}$$

where $G = 6.673\times10^{-11}$ $m^3/kg{\cdot}sec^2$ is the gravitational constant. m_e and m_p are, respectively, the mass of the electron and the mass of the proton. Equation (1.1.15) shows that the electrical force of attraction between an electron and a proton is about 2.2×10^{39} times greater than the gravitational force of attraction. Since the gravitational constant G is extremely small, the gravitational interaction can become considerable only for very large masses. For this reason gravitational forces are usually negligible in the mechanics of atoms and molecules.

From Eq.(1.1.14), it seems that the electrostatic forces induced by electrons are quite small due to very small charge of the electron. In fact, they are immensely strong. For example, electrostatic forces are responsible for the great strength of solids under compression. When neighboring atoms are close together, their electron clouds begin to overlap, and the mutual repulsion of these clouds opposes any compression force. Normally we do not notice that electrostatic forces are so powerful, because matter is usually electrically neutral, carrying equal amounts of positive and negative charges.

1.2 ELECTRIC AND MAGNETIC FIELDS

1.2.1 Electric Field

Coulomb's law was interpreted as a long-range interaction, that is, it was assumed that one body acts on another as if without intermediaries. Such a long-range interaction differs from a short-range interaction, which is defined according to bodies interacting only due to a continuous "transfer of forces" in the space between them.

To describe the interaction between material bodies, we may either formulate the action at a distance between the interacting bodies or separate the interaction process into the production of a field by one system and the action of the field on another system. These two descriptions are physically indistinguishable in the static case. However, if the bodies are in motion, it is both physically and mathematically advantageous to ascribe physical reality to the field itself. In what follows, we shall formulate electromagnetic and mechanical interaction as a field theory.

In classical electrodynamics, electric and magnetic fields are fundamental fields of forces that originate from charges. The electric field in free space is defined in terms of the force produced on a test charge q by the equation

$$\boldsymbol{E} = \lim_{q \to 0} \frac{\boldsymbol{F}}{q} \qquad (\mathrm{N/C}) \tag{1.2.1}$$

where $\boldsymbol{F}$ is the force on the test charge q. The limit $q \to 0$ is introduced here in order that the test charge will not influence the behavior of the sources of the field, which will then be independent of the presence of the test body. Practically, such a definition is entirely suitable only for macroscopic phenomena.

According to Coulomb's law, the electric field $\boldsymbol{E}$ at a position $\boldsymbol{r}$ due to a charge q_o at the origin of the radius vector may be found immediately by

$$\boldsymbol{E} = \frac{q_o}{4\pi\varepsilon_o} \times \frac{\boldsymbol{r}}{r^3} = -\frac{q_o}{4\pi\varepsilon_o}\nabla\left(\frac{1}{r}\right) \tag{1.2.2}$$

The force acting on a test charge q due to the interaction between the charge q_o and the charge q can now be expressed by $\boldsymbol{F} = q\boldsymbol{E}$.

Essentially, Coulomb's law describes quantitatively the interaction between two charges. Naturally, we may ask the question of how to describe interactions among large numbers of charges. To answer this question, a linear *superposition principle* of fields in free space was proved experimentally with sufficient accuracy in the classical domain of sizes and attainable field strengths. The superposition principle states that the total electric field produced by N charges q_α located respectively at $\boldsymbol{x}_\alpha$ ($\alpha = 1, 2,..., N$) in free space is simply the sum of the electric fields caused by individual charges:

$$\boldsymbol{E}(\boldsymbol{x}) = \sum_{\alpha=1}^{N} \frac{q_\alpha}{4\pi\varepsilon_o} \times \frac{(\boldsymbol{x}-\boldsymbol{x}_\alpha)}{|\boldsymbol{x}-\boldsymbol{x}_\alpha|^3} \tag{1.2.3}$$

For a continuous distribution of charges with charge density ρ_e in a volume V, the electric field can then be found by

$$\boldsymbol{E}(\boldsymbol{x}) = \int_V \frac{\rho_e(\boldsymbol{x}')}{4\pi\varepsilon_o} \times \frac{(\boldsymbol{x}-\boldsymbol{x}')}{|\boldsymbol{x}-\boldsymbol{x}'|^3} dV' \tag{1.2.4}$$

This expression for $\boldsymbol{E}$ is valid for the position point $\boldsymbol{x}$ outside the charged region as well as inside. If the charge density ρ_e and its first derivatives are continuous, the gradient of the electric field exists and is continuous at interior points of the charged region. Detailed discussion about the mathematics involved refers to the work of, for instance, Tiersten (1990).

1.2.2 Electric Potential

By taking the curl of Eq.(1.2.3) or (1.2.4), we may find $\nabla\times\boldsymbol{E} = 0$, which implies that the electrostatic field generated by charges at rest is irrotational. Thus, we may introduce a scalar electric potential ϕ defined by $\boldsymbol{E} = -\nabla\phi$, and can easily find

$$\phi(\boldsymbol{x}) = \sum_{\alpha=1}^{N} \frac{q_\alpha}{4\pi\varepsilon_o} \times \frac{1}{|\boldsymbol{x}-\boldsymbol{x}_\alpha|} \tag{1.2.5}$$

for the discrete distribution of N point charges q_α located respectively at $\boldsymbol{x}_\alpha$ (α

$= 1, 2, ..., N$), and

$$\phi(\boldsymbol{x}) = \int_V \frac{\rho_e(\boldsymbol{x}')}{4\pi\varepsilon_o|\boldsymbol{x}-\boldsymbol{x}'|}dV' \tag{1.2.6}$$

for the continuous distribution of charges with charge density ρ_e in the volume V. It can be shown that the expression (1.2.6) for the electric potential ϕ is valid for the position point $\boldsymbol{x}$ outside the charged region as well as inside. If the charge density ρ_e and its first derivatives are continuous, the second spatial derivative of ϕ exists and is continuous at interior points of the charged region.

The scalar electric potential ϕ has a physical interpretation when we consider the work done on a test charge q in transporting it from infinity to a position point $\boldsymbol{x}$ in a region of localized electric field described by ϕ (which vanishes at infinity). Since the force acting on the charge q at any point is $\boldsymbol{F} = q\boldsymbol{E}$, the work done in moving the charge from infinity to the position $\boldsymbol{x}$ is

$$W = -\int_\infty^{\boldsymbol{x}} \boldsymbol{F}\cdot d\boldsymbol{l} = \int_\infty^{\boldsymbol{x}} q(\nabla\phi\cdot d\boldsymbol{l}) = q\phi(\boldsymbol{x}) \tag{1.2.7}$$

which shows that $q\phi(\boldsymbol{x})$ can be interpreted as the potential energy of the test charge in the electrostatic field. The unit of the electric potential ϕ is the Volt (V). It is noticed here that we theoretically move the charge, but we did not release it and watch what happens.

If a charged particle is set free in an electric field, it will be accelerated by the electric force according to Newton's second law:

$$m_q\frac{d\mathbf{v}}{dt} = q\boldsymbol{E} \tag{1.2.8}$$

where m_q and q are, respectively, the mass and the charge of the particle. Integrating this force over the distance the particle moves from the position $\boldsymbol{x}_1$ to $\boldsymbol{x}_2$ yields the energy acquired by the particle. Thus, we may find

$$\frac{1}{2}m_q(\mathrm{v}_2^2-\mathrm{v}_1^2) = qU \tag{1.2.9}$$

where v_1 and v_2 are, respectively, the initial velocity and the final velocity of the particle. U denotes the magnitude of the potential difference between points $\boldsymbol{x}_1$ and $\boldsymbol{x}_2$. If the particle starts from rest, its initial velocity is zero, and the final velocity can be found by

$$\mathrm{v} = \sqrt{\frac{2qU}{m_q}} \tag{1.2.10}$$

For an electron charge, we may obtain $\mathrm{v}=5.9\times10^5U^{1/2}$, which is about 590 km/s if $U = 1$ V. It is apparent that a relatively small voltage may impart a very large velocity to an electron. The above relations are based on the assumption that the particle velocity is small compared with that of light. According to the relativistic effect (to be discussed in Section 1.10.5), the mass of a particle approaches an infinite value as the velocity approaches that of light, whereas the above relations are based on a constant mass. Actually, for most applications, the mass increase is of negligible consequence unless the velocity is at least 10 percent that of light.

1.2.3 Gauss′ Theorem for Electric Field

By the superposition principle, we may introduce the following *Gauss' theorem* for electric field as an integral form of Coulomb's law

$$\int_S \varepsilon_o \boldsymbol{E} \cdot \boldsymbol{n} dS = \int_V \rho_e dV \tag{1.2.11}$$

where V is a volume bounded by a closed surface S with $\boldsymbol{n}$ being its unit outward normal vector.

Equation (1.2.11) can be proved simply by writing

$$\int_S \varepsilon_o \boldsymbol{E} \cdot \boldsymbol{n} dS = \int_V \varepsilon_o \nabla \cdot \boldsymbol{E} dV = \int_V \varepsilon_o \nabla \cdot \left\{ \int_V \frac{\rho_e(\boldsymbol{x}')}{4\pi\varepsilon_o} \times \frac{(\boldsymbol{x}-\boldsymbol{x}')}{|\boldsymbol{x}-\boldsymbol{x}'|^3} dV' \right\} dV$$

$$= \iint_{VV} \frac{\rho_e(\boldsymbol{x}')}{4\pi} \nabla^2 \left(\frac{-1}{|\boldsymbol{x}-\boldsymbol{x}'|} \right) dV' dV = \iint_{VV} \rho_e(\boldsymbol{x}') \delta(\boldsymbol{x}-\boldsymbol{x}') dV' dV = \int_V \rho_e dV$$

where $\delta(\boldsymbol{x}-\boldsymbol{x}')$ is the Dirac delta function.

For a continuous distribution of volume charge density ρ_e, we have the following local differential form of Gauss′ theorem:

$$\nabla \cdot (\varepsilon_o \boldsymbol{E}) = \rho_e \tag{1.2.12}$$

This important result is one of Maxwell's equations in free space. The integral form of the Gauss theorem by Eq.(1.2.11) relates the electric field on any closed surface to the net amount of charge enclosed within the surface. Thus, by using the integral form of the Gauss theorem, the electric field $\boldsymbol{E}$ can be evaluated easily when the charge density has a distribution with a simple symmetry. The differential form of the Gauss theorem by Eq.(1.2.12) is, however, more useful in the case of analyzing the electric field $\boldsymbol{E}$ without a simple symmetry.

In the case of a discrete set of point charges q_α located at $\boldsymbol{x}_\alpha$ ($\alpha = 1, 2, ...$) in free space, we may write

$$\rho_e = \sum_\alpha q_\alpha \delta(\boldsymbol{x} - \boldsymbol{x}_\alpha) \tag{1.2.13}$$

and can find from Eq.(1.2.11)

$$\int_S \varepsilon_o \boldsymbol{E} \cdot \boldsymbol{n} dS = \sum_\alpha q_\alpha \tag{1.2.14}$$

where the sum is taken over only those charges inside the closed surface S.

It is shown that Gauss' theorem (1.2.11) [or (1.2.14)] expresses the physical fact that the total outward flux of the electric field over any closed surface in free space is equal to the total charge enclosed in the surface divided by the permittivity ε_0.

1.2.4 Electric Multipoles in Free Space

In this section, we shall introduce the concept of electric multipoles for localized charge distribution. The multipole concept is of interest since it may be used in modeling of electromagnetic materials, as we shall show in late sections. We consider an array of M point electric charges located in a small volume centered at $\boldsymbol{x}'$ as shown in Fig. 1.3, which ideally assumes that all charges are treated as geometrical points in space. In the laboratory, this approximation can be achieved by making any distances of separation that are involved very large compared to the dimensions of the charged particles. The electric potential ϕ at the position $\boldsymbol{x}$ for such an array of electric charges can be found by

$$\phi(x) = \sum_\alpha q^\alpha G^e(\boldsymbol{x}, \boldsymbol{x}' + \boldsymbol{d}^\alpha) \tag{1.2.15}$$

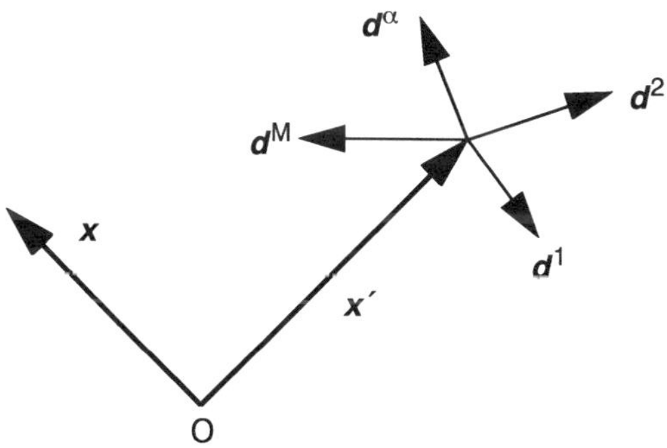

Figure 1.3 An array of point electric charges.

where $\boldsymbol{x}'+\boldsymbol{d}^\alpha$ is the position vector of the αth point charge q^α, and G^e is the electric Green's function in free space, defined by the following equation:

$$\nabla^2 G^e(\boldsymbol{x}, \boldsymbol{x}') + \frac{1}{\varepsilon_o}\delta(\boldsymbol{x} - \boldsymbol{x}') = 0 \tag{1.2.16}$$

with $\delta(\boldsymbol{x} - \boldsymbol{x}')$ being the Dirac delta function.

The fundamental solution of Eq.(1.2.16) can be found as

$$G^e(\boldsymbol{x}, \boldsymbol{x}') = \frac{1}{4\pi\varepsilon_o|\boldsymbol{x} - \boldsymbol{x}'|} \tag{1.2.17}$$

which can be explained physically as the electric potential at the position $\boldsymbol{x}$ produced by a unit point charge located at $\boldsymbol{x}'$ in free space.

By expending the electric Green's function in a Taylor series about $(\boldsymbol{x}, \boldsymbol{x}')$, Eq.(1.2.15) can be expressed as

$$\phi(\boldsymbol{x}) = \sum_{k=0}^{\infty} \frac{(-1)^k}{k!} p^e_{s_1 \ldots s_k} G^e_{,s_1 \ldots s_k}(\boldsymbol{x} - \boldsymbol{x}') \tag{1.2.18}$$

where we have introduced the point electric multipole moment of order k ($k = 0$, 1, 2, ...) defined by

$$p^e_{s_1 \ldots s_k} = \sum_{\alpha=1}^{M} q^\alpha d^\alpha_{s_1} \ldots d^\alpha_{s_k} \tag{1.2.19}$$

which is located at the position $\boldsymbol{x}'$. In Eq.(1.2.18), the following differential notation is used:

$$G^e_{,s_1 \ldots s_k}(\boldsymbol{x} - \boldsymbol{x}') = \frac{\partial^k G^e(\boldsymbol{x} - \boldsymbol{x}')}{\partial x_{s_1} \ldots \partial x_{s_k}}$$

For $k = 0$, we have the net charge of the array of point electric charges by

$$p^e = \sum_{\alpha=1}^{M} q^\alpha \tag{1.2.20}$$

which may also be called the electric monopole moment of the charge array. If the point $\boldsymbol{x}$ is far away from $\boldsymbol{x}'$ and if the net charge of the array of charges is not zero, the electric monopole term in Eq.(1.2.18) will be the dominant term in the electrical potential, which means that the whole electric charge distribution will act as if it were a point electric charge at $\boldsymbol{x}'$.

For $k = 1$, we arrive at the point electric dipole moment

$$p_k^e = \sum_{\alpha=1}^{M} q^\alpha d_k^\alpha \tag{1.2.21}$$

which involves only the properties of the charge distribution and does not involve the location of the field point $\boldsymbol{x}$. If the field point $\boldsymbol{x}$ is very far away and if the electric monopole moment p^e vanishes, the dipole term will then be the leading term in the expansion of the potential function ϕ. Thus the point electric dipole moment $\boldsymbol{p}^e$ will be the dominant feature of the charge array.

By introducing the electric center of gravity of the positive and negative charges, we can rewrite Eq.(1.2.21) in the following form:

$$\boldsymbol{p}^e = \sum_{positive} q^\alpha \boldsymbol{d}^\alpha + \sum_{negative} q^\alpha \boldsymbol{d}^\alpha = Q^+ \boldsymbol{x}^+ + Q^- \boldsymbol{x}^- \tag{1.2.22}$$

where $\boldsymbol{x}^+$ and $\boldsymbol{x}^-$ denotes the position vectors of the electric center of gravity of the positive charges and the negative charges, respectively. Q^+ and Q^- are the total positive charge and the total negative charge, respectively.

In the case of a zero net charge, Eq.(1.2.22) is reduced to

$$\boldsymbol{p}^e = Q\boldsymbol{d} \tag{1.2.23}$$

where Q denotes the total positive charge and $\boldsymbol{d} = \boldsymbol{x}^+ - \boldsymbol{x}^-$ represents the vectorial distance between the centers of gravity of the positive and negative charges.

In a simple case, the electric charge system consists of only two point charges $+q$ and $-q$ at a distance $\boldsymbol{d}$. Such a system is usually called a (physical) electric dipole, the moment of which is equal to $q\boldsymbol{d}$, where the vector $\boldsymbol{d}$ is pointing from the negative to the positive charge. A mathematical abstraction derived from the above defined physical dipole is the ideal or point dipole, which is defined as that obtained by replacing the distance $\boldsymbol{d}$ and the charge Q, respectively, by $\boldsymbol{d}/\mathrm{n}$ and $\mathrm{n}Q$, the limit approached as the number n tends to infinity is the ideal dipole.

It is worthwhile to mention that, apart from these permanent or intrinsic dipole moments, a temporary or induced dipole moment may arise in a molecule, having, for instance, originally a zero electric dipole moment, when it is brought into an external electric field, since in the presence of this field there are forces acting on those charges in the molecule which distort the molecular charge distribution from its originally symmetric form. In addition, molecules, especially when distorted by the field, may have higher-order electric multipoles, such as quadrupole moments [Stogryn (1966) and Böttcher (1973)]. By Eq.(1.2.18), we have shown that the electric potential at a field point $\boldsymbol{x}$ outside of a small volume containing the array of point electric charges could be described by a set of point electric multipole moments of the charge system located at $\boldsymbol{x}'$.

1.2.5 Interaction of Electric Multipoles with External Field

For a given array of M point electric charges located at $\boldsymbol{x}+\boldsymbol{d}^{\alpha}$ ($\alpha = 1, 2, ..., M$), the interaction energy of the array of electric charges with an external electric field $\boldsymbol{E}^{o}$ (its corresponding scalar potential is denoted by ϕ^{o}) reads

$$U^{eo} = \sum_{\alpha=1}^{M} q^{\alpha}\phi^{o}(\boldsymbol{x}+\boldsymbol{d}^{\alpha}) \tag{1.2.24}$$

If the scalar potential ϕ^{o} varies slowly in the region of the electric charge array, we can make a Taylor expansion of $\phi^{o}(\boldsymbol{x}+\boldsymbol{d}^{\alpha})$ around the point $\boldsymbol{x}$ and write

$$U^{eo} = p^{e}\phi^{o}(\boldsymbol{x}) + \sum_{k=1}^{\infty}\frac{1}{k!}\overset{e}{p}_{s_1\ldots s_k}\overset{o}{\phi}_{,s_1\ldots s_k}(\boldsymbol{x}) \tag{1.2.25}$$

where p^{e} is the total charge of the point charge array, and the point electric multipole moments defined in Eq.(1.2.19) is now located at $\boldsymbol{x}$.

The interaction energy of a point electric multipole moment of order k located at $\boldsymbol{x}$ in the external field can thus be introduced by

$$U^{eo(k)} = \frac{1}{k!}\overset{e}{p}_{s_1\ldots s_k}\overset{o}{\phi}_{,s_1\ldots s_k}(\boldsymbol{x}) \tag{1.2.26}$$

which, for a point electric dipole moment, reads

$$U^{eo(1)} = -\overset{e}{\boldsymbol{p}}\cdot\boldsymbol{E}^{o} \tag{1.2.27}$$

The total translational force acting on the array of electric charges in the external electric field $\boldsymbol{E}^{o}$ can then be found by

$$F^{e}_{n} = p^{e}E^{o}_{n}(\boldsymbol{x}) + \sum_{k=1}^{\infty}\frac{1}{k!}\overset{e}{p}_{s_1\ldots s_k}\overset{o}{E}_{n,s_1\ldots s_k}(\boldsymbol{x}) \tag{1.2.28}$$

It is shown that if the net charge of the array of electric charges is not zero ($p^{e}\neq 0$), the leading term of Eq.(1.2.28) will be the force on the point electric monopole moment. Obviously, there is no torque acting on the electric monopole moment. If, however, the net charge vanishes ($p^{e} = 0$), the electric force on the electric dipole moment $\boldsymbol{p}^{e}$ will be the dominant term in Eq.(1.2.28) and can be expressed by

$$\boldsymbol{F}^{e} = (\boldsymbol{p}^{e}\cdot\nabla)\boldsymbol{E}^{o} \tag{1.2.29}$$

This expression shows that the electric dipole force is proportional to the

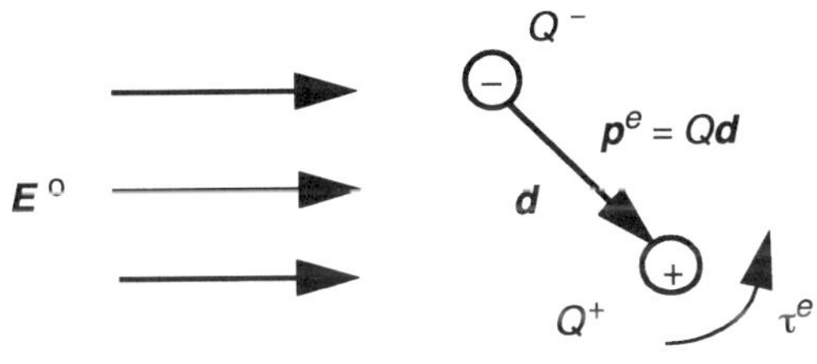

Figure 1.4 Dipole with a permanent dipole moment $\boldsymbol{p}^e$ in an external field $\boldsymbol{E}^{\mathrm{o}}$.

gradient of the external electric field and, therefore, vanishes in a uniform external electric field. There is, however, a torque acting on the electric dipole moment in a uniform field as well as in a non-uniform field. This torque may be derived by the negative derivative of the energy [see Eq.(1.2.27)] with respect to the angle between the external field and the dipole moment vector, that is,

$$\tau^e = \boldsymbol{p}^e \times \boldsymbol{E}^o \tag{1.2.30}$$

It is seen that the torque tends to orient the electric dipole moment in the direction of the external electric field, as shown in Fig. 1.4. From Eq.(1.2.27), we may notice that an unaligned dipole has a greater potential energy than an aligned one. The potential energy is minimum when the dipole moment vector $\boldsymbol{p}^e$ is parallel to $\boldsymbol{E}^{\mathrm{o}}$.

1.2.6 Magnetic Field and the Lorentz Force

So far, we have shown that interaction between fixed point charges is defined completely by Coulomb´s law. However, Coulomb´s law is incapable of describing the interaction between moving charges. Such a conclusion is based on relativistic properties of space and time and the relativistic equation of motion rather than on the specific features of Coulomb´s interaction. The interaction between moving charges is due not only to the Coulomb force but also to another kind of force, called the magnetic force.

When a small test charge q is moving in a magnetic field (to be defined), experiments show that it experiences a force that cannot be expressed in terms of $\boldsymbol{E}$, but can be expressed by defining a new vector field quantity, the magnetic (induction) field $\boldsymbol{B}$, such that

$$\boldsymbol{F}_m = q\mathbf{v} \times \boldsymbol{B} \tag{1.2.31}$$

where $\mathbf{v}$ is the velocity vector. It shows that the magnetic force always acts perpendicularly to the direction of motion of the charged particle, as shown in

Fig. 1.5. The unit of the magnetic field $\boldsymbol{B}$ is such that a charge of one coulomb moving with a speed of 1 m/s perpendicularly to a unit field experiences a force of one newton (N). This unit is called the tesla (1 tesla = 1 weber/meter^2 = 1 volt·second/meter^2). Often the strengths of magnetic fields are given in an alternative unit, called the gauss (G), which is ten thousand times smaller than the tesla, 1 G = 10^{-4} T. A magnetic field of one gauss is comparable with the earth magnetic field. Typical iron-cored electromagnets can produce steady fields of up to about 1 T. On a microscopic scale, fields as large as 10^4 T have been found to occur near the nuclei of certain highly ionized atoms.

In general, when both electric and magnetic fields are present, the total electromagnetic force on a charge q moving with a velocity $\mathbf{v}$ can be expressed by

$$\boldsymbol{F} = q\boldsymbol{E} + q\mathbf{v} \times \boldsymbol{B} \tag{1.2.32}$$

which is known as the Lorentz force law (equation). Its validity has been well established by experiments. In classical electrodynamics, the Lorentz force law may be considered as a fundamental postulate of our electromagnetic model. However, in Einstein's special theory of relativity, the Lorentz force expression may be obtained from the requirement of the invariance of the (special) relativistic equation of motion. The Lorentz force law is found experimentally to be true even for particles moving at speeds close to the speed of light. Electric charge is therefore relativistically invariant.

The Lorentz force for a charge moving in free space can be measured directly in a laboratory. A generalization of it to a charge moving in a material medium is not, however, subject to direct experimental confirmation, and therefore, it is considered as an assumption for the moment. Quantitatively, the magnetic interaction shown in Eq.(1.2.32) can be compared with electric interaction only at sufficiently high velocities of charged particles. If, however, Coulomb's interaction is absent for some reason, magnetic interaction can manifest itself at very low velocities, for instance, when an electric current flows in a conductor, which has an immensely high density of free electrons and

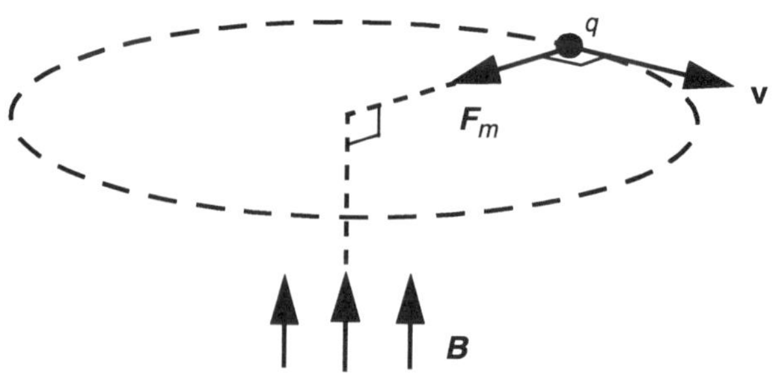

Figure 1.5 Force on a charged particle moving in a magnetic field.

where the electric field of moving charges is neutralized by the electric field of the opposite charges of the conductor, that is, it is screened. Thus the magnetic force in good conductors may become dominant since the product of charge density and velocity can be large even when the velocity is very low.

1.3 LAWS OF ELECTRODYNAMICS AND MAXWELL´S EQUATIONS IN FREE SPACE

1.3.1 Ampere´s Circuital Law in Free Space

In this section, we shall present the well-known laws of electrodynamics, based on which Maxwell's equations for free space are found. First of all, we introduce the integral form of the *Ampere circuital law* in free space by

$$\oint_L \left(\frac{\boldsymbol{B}}{\mu_o}\right) \cdot d\boldsymbol{l} = \int_S \boldsymbol{J} \cdot d\boldsymbol{S} + \frac{\partial}{\partial t}\int_S \varepsilon_o \boldsymbol{E} \cdot d\boldsymbol{S} \tag{1.3.1}$$

where $\mu_0 = 4\pi \times 10^{-7}$ H/m (henry per meter) is the permeability of free space. Here, L is any closed line that bounds a two-sided surface S in free space. The positive direction of the typical element $d\boldsymbol{S}$ may be taken to either side of S, but the positive integration sense about L must agree with the right-hand rule relative to $d\boldsymbol{S}$.

By applying Stokes' theorem to the line integral on the left-hand side of Eq.(1.3.1), we may derive the local differential form of Ampere´s circuital law in free space as

$$\nabla \times \left(\frac{\boldsymbol{B}}{\mu_o}\right) = \boldsymbol{J} + \frac{\partial(\varepsilon_o \boldsymbol{E})}{\partial t} \tag{1.3.2}$$

This equation states that the curl of $\boldsymbol{B}/\mu_0$ at any point in a region is the sum of the electric current density $\boldsymbol{J}$ and the displacement current density $\partial(\varepsilon_0 \boldsymbol{E})/\partial t$ at that point. The displacement current term is historically the contribution of Maxwell, who provided that missing link to unify the theories of electricity and magnetism. Apparently, Maxwell thought of the magnetic field as being caused by an actual displacement of electric charge, but such a picture is not needed to justify Eq.(1.3.2). This equation is justified by the far-reaching validity of the conclusions based on it. For instance, in an *RC* circuit, the displacement current may flow through a perfect capacitor and is numerically equal to the conduction current flowing through conductor wires in the complete ac circuit. Here, between the plates of a perfect capacitor (having a vacuum between the plates), there is no flow of real charges, but merely an oscillation of the electric field. Another well-known example is the prediction of propagation of electromagnetic waves in free space in the absence of charges and currents, which has become the base of today´s radio communication technologies.

In steady states of electromagnetic systems, the displacement current vanishes and the Ampere circuital law (1.3.1) is reduced to

$$\oint_L \left(\frac{\boldsymbol{B}}{\mu_o}\right) \cdot d\boldsymbol{l} = \int_S \boldsymbol{J} \cdot d\boldsymbol{S} \tag{1.3.3}$$

which states that the circulation of the magnetic field divided by μ_0 around any closed path is equal to the free current flowing through the surface bounded by the path. The integral form of the Ampere circuital law is very useful to determine the magnetic induction field $\boldsymbol{B}$ caused by a steady current I when there is a closed path around the current such that the magnitude of $\boldsymbol{B}$ is constant over the path.

It may be noticed that for steady-state current phenomena, we have $\nabla \cdot \boldsymbol{J} = 0$. In general, by noting Eqs.(1.2.12) and (1.3.2), we get the following equation for charge and current:

$$\nabla \cdot \boldsymbol{J} + \frac{\partial \rho_e}{\partial t} = 0 \tag{1.3.4}$$

which is consistent with the continuity equation (1.1.8) from the conservation law of charge.

As we shall show later, the displacement current is usually negligible in comparison with conduction currents in good conductors at microwave frequencies. But it may become important in many situations as the frequencies of the time-varying phenomena is increased.

1.3.2 Gauss′ Law for Magnetic Field

In contrast to Gauss′ theorem for electric field, there exists a hypothesis of non-existence of magnetic charges (also called magnetic monopoles), the mathematical expression of which is given by

$$\int_S \boldsymbol{B} \cdot d\boldsymbol{S} = 0 \tag{1.3.5}$$

for any closed surface S. This hypothesis is also often called Gauss′ law for magnetic field since, so far, there is no experimental evidence for the existence of magnetic charges or monopoles [Jackson (1975) and Johnk (1988)]. The differential form of Gauss′ law for magnetic field is

$$\nabla \cdot \boldsymbol{B} = 0 \tag{1.3.6}$$

This equation shows that we may introduce a vector field $\boldsymbol{A}$ defined by

$$B = \nabla \times A \tag{1.3.7}$$

so that Eq.(1.3.2) becomes

$$\nabla \times (\nabla \times A) = \mu_o J + \frac{1}{c^2}\frac{\partial E}{\partial t} \tag{1.3.8}$$

where $c = (\mu_0 \varepsilon_0)^{1/2}$ is the speed of light in free space, which has the numerical value of 2.998×10^8 m/s. The vector field $\boldsymbol{A}$ so defined is called the vector magnetic potential. Its SI unit is weber per meter (Wb/m). The vector magnetic potential $\boldsymbol{A}$ can be related to the magnetic flux Φ through a given surface S that is bounded by contour L

$$\Phi = \int_S B \cdot dS = \oint_L A \cdot dl \qquad \text{(Wb)} \tag{1.3.9}$$

with the aid of Stokes′ theorem.

1.3.3 The Biot-Savart Law

For steady currents, the displacement current term disappears in Eq.(1.3.8). Thus we may get

$$\nabla^2 A = \mu_o J \tag{1.3.10}$$

by using the Coulomb gauge $\nabla \cdot A = 0$. This is a vector Poisson's equation. Thus, for a steady current distribution confined in a finite volume V, the vector magnetic potential $\boldsymbol{A}$ can be found by

$$A(x) = \frac{\mu_o}{4\pi}\int_V \frac{J(x')}{|x - x'|} dV' \tag{1.3.11}$$

This expression for the magnetic vector potential $\boldsymbol{A}$ is valid for the position point $\boldsymbol{x}$ outside the current region as well as inside. If the current density $\boldsymbol{J}$ and its first derivative are continuous, the second derivative of the magnetic vector potential $\boldsymbol{A}$ exists and is continuous at interior points of the current region.

By taking the curl of Eq.(1.3.11), we may derive the following result, which is usually called the *Biot-Savart law* for a steady current distribution:

$$B(x) = \frac{\mu_o}{4\pi}\int_V \frac{J \times r}{r^3} dV' \tag{1.3.12}$$

with $\boldsymbol{r} = \boldsymbol{x} - \boldsymbol{x}'$ and $r = |\boldsymbol{x} - \boldsymbol{x}'|$. It is seen that the Biot-Savart law satisfies automatically the condition of the non-existence of magnetic charges by simply making the divergence of $\boldsymbol{B}$ for Eq.(1.3.12), the result of which gives $\nabla\cdot\boldsymbol{B} = 0$.

For a closed current circuit L' with current I', Eq.(1.3.12) may be written

$$\boldsymbol{B}(\boldsymbol{x}) = \frac{\mu_o}{4\pi}\oint_{L'} \frac{I'd\boldsymbol{l}' \times \boldsymbol{r}}{r^3} \tag{1.3.13}$$

In addition, for the steady current, we have $\nabla\cdot\boldsymbol{J} = 0$, which implies that any current flowing into an arbitrary volume must flow out. Historically, the Biot-Savart law was deduced from measurements on the fields due to currents in wires made by Ampere, Biot, and Savart in the years around 1820.

1.3.4 Magnetic Force on Current Circuit

Using the Lorentz force expression (1.2.31), the magnetic force acting on a line current element $d\boldsymbol{l}$ with the current I located at $\boldsymbol{x}$ can be calculated by

$$d\boldsymbol{F} = Id\boldsymbol{l} \times \boldsymbol{B}(\boldsymbol{x}) \tag{1.3.14}$$

Thus the total magnetic force acting on the current circuit L with the steady current I in the magnetic field $\boldsymbol{B}$ generated by another current circuit L' with the steady current I' can be found by

$$\boldsymbol{F}_{(L)} = \oint_L Id\boldsymbol{l} \times \boldsymbol{B}(\boldsymbol{x}) = \frac{\mu_o}{4\pi}\oint_L Id\boldsymbol{l} \times \oint_{L'} \frac{I'd\boldsymbol{l}' \times \boldsymbol{r}}{r^3} \tag{1.3.15}$$

We may also calculate the total magnetic force $\boldsymbol{F}_{(L')}$ acting on the current circuit L' due to the magnetic field generated by the current circuit L. It can be found that the force $\boldsymbol{F}_{(L')}$ has the same magnitude but opposite direction of the force $\boldsymbol{F}_{(L)}$, that is, $\boldsymbol{F}_{(L')} = -\boldsymbol{F}_{(L)}$, which means that the interaction forces between two steady current circuits satisfy Newton´s third law. Obviously, in order for the current circuit L (or L') to be in equilibrium, additional mechanical force has to be acted on it so that the net force acting on the current circuit L (or L') will be zero.

1.3.5 Magnetic Multipoles in Free Space

We shall now introduce the concept of point magnetic multipoles. It is known that although the existence of magnetic charges (monopoles) is compatible with the requirements of quantum mechanics [see Dirac (1931) and Amaldi (1968)], their reliable experimental discovery seems not yet to have been achieved. Therefore, we still admit here that the divergence of the magnetic field $\boldsymbol{B}$ is zero and that the sources of magnetic field $\boldsymbol{B}$ are only electric currents, until the magnetic monopole is discovered.

In analogy with electrostatics, we may consider the similar thing for an arbitrary localized current distribution. Suppose that there is a steady current distribution $\boldsymbol{J}$ within a small volume V in space. Denoting a certain point in the volume by a position vector $\boldsymbol{x}'$, which can be usually taken as the center of gravity of the volume "matter", shown in Fig. 1.6, the vector magnetic potential $\boldsymbol{A}$ (with the Coulomb gauge $\nabla\cdot\boldsymbol{A} = 0$) at the field point $\boldsymbol{x}$ can be found for the steady current distribution by

$$\boldsymbol{A}(\boldsymbol{x}) = \int_V \boldsymbol{J}(\boldsymbol{x}' + \boldsymbol{d})G^m(\boldsymbol{x}, \boldsymbol{x}' + \boldsymbol{d})dV(\boldsymbol{d}) \tag{1.3.16}$$

In this equation, G^m denotes the magnetic Green's function in free space, defined by the following equation:

$$\nabla^2 G^m(\boldsymbol{x}, \boldsymbol{x}') + \mu_o \delta(\boldsymbol{x} - \boldsymbol{x}') = 0 \tag{1.3.17}$$

which has the fundamental solution in the infinite space given by

$$G^m(\boldsymbol{x}, \boldsymbol{x}') = \frac{\mu_o}{4\pi|\boldsymbol{x} - \boldsymbol{x}'|} \tag{1.3.18}$$

If the current I flows in a closed loop L whose line element is $\boldsymbol{dl}$, Eq.(1.3.16) can be written

$$\boldsymbol{A}(\boldsymbol{x}) = \frac{\mu_o}{4\pi}\oint_L \frac{I\boldsymbol{dl}''}{|\boldsymbol{x} - \boldsymbol{x}''|} \tag{1.3.19}$$

where $\boldsymbol{x}''$ is the position point taken on the current loop L.

By expanding the magnetic Green's function $G^m(\boldsymbol{x}, \boldsymbol{x}'+\boldsymbol{d})$ in a Taylor series about $(\boldsymbol{x}, \boldsymbol{x}')$ for Eq.(1.3.16), the magnetic vector potential $\boldsymbol{A}$ can then be expressed in the following form:

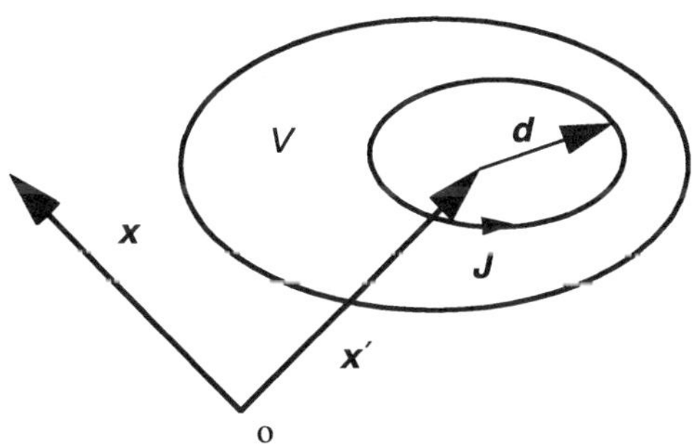

Figure 1.6 A volume distribution of currents in space.

$$A_i(\boldsymbol{x}) = \sum_{k=0}^{\infty} \frac{(-1)^k}{k!} p^m_{is_1\ldots s_k} G^m_{,s_1\ldots s_k}(\boldsymbol{x}-\boldsymbol{x}') \tag{1.3.20}$$

where we have formally introduced the point magnetic multipole moment of order k ($k = 0, 1, 2, \ldots$) located at $\boldsymbol{x}'$, defined by

$$p^m_{is_1\ldots s_k} = \int_V J_i d_{s_1} \ldots d_{s_k} dV(\boldsymbol{d}) \tag{1.3.21}$$

For $k = 0$, we have the point magnetic monopole term

$$\boldsymbol{p}^m = \int_V \boldsymbol{J} dV \tag{1.3.22}$$

which is zero since the localized current in the volume follows closed paths. This means that the leading term in the expansion of the magnetic potential (1.3.20) is always the magnetic dipole term under the constraint of nonexistence of magnetic charges (monopoles).

For $k = 1$, by noting the zero divergence of the electrical current density vector $\boldsymbol{J}$, the point magnetic dipole term may be expressed by

$$A_i(\boldsymbol{x}) = -p^m_{is} G^m_{,s}(\boldsymbol{x}-\boldsymbol{x}') = -\varepsilon_{iks} m_k G^m_{,s}(\boldsymbol{x}-\boldsymbol{x}') \tag{1.3.23}$$

or

$$\boldsymbol{A}(\boldsymbol{x}) = \boldsymbol{m} \times \nabla' G^m(\boldsymbol{x}-\boldsymbol{x}') \tag{1.3.24}$$

where ∇' denotes the gradient operator acting on $\boldsymbol{x}'$, and $\boldsymbol{m}$ is the point magnetic dipole moment for the localized current distribution by

$$\boldsymbol{m} = \frac{1}{2} \int_V \boldsymbol{d} \times \boldsymbol{J} dV(\boldsymbol{d}) \tag{1.3.25}$$

If the current is confined in an arbitrary current loop on a plane, and the current I flows in a closed circuit L whose element is $\boldsymbol{dl}$, Eq.(1.3.25) becomes

$$\boldsymbol{m} = \frac{1}{2} \oint_L \boldsymbol{d} \times \boldsymbol{dl} \tag{1.3.26}$$

the magnitude of which is $|\boldsymbol{m}| = I \times$ (area of the loop) regardless of the shape of the circuit. In analogy to electrostatics, it can also be shown that the magnetic dipole moment is a unique property of the current distribution and is independent of the choice of origin of coordinate systems since the magnetic monopole does not exist.

1.3.6 Interaction of Magnetic Multipoles with External Field

We now consider the magnetic interaction energy for a localized volume distribution of electric current which is subject to the influence of an external magnetic field B^o (– $\nabla \times A^o$). It can be found that the magnetic interaction energy reads

$$U^{mo} = \int_V J \cdot A^o(x + d)dV(d) \tag{1.3.27}$$

where x is a certain position point in the volume V.

If a steady current I flows in a closed loop L whose line element is dl and the magnetic vector potential A^o is produced by another current circuit L_o with its flowing steady current I_o, Eq.(1.3.27) becomes

$$U^{mo} = \frac{\mu_o}{4\pi} \oint_{L_o} \oint_{L} \frac{I_o I dl \cdot dl'}{|x - x'|} \tag{1.3.28}$$

which may also be identified as the magnetostatic energy of the system of the two current circuits.

In Eq.(1.3.27), if the volume V is supposed to be small and the magnetic vector potential A^o does not vary much over the current distribution, we can make a Taylor expansion of $A^o(x+d)$ around the point x and write

$$U^{mo} = \sum_{k=1}^{\infty} \frac{1}{k!} p^m_{is_1 \ldots s_k} A^o_{i,s_1 \ldots s_k}(x) \tag{1.3.29}$$

in which the leading term of this expansion is the magnetic interaction energy of the point magnetic dipole moment m in the external magnetic field B^o, which is

$$U^{mo(1)} = m \cdot (\nabla \times A^o) = m \cdot B^o \tag{1.3.30}$$

The total translational force acting on the electric current volume may be found from Eq.(1.3.29)

$$F^m_n = \sum_{k=1}^{\infty} \frac{1}{k!} p^m_{is_1 \ldots s_k} A^o_{i,ns_1 \ldots s_k}(x) \tag{1.3.31}$$

which, in the dipole approximation, reads

$$F^m = \nabla(m \cdot B^o) = (m \cdot \nabla) B^o \tag{1.3.32}$$

since m is a constant vector and the curl of the external magnetic field B^o at x is zero. It can be seen that the magnetic force on a magnetic dipole vanishes for a uniform external magnetic field.

The torque on the magnetic dipole moment $\boldsymbol{m}$ in the external magnetic field $\boldsymbol{B}^{o}$ is

$$\tau^{m} = \boldsymbol{m} \times \boldsymbol{B}^{o} \tag{1.3.33}$$

which exists even in a uniform external magnetic field.

We may notice that Eqs.(1.3.32) and (1.3.33) are completely analogous to the corresponding Eqs.(1.2.29) and (1.2.30) in electrostatics. However, the analogy breaks down in the energy Eqs.(1.3.30) and (1.2.27), where a sign difference appears. This is due to the fact that, to maintain the constant current distribution in applying the external magnetic field, certain amount of energy has to be supplied to the current system [Jackson (1975)]. It is common to define a quantity, called the potential energy of a magnetic dipole in an external magnetic field, by

$$U'^{m} = -\boldsymbol{m} \cdot \boldsymbol{B}^{o} \tag{1.3.34}$$

In such a case, we may write the expression for the force in complete analogy to that used in mechanics, namely, as the negative gradient of a potential energy. In other words, we may write

$$\boldsymbol{F}^{m} = -\nabla U'^{m} \tag{1.3.35}$$

which is completely in agreement with Eq.(1.3.32).

1.3.7 Faraday´s Law of Induction

The *Faraday law* of induction comes from the experimental results on the generation of electromotive forces (*emf*) by varying magnetic fields. It was observed by Faraday that a transient current is induced in a circuit if (a) the steady current flowing in an adjacent circuit is turned on or off, (b) the adjacent circuit with a steady current flowing is moved relative to the first circuit, (c) a permanent magnet is thrust into or out of the circuit. No current flows unless either the adjacent current changes or there is relative motion. Faraday interpreted the transient current flow as being due to a changing magnetic flux linked by the circuit. The changing flux induces an electric field around the circuit, the line integral of which is called the electromotive force. According to Faraday´s law, the electromotive force induced in a closed circuit at rest relative to the observer is equal to the negative rate of increase of the magnetic flux linking the circuit. Mathematically, the integral form of the Faraday law of induction can be expressed by

$$\varepsilon^{ind} = -\frac{\partial}{\partial t}\int_S \boldsymbol{B}\cdot d\boldsymbol{S} = \oint_L \boldsymbol{E}\cdot d\boldsymbol{l} \tag{1.3.36}$$

where ε^{ind} is the induced electromotive force. Since ε^{ind} is measured in volts, it is also called the induced voltage. If the closed path L is a metallic loop, an electric current is produced in the loop by an induced electromotive force. The negative sign appearing in Eq.(1.3.36) is due to the fact that the direction of the induced electromotive force is determined by the *Lenz law*, which states that induced current is in such a direction as to oppose the magnetic flux variation causing it. Thus, if the flux through a circuit in a certain direction is increasing, the induced current sets up a flux in the opposite direction, and if the flux is decreasing, this current sets up a flux in the same direction. The induced currents always seek to maintain the initial status of the magnetic field.

Faraday's law of induction expressed mathematically by Eq.(1.3.36) is a well-known experimental law, which expresses a new physical phenomenon: a varying electric field is created not only by electric charges but also by a varying magnetic field, and is also supposed to be valid even in the presence of material media. By applying Stokes' theorem to the line integral on the right-hand side of Eq.(1.3.36), we may derive the local differential form of Faraday's law

$$\nabla\times\boldsymbol{E} = -\frac{\partial\boldsymbol{B}}{\partial t} \tag{1.3.37}$$

which applies to every point in space, whether it is in free space or in a material medium. Equation (1.3.37) shows that the electric field in a region of time-varying magnetic induction field is non-conservative and cannot be expressed simply as the gradient of a scalar potential. However, for static fields, the term $\partial\boldsymbol{B}/\partial t = 0$ in Eq.(1.3.37) and, thus, Faraday's law states that the line integral of a static electric field $\boldsymbol{E}$ about any closed path is always zero. This means that all static electric fields are conservative.

1.3.8 Maxwell´s Equations in Free Space

According to Faraday's and Ampere's laws, electric and magnetic effects are not isolated phenomena; the variation in one must produce a change in the other. Thus we will speak of electromagnetic field, with the electric field and the magnetic field representing different aspects of the same phenomenon. The complete mathematical description of the electromagnetic phenomenon in free space was first proposed by Maxwell based on the experimental laws for isolated electric charges and electric current circuits in free space, discussed above. It is now known as Maxwell's equations in free space, which can be summarized as follows:

$$\nabla \times \boldsymbol{B} = \mu_o \boldsymbol{J} + \frac{1}{c^2}\frac{\partial \boldsymbol{E}}{\partial t} \tag{1.3.38}$$

$$\nabla \cdot \boldsymbol{B} = 0 \tag{1.3.39}$$

$$\nabla \times \boldsymbol{E} = -\frac{\partial \boldsymbol{B}}{\partial t} \tag{1.3.40}$$

$$\nabla \cdot (\varepsilon_o \boldsymbol{E}) = \rho_e \tag{1.3.41}$$

together with

$$\nabla \cdot \boldsymbol{J} + \frac{\partial \rho_e}{\partial t} = 0 \tag{1.3.42}$$

It is seen that, in free space, all electromagnetic phenomena with electric charges and electric currents as sources may be described by only two fields, the electric field $\boldsymbol{E}$ and the magnetic induction field $\boldsymbol{B}$. Here, $c = 1/(\varepsilon_o \mu_o)^{1/2}$ is the speed of light in free space (vacuum).

An important implication of Maxwell's equations is the prediction of electromagnetic waves that can carry energy and information across space, even the vacuum. Consider a source-free region, where ρ_e and $\boldsymbol{J}$ are zero. If we take the curl of Eq.(1.3.40), we may get

$$\nabla \times (\nabla \times \boldsymbol{E}) = \nabla(\nabla \cdot \boldsymbol{E}) - \nabla^2 \boldsymbol{E} = -\nabla \times \left(\frac{\partial \boldsymbol{B}}{\partial t}\right) \tag{1.3.43}$$

By noting that $\nabla \cdot \boldsymbol{E} = 0$ from Eq.(1.3.41) and that time and space partial derivatives may be taken in any order, we may find by using Eq.(1.3.38)

$$\nabla^2 \boldsymbol{E} = \frac{1}{c^2}\frac{\partial^2 \boldsymbol{E}}{\partial t^2} \tag{1.3.44}$$

which has the general form of a wave equation. This form applies as well to the magnetic field:

$$\nabla^2 \boldsymbol{B} = \frac{1}{c^2}\frac{\partial^2 \boldsymbol{B}}{\partial t^2} \tag{1.3.45}$$

We may consider a uniform plane wave solution for $\boldsymbol{E} = (E_x, 0, 0)$ with the positive traveling wave component of E_x, propagating along z-axis direction, by

$$E_x = F\left(t - \frac{z}{c}\right) \tag{1.3.46}$$

From Eq.(1.3.40), we may find a solution for the magnetic field $\boldsymbol{B} = (0, B_y, 0)$:

$$\frac{\partial B_y}{\partial t} = -\frac{\partial E_x}{\partial z} = \frac{1}{c}F'\left(t - \frac{z}{c}\right) \tag{1.3.47}$$

Integrating, and ignoring static constant of integration, we may get

$$B_y = \frac{1}{c}F\left(t - \frac{z}{c}\right) = \frac{1}{c}E_x \tag{1.3.48}$$

This equation may also be written

$$B_y = \frac{\mu_o}{\eta_o}E_x \tag{1.3.49}$$

by introducing a constant η_o, called the intrinsic impedance of the free space:

$$\eta_o = \sqrt{\frac{\mu_o}{\varepsilon_o}} \approx 377 \qquad \text{(ohms)} \tag{1.3.50}$$

We may notice that for such a uniform plane electromagnetic wave, the electric field $\boldsymbol{E}$ is perpendicular to the magnetic field $\boldsymbol{B}$, and the direction of the wave propagation is along the direction of the vector $\boldsymbol{E}\times\boldsymbol{B}$. Thus, no electric or magnetic field in the direction of propagation. Furthermore, we may show that the stored energy in the electric field per unit volume is

$$U_e = \frac{\varepsilon_o}{2}E_x^2 \tag{1.3.51}$$

and that in the magnetic field

$$U_m = \frac{1}{2\mu_o}B_y^2 \tag{1.3.52}$$

By Eq.(1.3.48), we find that U_e and U_m are equal, which means that the energy density at each point at each instant is equally divided between electric and magnetic energy for the uniform plane electromagnetic wave.

1.4 MAXWELL'S EQUATIONS FOR MATERIALS AT REST

Electromagnetic phenomena in the presence of material bodies are more complicated to describe than those in free space since materials in nature are composed of atoms or molecules, each made up of positively and negatively charged particles having various configurations in free space, varying states of motion, and their statistical character. The solution of Maxwell's equations for such complicated systems is almost impossible. This is not only because the number of individual sources is prohibitively large. One important aspect is that for macroscopic observations the detailed behavior of the fields, with their drastic variations in space over atomic distances, is not relevant. What is relevant is the average of a field or a source over a volume large compared to the volume occupied by a single atom or molecule. Such averaged quantities are called the macroscopic fields and macroscopic sources. The earliest work dealing with the electromagnetic phenomena in matter were given by Lorentz (1916), who used a volume average procedure to derive a set of macroscopic electromagnetic equations for the matter at rest. It was found that only two field variables were not sufficient to describe all electromagnetic phenomena in matter. Two more field variables, the polarization $\boldsymbol{P}$ and the magnetization $\boldsymbol{M}$, were then proposed to describe the behavior of materials. Subsequently, there have been many attempts to provide derivations of increasing rigor, such as the works of Van Vleck (1932), Rosenfeld (1951), Mazur and Nijboer (1953), De Groot and Vlieger (1965), Russakoff (1970), and Robinson (1973).

In this section, we shall give the derivation of macroscopic Maxwell's equations for materials at rest with the use of the concept of electric and magnetic multipoles. The material body considered here is assumed to be rigid. The effects of electromagnetic–mechanical interaction in deformable solids will, however, be treated in the next chapter.

1.4.1 Dipole Model of Electromagnetic Solids

We begin with the introduction of the macroscopic Maxwell's equations for materials at rest with the use of a dipole model by assuming that, on average, the dominant features of the material that are of interest to us are simply those associated with the electric and magnetic dipole moments.

We consider a material made up of a large number of atoms or molecules with bounded charges and free charges. In most materials, the electric monopole term (the net charge) from the bounded charges of the atom or molecule is zero in the absence of applied fields since the materials have equal amounts of positive and negative charges. In some simple materials, the multipole moments of each atom or molecule are also all zero in the absence of applied fields. However, permanent multipole moments may exist in some materials, even in the absence of applied fields. In the presence of an applied electric field, the materials are, in general, electrically polarized. Such an electric polarization in materials is mainly due microscopically to electronic polarization, ionic (or atomic) polarization and orientational polarization, which can be explained as follows.

The electronic polarization is due to the small displacement of negative electron cloud relative to the positive charge (the nucleus) of the atom in an external electric field since the field tends to shift the positive and negative charges in opposite directions. Such a polarization process does not change the true mass center of the atom because the nucleus is much heavier than electrons and, therefore, the predominant displacement is that of the electronic charges. Thus highest frequency phenomena, such as optical phenomena, are almost exclusively due to the electronic polarization since this process involves only the motion of light electrons.

The ionic polarization is caused in ionic crystals (such as NaCl) in which an external electric field displaces the positive (Na^+) ions relative to the negative (Cl^-) ions. This polarization process in general results in a certain deformation of the crystal lattice, and is essentially a low-frequency phenomenon because it involves the motion of ions with masses generally more than 10^4 greater than the mass of the electron. In addition, the ionic polarization may be produced by external forces which deform the crystal sublattices of positive ions and of negative ions differently so that the positive ions are also displaced relatively to the negative ions, and present, therefore, the piezoelectric effect of the ionic crystals.

The orientational polarization is due to the reorientation of electric dipole (permanent or induced) moments caused by the torques acting on the dipole moments in the presence of an external electric field, which make the dipole moments tend to orient themselves along the direction of the applied electric field since the stable state of the system is at its minimum energy. The orientational polarization is counteracted by the thermal movement of the atoms or molecules. Therefore, it is strongly temperature-dependent.

To describe macroscopically the effect of electrical polarization in the material, we shall assume here that the macroscopic electric behavior of a material body with a volume V bounded by a closed surface S can be modeled by a distribution of the macroscopic volume free charge density ρ_e, the macroscopic surface free charge density α_f and the macroscopic electric polarization (dipole) density $\boldsymbol{P}$ (measured in Coulomb/meter^2) in the material body. In this model, we do not consider the surface density of polarization since it plays no very significant part in the mathematical model of electromagnetism and it is difficult to devise experiments to involve it in an essential way [King and Prasad (1986)].

From a macroscopic point of view, the macroscopic electric potential field in the static case can be expressed, by noting Eq.(1.2.18), in the dipole model

$$\phi(\boldsymbol{x}) = \int_V \rho_e(\boldsymbol{x}')G^e(\boldsymbol{x}-\boldsymbol{x}')dV' + \int_S \alpha_f(\boldsymbol{x}')G^e(\boldsymbol{x}-\boldsymbol{x}')dS' - \int_V P_j(\boldsymbol{x}')G^e{}_{,j}(\boldsymbol{x}-\boldsymbol{x}')dV' \tag{1.4.1}$$

where the position vector $\boldsymbol{x}$ may be either inside the material body or outside the material body. It is noticed that, for the field point $\boldsymbol{x}$ being outside the material body, the electric field is well defined (by $\boldsymbol{E} = -\nabla\phi$) and can be checked by measuring the force acting on a test charge there. The electric field inside the material body is, however, not subject to direct measurement. The justification of the macroscopic model can thus be only based on its agreements with experiments.

1.4.2 Gauss´ Theorem in Material Medium

To examine how the Gauss theorem for electric field [Eq.(1.2.11)] is to be modified in the presence of a material medium, let us consider a larger closed surface S_o enclosing the material body V, as shown in Fig. 1.7, and make a surface integral of the electric field on S_o. We have then

$$\int_{S_o} \boldsymbol{E}\cdot\boldsymbol{n}dS = -\int_{S_o}\int_{V} \rho_e(\boldsymbol{x}')G^e{}_{,i}(\boldsymbol{x}-\boldsymbol{x}')n_i(\boldsymbol{x})dV'dS$$
$$-\int_{S_o}\int_{S} \alpha_f(\boldsymbol{x}')G^e{}_{,i}(\boldsymbol{x}-\boldsymbol{x}')n_i(\boldsymbol{x})dS'dS + \int_{S_o}\int_{V} P_j(\boldsymbol{x}')G^e{}_{,ij}(\boldsymbol{x}-\boldsymbol{x}')n_i(\boldsymbol{x})dV'dS \tag{1.4.2}$$

Equation (1.4.2) can further be written in the form

$$\int_{S_o} \boldsymbol{E}\cdot\boldsymbol{n}dS = -\int_{V_o}\int_{V} \rho_e(\boldsymbol{x}')\nabla^2 G^e(\boldsymbol{x}-\boldsymbol{x}')dV'dV$$
$$-\int_{V_o}\int_{S} \alpha_f(\boldsymbol{x}')\nabla^2 G^e(\boldsymbol{x}-\boldsymbol{x}')dS'dV$$
$$-\int_{V_o}\int_{S} \boldsymbol{P}^-(\boldsymbol{x}')\cdot\boldsymbol{n}(\boldsymbol{x}')\nabla^2 G^e(\boldsymbol{x}-\boldsymbol{x}')dS'dV + \int_{V_o}\int_{V} \nabla'\cdot\boldsymbol{P}(\boldsymbol{x}')\nabla^2 G^e(\boldsymbol{x}-\boldsymbol{x}')dV'dV \tag{1.4.3}$$

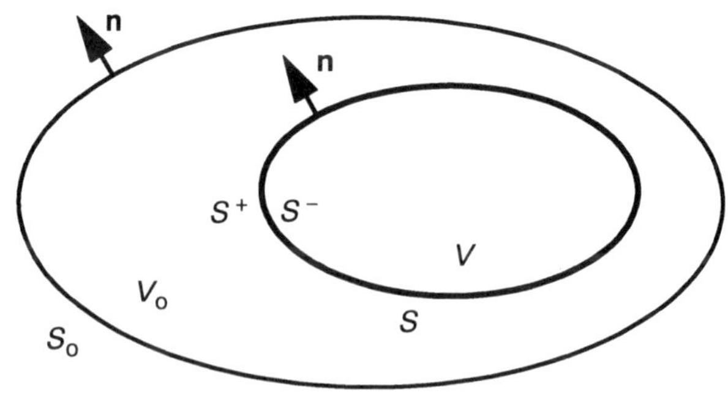

Figure 1.7 A material body with volume V is enclosed by a surface S_o.

by applying Gauss′ divergence theorem for the surface integrals on S_0 on the right-hand side of Eq.(1.4.2). Here the superscript "–" of $\boldsymbol{P}^-$ means its value being taken on the material side (S^-)
of the surface S. By noting Eq.(1.2.16), we can then arrive at

$$\int_{S_o} \varepsilon_o \boldsymbol{E} \cdot \boldsymbol{n} dS = \int_V \rho_e dV + \int_S \alpha_f \, dS + \int_S \boldsymbol{P}^- \cdot \boldsymbol{n} dS - \int_V \nabla \cdot \boldsymbol{P} dV \qquad (1.4.4)$$

We now write the surface integral on the left-hand side of Eq.(1.4.4) in the following form:

$$\int_{S_o} \varepsilon_o \boldsymbol{E} \cdot \boldsymbol{n} dS = \int_{V_o - V - S} \nabla \cdot (\varepsilon_o \boldsymbol{E}) dV + \int_{V-S} \nabla \cdot (\varepsilon_o \boldsymbol{E}) dV + \int_S \varepsilon_o (\boldsymbol{E}^+ - \boldsymbol{E}^-) \cdot \boldsymbol{n} dS \qquad (1.4.5)$$

with the use of Gauss′ divergence theorem for the regions V_0–V–S and V–S. By identifying Eq.(1.4.4) with Eq.(1.4.5), we can then find that, in the presence of the material medium, the local differential form of Gauss′ theorem (1.2.12) for electric field is modified to be

$$\nabla \cdot \boldsymbol{D} = \rho_e \quad \text{in } V \qquad (1.4.6)$$

with the introduction of an electric displacement vector $\boldsymbol{D}$, defined by

$$\boldsymbol{D} = \varepsilon_o \boldsymbol{E} + \boldsymbol{P} \quad \text{in } V \qquad (1.4.7)$$

where $\boldsymbol{P}$ is the electric polarization density vector, which vanishes in free space. The boundary condition at the interface between the material body and free space (or different media) may also be found

$$\boldsymbol{n} \cdot [\boldsymbol{D}] = \boldsymbol{n} \cdot (\boldsymbol{D}^+ - \boldsymbol{D}^-) = \alpha_f \qquad (1.4.8)$$

where α_f is the macroscopic surface free charge density.

1.4.3 Ampere´s Circuital Law in Material Medium

We shall now consider the problem where Ampere´s circuital law in free space is modified by the presence of a material body. We study the case when the material under the exertion of applied fields is in steady state. In the material, bounded charges are moving in closed paths, localized around the centers of atoms or molecules, and the free charges are moving continuously, forming the macroscopic current flow in the material. The localized microscopic current loops due to moving bounded charges result in the presence of magnetic multipole moments in the material as we have seen in Section 1.3.5, and thus lead to the magnetization of the material. In general, the magnetization of the material is due microscopically to the orbital current magnetization, the spin magnetization, and the orientational magnetization.

The orbital current magnetization appears in all atoms placed in an external magnetic field due to the effect of the interaction of the field on the electrons in orbits in the atom, explained qualitatively by Lenz's law of electricity. According to this law, atom's electronic motion, considered as a current loop, will be changed in such a sense that a magnetic moment will be induced in a direction opposite to that of the applied field. Such an induced magnetic moment is therefore called the diamagnetic moment. Obviously diamagnetism is a property of all matter.

The spin magnetization is due to the intrinsic angular momentums of electron and nucleus, that is, the electron spin and the nucleus spin. The nucleus spin is very small compared with electron spin and is normally neglected in the consideration of usual macroscopic magnetic properties of bulk materials.

The orientational magnetization comes from the preferred direction of orientation of magnetic moments due to torques acting on them caused by external magnetic fields or by anisotropic fields inherent to crystalline materials due to the exchange interactions among atomic moments.

To describe macroscopically the effect of magnetization of the material, we shall assume that the macroscopic magnetic behavior of the material body with a volume V bounded by a closed surface S may be modeled by a distribution of the macroscopic volume current density $\boldsymbol{J}$, the macroscopic surface current density $\boldsymbol{K}_f$, and the macroscopic magnetization (magnetic dipole density) $\boldsymbol{M}$ (measured in ampere/meter) in the material body.

The magnetic vector potential in the presence of the material body can thus be found, by noting Eqs.(1.3.20) and (1.3.24), in the dipole model by

$$\boldsymbol{A}(\boldsymbol{x}) = \int_V \boldsymbol{J}(\boldsymbol{x}')G^m(\boldsymbol{x}-\boldsymbol{x}')dV' + \int_S \boldsymbol{K}_f(\boldsymbol{x}')G^m(\boldsymbol{x}-\boldsymbol{x}')dS' + \int_V \boldsymbol{M}(\boldsymbol{x}') \times \nabla' G^m(\boldsymbol{x}-\boldsymbol{x}')dV' \tag{1.4.9}$$

This expression for the vector magnetic potential $\boldsymbol{A}$ is valid for the field point $\boldsymbol{x}$ outside the material volume V as well as inside. Similar to an electric field, it is noticed that for the field point $\boldsymbol{x}$ being outside the material body, the magnetic induction field $\boldsymbol{B}$ is well defined (by $\boldsymbol{B} = \nabla\times\boldsymbol{A}$) and can be checked by measuring the force acting on a test current element (loop) there. The magnetic field inside the material body is, however, not subject direct measurement. The justification of the macroscopic model could thus be only based on its agreement with experiment.

Consider now an arbitrary closed line loop L, which may be partly (L_0) outside and partly (L_1) inside the material volume V as shown in Fig. 1.8. We make a line integral of the magnetic induction field $\boldsymbol{B}$ from Eq.(1.4.9) on L (= $L_o + L_1$) and, after some manipulations, we can find

$$\oint_L \boldsymbol{B}\cdot d\boldsymbol{l} = \int_{S_{L1}} \mu_o(\boldsymbol{J}+\nabla\times\boldsymbol{M})\cdot\boldsymbol{n}_s dS + \int_{L_s} \mu_o(\boldsymbol{K}_f+\boldsymbol{M}^- \times\boldsymbol{n})\cdot\boldsymbol{n}_s dl_s \quad (1.4.10)$$

with the aid of Eq.(1.3.17). Here, L_s denotes the line from the intersect $S_L \cap S$ with S_L being the open surface on the closed line loop L. S_{L1} is the part of the open surface S_L and is on the closed line loop L_1+L_s. $\boldsymbol{n}_s$ is the unit normal vector on the open surface S_L, and $\boldsymbol{n}$ is the unit normal vector of the closed material surface S. We note that the unit direction vector $\boldsymbol{t}_s$ of the line element dl_s has its relation $\boldsymbol{t}_s = \boldsymbol{n}\times\boldsymbol{n}_s$ on the smoothly continuous line L_s (see Fig. 1.8). We write the line integral on the left-hand side of Eq.(1.4.10) in the following form:

$$\oint_L \boldsymbol{B}\cdot d\boldsymbol{l} = \int_{S_{Lo}} (\nabla\times\boldsymbol{B})\cdot\boldsymbol{n}_s dS + \int_{S_{L1}} (\nabla\times\boldsymbol{B})\cdot\boldsymbol{n}_s dS + \int_{L_s} (\boldsymbol{n}\times(\boldsymbol{B}^+ - \boldsymbol{B}^-))\cdot\boldsymbol{n}_s dl_s$$
(1.4.11)

with the use of Stokes′ theorem for the closed lines $L_0+L_s^+$ and $L_1+L_s^-$.

By identifying Eq.(1.4.10) with Eq.(1.4.11), we find that, in the presence of the material medium, Ampere′s circuital law becomes

$$\nabla\times\boldsymbol{H} = \boldsymbol{J} \quad \text{in } V \quad (1.4.12)$$

with the introduction of a magnetic intensity field $\boldsymbol{H}$ defined by

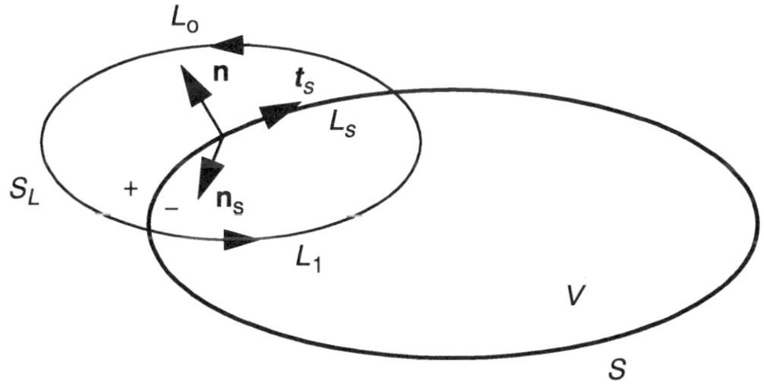

Figure 1.8 Configuration of a closed line loop and a material body.

$$H = \frac{1}{\mu_o}B - M \tag{1.4.13}$$

where M is zero in free space.

The boundary condition at the interface between the material body and the free space is found to be

$$n \times [H] = n \times (H^+ - H^-) = K_f \tag{1.4.14}$$

which may also be used as the interface condition at an interface between two different media.

In general, for dynamic cases, the local differential form of Ampere's circuital law (1.3.2) is found to be modified as

$$\nabla \times H = J + \frac{\partial D}{\partial t} \tag{1.4.15}$$

compatible with the equation of conservation law of charges (1.1.8) and Gauss' theorem for electric field (1.4.6). In Eq.(1.4.15), J can be either the conduction current density (motion of charges in conductors) or the convection current density (motion of charges in free space). The second term on the right-hand side of Eq.(1.4.15) is the displacement current density by Maxwell, as discussed in Section 1.3.1. The displacement current may appear not only in a dielectric in which there is an actual time-varying displacement of bound charges, but also in free space (vacuum). The displacement current term is now known to be essential for the existence of electromagnetic waves propagating either in material media, or in free space.

1.4.4 Maxwell's Equations for Materials at Rest

To summarize, we have obtained the following set of macroscopic Maxwell's equations for materials at rest:

$$\nabla \times H = J + \frac{\partial D}{\partial t} \tag{1.4.16}$$

$$\nabla \cdot B = 0 \tag{1.4.17}$$

$$\nabla \times E = -\frac{\partial B}{\partial t} \tag{1.4.18}$$

$$\nabla \cdot D = \rho_e \tag{1.4.19}$$

together with the equation of conservation of charges

$$\nabla \cdot \boldsymbol{J} + \frac{\partial \rho_e}{\partial t} = 0 \tag{1.4.20}$$

The interface conditions at an interface between two different media are

$$\boldsymbol{n} \cdot [\boldsymbol{D}] = \alpha_f \tag{1.4.21}$$

$$\boldsymbol{n} \times [\boldsymbol{E}] = 0 \tag{1.4.22}$$

$$\boldsymbol{n} \cdot [\boldsymbol{B}] = 0 \tag{1.4.23}$$

$$\boldsymbol{n} \times [\boldsymbol{H}] = \boldsymbol{K}_f \tag{1.4.24}$$

$$\boldsymbol{n} \cdot [\boldsymbol{J}] = -\frac{\partial \alpha_f}{\partial t} \tag{1.4.25}$$

where $[\ F\ ] = F^{+} - F^{-}$ denotes the jump of the quantity F across the interface S with $\boldsymbol{n}$ being the unit normal vector always drawn from S^{-} to S^{+}. Here, α_f is the surface density of free charge, and $\boldsymbol{K}_f$ is the surface density of free current. In particular, at an interface between two perfect dielectrics, the surface density of charge α_f vanishes and, consequently, the normal component of $\boldsymbol{D}$ is continuous across the interface. At an interface between two conductors of finite conductivity, the surface density of free current $\boldsymbol{K}_f$ cannot exist. Thus, the tangential component of $\boldsymbol{H}$ is continuous at an interface separating two materials with finite conductivity.

It may be worth notice that, in the absence of surface charges and surface currents, the interface conditions on normal components of $\boldsymbol{D}$ and $\boldsymbol{B}$ are not independent of those for tangential components of $\boldsymbol{E}$ and $\boldsymbol{H}$ for time-varying cases. The reason is that the former can be derived from the two curl equations (1.4.16) and (1.4.18) in the time-varying cases. However, the interface conditions on normal components may often be used as a check, especially when numerical solutions are concerned.

It can be of interest to look at the behavior of Maxwell´s equations under reflections. We consider the space reflection or parity P: $\boldsymbol{x} \rightarrow -\boldsymbol{x}$ and the time inversion T: $t \rightarrow -t$. From the Lorentz force law (1.2.32), we find that $\boldsymbol{E}$ and $\boldsymbol{B}$ transform differently under these reflections. Consider first the time reversal T, we know from mechanics that velocity is odd $\mathbf{v} \rightarrow -\mathbf{v}$, thus acceleration is even and so is the force $\boldsymbol{F}$. Then, Eq.(1.2.32) implies that $\boldsymbol{E}$ is even and $\boldsymbol{B}$ is odd:

$$\boldsymbol{E}(t, \boldsymbol{x}) \rightarrow \boldsymbol{E}(-t, \boldsymbol{x}), \qquad \boldsymbol{B}(t, \boldsymbol{x}) \rightarrow -\boldsymbol{B}(-t, \boldsymbol{x})$$

For parity, the behavior is the other way around; **v** is still odd, but acceleration is odd and so is the force $\boldsymbol{F}$. Thus, $\boldsymbol{E}$ is odd and $\boldsymbol{B}$ is even:

$$\boldsymbol{E}(t, \boldsymbol{x}) \to -\boldsymbol{E}(t, -\boldsymbol{x}), \qquad \boldsymbol{B}(t, \boldsymbol{x}) \to \boldsymbol{B}(t, -\boldsymbol{x})$$

A vector like $\boldsymbol{B}$ that does not change sign under parity is called a pseudovector or axial vector. Other vectors, $\boldsymbol{E}$, **v**, and $\boldsymbol{F}$ are true or polar vectors. Since temporal and spatial derivatives change sign under T and P, respectively, it is now easy to verify that Maxwell's equations (1.4.16)–(1.4.20) are invariant under time inversion T and space reflection P, provided that $\rho(t, \boldsymbol{x})$ is a scalar and $\boldsymbol{J}(t, \boldsymbol{x})$ is a true vector like **v** (which is consistent with $\boldsymbol{J} = \rho\mathbf{v}$). This is in accordance with their definitions.

So far, we have said nothing about the macroscopic polarization $\boldsymbol{P}$ and the macroscopic magnetization $\boldsymbol{M}$ except for their definitions respectively as the electric dipole moment per unit volume and the magnetic dipole moment per unit volume. Since microscopic electric and magnetic dipoles in materials are influenced by applied fields, we can expect that there must exist some macroscopic relations among the macroscopic polarization $\boldsymbol{P}$, the macroscopic magnetization $\boldsymbol{M}$ and the applied fields. In the case of neglecting the effects of deformation as well as temperature of the material, $\boldsymbol{P}$ and $\boldsymbol{M}$ are simply functions of the field quantities $\boldsymbol{E}$ and $\boldsymbol{H}$ (or $\boldsymbol{B}$). Such functional relations are called the constitutive relations, which are the heart of the matter. If such relations are known, the macroscopic Maxwell's equations (1.4.16)–(1.4.20) become a closed system and macroscopic electrodynamics can be worked out. In this section, we shall simply present some well-known linear constitutive relations used in classical electromagnetics without going into the details of constitutive theories which will be discussed in the next chapter.

For linear isotropic rigid materials at constant temperature, $\boldsymbol{P}$ and $\boldsymbol{M}$ are found experimentally to be directly proportional to $\boldsymbol{E}$ and $\boldsymbol{H}$ respectively, that is,

$$\boldsymbol{P} = \varepsilon_o \chi_e \boldsymbol{E} \qquad (\text{or } \boldsymbol{D} = \varepsilon \boldsymbol{E}) \tag{1.4.26}$$

and

$$\boldsymbol{M} = \chi_m \boldsymbol{H} \qquad (\text{or } \boldsymbol{B} = \mu \boldsymbol{H}) \tag{1.4.27}$$

where χ_e and χ_m are called the electric susceptibility and the magnetic susceptibility, respectively. $\varepsilon = \varepsilon_0(1+\chi_e)$ and $\mu = \mu_0(1+\chi_m)$ are called the dielectric permittivity and the magnetic permeability of the material, respectively. We may also introduce the relative permittivity ε_r, defined by $\varepsilon_r = \varepsilon/\varepsilon_0$ (or $\varepsilon_r = 1+\chi_e$), and the relative permeability μ_r, defined by $\mu_r = \mu/\mu_0$ (or $\mu_r = 1+\chi_m$) to describe the electromagnetic properties of some simple materials.

When conduction currents are present in the material medium, we have, in addition, Ohm's law (1827), expressed by

$$\boldsymbol{J} = \sigma \boldsymbol{E} \tag{1.4.28}$$

where $\boldsymbol{J}$ is the macroscopic current density, and σ is the electric conductivity of the material. According to the values of the conductivity σ, a simple classification of materials may be made. The dielectrics are the materials with low electric conductivity less than 10^{-5} S/m, semiconductors are materials having electric conductivity between 10^{-5} and 10^{3} S/m, and conductors are characterized by an electric conductivity higher than 10^{3} S/m. Mainly, conductors are metals. The best conductors among them are copper and silver which have electric conductivity of the order of 10^{7} S/m at room temperature. Shown in Table 1.1 are some data for several solid and fluid media.

Table 1.1 Conductivity and Dielectric Properties of Some Media

Material	Conductivity σ (S/m)	Relative Permittivity ε_r	$\tan\delta$ *
Air (atm pressure)	0	1.0006	0
Alumina (Al_2O_3)		9.5	10^{-4}
Aluminum	3.5×10^{7}	1	0
Copper	5.8×10^{7}	1	0
Glass (Corning 707)	10^{-13}	4	10^{-4}
Gold	4.3×10^{7}	1	0
$LaAlO_3$		24	3×10^{-5}
MgO		9.6	4×10^{-5}
$NdGaO_3$		24	3×10^{-4}
Oil (mineral)	10^{-14}	2.2	10^{-4}
Porcelain		5	0.004
Quartz	5	5	0.001
Silicon (lightly doped)	10	12	
Silicon dioxide (SiO_2)	10^{-13}	3.9	
Silver	6.3×10^{7}	1	
Soil (clay)	5×10^{-3}	14	
Sr_2AlTaO_6		12	4×10^{-5}
Teflon	10^{-15}	2.1	0.005
Water (fresh)	$10^{-2} \sim 10^{-3}$	80	
Water (sea)	4	80	

* $\tan\delta$ denotes the tangent of the dielectric loss angle of the medium.

Different from the electric properties of solid crystals, some crystals exhibit strong magnetic properties. Their difference may be understood from a difference in the behavior of microscopic charges and currents with respect to the time inversion: $t \to -t$, as discussed by Landau et al. (1984). Some simple classification of magnetic solids may be made according to the values of the magnetic susceptibility, χ_m. The diamagnetic material has a small negative temperature-independent susceptibility χ_m with a magnitude of order of 10^{-5}, and since it is negative, the induced magnetic moment in the diamagnetic material is directed oppositely to the applied magnetic field. This kind of magnetism is a direct consequence of Lenz's law applied to the motions of elementary charges of the system. All materials have diamagnetic contributions to their susceptibilities.

The paramagnetic material made of atoms or molecules having permanent magnetic moments has a positive temperature-dependent susceptibility χ_m with a magnitude of the order of 10^{-4} for aluminum and copper, and the order of 10^{-3} for $FeSO_4$, $NiSO_4$, and $CrCl_3$ at room temperature and varying with temperature approximately as $1/T$. This kind of magnetic behavior is explained as a consequence of two opposing effects: one, the tendency of the applied field to orient the magnetic moments in the direction of the field, and the other, the tendency of thermal agitation to preserve a random orientation of the magnetic moments. In most practical calculations for engineering applications, we often neglect such weak induced magnetization in diamagnetic and paramagnetic materials.

The soft ferromagnetic material has a positive temperature-dependent susceptibility χ_m with a magnitude of the order of 10^4 or higher for iron, nickel, cobalt, and their alloys. In general, ferromagnetic materials may present permanent macroscopic magnetization, and have nonlinear constitutive relations and hysteresis behaviors, which will be discussed in Section 1.8.

1.4.5 Electromagnetic Potentials and Gauge Transformation

Maxwell's equations are shown to be vector differential equations, each of which represents three scalar equations for each of three orthogonal components in general. It would be more convenient, therefore, if the vector problem were reduced to a scalar problem with a fewer number of equations. This has been done in electrostatics and magnetostatics by introducing electrostatic potential and vector magnetic potential to describe electric and magnetic fields, respectively. The concept of these potentials can also be extended to electrodynamic fields in the following manner.

For simplicity, let us assume that the medium is isotropic, homogeneous, and nondispersive. Thus, μ and ε are scalar and constant. Introducing the vector magnetic potential $\boldsymbol{A}$ by $\boldsymbol{B} = \nabla\times\boldsymbol{A}$, we find that the Maxwell equation (1.4.17) is satisfied, and we can write Eq.(1.4.18) as

$$\nabla \times \left(\boldsymbol{E} + \frac{\partial \boldsymbol{A}}{\partial t} \right) = 0 \tag{1.4.29}$$

Since the curl of the gradient of any scalar function is zero, the bracketed factor can be represented by the gradient of an arbitrary scalar function ϕ, which is called the scalar potential. Thus, we may write

$$\boldsymbol{E} = -\nabla\phi - \frac{\partial \boldsymbol{A}}{\partial t} \tag{1.4.30}$$

The Maxwell equations (1.4.16) and (1.4.19) can then be expressed as

$$\nabla^2 \boldsymbol{A} - \mu\varepsilon \frac{\partial^2 \boldsymbol{A}}{\partial t^2} = -\mu \boldsymbol{J} \tag{1.4.31}$$

and

$$\nabla^2 \phi - \mu\varepsilon \frac{\partial^2 \phi}{\partial t^2} = -\frac{\rho_e}{\varepsilon} \tag{1.4.32}$$

provided that the following *Lorentz gauge* (condition) is satisfied

$$\nabla \cdot \boldsymbol{A} + \mu\varepsilon \frac{\partial \phi}{\partial t} = 0 \tag{1.4.33}$$

Equations (1.4.31), (1.4.32), and (1.4.33) constitute the basic formulation of electromagnetic problems in terms of the vector and scalar potentials, from which the electromagnetic fields $\boldsymbol{E}$ and $\boldsymbol{B}$ can be determined. It can be shown that the electric field $\boldsymbol{E}$ and the magnetic field $\boldsymbol{B}$ are invariant with respect to the following gauge transformations:

$$\boldsymbol{A} \rightarrow \boldsymbol{A}' = \boldsymbol{A} + \nabla\Lambda \tag{1.4.34}$$

and

$$\phi \rightarrow \phi' = \phi - \frac{\partial \Lambda}{\partial t} \tag{1.4.35}$$

which implies that we can choose the Lorentz gauge with the aid of the freedom in the choice of $\boldsymbol{A}$ and ϕ. It should be noted that since more than one function Λ may be a solution of equation:

$$\nabla^2 \Lambda - \mu\varepsilon \frac{\partial^2 \Lambda}{\partial t^2} = 0 \tag{1.4.36}$$

the arbitrariness in the choice of $\boldsymbol{A}$ and ϕ may not be completely removed by imposing the Lorentz gauge. The advantage of introducing the Lorentz gauge is to give complete symmetry between the scalar potential ϕ and the vector potential $\boldsymbol{A}$, as shown by Eqs.(1.4.31) and (1.4.32). Sometimes, it is convenient to use the *Coulomb gauge* (condition) $\nabla \cdot \boldsymbol{A} = 0$, which is particularly useful when no sources ($\boldsymbol{J} = 0$ and $\rho_e = 0$) are present.

It is shown that, in the classical electrodynamics, potentials are merely a convenient mathematical tool for calculations of electromagnetic fields, which determine completely electromagnetism through the fundamental Maxwell's equations. In quantum mechanics, however, the potentials cannot be eliminated from the basic Schrödinger equation and consequently may be considered as physically effective. The early discussion on this subject was given in details by Aharonov and Bohm (1959). They showed that, contrary to the conclusion of classical electromagnetics and mechanics, there exist effects of potentials on charged particles, even in the region where all the fields vanish. The *Aharonov-Bohm effect* has been studied extensively both in theory and in experiments. Interested readers may refer to original works of, for instance, Aharonov and Bohm (1959), Peshkin and Tonomura (1989).

1.4.6 Time-Harmonic Fields and the Kramers-Kronig Relations

As we know, many engineering applications use sinusoidal fields. Besides, since the general behavior of a wave as a function of time can always be expressed as superposition of waves at different frequencies through the Fourier transformation, it is, therefore, of practical interest to investigate the characteristics of electromagnetic phenomena in the time-harmonic case. In such a case, it is most convenient to introduce the concept of the phasor field. Here, the field $\boldsymbol{E}(\boldsymbol{x}, t)$, a real function of position $\boldsymbol{x}$ and time t, is given by

$$\boldsymbol{E}(\boldsymbol{x}, t) = Re\{\boldsymbol{E}_{ph}(\boldsymbol{x})e^{i\omega t}\} \tag{1.4.37}$$

where $\boldsymbol{E}_{ph}(\boldsymbol{x})$ is a phasor field, which is, in general, complex. We shall use the convention that $\boldsymbol{E}_{ph}$ is the peak value rather than the *rms* value. We shall omit the subscript *ph* for the phasor whenever no confusion is expected to arise in the following discussions. Furthermore, in this section, we shall assume that the material medium involved is linear.

By using the phasor representation, the time derivative $\partial/\partial t$ may be replaced by the factor $i\omega$. Hence, Maxwell's equations in the time-harmonic case become

$$\nabla \times \boldsymbol{H} = \boldsymbol{J} + i\omega\boldsymbol{D} \tag{1.4.38}$$

$$\nabla \cdot \boldsymbol{B} = 0 \tag{1.4.39}$$

$$\nabla \times \boldsymbol{E} = -i\omega\boldsymbol{B} \tag{1.4.40}$$

$$\nabla \cdot \boldsymbol{D} = \rho_e \tag{1.4.41}$$

together with

$$\nabla \cdot \boldsymbol{J} + i\omega\rho_e = 0 \tag{1.4.42}$$

where field quantities must satisfy also certain boundary (or interface) conditions as discussed in Section 1.4.4. In addition to those conditions, if the region under consideration extends to infinity, the wave must be outgoing at infinity. This requirement at infinity is called the Sommerfeld radiation condition [Sommerfeld (1949)]. For a scalar field Ψ, the radiation condition is given by

$$\lim_{r \to \infty} r\left(\frac{\partial \Psi}{\partial r} + ik\Psi\right) = 0 \tag{1.4.43}$$

where $k = \omega/c$. For electromagnetic fields, they are given by

$$\lim_{r \to \infty} r\left(\sqrt{\frac{\mu}{\varepsilon}}\boldsymbol{r}^o \times \boldsymbol{H} + \boldsymbol{E}\right) = 0 \tag{1.4.44}$$

$$\lim_{r \to \infty} r\left(\boldsymbol{r}^o \times \boldsymbol{E} - \sqrt{\frac{\mu}{\varepsilon}}\boldsymbol{H}\right) = 0 \tag{1.4.45}$$

which mean that the field is outgoing and the field components E_i and H_i must decrease faster than r^{-1}. Besides, if the region includes a sharp edge, the field may become infinite, but the energy stored around the edge must be finite. Here, $\boldsymbol{r}^o$ denotes the unit directional vector of the vector $\boldsymbol{r}$.

For characterizing properties of electromagnetic materials in time-harmonic fields, the following linear constitutive relations of some isotropic material media are often used and can be expressed as

$$\boldsymbol{D} = (\varepsilon' - i\varepsilon'')\boldsymbol{E} \tag{1.4.46}$$

$$\boldsymbol{B} = (\mu' - i\mu'')\boldsymbol{H} \tag{1.4.47}$$

$$\boldsymbol{J} = (\sigma' - i\sigma'')\boldsymbol{E} \tag{1.4.48}$$

in the complex form. The introduction of the complex properties is due to the factor that, for instance, the electric displacement field $\boldsymbol{D}$ is generally not in phase with the electric field $\boldsymbol{E}$ in the time-harmonic case. For non-magnetic materials, we have simply $\mu' = \mu_0$ and $\mu'' = 0$. For normal conductors, it is often sufficient to consider only the real part σ' of the conductivity. For dielectric media, the imaginary part ε'' of the permittivity represents the dielectric loss in time-harmonic fields. Usually, the tangent of the loss angle $\tan\delta = \varepsilon''/\varepsilon'$ is introduced to characterize the dielectric loss of the material.

To see more clearly the physical significance of the real and imaginary part of the complex material constants, let us consider a time-harmonic wave propagating in an infinitely large linear and homogeneous dielectric medium in which there are no free charges ($\rho_e = 0$) and no source current ($\boldsymbol{J} = 0$). We suppose that the electric (phasor) field of the plane wave can be expressed as

$$\boldsymbol{E}(\boldsymbol{x}) = \boldsymbol{E}_o e^{-i(\boldsymbol{k}\cdot\boldsymbol{x})} \tag{1.4.49}$$

where $\boldsymbol{E}_0$ is a constant vector, perpendicular to the propagation vector $\boldsymbol{k}$. Thus we may find from Maxwell's equations (1.4.38)–(1.4.42) the magnetic field $\boldsymbol{H}$ by

$$\boldsymbol{H} = \frac{\boldsymbol{k}\times\boldsymbol{E}}{\omega\mu_o} \tag{1.4.50}$$

and the magnitude of the propagation vector $\boldsymbol{k}$ determined by

$$|\boldsymbol{k}|^2 = \omega^2\mu_o(\varepsilon' - i\varepsilon'') = \frac{\omega^2}{c^2}\varepsilon_r(1 - i\tan\delta) \tag{1.4.51}$$

It is shown that the propagating wave is a transverse (TEM) plane wave, where both the electric field and magnetic field of the wave are transverse to the direction of propagation. The phase velocity of the wave is determined approximately by $v = c/(\varepsilon_r)^{1/2}$, where c is the speed of light in free space. The wave has a damping factor of $e^{-\alpha x}$, if the direction of propagation is supposed to be along the x-axis, where α is the attenuation coefficient given approximately by $\alpha = \omega(\tan\delta)/2v$ for $\tan\delta \ll 1$. Thus, we see that the imaginary part of the complex permittivity ε'' characterizes effectively the damping of the wave with distance of propagation.

Frequently the inverse of the loss tangent, the quality factor $Q_d = 1/\tan\delta$ of the dielectric serves as the figure of merit, especially in waveguide problems. For power engineers interested in dielectric heating, the dielectric conductivity defined by

$$\sigma = \omega\varepsilon'' \tag{1.4.52}$$

may often be used. This dielectric conductivity may represent as well an actual ohmic conductivity caused by migrating charge carriers.

If the material medium is dispersive, these material parameters are also functions of frequency ω. In the linear approximation, we have, for instance, a linear relation between the temporal Fourier transformations:

$$D(\omega, x) = \varepsilon(\omega)E(\omega, x) \tag{1.4.53}$$

for an isotropic homogeneous dielectric medium, where

$$D(t, x) = \frac{1}{\sqrt{2\pi}} \int_{-\infty}^{+\infty} D(\omega, x)e^{-i\omega t}d\omega \tag{1.4.54}$$

This leads to a non-local relation in time between E and D:

$$\begin{aligned} D(t, x) &= \frac{1}{\sqrt{2\pi}} \int_{-\infty}^{+\infty} \varepsilon(\omega)E(\omega, x)e^{-i\omega t}d\omega \\ &= \frac{1}{2\pi} \int_{-\infty}^{+\infty} \varepsilon(\omega)\left(\int_{-\infty}^{+\infty} E(t', x)e^{i\omega t'}dt' \right) e^{-i\omega t}d\omega \end{aligned} \tag{1.4.55}$$

Noting $\varepsilon(\omega) \rightarrow \varepsilon_0$ for $\omega \rightarrow \infty$, due to the fact that the polarization cannot follow very rapid oscillations of the electric field, we may introduce the following function:

$$R(t) \equiv \frac{1}{2\pi} \int_{-\infty}^{+\infty} [\varepsilon(\omega) - \varepsilon_o]e^{-i\omega t}d\omega \tag{1.4.56}$$

which is usually called the response function. Now, instead of Eq.(1.4.55), we can write the following non-local time dependence of the fields

$$D(t, x) = \varepsilon_o E(t, x) + \int_{-\infty}^{+\infty} R(t - t')E(t', x)dt' \tag{1.4.57}$$

As a consequence of causality, the electric displacement $D(t, x)$ can only depend on the electric field $E(t', x)$ of earlier times $t' < t$. We find the response function $R(\tau) = 0$ for $\tau < 0$ ($\tau = t - t'$). Thus, by inversion of Eq.(1.4.56), we get

$$\varepsilon(\omega) = \varepsilon_o + \int_{0}^{+\infty} R(\tau)e^{i\omega\tau}d\tau \tag{1.4.58}$$

Since the lower limit of integration is zero, we can analytically continue this function to complex ω into the upper half-plane $\mathrm{Im}(\omega) > 0$. For real ω, we get by complex conjugation of Eq.(1.4.58)

$$\varepsilon(\omega)^* = \varepsilon(-\omega) \tag{1.4.59}$$

by noting that the response function $R(\tau)$ is real since $\boldsymbol{D}(t, \boldsymbol{x})$ and $\boldsymbol{E}(t, \boldsymbol{x})$ are real.

By Cauchy's theorem for a contour C in the upper half plane $\mathrm{Im}(z) > 0$, we may write

$$\varepsilon(z) = \varepsilon_o + \frac{1}{2\pi i}\int_C \frac{\varepsilon(z') - \varepsilon_o}{z' - z} dz' \tag{1.4.60}$$

from which we may derive the following relations for the real and imaginary parts of the complex permittivity $\varepsilon(\omega) = \varepsilon'(\omega) - i\varepsilon''(\omega)$:

$$\varepsilon'(\omega) = \varepsilon_o + \frac{2}{\pi}\int_0^\infty \frac{\omega'\varepsilon''(\omega')d\omega'}{\omega'^2 - \omega^2} \tag{1.4.61}$$

$$\varepsilon''(\omega) = -\frac{2\omega}{\pi}\int_0^\infty \frac{(\varepsilon'(\omega') - \varepsilon_o)d\omega'}{\omega'^2 - \omega^2} \tag{1.4.62}$$

where we have used the well-known Cauchy principle-value integration. Such relations between real and imaginary parts as a consequence of causality are called dispersion relations, known as the Kramers-Kronig relations. The Kramers-Kronig relations permit us to determine either the real or imaginary component of the permittivity if the other is known for all frequencies. This may be useful when it is easier to measure one part of $\varepsilon(\omega)$ rather than the other. Furthermore, these relations imply that a dispersive material must also be absorptive. A detailed derivation of the Kramers-Kronig relations may be found in the book of, for instance, Bartolo (1991) or Jackson (1975).

1.4.7 Dispersion in Materials

Electromagnetic properties of any material are generally dependent of frequency, and they can be considered constant only within a narrow frequency band. If a broadband pulse is propagated through such a material medium, the frequency dependence of the medium cannot be ignored. The variation of the material constants, such as dielectric constants, with frequency may cause the variation of phase velocity of a propagating wave in the medium. A material medium in which the phase velocity is a function of the frequency is said to be dispersive.

Let us consider the dispersion characteristics of dielectric materials. A simple mechanical model which led to a strikingly successful dispersion formula was proposed by Maxwell and independently by Sellmeyer. A further advance was accomplished by Lorentz, who extended the theory of the medium as a fine-grained assemble of molecular oscillators. By assuming that molecules with electrons bound elastically to the heavy nuclei, we may write the equation of motion for an electron as

$$m_e \frac{d^2 \boldsymbol{x}}{dt^2} = -m_e \omega_o^2 \boldsymbol{x} - m_e \alpha \frac{d\boldsymbol{x}}{dt} + \boldsymbol{F} \tag{1.4.63}$$

where m_e is the mass of the electron, $\boldsymbol{x}$ the displacement of the electron, α is a damping force factor, and ω_0 is a constant equal to the frequency of the free oscillations of the electron under the influence of restoring force alone. $\boldsymbol{F}$ is the Lorentz force acting on the electron, given by

$$\boldsymbol{F} = e\left(\boldsymbol{E} + \frac{\boldsymbol{P}}{3\varepsilon_o}\right) \tag{1.4.64}$$

according to the Mossotti approximation. Here, $\boldsymbol{P}$ is the polarization vector, given by

$$\boldsymbol{P} = Ne\boldsymbol{x} \tag{1.4.65}$$

assuming that there are N bound electrons per unit volume.

For time-harmonic fields ($\sim e^{i\omega t}$), noting that $\boldsymbol{D} = \varepsilon_0 \varepsilon_r \boldsymbol{E} = \varepsilon_0 \boldsymbol{E} + \boldsymbol{P}$, we may find the relative dielectric constant ε_r as a function of frequency ω:

$$\varepsilon_r = 1 + \frac{Ne^2}{m_e \varepsilon_o (\omega_1^2 - \omega^2 + i\omega\alpha)} \tag{1.4.66}$$

where

$$\omega_1 = \sqrt{\omega_o^2 - \frac{Ne^2}{3m_e \varepsilon_o}} \tag{1.4.67}$$

which implies that the effect of the polarization of the surroundings is to lower the resonance frequency of the individual oscillator from ω_0. From Eq.(1.4.66), we may notice that there exists a phase shift between the driving field and the resultant polarization because of the damping force factor α. This factor results in a complex permittivity of the dielectric material.

So far we have assumed that the dielectric contains only one oscillator type. In the more general case of s oscillator types which contribute to ε_r without mutual coupling, Eq.(1.4.66) may be generalized to be

$$\varepsilon_r = 1 + \sum_s \frac{N_s e^2}{m_s \varepsilon_o (\omega_s^2 - \omega^2 + i\omega\alpha_s)} \tag{1.4.68}$$

where N_s denotes the number of dispersion electrons per unit volume for the oscillator type s. Equation (1.4.68) is the well-known dispersion formula of classical physics.

In the absence of losses ($\alpha_s = 0$), Eq.(1.4.68) may be written

$$\varepsilon_r = 1 + \sum_s \frac{N_s e^2}{m_s \varepsilon_o (\omega_s^2 - \omega^2)} = 1 + \sum_s \frac{\lambda^2 C_s}{\lambda^2 - \lambda_s^2} \tag{1.4.69}$$

where $\lambda = 2\pi c/\omega$ and C_s are constants to be determined experimentally. For a non-magnetic dielectric medium, we may write $\varepsilon_r = n^2$, where n is the index of refraction of the medium. Equation (1.4.69), called the Sellmeier equation, is often used in the study of material dispersion of optical media. For example, the refractive index of pure fused silica (SiO_2) used for optical fibers in the wavelength $\lambda = 0.5$ to 2 μm can be given by Eq.(1.4.69) with $\lambda_1 = 0.1$ μm, $C_1 = 1.0955$ and $\lambda_2 = 9$ μm, $C_2 = 0.9$ [Marcuse (1982)].

To investigate further the behavior of oscillators of type p in the dispersion formula (1.4.68), let us consider the oscillator type which has the lowest resonance frequency. In the vicinity of ω_p, we have

$$\begin{aligned}\varepsilon_r &= 1 + \sum_s \frac{N_s e^2}{m_s \varepsilon_o (\omega_s^2 - \omega^2 + i\omega\alpha_s)} \\ &\approx 1 + \sum_{s \neq p} \frac{N_s e^2}{m_s \varepsilon_o \omega_s^2} + \frac{N_p e^2}{m_p \varepsilon_o (\omega_p^2 - \omega^2 + i\omega\alpha_p)}\end{aligned} \tag{1.4.70}$$

By noting

$$\omega_p^2 - \omega^2 = (\omega_p + \omega)(\omega_p - \omega) \approx 2\omega_p \Delta\omega \tag{1.4.71}$$

with $\Delta\omega = \omega_p - \omega$ in the vicinity of ω_p, we may obtain from Eq.(1.4.70)

$$\varepsilon_r = \varepsilon_r' - i\varepsilon_r'' \tag{1.4.72}$$

in which

$$\varepsilon_r' = 1 + \sum_{s \neq p} \frac{N_s e^2}{m_s \varepsilon_o \omega_s^2} + \frac{C\Delta\omega}{(\Delta\omega)^2 + (\alpha_p/2)^2} = \varepsilon_{rp}' + \frac{C\Delta\omega}{(\Delta\omega)^2 + (\alpha_p/2)^2} \tag{1.4.73}$$

$$\varepsilon_r'' = \frac{C\alpha_p/2}{(\Delta\omega)^2 + (\alpha_p/2)^2} \tag{1.4.74}$$

where $C = N_p e^2/(2m_p\varepsilon_0\omega_p)$, and ε_{rp}' is the real part of the relative dielectric constant due to the contributions from all oscillator types, which are different from the type p oscillator.

From Eq.(1.4.73), we may get

$$\frac{\partial\varepsilon_r'}{\partial\omega} = \frac{C(\omega_p - \omega - \alpha_p/2)(\omega_p - \omega + \alpha_p/2)}{[(\Delta\omega)^2 + (\alpha_p/2)^2]^2} \tag{1.4.75}$$

This results indicates that the real part of the relative dielectric constant ε_r' increases with increasing frequency ω for $\omega < \omega_p - \alpha_p/2$, which corresponds to the *normal dispersion*. However, in the frequency range: $\omega_p - \alpha_p/2 < \omega < \omega_p + \alpha_p/2$, we find that the real part of the relative dielectric constant ε_r' decreases with increasing frequency ω, which is known to be the *anomalous dispersion*.

In dielectric material, the resonant frequency ω_s is generally nonzero. In a normal conductor, there are free electrons that are not bound to molecules and, therefore, the restoring force $(-m_e\omega_0^2\boldsymbol{x})$ in Eq.(1.4.63) is absent. Besides, the interaction between the molecules can be ignored, and the Lorentz force is $\boldsymbol{F} = e\boldsymbol{E}$. Thus we have

$$m_e\frac{d^2\boldsymbol{x}}{dt^2} = -m_e\alpha\frac{d\boldsymbol{x}}{dt} + e\boldsymbol{E} \tag{1.4.76}$$

If there are N free electrons in unit volume, the current density is

$$\boldsymbol{J} = Ne\frac{d\boldsymbol{x}}{dt} \tag{1.4.77}$$

For time-harmonic fields ($\sim e^{i\omega t}$), noting that $\boldsymbol{J} = \sigma\boldsymbol{E}$, we may then find the conductivity σ as a function of frequency ω:

$$\sigma = \frac{Ne^2}{m_e(\alpha - i\omega)} \tag{1.4.78}$$

In many cases, the damping (collision) force is so strong in normal conductors, the conductivity σ can be treated as a real constant for frequency ω much lower than the collision frequency α, which lie usually in the infrared frequency range for metallic conductors.

1.4.8 Interfacial Polarization

Different from the electronic, atomic, and dipole orientation polarization, which are caused by the displacement or orientation of bound charge carriers, the interfacial polarization is caused by traveling charge carriers [Hippel (1954)]. The classical example of interfacial polarization is the Maxwell-Wagner two-layer condenser, as shown in Fig. 1.9*a*. The dielectric consists of two parallel sheets of materials, characterized by their dielectric constant, conductivity, and thickness (ε_1, σ_1, d_1) and (ε_2, σ_2, d_2), respectively. When a dc field is suddenly applied, the initial field distribution corresponds to the electrostatic requirement of constant flux density:

$$D_1 = D_2, \qquad \text{or} \qquad E_1/E_2 = \varepsilon_2/\varepsilon_1 \tag{1.4.79}$$

whereas the final distribution follows from the condition of current continuity:

$$J_1 = J_2, \qquad \text{or} \qquad E_1/E_2 = \sigma_2/\sigma_1 \tag{1.4.80}$$

The transient which links the initial and final state may be derived from the equivalent circuit, shown in Fig. 1.9*b*. In particular, the steady-state solution, when an ac voltage $V = V_0 e^{i\omega t}$ is applied, can be derived from the admittance of the circuit:

$$Y = \frac{I}{V} = \frac{(1 + i\omega\tau_1)(1 + i\omega\tau_2)}{(R_1 + R_2)(1 + i\omega\tau)} \tag{1.4.81}$$

in which τ_1 and τ_2 are the relaxation time of the two individual RC circuits, that is,

$$\tau_1 = R_1 C_1, \qquad \text{and} \qquad \tau_2 = R_2 C_2 \tag{1.4.82}$$

and τ is the relaxation time of the equivalent circuit given by

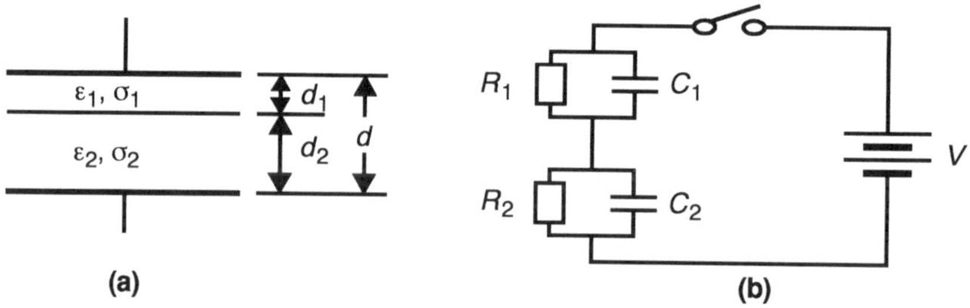

Figure 1.9 Maxwell-Wagner two-layer condenser and its equivalent circuit.

$$\tau = \frac{R_1 R_2 (C_1 + C_2)}{R_1 + R_2} = \frac{\varepsilon_1 d_2 + \varepsilon_2 d_1}{\sigma_1 d_2 + \sigma_2 d_1} \tag{1.4.83}$$

Note that we have made use of the following lumped elements of capacitors, given by

$$C_1 = \frac{A\varepsilon_1}{d_1}, \quad \text{and} \quad C_2 = \frac{A\varepsilon_2}{d_2} \tag{1.4.84}$$

and the lumped elements of resistors, given by

$$R_1 = \frac{d_1}{A\sigma_1}, \quad \text{and} \quad R_2 = \frac{d_2}{A\sigma_2} \tag{1.4.85}$$

where A is the surface area of the plate capacitor.

Noting that the admittance determines the complex permittivity ($\varepsilon = \varepsilon' - i\varepsilon''$) of the capacitor:

$$Y = i\omega C_o \frac{\varepsilon}{\varepsilon_o} \tag{1.4.86}$$

where $C_0 = A\varepsilon_0/d$, and using Eq.(1.4.81), we may find that the two-layer condenser appears to the outside observer as a dielectric having effectively the real part of the dielectric constant:

$$\varepsilon' = \varepsilon_\infty' \left(1 + \frac{k}{1 + \omega^2 \tau^2}\right) \tag{1.4.87}$$

with

$$\varepsilon_\infty' = \frac{\varepsilon_o}{C_o} \frac{C_1 C_2}{(C_1 + C_2)} = \frac{\varepsilon_1 \varepsilon_2 d}{\varepsilon_1 d_2 + \varepsilon_2 d_1} \tag{1.4.88}$$

being its optical value ($\omega \to \infty$), and

$$k = \frac{\varepsilon_s' - \varepsilon_\infty'}{\varepsilon_\infty'} \tag{1.4.89}$$

with

$$\varepsilon_s' = \frac{\varepsilon_o(\tau_1 + \tau_2 - \tau)}{C_o(R_1 + R_2)} = \varepsilon_\infty' \left[1 + d_1 d_2 \left(\frac{\sqrt{\varepsilon_1/\varepsilon_2}/\sigma_1 - \sqrt{\varepsilon_2/\varepsilon_1}/\sigma_2}{d_1/\sigma_1 + d_2/\sigma_2}\right)^2\right] \tag{1.4.90}$$

being the static value of ε' ($\omega \to 0$).

The dissipation factor of the two-layer dielectric becomes

$$\varepsilon'' = \varepsilon_\infty'\left(\frac{\tau}{\omega\tau_1\tau_2} + \frac{k\omega\tau}{1+\omega^2\tau^2}\right) = \frac{\sigma}{\omega} + \frac{(\varepsilon_s' - \varepsilon_\infty')\omega\tau}{1+\omega^2\tau^2} \tag{1.4.91}$$

in which σ characterizes the ohmic conductivity term given by

$$\sigma = \frac{d}{d_1/\sigma_1 + d_2/\sigma_2} \tag{1.4.92}$$

caused by the series resistor R_1+R_2.

It is shown by Eq.(1.4.88) that the optical dielectric constant of the two-layer condenser is determined by the real permittivities ε_1 and ε_2 of the two media. The static dielectric constant from Eq.(1.4.90) is larger than the optical one since these two media contain mobile charge carriers. Because of the conductivities σ_1 and σ_2 of these two media, charges pile up at the interface between two media until constant current transfer is established and the static conductivity results as given by Eq.(1.4.92). As the frequency increases, the interfacial polarization begins to lag. The dielectric constant ε' decreases to the midpoint between the static and optical value when $\omega = 1/\tau$.

1.4.9 Velocities of Wave Propagation

It is known that a single frequency sinusoid wave train of infinite duration can be expressed in its complex form by

$$F(z,t) = Ae^{i(\omega t - kz)} \tag{1.4.93}$$

where A is a constant denoting the amplitude of the wave propagating in the positive z direction, and k is the phase constant (or the wave number). The surface of constant state or phase defined by

$$\omega t - kz = \text{constant} \tag{1.4.94}$$

propagates with the velocity

$$v_p = \frac{dz}{dt} = \frac{\omega}{k} \tag{1.4.95}$$

This velocity is known as the *phase velocity*.

In general, a propagating wave of arbitrary shape may be expressed as a sum of sinusoidal (harmonic) waves according to Fourier analysis, by choosing the amplitude of the component harmonic waves as an appropriate function of the frequency or wave number:

$$F(z, t) = \int_{-\infty}^{+\infty} A(k) e^{i(\omega t - kz)} dk \tag{1.4.96}$$

If it happens that each frequency component of a "signal" wave has the same phase velocity v_p, and there is no attenuation of the wave amplitude, these component waves will add in proper phase to reproduce the original wave shape exactly, but delayed by the time of propagation z/v_p. In such a case, the phase velocity could be said to be the velocity of propagation of the signal.

In dispersive media, the phase velocity is generally frequency-dependent and, therefore, will vary for different frequency components. In this case, the individual sinusoidal components acting to make up a complex wave of the signal will shift in phase as they propagate along the propagation direction; the faster waves speeding ahead and the slower waves falling back, causing the phenomenon of dispersion of the wave. Thus the phase velocity of any particular frequency component does not represent the velocity of propagation of the signal. In general, it is then difficult to define a single velocity of propagation for the dispersive signal wave. However, when there is relatively little dispersion over the frequency band of interest, the concept of *group velocity* v_g may be introduced to characterize the velocity of propagation. This corresponds to the case where the amplitude function $A(k)$ in Eq.(1.4.96) is of negligible magnitude outside a narrow region:

$$k_o - \delta k \le k \le k_o + \delta k \tag{1.4.97}$$

so that from Eq.(1.4.96), we have

$$F(z, t) = \int_{k_o - \delta k}^{k_o + \delta k} A(k) e^{i(\omega t - kz)} dk \tag{1.4.98}$$

which represents a wave packet. Assuming that k is real, and ω is a known function of k, we may write $\omega(k)$ in a Taylor series:

$$\omega(k) = \omega(k_o) + \left.\frac{d\omega}{dk}\right|_{k = k_o} (k - k_o) + \dots \tag{1.4.99}$$

where only the first two terms are kept if we consider the case in which the function $\omega(k)$ deviates only slightly from its value at k_o in the small interval $2\delta k$. In such a case, the wave packet can be expressed by

$$F = F_o e^{i[\omega(k_o)t - k_o z]} \tag{1.4.100}$$

where F_0 is a mean amplitude defined by

$$F_o = \int_{k_o - \delta k}^{k_o + \delta k} A(k) \exp\left\{ i(k - k_o)\left[\left.\frac{d\omega}{dk}\right|_{k = k_o} t - z \right] \right\} dk \tag{1.4.101}$$

This amplitude is constant over surfaces defined by

$$\left.\frac{d\omega}{dk}\right|_{k = k_o} t - z = \text{constant} \tag{1.4.102}$$

from which we may introduce the group velocity v_g, defined by

$$v_g = \left.\frac{d\omega}{dk}\right|_{k = k_o} \tag{1.4.103}$$

which characterizes the velocity of propagation of the wave packet. It can be seen that if the medium is nondispersive, the group velocity v_g coincides with the phase velocity v_p. For dispersive media, both the group velocity v_g and the phase velocity v_p are functions of the wave number k_0. They have generally different values. A general relationship between these two velocities can be found in the following form:

$$v_g = v_p + k\frac{dv_p}{dk} = \frac{v_p}{1 - \dfrac{\omega}{v_p}\dfrac{dv_p}{d\omega}} \tag{1.4.104}$$

by noting $v_p = \omega/k$.

From above discussion, we may now understand that the group velocity is a proper concept to characterize the velocity of propagation of a signal wave only for media with small dispersions. For large dispersions, the spread in phase velocity of the harmonic components becomes significant; the wave packet is deformed rapidly, and the group velocity as a velocity of the whole loses its physical significance. It is then usually difficult to define a single velocity to describe the propagation of the wave in media with large dispersions. Some detailed discussions on this topic may be found in the work of, for instance, Stratton (1941).

1.5 ELECTROMAGNETIC ENERGY, MOMENTUM, AND VARIATIONAL PRINCIPLES

Electromagnetic energy can be transported through free space or within material media by means of electromagnetic waves or electric current flows. The determination of electromagnetic power losses during the energy transport in materials is of particular interest for the study of problems concerning Joule heating in normal conductors, ac losses in superconductors, and transmission losses in waveguides, such as electrical interconnects, optical fibers, and superconducting cables. In this section, problems concerning electromagnetic energies and their transport in material media will be discussed and formulated. Some variational principles in electromagnetism are introduced, which are often useful in numerical analysis of electromagnetic systems. Besides, the concept of momentum of electromagnetic fields will also be introduced.

1.5.1 Electric Field Energy for Charges

To study the electric field energy for charges, we first consider a test charge q in an electrostatic field $\boldsymbol{E}$. Since the force acting on the test charge q in the electric field is $\boldsymbol{F} = q\boldsymbol{E}$, the work done to displace the charge by $d\boldsymbol{l}$ is then

$$\delta W = -q\boldsymbol{E} \cdot d\boldsymbol{l} \tag{1.5.1}$$

where the minus sign appears because we are calculating the work done on the charge against the action of the field.

If the charge is displaced from point 1 to point 2 along a trajectory L, the work done per unit charge thus reads

$$W' = -\int_{(1)}^{(2)} \boldsymbol{E} \cdot d\boldsymbol{l} \tag{1.5.2}$$

Since the static electric field is irrotational, an electrostatic potential ϕ may be introduced such that $\boldsymbol{E} = -\nabla\phi$. Thus Eq.(1.5.2) becomes

$$W' = \phi(2) - \phi(1) \tag{1.5.3}$$

which shows that ϕ can be interpreted as the potential energy per unit charge in the electrostatic field. By defining $\phi(\infty) = 0$, we may also introduce an electric field potential for a unit point charge by

$$\phi(\boldsymbol{x}) = \int_{x}^{\infty} \boldsymbol{E} \cdot d\boldsymbol{l} \tag{1.5.4}$$

In particular, the electric potential at a distance r from a point charge q' is given by

$$\phi(r) = \frac{1}{4\pi\varepsilon_o}\frac{q'}{r} \tag{1.5.5}$$

Thus we can see that the interaction energy between two point charges q_1 and q_2 may be found by

$$W = \frac{1}{4\pi\varepsilon_o}\frac{q_1 q_2}{r} \tag{1.5.6}$$

which is equal to the work done by displacing the point charge q_2, in the field produced by the point charge q_1, from infinity to the position having a distance r from the point charge q_1.

Due to the symmetry of the problem, Eq.(1.5.6) can be written as

$$W = \frac{1}{2}\{\phi(r, q_1)q_2 + \phi(r, q_2)q_1\} \tag{1.5.7}$$

In general, the interaction energy associated with a collection of point charges $q^{(k)}$ ($k = 1, 2,..., N$) may be expressed by

$$W = \frac{1}{2}\sum_{k=1}^{N} q^{(k)}\phi^{(k)} \tag{1.5.8}$$

where $\phi(k)$ is the electric potential at the position of the kth charge due to all the other charges.

For a continuous distribution of charges in a space volume V, the work required to assemble the continuous charge distribution from infinitesimally small charge elements at infinity is given by

$$W^e = \frac{1}{2}\int_V \rho_e \phi dV \tag{1.5.9}$$

It is noticed that formula (1.5.9) is not simply a generalization of Eq.(1.5.8) for the continuous case. Since Eq.(1.5.9) takes into account the self-energy, which is infinity for a point charge, it is not allowable to use Eq.(1.5.9) for point charges [Böttcher (1973)].

With the use of the equation $\nabla \cdot \boldsymbol{D} = \rho_e$ and $\boldsymbol{E} = -\nabla\phi$, Eq.(1.5.9) may be expressed as

$$W^e = \frac{1}{2}\int_V \boldsymbol{E}\cdot\boldsymbol{D}dV + \frac{1}{2}\int_S \phi\boldsymbol{D}\cdot\boldsymbol{n}dS \tag{1.5.10}$$

where S is a closed surface enveloping the volume V.

If we assume that all charges are located in a finite region of space, at large distance r from the charges, the surface integral is of the order of $1/r$ and tends to zero as the surface of integration approaches infinity. Hence, for the entire space, we have

$$W^e = \frac{1}{2} \int_{V_\infty} \boldsymbol{E} \cdot \boldsymbol{D} dV \tag{1.5.11}$$

where V_∞ denotes the entire space.

1.5.2 Electrostatic Energy for Material Media

In the presence of a dielectric medium, Eq.(1.5.9) cannot be simply used due to the polarization effect in the dielectric medium. The reason is that with the dielectric medium work is done not only to bring real charge into position, but also to produce a certain state of polarization in the medium. If we consider a small change in energy δW^e due to some sort of change $\delta\rho_e$ in the charge density ρ_e existing in all space, we can find [Jackson (1975)] that the work done to accomplish this change is

$$\delta W^e = \int_{V_\infty} \phi \delta \rho_e dV \tag{1.5.12}$$

where ϕ is the potential due to the charge density ρ_e already present. Since $\nabla \cdot \boldsymbol{D} = \rho_e$, the change $\delta\rho_e$ can be related to a change in the displacement vector $\boldsymbol{D}$ such that Eq.(1.5.12) becomes

$$\delta W^e = \int_{V_\infty} \boldsymbol{E} \cdot \delta \boldsymbol{D} dV \tag{1.5.13}$$

extended over all space. Formally, this equation can be written as

$$\delta W^e = \delta \int_{V_\infty} \left(\int_0^D \boldsymbol{E} \cdot d\boldsymbol{D} \right) dV \tag{1.5.14}$$

which gives the electrostatic energy for dielectrics

$$W^e = \int_{V_\infty} \left(\int_0^D \boldsymbol{E} \cdot d\boldsymbol{D} \right) dV \tag{1.5.15}$$

For linear dielectric media, Eq.(1.5.15) becomes

$$W^e = \frac{1}{2} \int_{V_\infty} \boldsymbol{E} \cdot \boldsymbol{D} dV \qquad (1.5.16)$$

It is shown that Eq.(1.5.11) [or Eq.(1.5.9)] is valid macroscopically only for linear dielectric media (including vacuum). It should be noted that in deriving the energy expression, it is assumed that the medium is held at rest and hence no work is done in motion against forces. This implies that the virtual process of assembling the charges in the dielectric medium is a process with particular constraints. The resultant energy expression is nevertheless general, since no non-conservative forces are involved.

We shall now consider the energy of electrostatic field of conductors. It is known that a fundamental property of conductors in electrostatic cases is that the electric field inside a conductor must be zero. Hence, any charges in a conductor must be located on its surface. Thus the problem of the electrostatics of conductors amounts to determine the electric field in the vacuum outside the conductors and the distribution of charges on their surfaces. The total energy W^c of the electrostatic field of charged conductors is then

$$W^c = \frac{\varepsilon_o}{2} \int_{V_\infty} E^2 dV \qquad (1.5.17)$$

where the integral is taken over all space outside the conductors.

With the use of the result that the total charge on a conductor embedded in vacuum is

$$Q = \int_S \varepsilon_o \boldsymbol{E} \cdot d\boldsymbol{S} = -\int_S \varepsilon_o \frac{\partial \phi}{\partial n} dS \qquad (1.5.18)$$

where S is the surface of the conductor, we can write

$$W^c = \frac{1}{2} \sum_{k=1}^{N} Q_k \phi^{(k)} \qquad (1.5.19)$$

for a system of N conductors with charge Q_k on the kth conductor, where $\phi^{(k)}$ denotes the constant value of the electric potential on the surface of the kth conductor. Here, $\phi^{(k)}$ depends not only on Q_k, but depends linearly on all the Q_m as well. This follows from the general expression (1.2.6) of the electric potential. In general, we may write

$$Q_k = \sum_{m=1}^{N} C_{km} \phi^{(m)} \qquad (1.5.20)$$

where C_{km} are called the capacity coefficients, which fulfill the reciprocity relation: $C_{km} = C_{mk}$. It is noticed that if the conductor is embedded in a linear dielectric medium with $\boldsymbol{D} = \varepsilon \boldsymbol{E}$, then, the dielectric constant used in Eq.(1.5.18) should be replaced by ε (>0) instead of ε_0. It is often useful to find the capacitance by

$$C_{km} = \left.\frac{Q_k}{\phi^{(m)}}\right|_{\phi^{(k \neq m)} = 0} = \frac{1}{\phi^{(m)}} \int_{S_k} \varepsilon \frac{\partial \phi}{\partial n} dS \tag{1.5.21}$$

where S_k is the surface of the kth conductor. It is shown that the capacitance of the kth conductor (electrode) relative to the mth conductor (electrode) is the charge on the kth conductor per unit voltage on the mth conductor, with all other conductors (electrodes) held at zero voltage. The capacity coefficients play an important role in electronic systems where complicated systems of interacting conductors are of frequent occurrence. Theoretical calculation of these coefficients is generally not easy, requiring often approximations and/or numerical methods, which will be discussed later.

1.5.3 Variational Principle for Electrostatic System

An electrostatic system is described here by the following Maxwell equations:

$$\nabla \cdot \boldsymbol{D} = \rho_e, \qquad \nabla \times \boldsymbol{E} = 0 \ \ (\text{or } \boldsymbol{E} = -\nabla \phi) \tag{1.5.22}$$

with the interface condition:

$$\boldsymbol{n} \cdot [\boldsymbol{D}] = \alpha_f \tag{1.5.23}$$

where α_f denotes the surface charge density on the interface S. For such an electrostatic system, a *generalized variational principle* may be stated as follows:

Among all possible set of functions ϕ, E_i and D_i, the set that makes the functional U^e stationary is the solution of the electrostatic system (1.5.22)–(1.5.23), that is,

$$\delta U^e = 0 \tag{1.5.24}$$

with the condition of that the electrical potential ϕ is continuous everywhere. Here, the functional U^e is given by

$$U^e = \int_{V-S} \rho_e \phi dV - \int_{V-S} w^e(\boldsymbol{E}) dV - \int_S \alpha_f \phi dS + \int_{V-S} D_i(E_i + \phi_{,i}) dV \tag{1.5.25}$$

in which w^e is the electric field energy density of the system.

This variational principle can be proved as follows. By introducing the Lagrangian multipliers λ_i, we can define a functional:

$$U^e = \int_{V-S} \rho_e \phi dV - \int_{V-S} w^e(\boldsymbol{E}) dV - \int_S \alpha_f \phi dS + \int_{V-S} \lambda_i (E_i + \phi_{,i}) dV$$

Thus we have

$$\delta U^e = \int_{V-S} (\rho_e - \lambda_{i,i}) \delta\phi dV - \int_{V-S} \left(\lambda_i - \frac{\partial w^e}{\partial E_i}\right) \delta E_i dV$$

$$- \int_S (\alpha_f - [\lambda_i] n_i) \delta\phi dS + \int_{V-S} (E_i + \phi_{,i}) \delta\lambda_i dV$$

The stationary condition of $\delta U^e = 0$ gives

$$\rho_e - \lambda_{i,i} = 0 \quad \text{in } V-S$$

$$\lambda_i - \frac{\partial w^e}{\partial E_i} = 0 \quad \text{in } V-S$$

$$E_i + \phi_{,i} = 0 \quad \text{in } V-S$$

$$\alpha_f - [\lambda_i] n_i = 0 \quad \text{on } S$$

It can be seen that

$$\lambda_i = D_i = \frac{\partial w^e}{\partial E_i} \quad \text{in } V-S \tag{1.5.26}$$

For a linear isotropic dielectric medium, we have

$$w^e = \frac{1}{2}\varepsilon(\boldsymbol{E} \cdot \boldsymbol{E}) \tag{1.5.27}$$

and, therefore,

$$D_i = \frac{\partial w^e}{\partial E_i} = \varepsilon E_i \tag{1.5.28}$$

where ε is the permittivity of the dielectric medium.

1.5.4 Thomson's Theorem and Earnshaw's Theorem

Thomson's theorem states that charges placed on a system of fixed conductors embedded in a dielectric will distribute themselves on the surfaces of these conductors such that the energy of the resultant electrostatic field is a minimum.

The proof of this theorem may be given as follows. Suppose that there are N conductors bounded by the surfaces S_k ($k = 1, 2, ..., N$), each bearing a charge Q_k. The permittivity ε of the dielectric is supposed to be a continuous function of position. At every point within the dielectric, assumed to be linear, the field of the charges in equilibrium must satisfy Eq.(1.5.22) with the condition:

$$\phi_k = \text{constant}, \qquad \int_{S_k} \boldsymbol{D} \cdot \boldsymbol{n} dS = Q_k \tag{1.5.29}$$

over the surface of each conductor S_k.

Suppose that ϕ', $\boldsymbol{E}'$, $\boldsymbol{D}'$ is any other possible electrostatic field, which satisfies Eq.(1.5.22), but not necessarily the condition (1.5.29), and is known to differ somewhere, if not everywhere, from ϕ, $\boldsymbol{E}$, $\boldsymbol{D}$. Since the volume distribution ρ_e and the total charge on the conductors is fixed, we have

$$\nabla \cdot (\boldsymbol{D}' - \boldsymbol{D}) = 0, \qquad \int_{S_k} (\boldsymbol{D}' - \boldsymbol{D}) \cdot \boldsymbol{n} dS = 0 \tag{1.5.30}$$

Let U and U' be the electrostatic energies of the two fields. Their difference is then

$$U' - U = \frac{1}{2} \int \boldsymbol{E}' \cdot \boldsymbol{D}' dV - \frac{1}{2} \int \boldsymbol{E} \cdot \boldsymbol{D} dV \tag{1.5.31}$$

which becomes, since $\boldsymbol{D}' = \varepsilon \boldsymbol{E}'$,

$$U' - U = \frac{1}{2} \int (\boldsymbol{E}' - \boldsymbol{E}) \cdot (\boldsymbol{D}' - \boldsymbol{D}) dV + \int \boldsymbol{E} \cdot (\boldsymbol{D}' - \boldsymbol{D}) dV \tag{1.5.32}$$

The second term on the right-hand side vanishes. There remains

$$U' - U = \frac{1}{2} \int \varepsilon (\boldsymbol{E}' - \boldsymbol{E})^2 dV \tag{1.5.33}$$

which is an essentially positive quantity. It is therefore proved that if $\boldsymbol{E}'$ differs in any region of space from $\boldsymbol{E}$, the resultant energy U' will be greater than U.

Early in the history of electromagnetism some hopes emerged that all phenomena in the physical world could be explained by gravitational and electrostatic forces between charged particles. These hopes were extinguished, however, when Earnshaw (1805–1888), a mathematical physicist, concluded that inverse-square force laws between particles alone cannot account for the stability of a collection of particles.

The *Earnshaw theorem* states that a charged body placed in an electric field cannot be maintained in stable equilibrium under the influence of the electric forces alone.

Mathematically, this theorem is based on the fact that solutions of Laplace´s equation (or potential functions), which satisfied by electrostatic fields, cannot have any maxima or minima in a closed region except at the boundary of the region. Thus we can extend the theorem to say that it is impossible to find a stable equilibrium configuration of a body under any force potential satisfying Laplace´s equation, including gravity. Indeed, a more generalized version of Earnshaw´s theorem states that a body with steady charges, magnetization or currents placed in a steady electric or magnetic field cannot rest in stable equilibrium under the action of electric and magnetic forces alone [Jeans (1925)].

It should be noticed that Earnshaw´s theorem deals with potential energy only. In systems possessing both kinetic and potential energy, such as the gravitational system of the sun and its planets, stable equilibrium is possible. In electromagnetism, it is also possible to suspend objects in a stable position by the interaction of potential and kinetic energy. Various levitation systems are examples of such energy combinations. Earnshaw´s theorem is always true when all the sources of electromagnetism are held fixed in time and space as the test object is moved from equilibrium. The stable levitation of a permanent bar magnet in a superconducting bowl is however a classical counter example to Earnshaw´s theorem. Here, the induced superconducting currents in the bowl act to exclude the total flux from the superconductor. These currents act as repelling magnets and automatically adjust their strength and position as the bar magnet moves.

1.5.5 Magnetic Field Energy for Currents and Material media

We start with the consideration of magnetostatic energy for a set of elementary closed current loops L_k, with each loop carrying a steady current I_k ($k = 1, 2, \ldots, N$). They are confined to a finite region of space. We may find that the total magnetic field energy of the N current circuits can be expressed by

$$W^m = \frac{1}{2} \sum_{k=1}^{N} \sum_{m=1}^{N} L_{km} I_k I_m \tag{1.5.34}$$

where the self-energy of each particular circuit (for $k = m$) is taken into account. The factor 1/2 occurs because in the double summation each circuit is counted twice, and since $L_{km} = L_{mk}$. The coefficient L_{km} ($k \neq m$) is known as the coefficient of mutual inductance between the circuit k and circuit m, defined by

$$L_{km} = \frac{\mu_o}{4\pi} \oint_{l_k} \oint_{l_m} \frac{d\boldsymbol{l} \cdot d\boldsymbol{l}'}{|\boldsymbol{x} - \boldsymbol{x}'|} \qquad (H) \tag{1.5.35}$$

which is evidently symmetric. For $m = k$, the quantity L_{kk} (here k is not summed) is called the coefficient of self-inductance. Equation (1.5.35) is the Neumann formula for mutual inductance. It is seen that the coefficient of mutual (or self-) inductance is a property of the geometric shape and the physical arrangement of the coupled circuits, and it is independent of currents carried by the circuits.

By introducing the flux Φ_k threading the kth circuit due to all circuits

$$\Phi_k = \int_{S_k} \boldsymbol{B} \cdot d\boldsymbol{S} = \oint_{L_k} \boldsymbol{A} \cdot d\boldsymbol{l} \tag{1.5.36}$$

where S_k denotes the surface enclosed by the contour L_k followed by the current I_k, Eq.(1.5.34) may also be written as

$$W^m = \frac{1}{2} \sum_{k=1}^{N} I_k \Phi_k \tag{1.5.37}$$

Comparison of Eq.(1.5.37) and Eq.(1.5.34) shows

$$\Phi_k = \sum_{m=1}^{N} L_{km} I_m \tag{1.5.38}$$

It is often useful to find the inductance by

$$L_{km} = \left.\frac{\Phi_k}{I_m}\right|_{I_{k \neq m} = 0} = \frac{1}{I_m} \int_{S_k} \boldsymbol{B} \cdot d\boldsymbol{S} \tag{1.5.39}$$

which shows that the inductance L_{km} is the flux linked by the kth coil per unit current in the mth coil, with all other currents zero.

Equation (1.5.37) can be generalized to determine the magnetic field energy of a continuous distribution of current with a space volume V, which reads

$$W^m = \frac{1}{2} \int_V \boldsymbol{J} \cdot \boldsymbol{A}\, dV \tag{1.5.40}$$

This expression is valid for current distribution in linear conductive media (including free space). A more general form of the magnetic field energy for continuous current distribution in nonlinear magnetic media is

$$W^m = \int_V \left(\int_0^A \boldsymbol{J} \cdot d\boldsymbol{A} \right) dV \tag{1.5.41}$$

It is sometimes desirable to express the magnetic energy in terms of field quantities $\boldsymbol{B}$ and $\boldsymbol{H}$ instead of current density $\boldsymbol{J}$ and vector potential $\boldsymbol{A}$. By using the relations $\boldsymbol{B} = \nabla \times \boldsymbol{A}$ and $\boldsymbol{J} = \nabla \times \boldsymbol{H}$, we may find

$$W^m = \frac{1}{2} \int_{V_\infty} \boldsymbol{H} \cdot \boldsymbol{B} dV \tag{1.5.42}$$

for linear magnetic media, and

$$W^m = \int_{V_\infty} \left(\int_0^B \boldsymbol{H} \cdot d\boldsymbol{B} \right) dV \tag{1.5.43}$$

for nonlinear magnetic media. Here, the integration is taken over all space. It is seen that Eqs.(1.5.42) and (1.5.43) are analogous to those of electrostatic energy in respectively Eqs.(1.5.16) and (1.5.15).

1.5.6 Variational Principle for Magnetostatic System

A magnetostatic system is described here by the following Maxwell equations:

$$\nabla \times \boldsymbol{H} = \boldsymbol{J}, \qquad \nabla \cdot \boldsymbol{B} = 0 \ (\text{or } \boldsymbol{B} = \nabla \times \boldsymbol{A}) \tag{1.5.44}$$

with the interface condition:

$$\boldsymbol{n} \times [\boldsymbol{H}] = \boldsymbol{K}_f \tag{1.5.45}$$

where $\boldsymbol{K}_f$ is the surface current density on the interface S. For such a magnetostatic system, a *generalized variational principle* may be stated as follows:

Among all possible set of functions A_i, B_i and H_i, the set that makes the functional U^m stationary is the solution of the magnetostatic system (1.5.44)–(1.5.45), that is,

$$\delta U^m = 0 \tag{1.5.46}$$

with the condition of that the magnetic vector potential $\boldsymbol{A}$ is continuous everywhere. Here, the functional U^m is given by

$$U^m = \int_{V-S} J_i A_i dV - \int_{V-S} w^m(\boldsymbol{B}) dV - \int_S K_{fi} A_i dS + \int_{V-S} H_i(B_i - \varepsilon_{ijk} A_{k,j}) dV \tag{1.5.47}$$

in which w^m is the magnetic field energy density of the system.

This variational principle can be proved as follows. By introducing the Lagrangian multipliers λ_i, we can define a functional:

$$U^m = \int_{V-S} J_i A_i dV - \int_{V-S} w^m(\boldsymbol{B}) dV - \int_S K_{fi} A_i dS + \int_{V-S} \lambda_i(B_i - \varepsilon_{ijk} A_{k,j}) dV$$

It can be shown that the stationary condition of $\delta U^m = 0$ gives

$$J_i - \varepsilon_{ijk} \lambda_{k,i} = 0 \quad \text{in } V-S$$

$$\lambda_i - \frac{\partial w^e}{\partial B_i} = 0 \quad \text{in } V-S$$

$$B_i - \varepsilon_{ijk} A_{k,j} = 0 \quad \text{in } V-S$$

$$K_{fi} - \varepsilon_{ijk}[\lambda_k] n_j = 0 \quad \text{on } S$$

It can be seen that

$$\lambda_i = H_i = \frac{\partial w^m}{\partial B_i} \quad \text{in } V-S \tag{1.5.48}$$

For a linear isotropic magnetic medium, we have

$$w^m = \frac{1}{2\mu}(\boldsymbol{B} \cdot \boldsymbol{B}) \tag{1.5.49}$$

and, therefore, $\boldsymbol{B} = \mu \boldsymbol{H}$, where μ is the permeability of the medium.

1.5.7 Poynting's Theorem for Electrodynamic Systems

To formulate electromagnetic energy transport problems, we consider an arbitrary volume V enclosed by a surface ∂V at rest in free space. The volume V can be either an empty space volume or a material volume. Thus an energy integral can be obtained from Maxwell's equations (1.4.16)–(1.4.19)

$$\int_V \frac{\partial w}{\partial t} dV + \int_V \boldsymbol{J} \cdot \boldsymbol{E} dV = -\int_{\partial V} \boldsymbol{S} \cdot \boldsymbol{n} dS \tag{1.5.50}$$

where $\boldsymbol{S}$ is called the *Poynting vector* defined by

$$\boldsymbol{S} = \boldsymbol{E} \times \boldsymbol{H} \qquad (\text{W/m}^2) \tag{1.5.51}$$

which is usually interpreted as the electromagnetic field energy flux (flow) per unit time through a unit area of a given surface, oriented normally to the Poynting vector. It is the power flux density and its unit is W/m^2. $\partial w/\partial t$ denotes the rate of change of electromagnetic field energy density defined by

$$\frac{\partial w}{\partial t} = \boldsymbol{E} \cdot \frac{\partial \boldsymbol{D}}{\partial t} + \boldsymbol{H} \cdot \frac{\partial \boldsymbol{B}}{\partial t} \tag{1.5.52}$$

Equation (1.5.50) is known as the integral form of Poynting's theorem. It is a mathematical identity compatible with Maxwell's equations. In general, Eq.(1.5.50) does not represent the equation of conservation of energy since there might be other types of energy present, such as thermal and mechanical energies.

By using Eqs.(1.4.7) and (1.4.13), Eq.(1.5.52) may also be written

$$\frac{\partial w}{\partial t} = \frac{1}{2}\frac{\partial}{\partial t}(\varepsilon_o E^2 + \mu_o H^2) + \boldsymbol{E} \cdot \frac{\partial \boldsymbol{P}}{\partial t} + \mu_o \boldsymbol{H} \cdot \frac{\partial \boldsymbol{M}}{\partial t} \tag{1.5.53}$$

For linear electromagnetic solids with constitutive relations by Eqs.(1.4.26)–(1.4.27), we have

$$w = \frac{1}{2}(\varepsilon E^2 + \mu H^2) \tag{1.5.54}$$

which is the stored electromagnetic field energy density. Thus in the case of linearity, Poynting's equation (1.5.50) states that the net inward power-flux

$$P(t) = -\int_{\partial V} \boldsymbol{S} \cdot \boldsymbol{n} dS \tag{1.5.55}$$

supplied by the electromagnetic field over a closed surface ∂V, equals the sum of the time rate of increase of electromagnetic field energy inside the volume V, plus the total Ohmic loss (that may be converted into mechanical or heat energy) in V, assuming V contains no generators. If V contains a power generator, an additional term $\int \boldsymbol{J}_g \cdot \boldsymbol{E} dV$ with $\boldsymbol{J}_g$ being the source current density, independent of the field, should be added on the *l.h.s.*of Eq.(1.5.50).

1.5.8 Poynting's Theorem for Quasistatic Systems

At the quasistatic approximation, Poynting's theorem may be expressed by

$$\int_V \boldsymbol{H} \cdot \frac{\partial \boldsymbol{B}}{\partial t} dV + \int_V \boldsymbol{J} \cdot \boldsymbol{E} dV = -\int_{\partial V} \boldsymbol{S} \cdot \boldsymbol{n} dS \tag{1.5.56}$$

for magneto-quasistatic systems, and

$$\int_V \boldsymbol{E} \cdot \frac{\partial \boldsymbol{D}}{\partial t} dV + \int_V \boldsymbol{J} \cdot \boldsymbol{E} dV = -\int_{\partial V} \boldsymbol{S} \cdot \boldsymbol{n} dS \tag{1.5.57}$$

for electro-quasistatic systems.

If we consider ac fields of periodic variation with a time period t_p, the power dissipated per cycle in a conducting magnetizable medium can be obtained at the magneto-quasistatic approximation by

$$\int_0^{t_p} P(t)dt = -\int_0^{t_p}\int_{\partial V} (\boldsymbol{E} \times \boldsymbol{H}) \cdot \boldsymbol{n} dSdt = \int_0^{t_p}\int_V \left(\boldsymbol{J} \cdot \boldsymbol{E} + \boldsymbol{H} \cdot \frac{\partial \boldsymbol{M}}{\partial t}\right) dVdt \tag{1.5.58}$$

If the magnetization of the material is negligible, we have then

$$\int_0^{t_p} P(t)dt = \int_0^{t_p}\int_V (\boldsymbol{J} \cdot \boldsymbol{E}) dVdt \tag{1.5.59}$$

In such a case, we may take $\boldsymbol{J} \cdot \boldsymbol{E}$ to be the instantaneous power density dissipated in the medium.

For a static system carrying only direct currents, Eq.(1.5.50) is reduced to

$$P = -\int_{\partial V} \boldsymbol{S} \cdot \boldsymbol{n} dS = \int_V \boldsymbol{J} \cdot \boldsymbol{E} dV \tag{1.5.60}$$

assuming V contains no generators. It is shown that in a dc system the net power-flux entering a closed surface ∂V constructed about the current-carrying conductor is a measure of the Ohmic loss in the conductor.

1.5.9 Poynting's Theorem for Time-Harmonic Systems

In many practical problems, we deal with time-harmonic electromagnetic fields, and it is necessary to consider the Poynting theorem for the phasor field quantities, as discussed in Section 1.4.6. In a time-harmonic case, we may define the complex Poynting vector as

$$\boldsymbol{S}_c = \boldsymbol{E} \times \boldsymbol{H}^* \tag{1.5.61}$$

where $\boldsymbol{E}$ and $\boldsymbol{H}$ are phasors. Thus the complex form of Poynting's theorem can be introduced from Eqs.(1.4.38)–(1.4.41) by

$$\int_V \frac{i\omega}{2}(\boldsymbol{B} \cdot \boldsymbol{H}^* - \boldsymbol{E} \cdot \boldsymbol{D}^*)dV + \int_V \frac{1}{2}\boldsymbol{E} \cdot \boldsymbol{J}^* dV = -\int_{\partial V} \frac{1}{2}\boldsymbol{S}_c \cdot \boldsymbol{n} dS \tag{1.5.62}$$

where the quantity, for instance, $\boldsymbol{H}^*$ denotes the complex conjugate of the phasor $\boldsymbol{H}$.

To see the physical meanings of this equation, let us write first the complex phasors

$$\boldsymbol{E}(\boldsymbol{x}) = \boldsymbol{E}_R(\boldsymbol{x}) + i\boldsymbol{E}_I(\boldsymbol{x}) \quad \text{and} \quad \boldsymbol{H}(\boldsymbol{x}) = \boldsymbol{H}_R(\boldsymbol{x}) + i\boldsymbol{H}_I(\boldsymbol{x}) \tag{1.5.63}$$

We can then get

$$\boldsymbol{E}(\boldsymbol{x}, t) = Re\{\boldsymbol{E}(\boldsymbol{x})e^{i\omega t}\} = \boldsymbol{E}_R(\boldsymbol{x})\cos(\omega t) - \boldsymbol{E}_I(\boldsymbol{x})\sin(\omega t) \tag{1.5.64}$$

$$\boldsymbol{H}(\boldsymbol{x}, t) = Re\{\boldsymbol{H}(\boldsymbol{x})e^{i\omega t}\} = \boldsymbol{H}_R(\boldsymbol{x})\cos(\omega t) - \boldsymbol{H}_I(\boldsymbol{x})\sin(\omega t) \tag{1.5.65}$$

Substituting them into Eq.(1.5.51), we may find

$$\begin{aligned} \boldsymbol{S} = \boldsymbol{E} \times \boldsymbol{H} &= (\boldsymbol{E}_R \times \boldsymbol{H}_R)(\cos(\omega t))^2 + (\boldsymbol{E}_I \times \boldsymbol{H}_I)(\sin(\omega t))^2 \\ &\quad - (\boldsymbol{E}_R \times \boldsymbol{H}_I + \boldsymbol{E}_I \times \boldsymbol{H}_R)\sin(\omega t)\cos(\omega t) \end{aligned} \tag{1.5.66}$$

By taking the time-average of the Poynting vector

$$\bar{\boldsymbol{S}} = \frac{1}{t_p} \int_0^{t_p} \boldsymbol{S}(\boldsymbol{x}, t)dt \tag{1.5.67}$$

with $t_p = 2\pi/\omega$, we can obtain from Eq.(1.5.66) that

$$\bar{\boldsymbol{S}} = \frac{1}{2}(\boldsymbol{E}_R \times \boldsymbol{H}_R + \boldsymbol{E}_I \times \boldsymbol{H}_I) = \frac{1}{2}Re\{\boldsymbol{E} \times \boldsymbol{H}^*\} = \frac{1}{2}Re\{\boldsymbol{S}_c\} \tag{1.5.68}$$

which shows that the real part of the term on the right-hand side of Eq.(1.5.62) represents the time-average power flux entering the closed surface ∂V. It may be noticed that the real Poynting vector $\boldsymbol{S}$ and the complex Poynting vector $\boldsymbol{S}_c$

do not have a phasor relationship with each other. Instead, they are related by Eq.(1.5.68).

For isotropic linear electromagnetic materials with constitutive relations by Eqs.(1.4.46)–(1.4.48) (assuming a pure real values of the conductivity σ), we can obtain by taking the real part of Eq.(1.5.62)

$$-\int_{\partial V} \bar{S} \cdot \boldsymbol{n} dS = \int_V \frac{1}{2}\sigma \boldsymbol{E} \cdot \boldsymbol{E}^* dV + \int_V \frac{\omega\mu''}{2} \boldsymbol{H} \cdot \boldsymbol{H}^* dV + \int_V \frac{\omega\varepsilon''}{2} \boldsymbol{E} \cdot \boldsymbol{E}^* dV \quad (1.5.69)$$

which shows that the time-average power flux entering a closed surface ∂V equals the time-average power dissipated as the sum of the conduction loss (Joule heat), the magnetic loss, and the dielectric loss due to polarization damping forces inside the volume V bounded by ∂V, provided that there are no power sources in the volume V. Obviously, μ'' and ε'' must be positive in order to represent energy loss.

Similarly, we can get by taking the imaginary parts of Eq.(1.5.62)

$$-\int_{\partial V} \frac{1}{2} Im\{\boldsymbol{E} \times \boldsymbol{H}^*\} \cdot \boldsymbol{n} dS = \int_V 2\omega(w_m - w_e) dV \quad (1.5.70)$$

in which w_m and w_e denote respectively the time-average of the stored energy densities of the magnetic field and of the electric field in V, that is,

$$w_m = \frac{1}{4}\mu' \boldsymbol{H} \cdot \boldsymbol{H}^* \quad (1.5.71)$$

$$w_e = \frac{1}{4}\varepsilon' \boldsymbol{E} \cdot \boldsymbol{E}^* \quad (1.5.72)$$

for nondispersive media. If the material medium is dispersive, the time-averaged electric and magnetic stored energy densities are given by

$$w_m = \frac{1}{4} Re\left\{\frac{\partial}{\partial\omega}(\omega\mu) \boldsymbol{H} \cdot \boldsymbol{H}^*\right\} \quad (1.5.73)$$

and

$$w_e = \frac{1}{4} Re\left\{\frac{\partial}{\partial\omega}(\omega\varepsilon) \boldsymbol{E} \cdot \boldsymbol{E}^*\right\} \quad (1.5.74)$$

where the complex permeability μ and the complex permittivity ε are frequency-dependent [Landau and Lifshitz (1960)].

Equation (1.5.70) states that the imaginary part of the complex power flux entering the closed surface of the volume V is a measure of 2ω times the difference of the time-average energies stored in the magnetic and electric fields, provided that there is no power source in the volume V. In the general case where there is a power generator with the source current density $\boldsymbol{J}_g$ in the volume V, the complex Poynting theorem can be expressed as follows:

$$-\int_V \frac{1}{2}\boldsymbol{E}\cdot\boldsymbol{J}_g{}^* dV = \int_V \frac{i\omega}{2}(\boldsymbol{B}\cdot\boldsymbol{H}^* - \boldsymbol{E}\cdot\boldsymbol{D}^*)dV + \int_V \frac{1}{2}\boldsymbol{E}\cdot\boldsymbol{J}^* dV$$
$$+\int_{\partial V} \frac{1}{2}(\boldsymbol{E}\times\boldsymbol{H}^*)\cdot\boldsymbol{n}dS \qquad (1.5.75)$$

where the term on the left-hand side represents the net complex power supplied to the volume V.

The complex Poynting's theorem by Eq.(1.5.62) for a closed surface has a result of interest in the theory of equivalent circuits for waveguides. Consider a simple RLC circuit shown in Fig. 1.10. If the current in the circuit is I and applied voltage is V, the complex time-average power flow into the circuit is given by

$$P = \frac{1}{2}VI^* = \frac{1}{2}Z_{in}II^* = \frac{1}{2}\left(R + i\omega L - \frac{i}{\omega C}\right)II^* \qquad (1.5.76)$$

where Z_{in} denotes the input impedance to the circuit. This equation can also be written

$$\frac{1}{2}Z_{in}II^* = 2i\omega\left(\frac{1}{4}LII^* - \frac{1}{4}\frac{II^*}{\omega^2 C}\right) + \frac{1}{2}RII^* \qquad (1.5.77)$$

which has the similar interpretation as Eq.(1.5.62) by noting that the time-average reactive energy stored in the magnetic field around the inductor L is $W_m = LII^*/4$, while the time-average electric energy stored in the capacitor is $W_e = II^*/(4\omega^2 C)$ since the peak voltage across C is $I/(\omega C)$. The average energy dissipated in the resistor R per second is $P_L = RII^*/2$.

From above discussion, we may introduce a general definition of the input impedance for any one-port circuit by

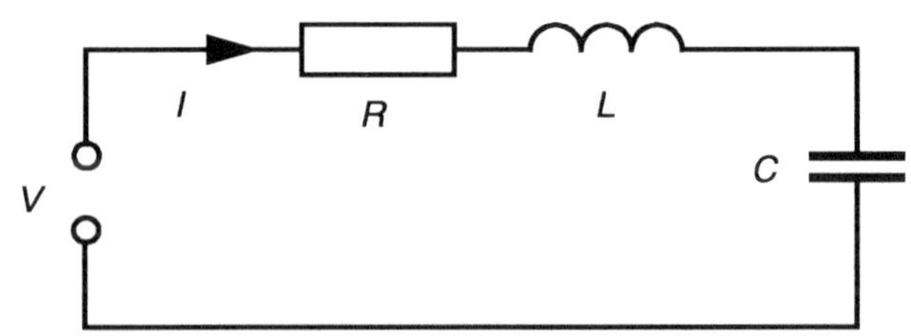

Figure 1.10 An RLC circuit.

$$Z_{in} = \frac{P_L + 2i\omega(W_m - W_e)}{II^*/2} \tag{1.5.78}$$

provided that a suitably defined equivalent terminal current I is used. Here, a one-port circuit (equivalent to a two-terminal network) is a circuit for which power can enter or leave through a single waveguide or transmission line. Some more detailed discussions on waveguide and transmission lines will be given in Chapter 6.

1.5.10 Variational Principle for Electrodynamic System

An electrodynamic system is described here by the following Maxwell equations:

$$\nabla \times \boldsymbol{H} = \boldsymbol{J} + \frac{\partial \boldsymbol{D}}{\partial t}, \qquad \nabla \cdot \boldsymbol{B} = 0 \ \ (\text{or } \boldsymbol{B} = \nabla \times \boldsymbol{A}) \tag{1.5.79}$$

$$\nabla \times \boldsymbol{E} = -\frac{\partial \boldsymbol{B}}{\partial t}, \qquad \nabla \cdot \boldsymbol{D} = \rho_e \tag{1.5.80}$$

with the interface condition:

$$\boldsymbol{n} \times [\boldsymbol{H}] = \boldsymbol{K}_f\,, \qquad [\boldsymbol{D}] \cdot \boldsymbol{n} = \alpha_f \tag{1.5.81}$$

where $\boldsymbol{K}_f$ is the surface current density and α_f the surface charge density on the interface S.

For such an electrodynamic system, a generalized variational principle may be stated as follows: Among all possible set of functions ϕ, A_i, E_i, D_i, B_i and H_i, the set that makes the functional U stationary is the solution of the electrodynamic system (1.5.79)–(1.5.81), that is,

$$\delta U = 0 \tag{1.5.82}$$

with the condition of that the magnetic vector potential $\boldsymbol{A}$ and the electric potential ϕ are continuous everywhere, and that $\boldsymbol{A}$ is not varied at the beginning and end of the process, that is, $\delta \boldsymbol{A}(t_o) = \delta \boldsymbol{A}(t_1) = 0$. Here, the functional U is

$$\begin{aligned} U = {} & \int_{t_o}^{t_1}\int_{V-S} (J_iA_i - \rho_e\phi - w^m(\boldsymbol{B}) + w^e(\boldsymbol{E}))\,dV\,dt + \int_{t_o}^{t_1}\int_S (\alpha_f\phi \quad K_{fi}A_i)\,dS\,dt \\ & + \int_{t_o}^{t_1}\int_{V-S} H_i(B_i - \varepsilon_{ijk}A_{k,j})\,dV\,dt - \int_{t_o}^{t_1}\int_{V-S} D_i\left(E_i + \phi_{,i} + \frac{\partial A_i}{\partial t}\right)dV\,dt \end{aligned} \tag{1.5.83}$$

in which w^m and w^e are respectively the magnetic field energy density and the electric field density of the system. This variational principle can be proved similarly as those for static systems (see Sections 1.5.3 and 1.5.6).

1.5.11 Uniqueness Theorem

Consider a typical electromagnetic boundary value problem, in which we have a volume of space V with interior surfaces S_k (k = 1, 2,..., n) and an exterior surface S_o, as shown in Fig. 1.11. Within this volume there may be regions occupied by different materials, such as dielectrics, magnetic materials, or conductors. These subvolumes are labeled V_o, V_1, V_2,..., V_n. Maxwell´s equations are to be solved for the electromagnetic field within each subvolume, and the field should satisfy the electromagnetic boundary (interface) conditions at each boundary (interface).

A *uniqueness theorem* states that for a volume V that contains only simple materials, specified by Eqs.(1.4.26)–(1.4.28), and for which the impressed currents $\boldsymbol{J}_i$ are specified, a solution to Maxwell´s equation [Eqs.(1.5.79)–(1.5.80)] is uniquely specified for all times $t > 0$ by the initial values of the field ($\boldsymbol{E}$ and $\boldsymbol{H}$ at time $t = 0$) throughout V and the values of the tangential component of either $\boldsymbol{E}$ or $\boldsymbol{H}$ over the boundary (interface) surfaces (S_k and S_o) of V for $t \geq 0$. Here, the tangential component of the electric field vector $\boldsymbol{E}$ can be specified on a portion of the boundary (interface) surface and the tangential component of the magnetic field vector $\boldsymbol{H}$ on the remainder.

The proof of this theorem may be given as follows: Suppose that we have two different solutions, denoted by $\boldsymbol{E}_a$, $\boldsymbol{D}_a$, $\boldsymbol{B}_a$, $\boldsymbol{H}_a$ and $\boldsymbol{E}_b$, $\boldsymbol{D}_b$, $\boldsymbol{B}_b$, $\boldsymbol{H}_b$, to Maxwell´s equations in the volume V for time $t > 0$. At time $t = 0$, the field $\boldsymbol{E}, \boldsymbol{B}$ is specified everywhere in V, and both solutions are equal to the specified field. Consider now the difference between two solutions, that is, $\delta\boldsymbol{E}=\boldsymbol{E}_a-\boldsymbol{E}_b$, $\delta\boldsymbol{D}=\boldsymbol{D}_a-\boldsymbol{D}_b$, $\delta\boldsymbol{B}=\boldsymbol{B}_a-\boldsymbol{B}_b$, $\delta\boldsymbol{H}=\boldsymbol{H}_a-\boldsymbol{H}_b$. Since Maxwell´s equations are linear, the difference field must be also a solution to Maxwell´s equations. Thus noting the Poynting theorem [Eq.(1.5.50)], we may obtain

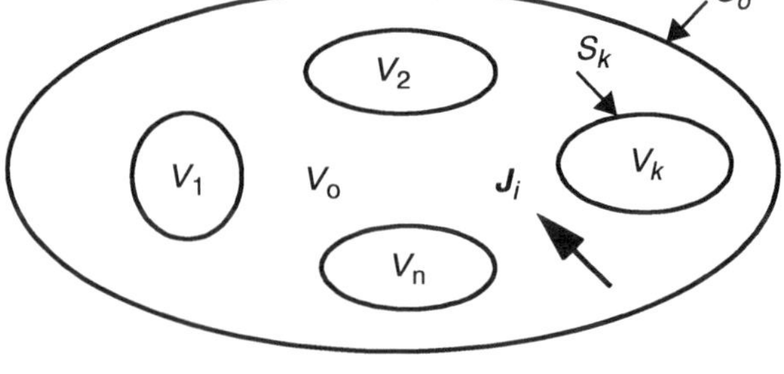

Figure 1.11 Geometrical scheme of an electromagnetic boundary value problem.

$$\int_V \left(\delta \boldsymbol{E} \cdot \frac{\partial(\delta \boldsymbol{D})}{\partial t} + \delta \boldsymbol{H} \cdot \frac{\partial(\delta \boldsymbol{B})}{\partial t}\right) dV + \int_V \delta \boldsymbol{J} \cdot \delta \boldsymbol{E} dV = 0 \tag{1.5.84}$$

where we have made use of the fact that both solutions satisfy the boundary (interface) conditions stated. For simple materials, Eq.(1.5.84) can be written in the following form:

$$\frac{\partial(\delta W)}{\partial t} = -\int_V \sigma |\delta \boldsymbol{E}|^2 dV \tag{1.5.85}$$

Here, we have introduced the notation:

$$\delta W = \frac{1}{2} \int_V (\varepsilon |\delta \boldsymbol{E}|^2 + \mu |\delta \boldsymbol{H}|^2) dV \tag{1.5.86}$$

which is always a positive number or zero for $t > 0$. Furthermore, we may see from Eq.(1.5.85) that the partial derivative of δW with respect to time must be a negative number or zero. By the initial conditions that we have assumed that $\delta \boldsymbol{E} = 0$, $\delta \boldsymbol{D} = 0$, $\delta \boldsymbol{B} = 0$, and $\delta \boldsymbol{H} = 0$, we have $\delta W = 0$ at $t = 0$. Thus we may conclude that $\delta W = 0$ for $t > 0$, which means that $\delta \boldsymbol{E} = 0$, $\delta \boldsymbol{H} = 0$ (and $\delta \boldsymbol{D} = 0$, $\delta \boldsymbol{B} = 0$) for $t \geq 0$. Therefore, there is only one solution, a unique solution, to Maxwell´s equations within V for stated conditions.

It may be noticed that although our proof of the uniqueness theorem is explicitly carried out for isotropic media, this theorem applies also to anisotropic media, for which electric and magnetic energies are then positive definite forms. Besides, the exterior surface S_o in the theorem can generally be extend to infinity. In such a case, the tangential components of the field $\boldsymbol{E}$, $\boldsymbol{H}$ on the exterior surface S_o are zero because the velocity of propagation for electromagnetic fields is finite.

Similarly, for static systems, we may also formulate the uniqueness theorem. For instance, consider an electrostatic system with the electric field $\boldsymbol{E} = -\nabla\phi$, formulated as follows:

1. The electrical potential ϕ satisfies the Laplace equation $\nabla^2\phi = 0$ at all points, but not on a boundary surface or within external sources,
2. The potential function ϕ is continuous everywhere, including boundaries of dielectrics or of conductors, and is finite except at external point charges introduced as primary sources. ϕ is also regular at infinity, provided all sources are within a finite distance of the origin,
3. The normal component of the displacement vector $(\varepsilon\partial\phi/\partial n)$ is continuous across an interface between two dielectrics, and $\varepsilon\partial\phi/\partial n = -\alpha_f$ at the interface of a conductor and a dielectric,
4. On the surface of a conductor either ϕ is a given constant ϕ_o, or ϕ is an unknown constant and charge Q_o is specified.

For such an electrostatic system, we may introduce a uniqueness theorem, which states that the solution of the electrostatic system is uniquely determined by the values either of the potential ϕ or of the charges specified on the surfaces of conductors of the system.

For magnetostatic systems, there exists also a similar uniqueness theorem [see, e.g., Stratton (1941)]. The proof of these uniqueness theorems is similar to the approach presented above, that is, by first assuming the contrary by introducing two possible solutions, and then show the assumption is false.

1.5.12 Momentum of Electromagnetic Fields

In above sections, we have discussed the energy of electromagnetic fields. Now, let us study the momentum of electromagnetic fields. The fact that the electromagnetic field can posses momentum and energy makes the field very real, and so, for better understanding, the original idea that there are just the forces between particles has to be modified to the idea that a particle makes a field, and a field acts on another particle, and the field itself has much familiar properties as energy content and momentum, just as particles can have. To formulate this idea mathematically, let us consider a system of charged particles located in a finite region V in free space. We assume that there are only the Lorentz force $\boldsymbol{f} = \rho_e\boldsymbol{E} + \boldsymbol{J}\times\boldsymbol{B}$ acting on the system of charges with the charge density ρ_e. By Maxwell's equations (1.3.38) and (1.3.41), we may write

$$\boldsymbol{f} = \varepsilon_o(\nabla\cdot\boldsymbol{E})\boldsymbol{E} + \frac{1}{\mu_o}(\nabla\times\boldsymbol{B})\times\boldsymbol{B} - \varepsilon_o\frac{\partial\boldsymbol{E}}{\partial t}\times\boldsymbol{B} \tag{1.5.87}$$

By Maxwell's equation (1.3.40), Eq.(1.5.87) can be expressed as

$$\boldsymbol{f} = \varepsilon_o(\nabla\cdot\boldsymbol{E})\boldsymbol{E} + \frac{1}{\mu_o}(\nabla\times\boldsymbol{B})\times\boldsymbol{B} + \varepsilon_o(\nabla\times\boldsymbol{E})\times\boldsymbol{E} - \frac{\partial}{\partial t}(\varepsilon_o\boldsymbol{E}\times\boldsymbol{B}) \tag{1.5.88}$$

After some mathematical manipulations for Eq.(1.5.88), we may write

$$\boldsymbol{f} + \frac{\partial\boldsymbol{p}^{em}}{\partial t} = \nabla\cdot\boldsymbol{t}^{em} \tag{1.5.89}$$

in which $\boldsymbol{p}^{em}$ denotes the density of momentum of the electromagnetic field, defined by

$$\boldsymbol{p}^{em} = \varepsilon_o\boldsymbol{E}\times\boldsymbol{B} \tag{1.5.90}$$

and $\boldsymbol{t}^{em}$ is often called the Maxwell stress tensor, defined by

$$t_{kl}^{em} = \varepsilon_o\left(E_kE_l - \frac{1}{2}(E_mE_m)\delta_{kl}\right) + \frac{1}{\mu_o}\left(B_kB_l - \frac{1}{2}(B_mB_m)\delta_{kl}\right) \tag{1.5.91}$$

which characterizes the flux density of momentum of the electromagnetic field. It can be seen that the Maxwell stresses are of purely electromagnetic origin; they are imagined to act on any surface lying in the electromagnetic field, irrespective of whether the surface passes through the charge system (medium) or is spread in free space. Consequently, Maxwell´s stress constitutes a peculiar kind of tension emerging from the electromagnetic field and acting in the surrounding space, even if the latter is empty. The well-known radiation pressure, which will be discussed in some details in Section 1.10.4, is obviously such an experimental evidence.

Since the Lorentz body force $\boldsymbol{f}$ equals to the rate of change of momentum density of the system of charges, we may introduce the density of mechanical (kinetic) momentum $\boldsymbol{p}^{me}$ of the system of charges by

$$\frac{\partial \boldsymbol{p}^{me}}{\partial t} = \boldsymbol{f} = \rho_e\boldsymbol{E} + \boldsymbol{J}\times\boldsymbol{B} \tag{1.5.92}$$

Thus, Eq.(1.5.89) can be written

$$\frac{\partial}{\partial t}(\boldsymbol{p}^{me} + \boldsymbol{p}^{em}) = \nabla\cdot\boldsymbol{t}^{em} \tag{1.5.93}$$

Integrating Eq.(1.5.93) over the volume V of the system of charges, we may get

$$\frac{\partial}{\partial t}\int_V (\boldsymbol{p}^{me} + \boldsymbol{p}^{em})dV = \int_{\partial V} \boldsymbol{t}^{em}\cdot\boldsymbol{n}dS \tag{1.5.94}$$

where $\boldsymbol{n}$ is the unit vector outward normal to the closed surface ∂V. Equation (1.5.94) represents the conservation of the momentum, in which the term on the right-hand side of this equation is the force per unit area transmitted across the surface ∂V and acting on the system of charges and fields inside the volume V. Equation (1.5.93) is the differential form of the conservation of momentum.

If we extend the integral over whole space, the term on the right-hand side of Eq.(1.5.94) vanishes, assuming all sources are located in a finite volume, and we obtain

$$\frac{\partial}{\partial t}(\boldsymbol{P}^{me} + \boldsymbol{P}^{em}) = 0 \tag{1.5.95}$$

where $\boldsymbol{P}^{em}$ and $\boldsymbol{P}^{me}$ denote the total momentum of the electromagnetic field and of the total mechanical momentum for the system of charges, respectively. Equation (1.5.95) implies the fact that the rate of change of the total momentum of the electromagnetic field and the system of charges is zero, which is the law of the conservation of momentum.

To understand better the physical meaning of the momentum of the electromagnetic field, we may recall that, in classical Newtonian mechanics, the momentum of an uncharged body, when changed by external forces, seeks to stay constant and resist such a change. So does an electromagnetic field. A steady magnetic field, for instance, is due to a steady current; if that field is changed by changing the current, it will, by Faraday´s law, induce an electric field that will seek to restore the current and its magnetic field to its previous value—its direction is given by Lenz´s law, and the entire effect is known as self-inductance. In effect, the magnetic field resists being changed.

Similarly, a steady electric field is due to a steady charge distribution. If the electric field is changed by moving the charges, the resulting displacement current $\partial \boldsymbol{D}/\partial t$ will give rise to a magnetic field $\boldsymbol{B}$ according to Ampere´s law by Eq.(1.3.38), and the change in magnetic field will induce an electric field $\boldsymbol{E}$ directed against the displacement of the charges. In effect, the electric field resists being changed.

By noting Eq.(1.5.51), we may find that the density of momentum $\boldsymbol{p}^{em}$ of the electromagnetic field and the density of energy flow $\boldsymbol{S}$ (the Poynting vector) of the electromagnetic field has the following relation:

$$\boldsymbol{p}^{em} = \varepsilon_o \boldsymbol{E} \times \boldsymbol{B} = \varepsilon_o \mu_o \boldsymbol{E} \times \boldsymbol{H} = \frac{1}{c^2}\boldsymbol{S} \tag{1.5.96}$$

Illustratively, for a plane electromagnetic wave, we have

$$\boldsymbol{B} = \frac{1}{c}\boldsymbol{n} \times \boldsymbol{E} \tag{1.5.97}$$

where $\boldsymbol{n}$ is the unit vector of the wave propagation direction. By substitution of Eq.(1.5.97) into Eq.(1.5.90), we may find

$$\boldsymbol{p}^{em} = \frac{\varepsilon_o}{c}(\boldsymbol{E} \cdot \boldsymbol{E})\boldsymbol{n} = \rho_{em} c\boldsymbol{n} \tag{1.5.98}$$

in which $\rho_{em} = \varepsilon_0 E^2/c^2$ may be thought formally as the electromagnetic mass density of the plane electromagnetic wave propagating (moving) with the velocity $c\boldsymbol{n}$.

The time average density of momentum of the plane electromagnetic wave in free space can be written

$$\overline{p^{em}} = \frac{1}{c^2}\bar{S} = \frac{\varepsilon_o}{2}Re(E \times B^*) = \frac{\varepsilon_o}{2c}|E|^2 n \tag{1.5.99}$$

by noting Eq.(1.5.68).

Besides, for the plane electromagnetic wave, we have $S = cwn$, where $w = \varepsilon_0 E^2$ is the energy density of the plane wave, and therefore, have

$$p^{em} = \frac{1}{c^2}S = \frac{w}{c}n \tag{1.5.100}$$

We may note that this relation is valid even when the electromagnetic field is quantized. The quantized electromagnetic field is composed of photons, each of which has the photon energy $\hbar\omega$, where $\hbar$ is the reduced Plank constant, and ω is the radian frequency. By Eq.(1.5.100), every photon carries the momentum $\hbar\omega n/c$.

We have shown that an electromagnetic wave can carry momentum with it. Its consequence on radiation pressure on an object impinged by the wave is obvious and will be discussed in more details in Section 1.10.4. Here, we consider briefly the phenomenon of radiation pressure in terms of photons. Consider a number of photons of the same frequency ω traveling in the same direction n in free space. All the photons in a volume c will strike a unit area perpendicular to their trajectory in the unit time. The number of photons in the volume c is given by $wc/(\hbar\omega)$. Since each photon carries a momentum $\hbar\omega$, the total momentum delivered to the unit area in the unit time is given by $\hbar\omega \times wc/(\hbar\omega) = w$, which gives the radiation pressure.

1.6 MAXWELL´S EQUATIONS FOR MOVING MEDIA

Classical electromagnetic theory for systems at rest can be used to deal with many engineering electromagnetic phenomena which involve no relative motion, but it may fail to describe electromagnetic phenomena involving relative motion. The correct formulation of electrodynamics of moving media has been a challenge to scientists in the pre-relativistic era as well as in modern times. In fact, the studies on the electrodynamics for moving bodies had led to the emergence of Einstein´s theory of relativity in the beginning of the twentieth century. Three years after Einstein´s enunciation of the special theory of relativity, Minkowski formulated a theory for moving media, which is widely used even nowadays. In this section, we shall give a short summary of some key points on the special theory of relativity relevant to the development of electrodynamics for moving media since there already exist many books and articles on this subject of relativity [see, e.g., Rosser (1964), Möller (1972) and Bladel (1984)].

1.6.1 Principles of Relativity and the Lorentz Transformation

The fundamental postulates of Einstein's special theory of relativity are:
1. The postulate of relativity: *All inertial systems of coordinates are equally suitable for the description of all physical phenomena.*
2. The postulate of the constancy of the speed of light: *The speed of light c in free space (vacuum) is the same for all observers and is independent of the motion of the source.*

From the first postulate, we may notice that Einstein had generalized the Galilean relativity principle, which states that the laws of mechanics formulated are absolutely identical in all inertial systems, to all physical phenomena, especially to electromagnetism. According to this postulate of relativity, all physical phenomena should have the same course of development in all systems of inertia, and observers installed in different systems of inertia should, as a result of their experiments, arrive at the establishment of the same laws of nature, especially, the same laws of electrodynamics as well as the same laws of mechanics. Obviously, if we accept Einstein's postulate of relativity, Newton's notion of absolute space loses its meaning since any system of inertia with equally good reason can claim to be the absolute system of reference. This simplifies certainly our description of nature since the transformation between the system of reference and the absolute system becomes unnecessary. However, this simplification has led to the abandonment of our usual notions of time and space due to the second postulate of Einstein's special theory of relativity.

The second postulate comes from the extension of the principle of relativity to electromagnetic phenomena, which are described fundamentally by Maxwell's equations. Since light waves, according to Maxwell's theory of light, are special electromagnetic waves, the velocity with which light is propagated in free space is equal to the universal constant c of the Maxwell equations, and must, therefore, be independent of the state of (uniform) motion of the light source. Obviously, the second postulate is in conflict with our usual kinematic concepts, and implies the fact that the velocity of propagation of interaction between two bodies is finite, which is contradictory to the assumption of instantaneous propagation of interaction used in classical Newtonian mechanics. Consequently, the acceptance of Einstein's special theory of relativity, which is firmly based on accurate experimental observations gathered in the process of the development of physical science, must necessarily lead to a revision of our ordinary concepts of space and time, based on which the classical Newtonian mechanics was founded.

In what follows, we shall show how the electrodynamics for rigid stationary media is generalized to study electromagnetic phenomena for moving rigid bodies with the aid of the special theory of relativity. We then introduce some approximations that are of sufficient accuracy and are of convenience for most of engineering applications.

To describe physical phenomena in nature, it is necessary to introduce a system of reference where a system of coordinates is used to indicate the position of, for instance, a particle in space, and clocks fixed in this system serve to indicate the time. According to the relativity principle of Einstein´s special theory of relativity, all the laws of nature are identical in all inertial systems of reference. In other words, equations expressing the laws of nature are invariant with respect to transformations of coordinates and time from one inertial system to another, that is, they should have the same form in all inertial systems. Electrodynamics as one of the fields of physics has also to satisfy the relativity principle. To formulate a relativistic electrodynamics which satisfies Einstein's special relativity principles, an elegant and convenient way is the four-dimensional formulation with the use of a transformation in a four-dimensional space, which was proposed by Minkowski.

The transformation of coordinates in 4-space, which is quantitatively in accordance with the second postulate, is called the Lorentz transformation. The most general Lorentz transformation can be interpreted as a rigid rotation of axes in the 4-space with coordinates $(x_1, x_2, x_3, x_4) = (x, y, z, ict)$. The rigidity expresses the basic condition for the Lorentz transformation between two coordinate systems K and K'

$$s^2 = \sum_{\nu=1}^{4} x_\nu x_\nu = \sum_{\nu=1}^{4} x'_\nu x'_\nu \tag{1.6.1}$$

where the summation is done over the subscript $\nu = 1, 2, 3, 4$.

In general, the transformation equations relating the two sets of coordinates are given by

$$x'_\nu = \sum_{\nu=1}^{4} a_{\nu\lambda} x_\lambda \tag{1.6.2}$$

in which the coefficients $a_{\nu\lambda}$ satisfy the relation $a_{\nu\mu}a_{\lambda\mu} = \delta_{\nu\lambda}$ with $\delta_{\nu\lambda}$ being the Kronecker delta in four-dimensional space. In particular, for the two coordinate systems K and K' in relative motion with a uniform velocity v along their common xx' direction, the coefficients $a_{\nu\lambda}$ of the Lorentz transformation may be written in the following matrix form:

$$[a_{\nu\lambda}] = \begin{bmatrix} \gamma & 0 & 0 & i\beta\gamma \\ 0 & 1 & 0 & 0 \\ 0 & 0 & 1 & 0 \\ -i\beta\gamma & 0 & 0 & \gamma \end{bmatrix} \tag{1.6.3}$$

with $\beta = \mathrm{v}/c$ and $\gamma = (1-\beta^2)^{-1/2}$.

1.6.2 Covariance of Maxwell's Equations

According to the special theory of relativity, the fundamental laws of electrodynamics, Maxwell's equations, must hold in any inertial frame of reference. In fact, the electromagnetism described by Maxwell's equations for free space is already covariant with respect to the Lorentz transformations. The invariance of form (or covariance) of Maxwell's equations may be shown by introducing the following four-tensor expressions:

$$[Z_{\nu\lambda}] = \begin{bmatrix} 0 & H_z & -H_y & -icD_x \\ -H_z & 0 & H_x & -icD_y \\ H_y & -H_x & 0 & -icD_z \\ icD_x & icD_y & icD_z & 0 \end{bmatrix} \tag{1.6.4}$$

and

$$[F_{\nu\lambda}] = \begin{bmatrix} 0 & E_z & -E_y & icB_x \\ -E_z & 0 & E_x & icB_y \\ E_y & -E_x & 0 & icB_z \\ -icB_x & -icB_y & -icB_z & 0 \end{bmatrix} \tag{1.6.5}$$

The pair of Maxwell's equations, $\nabla\times\boldsymbol{H} = \boldsymbol{J} + \partial\boldsymbol{D}/\partial t$ and $\nabla\cdot\boldsymbol{D} = \rho_e$, may now be expressed by

$$\sum_{\lambda=1}^{4} \frac{\partial Z_{\nu\lambda}}{\partial x_\lambda} = J_\nu \qquad (\nu = 1, 2, 3, 4) \tag{1.6.6}$$

with the four-current vector $J_\nu = (J_x, J_y, J_z, ic\rho_e)$, and the another pair of Maxwell's equations: $\nabla\times\boldsymbol{E} = -\partial\boldsymbol{B}/\partial t$ and $\nabla\cdot\boldsymbol{B} = 0$ may be expressed by

$$\sum_{\lambda=1}^{4} \frac{\partial F_{\nu\lambda}}{\partial x_\lambda} = 0 \qquad (\nu = 1, 2, 3, 4) \tag{1.6.7}$$

The equation of conservation of charge: $\nabla\cdot\boldsymbol{J} = -\partial\rho_e/\partial t$ reads

$$\sum_{\nu=1}^{4} \frac{\partial J_\nu}{\partial x_\nu} = 0 \tag{1.6.8}$$

We thus arrive at a set of covariant Maxwell's equations (1.6.6), (1.6.7), and

(1.6.8). It can be seen from Eq.(1.6.8) that relativistically charge density and current are simply different aspects of the same physical quantity.

With the use of the conventional transformation rules for the four-vectors and the four-tensors in the 4-space, that is,

$$J'_\nu = \sum_{\lambda=1}^{4} a_{\nu\lambda} J_\lambda \quad \text{and} \quad F'_{\nu\lambda} = \sum_{k=1}^{4} \sum_{n=1}^{4} a_{\nu k} a_{\lambda n} F_{kn} \tag{1.6.9}$$

we can find that the invariance of Maxwell's equations under the Lorentz transformation results in the following relations between the field quantities ($\boldsymbol{E}$, $\boldsymbol{B}$,...) defined in an inertial laboratory frame of reference $K\{(\boldsymbol{x}, t)\}$ (which may be supposed to be at rest) and the field quantities ($\boldsymbol{E}'$, $\boldsymbol{B}'$,...) defined in a primed inertial frame of reference $K'\{(\boldsymbol{x}', t')\}$ which is moving with a uniform velocity $\mathbf{v}$ relative to the laboratory frame K:

$$\boldsymbol{E}'_{\|} = \boldsymbol{E}_{\|}, \qquad \boldsymbol{E}'_{\perp} = \gamma(\boldsymbol{E} + \mathbf{v} \times \boldsymbol{B})_{\perp} \tag{1.6.10}$$

$$\boldsymbol{B}'_{\|} = \boldsymbol{B}_{\|}, \qquad \boldsymbol{B}'_{\perp} = \gamma(\boldsymbol{B} - c^{-2}\mathbf{v} \times \boldsymbol{E})_{\perp} \tag{1.6.11}$$

$$\boldsymbol{D}'_{\|} = \boldsymbol{D}_{\|}, \qquad \boldsymbol{D}'_{\perp} = \gamma(\boldsymbol{D} + c^{-2}\mathbf{v} \times \boldsymbol{H})_{\perp} \tag{1.6.12}$$

$$\boldsymbol{H}'_{\|} = \boldsymbol{H}_{\|}, \qquad \boldsymbol{H}'_{\perp} = \gamma(\boldsymbol{H} - \mathbf{v} \times \boldsymbol{D})_{\perp} \tag{1.6.13}$$

and

$$\boldsymbol{J}' = \boldsymbol{J} - \gamma\frac{\mathbf{v}}{\mathrm{v}^2}\left(\rho_e \mathrm{v} - \left(1 - \frac{1}{\gamma}\right)(\mathbf{v} \cdot \boldsymbol{J})\right) \tag{1.6.14}$$

$$\rho'_e = \gamma\left(\rho_e - \frac{1}{c^2}(\mathbf{v} \cdot \boldsymbol{J})\right) \tag{1.6.15}$$

where the subscripts $\|$ and $\perp$ are used, respectively, to represent directions parallel and perpendicular to the direction in which K' is moving relative to K with the uniform velocity $\mathbf{v}$.

1.6.3 Covariance of Potential Equations

The potential equations in the Lorentz gauge in free space can be given from Eqs.(1.4.31) and (1.4.32) with $\varepsilon = \varepsilon_o$ and $\mu = \mu_o$:

$$\Box^2 \boldsymbol{A} = -\mu_o \boldsymbol{J} \tag{1.6.16}$$

$$\square^2\phi = -\frac{\rho_e}{\varepsilon_o} \tag{1.6.17}$$

where we have used the D'Alembertian operator defined by

$$\square^2 = \nabla^2 - \frac{1}{c^2}\frac{\partial^2}{\partial t^2} \tag{1.6.18}$$

If we introduce a four-potential vector by

$$A_\nu \equiv \left(\boldsymbol{A}, A_4 = \frac{i}{c}\phi\right) \tag{1.6.19}$$

we may find that the Lorentz condition (1.4.33) in free space can be written

$$\sum_{\nu=1}^{4} \frac{\partial A_\nu}{\partial x_\nu} = 0 \tag{1.6.20}$$

Furthermore, by noting the four-current vector J_ν defined by

$$J_\nu \equiv (\boldsymbol{J}, J_4 = ic\rho_e) \tag{1.6.21}$$

we may write the potential equations (1.6.16) and (1.6.17) in the following form:

$$\square^2 A_\nu = -\mu_o J_\nu \qquad (\nu = 1, 2, 3, 4) \tag{1.6.22}$$

Equation (1.6.22) together with Eq.(1.6.20) are the Lorentz-covariant form of the potential equations.

1.6.4 Plane Wave and the Doppler Effect

From Eq.(1.6.22), we may find wave solutions if $J_\nu = 0$. These solutions are of the type

$$A_\nu(\boldsymbol{x}, t) = A_\nu^{(o)} e^{i(\boldsymbol{k}\cdot\boldsymbol{x} - \omega t)} \tag{1.6.23}$$

Since $\omega = kc$ with $k = |\boldsymbol{k}|$ in free space, we may express the phase part of A_ν as follows:

$$exp(i(\boldsymbol{k}\cdot\boldsymbol{x}-\omega t)) = exp\left(i\sum_{\nu=1}^{4}k_\nu x_\nu\right) \tag{1.6.24}$$

where we have set

$$k_\nu \equiv (\boldsymbol{k}, k_4 = ik) = \left(\boldsymbol{k}, i\frac{\omega}{c}\right) \tag{1.6.25}$$

Since A_ν is a four-vector, we can write

$$A_\nu(\boldsymbol{x}, t) = A_\nu^{(o)} exp\left(i\sum_{m=1}^{4}k_m x_m\right) = \sum_{\lambda=1}^{4}a_{\lambda\nu}A'_\lambda(\boldsymbol{x}', t')$$

$$= \left(\sum_{\lambda=1}^{4}a_{\lambda\nu}A'^{(o)}_\lambda\right) exp\left(i\sum_{m=1}^{4}k'_m x'_m\right)$$

from which we can obtain

$$A_\nu^{(o)} = \sum_{\lambda=1}^{4}a_{\lambda\nu}A'^{(o)}_\lambda \tag{1.6.26}$$

and

$$exp\left(i\sum_{\nu=1}^{4}k_\nu x_\nu\right) = exp\left(i\sum_{\nu=1}^{4}k'_\nu x'_\nu\right) \tag{1.6.27}$$

which shows that the phase factor is a Lorentz invariant, and k_ν by Eq.(1.6.25) is a four-vector.

Consider now a source at rest in a frame K', emitting radiation of frequency ω' in K'. Suppose that the frame K' is moving with respect to a laboratory frame K with velocity v in the x direction. By Eq.(1.6.27), we may write

$$\boldsymbol{k}\cdot\boldsymbol{x}-\omega t = \boldsymbol{k}'\cdot\boldsymbol{x}'-\omega' t' \tag{1.6.28}$$

Using the Lorentz transformation for the four-vector k_ν, we have

$$k'_x = \gamma\left(k_x - \frac{\mathrm{v}}{c^2}\omega\right), \quad k'_y = k_y, \quad k'_z = k_z, \quad \omega' = \gamma(\omega - \mathrm{v}k_x) \tag{1.6.29}$$

and by noting $k_x = kn_x = \omega n_x/c$, we may find

$$\frac{\omega}{\omega'} = \frac{\sqrt{1-\beta^2}}{1-\beta n_x} \tag{1.6.30}$$

where n_x denotes the direction cosine of the plane wave measured by the observer in the laboratory frame K. If K' moves with a velocity $\mathbf{v}$ in a general direction, βn_x is replaced by $\mathbf{v}\cdot\boldsymbol{n}/c$ with $\boldsymbol{n}$ being the direction of observation in K. Then, in general, we have

$$\frac{\omega}{\omega'} = \frac{\sqrt{1-\beta^2}}{1-\mathbf{v}\cdot\boldsymbol{n}/c} \tag{1.6.31}$$

This is the mathematical expression of the so-called *Doppler effect*, which indicates the shift in frequency that is observed at a detector when it moves with respect to a source of waves. In other words, it gives the connection between the frequency ω' in a moving frame K' and the frequency ω observed by an observer in the laboratory frame K. The Doppler effect has been observed in the spectra of the stars, the lines of the spectrum being shifted toward violet or red according as the earth, during its annual motion, moves nearer to or away from the observed star. When the speed of the moving source is low ($v/c \ll 1$), we have $\omega = \omega'(1 + \mathbf{v}\cdot\boldsymbol{n}/c)$, which gives the usual nonrelativistic longitudinal Doppler shift of the frequency. If $\mathbf{v}\cdot\boldsymbol{n} > 0$, the source moves toward the observer; if $\mathbf{v}\cdot\boldsymbol{n} < 0$, the source moves away from the observer. If the source moves in a direction perpendicular to the direction of observation: $\mathbf{v}\cdot\boldsymbol{n} = 0$, we have then $\omega = \omega'(1 - \beta^2)^{1/2}$, which is known to be the transversal Doppler shift of the frequency [Hay et al. (1960)].

1.6.5 Minkowski´s Electrodynamic Theory for Moving Media

The extension of Maxwell´s theory from media at rest to those in motion was a favorite problem of the older electrodynamics. As we probably know, an important characteristic of scientific theory aims at eliminating irrelevant subjective aspects from a field of human experience. In doing so it creates a possibility to focus on important observations. The covariance principle in physics, emphasized in Einstein´s theory of relativity, is a very typical example of this objective procedure which aims at an elimination of the subjective feature of the space-time frame of reference in the formulation of physical relations. In other words, the covariance principle (or the principle of form invariance of physical laws) emphasizes a formulation of physical laws so that observers in different positions and in different states of motion, including accelerated motion, can use these same laws and have unambiguous means of correlating their observations.

For a good understanding of covariance and electromagnetism, it is useful to pursue some details of covariance in the Maxwell equations (1.4.16)–(1.4.19) and the constitutive equations (1.4.26)–(1.4.28) for materials (media) at rest. The complete set of these equations determine the electromagnetic behavior of the media. One question may arise: are Maxwell´s equations (1.4.16)–(1.4.19) and the constitutive equations (1.4.26)–(1.4.28) both affected by a Lorentz transformation, or one of them only? The answer is immediately suggested by the fact that Maxwell´s equations (1.4.16)–(1.4.19) should be valid for any medium, including the vacuum (free-space), while equations (1.4.26)–(1.4.28) characterize the medium. Hence, the Maxwell (field) equations (1.4.16)–(1.4.19) should be form invariant under the Lorentz transformation, and the constitutive equations should be subject to certain specific changes to account for the properties and the state of motion of the medium [Post (1962)]. Following these considerations, we may get a better understanding of the problem of electrodynamic theory for moving media, which was first solved in 1908 by Minkowski for media moving with uniform velocities [Sommerfeld (1952)].

In Minkowski´s theory, to describe electrodynamics of a moving medium, a reference frame, called the laboratory frame, which is an inertial frame with a coordinate system K is introduced. When a material (rigid) body moves with a constant velocity $\mathbf{v}$, relative to the laboratory frame, we attach a reference frame with the coordinate system K' to the moving body, as shown in Fig. 1.12. This reference frame is called the rest frame because to an observer situated at the rest frame, the rigid body appears to be at rest. According to Einstein´s special theory of relativity, Maxwell´s equations in these two coordinate systems K and K' should be invariant. To formulate constitutive equations of moving media, Minkowski´s logic was simple: Firstly, the Maxwell equations formulated originally for stationary media are also valid for a medium moving with a constant (uniform) velocity $\mathbf{v}$ with respect to a “fixed” reference frame, the laboratory frame. Thus, in the rest frame K', Maxwell´s equations are

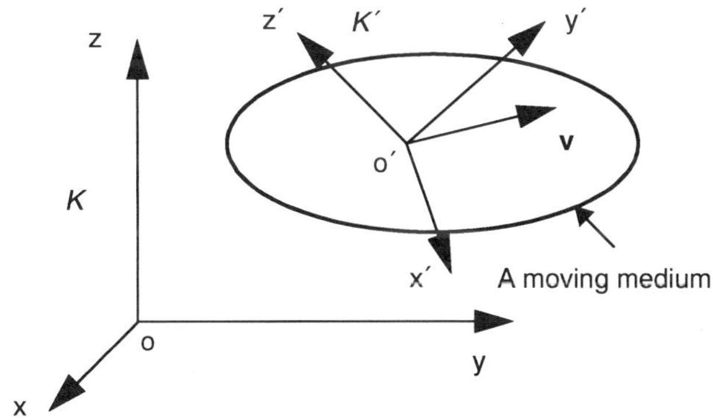

Figure 1.12 Scheme of two coordinate systems for a moving medium.

$$\nabla' \times \boldsymbol{H}' = \boldsymbol{J}' + \frac{\partial \boldsymbol{D}'}{\partial t'}, \qquad \nabla' \cdot \boldsymbol{B}' = 0 \tag{1.6.32}$$

$$\nabla' \times \boldsymbol{E}' = -\frac{\partial \boldsymbol{B}'}{\partial t'}, \qquad \nabla' \cdot \boldsymbol{D}' = \rho_e' \tag{1.6.33}$$

Secondly, suppose that the constitutive equations of a material body in the rest frame K' is known, for instance, by

$$\boldsymbol{D}' = \varepsilon \boldsymbol{E}', \qquad \boldsymbol{B}' = \mu \boldsymbol{H}', \qquad \boldsymbol{J}' = \sigma \boldsymbol{E}' \tag{1.6.34}$$

where $\boldsymbol{D}'$, $\boldsymbol{E}'$, $\boldsymbol{B}'$, $\boldsymbol{H}'$, and $\boldsymbol{J}'$ denote the dielectric displacement vector, the electric field vector, the magnetic induction field vector, the magnetic intensity vector, and the current density vector, respectively. Here, we have attached a prime to all variables in these equations to signify that they are defined in the rest-frame K'. ε, μ, and σ are, respectively, the permittivity, the permeability, and the conductivity of the medium. These material constants have the same values as if the material body were at rest with respect to the laboratory. The constitutive equations for the medium moving with the constant velocity $\mathbf{v}$ relative to the laboratory frame K are obtained by transforming the field variables from the K'–frame to the K–frame according to the well-known Lorentz transformation in the Minkowski theory. In particular, at the low-velocity approximation ($|\mathbf{v}| << c$ the velocity of light), we have

$$\boldsymbol{E}' = \boldsymbol{E} + \mathbf{v} \times \boldsymbol{B}, \qquad \boldsymbol{D}' = \boldsymbol{D} + \mathbf{v} \times \boldsymbol{H}/c^2 \tag{1.6.35}$$

$$\boldsymbol{H}' = \boldsymbol{H} - \mathbf{v} \times \boldsymbol{D}, \qquad \boldsymbol{B}' = \boldsymbol{B} - \mathbf{v} \times \boldsymbol{E}/c^2 \tag{1.6.36}$$

$$\boldsymbol{J}' = \boldsymbol{J} - \rho_e \mathbf{v}, \qquad \rho_e' = \rho_e \tag{1.6.37}$$

where ρ_e denotes the charge density. Substitution of these equations into Eq.(1.6.34) gives the following set of constitutive equations:

$$\boldsymbol{D} = \varepsilon \boldsymbol{E} + (\varepsilon\mu - \varepsilon_o\mu_o)\mathbf{v} \times \boldsymbol{H} \tag{1.6.38}$$

$$\boldsymbol{B} = \mu \boldsymbol{H} + (\varepsilon_o\mu_o - \varepsilon\mu)\mathbf{v} \times \boldsymbol{E} \tag{1.6.39}$$

$$\boldsymbol{J} = \rho_e \mathbf{v} + \sigma(\boldsymbol{E} + \mathbf{v} \times \boldsymbol{B}) \tag{1.6.40}$$

ignoring terms of the order v^2/c^2 and higher. This completes the Minkowski formulation for material bodies moving with the constant (uniform) velocity.

Minkowski's theory has been successfully used to analyze a number of interesting practical problems involving electromechanical devices, and has also been generalized to study moving deformable media. In general, the local velocity **v** of a moving deformable medium may vary from point to point and may also change with time. To study the moving deformable medium with acceleration, one often adopts the concept of the instantaneous (pointwise) rest-frame K' (an instantaneous co-moving inertial frame), in which the observer performs his measurements. Furthermore, one often assumes, as a working hypothesis, that the electrical properties of a continuous medium in the instantaneous rest-frame K' are unaffected by the acceleration for sufficiently low acceleration [see, e.g., Dunkin and Eringen (1963), and Bladel (1976)]. In such a case, the application of the Minkowski theory is straightforward. Indeed, such an approach has widely been used in modeling electromagnetic behaviors of many types of non-superconducting materials, which may be deformed dynamically and/or moving with accelerations. A question may arise whether or not the Minkowski theory could be used to model moving superconductors. This question will be discussed in Chapter 7 when we deal with electrodynamics for moving deformable superconductors.

1.7 ELECTROMAGNETO-QUASISTATIC APPROXIMATIONS

1.7.1 Galilean Approximation

It has been shown that Maxwell's equations of classical electrodynamics are invariant under the Lorentz transformation. However, Newton´s laws of classical mechanics are not invariant under the Lorentz transformation, but are invariant under the *Galilean transformation*, defined by

$$\boldsymbol{x}' = \boldsymbol{x} - \mathbf{v}t, \qquad t' = t \tag{1.7.1}$$

where **v** is the uniform velocity of the inertial frame $K'\{(\boldsymbol{x}', t')\}$ with respect to the inertial (laboratory) frame $K\{(\boldsymbol{x}, t)\}$. For simplicity, it is assumed here that the origins of the two systems of coordinates coincide at the time $t = 0$. The second relation in Eq.(1.7.1) states that the parameter describing the time is the same in all inertial systems, which means that the time is an absolute quantity in the Newtonian description of mechanics. Thus a consistent formulation of electromagneto–mechanics should be to modify the mechanical equations relativistically, which obviously will result in a set of relativistic equations being too complicated for many engineering applications, as we may have noticed from discussions in the above section.

However, if we realize that, for most problems of engineering applications, classical Newtonian mechanics is sufficiently accurate and relativistic effects are negligible in most cases where the velocity involved in considered kinematic processes is much smaller than the velocity of light, we may well introduce a so-called Galilean approximation in which the mechanical equations and Maxwell's equations are both treated classically. Indeed, from Eq.(1.6.3), we may notice that the Lorentz transformation becomes identical to the Galilean transformation when v << c so that $\beta = \mathrm{v}/c$ is very small and $\gamma = 1$. Besides, different from microscopic particles, for macroscopic material objects, it is practically very difficult, if it were not impossible, to force them to move in speeds near that of light, due to the fact that tremendous energy would be required to accelerate the macroscopic objects to such high speeds. Thus it is convenient and practically valid to introduce the Galilean approximation to study mechanical phenomena for macroscopic material objects. Furthermore, we may modify Maxwell's equations to their quasistatic forms which are then invariant under the Galilean transformation. In this way, we may consistently treat the classical mechanics and electrodynamics within the scope of *Galilean relativity* with sufficient accuracy for studying "low-frequency" electromagneto-mechanical phenomena. Actually, the so-called "low-frequency" approximation can often be used to describe time-varying phenomena with frequencies up to microwave frequencies in many cases.

1.7.2 Maxwell′s Equations at Magneto-Quasistatic Approximation

In this section, we shall introduce a set of Maxwell's equations at the magneto-quasistatic approximation expressed by

$$\nabla \times \boldsymbol{E} = 0 \tag{1.7.2}$$

$$\nabla \cdot \boldsymbol{B} = 0 \tag{1.7.3}$$

$$\nabla \cdot \boldsymbol{J} = 0 \tag{1.7.4}$$

$$\nabla \times \boldsymbol{E} + \frac{\partial \boldsymbol{B}}{\partial t} = 0 \tag{1.7.5}$$

with $\boldsymbol{B} = \mu_0(\boldsymbol{H} + \boldsymbol{M})$. Here, the displacement current $\boldsymbol{D}$ and free charge density ρ_e have been omitted in the magneto-quasistatic approximation. We may notice from this set of equations that, even with time-varying sources, the magnetic intensity field $\boldsymbol{H}$ and the magnetic induction field $\boldsymbol{B}$ are determined as if the system were magnetostatic. Then, the electric field $\boldsymbol{E}$ is found from the resulting flux density by using Eq.(1.7.5). This is the origin of the term magneto-quasistatic approximation. Such an approximation is of sufficient accuracy for describing "low-frequency" phenomena in good conductors.

By noting the following relations for the differential operators in the two coordinate systems K' and K

$$\nabla' = \nabla, \qquad \frac{\partial}{\partial t'} = \frac{\partial}{\partial t} + \mathbf{v} \cdot \nabla \tag{1.7.6}$$

a consistent set of field transformation relations, which satisfy the invariance of the magneto-quasistatic equations under the Galilean transformation, may be found

$$\boldsymbol{H}' = \boldsymbol{H} \tag{1.7.7}$$

$$\boldsymbol{B}' = \boldsymbol{B} \tag{1.7.8}$$

$$\boldsymbol{J}' = \boldsymbol{J} \tag{1.7.9}$$

$$\boldsymbol{E}' = \boldsymbol{E} + \mathbf{v} \times \boldsymbol{B} \tag{1.7.10}$$

from which we may also find the transformation relation for the magnetization $\boldsymbol{M}' = \boldsymbol{M}$.

The integral form of Faraday's law of induction for the magneto-quasistatic system may be of interest and can be given by

$$\oint_L \boldsymbol{E}' \cdot d\boldsymbol{l} = -\frac{d}{dt}\int_S \boldsymbol{B} \cdot d\boldsymbol{S} \tag{1.7.11}$$

where S is a material surface enclosed by a contour L, which is moving in space, and $\boldsymbol{E}'$ the electric field measured in the moving medium. The total time derivative of the flux

$$\frac{d}{dt}\int_S \boldsymbol{B} \cdot d\boldsymbol{S} = \lim_{\Delta t \to 0}\frac{1}{\Delta t}\left[\int_{S+\Delta S} \boldsymbol{B}(\boldsymbol{x} + \mathbf{v}\Delta t, t + \Delta t) \cdot d\boldsymbol{S} - \int_S \boldsymbol{B}(\boldsymbol{x}, t) \cdot d\boldsymbol{S}\right] \tag{1.7.12}$$

where ΔS denotes the change of the surface S as a result of the motion in a time interval Δt, can be given by the well-known Helmholtz´s formula:

$$\frac{d}{dt}\int_S \boldsymbol{B} \cdot d\boldsymbol{S} = \int_S \left[\frac{\partial \boldsymbol{B}}{\partial t} + \mathbf{v}(\nabla \cdot \boldsymbol{B}) + \nabla \times (\boldsymbol{B} \times \mathbf{v})\right] \cdot d\boldsymbol{S} \tag{1.7.13}$$

Thus, from Eqs.(1.7.11) and (1.7.13), we may obtain the following form of differential equation:

$$\nabla \times (\boldsymbol{E}' - \mathbf{v} \times \boldsymbol{B}) + \frac{\partial \boldsymbol{B}}{\partial t} = 0 \tag{1.7.14}$$

By noting Eq.(1.7.10), the argument of the curl in Eq.(1.7.14), $\boldsymbol{E}' - \mathbf{v} \times \boldsymbol{B}$ actually represents the electric field $\boldsymbol{E}$ measured by a stationary observer in the laboratory frame of reference K. This result implies the fact that the differential formulation of Faraday's law of induction is independent of the motion of the medium inside the field.

The interface conditions at an interface between two different material media which are moving in space can be found for the magneto-quasistatic system:

$$\boldsymbol{n} \times [\boldsymbol{E} + \mathbf{v} \times \boldsymbol{B}] = 0 \tag{1.7.15}$$

$$\boldsymbol{n} \cdot [\boldsymbol{B}] = 0 \tag{1.7.16}$$

$$\boldsymbol{n} \times [\boldsymbol{H}] = \boldsymbol{K}_f \tag{1.7.17}$$

$$\boldsymbol{n} \cdot [\boldsymbol{J}] = -\nabla_T \cdot \boldsymbol{K}_f \tag{1.7.18}$$

where $\nabla_T \cdot \boldsymbol{K}_f$ denotes a tangential (two-dimensional) surface divergence of the free surface current density $\boldsymbol{K}_f$. For example, if the interface coincides with the xy plane with $\boldsymbol{K}_f = (K_{fx}, K_{fy}, 0)$, we have then

$$\nabla_T \cdot \boldsymbol{K}_f = \frac{\partial K_{fx}}{\partial x} + \frac{\partial K_{fy}}{\partial y} \tag{1.7.19}$$

It is worth to mention that $\boldsymbol{K}_f$ vanishes at an interface between two conductors of finite conductivities. In such a case, Eq.(1.7.18) becomes $\boldsymbol{n} \cdot [\, \boldsymbol{J} \,] = 0$, that is, the normal component of current density is continuous across the interface between two conductors of finite conductivities.

1.7.3 Maxwell's Equations at Electro-Quasistatic Approximation

In the case of electro-quasistatics, Maxwell's equations become

$$\nabla \times \boldsymbol{E} = 0 \tag{1.7.20}$$

$$\nabla \cdot \boldsymbol{D} = \rho_e \tag{1.7.21}$$

$$\nabla \cdot \boldsymbol{J} = -\frac{\partial \rho_e}{\partial t} \tag{1.7.22}$$

$$\nabla \times \boldsymbol{H} = \boldsymbol{J} + \frac{\partial \boldsymbol{D}}{\partial t} \tag{1.7.23}$$

with $\boldsymbol{D} = \varepsilon_0 \boldsymbol{E} + \boldsymbol{P}$. Here, the magnetic induction term has been dropped from Faraday's law at the electro-quasistatic approximation. We may also notice that, even with time-varying sources, the electric field $\boldsymbol{E}$ and the electric displacement $\boldsymbol{D}$ are determined as if the system were static. Then, the current density $\boldsymbol{J}$ is determined from the equation of conservation of charges. The magnetic intensity field $\boldsymbol{H}$ (if it is of interest) is found from Eq.(1.7.23). Such an approximation is of sufficient accuracy for describing "low-frequency" phenomena in dielectrics.

Similarly, a consistent set of field transformation relations for the electro-quasistatic system, which satisfy the invariance of the electro-quasistatic equations under the Galilean transformation, may be found

$$\boldsymbol{E}' = \boldsymbol{E} \tag{1.7.24}$$

$$\boldsymbol{D}' = \boldsymbol{D} \tag{1.7.25}$$

$$\rho'_e = \rho_e \tag{1.7.26}$$

$$\boldsymbol{J}' = \boldsymbol{J} - \rho_e \mathbf{v} \tag{1.7.27}$$

$$\boldsymbol{H}' = \boldsymbol{H} - \mathbf{v} \times \boldsymbol{D} \tag{1.7.28}$$

from which we also get the transformation relation for the polarization $\boldsymbol{P}' = \boldsymbol{P}$.

For the electro-quasistatic system, the integral form of Ampere's circuital law may be written

$$\oint_L \boldsymbol{H}' \cdot d\boldsymbol{l} = \int_S \boldsymbol{J}' \cdot d\boldsymbol{S} + \frac{d}{dt}\int_S \boldsymbol{D} \cdot d\boldsymbol{S} \tag{1.7.29}$$

and the integral form of the conservation law of charge is

$$-\int_{\partial V} \boldsymbol{J}' \cdot d\boldsymbol{S} = \frac{d}{dt}\int_V \rho_e dV \tag{1.7.30}$$

where V is a material volume moving in space, and ∂V is the boundary surface of the volume V. $\boldsymbol{H}'$ and $\boldsymbol{J}'$ are, respectively, the magnetic intensity field and the free current density measured in the moving medium. Equations (1.7.29) and (1.7.30) may be shown similarly by carrying out the time derivatives of surface and volume integrals and by noting Eqs.(1.7.27) and (1.7.28). It should be noticed, however, that the given integral forms of Faraday's law of induction, Ampere's circuital law, and the conservation law of charge in Eqs.(1.7.11), (1.7.29) and (1.7.30) are valid only for quasistatic systems, in which the velocity of media are small as compared with the velocity of light.

The interface conditions at an interface between two different material media in the electro-quasistatic approximation are given by

$$\boldsymbol{n} \times [\boldsymbol{E}] = 0 \tag{1.7.31}$$

$$\boldsymbol{n} \cdot [\boldsymbol{D}] = \alpha_f \tag{1.7.32}$$

$$\boldsymbol{n} \times [\boldsymbol{H} - \mathbf{v} \times \boldsymbol{D}] = \boldsymbol{K}_f \tag{1.7.33}$$

$$\boldsymbol{n} \cdot [\boldsymbol{J} - \rho_e \mathbf{v}] = -\nabla_T \cdot \boldsymbol{K}_f - \frac{\partial \alpha_f}{\partial t} \tag{1.7.34}$$

For interfaces between two perfect dielectrics, the free surface charge α_f and the free surface current $\boldsymbol{K}_f$ vanish, provided that no excess charge is supplied to the interface by external agent (such as rubbing it with cat's fur). In such a case, the normal component of the dielectric displacement vector $\boldsymbol{D}$ is continuous across the interface separating two perfect dielectrics.

1.8 ELECTROMAGNETICS OF MAGNETIC SOLIDS

In this section, we shall introduce some basic phenomena, concepts, and theoretical models for some magnetic solids, such as ferromagnetic solids, gyromagnetic solids, and bianisotropic solids.

1.8.1 Saturation Magnetization and Magnetic Hysteresis

An important class of magnetic materials is ferromagnetic material. A ferromagnet is a solid that may exhibit a spontaneous magnetization in the absence of an applied magnetic field. The basic reason for the existence of ferromagnetism is that the magnetic moments of the individual atoms are ordered. These atomic magnetic moments in ferromagnets are almost entirely due to electron spin, with only a small contribution from the orbital motion of the electrons. In an external magnetic intensity field $\boldsymbol{H}$, a distinguishing feature of macroscopic behaviors of ferromagnetic materials may be seen from the shape of its magnetization curve $M(H)$ as shown in Fig. 1.13 [Stoletov (1873a, b)]. It is shown that starting from the initial state ($H = M = 0$), the magnetization of the ferromagnetic sample increases sharply, following the curve *a*, with the increase in the field H and eventually reaches a constant value M_s, called the saturation magnetization, which is the maximum value of the magnetization for the ferromagnet. Some values of $\mu_0 M_s$ for typical ferromagnetic materials are, for instance, 0.79 tesla for supermalloy, 0.69 tesla for nickel, 1.79 tesla for cobalt, and 2.15 tesla for iron [Chikazumi (1964)]. It may be noticed that, since $\boldsymbol{B} = \mu_0(\boldsymbol{H} + \boldsymbol{M})$, the magnetic induction field B cannot in any event reach a saturation value because the component $\mu_0 H$ increases continuously with increasing H. The saturation magnetization of a

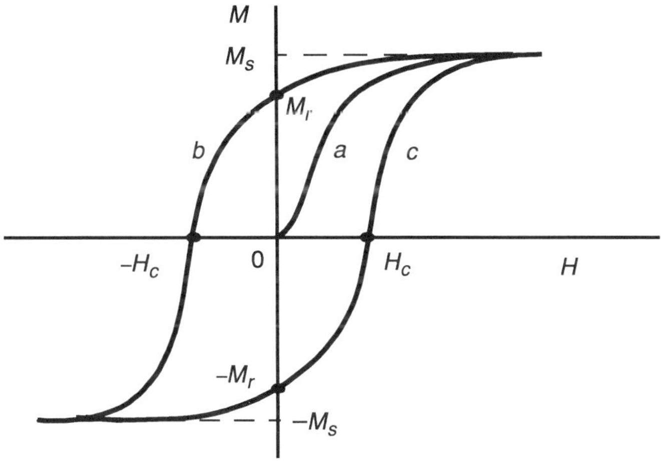

Figure 1.13 Magnetic hysteresis loop.

ferromagnetic body is a structure-insensitive property. However, the approach to the saturation is a structure-sensitive process. The magnitude of the saturation magnetization depends on temperature. It decreases with the increase of temperature and vanishes at a certain (critical) temperature T_{cu}, called the Curie temperature. The Curie temperature separates the ordered ferromagnetic phase (at $T < T_{cu}$) from a disordered paramagnetic phase (at $T > T_{cu}$). This is because at temperatures above T_{cu}, thermal energy destroys the long-range order of the parallel magnetic moments of the ferromagnetic phase. However, the atomic magnetic moments may still be aligned by an applied external magnetic field, so the material is paramagnetic at temperatures above T_{cu}, but there is no spontaneous magnetic moment. Some typical values of the Curie temperature T_{cu} are, for instance, 673 K for supermalloy, 631 K for nickel, 1400 K for cobalt, and 1043 K for iron [Chikazumi (1964) and Brailsford (1966)].

From Fig. 1.13, we may observe that, during the process of magnetization, if we decrease the field H, we find that the magnetization curve does not retrace its path, and M does not decrease as rapidly as it originally increased as shown by the curve b in the figure. Such a phenomenon is known as hysteresis behavior. If we continue decreasing H until $H = 0$, we find that $M = M_r$ ($\neq 0$). This value M_r is called the remnant magnetization of the sample, with a value somewhat lower than M_s. Typical values of $\mu_0 M_r$ are, for instance, 0.5 tesla for supermalloy, 1 tesla for cobalt steel, and 0.5 tesla for iron.

In order to reduce the magnetization of a ferromagnet to zero, a reverse magnetic field must be applied so that $M = 0$ when $H = -H_c$. This value H_c is called the coercive force, which has typical values of, for example, 0.48 $A{\cdot}m^{-1}$ for supermalloy, 18500 $A{\cdot}m^{-1}$ for cobalt steel, and 48000 $A{\cdot}m^{-1}$ for iron. The remnant magnetization and the coercive force are both structure-sensitive properties. The coercive force is often used as the most important single criterion for determining whether a ferromagnetic material is soft or hard. Although the classification is not sharp, materials with the maximum H_c values (coercivity) less than 400 A·m^{-1} are definitely considered soft, and those with coercitivity values greater than 8000 A·m^{-1} are labeled hard. For materials with coercivity values between 400 and 8000 A·m^{-1}, other criteria, such as the energy product $\mu_0 M_r H_c$, must be considered to have a clear classification.

As shown in Fig. 1.13, further strengthening of the negative field may finally bring the sample to the magnetization in the opposite direction $-M_S$. Returning backward, we reach $M = -M_r$ at $H = 0$, then $M = 0$ at $H = +H_c$, and finally the saturation $M = M_S$ again, as shown by following the curve c in Fig. 1.13. Thus, over a complete cycle of H, a closed curve is traced out. This closed curve is called the maximum loop of magnetic hysteresis. If, in a cycle, the maximum values of the magnetization are lower than M_S but the same, the loop will describe a similar symmetrical partial cycle within the maximum loop. If the maximum values of the magnetization are not equal, the loop will have the shape of unsymmetrical partial cycle.

To formulate the magnetic hysteresis behavior of ferromagnetic solids, we may use sometime simple mathematical models without going to details about the structures of ferromagnetic domains and their orientational configurations influenced by external magnetic fields [see, e.g., Potter and Schmulian (1971) and Zhou (1991a)].

Since the process of magnetization involves variations in magnetic moments and movements of domain walls, which are resisted by the crystals, certain energy must be required to flow into the ferromagnetic specimen to provide for the hysteresis loss. If we consider ac fields of periodic variation with a time period t_p, the energy supplied (power loss) per cycle from the magnetic hysteresis is given by

$$W_m = \int_0^{t_p} P(t)dt = \int_0^{t_p} \int_V \boldsymbol{H} \cdot \frac{\partial \boldsymbol{M}}{\partial t} dV dt \tag{1.8.1}$$

If $\boldsymbol{H}$ and $\boldsymbol{M}$ (or $\boldsymbol{B}$) are parallel, and if each is uniform throughout V, Eq.(1.8.1) reduces to

$$W_m = V \oint_C H dM = V \oint_C H dB \tag{1.8.2}$$

in which C denotes the contour of the hysteresis loop. The integral in Eq.(1.8.2) can be recognized as the area enclosed by the hysteresis loop in the H-M plane, as shown in Fig. 1.13. The time average power supplied to account for the hysteresis loss is then simply $1/t_p$ times the value of W_m computed from Eq.(1.8.2).

It should be noted that this power loss from Eq.(1.8.1) [or Eq.(1.8.2)] is unrelated to the losses associated with eddy currents in metallic ferromagetic materials, as shown by Eq.(1.5.58). Eddy current losses, which are caused by changes in the magnetic field, can be minimized by using laminated materials as in transformers, or by using non-conducting materials. Hysteresis losses, however, can be minimized only by selecting a material with a narrow (small) hysteresis loop.

1.8.2 Gyromagetic Media

Among magnetic materials, there exists also a group of materials, called ferrimagnetic materials, which has one essential difference from ferromagnetic materials. Whereas the exchange integral has a positive value for ferromagnetic materials, causing parallel alignment of the net spins of all the atoms, ferrimagnetism corresponds to a negative value of the exchange integral and to anti-parallel spin alignment. (Antiferromagnetism may be viewed as a special case of ferrimagnetism, occurring when the opposed spins are of equal strength.) Prominent among ferrimagnetic materials are ferrites, which have the important characteristics of low loss and strong magnetic effects at microwave frequencies. These are solid materials with a particular type of crystal structure made up of oxygen, iron, and another element, such as lithium, magnesium, zinc, and so forth. In the macroscopic model for ferrites, the magnetization $\boldsymbol{M}$ obeys the following relation:

$$\frac{\partial \boldsymbol{M}}{\partial t} = -\gamma_e \boldsymbol{M} \times \boldsymbol{B} \tag{1.8.3}$$

where γ_e is called the gyromagnetic ratio ($\gamma_e = |e|/m_e = 1.7592\times10^{11}$ C/kg).

As usual, the constitutive relation for $\boldsymbol{B}$ and $\boldsymbol{H}$ is written

$$\boldsymbol{B} = \mu_o(\boldsymbol{H} + \boldsymbol{M}) \tag{1.8.4}$$

Thus, noting that $\boldsymbol{M}\times\boldsymbol{M} = 0$, Eq.(1.8.3) may be expressed as [Elliott (1993)]

$$\frac{\partial \boldsymbol{M}}{\partial t} = -\gamma_e \mu_o \boldsymbol{M} \times \boldsymbol{H} \tag{1.8.5}$$

We now consider a case in which there is a large dc magnetic intensity field $\boldsymbol{H}_o$ is present in the ferrite with a small-signal ac component $\boldsymbol{h}$, added to it, so that we can write

$$\boldsymbol{H} = \boldsymbol{H}_o + \boldsymbol{h}, \qquad \boldsymbol{M} = \boldsymbol{M}_o + \boldsymbol{m} \tag{1.8.6}$$

We then obtain the linearized equation:

$$\frac{\partial \boldsymbol{m}}{\partial t} = -\gamma_e \mu_o (\boldsymbol{M}_o \times \boldsymbol{h} + \boldsymbol{m} \times \boldsymbol{H}_o) \tag{1.8.7}$$

assuming the ac component is much small compared with the dc component. It should be noticed here that the internal dc magnetic intensity field $\boldsymbol{H}_0$ is generally not equal to an externally applied field $\boldsymbol{H}_e$ since $\boldsymbol{H}_o$ may depend on the shape of the ferrite body.

If we consider the time-harmonic case with $\exp(i\omega t)$ dependence, and choose the z axis along the direction of the internal dc magnetic field $\boldsymbol{H}_0$ and $\boldsymbol{M}_0$, we may express Eq.(1.8.7) in the following matrix form:

$$\begin{bmatrix} m_x \\ m_y \\ m_z \end{bmatrix} = \frac{\omega_M}{\omega_o^2 - \omega^2} \begin{bmatrix} \omega_o & i\omega & 0 \\ -i\omega & \omega_o & 0 \\ 0 & 0 & 0 \end{bmatrix} \begin{bmatrix} h_x \\ h_y \\ h_z \end{bmatrix} \tag{1.8.8}$$

in which $\omega_0 = \gamma_e \mu_0 H_0$ is the gyromagnetic response frequency, and $\omega_M = \gamma_e \mu_0 M_0$ is the saturation magnetization frequency. Thus, by noting the ac component of the magnetic induction field

$$b_j = \mu_o (h_j + m_j) = \mu_{jl} h_l \tag{1.8.9}$$

we may find the effective permeability tensor μ_{jl} given by

$$[\mu_{jl}] = \begin{bmatrix} \mu_{11} & \mu_{12} & 0 \\ \mu_{21} & \mu_{22} & 0 \\ 0 & 0 & \mu_o \end{bmatrix} \tag{1.8.10}$$

in which

$$\mu_{11} = \mu_{22} = \mu_o \left(1 + \frac{\omega_o \omega_M}{\omega_o^2 - \omega^2} \right) \tag{1.8.11}$$

$$\mu_{12} = \frac{i\mu_o \omega \omega_M}{\omega_o^2 - \omega^2} \quad \text{and} \quad \mu_{21} = \frac{-i\mu_o \omega \omega_M}{\omega_o^2 - \omega^2} \tag{1.8.12}$$

where we may notice $\mu_{12} = \mu_{21}^*$. The permeability matrix shown by Eq.(1.8.10) is known to be a gyrotropic permeability tensor, which indicates that ferrites exhibit gyromagnetism.

1.8.3 The Faraday Rotation

In 1845, Faraday discovered the rotation of the plane of polarization of light when propagating through some materials under the influence of a dc magnetic field, now called the Faraday rotation. With the discovery of low-loss ferrite materials, the effect of the Faraday rotation has found a variety of applications in microwave engineering.

To understand the phenomenon, let us first consider the characteristics of a plane wave propagating in a gyromagntic medium, characterized generally by a constitutive equation of the form: $B_j = \mu_{jl}H_l$ with μ_{jl} being a gyrotropic permeability tensor, as shown by Eq.(1.8.10). For a plane wave of the form: $\exp(i(\omega t - \boldsymbol{k}\cdot\boldsymbol{x}))$, from Maxwell's equations, we may find

$$\boldsymbol{k} \times \boldsymbol{E} = \omega \boldsymbol{B} \tag{1.8.13}$$

$$\boldsymbol{k} \times \boldsymbol{H} = -\omega \boldsymbol{D} \tag{1.8.14}$$

We assume that the dielectric constant of the gyromagnetic medium considered is a scalar so that $\boldsymbol{D} = \varepsilon \boldsymbol{E}$, which is the case for the ferrites. Thus we may write

$$\boldsymbol{E} = -\frac{1}{\omega\varepsilon}\boldsymbol{k} \times \boldsymbol{H} \tag{1.8.15}$$

and, therefore,

$$\boldsymbol{k} \times (\boldsymbol{k} \times \boldsymbol{H}) = -\varepsilon\omega^2 \boldsymbol{B} = -\varepsilon\omega^2 \mu \cdot \boldsymbol{H} \tag{1.8.16}$$

Suppose now that the plane wave is propagating along z axis direction, that is, $\boldsymbol{k} = k z_o$ with $k = |\boldsymbol{k}|$, and $\boldsymbol{H} = (H_x, H_y, 0)$. We may find the following matrix equation:

$$k^2 \begin{bmatrix} H_x \\ H_y \end{bmatrix} = \varepsilon\omega^2 \begin{bmatrix} \mu_1 & i\mu_2 \\ -i\mu_2 & \mu_1 \end{bmatrix} \begin{bmatrix} H_x \\ H_y \end{bmatrix} \tag{1.8.17}$$

where we have made use of the relations (1.8.11) and (1.8.12). Thus, μ_1 (= μ_{11} = μ_{22}) and μ_2 (= Im(μ_{12}) for the ferrite discussed in above section) are both real constants. It can be shown that Eq.(1.8.17) has a solution determined by

$$k^2 = \varepsilon\omega^2(\mu_1 + \mu_2) \tag{1.8.18}$$

for the clockwise circularly polarized wave (field vectors rotate clockwise as a function of time at a fixed plane in space, looking in the direction of propagation) of the form:

$$\boldsymbol{H}^{cw} = (H_{cw}, -iH_{cw}, 0) \tag{1.8.19}$$

This wave has the phase velocity

$$v_{cw} = \frac{\omega}{k} = \frac{1}{\sqrt{\varepsilon(\mu_1 + \mu_2)}} \tag{1.8.20}$$

Equation (1.8.17) has also a solution determined by

$$k^2 = \varepsilon\omega^2(\mu_1 - \mu_2) \tag{1.8.21}$$

for the counterclockwise circularly polarized wave of the form:

$$\boldsymbol{H}^{ccw} = (H_{ccw}, iH_{ccw}, 0) \tag{1.8.22}$$

This wave has the phase velocity

$$v_{ccw} = \frac{1}{\sqrt{\varepsilon(\mu_1 - \mu_2)}} \tag{1.8.23}$$

Let us now study the behavior of a linearly polarized wave propagating in $+z$ direction, passing through the gyromagnetic medium. We first decompose the wave into two orthogonal circular polarization as

$$\boldsymbol{H} = H_o\begin{bmatrix}1\\0\\0\end{bmatrix} = \frac{H_o}{2}\begin{bmatrix}1\\i\\0\end{bmatrix} + \frac{H_o}{2}\begin{bmatrix}1\\-i\\0\end{bmatrix} \tag{1.8.24}$$

which is supposed to be the initial form of the wave. After the wave has propagated a distance d into the gyromagnetic medium, the magnetic field can be written

$$\boldsymbol{H} = \frac{H_o}{2}\begin{bmatrix}1\\i\\0\end{bmatrix}e^{-ik_{ccw}d} + \frac{H_o}{2}\begin{bmatrix}1\\-i\\0\end{bmatrix}e^{-ik_{cw}d} \tag{1.8.25}$$

in which k_{ccw} is given by

$$k_{ccw} = \omega\sqrt{\varepsilon(\mu_1 - \mu_2)} \tag{1.8.26}$$

and k_{cw} by

$$k_{cw} = \omega\sqrt{\varepsilon(\mu_1 + \mu_2)} \tag{1.8.27}$$

By writing

$$k_{cww} = \frac{k_{cww} + k_{cw}}{2} + \frac{k_{cww} - k_{cw}}{2} \tag{1.8.28}$$

and

$$k_{cw} = \frac{k_{cww} + k_{cw}}{2} - \frac{k_{cww} - k_{cw}}{2} \tag{1.8.29}$$

we may express

$$e^{-ik_{ccw}d} + e^{-ik_{cw}d} = 2e^{-i(k_{cww} + k_{cw})d/2}\cos\left(\frac{(k_{cww} - k_{cw})}{2}d\right) \tag{1.8.30}$$

and

$$i\left(e^{-ik_{ccw}d} - e^{-ik_{cw}d}\right) = 2e^{-i(k_{cww} + k_{cw})d/2}\sin\left(\frac{(k_{cww} - k_{cw})}{2}d\right) \tag{1.8.31}$$

Thus we may express Eq.(1.8.25) in the following form:

$$\boldsymbol{H} = H_o e^{-i(k_{cww} + k_{cw})d/2}\cos\left(\frac{(k_{cww} - k_{cw})}{2}d\right)\begin{bmatrix} 1 \\ \tan\left(\frac{(k_{cww} - k_{cw})}{2}d\right) \\ 0 \end{bmatrix} \tag{1.8.32}$$

which represents a linearly polarized wave that propagates with the propagation constant

$$k = \frac{k_{cww} + k_{cw}}{2} \tag{1.8.33}$$

and whose plane of polarization rotates with the angle

$$\theta = \frac{(k_{cww} - k_{cw})}{2}d \tag{1.8.34}$$

This angle θ is proportional to the propagation distance d. The rotation of a linearly polarized field to another linear polarization of different orientation by a gyrotropic medium (gyromagnetic medium in this example) is known to be the Faraday rotation. The effect of the Faraday rotation has found use of microwave applications for the construction of non-reciprocal microwave networks among others. The first and simplest of the microwave devices using the Faraday rotation effect was the gyrator, which serves to produce a phase shift of 180 degrees in one direction and no phase shift in the opposite direction. Another well-known application of the Faraday rotation effect is the circulator. Some details on these devices may be found in the work of, for instance, Collin (1968).

1.8.4 Bianisotropic Media and Chiral Media

In preceding sections, we have presented an example of analyzing the effect of anisotropy of the gyromagnetic media. In general, the electric and magnetic constitutive relations of non-conducting anisotropic media may be expressed by

$$D_j = \varepsilon_{jl} E_l \qquad \text{and} \qquad B_j = \mu_{jl} H_l \tag{1.8.35}$$

where ε_{jl} and μ_{jl} are the dielectric permittivity tensor and the magnetic permeability tensor respectively. We can further generalize these relations to the following form:

$$\begin{bmatrix} \boldsymbol{D} \\ \boldsymbol{H} \end{bmatrix} = \begin{bmatrix} \boldsymbol{R} & \boldsymbol{L} \\ \boldsymbol{N} & \boldsymbol{Q} \end{bmatrix} \begin{bmatrix} \boldsymbol{E} \\ \boldsymbol{B} \end{bmatrix} \tag{1.8.36}$$

where the coefficients $\boldsymbol{R}$, $\boldsymbol{Q}$, $\boldsymbol{L}$, and $\boldsymbol{N}$ are, in general, 3×3 matrices for bianisotropic media. It may be noticed that, in Eq.(1.8.36), $\boldsymbol{D}$ and $\boldsymbol{H}$ are given in terms of both $\boldsymbol{E}$ and $\boldsymbol{B}$. Here, we consider $\boldsymbol{E}$ and $\boldsymbol{B}$ as the fundamental field quantities since the electromagnetic force depends directly on $\boldsymbol{E}$ and $\boldsymbol{B}$. $\boldsymbol{D}$ and $\boldsymbol{H}$ are thus considered as the derived fields through the constitutive relations.

The set of coefficient matrices $\boldsymbol{R}$, $\boldsymbol{Q}$, $\boldsymbol{L}$, and $\boldsymbol{N}$ satisfies certain symmetric conditions as discussed by Kong (1972). Under time harmonic excitations, these coefficient matrices are usually complex and frequency dependent. From consideration of energy conservation, it is known that $\boldsymbol{R}$, $\boldsymbol{Q}$, $\boldsymbol{L}$, and $\boldsymbol{N}$ satisfy the following symmetric conditions:

$$\boldsymbol{R} = \boldsymbol{R}^+ , \quad \boldsymbol{Q} = \boldsymbol{Q}^+ , \quad \boldsymbol{N} = -\boldsymbol{L}^+ \tag{1.8.37}$$

where the superscript + denotes transpose and complex conjugate. It is shown from Eq.(1.8.37) that $\boldsymbol{N}$ is completely determined by $\boldsymbol{L}$. $\boldsymbol{R}$ and $\boldsymbol{Q}$ are Hermitian, and are symmetrical when they are real.

The constitutive relations for bianisotropic media provide a phenomenological description of a class of materials. The bianisotropic media that received the earliest attention are media in motion. Almost any medium, when it is in motion, is a bianisotropic medium. In Section 1.6.5, we have shown that the electromagnetic properties of a simple electromagnetic medium moving with velocity **v** with respect to a laboratory frame can be described by the constitutive relations (1.6.38)–(1.6.40) at the magneto-quasistatic approximation according to Minkowski's theory, where we may notice that a moving isotropic dielectric/magnetic medium appears to be bianisotropic.

From Eqs.(1.8.35) and (1.8.36), we may say that the anisotropic media is a special case of the bianisotropic media with $\boldsymbol{L} = \boldsymbol{N} = 0$. If $\boldsymbol{R}$, $\boldsymbol{Q}$, $\boldsymbol{L}$, and $\boldsymbol{N}$ are scalar, that is,

$$R_{kl} = R\delta_{kl}, \qquad Q_{kl} = Q\delta_{kl}, \qquad L_{kl} = L\delta_{kl}, \qquad N_{kl} = N\delta_{kl} \tag{1.8.38}$$

the corresponding medium is called the biisotropic medium, which is also called the chiral medium. In particular, for a lossless chiral medium in time-harmonic fields, we may write its constitutive relations by

$$\boldsymbol{D} = \varepsilon \boldsymbol{E} - i\zeta \boldsymbol{B} \tag{1.8.39}$$

$$\boldsymbol{H} = -i\zeta \boldsymbol{E} + \frac{1}{\mu}\boldsymbol{B} \tag{1.8.40}$$

where ε, μ, and ζ are real constants.

Using the Maxwell equations $\nabla\times\boldsymbol{H} = i\omega\boldsymbol{D}$ and $\nabla\times\boldsymbol{E} = -i\omega\boldsymbol{B}$ for the non-conducting chiral medium, we may rewrite Eqs.(1.8.39) and (1.8.40) in the following form:

$$\boldsymbol{D} = \varepsilon(\boldsymbol{E} + \beta\nabla\times\boldsymbol{E}) \tag{1.8.41}$$

$$\boldsymbol{B} = \mu_c(\boldsymbol{H} + \beta\nabla\times\boldsymbol{H}) \tag{1.8.42}$$

with

$$\mu_c = \frac{\mu\varepsilon}{\varepsilon + \mu\zeta^2} \qquad \text{and} \qquad \beta = \frac{\zeta}{\omega\varepsilon} \tag{1.8.43}$$

which shows that, for the biisotropic (chiral) medium, $\boldsymbol{D}$ depends not only on $\boldsymbol{E}$ at a point, but also on the behavior of $\boldsymbol{E}$ in the neighborhood of that point represented by $\nabla\times\boldsymbol{E}$. Such a nonlocal behavior of $\boldsymbol{D}$ is called the spatial dispersion.

We now consider an example of finding the propagation constant k for a plane wave of the form: $\boldsymbol{E} = (E_x, E_y, 0)$ (or $\boldsymbol{H}$ field) with the common factor $\exp(i(\omega t - kz))$ in a non-conducting chiral medium. By Maxwell's equations and Eqs.(1.8.41) and (1.8.42) for the chiral medium, we have

$$\nabla\times\boldsymbol{H} = i\omega\boldsymbol{D} = i\omega\varepsilon(\boldsymbol{E} + \beta\nabla\times\boldsymbol{E}) \tag{1.8.44}$$

$$\nabla \times \boldsymbol{E} = -i\omega \boldsymbol{B} = -i\omega\mu_c(\boldsymbol{H} + \beta\nabla \times \boldsymbol{H}) \tag{1.8.45}$$

By taking the curl of Eq.(1.8.45) and using Eq.(1.8.44), we may find an equation for $\boldsymbol{E}$:

$$(\omega^2\varepsilon\mu_c\beta^2 - 1)\nabla \times (\nabla \times \boldsymbol{E}) + 2\omega^2\varepsilon\mu_c\beta\nabla \times \boldsymbol{E} + \omega^2\varepsilon\mu_c\boldsymbol{E} = 0 \tag{1.8.46}$$

For the plane wave under consideration, this equation becomes

$$\begin{bmatrix} \omega^2\varepsilon\mu_c + k^2(\omega^2\varepsilon\mu_c\beta^2 - 1) & i2k\omega^2\varepsilon\mu_c\beta \\ -i2k\omega^2\varepsilon\mu_c\beta & \omega^2\varepsilon\mu_c + k^2(\omega^2\varepsilon\mu_c\beta^2 - 1) \end{bmatrix} \begin{bmatrix} E_x \\ E_y \end{bmatrix} = 0 \tag{1.8.47}$$

A non-zero solution to this equation is obtained by letting the determinant of the coefficient matrix be zero, that is,

$$[\omega^2\varepsilon\mu_c + k^2(\omega^2\varepsilon\mu_c\beta^2 - 1)]^2 - (2k\omega^2\varepsilon\mu_c\beta)^2 = 0 \tag{1.8.48}$$

from which we may obtain two propagation constants:

$$k_1 = \frac{\omega\sqrt{\varepsilon\mu_c} - \beta\omega^2\varepsilon\mu_c}{1 - \beta^2\omega^2\varepsilon\mu_c} \tag{1.8.49}$$

and

$$k_2 = \frac{\omega\sqrt{\varepsilon\mu_c} + \beta\omega^2\varepsilon\mu_c}{1 - \beta^2\omega^2\varepsilon\mu_c} \tag{1.8.50}$$

It can be seen that if $\beta = 0$, we get $k_1 = k_2 = \omega(\varepsilon\mu)^{1/2}$, which recovers the well-known solution for a plane wave propagating in an isotropic medium. For the biisotropic (chiral) medium, we have two solutions, one of which represents a counterclockwise circularly polarized wave,

$$E_x = -iE_y \tag{1.8.51}$$

with the propagation constant k_1. The another one represents a clockwise circularly polarized wave,

$$E_x = iE_y \tag{1.8.52}$$

with the propagation constant k_2. It may be noticed that, if $\beta > 0$, the phase velocity of the clockwise circularly polarized wave is slower than the phase velocity of the counterclockwise circularly polarized wave. If $\beta < 0$, the phase velocity of the clockwise circularly polarized wave is then higher than the phase velocity of the counterclockwise circularly polarized wave.

1.9 ELECTROMAGNETICS OF CIRCUITS

In electrical and electronic industries, much of engineering design and analysis of electromagnetic interactions is carried out by using the so-called lumped-element circuit models so long as those physical elements modeled have their geometric sizes small compared with the electromagnetic wavelength in question. In this section, we shall introduce some useful concepts in circuit models from the view point of electromagnetic fields. This subject has been discussed in the work of, for instance, Ramo et al. (1984). Here, we shall present the subject in a slightly different way and include some additional materials of possible interest.

1.9.1 Lumped Circuit Elements and Source Functions

Before discussion of circuits, it is helpful to define a standard conventional direction for electric current in a circuit. This is essential if we are to be able to calculate the algebraic sum of the currents at a junction point in the circuit. As we nowadays know, the charge carries in metallic conductors are free electrons which flow from the negative terminal of a voltage source (that is a battery) toward the positive terminal. Since most electric circuits use metallic conductors, it would have been a very happy arrangement to select the direction of electron flow in such conductors as the mathematically conventional current direction. However, experiments in the days of Benjamin Franklin and Michael Faraday had no knowledge of the atomic structure of material. They selected a conventional direction for electric current on the basis of observation. The clearly observable result of electric current through Faraday´s electrolytic suggested that electric current consists of a flow of positively charged particles. Since positive charge carriers flow in the opposite direction to negative charge carriers such as free electrons, all the laws and rules for electric circuits were based on a direction from the positive terminal of the source, through the external circuit, and back to the negative terminal of the source. In this text, we shall follow the conventional rules of associating the current in our circuit diagrams with conventional direction, that is the current in the external circuit flows from the positive terminal to the negative terminal of the voltage source. As long as we keep in mind the purpose of a mathematical convention for current direction, the fact that physical direction of electron flow is just opposite to the mathematical convention should not present any problem.

Let us now start the discussion on lumped circuit elements. The so-called lumped elements of a passive network is built are quantities that are mathematical abstraction of physical reality. They exist by mathematical definition. A simple example may be given to explain this point. Consider a peace of a conductor coil. This coil has not only inductance, but also resistance. To model the property of the coil, we often use two lumped elements, a resistor to model the resistive property of the coil, and an inductor to model the inductive property of the coil. Clearly, neither of them represents the physical reality of the coil.

The resistance R (or conductance G), as shown in Fig. 1.14(a), is the simplest lumped element. The resistance R (or the conductance G) is defined in the relationship between voltage V and current I at its terminals:

$$V(t) = RI(t) \quad \text{or} \quad I(t) = GV(t) \tag{1.9.1}$$

in which the resistance R and conductance G have reciprocal values, that is, $R = 1/G$ for the element. The voltage $V(t)$ and current $I(t)$ are arbitrary functions of the time t, except that they are related to each other by Eq.(1.9.1). Such an expression of the approximate proportionality between the voltage and current is expected to be independent of the frequency of variation of the current and voltage only if the frequency is not too high. The linearity of relation (1.9.1) is approximate for real conducting materials.

The resistance element R may be understood from a peace of a real conductive material in which an electric current is passing through its two ends. From the view point of electromagnetic fields, the voltage V_{ba} across two reference points at the ends of the conductive material can be expressed by

$$V_{ba} = -\int_a^b \boldsymbol{E} \cdot dl = -\int_a^b \frac{1}{\sigma}\boldsymbol{J} \cdot dl \tag{1.9.2}$$

by using Ohm´s law: $\boldsymbol{J} = \sigma\boldsymbol{E}$ with σ being the conductivity. Here, $\boldsymbol{E}$ is the electric field and $\boldsymbol{J}$ is the current density vector inside the conductive material.

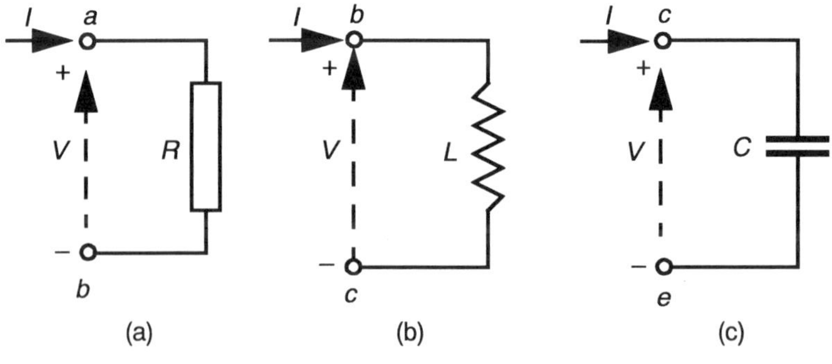

Figure 1.14 Passive lumped circuit elements.

When frequencies are not too high so that the current density $\boldsymbol{J}$ is approximately uniform over the cross section A of the material, Eq.(1.9.2) can be written

$$V_{ba} = I\int_a^b \frac{1}{\sigma A}dl = -IR \tag{1.9.3}$$

where $I = JA$ is the current, and R is the resistance defined by

$$R = \int_a^b \frac{1}{\sigma A}dl \tag{1.9.4}$$

Obviously, the result derived here is only valid at low frequencies. The situation becomes more complicated at high frequencies, where the current density distribution is not uniform over the cross section of the conductive material because of the skin effect, as we shall discussed in later chapters. By comparing Eq.(1.9.3) with Eq.(1.9.1), we find $V = -V_{ba} = V_{ab}$, which may be understood as the voltage (potential) drop $(V_a - V_b)$ across the resistor.

The resistance element R (or the conductance element G) is a circuit element which causes energy dissipation in the circuit. The instantaneous rate of energy dissipation $P(t)$ (instantaneous power dissipation) by the resistance element R (or the conductance element G) is

$$P(t) = V(t)I(t) = RI^2(t) = GV^2(t) \tag{1.9.5}$$

For time-harminic fields, the time-average power dissipation P_{av} in the resistance element R (or the conductance element G) is then given by

$$P_{av} = \frac{1}{2}R|I|^2 = \frac{1}{2}G|V|^2 \tag{1.9.6}$$

which is familiar to us from basic studies in physics. It can be seen that the peak value of the instantaneous power dissipation equals just twice the average value, that is, $P_{peak} = 2P_{av}$.

The second lumped element is the inductance L, as shown in Fig. 14(b). The inductance L of an element is defined in the relationship between voltage V and time-derivative of current I at its terminals:

$$V(t) = L\frac{dI(t)}{dt} \tag{1.9.7}$$

where L is the inductance of the element, which is assumed to be independent of time t in low frequency circuits.

From the view point of electromagnetic fields, the voltage V_{bc} across the terminals of the inductive element comes from the time rate of change of magnetic flux within the inductor. Consider a closed line integral of electric field along the conductor of the coil, returning by the path across the terminals, as shown in Fig. 1.14(*b*). In the calculation of the inductance, we assume that the conductor is perfect (having no resistance). The resistive property of the coil can be taken care of by introducing effectively a separate resistor in series with the inductor. Thus we may write

$$V = V_{bc} = -\int_c^b \mathbf{E} \cdot d\mathbf{l} = -\oint \mathbf{E} \cdot d\mathbf{l} = \frac{\partial}{\partial t}\int_S \mathbf{B} \cdot d\mathbf{S} = L\frac{dI}{dt} \tag{1.9.8}$$

where the inductance L is defined as the magnetic flux linkage per unit of current

$$L = \frac{1}{I}\int_S \mathbf{B} \cdot d\mathbf{S} \tag{1.9.9}$$

Note that we assume here that there is only one turn of the coil so that S is the surface bounded by the coil loop. In the case of a coil having many turns, we may add a contribution each time we follow another turn around the flux. Thus, for N turns, the contribution to the inductance is just N times that of one turn, provided the same flux links each turn. This corresponds to the case of, for instance, a long solenoid.

As we may know from basic physics, the inductance element L is a circuit element which stores magnetic energy in the circuit. The instantaneous stored magnetic energy by the inductance element L is

$$W^m(t) = \frac{1}{2}LI^2(t) \tag{1.9.10}$$

For time-harminic fields, the time-average stored magnetic energy in the inductance element L is

$$W_m = \frac{1}{4}L|I|^2 \tag{1.9.11}$$

It can be seen that the peak value of the instantaneous stored magnetic energy equals just twice the average value, that is, $W^m{}_{peak} = 2W_m$.

The third lumped element is the capacitance C, as shown in Fig. 1.14(*c*). A capacitor consists of a pair of conducting plates from which two wires are brought out to suitable terminals. The plates may be of any shape whatsoever, and are often separated by some dielectric material. So long as the capacitance of the capacitor is concerned, we assume that the plates and wires are perfect conductor, and the insulator between the plates is also perfect so that there no charges can flow across the insulator from one plate to the other. We also assume that the two conductor plates are close to each other so that all field lines which leave one plate end up on the other. There are then always equal and opposite charges on the two plates. Furthermore, we assume that the magnetic fields are negligible so there is no contribution to voltage from changing magnetic fields, but only from the charges on plates of the capacitor.

We now consider the line integral of $\boldsymbol{E}$ around a closed loop starts at terminal c, goes along inside the wire to the top plate of the capacitor, jumps across the space between the plates, passes from the lower plate to terminal e through the wire, and returns to terminal c in the space outside the capacitor, as shown in Fig. 1.14(c). Thus we have

$$\oint \boldsymbol{E}\cdot d\boldsymbol{l} = \int_{\text{along wires}} \boldsymbol{E}\cdot d\boldsymbol{l} + \int_{\text{between plates}} \boldsymbol{E}\cdot d\boldsymbol{l} + \int_e^c \boldsymbol{E}\cdot d\boldsymbol{l} = 0 \qquad (1.9.12)$$

since magnetic fields are ignored. The integral along the wires is zero because there are no electric fields inside perfect conductors. Thus we have from Eq.(1.9.12)

$$V_{ce} = -\int_e^c \boldsymbol{E}\cdot d\boldsymbol{l} = \int_{\text{between plates}} \boldsymbol{E}\cdot d\boldsymbol{l} = \frac{Q}{C} \qquad (1.9.13)$$

where Q is the charge on the upper plate, and C is called the capacitance, as defined in electrostatics.

By noting that the current I entering the capacitor through terminal c (and leaving through terminal e) is equal to dQ/dt, the rate of change of the charge on the plates, we may find the following voltage-current relation:

$$I(t) = C\frac{dV(t)}{dt} \qquad (1.9.14)$$

where $V = V_{ce}$ is the voltage drop $(V_c - V_e)$ across terminals of the capacitor. Here, we have assumed that the capacitance C is not changing with time.

One may wonder how current can flow through such a capacitor with the perfect insulator. The answer is that current in the conventional sense (that associates with charge carries, such as electrons) does not flow through a perfect insulator. The current expressed by Eq.(1.9.14) is the displacement current, which has numerically the same magnitude as the conduction current through the wire to the capacitance element. Obviously, the displacement current is zero in dc cases. Usually, both the conductor plates and the dielectric material used inside a capacitor are not perfect, and have losses due to normal conduction currents. In such a case, these losses are usually taken care of by introducing effectively a conductance element G, in parallel with the capacitance element C.

The capacitance element C is a circuit element, which stores electric energy in the circuit. The instantaneous stored electric energy by the capacitance element C is

$$W^e(t) = \frac{1}{2}CV^2(t) \tag{1.9.15}$$

For time-harmonic fields, the time-average stored magnetic energy in the inductance element L is

$$W_e = \frac{1}{4}C|V|^2 \tag{1.9.16}$$

It can be seen that the peak value of the instantaneous stored electric energy equals just twice the average value, i.e., $W^e{}_{peak} = 2W_e$.

Based on these basic elements R, G, L, and C, linear passive networks can be built by proper interconnection of these elements. In real problems, the passive network must in some way be excited, that is a source of some sort must be present. In the consideration of resistive networks, there are commonly two types of sources, which supply energy, a voltage source and a current source. These are shown illustratively in Fig. 1.15. Both are fictitious in circuit theory since they exist only by definition; but as in the case of the network elements, they suffice, together with appropriate elements, for the representation of actual physical sources.

The voltage source produces at its terminals the voltage $V_s(t)$, the value of which cannot be altered through any circumstances of environment in which the source may find itself. That is to say, the terminal voltage remains precisely $V_s(t)$ regardless of what may be placed across these terminals. An important factor is that the voltage source itself is a short circuit, which means that if a voltage source is bridged across two points in a network, then, except for the voltage $V_s(t)$ existing between them, these points are short-circuited.

In analogous with the voltage source, the current source shown in Fig. 1.15 is regarded as a device that provides a stated current $I_s(t)$ at its terminals.

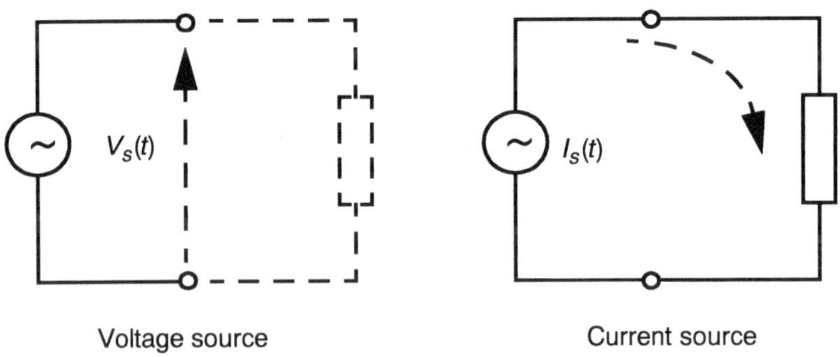

Figure 1.15 Scheme of voltage and current sources.

Since the terminal current is $I_s(t)$ by definition, it cannot be altered through any circumstances of environment in which the source may find itself. That is to say, the terminal current remains precisely $I_s(t)$, regardless of what may be placed across these terminals. Like an open circuit, the current source is a constraint; but it is more general in that it constrains the current at its terminals to any desired value. In this sense, one may say that a current source is a generalized open circuit. Some more detailed discussion about the source functions may be found in the work of Guillemin (1953).

1.9.2 Kirchhoff´s Voltage and Current Laws

Kirchhoff´s voltage law states that for any closed loop of a circuit, the algebraic sum of the voltages for the individual branches of the loop is zero:

$$\sum_k V_k = 0 \tag{1.9.17}$$

To illustrate the Kirchhoff voltage law, let us consider a circuit loop shown in Fig. 1.16. The interpretation of Eq.(1.9.17) by the Kirchhoff voltage law for the circuit is then

$$V_s(t) - RI(t) - L\frac{dI(t)}{dt} - \frac{1}{C}\int I(t)dt = 0 \tag{1.9.18}$$

according to our results about lumped circuit elements. This equation may be understood from Faraday´s law for the closed loop of the circuit:

$$-\oint \mathbf{E} \cdot d\mathbf{l} = -\int_a^b \mathbf{E} \cdot d\mathbf{l} - \int_b^c \mathbf{E} \cdot d\mathbf{l} - \int_c^e \mathbf{E} \cdot d\mathbf{l} - \int_e^a \mathbf{E} \cdot d\mathbf{l} = 0 \tag{1.9.19}$$

Here, we have assumed that there is no magnetic field in the region outside the individual circuit element. This implies also that the contribution to *emf*

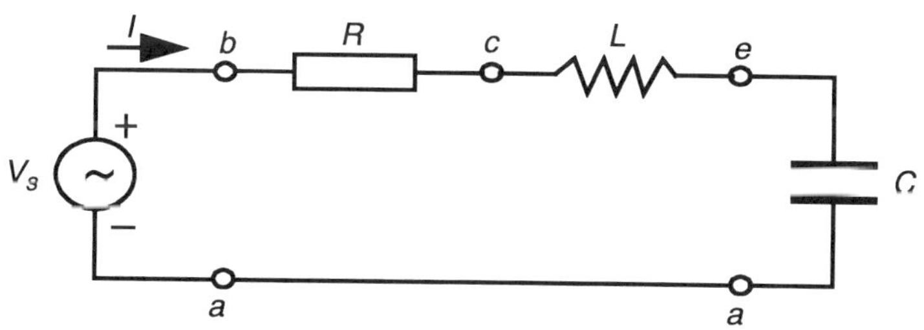

Figure 1.16 A series *RLC* circuit.

generated by any rate of change of magnetic flux within the path defined as the circuit is either negligible, or taken into account effectively in the lumped inductance element in the circuit.

We now consider an analogous law in terms of branch currents: the so-called *Kirchhoff current law*. Since the electric current in a branch of a circuit is the time rate at which charge flows through that branch, the algebraic sum of currents for a group of branches confluent in the same node must be zero, otherwise, electric charges would be either created or destroyed at that node. Kirchhoff´s current law, which in essence expresses the principle of the conservation of charge, states therefore that an algebraic summation of branch currents confluent in the same node must equal zero. Symbolically, this fact is expressed by

$$\sum_k I_k = 0 \tag{1.9.20}$$

as we have discussed in Section 1.1.3 for electrostatic cases. In time-varying cases, from the continuity equation implicit in Maxwell´s equations:

$$\int_{\partial V} \boldsymbol{J} \cdot d\boldsymbol{S} = -\frac{\partial}{\partial t} \int_V \rho dV \tag{1.9.21}$$

we find that there exists possible time rate of change of charge at the junction. In such a case, Eq.(1.9.20) is understood that the time rate of change of charge, if any, accumulating at the junction is either negligible, or taken into account effectively in the lumped capacitance element in the circuit.

1.9.3 Electromagnetic Energy and Power in a Circuit

To study electromagnetic energy and power in a circuit, let us consider again the simple series *RLC* circuit, as shown in Fig. 1.16, which is excited by a voltage source $V_s(t)$. For this circuit, we have Eq.(1.9.18) from Kirchhoff´s voltage law. Suppose each term in this equation is multiplied by $I(t)$. We then have

$$V_s \times I = RI^2 + LI\frac{dI}{dt} + \frac{I}{C}\int Idt \tag{1.9.22}$$

Using Eqs.(1.9.10) and (1.9.15), Eq.(1.9.22) can be written

$$V_s \times I = RI^2 + \frac{d}{dt}(W^m + W^e) \tag{1.9.23}$$

where RI^2 represents the instantaneous rate of energy dissipation by the resistive element in the circuit; $d(W^m + W^e)/dt$ is the instantaneous rate of energy absorption by the energy storage elements; and $V_s \times I$ is the instantaneous rate of energy supply from the source. Thus, Eq.(1.9.23) represents the conservation of energy in the circuit.

In time-harmonic cases, by letting $V_s(t) = V_s e^{i\omega t}$ and $I(t) = Ie^{i\omega t}$, we may get from Eq.(1.9.18)

$$V_s = RI + i\omega\left(LI - \frac{1}{C\omega^2}I\right) \tag{1.9.24}$$

or its conjugate

$$V_s^* = RI^* + i\omega\left(\frac{1}{C\omega^2}I^* - LI^*\right) \tag{1.9.25}$$

Multiplying by I and taking one-half, we obtain

$$\frac{1}{2}IV_s^* = \frac{1}{2}R|I|^2 + i2\omega\left(\frac{1}{4C\omega^2}|I|^2 - \frac{1}{4}L|I|^2\right) \tag{1.9.26}$$

By Eqs.(1.9.11), (1.9.14), and (1.9.16), Eq.(1.9.26) can be written as

$$\frac{1}{2}IV_s^* = \frac{1}{2}R|I|^2 + i2\omega(W_e - W_m) \tag{1.9.27}$$

The complex quantity $IV_s^*/2$ is often called the vector power, which may be written as

$$\frac{1}{2}IV_s^* = P_{av} + iQ_{av} = \frac{1}{2}R|I|^2 + i2\omega(W_e - W_m) \tag{1.9.28}$$

It is shown that the real part of the vector power is the actual average power supplied by the source (also referred to as the active power), while the imaginary part Q_{av} is the reactive power, given by the definition:

$$Q_{av} = 2\omega(W_e - W_m) \tag{1.9.29}$$

It can be seen that the active power is equal to the time average power dissipation in the resistive element. The reactive power is proportional to the difference between the time average energy stored in the electric field and that stored in the magnetic field. It may be noticed that the stored electric energy in the capacitance element C and the stored magnetic energy in the inductance element L are not lost within these elements. Any energy that, may, throughout a given time interval, be absorbed by an inductance or capacitance element must ultimately be wholly returned. Although derived here for the simple series

RLC circuit only, the expression (1.9.29) holds true for all linear passive networks, however complicated (see, for instance, Guillemin (1953)).

A significant feature about the power supplied to a circuit in the time-harmonic (sinusoidal) steady state is that it is not represented by a uniform flow of energy, but rather that this flow has a pulsating character and that energy flows generally in both directions; that is to say, it flows from the circuit back into the source as well as from the source into the circuit. If the circuit contains lossy elements (resistors), then on the average more energy flows into the circuit than is returned to the source. It can be seen from Eq.(1.9.29) that, if the two (electric and magnetic) fields associated with a given circuit store, on the average, equal amounts of energy, then they merely swap a certain amount of energy back and forth between them, and the source is not called upon to enter into this interplay once it has reached a steady state. It is only when $W_e \neq W_m$ that some of the stored energy is continuously played back and forth between the source and the circuit. The reactive power Q_{av} is thus seen to be a measure of the extent to which the source participates in the interplay of stored energy because it is proportional to the excess in the average value of electric as compared with the magnetic stored energy.

1.10 SOME EXAMPLES IN CLASSICAL ELECTRODYNAMICS

In this section, we shall give some examples to illustrate how the classical electromagnetic theory developed in this chapter can be used to analyze problems of practical interest. In particular, examples are selected to illustrate some methods in electromagnetic field analyses, and to introduce some concepts and formulas which can be useful in engineering applications.

1.10.1 Capacitance Calculations of Some Capacitors

We first consider a cylindrical capacitor, which is supposed to be composed of a pair of concentric circular conducting cylinders of infinite length, the inner, of radius a, carrying a charge Q per unit length and the outer, of radius b, carrying a charge $-Q$ per unit length, as shown in Fig. 1.17. Let the region between the two conductors be filled with a homogeneous isotropic dielectric medium of permittivity ε.

From symmetry, the electric displacement vector $\boldsymbol{D}$ ($= \varepsilon \boldsymbol{E}$) must be directed radially outward from the axis and lie in a plane normal to the axis, and its magnitude must depend only on the coordinate r in a cylindrical coordinate system (r, θ, z). Applying the Gauss theorem to the volume enclosed by a closed surface S, that is, two planes, normal to the axis and one unit length apart, and the concentric circular cylinder of radius r when $a < r < b$. Since the plane walls contribute nothing to the surface integral, we have

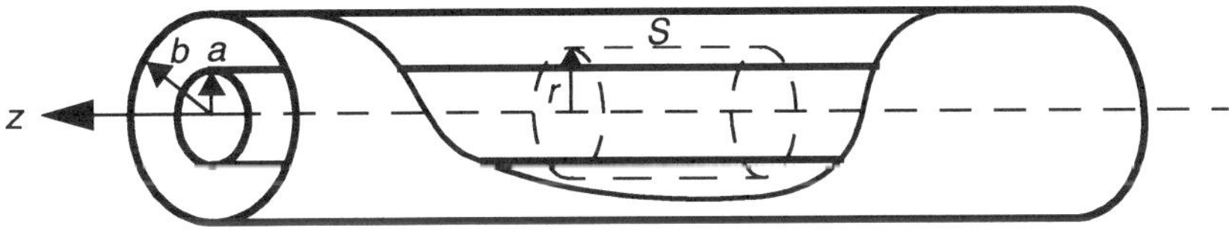

Figure 1.17 Scheme of an infinitely long coaxial cylindrical capacitor.

$$\int_S \varepsilon \boldsymbol{E} \cdot d\boldsymbol{S} = 2\pi r \varepsilon E_r = Q \tag{1.10.1}$$

from which we may obtain the potential difference between the conducting cylinders by

$$V_a - V_b = -\int_b^a E_r dr = \int_a^b \frac{Q}{2\pi\varepsilon r} dr = \frac{Q}{2\pi\varepsilon} \ln\left(\frac{b}{a}\right) \tag{1.10.2}$$

Thus the capacitance per unit length of a long cylindrical capacitor is

$$C = \frac{Q}{V_a - V_b} = \frac{2\pi\varepsilon}{\ln(b/a)} \tag{1.10.3}$$

Similarly, for two infinite parallel conducting plates, carrying surface charge densities σ and $-\sigma$ and being a distance a apart, we may find the potential difference (voltage) V between the plates by

$$V = -\int_0^a E_z dz = \int_0^a \frac{\sigma}{\varepsilon} dz = \frac{\sigma a}{\varepsilon} \tag{1.10.4}$$

from which we may obtain the capacitance of an area A of the plate by

$$C = \frac{\sigma A}{V} = \frac{A\varepsilon}{a} \tag{1.10.5}$$

It may be noticed that the derived formulas (1.10.3) and (1.10.5) both involve the hypothesis of conductors of infinite dimensions. In practice, these two formulas can be used in good approximation if the distance a between the two parallel plates (or b a between the two cylindrical surfaces) is sufficiently small as compared with the lateral dimension of the capacitors so that the fringing effect is negligible. The use of so-called guard rings are such examples [Smythe (1950)].

Finally, let us consider a spherical capacitor, which consists of a solid conductor sphere with the radius a surrounded by a concentric conducting spherical shell with its inner radius b. From symmetry of the problem, we could also apply the Gauss theorem to find a solution. However, to illustrate a different method, we study the problem based on the solution of a partial differential equation. As we know, in electrostatics, the electric potential ϕ ($\boldsymbol{E} = -\nabla\phi$) satisfies the Laplace equation in dielectric regions. For this problem, due to symmetry, the electric potential ϕ satisfies the following equation in a spherical coordinate system:

$$\nabla^2\phi = \frac{\partial^2\phi}{\partial r^2} + \frac{2}{r}\frac{\partial\phi}{\partial r} = 0 \tag{1.10.6}$$

which has a general solution of the form:

$$\phi = \frac{A_1}{r} + A_2 \tag{1.10.7}$$

where A_1 and A_2 are two constants to be determined by boundary conditions.

Suppose that V_a is the potential value at $r = a$ and V_b is the potential value at $r = b$. we may find the potential function ϕ as

$$\phi = \frac{ab}{b-a}\left(\frac{V_a - V_b}{r} + \frac{V_b}{a} - \frac{V_a}{b}\right) \tag{1.10.8}$$

and the electric field E_r by

$$E_r = -\frac{\partial\phi}{\partial r} = \frac{ab(V_a - V_b)}{(b-a)r^2} \tag{1.10.9}$$

Thus the total charge on the inner conductor is given by

$$Q = \int_{\theta=0}^{\pi}\int_{\varphi=0}^{2\pi} \varepsilon E_r(a)a^2\sin(\theta)d\theta d\varphi = \frac{4\pi\varepsilon ab}{b-a}(V_a - V_b) \tag{1.10.10}$$

and the capacitance is given accordingly by

$$C = \frac{Q}{V_a - V_b} = \frac{4\pi\varepsilon ab}{b-a} \tag{1.10.11}$$

In general, for two conductors of arbitrary shapes embedded in a dielectric medium with the permittivity ε, as shown in Fig. 1.18, the capacitance can be found by

$$C = \frac{1}{V_1 - V_2}\int_{S_1} \varepsilon\boldsymbol{E}\cdot d\boldsymbol{S} \tag{1.10.12}$$

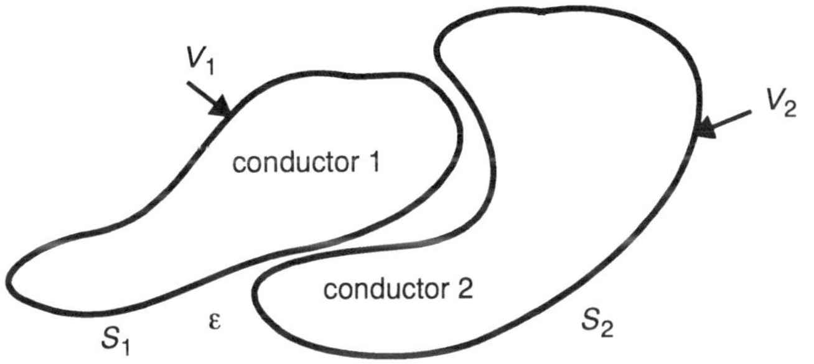

Figure 1.18 Scheme of a capacitor formed by two conductors of arbitrary shapes.

where V_1 and V_2 are, respectively, the electric potential on conductor 1 with surface S_1 and conductor 2 with surface S_2. Formula (1.10.12) is often useful when numerical results can be obtained for the electric field. Nowadays, commercial programs based on the finite element method are available for solving electrostatic field problems numerically, and can, therefore, be used here for the calculation of capacitance of some capacitors with complicated three-dimensional geometries.

1.10.2 Inductance Calculations of Some Inductors

In calculation of inductance, we have learned in Section 1.9.1 that the inductance L for one turn of a coil loop can be found as the magnetic flux linkage per unit of current from Eq.(1.9.9). This formula is useful in many cases when the finite cross-sectional area of the wire is small so that the internal inductance is negligible as compared with the external inductance. In many cases when the internal inductance is of importance, it is more convenient to use the following formula for the calculation of inductance:

$$L = \frac{2W^m}{I^2} = \frac{1}{I^2}\int_{V_\infty} \boldsymbol{B} \cdot \boldsymbol{H} dV \tag{1.10.13}$$

based on the consideration of stored magnetic energy, W^m, in the inductance element. In Eq.(1.10.13), $\boldsymbol{B}$ and $\boldsymbol{H}$ are, respectively, the static magnetic induction field and the static magnetic intensity field, which are induced by the steady current I flowing in the inductance element. For linear magnetic media, we have simply $\boldsymbol{B} = \mu\boldsymbol{H}$ with μ being the permeability. Here, V_∞ denotes the volume extended to infinity.

If we consider the case in which the current flows in a finite region, as shown in Fig. 1.19, Eq.(1.10.13) can also be written in the form:

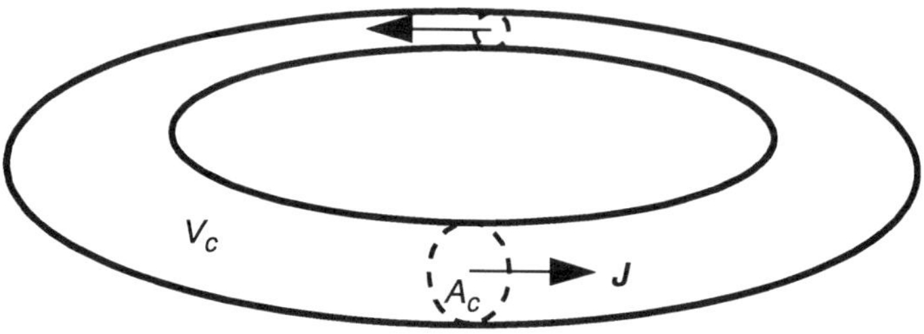

Figure 1.19 Scheme of a three-dimensional inductance element.

$$L = \frac{1}{I^2}\int_{V_c} \boldsymbol{A}\cdot\boldsymbol{J}dV \tag{1.10.14}$$

where $\boldsymbol{A}$ is the magnetic vector potential ($\boldsymbol{B} = \nabla \times \boldsymbol{A}$), $\boldsymbol{J}$ the current density, and V_c the finite volume in which current flows. The total current I in Eq.(1.10.14) can be obtained from

$$I = \int_{A_c} \boldsymbol{J}\cdot\boldsymbol{n}dS \tag{1.10.15}$$

where A_c is the cross-sectional area of the current tube, as shown in Fig. 1.19.

In time-harmonic cases ($\sim e^{i\omega t}$), we often deal with time-average stored magnetic energy. In such cases, we have

$$L = \frac{4W_m}{|I|^2} = \frac{1}{|I|^2}Re\left\{\int_{V_c} \boldsymbol{A}\cdot\boldsymbol{J}^*dV\right\} \tag{1.10.16}$$

where W_m denotes the time-average stored magnetic energy, $\boldsymbol{A}$ and $\boldsymbol{J}$ are the corresponding phasor fields. The superscript * indicates the complex conjugate. This set of formulas is particularly useful when numerical results on the magnetic vector potential $\boldsymbol{A}$ and the distribution of the current density $\boldsymbol{J}$ can be obtained by, for instance, commercially available electromagnetic field solvers.

In many cases, we are interested in finding the inductance for an open current path, such as a peace of bonding wire, and a spiral inductor, etc. In such cases, we may introduce the concept of partial inductance, which is calculated for the part of a current loop of interest. The inductance of the total current loop can be considered as the sum of the partial inductances of each section of the current loop, that is,

$$L = \frac{1}{I^2}\int_{V_c} \boldsymbol{A}\cdot\boldsymbol{J}dV = \sum_{k=1}^{N} L_k \tag{1.10.17}$$

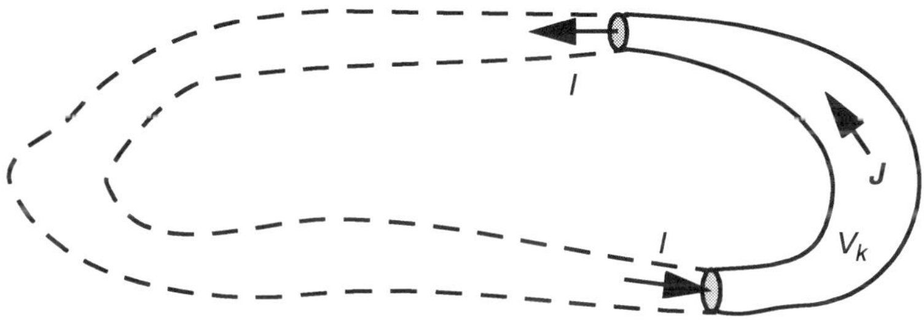

Figure 1.20 Scheme of a section of current loop.

in which the partial inductance L_k of the kth section is defined by

$$L_k = \frac{1}{I^2}\int_{V_k} \boldsymbol{A} \cdot \boldsymbol{J} dV \tag{1.10.18}$$

where V_k denotes the volume of the kth section ($k = 1, 2, ..., N$), as shown in Fig. 1.20.

Let us now present a typical example for the inductance calculation of a long solenoid, as shown in Fig. 1.21. We assume that the solenoid is tightly wound so that the current may be modeled approximately to be circumferential though the wires makes a small helical angle with a cross-sectional plane. At first, we suppose that the solenoid is infinitely long, and the current flowing around the solenoid per unit length is nI, where n is the number of turns per unit length and I is the current flowing in the wire (in each turn).

By Ampere´s circuital law for the loop l_c ($CDEFC$) with CD a unit length, we have

$$\oint_{l_c} \boldsymbol{H} \cdot d\boldsymbol{l} = nI \tag{1.10.19}$$

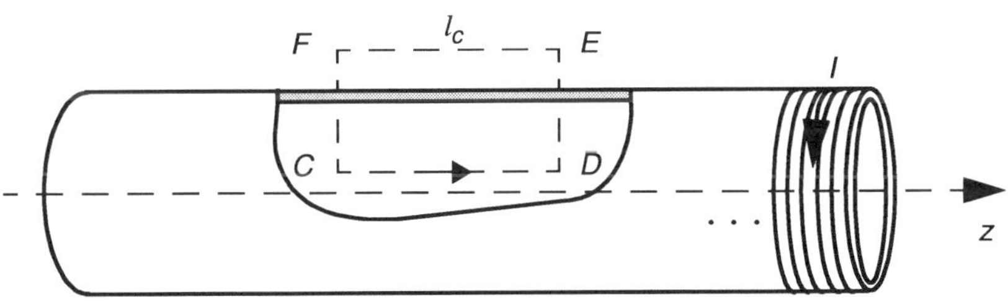

Figure 1.21 Scheme of a section of an infinitely long solenoid.

By noting that the magnetic field outside the infinitely long solenoid is zero, we may find that the magnetic field inside the solenoid is everywhere z-directed and has the value

$$H_z = nI \tag{1.10.20}$$

Thus the inductance L per unit length of the solenoid can be obtained by Eq.(1.9.9) as

$$L = \frac{n}{I}(\mu_o nI\pi a^2) = \mu_o \pi n^2 a^2 \tag{1.10.21}$$

if the medium inside the solenoid is non-magnetic, or free space. Here, a is the radius of the solenoid (the current sheet in the model).

For a solenoid of finite length l, the field inside the solenoid may be assumed to be uniform across any section taken normal to the axis if the solenoid is long compared with its diameter ($l>a$). In such a case, the inductance L of the solenoid can be given approximately by [Smythe (1950)]

$$L = \mu_o \pi n^2 [\sqrt{l^2 + a^2} - a] \tag{1.10.22}$$

Next, we consider two parallel cylindrical wires embedded in a medium with a permeability μ, as shown in Fig. 1.22. The axis of each wire is parallel to the z axis. One wire carries a total current I, and its axis passes through a coordinate point $(-b, 0, 0)$. The other wire carries the return current, which flows in the opposite direction, and its axis passes through the coordinate point $(b, 0, 0)$. Each wire is assumed to have a radius of r_o ($<< b$). We assume that these two wires are far enough apart so that the current distribution in either wire is not much affected by the presence of the other. Thus the magnetic field on the plane $y = 0$ can be expressed approximately by

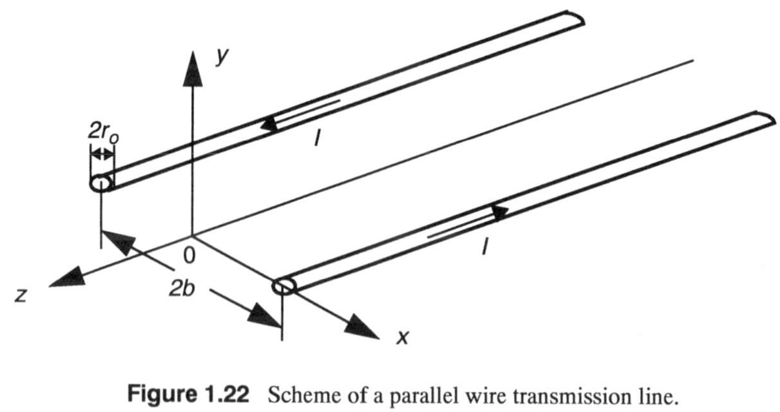

Figure 1.22 Scheme of a parallel wire transmission line.

$$H_y(x, 0) \approx \frac{I}{2\pi(b-x)} + \frac{I}{2\pi(b+x)} \tag{1.10.23}$$

Thus, by Eq.(1.9.9), we may find the external inductance L_{ext} per unit length of the two parallel wire transmission line by

$$L_{ext} = \frac{\mu}{2\pi} \int_{-(b-r_o)}^{b-r_o} \left(\frac{1}{b-x} + \frac{1}{b+x}\right) dx = \frac{\mu}{\pi} \ln\left(\frac{2b}{r_o} - 1\right) \tag{1.10.24}$$

When spacing between two wires is comparable with wire radius, current distribution in the wires is affected. A more accurate result for the external inductance of the parallel wire transmission line can be found by [Ramo et al. (1984)]

$$L_{ext} = \frac{\mu}{\pi} \cosh^{-1}(b/r_o) \tag{1.10.25}$$

Finally, let us consider the calculation of the internal inductance L_{int} per unit length of a infinitely long wire with the radius r_o. By using Ampere´s circuital law, it is easy to find that the magnetic field inside the round wire ($r < r_o$), assuming a uniform distribution of the current density $I/(\pi a^2)$, can be expressed by

$$H_\theta = \frac{Ir}{2\pi r_o^2} \tag{1.10.26}$$

Thus the magnetic energy $W^m{}_{int}$ per unit length inside the wire with a permeability μ can be given by

$$W^m_{int} = \frac{\mu}{2} \int_{\theta=0}^{2\pi} \int_{r=0}^{r_o} H_\theta^2 r\,dr\,d\theta = \frac{\mu I^2}{16\pi} \tag{1.10.27}$$

from which we may get the internal inductance L_{int} per unit length of the wire:

$$L_{int} = \frac{2W^m_{int}}{I^2} = \frac{\mu}{8\pi} \tag{1.10.28}$$

Calculations of inductances for some conductors with more complicated geometries, such as coaxial lines, microstrip lines, and composite striplines, will be given in Chapter 6.

1.10.3 Reflection and Refraction of Electromagnetic Wave at Dielectric Interface

It is known that plane electromagnetic waves may propagate in an infinite and homogeneous dielectric medium. In this section, we shall study the effect of a discontinuity in the medium of propagation. We assume an ideally infinite plane interface between two linear, homogeneous, isotropic dielectric media. Consider now a plane electromagnetic wave of the form:

$$\boldsymbol{E}^{(i)}(\boldsymbol{x}, t) = \boldsymbol{E}_o^{(i)} \exp\left[i\omega^{(i)}\left(t - \frac{\boldsymbol{n}^{(i)} \cdot \boldsymbol{x}}{\mathrm{v}_1}\right)\right] \tag{1.10.29}$$

incident on the interface. Here, $\omega^{(i)}$ is the radian frequency of the incident wave, and $\boldsymbol{n}^{(i)}$ is the unit direction vector of propagation of the incident wave. v_1 is generally complex for a lossy medium. For some simplicity, we shall assume here that the dielectric media are lossless. Thus v_1 is the phase velocity of the wave in medium 1 with the permittivity ε_1 and permeability μ_1. In general, the incident wave along $\boldsymbol{n}^{(i)}$ will give rise to both a reflected wave along $\boldsymbol{n}^{(r)}$ and a transmitted (refracted) wave along $\boldsymbol{n}^{(t)}$, as shown in Fig. 1.23. The three waves have to satisfy the conditions of continuity for the tangential components of the electric field $\boldsymbol{E}$ and of the magnetic field $\boldsymbol{H}$ at the interface.

Suppose that the electric fields for the reflected and the transmitted waves are of the form

$$\boldsymbol{E}^{(r)}(\boldsymbol{x}, t) = \boldsymbol{E}_o^{(r)} \exp\left[i\omega^{(r)}\left(t - \frac{\boldsymbol{n}^{(r)} \cdot \boldsymbol{x}}{\mathrm{v}_1}\right)\right] \tag{1.10.30}$$

and

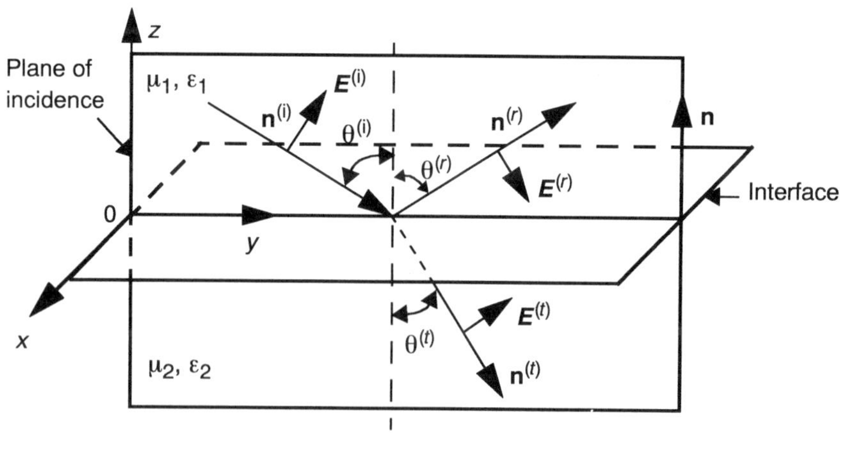

Figure 1.23 Scheme of incident, reflected, and transmitted waves at an interface.

$$\boldsymbol{E}^{(t)}(\boldsymbol{x}, t) = \boldsymbol{E}_o^{(t)} \exp\left[i\omega^{(t)}\left(t - \frac{\boldsymbol{n}^{(t)} \cdot \boldsymbol{x}}{\mathrm{v}_2}\right)\right] \tag{1.10.31}$$

where v_2 is the phase velocity of the transmitted wave in medium 2 with the permittivity ε_2 and permeability μ_2. Generally, the time t and the origin $\boldsymbol{x} = 0$ can be chosen arbitrarily. Here, we shall choose the origin at some convenient point on the interface. The characteristics of both the reflected and the transmitted waves can be determined by the conditions that the sum of the tangential components of $\boldsymbol{E}^{(i)}$ and $\boldsymbol{E}^{(r)}$ just above the interface must be equal to the tangential component of $\boldsymbol{E}^{(t)}$ just below the interface, and that a similar situation holds for magnetic field $\boldsymbol{H}$.

Since the conditions of continuity of the tangential components of both $\boldsymbol{E}$ and $\boldsymbol{H}$ at the interface with its normal unit vector $\boldsymbol{n}$ must be satisfied for all time t and for all points on the interface, we may find that all three vectors $\boldsymbol{E}^{(i)}$, $\boldsymbol{E}^{(r)}$, $\boldsymbol{E}^{(t)}$ must be identical functions of time t, which implies that all three waves must be of the same frequency, that is,

$$\omega^{(i)} = \omega^{(r)} = \omega^{(t)} \tag{1.10.32}$$

Furthermore, we have, at any point $\boldsymbol{x}^{(int)}$ on the interface,

$$\frac{\boldsymbol{n}^{(i)} \cdot \boldsymbol{x}^{(int)}}{\mathrm{v}_1} = \frac{\boldsymbol{n}^{(r)} \cdot \boldsymbol{x}^{(int)}}{\mathrm{v}_1} = \frac{\boldsymbol{n}^{(t)} \cdot \boldsymbol{x}^{(int)}}{\mathrm{v}_2} \tag{1.10.33}$$

Thus we find that the vector $\boldsymbol{n}^{(i)} - \boldsymbol{n}^{(r)}$ must be normal to the interface and, therefore,

$$\theta^{(i)} = \theta^{(r)} \tag{1.10.34}$$

which means that the angle of reflection is equal to the angle of incidence. Besides, since the vector $\boldsymbol{n}^{(i)} - \boldsymbol{n}^{(r)}$ is parallel to the normal $\boldsymbol{n}$, the three vectors $\boldsymbol{n}^{(i)}$, $\boldsymbol{n}^{(r)}$, $\boldsymbol{n}$ are coplanar. The plane of these three vectors is normal to the interface and is called the plane of incidence. These are the *laws of reflection.*

From Eq.(1.10.33), we have

$$\left(\frac{\boldsymbol{n}^{(i)}}{\mathrm{v}_1} - \frac{\boldsymbol{n}^{(t)}}{\mathrm{v}_2}\right) \cdot \boldsymbol{x}^{(int)} = 0 \tag{1.10.35}$$

which indicates the vector $\boldsymbol{n}^{(t)}$ is also in the plane of incidence. Moreover, the tangential components of $\boldsymbol{n}^{(i)}/\mathrm{v}_1$ and $\boldsymbol{n}^{(t)}/\mathrm{v}_2$ must be equal, that is,

$$\frac{\sin\theta^{(i)}}{\mathrm{v}_1} = \frac{\sin\theta^{(t)}}{\mathrm{v}_2} \tag{1.10.36}$$

which may also be written as

$$\frac{\sin\theta^{(t)}}{\sin\theta^{(i)}} = \frac{n_1}{n_2} \tag{1.10.37}$$

where n_1 $(= (\mu_{r1}\varepsilon_{r1})^{1/2})$ and n_2 $(= (\mu_{r2}\varepsilon_{r2})^{1/2})$ are the indices of refraction of medium 1 and medium 2, respectively. Here, μ_{r1}, ε_{r1} and μ_{r2}, ε_{r2} are the relative permeability and the relative permittivity of medium 1 and of medium 2, respectively. Obviously, the refractive index of free space is 1. Equation (1.10.37) is the well-known *Snell's law* of refraction in optics.

By Snell's law, it is noticed that if $n_1 > n_2$, there exists a critical angle $\theta_c^{(i)}$ of incidence, for which $\sin\theta^{(t)} = 1$ and $\theta^{(t)} = 90^{\circ}$. This critical angle of incidence is determined by

$$\sin\theta_c^{(i)} = \frac{n_2}{n_1} \tag{1.10.38}$$

Thus, when $\theta^{(i)} \geq \theta_c^{(i)}$, the wave originating in medium 1 and incident on the interface is totally reflected back into medium 1. Correspondingly, the energy is also totally reflected. This phenomenon, called total reflection, has been observed experimentally. The phenomenon of total reflection is very important at optical frequencies since it provides reflection with less loss than from conducting mirrors. The use in total reflecting prisms is a well-known example. It may be worth notice that, in the case of total reflection, there exists also a transmitted surface wave propagating along the interface, otherwise the interface conditions would not be satisfied with only the incident and reflected waves. A physical explanation of this phenomenon is that a given incident (wave) ray, instead of being reflected abruptly at the interface, penetrates into medium 2, where it is bent back into medium 1, which gives rise to the transmitted surface wave. A detailed analysis on this phenomenon can be found in the book of Lorrain and Corson (1970).

As we have known, the $\boldsymbol{E}$ and $\boldsymbol{H}$ field vectors in a plane electromagnetic wave are always perpendicular to the direction of propagation and to each other. If we suppose that the $\boldsymbol{E}^{(i)}$ vector of the incident wave is normal to the plane of incidence, we may find

$$\boldsymbol{H}^{(i)} = \frac{1}{\mu_1 v_1}\boldsymbol{n}^{(i)} \times \boldsymbol{E}^{(i)} \tag{1.10.39}$$

which is parallel to the plane of incidence. By the conditions of continuity for the tangential components of $\boldsymbol{E}$ and $\boldsymbol{H}$ at the interface, we have

$$\boldsymbol{n} \times (\boldsymbol{E}^{(i)} + \boldsymbol{E}^{(r)}) = \boldsymbol{n} \times \boldsymbol{E}^{(t)} \tag{1.10.40}$$

and

$$\boldsymbol{n} \times \left(\frac{1}{\mu_1 v_1}\boldsymbol{n}^{(i)} \times \boldsymbol{E}^{(i)} + \frac{1}{\mu_1 v_1}\boldsymbol{n}^{(r)} \times \boldsymbol{E}^{(r)}\right) = \boldsymbol{n} \times \left(\frac{1}{\mu_2 v_2}\boldsymbol{n}^{(t)} \times \boldsymbol{E}^{(t)}\right) \tag{1.10.41}$$

which may also be expressed as

$$E_o^{(i)} + E_o^{(r)} = E_o^{(t)} \tag{1.10.42}$$

and

$$\frac{1}{\mu_1 v_1}(E_o^{(i)} - E_o^{(r)})\cos\theta^{(i)} = \frac{1}{\mu_2 v_2}E_o^{(t)}\cos\theta^{(t)} \tag{1.10.43}$$

if we assume that $\boldsymbol{E}^{(i)}$, $\boldsymbol{E}^{(r)}$, $\boldsymbol{E}^{(t)}$ are all along the positive x-axis direction in the coordinate system, shown in Fig. 1.23. From this set of equations, we may find the following relations:

$$\left(\frac{E_o^{(r)}}{E_o^{(i)}}\right)_\perp = \frac{\dfrac{n_1}{\mu_{r1}}\cos\theta^{(i)} - \dfrac{n_2}{\mu_{r2}}\cos\theta^{(t)}}{\dfrac{n_1}{\mu_{r1}}\cos\theta^{(i)} + \dfrac{n_2}{\mu_{r2}}\cos\theta^{(t)}} \tag{1.10.44}$$

and

$$\left(\frac{E_o^{(t)}}{E_o^{(i)}}\right)_\perp = \frac{\dfrac{2n_1}{\mu_{r1}}\cos\theta^{(i)}}{\dfrac{n_1}{\mu_{r1}}\cos\theta^{(i)} + \dfrac{n_2}{\mu_{r2}}\cos\theta^{(t)}} \tag{1.10.45}$$

where the subscript $\perp$ indicates that $\boldsymbol{E}^{(i)}$ is normal to the plane of incidence. For non-magnetic dielectric media, Eqs.(1.10.44) and (1.10.45) become

$$\left(\frac{E_o^{(r)}}{E_o^{(i)}}\right)_\perp = \frac{n_1\cos\theta^{(i)} - n_2\cos\theta^{(t)}}{n_1\cos\theta^{(i)} + n_2\cos\theta^{(t)}} \tag{1.10.46}$$

and

$$\left(\frac{E_o^{(t)}}{E_o^{(i)}}\right)_\perp = \frac{2n_1\cos\theta^{(i)}}{n_1\cos\theta^{(i)} + n_2\cos\theta^{(t)}} \tag{1.10.47}$$

Similarly, if we suppose that the $\boldsymbol{E}^{(i)}$ vector of the incident wave is parallel to the plane of incidence, as shown in Fig. 1.23, we may then find the following relations:

$$\left(\frac{E_o^{(r)}}{E_o^{(i)}}\right)_{\|} = \frac{-\dfrac{n_2}{\mu_{r2}}\cos\theta^{(i)} + \dfrac{n_1}{\mu_{r1}}\cos\theta^{(t)}}{\dfrac{n_2}{\mu_{r2}}\cos\theta^{(i)} + \dfrac{n_1}{\mu_{r1}}\cos\theta^{(t)}} \tag{1.10.48}$$

and

$$\left(\frac{E_o^{(t)}}{E_o^{(i)}}\right)_{\|} = \frac{\dfrac{2n_1}{\mu_{r1}}\cos\theta^{(i)}}{\dfrac{n_2}{\mu_{r2}}\cos\theta^{(i)} + \dfrac{n_1}{\mu_{r1}}\cos\theta^{(t)}} \tag{1.10.49}$$

where the subscript || indicates that $\boldsymbol{E}^{(i)}$ is parallel to the plane of incidence. For non-magnetic dielectric media, Eqs.(1.10.48) and (1.10.49) become

$$\left(\frac{E_o^{(r)}}{E_o^{(i)}}\right)_{\|} = \frac{-n_2\cos\theta^{(i)} + n_1\cos\theta^{(t)}}{n_2\cos\theta^{(i)} + n_1\cos\theta^{(t)}} \tag{1.10.50}$$

and

$$\left(\frac{E_o^{(t)}}{E_o^{(i)}}\right)_{\|} = \frac{2n_1\cos\theta^{(i)}}{n_2\cos\theta^{(i)} + n_1\cos\theta^{(t)}} \tag{1.10.51}$$

These relations (1.10.44)–(1.10.51) are the well-known *Fresnel's equations*. In what follows, we shall discuss in more details about the case for incident wave polarized with its $\boldsymbol{E}^{(i)}$ vector parallel to the plane of incidence. The other case for incident wave polarized with its $\boldsymbol{E}^{(i)}$ vector perpendicular to the plane of incidence can be analyzed similarly. We shall consider only non-magnetic dielectric media for simplicity.

It can be seen from Eq.(1.10.51) that at the interface, the transmitted wave is always in phase with the incident wave. At the interface, the reflected wave may, however, be either in phase, or out of phase with the incident wave. If the following condition:

$$-n_2\cos\theta^{(i)} + n_1\cos\theta^{(t)} > 0 \tag{1.10.52}$$

or, by using Snell´s law,

$$\sin(\theta^{(t)} - \theta^{(i)})\cos(\theta^{(t)} + \theta^{(i)}) > 0 \tag{1.10.53}$$

is satisfied, the reflected wave is in phase with the incident wave from Eq.(1.10.50).

In particular, for

$$\theta^{(t)} + \theta^{(i)} = \frac{\pi}{2} \tag{1.10.54}$$

there is no reflected wave for the incident wave polarized with its $\boldsymbol{E}^{(i)}$ vector parallel to the plane of incidence. This is rather remarkable because it involves the passage of a wave through a discontinuity in the medium of propagation without the production of a reflected wave. This angle of incidence is called the *Brewster angle*, which may be found by using Snell´s law:

$$\frac{\sin\theta_B^{(i)}}{\sin(\pi/2 - \theta_B^{(i)})} = \frac{\sin\theta_B^{(i)}}{\cos\theta_B^{(i)}} = \tan\theta_B^{(i)} = \frac{n_2}{n_1} \tag{1.10.55}$$

For instance, for a typical ratio $n_2/n_1 = 1.5$, we find that $\theta_B{}^{(i)}$ is about 56°. Thus, if a plane wave of mixed polarization is incident on a plane interface with the Brewster angle, the reflected wave is completely plane polarized with $\boldsymbol{E}^{(r)}$ perpendicular to the plane of incidence.

1.10.4 Radiation Pressure on a Good Conductor

In this section, we shall study a phenomenon that is related to reflection of an electromagnetic wave from conductors, namely radiation pressure. We consider an electromagnetic wave incident on a good conductor, which is assumed here to be non-magnetic. It can be shown that the index of refraction n_2 of a good conductor is given by

$$n_2 = \sqrt{\frac{\sigma}{2\omega\varepsilon_o}}(1 - i) \tag{1.10.56}$$

which is complex. Here, σ is the conductivity of the conductor. By Snell´s law, which is also valid if we disregard the fact that n_2 is complex, we have

$$\cos\theta^{(t)} = \sqrt{1 - \left(\frac{n_1}{n_2}\right)^2 \sin^2\theta^{(i)}} \tag{1.10.57}$$

Since $|n_2| >> n_1$ ($n_1 = 1$ in free space) for good conductors in a wide frequency range (e.g., up to the ultraviolet for copper), we find

$$\cos\theta^{(t)} \approx 1, \qquad \theta^{(t)} \approx 0 \tag{1.10.58}$$

which implies that the wave penetrates into the conductor essentially along the normal to the surface of the conductor. Besides, we may find that this transmitted wave is attenuated by a factor of e over one skin depth δ_n, defined by

$$\delta_n = \sqrt{\frac{2}{\omega\mu_o\sigma}} \tag{1.10.59}$$

Thus, whatever the angle of incidence $\theta^{(i)}$, the transmitted wave can be considered to be a plane wave propagating along the normal to the surface, with the enormous damping which is characteristic of electromagnetic waves in good conductors.

For simplicity, let us now study the radiation pressure of an electromagnetic wave incident normally on a good conductor (i.e., $\theta^{(i)} = 0$). We suppose that the incident wave is polarized with its $\boldsymbol{E}^{(i)}$ vector normal to the plane of incidence. Thus, from Fresnel´s equation (1.10.44)–(1.10.45), we have

$$\frac{E_o^{(r)}}{E_o^{(i)}} = \frac{n_1 - n_2}{n_1 + n_2} \tag{1.10.60}$$

and

$$\frac{E_o^{(t)}}{E_o^{(i)}} = \frac{2n_1}{n_1 + n_2} \tag{1.10.61}$$

from which we may, therefore, obtain

$$E^{(t)} = n_1\sqrt{\frac{2\omega\varepsilon_o}{\sigma}}(1+i)E_o^{(i)}e^{z/\delta_n}\exp\left[i\left(\omega t + \frac{z}{\delta_n}\right)\right] \tag{1.10.62}$$

$$H^{(t)} = 2H_o^{(i)}e^{z/\delta_n}\exp\left[i\left(\omega t + \frac{z}{\delta_n}\right)\right] \tag{1.10.63}$$

by Eq.(1.10.56) and for $|n_2| >> n_1$. We notice here that the direction of the $\boldsymbol{E}^{(t)}$ field is along the positive y-axis, and that of the $\boldsymbol{H}^{(t)}$ field is along the positive x-axis in the coordinate system defined in Fig. 1.23.

The time-average electromagnetic force per unit area, p_{av}, on the conductor is then given by

$$p_{av} = \frac{1}{2}\int_{-\infty}^{0} Re\{\sigma\boldsymbol{E}^{(t)} \times (\mu_o\boldsymbol{H}^{(t)})\}dz \tag{1.10.64}$$

which, by using Eqs.(1.10.62) and (1.10.63), becomes

$$p_{av} = \frac{2n_1 E_o^{(i)} H_o^{(i)}}{c\delta_n} \int_{-\infty}^{0} e^{2z/\delta_n} dz = \frac{n_1}{c} E_o^{(i)} H_o^{(i)} \tag{1.10.65}$$

where c is the speed of light in free space. The direction of the time-average force p_{av} is in the direction of propagation of the incident wave. Thus we may call the quantity p_{av} as the time-average pressure exerted on the conductor by the incident radiation, or the radiation pressure.

If the incident wave propagates in free space, we have $n_1 = 1$, and

$$p_{av} = \frac{2\bar{S}^{(i)}}{c} \tag{1.10.66}$$

where $S^{(i)}$ is the time-average of the absolute value of the Poynting vector for the incident wave (see also Section 1.5.9). Thus we may ascribe the radiation pressure to a change in momentum of $2S^{(i)}/c$ per unit time and per unit area in the incident wave. The factor 2 is required because the wave is reflected with a momentum equal to its initial momentum, but of opposite sign.

The radiation pressure of electromagnetic waves has been observed experimentally, and it has been found to agree with the above theory. Usually, radiation pressure is small. In Sunlight, the Poynting vector is approximately 1.4 kw/m^2 at the surface of the earth, giving a radiation pressure of about 9.3×10^{-6} N/m^2 on a metallic reflector. In a laser beam with a power density of 10^{11} watts/m^2, the radiation pressure may reach 660 N/m^2.

1.10.5 Charged Particle in Electromagnetic Fields

In Section 1.6, we have introduced Einstein´s special theory of relativity, with particular emphasis on relativistic formulations of fundamental laws of electrodynamics. Einstein´s special theory of relativity is general and also valid in mechanics. As we may know, in many engineering applications, classical Newtonian mechanics has been very successful in studying mechanical phenomena when the velocity of motion involved is much less than the velocity of light. However, when we deal with fast moving charged particles, such as electrons and protons etc., in electromagnetic fields, relativistic effects may become important. To study behaviors of charged particles in electromagnetic fields, we shall therefore introduce here a relativistic formulation of mechanics for the charged particles, and discuss some consequences of interest.

By defining a four-velocity vector by

$$V_\mu = \left(\frac{\mathbf{v}}{\sqrt{1-\beta^2}}, \frac{ic}{\sqrt{1-\beta^2}} \right) \tag{1.10.67}$$

where $\mathbf{v}$ denotes the instantaneous velocity of the charged particle, and $\beta = \mathrm{v}/c$ with c being the velocity of light, we may introduce the covariance form of mechanical equation of motion:

$$m_o \frac{dV_\mu}{d\tau} = F_\mu \tag{1.10.68}$$

where F_μ is a four-vector, often called the Minkowski force. m_o is the rest mass (a constant) of the charged particle, and τ is the proper time, defined by $d\tau = dt/\gamma$ with $\gamma = (1-\beta^2)^{-1/2}$. The Minkowski force F_μ can be expressed by

$$F_\mu = \gamma\left(\boldsymbol{f}, \frac{i}{c}(\boldsymbol{f}\cdot\mathbf{v})\right) \tag{1.10.69}$$

where $\boldsymbol{f}$ is the Lorentz force on the particle with charge q in electromagnetic fields $\boldsymbol{E}$ and $\boldsymbol{B}$ by

$$\boldsymbol{f} = q(\boldsymbol{E} + \mathbf{v} \times \boldsymbol{B}) \tag{1.10.70}$$

The covariant mechanical equation of motion (1.10.68) describes the general motion of a charged particle in external electromagnetic fields (neglecting the emission of radiation).

We may also write Eq.(1.10.68) in the following form:

$$\frac{dP_\mu}{d\tau} = F_\mu \tag{1.10.71}$$

by introducing a four-momentum vector P_μ by

$$P_\mu = \left(\boldsymbol{p}, \frac{i}{c}W\right) \tag{1.10.72}$$

in which $\boldsymbol{p} = m\mathbf{v}$ with m being the relativistic mass of the particle, defined by

$$m = \gamma m_o = \frac{m_o}{\sqrt{1-(\mathrm{v}/c)^2}} \tag{1.10.73}$$

It may be noticed that if an observer follows the particle, $\mathrm{v} = 0$ and $m = m_o$, whatever be the velocity of the particle with respect to some other observer in a laboratory frame of reference. In the laboratory frame in which the particle is moving with the velocity v, we may however find that the relativistic mass of the particle increases for increasing v. The relativistic mass would become infinite if the velocity v of the particle were equal to the velocity c of light. The fact that the velocity of a particle is always smaller than the velocity of light is a crucial point in relativity.

From the property of the relativistic mass, one might wonder if the electric charge of a particle would also vary with the velocity of the particle. Experimentally, it has been proven that the electric charge is invariant. Possibly the most direct demonstration of the invariance of electric charge is the fact that the charge-to-mass ratio e/m for a charged particle moving at a velocity v is found experimentally to agree with the law:

$$\frac{e}{m} = \frac{e}{m_o}\sqrt{1-(\mathrm{v}/c)^2} \tag{1.10.74}$$

The elementary charge e therefore remains constant, irrespective of the velocity v of the particle, while the mass m varies with the velocity, as shown by Eq.(1.10.73). This relation is found to apply in particle accelerators up to the highest energies attained to date [Lorrain and Corson (1970)].

The quantity W in Eq.(1.10.72) is the relativistic energy of the particle, given by

$$W = \gamma m_o c^2 = \frac{m_o c^2}{\sqrt{1-(\mathrm{v}/c)^2}} \tag{1.10.75}$$

Since the kinetic energy of the particle is zero for v = 0, we may write $W = T + m_o c^2$, in which the term of $m_o c^2$ is called the rest energy, and T is the kinetic energy of the particle, given by

$$T = \frac{m_o c^2}{\sqrt{1-(\mathrm{v}/c)^2}} - m_o c^2 \tag{1.10.76}$$

The existence of the rest energy has been demonstrated by, for instance, the annihilation of electrons, which is a well-known phenomenon to nuclear physicists, that is, a positive electron combines with a negative electron to give two gamma rays, each of which has an energy equal to the rest energy of one electron (0.511 MeV).

By Eqs.(1.10.73) and (1.10.75), we may obtain the following relation:

$$W = \sqrt{p^2 c^2 + m_o^2 c^4} \tag{1.10.77}$$

which is an important relation among the energy, momentum and mass of the particle. An interesting example is for a photon with $m_0 = 0$ and energy $\hbar\omega$. We may find from Eq.(1.10.77) that the photon has a momentum of $p = \hbar\omega/c$, where ω is the radian frequency of the photon.

From the covariant mechanical equation of motion (1.10.71), we may find

$$\frac{d\boldsymbol{p}}{dt} = q(\boldsymbol{E} + \mathbf{v} \times \boldsymbol{B}) \tag{1.10.78}$$

$$\frac{dW}{dt} = q(\boldsymbol{E} + \mathbf{v} \times \boldsymbol{B}) \cdot \mathbf{v} \tag{1.10.79}$$

This is a set of mechanical equations, valid in any inertial system of reference, for a particle with charge q, which may move with a high velocity $\mathbf{v}$ in external electromagnetic fields, $\boldsymbol{E}$ and $\boldsymbol{B}$.

As an example, let us now study the motion of a charged particle with its rest mass m_0 and charge q in a uniform and static magnetic field $\boldsymbol{B}$. In such a case, Eqs.(1.10.78)–(1.10.79) become

$$\frac{d\boldsymbol{p}}{dt} = q\mathbf{v} \times \boldsymbol{B} \tag{1.10.80}$$

$$\frac{dW}{dt} = q(\mathbf{v} \times \boldsymbol{B}) \cdot \mathbf{v} = 0 \tag{1.10.81}$$

From Eq.(1.10.81), we find that the relativistic energy W of the particle is constant in time. Thus the magnitude of the velocity v of the particle is also constant and so are the parameter γ and the relativistic mass $m = \gamma m_o$. Then, Eq.(1.10.80) can be written

$$m\frac{d\mathbf{v}}{dt} = q\mathbf{v} \times \boldsymbol{B} \tag{1.10.82}$$

The motion of the particle described by this equation is a circular motion perpendicular to the magnetic field $\boldsymbol{B}$ and a uniform translation parallel to $\boldsymbol{B}$. By setting $\mathbf{v} = \mathbf{v}_{||} + \mathbf{v}_{\perp}$, where $\mathbf{v}_{||}$ is the velocity component parallel to $\boldsymbol{B}$ and $\mathbf{v}_{\perp}$ is the velocity component perpendicular to $\boldsymbol{B}$, we may find the radius a, called the gyration radius, of the circular motion by

$$a = \frac{m\mathrm{v}_{\perp}}{qB} = \frac{p_{\perp}}{qB} \tag{1.10.83}$$

and the gyration or precession frequency ω_B by

$$\omega_B = \frac{\mathrm{v}_{\perp}}{a} = \frac{qB}{m} = \frac{qB}{m_o}\sqrt{1-(\mathrm{v}/c)^2} \tag{1.10.84}$$

where $\mathrm{v}_{\perp}$ is the magnitude of the velocity component $\mathbf{v}_{\perp}$. Since $\mathrm{v}_{||}$ is constant in time, $\mathrm{v}_{\perp}$ is also constant. It is shown that ω_B decreases with increasing velocity of the particle. The path of the particle is a helix of radius a and pitch angle $\alpha = \tan^{-1}(\mathrm{v}_{||}/\omega_B a)$. Equation (1.10.83) is of a convenient form often used in practice for the determination of particle momenta.

2

Continuum Electrodynamics of Deformable Solids

Continuum electrodynamics of deformable solids is a subject covering electromagnetic phenomena as well as their interaction with mechanical phenomena in material solids which may be deformed or in motion under electromagnetic and/or mechanical loadings. The purpose of this chapter is to give a brief introduction to this subject, its basic concepts, and theoretical principles for the study of phenomena of electromagneto-thermoelastic interaction in material solids. Particular emphasis will be put on electro-quasistatic and magneto-quasistatic problems, which are of practical interest for many engineering applications of electromagnetic–mechanical devices involving velocities, small compared to the velocity of light. Starting from elementary knowledge of continuum mechanics, the subject will be presented systematically and concisely in a way that both electrical and mechanical engineers who have just begun this subject may benefit and can get a quick and global view about this seemingly complicated subject of electromagnetic-mechanical interaction in material solids. Illustratively, continuum models for thermoelastic conductors and dielectrics are formulated in some detail to demonstrate the methodology used in continuum electrodynamics. Continuum models for photothermoelastic solids and elastic dielectric composites are also introduced, which are useful for a variety of engineering problems of practical interest. Some boundary value problems are provided and solved as examples.

2.1 MASS AND MOTION OF CONTINUOUS MEDIA

2.1.1 Mass and Mass Density

It is known that, besides electric charges, a material body has mass which characterizes the property of inertia, that is, the measure of the material body´s resistance to having its momentum changed. In classical mechanics, the mass is assumed to be conserved, that is, the mass of a material body is the same at all time. In continuum mechanics it is further assumed that the mass is an absolutely continuous function of volume. In other words, it is assumed that a positive quantity ρ, the mass density, can be defined at every point in the material body as the limit

$$\rho(X) = \lim_{\Delta V \to 0} \frac{\Delta m}{\Delta V} \tag{2.1.1}$$

where Δm is the total mass contained in the small volume ΔV which is shrinking down upon the position point $\boldsymbol{X}$. It is seen that the definition of the mass density is similar to the definition of charge density presented in Section 1.1.1. Such a concept of the material continuum as a mathematical idealization of the real material is applicable to problems in which fine structures, such as atomic or molecular structures of the matter can be ignored. If the mass density is known at every point in a material body, the total mass of the body can be obtained by means of an integral taken over the volume of the body, that is,

$$m = \int_V \rho dV \tag{2.1.2}$$

It can be understood that the inertial properties of a point mass are fully defined by the value of its mass. The inertial properties of a body of finite dimensions are, however, defined by the law for the distribution of the density throughout the volume of the body. In the case of a rigid body, the inertial characteristics can be completely defined by the total mass of the body, the position of the center of mass (center of gravity), and the moment of inertia tensor at the center of mass.

2.1.2 Motion of Continuum Media

In continuum mechanics, the material body are described mathematically by specifying the positions of the material points of the idealized continuous material medium at time t in a region V of space with the use of a chosen frame of reference. Here, for simplicity, we shall choose the same Cartesian coordinate system to describe the reference configuration of a continuum body at time $t = t_0$ with conventional right-handed set of orthogonal axes O–XYZ, and the current configuration of the body at the current time t with the same set of orthogonal axes but denoted by O–xyz. A general analysis of deformation based on different coordinate systems refers to the classical work of continuum mechanics [see, e.g., Eringen (1967)].

Let us now suppose that, at time $t = t_0$ (for convenience, we may set $t_0 = 0$), the material body occupies a region of space V_0 bounded by the surface ∂V_0, and the material points are denoted by the position vector $\boldsymbol{X}(X, Y, Z)$. After motion and deformation have taken place, at time $t > 0$, the material body occupies a region V bounded by ∂V in space, and the material particle (point) labeled X is moved to a position x as shown in Fig. 2.1. Thus one possible description of the motion of the material body can be given by the vector equation

$$x = x(X, t) \tag{2.1.3}$$

where, for each time t, $x(X, t)$ is a continuously differentiable function if no discontinuity of the shock-wave type occur. Obviously, we have $x(X, t_o) = X$. Equation (2.1.3) defines the material (or Lagrangian) description of the motion of a continuum.

The velocity and acceleration of the material particle labelled X in the material description are given, respectively, by

$$\mathbf{V}(X, t) = \frac{\partial x(X, t)}{\partial t} \tag{2.1.4}$$

and

$$A(X, t) = \frac{\partial^2 x(X, t)}{\partial t^2} \tag{2.1.5}$$

It is important to notice that the spatial arguments of $\mathbf{V}$ and A do not denote the current position of the material particle (which is, of course, x), but the initial position. For many purposes it is more convenient to express velocity and acceleration in terms of x rather than X. This can be accomplished by using a spatial (Eulerian) description of motion.

We shall assume that the particle located initially at X moves to one and only one point x, and conversely that no two separated particles in the initial configuration V_o arrive at the same point x. Thus the mapping of $x \to X$ is one to one and Eq.(2.1.3) can be solved to obtain X as a function of x and t by

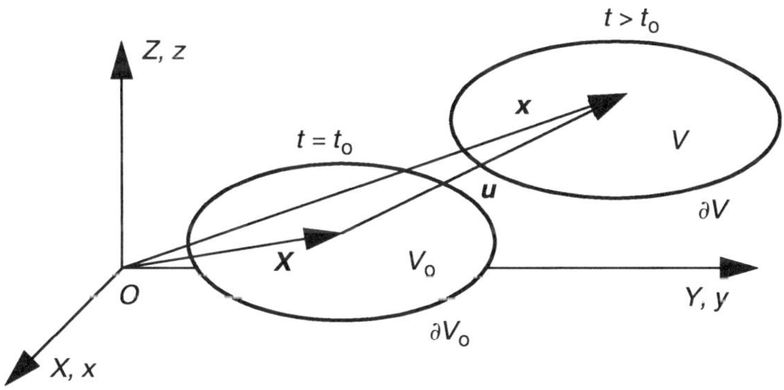

Figure 2.1 Configurations of a moving material body.

$$X = X(x, t) \tag{2.1.6}$$

The necessary condition for the existence of the unique solution (2.1.6) of Eq.(2.1.3) is the non-vanishing of the Jacobian J defined by

$$J = \det\left(\frac{\partial x_i}{\partial X_K}\right) \equiv |x_{i,K}| \tag{2.1.7}$$

where we have used the denotation $x_{i,K} \equiv \partial x_i/\partial X_K$.

In a spatial description, the instantaneous motion of the material body is described by a velocity vector field $\mathbf{v}(x, t)$ associated with the instantaneous location of each material particle

$$\mathbf{v}(x, t) = \mathbf{V}(X(x, t), t) \tag{2.1.8}$$

It is important to recognize that $\mathbf{v}$ as a function of x and t is a totally different function from $\mathbf{V}$ as a function of X and t. The velocity vector field $\mathbf{v}(x, t)$ may be interpreted as follows: Consider some fixed point x in space. At this location suppose that an apparatus is installed to record the velocity of the particles passing through x as a function of time. The measurements yield $\mathbf{v}(x, t)$.

The acceleration of the particle currently passing through x is given by the material time derivative of $\mathbf{v}(x, t)$

$$\frac{d\mathbf{v}(x, t)}{dt} = \frac{\partial \mathbf{v}}{\partial t} + (\mathbf{v} \cdot \nabla)\mathbf{v} \tag{2.1.9}$$

The first term in Eq.(2.1.9) is interpreted as the local part of the acceleration arising from the time dependence of the velocity field while the second term represents the convective part of the acceleration as the contribution of the motion of the material particle in the instantaneous velocity field.

2.1.3 Conservation Law of Mass

The conservation law of mass, which states mass can be neither created nor destroyed, may now be expressed mathematically by

$$\int_{V_o} \rho_o(X) dV_o = \int_V \rho(x, t) dV = \int_{V_o} \rho(x(X, t), t) J dV_o \tag{2.1.10}$$

Since this relation must hold for any volume element of a continuous body, we then have

$$\rho_o(X) = \rho(x(X, t), t) J \tag{2.1.11}$$

which expresses the mass density in the initial configuration in terms of the mass density in the current configuration. Since physically ρ and ρ_0 are both positive, the Jacobian J is then also a positive quantity.

With the use of the material time derivative, the conservation law of mass may also be written alternatively by the following mathematical expression

$$\frac{d}{dt}\int_V \rho(x, t)dV = 0 \tag{2.1.12}$$

where V is a moving volume of space occupied by particles of the body. The integral form of the conservation of mass (2.1.12) can be applied to arbitrary motions of a material medium, in which the field of the density, velocity, and other mechanical characteristics can be discontinuous. For continuous motion described by smooth functions, Eq.(2.1.12) can be replaced by a partial differential equation from the material time derivative of the volume integral

$$\frac{d}{dt}\int_V \rho(x, t)dV = \int_V \left(\frac{d\rho}{dt} + \rho(\nabla \cdot \mathbf{v})\right)dV = 0 \tag{2.1.13}$$

from which we may get the local equation of conservation of mass by

$$\frac{d\rho}{dt} + \rho(\nabla \cdot \mathbf{v}) = 0 \tag{2.1.14}$$

This equation is also called the equation of continuity.

2.2 CONTINUUM DEFORMATION AND STRAIN ANALYSIS

2.2.1 Deformation and Strain Tensors

In this section, a brief description of the geometric aspects of the deformation of a continuum and the strain analysis will be given. The nature of the deformation of material in the immediate vicinity of a particle labelled X is made by examining the behavior of all line vector elements connecting the material particle X with an arbitrary neighboring material particle $X+dX$. In the reference configuration the line vector element is dX. At time t, the line vector element connecting the same particles now located at $x(X, t)$ and $x(X+dX, t)$ is

$$dx = x(X + dX, t) - x(X, t) \tag{2.2.1}$$

The effect of the motion is to carry the line vector element from X to x and concomitantly to transform the line vector element from dX to dx.

With the use of the assumption that x is a differentiable function, Eq.(2.2.1) may be written at the first order approximation as

$$dx_i = x_{i,K}dX_K \equiv F_{iK}dX_K \tag{2.2.2}$$

where F_{iK} is called the deformation gradient tensor defined by $F_{iK} \equiv x_{i,K}$. The squares of the length of the line vector elements $\boldsymbol{dX}$ and $\boldsymbol{dx}$ are, respectively,

$$dS^2 = \boldsymbol{dX} \cdot \boldsymbol{dX} = \delta_{KL}dX_K dX_L \tag{2.2.3}$$

$$ds^2 = \boldsymbol{dx} \cdot \boldsymbol{dx} = \delta_{ij}dx_i dx_j \tag{2.2.4}$$

where δ_{KL} and δ_{ij} denote the Euclidean metric tensors for their respective coordinate systems, which, in the case of choosing the same Cartesian coordinate system, both become the Kronecker delta. The difference between the squares of the length of line elements may now be expressed as

$$ds^2 - dS^2 = (\delta_{ij}F_{iK}F_{jL} - \delta_{KL})dX_K dX_L \tag{2.2.5}$$

or

$$ds^2 - dS^2 = 2E_{KL}dX_K dX_L \tag{2.2.6}$$

in which we have introduced the Lagrangian strain tensor E_{KL} by

$$E_{KL} = \frac{1}{2}(C_{KL} - \delta_{KL}) \tag{2.2.7}$$

where $C_{KL} \equiv F_{iK}F_{iL}$ is called the Cauchy deformation tensor. The tensors C_{KL} and E_{KL} are obviously symmetric. From Eq.(2.2.6), we find a fundamental result that a necessary and sufficient condition for the motion of a body being a rigid-body motion is that all the components of the strain tensor E_{KL} are zero throughout the body.

By introducing further a displacement vector $\boldsymbol{u}$ defined by (see Fig. 2.1)

$$\boldsymbol{u} = \boldsymbol{x} - \boldsymbol{X} \qquad (\text{or } u_K = (\delta_{Ki}x_i - X_K)) \tag{2.2.8}$$

with δ_{Ki} being the so-called shifter which plays a similar role as the Kronecker delta, Eq.(2.2.7) can then be expressed as

$$E_{KL} = \frac{1}{2}(u_{K,L} + u_{L,K} + u_{I,K}u_{I,L}) \tag{2.2.9}$$

For infinitesimal strains, Eq.(2.2.9) is reduced to be

$$E_{KL} = \varepsilon_{kl} = \frac{1}{2}(u_{k,l} + u_{l,k}) \tag{2.2.10}$$

where ε_{ij} is called the infinitesimal strain tensor. The physical meaning of the Lagrangian strain tensor (and the infinitesimal strain tensor) can be illustrated by considering a simple stretch of a line vector element $dX = (0, 0, dX)$ along the X-axis, the result of which gives

$$E_{XX} = \frac{(ds - dS)(ds + dS)}{2dS^2} \tag{2.2.11}$$

In the case of an infinitesimal strain, Eq.(2.2.11) is reduced to be

$$E_{XX} \approx \varepsilon_{xx} = \frac{ds - dS}{dS} \tag{2.2.12}$$

which shows that ε_{xx} represents the extension, or change of length per unit length of a line vector element parallel to the x-axis.

2.2.2 Rate of Deformation and Rigid Body Rotation

Based on above knowledge, we can now derive the following useful relations for the study of the time-rate of change of deformation fields:

$$\frac{d}{dt}(x_{i,K}) = \mathrm{v}_{i,n} x_{n,K} \tag{2.2.13}$$

$$\frac{d}{dt}|x_{i,K}| = (\nabla \cdot \mathbf{v})|x_{i,K}| \tag{2.2.14}$$

$$\frac{d}{dt}(E_{KL}) = D_{ij} x_{i,K} x_{j,L} \tag{2.2.15}$$

in which $\mathbf{v}$ is the velocity field defined in Eq.(2.1.8) and $\mathrm{v}_{i,j}$ is called the velocity gradient. This velocity gradient $\mathrm{v}_{i,j}$ can be decomposed into a symmetric part and an antisymmetric part

$$\mathrm{v}_{i,j} = D_{ij} + \Omega_{ij} \tag{2.2.16}$$

where D_{ij} is called the rate of strain tensor defined by

$$D_{ij} = \frac{1}{2}(\mathrm{v}_{i,j} + \mathrm{v}_{j,i}) \tag{2.2.17}$$

which, in the case of infinitesimal strains, reads $D_{ij} = d(\varepsilon_{ij})/dt$. The quantity Ω_{ij} in Eq.(2.2.16) is called the spin tensor, defined by

$$\Omega_{ij} = \frac{1}{2}(\mathrm{v}_{i,j} - \mathrm{v}_{j,i}) \tag{2.2.18}$$

which can be associated uniquely with an axial vector Ω, called the vorticity vector defined by

$$\Omega_i = -\frac{1}{2}\varepsilon_{ijk}\Omega_{jk} \qquad \left(\text{or } \Omega = \frac{1}{2}\nabla \times \mathbf{v}\right) \tag{2.2.19}$$

where ε_{ijk} is the permutation symbol (see Appendix). This vorticity vector Ω measures the instantaneous angular velocity of a rigid body rotation. The infinitesimal rigid body rotation experienced by a continuum at an infinitesimal interval of time Δt may thus be expressed as

$$\omega_i = \Omega_i \Delta t \tag{2.2.20}$$

where ω_i is called the rotation vector defined by

$$\omega = \frac{1}{2}\nabla \times \boldsymbol{u} \qquad \left(\text{or } \omega_k = \frac{1}{2}\varepsilon_{kij}\omega_{ij}\right) \tag{2.2.21}$$

in which ω_{ij} is called the rotation tensor of the displacement field $\boldsymbol{u}$, defined by

$$\omega_{ij} = \frac{1}{2}(u_{j,i} - u_{i,j}) \tag{2.2.22}$$

By noting Eq.(2.2.15), the rigid body motion of a continuum body may now be stated as the motion of the body with $\boldsymbol{D}(\boldsymbol{x}, t) = 0$ throughout the body for all time t. In addition, motions in which $\Omega = 0$ throughout the body for all time t are said to be irrotational.

2.2.3 Compatibility Conditions

For some problems, a question of how to determine the displacement field $\boldsymbol{u}$ may arise when the strain field has been given. Since the strain tensor is symmetric, we have then six equations for the determination of three unknown functions u_i ($i = 1, 2, 3$). Generally, the system of Eq.(2.2.10) [or (2.2.9)] will not have a single-valued solution for u_i if the functions ε_{ij} were arbitrarily assigned. A single-valued solution may exist only if the functions ε_{ij} satisfy certain conditions. Besides, since strain components only determine the relative positions of points in the material body, and since any rigid-body motion corresponds to zero strain, we expect that the solution u_i can be determined only up to an arbitrary rigid-body motion.

Thus, to have a single-valued continuous displacement field $\boldsymbol{u}$, the so-called compatibility conditions have been introduced, which, in the case of infinitesimal strains, can be expressed by

$$\varepsilon_{ij,kl} + \varepsilon_{kl,ij} - \varepsilon_{ik,jl} - \varepsilon_{jl,ik} = 0 \tag{2.2.23}$$

This is the equation of compatibility of St. Venant, first obtained by him in 1860.

Of the 81 equations represented by Eq.(2.2.23), only six are essential and they can be written in unabridged notation

$$\frac{\partial^2 \varepsilon_{xx}}{\partial y \partial z} = \frac{\partial}{\partial x}\left(-\frac{\partial \varepsilon_{yz}}{\partial x} + \frac{\partial \varepsilon_{zx}}{\partial y} + \frac{\partial \varepsilon_{xy}}{\partial z}\right) \tag{2.2.24}$$

$$\frac{\partial^2 \varepsilon_{yy}}{\partial z \partial x} = \frac{\partial}{\partial y}\left(-\frac{\partial \varepsilon_{zx}}{\partial y} + \frac{\partial \varepsilon_{xy}}{\partial z} + \frac{\partial \varepsilon_{yz}}{\partial x}\right) \tag{2.2.25}$$

$$\frac{\partial^2 \varepsilon_{zz}}{\partial x \partial y} = \frac{\partial}{\partial z}\left(-\frac{\partial \varepsilon_{xy}}{\partial z} + \frac{\partial \varepsilon_{yz}}{\partial x} + \frac{\partial \varepsilon_{zx}}{\partial y}\right) \tag{2.2.26}$$

$$2\frac{\partial^2 \varepsilon_{xy}}{\partial x \partial y} = \frac{\partial^2 \varepsilon_{xx}}{\partial y^2} + \frac{\partial^2 \varepsilon_{yy}}{\partial x^2} \tag{2.2.27}$$

$$2\frac{\partial^2 \varepsilon_{yz}}{\partial y \partial z} = \frac{\partial^2 \varepsilon_{yy}}{\partial z^2} + \frac{\partial^2 \varepsilon_{zz}}{\partial y^2} \tag{2.2.28}$$

$$2\frac{\partial^2 \varepsilon_{zx}}{\partial z \partial x} = \frac{\partial^2 \varepsilon_{zz}}{\partial x^2} + \frac{\partial^2 \varepsilon_{xx}}{\partial z^2} \tag{2.2.29}$$

These conditions are derived for infinitesimal strains referred to rectangular Cartesian coordinates. In general cases, one requires that the Riemannian-Christoffel curvature tensor formed from a metric tensor for the Euclidean material manifold vanishes [Eringen (1971)].

It can be proved that the satisfaction of the compatibility conditions (2.2.23) is necessary and sufficient for the displacement to be single-valued in a simply connected region. For a multiply connected region, Eq.(2.2.23) is necessary, but no longer sufficient. To guarantee single-valuedness of the displacement, some additional conditions are required [see Fung (1965)].

2.3 THE LAWS OF MOTION AND STRESS HYPOTHESIS

2.3.1 Objective Tensors

In the formulation of physical laws, it is often desirable to employ quantities that are independent of the motion of the observer. Such quantities are called objective. In classical mechanics, a tensorial quantity is said to be objective or frame-indifferent if it obeys an appropriate tensor transformation law for all times in any two objectively equivalent (rigid body) motions $\boldsymbol{x}(X, t)$ and $\boldsymbol{x}'(X, t')$, that is,

$$x'_k(X, t') = Q_{kl}(t)x_l(X, t) + b_k(t), \qquad t' = t - a \tag{2.3.1}$$

with $b_k(t)$ being a time-dependent vector, a being a constant characterizing a shift of the origin of time on the clocks in reference frames, and $Q_{kl}(t)$ a time-dependent orthogonal tensor satisfying the proper condition:

$$Q_{kl}Q_{ml} = Q_{lk}Q_{lm} = \delta_{km} \tag{2.3.2}$$

For example, if a vector A_k and a second-order tensor S_{kl} are objective, then they must obey the following transformation law appropriate to two objectively equivalent motions:

$$A'_k(X, t') = Q_{kl}(t)A_l(X, t) \tag{2.3.3}$$

$$S'_{kl}(X, t') = Q_{km}(t)Q_{ln}(t)S_{mn}(X, t) \tag{2.3.4}$$

For vectors and tensors that are independent of time, objectivity readily applies. For time-dependent quantities, this is not always the case. For instance, the velocity vector v_k and the spin tensor Ω_{kl} are not objective while the rate of strain tensor D_{kl} is objective [see Eringen (1967) for more detailed discussions].

The transformation relation (2.3.1) between $\boldsymbol{x}'$ and $\boldsymbol{x}$ is also called the Euclidean transformation. In particular, a Euclidean transformation with Q_{kn} being time-independent (for which we may take particular coordinate systems such that we have $Q_{kn} = \delta_{kn}$), $a = 0$ and $\boldsymbol{b} = -\mathbf{v}t$ is called the Galilean transformation [see Eq.(1.7.1)].

2.3.2 The Laws of Motion

The motion of a continuum in space is governed by the laws of continuum mechanics based on Newton's laws of motion and laws of thermodynamics. Newton's laws of motion state that, in an inertial frame of reference, the material rate of change of the linear momentum of a body is equal to the resultant of applied forces acting on the body, and that the material rate of change of the angular momentum of the body with respect to the coordinate origin is equal to the resultant moment of applied forces and couples about the same origin.

Mathematically, the balance law of linear momentum for a continuum medium with volume V bounded by ∂V may be written

$$\frac{d}{dt}\int_V \rho \mathbf{v} dV = \int_{\partial V} \mathbf{t}^{(n)} dS + \int_V \mathbf{f} dV \tag{2.3.5}$$

where $\mathbf{t}^{(n)}$ denotes the surface traction, which are, for instance, aerodynamic pressure, mechanical contact pressure and magnetic pressure. $\mathbf{f}$ is the body force, such as the earth gravitational force and electromagnetic body forces.

The balance law of angular momentum is

$$\frac{d}{dt}\int_V \mathbf{x} \times \rho \mathbf{v} dV = \int_{\partial V} \mathbf{x} \times \mathbf{t}^{(n)} dS + \int_V (\mathbf{x} \times \mathbf{f} + \mathbf{c}) dV \tag{2.3.6}$$

in which $\mathbf{c}$ denotes the body couple which may result from the presence of electromagnetic fields in magnetized or electrically polarized materials. No surface couple effect is being taken into account here [Eringen (1964)]. The integral form of the balance laws of motion (2.3.5) and (2.3.6) are valid for any motion including discontinuous motion, in which the distributions of the characteristics of the motion and state within the volume V can be step functions of the time t (shock processes).

From a continuum point of view, the forces and couples acting on the material body may be classified into three categories, that is, the extrinsic body loads, the extrinsic surface loads, and the internal loads. The extrinsic body loads are the forces and couples that arise from the external effects, such as the force of gravity. They act on the mass point of the body. In continuum mechanics, a load density per unit mass is assumed to exist. The extrinsic surface loads may arise from the action of one body on another through the bounding surface. The surface density of these loads is also assumed to exist, and the extrinsic surface force per unit area is called the surface traction. The internal loads are the result of the mutual action of pairs of particles that are located in the interior of the body. The effect of interparticle forces in a continuum appears in the form of a resultant effect of one part of the body on another part through the latter's bounding surface. This concept gives rise to the stress hypothesis as follows.

2.3.3 Stress Tensor

The internal loads and their connection to surface loads may be understood by applying the balance law of momenta on a region v with its surface ∂v fully or partially contained in the body. In particular, we may consider a small tetrahedron with its vertex in the interior of v and having three of its faces on the coordinate surfaces and the fourth face on ∂v, as shown in Fig. 2.2.

We introduce the stress vector (or traction) $-\boldsymbol{t}^k$ acting on the coordinate surface $x_k =$ const. ($k = 1, 2, 3$) and apply Eq.(2.3.5) to this tetrahedron. We may obtain, by using the mean value theorem,

$$\frac{d}{dt}(\rho^* \mathbf{v}^* \Delta v) = \boldsymbol{t}^{(n)*} \Delta S - \boldsymbol{t}^{k*} \Delta S_k + \rho^* \mathbf{f}^* \Delta v \tag{2.3.7}$$

where ρ^*, $\mathbf{v}^*$ and $\mathbf{f}^*$ denote, respectively, the values of ρ, $\mathbf{v}$ and $\mathbf{f}$ at some interior point of the volume element v and $\boldsymbol{t}^{(n)*}$ and $\boldsymbol{t}^{k*}$ are the values of $\boldsymbol{t}^{(n)}$ and $\boldsymbol{t}^k$ at some point on the surfaces ΔS and ΔS_k, respectively. Using the principle of conservation of mass, and letting $\Delta S \to 0$, we can find, by noting that ρ^*, $d\mathbf{v}^*/dt$ and $\mathbf{f}^*$ are bounded quantities,

$$\boldsymbol{t}^{(n)} dS = \boldsymbol{t}^k dS_k \tag{2.3.8}$$

and

$$dS_k = n_k dS \tag{2.3.9}$$

where $\boldsymbol{n}$ is the unit outward normal vector of the surface ΔS.

Substituting Eq.(2.3.9) into (2.3.8), we have

$$\boldsymbol{t}^{(n)} = \boldsymbol{t}^k n_k \tag{2.3.10}$$

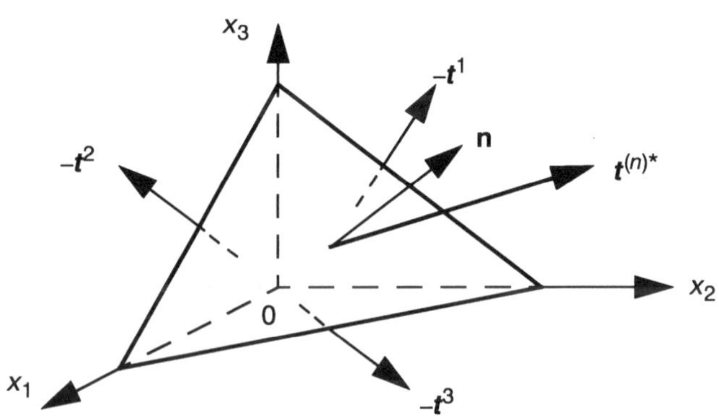

Figure 2.2 Surface tractions on a tetrahedron.

from which, since the stress vectors $\boldsymbol{t}^k$ are, by definition, independent of $\boldsymbol{n}$, we can get immediately the following *Cauchy's lemma*: The stress vector $\boldsymbol{t}^{(n)}$ acting upon opposite sides of the same surface at a given point are equal in magnitude and opposite in direction, that is,

$$\boldsymbol{t}^{(n)} = -\boldsymbol{t}^{(-n)} \tag{2.3.11}$$

We may now introduce the stress tensor with its component t_{kj} defined as the jth component of the stress vector $\boldsymbol{t}^k$ acting on the positive side of the kth coordinate surface, that is,

$$\boldsymbol{t}^k = t_{kj}\mathbf{i}_j \qquad (k = 1, 2, 3) \tag{2.3.12}$$

with $\mathbf{i}_j$ (j = 1, 2, 3) being the unit base vectors of a rectangular Cartesian coordinate system. For example, t_{13} is the x_3-component of the stress vector $\boldsymbol{t}^1$ acting on the surface x_1 = const. The stress tensor t_{kj} may be arranged in a matrix form:

$$[t_{kj}] = \begin{bmatrix} t_{11} & t_{12} & t_{13} \\ t_{21} & t_{22} & t_{23} \\ t_{31} & t_{32} & t_{33} \end{bmatrix} \tag{2.3.13}$$

in which the components t_{11}, t_{22}, and t_{33} are usually called normal stresses, and the mixed components t_{12}, t_{23}, etc., are called shearing stresses.

By Eqs.(2.3.10) and (2.3.12), we may further get the following *Cauchy´s theorem*: From the stress vectors acting across three mutually perpendicular planes at a point, all stress vectors at that point are determinated; they are given by

$$t_k^{(n)} = t_{jk}n_j \tag{2.3.14}$$

as linear functions of the stress tensor t_{jk}. The Cauchy formula assures us that the nine components of stresses t_{jk} are necessary and sufficient to define the traction across any surface element in a body. Hence, the stress state in a body is characterized completely by the stress tensor t_{jk}.

In particular, at the boundary surface of a body with the prescribed surface loading $\boldsymbol{t}^{(n)o}$, we have the following stress boundary condition:

$$t_{jk}n_j = t_k^{(n)o} \tag{2.3.15}$$

with $\boldsymbol{n}$ being the unit outward normal vector of the boundary surface.

It can be shown that the Cauchy stress tensor is objective since it obeys the appropriate tensor transformation rule

$$t'_{kl}(X, t') = Q_{km}(t)Q_{ln}(t)t_{mn}(X, t) \tag{2.3.16}$$

in any two objectively equivalent motions described by Eq.(2.3.1). The objectivity of the Cauchy stress tensor has important consequences on constitutive equations which will be discussed later.

With the aid of the stress tensor t_{jk}, we may get the following local balance equation of linear momentum

$$\rho\frac{dv_i}{dt} = t_{ji,j} + f_i \tag{2.3.17}$$

and the local balance equation of angular momentum

$$c_i + \varepsilon_{ijk}t_{jk} = 0 \tag{2.3.18}$$

in the region of continuous smooth motions, where Eqs.(2.3.5) and (2.3.6) are valid for every volume element. It can be seen from Eq.(2.3.18) that, in the absence of body couples, Cauchy's stress tensor t_{jk} is symmetric, that is, $t_{[jk]} = t_{jk} - t_{kj} = 0$. Equation (2.3.17) is often called the Eulerian equation of motion of a continuum.

2.4 THE LAWS OF CONTINUUM THERMODYNAMICS

The actual motion and physical states of material bodies in contact with their surroundings at a certain temperature also have to be subject to the restrictions of classical laws of thermodynamics. In the classical theory of thermodynamics, we study thermodynamic systems which are characterized by a set of independent macroscopic state variables, such as the temperature, the strain tensor, the electromagnetic fields, and so on. The thermodynamic system which does not exchange matter with its surroundings is called the closed system. Here, we shall only study the closed systems. The closed system of a continuous body can interact with its surrounding by exchanging heat and work performed by body forces and surface forces on it. A system having no interaction with its surrounding is called an isolated system.

It is worth mentioning that where electric conduction in materials is concerned, we shall only consider the electric conduction in metallic solids so that the electric current is due to the flow of free electrons in the metallic solids and, thus, the effect of matter transfer of electrons may be ignored since the mass of the electron is very small compared with the mass of the atoms composing the material. In contrast to the metallic solids, the electric current in electrolytic media is due to the flow of ions which form the substance, and, therefore, a passage of an electric current due to the migration of ions under electric field is always associated with transport of matter. In addition, chemical reactions are always present during the passage of an electric current from a metallic to an electrolytic conductor or vice versa [Kortum (1965)].

In each problem of a thermodynamic system, it is important to select a particular set of independent state variables. However, the choice is to a certain extent arbitrary. We shall see this point more clearly in later discussion of a concrete problem in Section 2.6.2. If a certain state variable can be expressed as a single-valued function of a set of other state variables, it is called a state function, and the functional relationship is said to be an equation of state. For a given system, if the state variables does not depend on space coordinates, the system is said to be homogeneous. If the values of the state variables are independent of time, the system is said to be in thermodynamic equilibrium, otherwise, if the state variables vary with time, the system is said to undergo a thermodynamic process. The thermodynamic process can be either irreversible or reversible, depending on whether the energy of the system is dissipative or not during the process.

In the theory of thermodynamics for continuous media, the first law of thermodynamics is the law of energy conservation. The mathematical expression of the law of energy conservation for an electromagnetic continuum body in electromagnetic fields can be given by

$$\frac{d}{dt}\int_V \rho\left(\frac{1}{2}\mathbf{v}\cdot\mathbf{v}+u\right)dV = \int_{\partial V}(\boldsymbol{t}^{me(n)}\cdot\mathbf{v}-\boldsymbol{q}\cdot\boldsymbol{n})dS + \int_V \boldsymbol{f}^{me}\cdot\mathbf{v}dV + W^{em} \tag{2.4.1}$$

where u denotes the internal stored energy per unit mass, $\boldsymbol{t}^{me(n)}\cdot\mathbf{v}$ the power per unit volume by mechanical surface traction, $\boldsymbol{f}^{me}\cdot\mathbf{v}$ the power per unit volume by mechanical body force, and $\boldsymbol{q}$ the heat flux. W^{em} is the power supply of electromagnetic origin, depending on materials, some examples of which will be given explicitly in Section 2.6.

The first law of thermodynamics only states that energy is conserved in any process. We may naturally ask the question what kind of process is physically possible for a given thermodynamic system with a certain amount of energy. For instance, our common experience tells us that heat flow is always from a body with high temperature to one at lower temperature. What determines the direction of the process? The second law of thermodynamics is such a law which gives a restriction on the possible thermodynamic processes by using the concepts of absolute temperature and entropy. For thermodynamics of continuum systems, the entropy is postulated to be a function of state in irreversible processes as well as in reversible processes, and the balance equation of entropy can be expressed by

$$\frac{d}{dt}\int_V \rho\eta dV = -\int_{\partial V}\mathbf{S}\cdot\boldsymbol{n}dS + \int_V \rho\sigma_s dV \tag{2.4.2}$$

where η denotes the specific entropy (entropy per unit mass), $\mathbf{S}$ the entropy flux, and T the local absolute temperature being a strictly positive scalar with the unit K. The scale of the absolute temperature is fixed by defining the temperature at equilibrium between liquid water and ice at a pressure of 1 atm at 273.16 K, that is, 273.16 on Kelvin's scale. σ_s is the rate of internal entropy production per unit mass. It is important to realize that entropy is an attribute of a material body, just as its mass or its electric charges. Obviously, the entropy of a system is equal to the sum of entropies of its parts.

The second law of thermodynamics can now be stated by the simple mathematical expression:

$$\int_V \rho\sigma_s dV \geq 0, \qquad \textit{for all possible thermodynamic processes} \tag{2.4.3}$$

where the equality sign is true only for reversible processes, and the inequality sign is valid for irreversible processes.

When Eq.(2.4.2) is valid for any part of a continuous body, we obtain the local production of entropy

$$\rho\sigma_s = \rho\frac{d\eta}{dt} + \nabla \cdot \mathbf{S} \geq 0 \tag{2.4.4}$$

and the interface condition

$$[\mathbf{S}] \cdot \boldsymbol{n} \geq 0 \tag{2.4.5}$$

in the absence of surface heat sources.

The local expression (2.4.4) of the second law of thermodynamics is called the local *Clausius-Duhem inequality*. We may now introduce the definition for a thermodynamically admissible process by stating that a process is called thermodynamically admissible if and only if it obeys the local Clausius-Duhem inequality and possesses a non-negative finite temperature. In addition, we may also introduce the definition that a thermodynamic process is called mechanically admissible if it obeys the laws of motion, namely, the conservation law of mass, the balance laws of momenta, and the conservation law of energy.

For a simple thermodynamic process, in which there are no chemical reactions and matter transfers, the entropy flux can be expressed simply by

$$\mathbf{S} = \frac{\boldsymbol{q}}{T} \tag{2.4.6}$$

where $\boldsymbol{q}$ is the heat flux and T the local absolute temperature.

It can be seen from Eq.(2.4.4) that, in an isolated system, the second law of thermodynamics tells us that the possible spontaneous processes occur only in the direction of increasing entropy. Further development of irreversible thermodynamics for various thermodynamic systems requires a detailed description of the entropy production σ_S. In Section 2.6, we shall present an example of describing the entropy production σ_S.

2.5 ON FORMULATION OF ELECTRODYNAMICS FOR DEFORMABLE MEDIA

During the past three decades, the macroscopic continuum theory of electromagnetism for moving and deformable solids has been studied extensively because of the rapid development of technology and applied physical sciences. Many formulations of the macroscopic Maxwell's equations were proposed, but none of them has so far been accepted as a universal law, owing to the complication of the electromagnetic phenomena for moving deformable solids, especially for high velocity and highly accelerated motion of the solids. The difficulties in formulating the electromagneto-mechanical interaction phenomena are mainly due to the following facts:

(*a*) *The invariance properties of electromagnetic field equations and mechanical equations.* In classical electrodynamics and classical mechanics, it is well known that Maxwell's equations are invariant under Lorentz transformation but the mechanical balance laws are invariant under Galilean transformations. Therefore a consistent formulation of electromagneto–mechanical interactions should be treated from the relativity point of view, especially, for high velocity and highly accelerated motion of deformable solids, where the formulation has to be based on the general principle of relativity. So far, the formulations with less controversy are those for electromagneto-mechanical interaction phenomena in the electro-quasistatic and the magneto-quasistatic approximations as we have discussed in Section 1.7. In these cases, with the neglect of relativistic effects, the electromagneto-mechanical interactions may be treated consistently in the classical Galilean relativity. According to Beckmann (1987), due probably to today´s technology limits, it seems that almost all experimental evidences on relativistic effects obtained so far could be treated consistently within the Galilean relativity, provided that the velocity of motion is properly defined.

(*b*) *The definitions of electromagnetic body force, body couple, and the energy supply.* This difficulty mainly lies in the separation of the electromagneto-mechanical interactions in near and far field effects. As we may know, in the classical continuum theory of elasticity, the introduction of a stress vector implies that the range of action of inter-particle forces may be taken to be effectively zero, that is, the inter-particle forces, which are dealt with by the stress vector, are considered as short-range forces acting only between neighboring points. The mechanical body force is, for instance, the gravitational force between the body and the earth, which is a long-range force. The gravitational force also acts at short distances, but no two parts of the body and the earth, respectively, are separated by a distance comparable with the microscopic dimensions of the elastic body. Therefore, in the classical elasticity, the surface (stress) force and the body force are uniquely well-defined quantities. However, in an electromagnetic body, the electromagnetic inter-particle forces are partly of short-range character (e.g., exchange forces) and partly of long-range character (e.g., electric and magnetic dipole–dipole forces). In addition, the long-range part of the electromagnetic inter-particle forces also act at short distances and these forces act on some of the particles which are separated from each other by short distances comparable to the microscopic dimensions of the solid. So far, no unique separation between the electromagnetic stress force and the electromagnetic body force seems to be possible.

(c) The forms of Maxwell's equations for deformable solids, especially, the definitions of the macroscopic quantities of the electric polarization and the magnetization. It has been shown in the theory of electrodynamics for free space that the electric and magnetic fields may be well defined from the measurable forces acting on testing charges in free space. The electromagnetic fields inside materials, however, cannot be measured directly in usual ways. They have to be inferred usually from the measurable fields adjacent to the material in free space. Different forms of the polarization and the magnetization vectors then appear as a consequence of different choices of fundamental field variables in the material, which are somewhat abstract quantities. The differences between these formulations are mainly in their different expressions for electromagnetic body forces, body couples, and power supply.

At present, there are essentially four types of formulations, that is, the Maxwell-Minkowski formulation, the Chu formulation (two-dipole model), the Lorentz-Ampere formulation (dipole–current circuit model), and the statistical formulation. A comparison of these four types of formulations has been made by Pao (1978) and Hutter and van de Ven (1978).

In a recent work of Zhou (1991a), a complete set of expressions for the electromagnetic body forces, body couples and powers has been formulated for some non-magnetizable conductors, magnetizable insulators, magnetizable conductors, and dielectric solids based on the dipole model being consistent with its use in the derivation of the macroscopic Maxwell equations for materials, as shown in Section 1.4. These expressions may then be used in the formulation of constitutive equations for these materials.

As we may know, the basic laws of electromagneto–mechanical phenomena presented in above sections are valid for all types of continuous media irrespective of their internal constitutions. To complete the study of material responses of the continuous media under mechanical and electromagnetic loadings, material models characterizing the internal constitution of the continuous media are, therefore, required. In continuum physics, we are not concerned with the atomic or molecular structures of the material, rather, we are interested in the global macroscopic behavior of the material, which can be modeled phenomenologically by a set of equations called the constitutive equations. To formulate a constitutive theory which can represent a material adequately, certain physical and mathematical rules (axioms) have to be satisfied [see, e.g., Eringen (1967)]. Here, we present briefly several axioms which are of direct relevance to our material models.

The axiom of determinism states that past and present "causes" (independent constitutive (state) variables) determine the present "effects" (dependent constitutive (state) variables).

This axiom excludes essentially the dependence of the material behavior on any future events. Accordingly, the future phenomena concerning the behavior of the material body are fully determined and observable once all the past events of the body are known.

The axiom of memory states that the values of the independent constitutive variables at distant past from the present do not affect appreciably the values of the constitutive functions (dependent constitutive variables).

This axiom implies essentially that the memory of past events of the material body decays rapidly in the time domain. In particular, when hereditary phenomena, such as viscoelasticity, are not of interest, we may adopt a simpler statement that the present "causes" determine the present "effects", which means that both independent and dependent constitutive variables are now defined for the same event $(\boldsymbol{x}, t)$. Here, the independent constitutive variables are, for instance, the motion of the material points of a continuum body, the temperature, the electric and magnetic fields. The dependent constitutive variables are, for instance, the stress tensor, the heat flux vector, the electric polarization, the magnetization, and the internal energy.

The axiom of equipresence states that at the outset all constitutive functionals should be expressed in terms of the same list of independent constitutive variables until the contrary is deduced.

This axiom is a precautionary measure. It helps us not to forget or be prejudiced against a certain class of variables and favour others in the expression of constitutive functionals, though some of the variables may be eliminated finally by the basic laws of continuum physics and various approximations.

The axiom of neighborhood states that the values of the independent constitutive variables at distant material points from $\boldsymbol{X}$ do not affect appreciably the values of the dependent constitutive variables at $\boldsymbol{X}$.

In this book, we are not going to discuss the general nonlocal theory. Only local first-order gradient theories will be considered. Some discussion about Pippard´s nonlocal theory for superconductors will, however, be presented in the next chapter.

The axiom of objectivity states that constitutive equations must be form-invariant with respect to rigid motions of the spatial frame of reference.

This axiom has been considered as self-evident in classical continuum mechanics. To illustrate the objectivity, a simple example was cited by Truesdall (1961), who considered a spring mounted on a rotating table with one end attached to the table and the other fixed to a mass. As the table is rotating the spring extends. From the viewpoint of an external observer, the force in the spring is merely that required to satisfy Newton's law for the accelerated motion of the mass. For an observer utilizing a reference frame fixed in the rotating table the spring force would derive seemingly from a counterbalancing body force exerted on the mass by an external field. Nevertheless, for both observers the force in the spring is the same, and hence so are the contact forces exerted on both the table and mass by the spring, which means that the contact forces are frame-independent (objective) quantities.

However, the axiom of objectivity is not self-evident in dealing with the modelling of electrodynamic properties of deformable media. Nevertheless, the axiom of objectivity has, so far, been used widely and successfully in the modelling of continuum electromagnetic media in many cases.

Other self-evident axioms are the axiom of material invariance and the axiom of admissibility. *The axiom of material invariance* states that constitutive equations must be form-invariant with respect to a group of orthogonal transformations $\{\mathbf{S}\}$ and translations $\{\mathbf{C}\}$ of the material coordinates, defined by

$$\bar{X} = \mathbf{S} \cdot X + \mathbf{C} \tag{2.5.1}$$

These restrictions express the geometric symmetry, represented by $\{\mathbf{S}\}$, and inhomogeneity represented by $\{\mathbf{C}\}$, at X, in the physical properties of the material body. When $\{\mathbf{S}\}$ is the proper orthogonal group ($\det S = 1$), the material is called hemitropic. If $\{\mathbf{S}\}$ is the full orthogonal group ($\det \mathbf{S} = \pm 1$), the material is called isotropic. A material that is not hemitropic is called anisotropic. When the constitutive response functions do not depend on the translations $\{\mathbf{C}\}$ of the origin of material coordinates, the material is said to be homogeneous. Otherwise, the material is called inhomogeneous. Obviously, constitutive equations of a homogeneous material are independent of the material coordinate X explicitly. It should be noticed that a material may posses different types of material symmetries for its different physical and mechanical properties. For example, a material isotropic with respect to mechanical property may not be isotropic with respect to other physical properties, such as magnetization.

The axiom of admissibility states that all constitutive equations must be consistent with the basic laws of continuum physics.

This axiom is a simple statement of the fact that in the formulation of constitutive equations, we must not violate the basic laws of continuum physics. The conservation of mass, the balance of momenta, the conservation of energy and the second law of thermodynamics introduced in above sections are such basic laws in continuum mechanics. The axiom of admissibility may help us to eliminate the dependence on certain constitutive variables whenever they are obtainable through, for instance, the equations of motion.

With the aid of the above knowledge, we may formulate the constitutive equations for electromagnetic deformable solids [Zhou (1991a)]. Illustratively, in the following sections of this chapter, we shall introduce some continuum theories for thermoelastic conductors and thermoelastic dielectrics to demonstrate the methodology of formulating constitutive equations often used in continuum electrodynamics. Comprehensive studies on other types of electromagnetic deformable media may be found in the work of, for instance, Maugin (1988), Eringen and Maugin (1990), and Zhou (1991a).

2.6 CONTINUUM THEORY OF THERMOELASTIC CONDUCTORS

2.6.1 Constitutive Equations for Non-Magnetizable Conductors

To formulate constitutive equations for deformable electromagnetic solids, we first formulate the electromagnetic forces, couples, and powers in the solids. For a non-magnetizable conductor at the magneto-quasistatic approximation, the electromagnetic body force in the Lorentz force form, acting on the non-magnetic conductor, may be expressed by

$$\boldsymbol{f}^{em} = \boldsymbol{J} \times \boldsymbol{B} \tag{2.6.1}$$

where $\boldsymbol{J}$ is the macroscopic current density vector, and $\boldsymbol{B}$ the magnetic induction field. The free charge density ρ_e is assumed to be zero in conductors at the magneto-quasistatic approximation.

By introducing an electromagnetic stress tensor $\boldsymbol{t}^{em}$ by

$$t_{ji}^{em} = \frac{1}{\mu_o}\left\{B_j B_i - \frac{1}{2}(B_k B_k)\delta_{ji}\right\} \tag{2.6.2}$$

we may write Eq.(2.6.1) in the form:

$$\boldsymbol{f}^{em} = \boldsymbol{J} \times \boldsymbol{B} = \nabla \cdot \boldsymbol{t}^{em} \tag{2.6.3}$$

where we has introduced the notation $(\nabla \cdot \boldsymbol{t}^{em})_i = t_{ji,j}^{em}$.

There is no electromagnetic body couple acting on the non-magnetizable conductors. The electromagnetic power can be written as

$$W^{em} = \int_V (\boldsymbol{J} \times \boldsymbol{B}) \cdot \mathbf{v}\, dV + \int_V \boldsymbol{J} \cdot \boldsymbol{E}' dV \tag{2.6.4}$$

with $\boldsymbol{E}' = \boldsymbol{E} + \mathbf{v} \times \boldsymbol{B}$. Here, $\boldsymbol{J} \cdot \boldsymbol{E}'$ represents the Joule heating of the conductor.

Next, we study the energy balance equation for the deformable electromagnetic solid. In the case of non-magnetizable conductors, by using Eq.(2.4.1) and Eq.(2.6.4), we may find the following energy balance equation:

$$\frac{d}{dt}\int_V \rho\left(\frac{1}{2}\mathbf{v}\cdot\mathbf{v}+u\right)dV = \int_{\partial V}(\boldsymbol{t}^{(n)}\cdot\mathbf{v}-\boldsymbol{q}\cdot\boldsymbol{n})dS+\int_V \boldsymbol{f}\cdot\mathbf{v}dV+\int_V \boldsymbol{J}\cdot\boldsymbol{E}'dV \tag{2.6.5}$$

in which $\boldsymbol{f}$ is the sum of the mechanical body force $\boldsymbol{f}^{me}$ and the electromagnetic body force $\boldsymbol{f}^{em}$ given by Eq.(2.6.3). The stress traction $\boldsymbol{t}^{(n)}$ is related to the Cauchy stress tensor by Eq.(2.3.14).

Equation (2.6.5) represents the global conservation of energy, valid for any motion of the material body, including discontinuous distribution of material characteristics. In the region of continuous smooth motions of a continuous body, where Eqs.(2.6.5) is valid for every volume element, we may then obtain the following local balance equation of energy:

$$\rho\dot{u} = \boldsymbol{t}:(\nabla\mathbf{v})-\nabla\cdot\boldsymbol{q}+\boldsymbol{J}\cdot\boldsymbol{E}' \tag{2.6.6}$$

with the aid of the balance law of linear momentum (2.3.5) and the conservation law of mass (2.1.14). Here, the notations of $\boldsymbol{t}:(\nabla\mathbf{v}) = t_{ji}\mathrm{v}_{i,j}$ and $\dot{u} = du/dt$ have been used.

By the second law of thermodynamics (2.4.4), (2.4.6), and Eq.(2.6.6), we may get

$$\rho T\sigma_s = -\rho(\dot{\Psi}+\eta\dot{T})+\boldsymbol{t}:(\nabla\mathbf{v})-\frac{1}{T}(\boldsymbol{q}\cdot\nabla T)+\boldsymbol{J}\cdot\boldsymbol{E}' \geq 0 \tag{2.6.7}$$

where Ψ is a thermodynamic free energy function defined by

$$\Psi = u-\eta T \tag{2.6.8}$$

Since no hereditary phenomena and nonlocality are considered here for the non-magnetizable thermoelastic conductor, Ψ may be taken in the following form:

$$\Psi = \Psi(E_{KL}, T) \tag{2.6.9}$$

by noting the fact that Ψ must be an objective scalar functional under the full Euclidean transformation group (2.3.1) according to the axiom of objectivity. Here, E_{KL} is the Lagrangian strain, defined by Eq.(2.2.7), and T is the local temperature. The free energy function Ψ may also depend on the position vector $\boldsymbol{X}$ if the material is inhomogeneous.

To satisfy the inequality expressed by Eq.(2.6.7) according to the axiom of admissibility, discussed in Section 2.5, we must have

$$\eta = -\frac{\partial \Psi}{\partial T} \tag{2.6.10}$$

$$t_{ji} = \rho \frac{\partial \Psi}{\partial E_{KL}} x_{i,K} x_{j,L} \tag{2.6.11}$$

$$t_{[ji]} = t_{ij} - t_{ji} = 0 \tag{2.6.12}$$

and

$$-\frac{1}{T}(\boldsymbol{q} \cdot \nabla T) + \boldsymbol{J} \cdot \boldsymbol{E}' \geq 0 \tag{2.6.13}$$

where the heat flux $\boldsymbol{q}$ and the current density $\boldsymbol{J}$ can be, in general, functions of the Lagrangian strain E_{KL}, the temperature T, the temperature gradient $(\nabla_R T)_L = T,_i x_{i,L}$ and the electric field $W'_L = E'_i x_{i,L}$ according to the axiom of equipresence. It is noticed that the disappearance of the dependence of the thermodynamic free energy function Ψ on the temperature gradient $\nabla_R T$ and the electric field W'_L is due to the fact that Ψ is independent of the rate of the independent constitutive variables in the non-hereditary material model.

It can be seen that Eq.(2.6.12) implies that the balance law of angular momentum (2.3.18) is identically satisfied in the absence of body couple for the non-magnetizable conductors. As to the inequality (2.6.13) which characterizes the irreversibility of thermodynamic processes of the system, the following set of phenomenological equations may be introduced to describe the irreversible thermodynamic process of the system which is supposed to be not far from its equilibrium state by

$$q_i = -\kappa_{ij} T,_j + \Gamma^Q_{ij} J_j \tag{2.6.14}$$

$$E'_i = \sigma^{-1}_{ij} J_j + \Gamma^E_{ij} T,_j \tag{2.6.15}$$

with $\boldsymbol{E}' = \boldsymbol{E} + \mathbf{v} \times \boldsymbol{B}$. This set of phenomenological relations is usually called the generalized Fourier-Ohm´s law. Here, κ_{ij}, σ_{ij}, Γ^Q_{ij} and Γ^E_{ij} are, respectively, the thermal conductivity, the electric conductivity and the thermoelectric coefficient tensors. In general, these coefficient tensors may still be functions of the elastic strain, and temperature according to the axiom of equipresence. To make the Clausius-Duhem inequality (2.6.13) be satisfied for all possible thermodynamic processes, the thermal conductivity tensor κ_{ij} and the electric conductivity tensor σ_{ij} have to be both positive definite.

In many engineering applications, if the material is homogeneous, we often assume that these coefficient tensors are constant tensors at a given reference temperature, provided that those neglected effects are not important in the problems considered. In such a case, the following relations of the Onsager symmetry apply

$$\kappa_{ij} = \kappa_{ji} \,, \qquad \sigma_{ij} = \sigma_{ji} \,, \qquad \Gamma^{Q}_{ij} = T_o \Gamma^{E}_{ji} \tag{2.6.16}$$

where T_0 is a constant uniform reference temperature. It should be noticed that, in an anisotropic solid, the thermoelectric coefficient tensors Γ^{Q}_{ij} and Γ^{E}_{ij} are generally not symmetrical. For isotropic solids, according to the axiom of material invariance, we simply have

$$\kappa_{ij} = \kappa\delta_{ij} \,, \qquad \sigma_{ij} = \sigma\delta_{ij} \,, \qquad \Gamma^{Q}_{ij} = T_o\Gamma\delta_{ij} \,, \qquad \Gamma^{E}_{ij} = \Gamma\delta_{ij} \tag{2.6.17}$$

with the thermal conductivity κ, which gives the heat flux in the absence of an electric current, and the electric conductivity σ being positive ($\kappa > 0$ and $\sigma > 0$). The thermoelectric coefficient Γ can be positive or negative, depending on materials. Interesting thermoelectric phenomena include, for instance, the *Thomson effect*, which shows the Thomson heat (emitted or absorbed) when an electric current flows in the direction of temperature gradient, and the *Peltier effect*, which occurs when a current passes through a junction of two different metals [see, e.g., Landau et al. (1984)].

The local balance equation of entropy (2.4.4) becomes now

$$\rho T\dot{\eta} = -\nabla \cdot \boldsymbol{q} + \boldsymbol{J} \cdot \boldsymbol{E}' \tag{2.6.18}$$

which is the equation of heat conduction with Joule heating being the heat source. The derived set of equations (2.6.10), (2.6.11), (2.6.14), and (2.6.15) form the complete set of constitutive equations for the non-magnetizable thermoelastic conductors.

2.6.2 Constitutive Equations for Magnetizable Conductors

Let us now consider a soft-ferromagnetic thermoelastic conductor, for which no hysteresis is exhibited, and gyromagnetic and exchange effects are neglected. At the magneto-quasistatic approximation, the total electromagnetic force on the magnetizable conductor in the absence of surface currents can be expressed in the dipole model by Alblas (1979)

$$\boldsymbol{F}^{em} = \int_V \{\boldsymbol{J} \times \boldsymbol{B}^o + (\boldsymbol{M} \cdot \nabla)\boldsymbol{B}^o\} dV \tag{2.6.19}$$

which can be further transformed into the following form

$$\boldsymbol{F}^{em} = \int_V \boldsymbol{f}^{em} dV + \int_{\partial V} \frac{1}{2}\mu_o(\boldsymbol{M}\cdot\boldsymbol{n})^2 \boldsymbol{n} dS \tag{2.6.20}$$

where we have introduced an electromagnetic body force defined by

$$\boldsymbol{f}^{em} = \boldsymbol{J}\times\mu_o\boldsymbol{H} + \mu_o(\boldsymbol{M}\cdot\nabla)\boldsymbol{H} = \boldsymbol{J}\times\boldsymbol{B} + \mu_o(\nabla\boldsymbol{H})\cdot\boldsymbol{M} \tag{2.6.21}$$

It is shown from Eq.(2.6.20) that the presence of the magnetic surface traction means that the long-range magnetic force cannot be expressed simply as a form per unit volume; it depends also on the shape of the volume considered. However, we may notice that, for practical magnetic materials, the magnetic surface traction is usually much smaller than the yield stress traction of the material so that, for some engineering problems, this magnetic surface traction may be ignored at the first-order approximation [Moon (1984)].

By introducing the electromagnetic stress tensor $\boldsymbol{t}^{em}$, defined by

$$t_{ji}^{em} = B_j H_i - \frac{1}{2}\mu_o(H_k H_k)\delta_{ji} \tag{2.6.22}$$

we may write Eq.(2.6.21) in the form:

$$\boldsymbol{f}^{em} = \boldsymbol{J}\times\mu_o\boldsymbol{H} + \mu_o(\boldsymbol{M}\cdot\nabla)\boldsymbol{H} = \nabla\cdot\boldsymbol{t}^{em} \tag{2.6.23}$$

For the magnetizable conductor, the total magnetic torque on the material body in the presence of an external magnetic field $\boldsymbol{B}^o$ may be expressed by

$$\boldsymbol{L}^m = \int_V \{\boldsymbol{M}\times\boldsymbol{B}^o + \boldsymbol{x}\times((\boldsymbol{M}\cdot\nabla)\boldsymbol{B}^o + \boldsymbol{J}\times\boldsymbol{B}^o)\}dV \tag{2.6.24}$$

from which we may find the electromagnetic body couple $\boldsymbol{c}^m$ acting on the magnetizable conductor in the dipole model by [Zhou (1991a)]

$$\boldsymbol{c}^m = \mu_o\boldsymbol{M}\times\boldsymbol{H} = \boldsymbol{M}\times\boldsymbol{B} \tag{2.6.25}$$

which can be related to the electromagnetic stress from Eq.(2.6.22) by $t_{[ji]}^{em} = \varepsilon_{jik} c_k^m$.

Here we emphasis that the electromagnetic field quantities $\boldsymbol{E}$, $\boldsymbol{H}$, and $\boldsymbol{B}$, defined in this section, are still governed by the set of Maxwell's equations at the magneto-quasistatic approximation given by Eqs.(1.7.2)–(1.7.5) with the transformation relations by Eqs.(1.7.7)–(1.7.10) rather than Chu's form of Maxwell's equations [see Hutter and van de Ven (1978)]. The present formulation can, however, be used consistently only at the magneto-quasistatic approximation.

We may now introduce the electromagnetic power for the magnetizable conductor by

$$W^{em} = \int_V \boldsymbol{J}\cdot\boldsymbol{E}'dV + \int_V (\boldsymbol{J}\times\mu_o\boldsymbol{H})\cdot\mathbf{v}dV + \int_V \mu_o((\boldsymbol{M}\cdot\nabla)\boldsymbol{H})\cdot\mathbf{v}dV$$
$$+ \int_{\partial V} \frac{1}{2}\mu_o(\boldsymbol{M}\cdot\boldsymbol{n})^2\boldsymbol{n}\cdot\mathbf{v}dS + \int_V \rho\mu_o\boldsymbol{H}\cdot\frac{d}{dt}\left(\frac{\boldsymbol{M}}{\rho}\right)dV \qquad (2.6.26)$$

in which the first term on the right-hand side is the Joule heating of the conductor. The energy balance equation for the magnetizable conductor in electromagnetic fields may thus be expressed from Eqs.(2.4.1) and (2.6.26) at the magneto-quasistatic approximation by

$$\frac{d}{dt}\int_V \rho\left(\frac{1}{2}\mathbf{v}\cdot\mathbf{v}+u\right)dV = \int_{\partial V} (\boldsymbol{t}^{(n)}\cdot\mathbf{v}-\boldsymbol{q}\cdot\boldsymbol{n})dS + \int_V \boldsymbol{f}\cdot\mathbf{v}dV$$
$$+\int_V \boldsymbol{J}\cdot\boldsymbol{E}'dV + \int_V \rho\mu_o\left(\boldsymbol{H}\cdot\frac{d\boldsymbol{M}'}{dt}dV\right)dV \qquad (2.6.27)$$

in which $\boldsymbol{M}' = \boldsymbol{M}/\rho$ denotes the magnetization per unit mass, and u the specific internal energy. Here, $\boldsymbol{f}$ is the sum of the mechanical body force $\boldsymbol{f}^{me}$ and the electromagnetic body force $\boldsymbol{f}^{em}$ given by Eq.(2.6.21). The stress traction $\boldsymbol{t}^{(n)}$, including both the mechanical and magnetic surface tractions, is related to Cauchy's stress tensor by Eq.(2.3.14).

Equation (2.6.27) represents the global conservation of energy, valid for any motion of the material body, including discontinuous distribution of material characteristics. In the region of continuous smooth motions of a continuous body, where Eqs.(2.6.27) is valid for every volume element, we can then get the following local balance equation of energy:

$$\rho\dot{u} = \boldsymbol{t}:(\nabla\mathbf{v}) - \nabla\cdot\boldsymbol{q} + \boldsymbol{J}\cdot\boldsymbol{E}' + \rho\mu_o(\boldsymbol{H}\cdot\dot{\boldsymbol{M}}') \qquad (2.6.28)$$

with the aid of the balance law of linear momentum (2.3.5) and the conservation law of mass (2.1.14). Here, the notations of $\boldsymbol{t} : (\nabla\mathbf{v}) = t_{ji}\mathrm{v}_{i,j}$ and $\dot{u} = du/dt$ have been used.

By the second law of thermodynamics (2.4.4) and Eq.(2.6.28), we find

$$\rho T\sigma_s = -\rho(\dot{\Psi}+\eta\dot{T}) + \boldsymbol{t}:(\nabla\mathbf{v}) + \rho\mu_o(\boldsymbol{H}\cdot\dot{\boldsymbol{M}}') - \frac{1}{T}(\boldsymbol{q}\cdot\nabla T) + \boldsymbol{J}\cdot\boldsymbol{E}' \geq 0$$
$$(2.6.29)$$

where Ψ is a thermodynamic free energy function defined by

$$\Psi = u - \eta T \tag{2.6.30}$$

Since no hereditary phenomena and nonlocality are considered for the soft ferromagnetic conducting thermoelastic solid, Ψ may be taken to be of the following form:

$$\Psi = \Psi(E_{KL}, G_L, T) \tag{2.6.31}$$

with the quantities G_L defined by $G_L = M'_i x_{i,L}$ due to the fact that Ψ must be an objective scalar functional under the full Euclidean transformation group (2.3.1) according to the axiom of objectivity. The free energy function Ψ may also depend on the position vector $\boldsymbol{X}$ if the material is inhomogeneous.

To satisfy the inequality expressed by Eq.(2.6.29) according to the axiom of admissibility, we must have

$$\eta = -\frac{\partial \Psi}{\partial T} \tag{2.6.32}$$

$$t_{ji} = \rho \frac{\partial \Psi}{\partial E_{KL}} x_{i,K} x_{j,L} + \rho \frac{\partial \Psi}{\partial G_L} M'_i x_{j,L} = \rho \frac{\partial \Psi}{\partial E_{KL}} x_{i,K} x_{j,L} + \mu_o M_i H_j \tag{2.6.33}$$

$$t_{[ji]} = \rho \frac{\partial \Psi}{\partial G_L} M'_{[i} x_{j],L} = \mu_o H_{[j} M_{i]} = \mu_o (H_j M_i - H_i M_j) \tag{2.6.34}$$

$$\mu_o H_i = \frac{\partial \Psi}{\partial G_L} x_{i,L} \tag{2.6.35}$$

and

$$-\frac{1}{T}(\boldsymbol{q} \cdot \nabla T) + \boldsymbol{J} \cdot \boldsymbol{E}' \geq 0 \tag{2.6.36}$$

where the heat flux $\boldsymbol{q}$ and the current density $\boldsymbol{J}$ can be, in general, functions of the Lagrangian strain E_{KL}, the magnetization G_L, the temperature T, the temperature gradient $(\nabla_R T)_L = T_{,i}\, x_{i,L}$ and the electric field $W'_L = E'_i x_{i,L}$ according to the axiom of equipresence. It is noticed that the disappearance of the dependence of the thermodynamic free energy function Ψ on the temperature gradient $\nabla_R T$ and the electric field W'_L is due to the fact that Ψ is independent of the rate of the independent constitutive variables in the non-hereditary material model.

It can be seen that Eq.(2.6.34) implies that the balance law of angular momentum (2.3.18) is identically satisfied by noting Eq.(2.6.25). As to the inequality (2.6.36) which characterizes the irreversibility of thermodynamic processes of the system, the same set of phenomenological equations as

Eqs.(2.6.14) and (2.6.15) may be introduced to describe the irreversible thermodynamic process of the system which is supposed to be not far from its equilibrium state. The local balance equation of entropy (2.4.4) has now the same form as Eq.(2.6.18), which is the equation of heat conduction with Joule heating being the heat source. The derived set of equations (2.6.32), (2.6.33), (2.6.35), (2.6.14), and (2.6.15) form the complete set of constitutive equations for the soft ferromagnetic thermoelastic conductors.

It is sometimes useful to make an inversion of the independent constitutive (state) variable of the magnetization $\boldsymbol{M}'$ to the magnetic intensity field $\boldsymbol{H}$ as the independent constitutive (state) variable so that the formulation may be more convenient to be used than those presented in Eqs.(2.6.32)–(2.6.35) for practical applications, especially for studying nonlinear magnetic behaviors. To do so we can introduce a thermodynamic Gibbs function Φ defined by

$$\Phi = \Psi - \mu_o(\boldsymbol{H} \cdot \boldsymbol{M}') \tag{2.6.37}$$

Thus equation (2.6.29) may now be expressed as

$$\rho T \sigma_s = -\rho(\dot{\Phi} + \eta \dot{T}) + \boldsymbol{t} : (\nabla \mathbf{v}) - \frac{1}{T}(\boldsymbol{q} \cdot \nabla T) - \mu_o \boldsymbol{M} \cdot \dot{\boldsymbol{H}} + \boldsymbol{J} \cdot \boldsymbol{E}' \geq 0 \tag{2.6.38}$$

from which we can arrive at the following form of constitutive equations:

$$\eta = -\frac{\partial \Phi}{\partial T} \tag{2.6.39}$$

$$t_{ji} = \rho \frac{\partial \Phi}{\partial E_{KL}} x_{i,K} x_{j,L} + \rho \frac{\partial \Phi}{\partial H_L} H_i x_{j,L} = \rho \frac{\partial \Phi}{\partial E_{KL}} x_{i,K} x_{j,L} - \mu_o M_j H_i \tag{2.6.40}$$

$$t_{[ji]} = \rho \frac{\partial \Phi}{\partial H_L} H_{[i} x_{j],L} = -\mu_o M_{[j} H_{i]} \tag{2.6.41}$$

$$\mu_o M_i = -\rho \frac{\partial \Phi}{\partial H_L} x_{i,L} \tag{2.6.42}$$

with $H_L = H_i x_{i,L}$. A consequence of such an inversion may now be seen from Eq.(2.6.35) and Eq.(2.6.42).

2.6.3 Linearized Model for Thermoelastic Conductors

The constitutive equations formulated in the above section is, in general, highly nonlinear. Thus solutions of such a general set of nonlinear equations, together with field equations and boundary equations, are hardly possible to be found exactly. Some further simplification has to be made in order to deal with engineering problems for various application purposes. For some simplicity, in this section and the following section, we shall only discuss non-magnetizable conductors. Readers who are interested in linearizaed models for soft-ferromagnetic conductors may refer to the work of Zhou (1991a).

In the case of non-magnetizable thermoelastic conductors, in many cases of practical interest, we may introduce the following set of linearized constitutive equations by

$$t_{ij} = C_{ijkl}\varepsilon_{kl} - \beta_{ij}\theta \tag{2.6.43}$$

$$\eta = \eta_o + \frac{C_v}{T_o}\theta + \frac{1}{\rho_o}\beta_{ij}\varepsilon_{ij} \tag{2.6.44}$$

$$q_i = -\kappa_{ij}\theta,_j + \Gamma^Q_{ij}J_j \tag{2.6.45}$$

$$E'_i = \sigma^{-1}_{ij}J_j + \Gamma^E_{ij}\theta,_j \tag{2.6.46}$$

with $\boldsymbol{E}' = \boldsymbol{E} + \mathbf{v} \times \mathbf{B}$, and ε_{kl} being the infinitesimal strain defined in Eq.(2.2.10). θ is the temperature deviation defined by $\theta = T - T_0$ with the condition $|\theta| << T_0$, where T_0 (> 0) is the reference temperature. η_0 is the specific entropy (per unit mass) of the material at a strain-free reference configuration with the reference temperature T_0. ρ_0 is the mass density of the material, which is supposed to be a constant in the infinitesimal deformation approximation, and C_V is the specific heat per unit mass. C_{ijkl} and β_{ij} are, respectively, the elastic moduli and the thermal moduli. All material coefficients defined here are only functions of the position vector $\boldsymbol{x}$ if the material is inhomogeneous at a given temperature. Otherwise, they are all constants at T_0.

In addition, the material coefficients C_{ijkl}, β_{ij}, κ_{ij} and σ_{ij} satisfy the following symmetric conditions:

$$C_{ijkl} = C_{klij} = C_{jikl} = C_{ijlk}\ , \qquad \beta_{ij} = \beta_{ji}\ , \qquad \kappa_{ij} = \kappa_{ji}\ , \qquad \sigma_{ij} = \sigma_{ji} \tag{2.6.47}$$

These symmetric conditions may reduce the number of independent material coefficients. For instance, the elastic modulus tensor C_{ijkl} have generally 36 coefficients. Due to the symmetric conditions, it has, however, only 21 independent coefficients in the most general anisotropic case. In particular, for isotropic materials, we have then

$$C_{ijkl} = \lambda\delta_{ij}\delta_{kl} + G(\delta_{ik}\delta_{jl} + \delta_{il}\delta_{jk}) \tag{2.6.48}$$

where λ and G are called Lamé´s constants. Often, G is also called the elastic shear modulus. Thus only two independent elastic coefficients are left for isotropic materials. In engineering applications, we sometimes use other equivalent parameters, such as $K = \lambda+2G/3$, called the elastic bulk modulus, and $E = 2G(1 + \nu)$, called Young´s modulus, where ν is called Poisson´s ratio, defined by $\nu = \lambda/(2(\lambda+G))$. Shown in Table 2.1 are the elastic constants for some isotropic polycrystalline materials at room temperature.

For other material coefficients, we have, in the isotropic case, Eq.(2.6.17), and

$$\beta_{ij} = \beta\delta_{ij} \tag{2.6.49}$$

In Table 2.2, the thermal coefficient α of linear expansion ($\alpha = \beta/(3\lambda+2G)$) and the thermal conductivity κ are shown for some materials [Khazan (1994)].

It is known that isotropy is the material symmetry assumed on statistical basis for polycrystalline materials. In general, material properties of crystals are anisotropic. To describe the anisotropic material properties, especially the elastic properties, we often use the matrix presentation instead of the tensor presentation. For some simplicity, let us, for the moment, consider only the pure elastic case, assuming the material is in a constant and uniform temperature field so that thermal effects can be ignored. Since the stress tensor t_{ij} and the strain tensor ε_{kl} are both symmetric tensors, there are only six independent variables for each of them. Thus we may write

$$(t_{11}, t_{22}, t_{33}, t_{23}, t_{13}, t_{12}) = (t_1, t_2, t_3, t_4, t_5, t_6) \tag{2.6.50}$$

Table 2.1 Elastic Properties of Some Polycrystalline Materials

Materials	Density ρ (g/cm^3)	E (10^{10} N/m^2)	Poisson´s Ratio ν
Aluminum	2.7	6.8–7.1	0.355
Copper	8.3–8.93	12.1–12.8	0.37
Lead	11.4	1.5–1.7	0.43
Stainless steel	7.91	19.6	0.30
Tin	7.3	5.5	0.34
Titanium	4.5	36.2	0.35
Zinc	7.04–7.18	10.5	0.25

Source: American Institute of Physics Handbook, 2nd ed., McGraw-Hill, New York, 1951.

Table 2.2 Thermal Conductivity and Expansion Coefficients of Materials

Materials	Density ρ (g/cm^3)	κ (W/cm·°C)	α ($\times 10^{-6}$ /°C)
Aluminum	2.7	2.36	25.0
Stainless steel	7.9	0.329	17.3
Iron	7.8	0.803	12.0
Diamond*	3.5	20	1.0
Al_2O_3*	4.0	0.5	5.4
Si*	2.3	1.57	2.33
SiO_2*	2.5	0.014	0.55

* Single crystal.

and

$$(\varepsilon_{11}, \varepsilon_{22}, \varepsilon_{33}, 2\varepsilon_{23}, 2\varepsilon_{13}, 2\varepsilon_{12}) = (\varepsilon_1, \varepsilon_2, \varepsilon_3, \varepsilon_4, \varepsilon_5, \varepsilon_6) \tag{2.6.51}$$

We may then express the elastic stress-strain relation: $t_{ij} = C_{ijkl}\varepsilon_{kl}$ in the following matrix form:

$$\begin{bmatrix} t_1 \\ t_2 \\ t_3 \\ t_4 \\ t_5 \\ t_6 \end{bmatrix} = \begin{bmatrix} C_{11} & C_{12} & C_{13} & C_{14} & C_{15} & C_{16} \\ & C_{22} & C_{23} & C_{24} & C_{25} & C_{26} \\ & & C_{33} & C_{34} & C_{35} & C_{36} \\ & & & C_{44} & C_{45} & C_{46} \\ & & & & C_{55} & C_{56} \\ & & & & & C_{66} \end{bmatrix} \begin{bmatrix} \varepsilon_1 \\ \varepsilon_2 \\ \varepsilon_3 \\ \varepsilon_4 \\ \varepsilon_5 \\ \varepsilon_6 \end{bmatrix} \tag{2.6.52}$$

where the unwritten members are identical to those located symmetrically with respect to the main diagonal of the matrix. It is seen that there are generally 21 independent elastic coefficients.

For single crystals, the simplest case is for a crystal with cubic symmetry. For crystals with cubic symmetry, the elastic coefficient matrix C_{mn} becomes

$$[C_{mn}] = \begin{bmatrix} C_{11} & C_{12} & C_{12} & 0 & 0 & 0 \\ C_{12} & C_{11} & C_{12} & 0 & 0 & 0 \\ C_{12} & C_{12} & C_{11} & 0 & 0 & 0 \\ 0 & 0 & 0 & C_{44} & 0 & 0 \\ 0 & 0 & 0 & 0 & C_{44} & 0 \\ 0 & 0 & 0 & 0 & 0 & C_{44} \end{bmatrix} \tag{2.6.53}$$

where there are only three independent elastic coefficients, C_{11}, C_{12} and C_{44}. In Table 2.3, these elastic coefficients are shown for some crystals with cubic symmetry at room temperature [Maugin (1988) and Khazan (1994)].

In particular, for isotropic materials, we have the following relationships between the elastic coefficients C_{mn} from Eq.(2.6.53) and Lamé´s constants λ and G:

$$C_{11} = \lambda + 2G, \qquad C_{12} = \lambda \qquad \text{and} \qquad C_{44} = G \tag{2.6.54}$$

Table 2.3 Elastic Properties of Some Crystals at Room Temperature

Materials	C_{11}	C_{12} ($\times 10^9$ N/m^2)	C_{44}
Al	106.8	60.7	28.2
Cu	168.4	121.4	75.4
Ag	124.0	93.7	46.1
Ge	129.2	47.9	67.0
Si	165.7	63.9	79.6
GaAs	117.6	52.7	59.6
Diamond	1020	250	492

2.6.4 Field Equations and Boundary Conditions

In general, the set of the mechanical motion equation, the heat conduction equation, Maxwell's equations and boundary conditions for non-magnetizable thermoelastic conductors can be summarized as follows.

The mechanical motion equation is

$$\nabla \cdot \boldsymbol{t} + \boldsymbol{f} = \rho \frac{d\mathbf{v}}{dt} \tag{2.6.55}$$

which may also be written as

$$\nabla \cdot (\boldsymbol{t} + \boldsymbol{t}^{em}) + \boldsymbol{f}^{me} = \rho \frac{d\mathbf{v}}{dt} \tag{2.6.56}$$

where $\boldsymbol{f}^{me}$ denotes the mechanical body force. $\boldsymbol{t}^{em}$ is the electromagnetic stress tensor defined by Eq.(2.6.2) for the non-magnetizable conductors.

The stress interface condition at the interface S between two material media is

$$\boldsymbol{n} \cdot [\boldsymbol{t} + \boldsymbol{t}^{em}] = 0 \tag{2.6.57}$$

where $[\ \mathrm{F}\ ] = \mathrm{F}^{+} - \mathrm{F}^{-}$ denotes the jump of the quantity F across the interface S with $\boldsymbol{n}$ being the unit normal vector always drawn from S^{-} to S^{+}. The stress interface condition (2.6.57) can be reduced to

$$\boldsymbol{n} \cdot [\boldsymbol{t}] = 0 \tag{2.7.58}$$

if no free surface currents are present at the material interface since, in such a case, the electromagnetic traction is continuous across the material interface (including the boundary surface between the material body and free space). On the boundary ∂V (= $\partial V_t + \partial V_u$) of the material body, the mechanical boundary conditions can be given from Eq.(2.6.57) by

$$\boldsymbol{n} \cdot \boldsymbol{t} = \boldsymbol{t}^{(n)o} + \boldsymbol{n} \cdot [\boldsymbol{t}^{em}] \quad \text{on } \partial V_t \tag{2.6.59}$$

$$\boldsymbol{u} = \boldsymbol{u}^{o} \quad \text{on } \partial V_u \tag{2.6.60}$$

where $\boldsymbol{t}^{(n)o}$ are the prescribed surface traction on ∂V_t and $\boldsymbol{u}^{o}$ the prescribed displacement on ∂V_u. Here, we have simply replaced $\boldsymbol{t}^{-}$ by $\boldsymbol{t}$ and $\boldsymbol{u}^{-}$ by $\boldsymbol{u}$ on the boundary of the material body, where the quantity with superscript "−" means its value being taken on the material side of the body at the boundary surface. The second term on the right-hand side of Eq.(2.6.59) vanishes in the absence of free surface currents at the boundary of the material body.

The heat conduction equation with Joule heating is given by Eq.(2.6.18) with the energy interface condition at the interface S between two material media by

$$\boldsymbol{n} \cdot [\boldsymbol{q} + \boldsymbol{E}' \times \boldsymbol{H} - \mathbf{v} \cdot (\boldsymbol{t} + \boldsymbol{t}^{em})] = 0 \qquad \text{on } S \tag{2.6.61}$$

which states that the jump in the energy of traction across a material interface is balanced by that of the sum of the normal components of the heat flux vector and of Poynting's vector. This interface condition is derived from the global balance equation of energy (2.4.1) by noting the possible discontinuity of the electromagnetic traction and of Poynting's vector across the material interface.

With the use of Eq.(2.6.57), Eq.(2.6.61) may be reduced to be

$$\boldsymbol{n} \cdot [\boldsymbol{q} + \boldsymbol{E}' \times \boldsymbol{H}] = 0 \qquad \text{on } S \tag{2.6.62}$$

if the material velocity $\mathbf{v}$ is continuous across the material interface, which is the case when two material media are supposed to be perfectly bonded at the interface. Furthermore, if no surface currents exist at the interface of the material media which are also supposed to be non-magnetic, we then have

$$\boldsymbol{n} \cdot [\boldsymbol{q}] = 0 \qquad \text{on } S \tag{2.6.63}$$

since, in such a case, the magnetic intensity field $\boldsymbol{H}$ and the tangential component of the field $\boldsymbol{E}'$ ($= \boldsymbol{E} + \mathbf{v} \times \boldsymbol{B}$) are both continuous across the material interface. Equation (2.6.63) simply states that the normal component of the heat flux vector $\boldsymbol{q}$ is continuous across the material interface.

On the boundary ∂V ($= \partial V_q + \partial V_\theta$) of the material body where no free surface currents present, the boundary conditions for heat flux and temperature may be written as

$$\boldsymbol{q} \cdot \boldsymbol{n} = q^{(n)} \qquad \text{on } \partial V_q \tag{2.6.64}$$

$$\theta = \theta^o \qquad \text{on } \partial V_\theta \tag{2.6.65}$$

where $q^{(n)}$ and θ^o are both prescribed on the part of the boundary ∂V_q and ∂V_θ, respectively.

Another important condition, replacing any one of (2.6.64) and (2.6.65), is concerned with heat convection from a part ∂V_r of the boundary surface ∂V, which may be expressed by

$$\boldsymbol{q} \cdot \boldsymbol{n} + h(\theta - \theta_1) = 0 \qquad \text{on } \partial V_r \tag{2.6.66}$$

where h (≥ 0) is termed the boundary conductance (or the surface heat transfer coefficient) being an appropriate function of the surface coordinates and θ_1 the known temperature outside of the body near the boundary ∂V_r.

In many cases, for studying non-magnetizable thermoelastic conductors with small deformations, the mechanical motion equation (2.6.55) may be expressed approximately by

$$\nabla \cdot \boldsymbol{t} + \boldsymbol{J} \times \boldsymbol{B} = \rho_o \frac{\partial^2 \boldsymbol{u}}{\partial t^2} \tag{2.6.67}$$

in the absence of mechanical body forces. The heat conduction equation (2.6.18) may be expressed by

$$\rho_o C_{\mathrm{v}} \frac{\partial \theta}{\partial t} = \frac{\partial}{\partial x_i}\left(\kappa_{ij} \frac{\partial \theta}{\partial x_j}\right) - \frac{\partial}{\partial x_i}(\Gamma^Q_{ij} J_j) - T_o \beta_{ij} \frac{\partial \varepsilon_{ij}}{\partial t} + J_i E'_i \tag{2.6.68}$$

in the linearized constitutive model.

Maxwell's equations for the study of non-magnetizable conductors at the magneto-quasistatic approximation are given by Eqs.(1.7.2)–(1.7.5) with $\boldsymbol{B} = \mu_0 \boldsymbol{H}$ and with the electromagnetic interface conditions by Eqs.(1.7.15)–(1.7.18).

In addition to boundary (interface) conditions, initial conditions have to be given for general dynamic cases. For instance, initial values of the displacement and velocity fields, the temperature field, and the electromagnetic fields may be specified according to concrete problems. No need, however, arises for the initial conditions in the case of steady-state problems.

2.7 CONTINUUM THEORY OF THERMOELASTIC DIELECTRICS

2.7.1 Constitutive Equations for Thermoelastic Dielectrics

In this section, we shall formulate constitutive equations for thermoelastic dielectric solids at the electro-quasistatic approximation. For a dielectric solid in the absence of free charges, the total electric force acting on an electrically polarized body in the presence of an external electric field $\boldsymbol{E}^{\mathrm{o}}$ may be expressed in our dipole model [see Eq.(1.2.29)] by

$$\boldsymbol{F}^e = \int_V (\boldsymbol{P} \cdot \nabla) \boldsymbol{E}^o dV \tag{2.7.1}$$

This force expression can also be transformed into the following form

$$\boldsymbol{F}^e = \int_V \boldsymbol{f}^e dV + \int_{\partial V} \frac{1}{2\varepsilon_o} (\boldsymbol{P} \cdot \boldsymbol{n})^2 \boldsymbol{n} dS \tag{2.7.2}$$

where we have introduced an electric body force defined by

$$\boldsymbol{f}^e = (\boldsymbol{P} \cdot \nabla)\boldsymbol{E} \tag{2.7.3}$$

with $\boldsymbol{P}$ being the polarization vector of the dielectric solid. By introducing an electric stress tensor $\boldsymbol{t}^e$, defined by

$$t_{ji}^e = D_j E_i - \frac{1}{2}\varepsilon_o (E_k E_k)\delta_{ji} \tag{2.7.4}$$

we may write Eq.(2.7.3) in the form:

$$\boldsymbol{f}^e = (\boldsymbol{P} \cdot \nabla)\boldsymbol{E} = \nabla \cdot \boldsymbol{t}^e \tag{2.7.5}$$

For the dielectric solid, the total electric torque on the material body in our dipole model [see Eq.(1.2.30)] can be found by

$$\boldsymbol{L}^e = \int_V \{\boldsymbol{P} \times \boldsymbol{E}^o + \boldsymbol{x} \times (\boldsymbol{P} \cdot \nabla)\boldsymbol{E}^o\} dV \tag{2.7.6}$$

which may also be expressed as

$$\boldsymbol{L}^e = \int_V \{\boldsymbol{c}^e + \boldsymbol{x} \times (\boldsymbol{P} \cdot \nabla)\boldsymbol{E}\} dV + \int_{\partial V} \frac{1}{2\varepsilon_o} \boldsymbol{x} \times (\boldsymbol{P} \cdot \boldsymbol{n})^2 \boldsymbol{n} dS \tag{2.7.7}$$

Here, we have introduced the electric body couple $\boldsymbol{c}^e$ defined by

$$\boldsymbol{c}^e = \boldsymbol{P} \times \boldsymbol{E} \tag{2.7.8}$$

which can be related to the electric stress from Eq.(2.7.4) by $t_{[ji]}^e = \varepsilon_{jik} c_k^e$.

The electromagnetic power on the dielectric solid may be written in the dipole model as

$$W^{em} = \int_V ((\boldsymbol{P} \cdot \nabla)\boldsymbol{E}) \cdot \mathbf{v} dV + \int_{\partial V} \frac{1}{2\varepsilon_o} (\boldsymbol{P} \cdot \boldsymbol{n})^2 \boldsymbol{n} \cdot \mathbf{v} dS + \int_V \rho \boldsymbol{E} \cdot \frac{d}{dt}\left(\frac{\boldsymbol{P}}{\rho}\right) dV \tag{2.7.9}$$

in which the last term on the right-hand side represents the rate of work done by the electric couple. No surface electric torques are considered in this model.

We may now introduce the energy balance equation for the thermoelastic dielectrics by

$$\frac{d}{dt}\int_V \rho\left(\frac{1}{2}\mathbf{v} \cdot \mathbf{v} + u\right) dV = \int_{\partial V} (\boldsymbol{t}^{(n)} \cdot \mathbf{v} - \boldsymbol{q} \cdot \boldsymbol{n}) dS + \int_V \boldsymbol{f} \cdot \mathbf{v} dV + \int_V \rho \boldsymbol{E} \cdot \frac{d}{dt}\left(\frac{\boldsymbol{P}}{\rho}\right) dV \tag{2.7.10}$$

by noting Eqs.(2.4.1) and (2.7.9). Here, f is the total body force being the sum of the mechanical body force f^{me} and the electric body force f^{e} given by Eq.(2.7.3). The stress traction $t^{(n)}$, including both the mechanical and electric surface tractions, is related to Cauchy's stress tensor by Eq.(2.3.14).

Equation (2.7.10) represents the global conservation of energy, valid for any motion of the thermoelastic dielectric body, including discontinuous distribution of material characteristics. In the region of continuous smooth motions of a continuous body, where Eq.(2.7.10) is valid for every volume element, we can get the following local balance equation of energy

$$\rho \dot{u} = \boldsymbol{t}:(\nabla \mathbf{v}) - \nabla \cdot \boldsymbol{q} + \rho \boldsymbol{E} \cdot \frac{d}{dt}\left(\frac{\boldsymbol{P}}{\rho}\right) \tag{2.7.11}$$

with the aid of the balance law of linear momentum (2.3.5) and the conservation law of mass (2.1.14). By Eq.(2.7.11) and the second law of thermodynamics (2.4.4), we may obtain

$$\rho T \sigma_s = -\rho(\dot{\Phi}^e + \eta \dot{T}) + \boldsymbol{t}:(\nabla \mathbf{v}) + \boldsymbol{P} \cdot \dot{\boldsymbol{E}} - \frac{1}{T}\boldsymbol{q} \cdot \nabla T \geq 0 \tag{2.7.12}$$

in which the thermodynamical function Φ^e is defined by

$$\Phi^e = u - \eta T - (\boldsymbol{E} \cdot \boldsymbol{P})/\rho \tag{2.7.13}$$

Since no hereditary phenomena and nonlocality are considered for the thermoelastic dielectrics, Φ^e may be taken to be of the following form:

$$\Phi^e = \Phi^e(E_{KL}, W_L, T) \tag{2.7.14}$$

with the quantities W_L defined by $W_L = E_i x_{i,L}$ due to the fact that Φ^e must be an objective scalar functional under the full Euclidean transformation group (2.3.1) according to the axiom of objectivity. The free energy function Φ^e may also depend on the position vector $\boldsymbol{X}$ if the material is inhomogeneous.

To satisfy the inequality expressed by Eq.(2.7.12) according to the axiom of admissibility, we must have

$$\eta = -\frac{\partial \Phi^e}{\partial T} \tag{2.7.15}$$

$$t_{ji} = \rho \frac{\partial \Phi^e}{\partial E_{KL}} x_{i,K} x_{j,L} + \rho \frac{\partial \Phi^e}{\partial W_L} E_i x_{j,L} = \rho \frac{\partial \Phi^e}{\partial E_{KL}} x_{i,K} x_{j,L} - P_j E_i \tag{2.7.16}$$

$$t_{[ji]} = \rho \frac{\partial \Phi^e}{\partial W_L} E_{[i} x_{j],L} = -P_{[j} E_{i]} \tag{2.7.17}$$

$$P_i = -\rho \frac{\partial \Phi^e}{\partial W_L} x_{i,L} \tag{2.7.18}$$

and

$$-\frac{1}{T}(\boldsymbol{q} \cdot \nabla T) \geq 0 \tag{2.7.19}$$

It can be seen that Eq.(2.7.17) implies that the balance law of angular momentum (2.3.18) is identically satisfied by noting Eq.(2.7.8). It can also be seen that the Clausius–Duhem inequality (2.7.19) implies simply the fact that the heat can only flow from the place of high temperature to the place of lower temperature consistent with our common experience.

A generalized Fourier's law may be expressed by

$$q_i = -\kappa_{ij} T_{,j} \tag{2.7.20}$$

where κ_{ij} is the thermal conductivity tensor, which must be non-negative definite for making the Clausius–Duhem inequality be satisfied generally. In general, the thermal conductivity tensor κ_{ij} can be a function of the elastic strain, the electric field, and the local temperature. In many practical cases, if the material is homogeneous, we may assume that κ_{ij} is a constant tensor at a given reference temperature. In such a case, the thermal conductivity tensor is a constant tensor being positive definite and satisfying the symmetric relation $\kappa_{ij} = \kappa_{ji}$. For isotropic thermoelastic dielectrics, according to the axiom of material invariance, we simply have

$$\kappa_{ij} = \kappa \delta_{ij} \qquad (\kappa > 0) \tag{2.7.21}$$

The local balance equation of entropy (2.4.4) now reads

$$\rho T \dot{\eta} = -\nabla \cdot \boldsymbol{q} \tag{2.7.22}$$

which is the equation of heat conduction in the dielectrics in the absence of heat sources. The derived equations (2.7.15), (2.7.16), (2.7.18), and (2.7.20) form the complete set of constitutive equations for thermoelastic dielectric solids.

2.7.2 Linearized Model for Thermopiezoelectric Solids

An important class of thermoelastic dielectrics of practical interest is the thermopiezoelectric solid. In this section, we shall introduce the following set of linearized constitutive equations for thermopiezoelectric solids:

$$t_{ij} = C_{ijkl}\varepsilon_{kl} - e_{k \cdot ij} E_k - \beta_{ij}\theta \tag{2.7.23}$$

$$P_i = \chi_{ij}^e E_j + e_{i \cdot kl}\varepsilon_{kl} + \pi_i^e \theta \tag{2.7.24}$$

$$\eta = \eta_o + \frac{C_v}{T_o}\theta + \frac{1}{\rho_o}\beta_{ij}\varepsilon_{ij} + \frac{1}{\rho_o}\pi_i^e E_i \tag{2.7.25}$$

together with the Fourier law by

$$q_i = -\kappa_{ij}\theta,_j \tag{2.7.26}$$

Here, π_i^e denote the pyroelectric coefficients. χ_{ij}^e and $e_{i \cdot kl}$ are, respectively, the electric susceptibilities and the piezoelectric coefficients, which satisfy the following symmetrical conditions:

$$\chi_{ij}^e = \chi_{ji}^e \quad \text{and} \quad e_{i \cdot kl} = e_{i \cdot lk} \tag{2.7.27}$$

All other coefficients have been already introduced in Section 2.6.3. These material coefficients are all supposed to be constants if the material is homogeneous and at a constant reference temperature. Due to the symmetric condition by Eq.(2.7.27), there are 18 independent piezoelectric coefficients in the most general case of material anisotropy. Obviously, the piezoelectric phenomena occur only in dielectric solids presenting no center of symmetry. In contrast to piezoelectricity, piezomagnetism is seldom encountered in nature since in such materials certain types of magnetic structures have to exist and adequate magnetic symmetric conditions must be met, which is rarely the case [see Landau et al. (1984)].

Similar to the case of anisotropic elasticity discussed in Section 2.6.3, we often use the matrix presentation instead of the tensor presentation to describe piezoelectric phenomena which appear only in anisotropic materials. For simplicity, let us, for the moment, ignore thermal effects. Using the matrix presentation discussed in Section 2.6.3, we may write the following matrix expressions for piezoelectric materials:

$$t_m = C_{mn}\varepsilon_n - e_{km}E_k \tag{2.7.28}$$

$$P_i = \chi_{ik}^e E_k + e_{in}\varepsilon_n \tag{2.7.29}$$

with $i, k = 1, 2, 3$, and $m, n = 1, 2, \ldots, 6$. The elastic coefficient matrix C_{mn} has been discussed in Section 2.7.2. Here, we may write explicitly the piezoelectric coefficient matrix e_{in} in the general form:

$$[e_{in}] = \begin{bmatrix} e_{11} & e_{12} & e_{13} & e_{14} & e_{15} & e_{16} \\ e_{21} & e_{22} & e_{23} & e_{24} & e_{25} & e_{26} \\ e_{31} & e_{32} & e_{33} & e_{34} & e_{35} & e_{36} \end{bmatrix} \tag{2.7.30}$$

which has 18 independent coefficients.

For crystals with hexagonal symmetry, for instance, for some polarized ferroelectric ceramics in class $C_{6v} = 6mm$, the piezoelectric coefficient matrix (2.7.30) becomes

$$[e_{in}] = \begin{bmatrix} 0 & 0 & 0 & 0 & e_{15} & 0 \\ 0 & 0 & 0 & e_{15} & 0 & 0 \\ e_{31} & e_{31} & e_{33} & 0 & 0 & 0 \end{bmatrix} \tag{2.7.31}$$

which has only three independent piezoelectric coefficients. Besides, the electric susceptibility matrix χ_{ik}^{e} becomes

$$[\chi_{ik}^{e}] = \begin{bmatrix} \chi_{11}^{e} & 0 & 0 \\ 0 & \chi_{11}^{e} & 0 \\ 0 & 0 & \chi_{33}^{e} \end{bmatrix} \tag{2.7.32}$$

and the elastic coefficient matrix C_{mn} becomes

$$[C_{mn}] = \begin{bmatrix} C_{11} & C_{12} & C_{13} & 0 & 0 & 0 \\ C_{12} & C_{11} & C_{13} & 0 & 0 & 0 \\ C_{13} & C_{13} & C_{33} & 0 & 0 & 0 \\ 0 & 0 & 0 & C_{44} & 0 & 0 \\ 0 & 0 & 0 & 0 & C_{44} & 0 \\ 0 & 0 & 0 & 0 & 0 & C_{66} \end{bmatrix} \tag{2.7.33}$$

with $C_{66} = (C_{11} - C_{12})/2$. Thus, in this case, there are altogether 10 independent coefficients to characterize the piezoelectric properties of this type of crystals with hexagonal symmetry (6*mm*). Detailed numerical values of the piezoelectric coefficients for various types of piezoelectric crystals may be found in books and papers specialized in piezoelectricity [Mason (1950), Nye (1960), and Auld (1973)].

2.7.3 Field Equations and Boundary Conditions

The set of field equations and boundary conditions for the linear thermopiezoelectric solid may be summarized as follows.

The linearized mechanical equation of motion for the thermopiezoelectric solid can be given by

$$\frac{\partial}{\partial x_j}(C_{ijkl}u_{k,l} - e_{k \cdot ij}E_k - \beta_{ij}\theta) = \rho_o \frac{\partial^2 u_i}{\partial t^2} \tag{2.7.34}$$

in the absence of mechanical body force. The stress interface condition can be expressed as

$$\boldsymbol{n} \cdot [\boldsymbol{t}] = 0 \tag{2.7.35}$$

On the boundary ∂V (= ∂V_t + $\partial \mathrm{V}_u$) of the thermopiezoelectric body, the mechanical boundary conditions can be expressed by

$$\boldsymbol{n} \cdot \boldsymbol{t} = \boldsymbol{t}^{(n)o} \qquad \text{on } \partial V_t \tag{2.7.36}$$

$$\boldsymbol{u} = \boldsymbol{u}^o \qquad \text{on } \partial V_u \tag{2.7.37}$$

where $\boldsymbol{t}^{(n)\mathrm{o}}$ and $\boldsymbol{u}^{\mathrm{o}}$ are, respectively, the prescribed surface traction on ∂V_t and the prescribed displacement on ∂V_u.

The linearized heat conduction equation can be found as

$$\rho_o C_{\mathrm{v}} \frac{\partial \theta}{\partial t} = \frac{\partial}{\partial x_i}\left(\kappa_{ij}\frac{\partial \theta}{\partial x_j}\right) - T_o \beta_{ij}\frac{\partial \varepsilon_{ij}}{\partial t} - T_o \pi_i^e \frac{\partial E_i}{\partial t} \tag{2.7.38}$$

in the absence of internal heat sources. The interface condition for heat flux is given by

$$\boldsymbol{n} \cdot [\boldsymbol{q}] = 0 \tag{2.7.39}$$

where we have assumed that the material velocity is continuous across the material interface.

On the boundary ∂V (= ∂V_q+$\partial \mathrm{V}_\theta$) of the material body, the heat flux boundary conditions is

$$\boldsymbol{q} \cdot \boldsymbol{n} = q^{(n)} \qquad \text{on } \partial V_q \tag{2.7.40}$$

$$\theta = \theta^o \qquad \text{on } \partial V_\theta \tag{2.7.41}$$

where $q^{(n)}$ and θ^o are both prescribed on the part of the boundary ∂V_q and ∂V_θ, respectively. Similarly, we may also have the boundary condition of heat convection as shown by Eq.(2.6.66).

The Maxwell equations at the electro-quasistatic approximation are given by Eqs.(1.7.20) and (1.7.21) with $\boldsymbol{D} = \varepsilon_0 \boldsymbol{E} + \boldsymbol{P}$ and with the electric interface conditions (1.7.31) and (1.7.32) where the free surface and volume charges are supposed to be absent in the thermopiezoelectric body considered.

Furthermore, we may introduce a scalar electric potential function ϕ by

$$\boldsymbol{E} = -\nabla\phi \tag{2.7.42}$$

so that Eq.(1.7.20) is identically satisfied, and Eq.(1.7.21) may now be expressed by using the constitutive equation (2.7.24) as

$$\frac{\partial}{\partial x_i}\left(\overset{e}{\varepsilon}_{ij}\frac{\partial\phi}{\partial x_j} - e_{i\cdot kl}\frac{\partial u_k}{\partial x_l} - \overset{e}{\pi}_i\theta\right) = 0 \tag{2.7.43}$$

where $\overset{e}{\varepsilon}_{ij} = \varepsilon_o\delta_{ij} + \overset{e}{\chi}_{ij}$ is the permittivity tensor of the dielectric solid.

The set of five linearized equations (2.7.34), (2.7.38), and (2.7.43), with the aid of Eq.(2.7.42), constitute the complete set of field equations of thermopiezoelectricity for the determination of the five unknowns u_i ($i = 1, 2, 3$), θ and ϕ. Of course, in all above theories, initial conditions have to be specified if non-steady–state problems are involved.

In above sections, we have demonstrated how the constitutive equations for thermoelastic conductors and thermoelastic dielectrics can be formulated in the theory of continuum electrodynamics. In the following section, we shall introduce a continuum model for studying phothothermoelasticity, which is of practical interest in proper utilization of optical materials for applications in, for instance, fiber optical communications and various types of optical sensors.

2.8 PHOTOTHERMOELASTICITY

It is known that optical characteristics of a dielectric may be changed by the application of a force field, which may be of mechanical, thermal, electric or magnetic origin. These effects have applications both in mechanically static arrangements, where the phenomenon is referred to as photoelasticity, and at high frequencies, where the effect of stress wave travelling through the medium is called acousto-optics. The photoelastic phenomenon was discovered early in 1816 by Brewster, who found that transparent isotropic glasses under anisotropic stress fields become optically anisotropic and cause birefringence in general. In this section, we shall introduce the theoretical basis for analyzing these phenomena.

2.8.1 Principal Stresses and Principal Refractive Indices

The concepts of principal stresses and principal refractive indices are useful for the study of general stress and optical behaviors of anisotropic media. In solid mechanics, we introduce the concept of the principal stress, which is defined as the stress acting on a principal plane on which shearing stress vanishes. The values of the principal stresses at each point of the solid medium are determined as the roots of the following equation:

$$\det(t_{ij} - \sigma\delta_{ij}) = 0 \tag{2.8.1}$$

which is a cubic equation in σ. The obtained principal stresses σ_i ($i = 1, 2, 3$) are independent of any coordinates of reference since they characterize the physical state of stress at a point. For a symmetric stress tensor $t_{ij} = t_{ji}$, the three principal stresses are all real and their corresponding three principal planes are mutually orthogonal. The unit normal vector $\boldsymbol{s}$ for each value of the principal stress is determined by the equation:

$$(t_{ij} - \sigma\delta_{ij})s_i = 0 \tag{2.8.2}$$

with the condition $s_i s_i = 1$. The three mutually orthogonal vectors $\boldsymbol{s}^{(k)}$ ($k = 1, 2, 3$) may be used to constitute a coordinate system, called the principle coordinate system. In the principal coordinate system with the coordinate axes x_1, x_2, x_3 being chosen to coincide with the principal axes, the matrix of stress components becomes simply

$$[t_{ij}] = \begin{bmatrix} \sigma_1 & 0 & 0 \\ 0 & \sigma_2 & 0 \\ 0 & 0 & \sigma_3 \end{bmatrix} \tag{2.8.3}$$

On any surface element with a unit outer normal vector $\boldsymbol{s}$, there acts a traction $\boldsymbol{t}^{(s)}$, which, in the principal coordinate system, may be expressed (by $t_i^{(s)} = t_{ji}s_j$) as

$$t_1^{(s)} = \sigma_1 s_1, \qquad t_2^{(s)} = \sigma_2 s_2, \qquad t_3^{(s)} = \sigma_3 s_3 \tag{2.8.4}$$

By noting the fact that $\boldsymbol{s}$ is a unit vector, we have

$$\frac{(t_1^{(s)})^2}{(\sigma_1)^2} + \frac{(t_2^{(s)})^2}{(\sigma_2)^2} + \frac{(t_3^{(s)})^2}{(\sigma_3)^2} = 1 \tag{2.8.5}$$

which is an equation of an ellipsoid with reference to a system of rectangular Cartesian coordinates with axes labelled $t_1^{(s)}$, $t_2^{(s)}$, $t_3^{(s)}$. This ellipsoid is often called the *Lamé stress ellipsoid.*

In addition to the Lamé stress ellipsoid, Cauchy considered the following quadratic surface:

$$t_{ij}x_ix_j = \pm c_o^2 \tag{2.8.6}$$

which, in the principal coordinate system, becomes

$$\sigma_1(x_1)^2 + \sigma_2(x_2)^2 + \sigma_3(x_3)^2 = \pm c_o^2 \tag{2.8.7}$$

where the ambiguous sign is to be chosen so that Eq.(2.8.6) or Eq.(2.8.7) represents a real surface. It can be seen that Eq.(2.8.7) represents an ellipsoid, called the *Cauchy stress ellipsoid*, if the principal stresses are all of the same sign.

In classical optics, an electrically anisotropic medium is characterized by a dielectric coefficient (permittivity) tensor ε^e_{ij} instead of a scalar permittivity. This permittivity tensor is symmetric, that is, $\varepsilon^e_{ij} = \varepsilon^e_{ji}$. Thus, we may also introduce a principal coordinate system, in which the permittivity tensor becomes simply

$$[\varepsilon^e_{ij}] = \begin{bmatrix} \varepsilon^e_1 & 0 & 0 \\ 0 & \varepsilon^e_2 & 0 \\ 0 & 0 & \varepsilon^e_3 \end{bmatrix} \tag{2.8.8}$$

and the electric displacement vector $\boldsymbol{D}$ can be expressed in the principal coordinate system as

$$D_1 = \varepsilon^e_1 E_1, \quad D_2 = \varepsilon^e_2 E_2, \quad D_3 = \varepsilon^e_3 E_3 \tag{2.8.9}$$

with ε^e_1, ε^e_2, ε^e_3 being called the principal dielectric coefficients. $\boldsymbol{E}$ is the electric field vector.

Similar to the Cauchy stress ellipsoid, we may construct an index ellipsoid of the form:

$$\frac{(x_1)^2}{\varepsilon^e_1} + \frac{(x_2)^2}{\varepsilon^e_2} + \frac{(x_3)^2}{\varepsilon^e_3} = 1 \tag{2.8.10}$$

which can also be written in the following form:

$$\frac{(x_1)^2}{(n_1)^2} + \frac{(x_2)^2}{(n_2)^2} + \frac{(x_3)^2}{(n_3)^2} = 1 \qquad (2.8.11)$$

by introducing the principal refractive indices, defined by

$$n_1 = \sqrt{\varepsilon_{r1}}, \qquad n_2 = \sqrt{\varepsilon_{r2}}, \qquad n_3 = \sqrt{\varepsilon_{r3}} \qquad (2.8.12)$$

with $\varepsilon_{ri} = \varepsilon^e{}_i/\varepsilon_o$ $(i = 1, 2, 3)$. The index ellipsoid described by Eq.(2.8.11) is often called the *Fresnell ellipsoid.*

2.8.2 Optical Birefringence

Using Maxwell's equations, and the constitutive equation, $D_i = \varepsilon^e{}_{ij}E_j$ for simple non-magnetic dielectrics, we may derive the following electromagnetic wave equation:

$$\nabla^2 \boldsymbol{E} - \nabla(\nabla \cdot \boldsymbol{E}) = \frac{1}{c^2}\frac{\partial^2}{\partial t^2}(\varepsilon_r \cdot \boldsymbol{E}) \qquad (2.8.13)$$

where c is the velocity of light in vacuum $[c^2 = 1/(\varepsilon_o\mu_o)]$.

For a homogeneous dielectric medium in a uniform temperature and stress field so that ε_r is constant in space, Eq.(2.8.13) may be reduced to be

$$\nabla^2 \boldsymbol{E} = \frac{1}{c^2}\frac{\partial^2}{\partial t^2}(\varepsilon_r \cdot \boldsymbol{E}) \qquad (2.8.14)$$

In practice, this wave equation may be used approximately when ε_r varies in space, provided that its variation is slight over the distance of the light wavelength. More specifically, the relative change of ε_r over the distance of one light wavelength must be less than unit [Marcuse (1982)]. This condition is often satisfied in inhomogeneous optical media with continuous variation of ε_r in space, such as graded index optical fibers.

Consider now a plane wave of the form:

$$\boldsymbol{E} = \boldsymbol{E}_o \exp\left[\frac{\omega}{\text{v}}(\boldsymbol{x} \cdot \boldsymbol{l} - \text{v}t)\right] \qquad (2.8.15)$$

where $\boldsymbol{x}$ is the position vector, $\boldsymbol{l}$ the unit vector along the direction of light propagation, ω the circular frequency, v the phase velocity, and t the time. We may find

$$D_i l_i = 0, \qquad \varepsilon_{rij} E_j l_i = 0 \tag{2.8.16}$$

which implies that the electric displacement vector $\boldsymbol{D}$ is at right angle to the direction of the wave propagation, while $\boldsymbol{E}$ is not, in general, perpendicular to the direction of wave propagation, as in the case of isotropy. Furthermore, by using Maxwell´s equations, we may find

$$\mathrm{v}\boldsymbol{D} = \boldsymbol{l} \times \boldsymbol{H}, \qquad \mu_o \mathrm{v}\boldsymbol{H} = \boldsymbol{l} \times \boldsymbol{E} \tag{2.8.17}$$

which shows that the magnetic field vector $\boldsymbol{H}$ (and $\boldsymbol{B}$) is at right angle to the direction of the wave propagation, and is perpendicular to the electric field vector $\boldsymbol{E}$ as well as the electric displacement vector $\boldsymbol{D}$. Thus, vectors $\boldsymbol{E}$, $\boldsymbol{D}$, and $\boldsymbol{l}$ are coplanar, and the direction of energy flow given by the Poynting vector $\boldsymbol{E}\times\boldsymbol{H}$ is not generally colinear with the direction of the wave propagation.

From Eq.(2.8.17), we may also find

$$\boldsymbol{D} = \frac{1}{\mu_o \mathrm{v}^2}[\boldsymbol{E} - (\boldsymbol{l} \cdot \boldsymbol{E})\boldsymbol{l}] \tag{2.8.18}$$

It can be seen that, in an unloaded isotropic dielectric medium at reference temperature, we may get the phase velocity v of the plane electromagnetic wave in the medium

$$\mathrm{v} = \frac{c}{\sqrt{\varepsilon_r}} = \frac{c}{n} \tag{2.8.19}$$

where $n = (\varepsilon_r)^{1/2}$ is the refractive index of the isotropic dielectric medium, and ε_r the relative permittivity.

For anisotropic media, by choosing a principal coordinate system, we may get from Eqs.(2.8.9) and (2.8.18)

$$E_i = \frac{(\boldsymbol{l} \cdot \boldsymbol{E}) l_i}{1 - \varepsilon_{ri} \mathrm{v}^2 / c^2} \qquad (i = 1, 2, 3) \tag{2.8.20}$$

Multiplying this equation by l_i and adding the resulting three equations, we obtain

$$\frac{(l_1)^2}{1 - \varepsilon_{r1} \mathrm{v}^2 / c^2} + \frac{(l_2)^2}{1 - \varepsilon_{r2} \mathrm{v}^2 / c^2} + \frac{(l_3)^2}{1 - \varepsilon_{r3} \mathrm{v}^2 / c^2} = 1 \tag{2.8.21}$$

which may also be deduced, by using the relation $l_1^2 + l_2^2 + l_3^2 = 1$, as

$$\frac{\varepsilon_{r1}(l_1)^2}{1-\varepsilon_{r1}\mathrm{v}^2/c^2}+\frac{\varepsilon_{r2}(l_2)^2}{1-\varepsilon_{r2}\mathrm{v}^2/c^2}+\frac{\varepsilon_{r3}(l_3)^2}{1-\varepsilon_{r3}\mathrm{v}^2/c^2} = 0 \tag{2.8.22}$$

Equation (2.8.22) is a quadratic equation in v^2. Therefore, for each direction of propagation, it yields two solutions for the phase velocity. To each of these two velocities, Eq.(2.8.20) gives the ratios E_1:E_2:E_3, and then Eq.(2.8.9) gives the corresponding ratios D_1:D_2:D_3. Since these ratios are real, the fields $\boldsymbol{E}$ and $\boldsymbol{D}$ are linearly polarized. It is thus shown that when a light ray enters an anisotropic dielectric medium, it may split into two rays that are linearly polarized and propagate with different phase velocities. Such a phenomenon is called *optical birefringence* or simply birefringence.

2.8.3 Photothermoelastic Law

Consider dielectrics which are isotropic in a stress (strain) free state. We ignore possible small effects on mechanical behavior and temperature of the dielectrics from applied electromagnetic fields by assuming that the involved electromagnetic fields are weak in our problems. Thus, we are dealing with an uncoupled problem, in which we may study the effect of material deformation as well as temperature variation on dielectric properties of the dielectric media by perturbation. For this purpose, we may introduce the constitutive equation:

$$D_i = \varepsilon^e_{ij}E_j \tag{2.8.23}$$

with

$$\varepsilon^e_{ij} = \varepsilon_o\varepsilon_r\delta_{ij}+\pi^e\theta\delta_{ij}+\Gamma_{ijkl}\varepsilon_{kl} \tag{2.8.24}$$

where π^e denotes the second order pyroelectric coefficient, and Γ_{ijkl} the photoelastic coefficient tensor. Besides, we have the following thermoelastic constitutive equation:

$$t_{ij} = \lambda\varepsilon_{kk}\delta_{ij}+2G\varepsilon_{ij}-\beta\theta\delta_{ij} \tag{2.8.25}$$

where λ, G are the Lamé elastic coefficients, and β the thermoelastic modulus. And, for heat conduction, we have the Fourier law: $\boldsymbol{q} = -\kappa\nabla\theta$ with κ being the thermal conductivity.

In analogy with Pockel´s photoelastic law [Narasimhamurty (1981)], by using above constitutive equations, we may introduce a photothermoelastic law by

$$\left(\frac{1}{n^2}\right)_{ij}-\frac{1}{n^2}\delta_{ij} = (\pi'\theta+B'_2t_{kk})\delta_{ij}+2B'_1t_{ij} \tag{2.8.26}$$

where n denotes the refractive index of the unloaded dielectric at reference temperature.

It can be seen that the refractive index matrix and the stress matrix have the same principal axes, and the principal refractive indices n_i $(i = 1, 2, 3)$ are related to the principal stresses σ_i $(i = 1, 2, 3)$ by the following equations:

$$\frac{1}{n_1^2} - \frac{1}{n^2} = \pi'\theta + (B'_2 + 2B'_1)\sigma_1 + B'_2(\sigma_2 + \sigma_3) \tag{2.8.27}$$

$$\frac{1}{n_2^2} - \frac{1}{n^2} = \pi'\theta + (B'_2 + 2B'_1)\sigma_2 + B'_2(\sigma_1 + \sigma_3) \tag{2.8.28}$$

$$\frac{1}{n_3^2} - \frac{1}{n^2} = \pi'\theta + (B'_2 + 2B'_1)\sigma_3 + B'_2(\sigma_1 + \sigma_2) \tag{2.8.29}$$

in the same principal coordinate system.

Since the difference $|n - n_i|$ is usually small, compared with n or n_i, we may use the approximation:

$$\frac{1}{n_i^2} - \frac{1}{n^2} \approx \frac{2}{n^3}(n - n_i), \qquad (i = 1,2,3) \tag{2.8.30}$$

and write Eqs.(2.8.27)–(2.8.29) into the following form:

$$n_1 - n = \pi^*\theta + B_1\sigma_1 + B_2(\sigma_2 + \sigma_3) \tag{2.8.31}$$

$$n_2 - n = \pi^*\theta + B_1\sigma_2 + B_2(\sigma_1 + \sigma_3) \tag{2.8.32}$$

$$n_3 - n = \pi^*\theta + B_1\sigma_3 + B_2(\sigma_1 + \sigma_2) \tag{2.8.33}$$

with relevant coefficients defined by

$$\pi^* = -\frac{n^3}{2}\pi', \quad B_1 = -\frac{n^3}{2}(B'_2 + 2B'_1), \quad B_2 = -\frac{n^3}{2}B'_2 \tag{2.8.34}$$

where π^* may be called the photothermal coefficient, and B_1 and B_2 the stress-optical coefficients. Obviously, if we can determine experimentally the coefficients π^*, B_1 and B_2, the coefficients π', B'_1, B'_2 can be easily obtained by Eq.(2.8.34). As an example, the stress-optical coefficients measured by Primak and Post (1959) for fused silica are: $B_1 = -6.5\times10^{-7}$/Mpa and $B_2 = -4.2\times10^{-6}$/Mpa. The negative sign of the stress-optical coefficients implies that the refractive index decreases when the stresses are tensile, and increases when the stresses are compressive.

Using the thermoelastic relation (2.8.25), we may find

$$\sigma_1 = -\beta\theta + (\lambda + 2G)\varepsilon_1 + \lambda(\varepsilon_2 + \varepsilon_3) \tag{2.8.35}$$

$$\sigma_2 = -\beta\theta + (\lambda + 2G)\varepsilon_2 + \lambda(\varepsilon_1 + \varepsilon_3) \tag{2.8.36}$$

$$\sigma_3 = -\beta\theta + (\lambda + 2G)\varepsilon_3 + \lambda(\varepsilon_1 + \varepsilon_2) \tag{2.8.37}$$

where ε_1, ε_2, and ε_3 represent the principal strains.

Inserting Eqs.(2.8.35)–(2.8.37) into Eqs.(2.8.31)–(2.8.33), we can obtain

$$n_1 - n = \Pi^*\theta + C_1\varepsilon_1 + C_2(\varepsilon_2 + \varepsilon_3) \tag{2.8.38}$$

$$n_2 - n = \Pi^*\theta + C_1\varepsilon_2 + C_2(\varepsilon_1 + \varepsilon_3) \tag{2.8.39}$$

$$n_3 - n = \Pi^*\theta + C_1\varepsilon_3 + C_2(\varepsilon_1 + \varepsilon_2) \tag{2.8.40}$$

with the coefficients given by

$$\Pi^* = \pi^* - (B_1 + 2B_2)\beta \tag{2.8.41}$$

$$C_1 = (\lambda + 2G)B_1 + 2\lambda B_2 \tag{2.8.42}$$

$$C_2 = \lambda B_1 + 2(\lambda + G)B_2 \tag{2.8.43}$$

where C_1 and C_2 may be called the strain-optical coefficients.

If we are interested only in the difference between the refractive indices along the principal stress directions, we may obtain from Eqs.(2.8.31)–(2.8.33)

$$n_1 - n_2 = B(\sigma_1 - \sigma_2) \tag{2.8.44}$$

$$n_2 - n_3 = B(\sigma_2 - \sigma_3) \tag{2.8.45}$$

$$n_3 - n_1 = B(\sigma_3 - \sigma_1) \tag{2.8.46}$$

with $B = B_1 - B_2$. Similarly, we may find

$$n_1 - n_2 = C(\varepsilon_1 - \varepsilon_2) \tag{2.8.47}$$

$$n_2 - n_3 = C(\varepsilon_2 - \varepsilon_3) \tag{2.8.48}$$

$$n_3 - n_1 = C(\varepsilon_3 - \varepsilon_1) \tag{2.8.49}$$

with $C = C_1 - C_2$. The relations (2.8.44)–(2.8.49) were first experimentally found by Werteim in 1850s, and known as the Wertheim stress or strain optical laws. It is seen that the photothermal effect accounted by the term with the coefficient π^* or Π^* is absent if only the differences between refractive indices along principal stress directions are considered.

2.9 ON CONTINUUM MODELS AND THEIR LIMITATIONS

In above sections, we have presented some continuum models for describing macroscopic electromagnetic and mechanical behaviors of solid materials, where the intrinsic structures of the materials are not discussed. At early time, material utilization was indeed a simple selection process, that is, deciding from a given, rather limited set of materials that was best suited for an application by virtue of its characteristics. It was not until relatively recent times that the relationships between the structure elements of materials and their properties came to be understood. Thus, tens of thousands of different materials have evolved with rather specialized characteristics, which may meet the needs of our modern and complex society. Shown in Fig. 2.3 is the material structures that play an important role in the macroscopic behaviors of solids range dimensionally from extremely small values on the order of atomic dimensions to relatively large values.

At subatomic level, solid materials are known to be composed of immense number of elementary particles which obey the basic laws of physics. Obviously, it is impossible to treat macroscopic solids in the full generality and rigor of the pertaining physical laws. A number of simplifications are thus introduced.

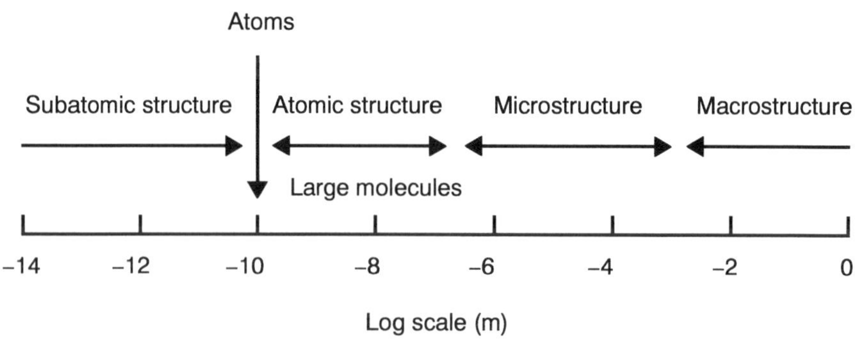

Figure 2.3 Dimensional comparison between certain structures in materials.

If nuclear reactions as well as inter-particle gravitation are neglected, then the strong, weak and gravitational interactions can be omitted, thus reducing the problem to one of quantum electrodynamics. In many situations, also electromagnetic radiation effects are not important or can be included in some global manner. If also special relativistic effects play no role, then the problem reduces to one of non-relativistic quantum mechanics in which the particles interact with electromagnetic forces. Although it may be easy to formulate the Schrödinger equation for systems of interacting particles, the resulting mathematical problem is still far too complex for exact solution. It is common at this point to introduce a further basic approximation known as the Born-Oppenheimer approximation or adiabatic approximation [Born and Oppenheimer (1927) and Atkins (1983)]. However, exact solutions are still rare. Therefore, one usually treats the materials as aggregations of atoms bonded together by certain semi-empirical bonding mechanisms without looking for exact solution at the subatomic level [Born and Huang (1954)]. The well-known bonds are respectively the ionic, the covalent, the metallic and the Van der Waals bond. Relationships between the bonds and properties are listed in Table 2.4.

Table 2.4 Classification of Atomic Bonds and Relevant Material Properties

Classification of Bonds	Subatomic Structures	Examples	Properties
Ionic bond	Electrostatic attraction between positive and negative ions, formed by complete transfer of electron between atoms	NaCl Al_2O_3 MgO	Transparent, hard, brittle crystals; Poor conductors of electricity; Melting temperature (and strength) increases with bond energy; Low coefficient of thermal expansion
Covalent bond	Shearing electrons between adjacent atoms	H_2 C O_2	Brittle, hard crystals; Insulator; High melting temperature; Low coefficient of thermal expansion
Metallic bond	Combination of the positive charged ions and the surrounding electrons cloud	Cu Ni Fe	Opague, ductile crystals; Good conductors of heat and electricity; Various melting temperatures and thermal expansion coefficients
Van der Waals bond	Electrostatic dipole attraction due to unsymmetric electrical charges in electrically neutral atoms or molecules	H_2O Ar He	Weak, soft crystals; Insulators; Low melting temperature; Large coefficient of thermal expansion

Solid materials at the atomic level may be classified as crystalline materials and noncrystalline (or amorphous) materials according to the regularity of the arrangement of the atoms or ions. Crystalline materials consist of crystals which may have different type of crystal structures that are atoms situated in an orderly, repetitive arrangements in space. In crystalline solids, there are seven crystal systems (cubic, tetragonal, orthorhombic, monoclinic, triclinic, hexagonal, and rhombohedral). A spatial arrangement of equivalent points in a crystal system is called a space lattice. When this arrangement comprises the smallest repeating network of sites in the crystal, it is called a unit cell. Only 14 space lattices may be constructed from the seven crystal systems. Mechanical and physical properties such as density, strength, hardness, ductility, electrical conductivity, thermal conductivity, thermal expansion, and magnetic behavior of a macroscopic solid material are strongly affected by the arrangement of atoms in the solid. Specific arrangements of certain atoms or molecules are also responsible for the phase transformations of solid materials [see, e.g., Aifantis and Gittus (1986)].

Crystals occurring in nature and in engineering materials, more or less, always deviate from perfect periodicity. They contain numerous imperfections (crystalline defects, such as vacancies, interstitials, impurity atoms, dislocations, grain boundaries, etc.). Although such imperfections exist on an atomic scale, in other words, their individual proportions are extremely small, there can be an extremely large number of such defects in ordinary crystals. For this reason, they are responsible for a substantial part of observed mechanical and electromagnetic behaviors of the materials. Theories concerning with the study of the imperfections at atomic scales are listed in Table 2.5 and are compared with other theories on defects at larger scales.

At a relative large scale, solid materials often display another type of internal structure, the microstructure. The microstructure is referred to the physical appearance of the internal structure of engineering materials. This type of structure results from various combinations of chemical elements in the solid state and from various microdefects distributed in the materials. In essence, microstructure is the architecture of the materials, that is, how the various phases and crystals (units) are put together, their particular patterns and arrangements, and how these are joined. Such type of structures are usually revealed under the microscope in specimens that have been prepared for microscopy by some special treatments like polishing, etching, etc.

It has been experimentally found that many macroscopic properties of solid materials depend on their microstructures. An example of the effect of microstructures on the macroscopic properties of solids is the macroscopic isotropic behavior of a polycrystalline solid which consists of grains that may behave anisotropically. Such a phenomenon is due to the random crystallographic orientations of the individual grains forming the microstructure of the solid. Microstructure effects of two-phase alloys such as iron-carbon alloys having pearlitic and spheroiditic microstructures are also known. Other examples are, for instance, a permanent magnetic material should be very fine

Table 2.5 Structures, Imperfections and Material Theories

Structures	Elements	Theories
Atomic defects & large molecules (0.1–100 nm)	Vacancies, interstitials, substitutionals, impurity atoms, dislocations, grain boundaries, single domain ferroparticles	Classical or quantum mechanics Theory of solid state physics Atomic lattice theory Crystal dislocation theory Molecular field theory
Microstructures (0.1–100 μm)	Grains, voids, microcracks, inhomogeneities, (second phase particles), inclusions	Classical mechanics Crystal elasticity and plasticity Metallurgy and Metallography Continuum dislocation theory Micromechanics of composites
Macrostructures (0.1–100 mm)	Cracks, cavities, notches, pits, inhomogeneities, inclusions	Classical continuum theory Elastic theory, Plastic theory Fracture mechanics Elastic-plastic theory

grained (under one micron) to prevent the formation of domain walls, whereas high-permeability materials should have large uniform grains because magnetization rotation requires more energy than domain-wall motion.

With the use of effects of microstructures, various types of materials with synthetic microstructures, such as composite materials, are now being produced in materials industry. The microstructures in materials can often be controlled by certain designed processes. For instance, microstructures of single-phase metals can be adjusted by plastic deformation, by recrystallization, by appropriate selection of solid solutions, or by crystal and molecular orientations. Multi-phase solid materials can have properties modified and controlled through the same procedures availablc for single-phase materials. There are also additional means of control of microstructures in multi-phase materials such as the ways of varying the relative amounts of the phases, the size of the phase grains, the shape and the distribution of the phases. Examples of the microstructures of some composite materials composed of reinforced particles with various sizes and shapes in a homogeneous matrix medium are shown in Fig. 2.4.

To study quantitatively the effects of microstructures, we may require theories of studying imperfections at different levels, which are summarized briefly in Table 2.5. Some details about material microstructures and their effects on macroscopic properties of solid materials are referred to in the work of, for instance, Van Vlack (1975), Hale (1976), and Bradley (1971).

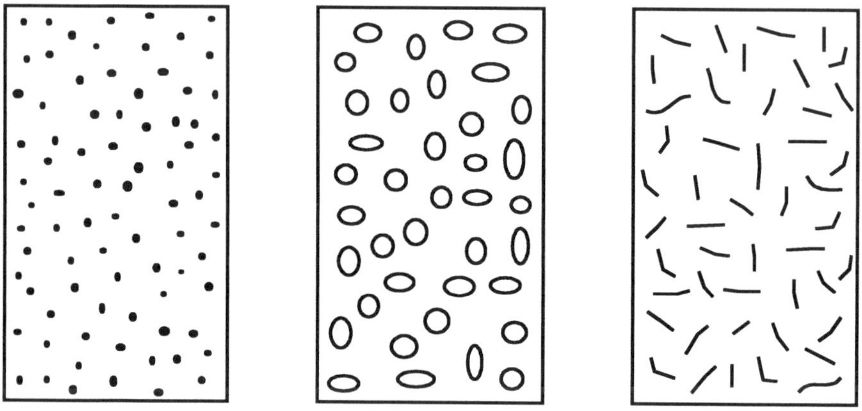

(a) (b) (c)

Figure 2.4 (*a*) Dispersion-strengthened, (*b*) particle-reinforced, (*c*) fiber-reinforced composites.

We have shown that real material solids are extremely complicated in their internal structures, and they are, in particular, inhomogeneous. Since it is impossible to follow up the motion of each internal particles due to their large numbers and a lack of knowledge regarding their interacting forces, the question is then whether a continuum theory would permit us to obtain approximate solutions to study quantitatively the macroscopic behaviors of solids. This is undoubtedly the case in many situations and, among other reasons, provides a strong motivation for the development of such theories. Here, we shall give a brief discussion on some well-known continuum theories and their limitations.

As we know, the development of continuum theories is relevant to the internal structures of the materials. In the case of defect-free Bravais lattices, in which the elementary cell contains only one atom, and with atomic interaction of a range that is small compared to the relevant dimensions, for instance, the wave length of a propagation mechanical wave, we arrive at the common structureless elastic theory. Because of continuum hypothesis, it is expected that the lower limit of interesting length scale in the structureless crystal elasticity theory would be about 10 nm if the average may be made on about 100 atomic distances.

For macroscopic engineering materials, a well-known empirical proportional relation between loads and deflections of macro-homogeneous material specimens was first found by Robert Hooke in 1678. This relation is referred to as Hooke´s law. Later, with the introduction of the concepts of stress and strain, Cauchy generalized Hooke´s law to that the components of stress are linearly related to the components of strain. Such a phenomenological approach provides effective mean of the solution of practically important problems and results achieved with their aid agree with experiments. However, the validity of Hooke´s law is relevant to the scales used in continuum treatments of the elastic internal structures in practical materials such that local variations in, for instance, polycrystalline solids may be averaged out. Therefore, when one studies problems of defects having a scale comparable with the size of the internal structural features in a solid, Hooke´s theory of elasticity for the solid cannot be applied. In addition, limitations of the phenomenological approach to material science should also be noted. On one hand, the macroscopic approach based upon small laboratory specimens does not provide an adequate method of studying the influence of imperfections in materials and the reasons why materials behave as they do. Nor is it possible to completely determine by macroscopic tests what can be done to a material by changes in chemical composition and metallurgical treatment. On the other hand, the predictions based on macroscopic properties are subject to error due to the difference between the laboratory specimen and the prototype. These factors include influences of stress gradients, volume effects, anisotropy, and inhomogeneities.

However, because of the complexity of interaction problems, the lack of statistical informations concerning with the material structures, and the various special characteristics of imperfections in the materials, up to this century, most of the studies at the macroscopic level and that at the microscopic level are investigated independently. Only recently, the necessity of the bridging of the two levels is being realized, and more sophisticated continuum theories which can account for microstructure effects in materials are being required, and various attempts have being made by more and more researchers.

The self-consistent scheme (SCS), as one of the well-known approaches, has been widely used by many researchers to develop an effective medium theory [Böttcher (1945), Hershey (1954), Kröner (1958, 1978), Hill (1965), Hashin (1968), and Willis (1977)]. The SCS approximates the interaction between any element of an aggregate and all the others by the interaction between this element and a homogeneous matrix, the behavior of which is the overall behavior that is to be determined. In this scheme, the matrix may be considered infinite with uniform boundary conditions in the case of macro-homogeneity. One is thus left with a classical inclusion/matrix problem, the solution of which may, in principle, be obtained as a function of the characteristic parameters of the unknown effective behavior (i.e., the behavior of the "homogeneous effective medium"). Finally, these parameters can be identified when the inclusion/matrix problem has been solved for each element by performing an average stress and strain of the elements with the macroscopic ones.

Though SCS may be considered as a rigorous model for "perfectly disordered" materials in the sense defined by Kröner (1978), that is, all (order n = 1, 2, ..., ∞) of correlation functions are statistically homogeneous, isotropic and disordered, there are some problems for multiphase composite materials since these materials are often not disordered in this sense [Kröner (1980)]. A well-known symptom of the difficulty can be observed in the case of rigid inclusions and cavities, in which unrealistic results were predicted with the application of the SCS [see Budiansky (1965) and Christensen (1979)]. One of the main reasons leading to erroneous results is that the SCS method takes too many liberties with the geometry of the material combination to be able to reflect the reality of various materials. A consequential difficulty is that there is uncertainty about the degree of the approximation involved in using SCS results. Besides, in order to apply the SCS, the overall effective constitutive relation for the composite under study has to be preassumed as a known form with some unknown effective material constants to be determined. Such a procedure may, however, fail if one does not even know the form of the overall macroscopic constitutive relation of the composite, which is generally the case for electromagnetic deformable composites. An example for elastic dielectric composites may be found in, for instance, the work of Zhou (1991b), which will be discussed in the next section.

Homogenization method, another useful continuum approach to study the global behavior of materials, was introduced and developed by Bensoussan, Lions, and Papanicolaou (1978), Lions (1981), Sanchez-Palencia (1980), and Levy and Sanchez-Palencia (1983). The material is assumed to have local periodic structures and the method is based on some knowledge of asymptotic methods (two-scale and matching asymptotic expansions) and boundary value problems for partial differential equations. For instance, one deals with a set of elastic field equations in which the elastic moduli $C_{ijkl}(\boldsymbol{x})$ of the elastic material are supposed to be periodic functions of the position vector $\boldsymbol{x}$. One then searches for an asymptotic expansion of the displacement field $\boldsymbol{u} = \boldsymbol{u}^{\mathrm{o}}(\boldsymbol{x}) + \varepsilon \boldsymbol{u}^1(\boldsymbol{x}, \boldsymbol{y}) + o(\varepsilon^2)$, where $\boldsymbol{y} = \boldsymbol{x}/\varepsilon$, as a function of the small parameter ε (the ratio of the micro- to macro-lengths) for $\varepsilon \to 0$. When the length of the period is very small with respect to the other lengths appearing in the problem, one may think that the solution $\boldsymbol{u}$ of the elastic field equations may be approximately the same as the corresponding solution for a "homogenized" medium with elastic constants $C^{\mathrm{h}}{}_{ijkl}$, which only depend on the local structure of the medium and may be obtained by the numerical solution of some boundary value problems in a period of the structure, details of which are referred to in the original work of, for instance, Sanchez-Palencia (1980). Some interesting and useful results have been obtained by this method. The best results are, however, only obtainable under hypotheses of periodicity of the local structure since, in this case, this method leads to a "rigorous" deduction of the macroscopic behavior. In addition, there is a problem on the mathematical proof of the convergence of solutions as the parameter ε tends to zero. Nevertheless, the interest of this method is very qualitative and it gives relevant information on the relation between the local and global behavior.

The first phenomenological continuum theory which deviated from the classical local theory of elasticity is the micropolar continuum theory [Eringen (1965, 1966)]. Such a theory was originally from Cosserat (1909) and its generalized models developed intensively from the 1960s [see, e.g., Mindlin and Tiersten (1962), Green and Rivlin (1964), Nowacki (1970), Brulin and Hjalmars

(1981)]. In this theory, the motion of material medium is described not only by a deformation but also a micro-rotation giving six degrees of freedom, and the interaction between two parts of a body is transmitted not only by a force but also by a torque, resulting in asymmetric force stresses and coupled stresses. In addition, on the volume element, both body forces and body couples might act. Interesting possible applications of this theory to liquid crystals, composite materials, and ferromagnetic materials have been studied by many researchers. There are, however, some problems in the micropolar continuum theory on the physical boundary conditions, the new material constants, and the natural principle of correspondence [Kunin (1983)].

Independent of the micropolar theory, the nonlocal continuum theory was also developed by Eringen (1965), and Kröner and Datta (1966). The specific features of the nonlocal theory are that the long-range inter-particle forces and their effects are taken into account, which results in an integrodifferential form of the field equations. Obviously, the nonlocality is the characteristic feature of all consistent theories for materials with microstructures. Applications of this theory have found some interesting results, such as the disappearance of stress and energy singularities around crack-tips [Eringen and Kim (1974)], the agreement of dispersion curve of the Rayleigh surface wave with the calculation of atomic lattice dynamics [Eringen (1978)], and the explanation of the anomalous skin effect [Chambers (1952)]. The nonlocal theory is now still in progress though there are some mathematical difficulties in applying this theory [Kunin (1983)].

2.10 CONTINUUM THEORY OF MATERIAL COMPOSITES

In this section, we shall introduce a continuum theory for studying materials with internal structures, such as material composites. In accordance with our derivation of Maxwell´s equations for materials based on the dipole model, as shown in Section 1.4, we shall show here how the overall mechanical and electric properties of elastic dielectric composites can be derived with the aid of the concept of statistical continuum material multipoles [Zhou and Hsieh (1986)]. In particular, different from the classical effective medium theory (SCS), we shall demonstrate that the continuum material multipole theory may provide us a method of studying some complex electromagnetoelastic coupling problems in material composites. Here, a material composite is supposed to be a medium composed of a large number of inhomogeneous material particles distributed in a uniform material matrix, and the material particles are assumed to be firmly bonded with the material matrix at their interfaces. The microstructure of the composite is supposed to be of random nature so that a statistical approach can be adopted.

2.10.1 Solution of an Ellipsoidal Inhomogeneity in an Elastic Dielectric

It is well known that classical Eshelby´s solutions on elastic inhomogeneous inclusions have fundamental importance in studying overall properties of various elastic material composites [see, e.g., Eshelby (1961) and Mura (1982)]. In this section, we shall introduce a method of using Green´s function to find a basic solution for the problem of an electrically polarized ellipsoidal inhomogeneous elastic particle embedded in an infinite elastic dielectric medium [Zhou (1991b)].

We start with the following mechanical equilibrium equation:

$$\nabla \cdot \boldsymbol{t} + \boldsymbol{f} = 0 \tag{2.10.1}$$

where $\boldsymbol{t}$ denotes the Cauchy stress tensor and $\boldsymbol{f}$ the body force. If the medium is purely elastic and can be characterized by the following linear Hooke´s law

$$t_{ij}(\boldsymbol{x}) = C_{ijkl} u_{k,l}(\boldsymbol{x}) \tag{2.10.2}$$

with C_{ijkl} being the elastic moduli and $\boldsymbol{u}$ the mechanical displacement vector field, we may find with the aid of the method of Green´s function that the mechanical displacement field $\boldsymbol{u}$ due to the body force $\boldsymbol{f}$ may be obtained by

$$u_i = u_i^o + \int_{V^\infty} f_j(\boldsymbol{x}') G_{ij}(\boldsymbol{x} - \boldsymbol{x}') d\boldsymbol{x}' \tag{2.10.3}$$

where G_{ij} is the elastic Green function, which is given by

$$G_{ij}(\boldsymbol{x} - \boldsymbol{x}') = \frac{1}{16\pi G(1-\nu)|\boldsymbol{x} - \boldsymbol{x}'|}\left[(3-4\nu)\delta_{ij} + \frac{(x_i - x'_i)(x_j - x'_j)}{|\boldsymbol{x} - \boldsymbol{x}'|^2}\right] \tag{2.10.4}$$

for an isotropic elastic medium, where G and ν are, respectively, the elastic shear modulus and Poisson´s ratio of the elastic medium.

If the material medium is not elastic and it is characterized by the following relation:

$$t_{ij}(\boldsymbol{x}) = C_{ijkl} u_{k,l}(\boldsymbol{x}) + t_{ij}^{ine}(\boldsymbol{x}) \tag{2.10.5}$$

where t_{ij}^{ine} denotes the nonelastic part of the stress, we may also find the displacement field $\boldsymbol{u}$ that can be expressed formally with the use of the elastic Green´s function as

$$u_i(\boldsymbol{x}) = u_i^o + \int_{V^\infty} f_j(\boldsymbol{x}') G_{ij}(\boldsymbol{x} - \boldsymbol{x}') d\boldsymbol{x}' + \int_{V^\infty} t_{jk}^{ine}(\boldsymbol{x}') G_{ij,k}(\boldsymbol{x} - \boldsymbol{x}') d\boldsymbol{x}' \tag{2.10.6}$$

With the aid of above results, we may now study the problem of elastic deformation at point $\boldsymbol{x}$ due to the electric forces acting on an electric dipole centered at $\boldsymbol{x}'$ in an infinite elastic homogeneous medium under the exertion of an electric field $\boldsymbol{E}$. By the method of Green's functions, the elastic displacement field may be found by

$$u_i(\boldsymbol{x}) = G_{ij}\left(\boldsymbol{x}-\boldsymbol{x}'-\frac{d}{2}\boldsymbol{n}'\right)f_j^{\,+} + G_{ij}\left(\boldsymbol{x}-\boldsymbol{x}'+\frac{d}{2}\boldsymbol{n}'\right)f_j^{\,-} \tag{2.10.7}$$

in which $\boldsymbol{n}'$ is a unit vector of the direction of the electric dipole and d is the distance between two point electric charges which constitute the electric dipole. $\boldsymbol{f}^+$ and $\boldsymbol{f}^-$ are the point electric body forces given by

$$\boldsymbol{f}^{\,+} = q\boldsymbol{E}\left(\boldsymbol{x}'+\frac{d}{2}\boldsymbol{n}'\right), \quad \text{and} \quad \boldsymbol{f}^{\,-} = -q\boldsymbol{E}\left(\boldsymbol{x}'-\frac{d}{2}\boldsymbol{n}'\right) \tag{2.10.8}$$

where q denotes the point body charge. For an ideal point electric dipole $\boldsymbol{p}$, resulting from the limiting process of letting the distance between the two charges decrease indefinitely and at the same time let the amount of charge increase in such a way that the product $\boldsymbol{p} = qd\boldsymbol{n}'$ remains a finite (constant) vector, the displacement field by Eq.(2.10.7) becomes then

$$u_i(\boldsymbol{x}) = G_{ij}(\boldsymbol{x}-\boldsymbol{x}')p_kE_{j,k}(\boldsymbol{x}') - G_{ij,k}(\boldsymbol{x}-\boldsymbol{x}')p_kE_j(\boldsymbol{x}') \tag{2.10.9}$$

where the electric field $\boldsymbol{E}$ has been assumed to be smooth enough such that all the high order terms $O(d^2)$ vanish during the limiting process.

Equation (2.10.9) shows that the elastic field caused by such an electric dipole may be modeled as the one generated by an induced point elastic monopole at $\boldsymbol{x}'$ defined by

$$P_{jk} = p_kE_j(\boldsymbol{x}') \tag{2.10.10}$$

and an induced point body force at $\boldsymbol{x}'$ defined by

$$f_j = p_kE_{j,k}(\boldsymbol{x}') \tag{2.10.11}$$

which vanishes for a uniform electric field.

Consider now an inhomogeneous particle with the continuum dielectric polarization $\boldsymbol{P}$ embedded in the elastic dielectric matrix medium, which is under the exertion of certain external electric and mechanical loads (denoted by $\boldsymbol{E}^o$ and $\boldsymbol{u}^o$). We may find

$$E_i = E_i^o + \int_{V_I} P_k^e(\boldsymbol{x}')G_{,ki}^e(\boldsymbol{x}-\boldsymbol{x}')d\boldsymbol{x}' \tag{2.10.12}$$

and

$$u_i = u_i^o + \int_{V_I} f_j(x')G_{ij}(x-x')dx' - \int_{V_I} P_{jk}(x')G_{ij,k}(x-x')dx' \tag{2.10.13}$$

where V_I is the volume of the particle, G^e and G_{ij} are, respectively, the electric and elastic Green's functions for the infinite matrix medium. The superscript e is used here to identify quantities which are related to electricity.

Here, we have assumed that the size of the particle is large enough compared with molecular size so that it can be treated as a continuum. The dielectric polarization of the particle has the form

$$\boldsymbol{P} = \varepsilon_o \chi^* \boldsymbol{E} + \boldsymbol{P}^o \tag{2.10.14}$$

where $\boldsymbol{P}^o$ denotes the spontaneous polarization, and χ^* the dielectric susceptibility of the particle. It is also assumed that the dielectric polarization of the matrix material is considered to be small such that the electric body force acting on the matrix may be neglected.

It can be seen from Eqs.(2.10.12) and (2.10.13) that the effect of such an inhomogeneous particle embedded in the elastic dielectric matrix medium may be modeled by a distribution of continuum electric dipole defined by

$$P_k^e = P_k^o + \Delta\varepsilon_{kj}^e E_j \tag{2.10.15}$$

and the body force by

$$f_j = P_k E_{j,k} \tag{2.10.16}$$

and the continuum elastic monopole by

$$P_{jk} = P_k E_j - \Delta C_{jkmn} u_{m,n} \tag{2.10.17}$$

where $\Delta\varepsilon^e$ and $\Delta\boldsymbol{C}$ denote the perturbation values of the material properties between the particle and the matrix, which, for isotropic materials, can be expressed by

$$\Delta\varepsilon_{ij}^e = (\varepsilon^* - \varepsilon^e)\delta_{ij} \tag{2.10.18}$$

where ε^e (= $\varepsilon_o(1+\chi)$) is the permittivity of the matrix, ε^* (= $\varepsilon_o(1+\chi^*)$) is the permittivity of the particle, and

$$\Delta C_{ijkl} = (\lambda^* - \lambda)\delta_{ij}\delta_{kl} + (G^* - G)(\delta_{ik}\delta_{jl} + \delta_{il}\delta_{jk}) \tag{2.10.19}$$

where λ, G and λ^*, G^* are the Lamé constants of the matrix and that of the particle, respectively.

In order to determine these induced continuum electric and elastic multipoles, we have to solve Eqs.(2.10.12) and (2.10.13), which are generally coupled since the elastic deformation depends on the electric field that is dependent of the orientation of the particle which is, in turn, affected by the elastic deformation. At the first approximation, we may ignore the influence of the small change of the orientation of the particle, due to small elastic deformation, on electric field. Thus, for uniform external electric and elastic strain fields, we can find that the electrically induced body force vanishes and that the electric and elastic fields within the particle are also uniform and they become, respectively,

$$E_i = \left[\delta_{ik} + \frac{\Delta\varepsilon^e}{\varepsilon^e} L^e_{ik}\right]^{-1} \left[E^o_k - \frac{1}{\varepsilon^e} L^e_{km} P^o_m\right] \tag{2.10.20}$$

and

$$e_{ij} = [\delta_{im}\delta_{jn} + L_{ijkl}\Delta C_{lkmn}]^{-1}[e^o_{mn} + L_{mnkl}E_l P_k] \tag{2.10.21}$$

where e_{ij} denotes the infinitesimal elastic strain tensor, and $\boldsymbol{L}^e$ and $\boldsymbol{L}$ denote, respectively, the electric and elastic depolarization tensors defined by

$$L^e_{ik} = \frac{1}{4\pi} \int_{S_I} \frac{x'_i n'_k}{|\boldsymbol{x}'|^{3/2}} dS' \tag{2.10.22}$$

and

$$L_{ijkl} = -\frac{1}{2} \int_{S_I} (G_{il,j'} + G_{jl,i'}) n'_k dS' \tag{2.10.23}$$

which are constant tensors in the case of that the particle has an ellipsoidal shape. Here, S_I is the surface of the particle, and $\boldsymbol{n}'$ the outer normal vector of the surface S_I. For an isotropic spherical particle, the electric and elastic depolarization tensors may be found explicitly as

$$L^e_{ik} = \frac{1}{3}\delta_{ik} \tag{2.10.24}$$

and

$$L_{ijkl} = \frac{1}{30G}\left[\frac{9K + 18G}{3K + 4G}(\delta_{kj}\delta_{il} + \delta_{ik}\delta_{jl}) - \frac{6K + 2G}{3K + 4G}\delta_{ij}\delta_{kl}\right] \tag{2.10.25}$$

and the electric and elastic strain fields in the spherical particle may be derived explicitly as

$$E_i = \frac{3\varepsilon^e}{3\varepsilon^e + \Delta\varepsilon^e}\left(E_i^o - \frac{1}{3\varepsilon^e}P_i^o\right) \tag{2.10.26}$$

and

$$e_{kk} = \frac{3K+4G}{3K+4G+3\Delta\lambda+2\Delta G}e_{kk}^o + \frac{1}{3K+4G+3\Delta\lambda+2\Delta G}E_k P_k \tag{2.10.27}$$

where sum is made over the suffix k, and for $i \neq j$

$$e_{ij} = \frac{5G(3K+4G)}{5G(3K+4G)+\Delta G(6K+12G)}e_{ij}^o$$
$$+ \frac{(3K+6G)(E_i P_j + E_j P_i)}{10G(3K+4G)+\Delta G(12K+24G)} \tag{2.10.28}$$

in which $K = \lambda + 2G/3$ is the elastic bulk modulus of the matrix, $\boldsymbol{P}$ the dielectric polarization in the particle given by Eq.(2.10.14) and $\boldsymbol{E}$ the electric field in the particle given by Eq.(2.10.26). It is shown that for an ellipsoidal elastic inhomogeneous particle with electric polarization embedded in an infinite elastic dielectric medium, Eshelby's classical solution is modified by the presence of the additional term on the right-hand side of Eq.(2.10.21) due to the electroelastic coupling effect.

The stress fields within or outside the particle may be found by noting Eqs.(2.10.5), (2.10.6), and (2.10.13) as

$$t_{ij}(\boldsymbol{x}) = C_{ijkl}u_{k,l}(\boldsymbol{x}) + [\Delta C_{ijkl}u_{k,l}(\boldsymbol{x}) - P_j(\boldsymbol{x})E_i(\boldsymbol{x})]\gamma(\boldsymbol{x}) \tag{2.10.29}$$

where the indicative function $\gamma(\boldsymbol{x})$ is defined by

$$\gamma(\boldsymbol{x}) = \begin{cases} 1, & \boldsymbol{x} \text{ in } V_I \\ 0, & \text{otherwise.} \end{cases} \tag{2.10.30}$$

The antisymmetric part of the elastic deformation field in the particle may be studied by introducing the antisymmetric deformation tensor by $2\omega = \nabla\boldsymbol{u} - (\nabla\boldsymbol{u})^{\mathrm{T}}$. After some calculations, we can get

$$\omega_{ij} = \omega_{ij}^o + L_{ijkl}^{\omega}(\Delta C_{lkmn}u_{m,n} - P_k E_l) \tag{2.10.31}$$

where the tensor $\boldsymbol{L}^{\omega}$ is defined by

$$L_{ijkl}^{\omega} = \frac{1}{2}\int_{S_I}(G_{il,j'} - G_{jl,i'})n'_k dS' \tag{2.10.32}$$

which, in the case of a spherical particle, reads

$$L^{\omega}_{ijkl} = \frac{1}{6G}(\delta_{ik}\delta_{lj} - \delta_{il}\delta_{kj}) \quad \text{in } V_I \tag{2.10.33}$$

For a spherical particle, we can deduce explicitly from Eq.(2.10.31) that

$$\nabla \times \boldsymbol{u} = \nabla \times \boldsymbol{u}^o + \frac{\varepsilon^e}{G(3\varepsilon^e + \Delta\varepsilon^e)}\boldsymbol{P}^o \times \boldsymbol{E}^o \tag{2.10.34}$$

which gives the proportional relation between the rotation vector and the electric moment caused by the spontaneous polarization of the particle. It is shown that the rotation vector of the spherical particle is independent of the elastic properties of the particle, and it is also a constant vector within the particle under uniform external loads. It should be noticed that the formula (2.10.34) may only be used to predict small rotation of the particle since the solution is derived from an infinitesimal deformation theory. For large rotation of the particle, we need a finite deformation theory.

It is shown that a basic solution of an ellipsoidal elastic inhomogeneity with electric polarization embedded in an infinite elastic dielectric medium can be obtained by using the Green function (multipole) approach. It is found that, under the exertion of uniform electric and elastic loadings, the elastic strain and electric field in the ellipsoidal elastic inhomogeneity with electric polarization are both uniform and can be determined generally by Eqs.(2.10.20) and (2.10.21). The inhomogeneous particle can also be rotated uniformly due to the electric moment acting on the inhomogeneity with spontaneous polarization as shown in Eq.(2.10.34). Eshelby's classical results on an ellipsoidal elastic inhomogeneity embedded in an infinite elastic medium are found to be modified in the case of elastic dielectrics due to the electric-elastic coupling effect.

2.10.2 Statistical Continuum Material Multipoles

It has been shown that an inhomogeneity embedded in a matrix medium may be modeled as continuum material multipoles in the uniform matrix medium. Material composites may, however, be composed of many such inhomogeneities. In addition, though some of the composites may have a regular structure such as laminated media in which the material properties can be well defined periodically, there is, however, a large class of material composites in which the microstructure is so complex that it is hardly feasible to define its material properties at each point. It is more likely, instead, that only a certain amount of statistical information on the microstructure of the composites is available. Such composites are, for instance, dispersion strengthened, particle reinforced, and chopped fiber reinforced materials. To study this class of material composites with large numbers of randomly distributed inhomogeneous particles, a statistical approach will, therefore, be adopted and the concepts of the statistical continuum material multipoles will be introduced in this section. We shall start with the generalization of the concept of statistical continuum elastic multipoles [Zhou and Hsieh (1986)] to the case where some random parameters accounting for the microstructure of the body-force array are required. Then, concepts of statistical continuum electric and magnetic multipoles may be considered similarly.

In analogy with the Gibbs ensemble used in classical statistical mechanics, we may imagine a great number of independent samples identical in the sense of having the same macroscopic elastic property, the same geometric shape and subject to the same number of point body-force arrays, but varying in an undetermined manner in the distributions and orientations of these point body-forces from sample to sample (see Fig. 2.5).

The total elastic displacement fields in a given sample due to these M body-force arrays can be written in general as

$$u_i(\boldsymbol{x};\ \boldsymbol{x}^1,\ldots,\ \boldsymbol{x}^M,\omega^1,\ldots,\omega^M) = \sum_{\alpha=1}^{M}\sum_{k=0}^{\infty}\frac{(-1)^k}{k!}P^{\alpha}_{js_1\ldots s_k}G_{ij,s_1\ldots s_k}(\boldsymbol{x},\ \boldsymbol{x}^{\alpha}) \tag{2.10.35}$$

where $P^{\alpha}_{js_1\ldots s_k}$ is the elastic multipole moment of order k for the αth point body-force array. $\boldsymbol{x}^{\alpha}$ is a random variable (varying with different samples) which indicates the center position of the αth point body-force array, and ω^{α} is a random parameter (probably several parameters) characterizing the microstructure of the body-force array. G_{ij} is the elastic Green function of the

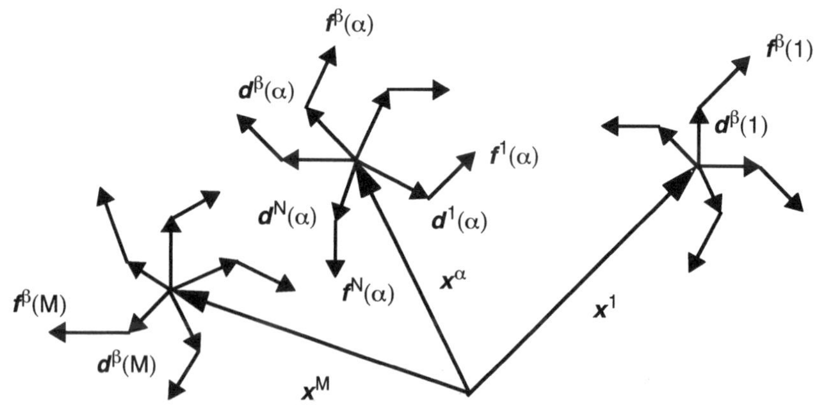

Figure 2.5 A distribution of point body-force arrays.

sample, which is supposed to be an infinite elastic medium.

For force arrays in self-equilibrium, we have $P_j^\alpha = 0$. If the point body forces are permanent in the sense of that they are independent of each other, we may write

$$P^\alpha_{js_1\ldots s_k} = P_{js_1\ldots s_k}(\boldsymbol{x}^\alpha,\omega^\alpha) = \sum_{\beta=1}^{N(\alpha)} f_j^\beta(\boldsymbol{x}^\alpha,\omega^\alpha)d_{s_1}^\beta(\boldsymbol{x}^\alpha,\omega^\alpha)\ldots d_{s_k}^\beta(\boldsymbol{x}^\alpha,\omega^\alpha) \tag{2.10.36}$$

where $N(\alpha)$ denotes the number of the point body-forces in the αth point body-force array and $\boldsymbol{d}^\beta(\alpha)$ denotes the βth position vector from $\boldsymbol{x}^\alpha$ to its corresponding position of the point body-force $\boldsymbol{f}^\beta(\alpha)$. If the point body-forces are of induced type so that they are dependent of each other due to interaction among them, we shall, in general, have

$$P^\alpha_{js_1\ldots s_k} = \sum_{\beta=1}^{N(\alpha)} f_j^\beta(\boldsymbol{x}^\alpha;\ \boldsymbol{x}^1,\ldots,\ \boldsymbol{x}^M,\omega^1,\ldots,\omega^M)d_{s_1}^\beta\ldots d_{s_k}^\beta \tag{2.10.37}$$

which means that the elastic multipoles modeling the αth point body-force array are dependent of other body-force arrays, their positions, and distributions.

With the aid of the joint-probability density function $f(\Omega^1, \ldots, \Omega^M)$ with $\Omega^\alpha = (\mathbf{x}^\alpha, \omega^\alpha)$, the ensemble average of elastic fields due to the randomly distributed point body-force arrays may be expressed by

$$\langle u_i\rangle(\boldsymbol{x}) = \sum_{k=0}^{\infty} \frac{(-1)^k}{k!}\int_V \bar{P}_{js_1\ldots s_k}(\boldsymbol{x}')G_{ij,s_1\ldots s_k}(\boldsymbol{x},\ \boldsymbol{x}')d\boldsymbol{x}' \tag{2.10.38}$$

where we have introduced the statistical continuum elastic multipole of order k, defined by

$$\bar{P}_{js_1\ldots s_k}(\boldsymbol{x}') = \sum_{\alpha=1}^{M}\int_\Gamma F^\alpha_{js_1\ldots s_k}(\Omega')d\Gamma' \tag{2.10.39}$$

in which

$$F^\alpha_{js_1\ldots s_k} = \int_\Omega\ldots\int_\Omega f(\Omega^1,\ldots,\Omega^M)P^\alpha_{js_1\ldots s_k}d\Omega^1\ldots d\Omega^{\alpha-1}d\Omega^{\alpha+1}\ldots d\Omega^M \tag{2.10.40}$$

where Ω is a defined space, $\Omega = V \times \Gamma$, in which V is the volume of the material body and Γ is a parameter space, $\omega^\alpha \in \Gamma$.

In the case of permanent point body-force arrays, Eq.(2.10.39) may be reduced to

$$\bar{P}_{js_1\ldots s_k}(\boldsymbol{x}') = \int_\Gamma \rho(\Omega')P_{js_1\ldots s_k}(\Omega')d\Gamma' \tag{2.10.41}$$

where

$$\rho(\Omega') = \sum_{\alpha=1}^{M} W_\alpha(\Omega') \tag{2.10.42}$$

with

$$W_\alpha = \int_\Omega \ldots \int_\Omega f(\Omega^1,\ldots,\Omega^M)d\Omega^1\ldots d\Omega^{\alpha-1}d\Omega^{\alpha+1}\ldots d\Omega^M \tag{2.10.43}$$

The physical meaning of the function, defined by

$$\int_\Omega \rho(\Omega')d\Omega' = \int_V\left(\int_\Gamma \rho(\Omega')d\Gamma'\right)d\boldsymbol{x}' = M \tag{2.10.44}$$

may be explained as the number of the elastic multipoles per unit volume. It is shown that the ensemble average elastic fields due to a statistically discrete distribution of point body-force arrays may be modeled as the elastic fields due to a distribution of statistical continuum elastic multipoles defined in Eq.(2.10.39) or Eq.(2.10.41).

Similarly, for M randomly distributed point body-charge arrays in a dielectric body, we find that the ensemble average of the electric field due to the randomly distributed point body-charge arrays may be expressed as

$$\langle E_i\rangle(\boldsymbol{x}) = \sum_{k=0}^{\infty}\frac{(-1)^{k+1}}{k!}\int_V \bar{P}^e_{s_1\ldots s_k}(\boldsymbol{x}')G^e{}_{,is_1\ldots s_k}(\boldsymbol{x},\ \boldsymbol{x}')d\boldsymbol{x}' \tag{2.10.45}$$

where we have introduced the statistical continuum electric multipole of order k, defined by

$$\bar{P}^e_{s_1\ldots s_k}(\boldsymbol{x}') = \sum_{\alpha=1}^{M}\int_\Gamma F^{e\alpha}_{s_1\ldots s_k}(\Omega')d\Gamma' \tag{2.10.46}$$

in which

$$F^{e\alpha}_{s_1\ldots s_k} = \int_\Omega\ldots\int_\Omega f(\Omega^1,\ldots,\Omega^M)P^{e\alpha}_{s_1\ldots s_k}d\Omega^1\ldots d\Omega^{\alpha-1}d\Omega^{\alpha+1}\ldots d\Omega^M \tag{2.10.47}$$

with $P^{e\alpha}_{s_1\ldots s_k}$ being the electric multipole moment of order k for the αth point body-charge array, defined by

$$P^{e\alpha}_{s_1 \ldots s_k} = \sum_{\beta=1}^{N(\alpha)} q^{\beta}(x^{\alpha};\ x^1, \ldots,\ x^M, \omega^1, \ldots, \omega^M) d^{\beta}_{s_1} \ldots d^{\beta}_{s_k} \tag{2.10.48}$$

If the αth point body-charge array is self-electrically neutral, we have $P^{e\alpha} = 0$.

In particular, if the point body-charge arrays are of permanent type, that is,

$$P^{e\alpha}_{s_1 \ldots s_k} = P^{e}_{s_1 \ldots s_k}(x^{\alpha}, \omega^{\alpha}) = \sum_{\beta=1}^{N(\alpha)} q^{\beta}(x^{\alpha}, \omega^{\alpha}) d^{\beta}_{s_1}(x^{\alpha}, \omega^{\alpha}) \ldots d^{\beta}_{s_k}(x^{\alpha}, \omega^{\alpha}) \tag{2.10.49}$$

then the statistical continuum electric multipole of order k may be expressed simply by

$$\overline{P}^{e}_{s_1 \ldots s_k}(x') = \int_{\Gamma} \rho(\Omega') P^{e}_{s_1 \ldots s_k}(\Omega') d\Gamma' \tag{2.10.50}$$

where $\rho(\Omega')$ is given by Eq.(2.10.42).

2.10.3 Statistical Continuum Multipole Modeling of Composites

To illustrate the use of statistical continuum material multipoles, we shall consider, in this section, the statistical continuum electric and elastic multipole modeling of elastic dielectric material composites with a large number of M statistically distributed inhomogeneous particles with electrical polarization. All particles will be assumed to have the same shape, the same size, and the same strength of spontaneous electric polarization but they can have different orientations. Other types of material composites may be studied similarly. To formulate the problem mathematically, we have several possibilities in regard to the composite specimen shape and boundary conditions. To avoid using the concept of an "infinite" specimen as well as the convergence difficulty [Jeffrey (1973)], we shall consider a finite spherical specimen, which is assumed to be perfectly embedded into an infinite homogeneous elastic dielectric medium with the same material properties as the matrix of the composite (see Fig. 2.6).

The problem of interest to us is to find the overall properties of such an elastic dielectric composite. The geometrical arrangement of the inhomogeneities in a given sample of the composite is specified by the indicative functions

$$\gamma^{\alpha}(x) = \begin{cases} 1, & x \text{ in } V_{\alpha} \\ 0, & \text{otherwise.} \end{cases} \tag{2.10.51}$$

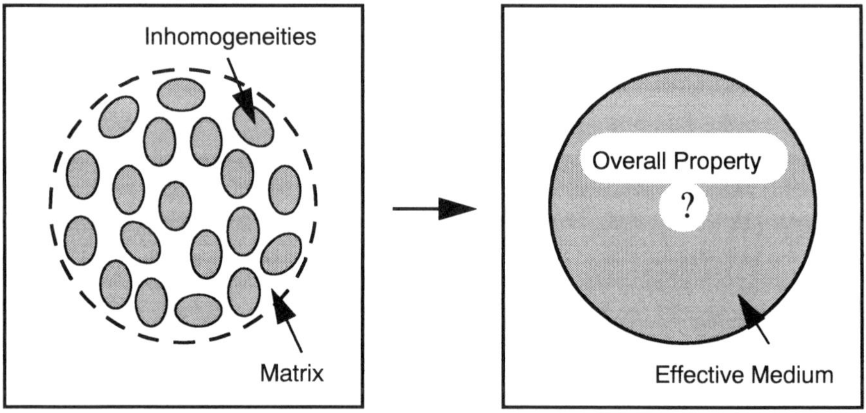

Figure 2.6 A material composite of spherical shape.

where V_α is the volume of the αth inhomogeneity ($\alpha = 1, 2, ..., M$).

The dimension of the inhomogeneities will be always assumed to be much smaller than the dimension of the composite specimen, but much larger than the size of the molecule (or atom) such that the inhomogeneous particle may be treated as a continuum with the permittivity ε^* and the elastic moduli $\boldsymbol{C}^*$.

With the aid of results given in Section 2.9.1, the microscopic electric and elastic fields in the composite under the exertion of external fields ($\boldsymbol{E}^o$ and $\boldsymbol{u}^o$) may be written as

$$E_i = E_i^o + \sum_{\alpha=1}^{M} \int_{V_\alpha} (P_k^{o(\alpha)} + \Delta\varepsilon_{kj}^e E_j) G^e{}_{,ki}(\boldsymbol{x}-\boldsymbol{x}')d\boldsymbol{x}' \tag{2.10.52}$$

and

$$u_i = u_i^o + \sum_{\alpha=1}^{M} \left[\int_{V_\alpha} f_j^{(\alpha)} G_{ij}(\boldsymbol{x}-\boldsymbol{x}')d\boldsymbol{x}' - \int_{V_\alpha} P_{jk}^{(\alpha)} G_{ij,k}(\boldsymbol{x}-\boldsymbol{x}')d\boldsymbol{x}' \right] \tag{2.10.53}$$

By noting the geometrical arrangements and the orientations of the inhomogeneities, Eqs.(2.10.52) and (2.10.53) can be further expressed as

$$E_i(\boldsymbol{x}; \Omega^1,\ldots,\Omega^M) = E_i^o + \sum_{\alpha=1}^{M} \int_{V_I} (P_k^{o(\alpha)} + \Delta\varepsilon_{kj}^e E_j^{(\alpha)}) G^e{}_{,ki}(\boldsymbol{x}-\boldsymbol{x}^\alpha - \boldsymbol{y}^\alpha)d\xi \tag{2.10.54}$$

and

$$u_i(\boldsymbol{x};\Omega^1,\ldots,\Omega^M) = u_i^o + \sum_{\alpha=1}^{M}\left[\int_{V_I} f_j^{(\alpha)}\, G_{ij}(\boldsymbol{x}-\boldsymbol{x}^\alpha-\boldsymbol{y}^\alpha)d\xi - \int_{V_I} P_{jk}^{(\alpha)} G_{ij,k}(\boldsymbol{x}-\boldsymbol{x}^\alpha-\boldsymbol{y}^\alpha)d\xi\right] \tag{2.10.55}$$

in which $\boldsymbol{P}^{o(\alpha)}$, $\boldsymbol{E}^{(\alpha)}$, $\boldsymbol{f}^{(\alpha)}$ and $\boldsymbol{P}^{(\alpha)}$ are, respectively, the permanent electric polarization, the electric field, the induced electric body force and the induced elastic dipole density defined in the αth inhomogeneity, which are, in general, dependent of the geometrical arrangements of all other inhomogeneities due to interaction among them.

Here, we have introduced the notation $\Omega^\alpha = (\boldsymbol{x}^\alpha, \theta^\alpha, \psi^\alpha, \omega^\alpha)$ and

$$\boldsymbol{y}^\alpha = \boldsymbol{Q}^\alpha(\theta^\alpha,\psi^\alpha,\omega^\alpha)\cdot\xi \tag{2.10.56}$$

where $\boldsymbol{x}^\alpha$ is the position vector of the gravitational center and $\boldsymbol{Q}^\alpha$ the orientation tensor of the αth inhomogeneous ellipsoid, which is an orthogonal tensor, defined by

$$[Q_{ij}^\alpha] = \begin{bmatrix} \cos\psi^\alpha\cos\omega^\alpha - \cos\theta^\alpha\sin\psi^\alpha\sin\omega^\alpha & -\cos\psi^\alpha\sin\omega^\alpha - \cos\theta^\alpha\sin\psi^\alpha\cos\omega^\alpha & \sin\psi^\alpha\sin\theta^\alpha \\ \cos\omega^\alpha\sin\psi^\alpha + \cos\theta^\alpha\cos\psi^\alpha\sin\omega^\alpha & -\sin\psi^\alpha\sin\omega^\alpha + \cos\theta^\alpha\cos\psi^\alpha\cos\omega^\alpha & -\cos\psi^\alpha\sin\theta^\alpha \\ \sin\theta^\alpha\sin\omega^\alpha & \cos\omega^\alpha\sin\theta^\alpha & \cos\theta^\alpha \end{bmatrix} \tag{2.10.57}$$

where θ^α, ψ^α and ω^α are the Euler angles of the αth inhomogeneity, as shown in Fig. 2.7.

In particular, for inhomogeneities with ellipsoidal shape of revolution, we may let $\omega^\alpha = 0$ and the orientation tensor is reduced to be

$$[Q_{ij}^\alpha] = \begin{bmatrix} \cos\psi^\alpha & -\cos\theta^\alpha\sin\psi^\alpha & \sin\theta^\alpha\sin\psi^\alpha \\ \sin\psi^\alpha & \cos\theta^\alpha\cos\psi^\alpha & -\sin\theta^\alpha\cos\psi^\alpha \\ 0 & \sin\theta^\alpha & \cos\theta^\alpha \end{bmatrix} \tag{2.10.58}$$

For spherical inhomogeneous particles, we have simply $Q^\alpha{}_{ij} = \delta_{ij}$ which is a

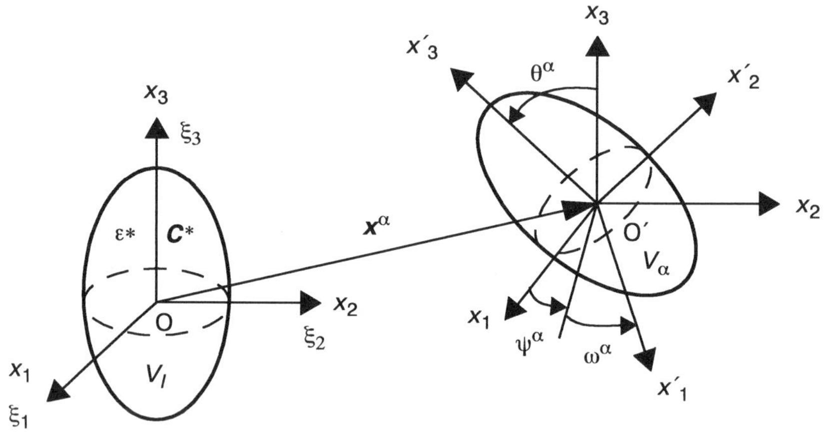

Figure 2.7 Phase geometry of the composite.

Kronecker delta. By a Taylor expansion of the Green function, Eqs.(2.10.54) and (2.10.55) may be written as

$$E_i(\boldsymbol{x};\,\Omega^1,\ldots,\Omega^M) = E_i^o - \sum_{\alpha=1}^{M}\sum_{k=1}^{\infty}\frac{(-1)^k}{k!}P^{e(\alpha)}_{s_1\ldots s_k}(\Omega^1,\ldots,\Omega^M)G^e{}_{,is_1\ldots s_k}(\boldsymbol{x}-\boldsymbol{x}^\alpha) \tag{2.10.59}$$

and

$$u_i(\boldsymbol{x};\,\Omega^1,\ldots,\Omega^M) = u_i^o + \sum_{\alpha=1}^{M} f_j^{(\alpha)}(\Omega^1,\ldots,\Omega^M)G_{ij}(\boldsymbol{x}-\boldsymbol{x}^\alpha)$$

$$+\sum_{\alpha=1}^{M}\sum_{k=1}^{\infty}\frac{(-1)^k}{k!}P^{(\alpha)}_{js_1\ldots s_k}(\Omega^1,\ldots,\Omega^M)G_{ij,s_1\ldots s_k}(\boldsymbol{x}-\boldsymbol{x}^\alpha) \tag{2.10.60}$$

which shows that a given discretely distributed inhomogeneities may be modeled by a discrete distribution of the induced electric and elastic multipoles defined, respectively, by

$$P^{e(\alpha)}_{s_1\ldots s_k} = k\int_{V_I}(P^{o(\alpha)}_{s_1} + \Delta\varepsilon^e_{js_1}E^{(\alpha)}_j)y^\alpha_{s_2}\ldots y^\alpha_{s_k}d\xi \qquad (k = 1,2,\ldots) \tag{2.10.61}$$

and

$$f_j^{(\alpha)} = \int_{V_I} P_n^{(\alpha)}E^{(\alpha)}_{j,n}d\xi \tag{2.10.62}$$

$$P^{(\alpha)}_{js_1\ldots s_k} = -k\int\limits_{V_I}\left[\Delta C_{mnjs_1}u^{(\alpha)}_{m,n} - P^{(\alpha)}_{s_1}E^{(\alpha)}_j - \frac{1}{k}P^{(\alpha)}_n E^{(\alpha)}_{j,n}y^{\alpha}_{s_1}\right]y^{\alpha}_{s_2}\ldots y^{\alpha}_{s_k}d\xi \quad (2.10.63)$$

for $k = 1, 2, \ldots$, where $y^{\alpha}_{s_2}\ldots y^{\alpha}_{s_k} = 1$ when $k = 1$. It is also shown that the total resultant charge (the zeroth order of the electric multipole) of the point charge arrays modeling the inhomogeneities vanishes since these inhomogeneities are self-electrically neutral. The induced body force term due to the electric and elastic interaction given by Eq.(2.10.62), in general, does not vanishes. If, however, the microscopic electric field inside particles is uniform, this term will be zero.

Considering now an ensemble of large numbers of such samples, the ensemble average of the electric field and the elastic displacement field in the composite may be expressed by

$$\langle E_i\rangle(\boldsymbol{x}) = \int\limits_{\Omega}\ldots\int\limits_{\Omega} f(\Omega^1,\ldots,\Omega^M)E_i(\boldsymbol{x};\Omega^1,\ldots,\Omega^M)d\Omega^1\ldots d\Omega^M$$

$$= E^o_i - \sum_{k=1}^{\infty}\frac{(-1)^k}{k!}\int\limits_{V_R}\bar{P}^e_{s_1\ldots s_k}(\boldsymbol{x}')G^e_{,is_1\ldots s_k}(\boldsymbol{x},\boldsymbol{x}')d\boldsymbol{x}' \quad (2.10.64)$$

and

$$\langle u_i\rangle(\boldsymbol{x}) = \int\limits_{\Omega}\ldots\int\limits_{\Omega} f(\Omega^1,\ldots,\Omega^M)u_i(\boldsymbol{x};\Omega^1,\ldots,\Omega^M)d\Omega^1\ldots d\Omega^M$$

$$= u^o_i + \int\limits_{V_R}\bar{f}_j(\boldsymbol{x}')G_{ij}(\boldsymbol{x},\boldsymbol{x}')d\boldsymbol{x}' + \sum_{k=1}^{\infty}\frac{(-1)^k}{k!}\int\limits_{V_R}\bar{P}_{js_1\ldots s_k}(\boldsymbol{x}')G_{ij,s_1\ldots s_k}(\boldsymbol{x},\boldsymbol{x}')d\boldsymbol{x}' \quad (2.10.65)$$

where $f(\Omega^1,\ldots,\Omega^M)$ is the M-point joint probability density function, and $\Omega^\alpha \in \Omega$ which is a defined space, $\Omega = V_R \times \Gamma$, in which $\Gamma = (0, \pi)\times(0, 2\pi)\times(0, 2\pi)$ is a parameter space and V_R the volume of the composite specimen.

Equations (2.10.64) and (2.10.65) show that the macroscopic ensemble average behavior of such M statistically distributed inhomogeneous particles may be modeled overall by a distribution of induced statistical continuum electric and elastic multipoles defined, respectively, by

$$\bar{P}^e_{s_1\ldots s_k}(\boldsymbol{x}') = \sum_{\alpha=1}^{M}\int\limits_{\Gamma} Z^{e(\alpha)}_{s_1\ldots s_k}(\Omega')d\Gamma' \quad (2.10.66)$$

and

$$\bar{f}_j(\boldsymbol{x}') = \sum_{\alpha=1}^{M} \int_{\Gamma} Z_j^{(\alpha)}(\Omega')d\Gamma' \tag{2.10.67}$$

$$\bar{P}_{js_1\ldots s_k}(\boldsymbol{x}') = \sum_{\alpha=1}^{M} \int_{\Gamma} Z_{js_1\ldots s_k}^{(\alpha)}(\Omega')d\Gamma' \tag{2.10.68}$$

where

$$Z_{s_1\ldots s_k}^{e(\alpha)} = \int_{\Omega}\ldots\int_{\Omega} f(\Omega^1,\ldots,\Omega^M)P_{s_1\ldots s_k}^{e(\alpha)}d\Omega^1\ldots d\Omega^{\alpha-1}d\Omega^{\alpha+1}\ldots d\Omega^M \tag{2.10.69}$$

and

$$Z_j^{(\alpha)} = \int_{\Omega}\ldots\int_{\Omega} f(\Omega^1,\ldots,\Omega^M)f_j^{(\alpha)}d\Omega^1\ldots d\Omega^{\alpha-1}d\Omega^{\alpha+1}\ldots d\Omega^M \tag{2.10.70}$$

$$Z_{js_1\ldots s_k}^{(\alpha)} = \int_{\Omega}\ldots\int_{\Omega} f(\Omega^1,\ldots,\Omega^M)P_{js_1\ldots s_k}^{(\alpha)}d\Omega^1\ldots d\Omega^{\alpha-1}d\Omega^{\alpha+1}\ldots d\Omega^M \tag{2.10.71}$$

in which $d\Omega^{\alpha-1} = 1$ when $\alpha = 1$ and $d\Omega^{\alpha+1} = 1$ when $\alpha = M$.

Some special cases may be of interest. If the electric and elastic multipoles modeling the αth inhomogeneity do not explicitly depend on the geometrical positions and orientations of its surrounding inhomogeneities, that is,

$$P_{s_1\ldots s_k}^{e(\alpha)} = P_{s_1\ldots s_k}^{e}(\Omega^\alpha) \tag{2.10.72}$$

and

$$f_j^{(\alpha)} = f_j(\Omega^\alpha) \tag{2.10.73}$$

$$P_{js_1\ldots s_k}^{(\alpha)} = P_{js_1\ldots s_k}(\Omega^\alpha) \tag{2.10.74}$$

Eqs.(2.10.66), (2.10.67), and (2.10.68) may be reduced to be

$$\bar{P}^e_{s_1\ldots s_k}(\boldsymbol{x}') = \int_{\Gamma} \rho(\Omega')P_{s_1\ldots s_k}^{e}(\Omega')d\Gamma' \tag{2.10.75}$$

and

$$\bar{f}_j(\boldsymbol{x}') = \int_{\Gamma} \rho(\Omega')f_j(\Omega')d\Gamma' \tag{2.10.76}$$

$$\bar{P}_{js_1\ldots s_k}(\boldsymbol{x}') = \int_{\Gamma} \rho(\Omega')P_{js_1\ldots s_k}(\Omega')d\Gamma' \tag{2.10.77}$$

where function $\rho(\Omega')$ is defined in Eq.(2.10.42). Examples of such cases may be dilute suspensions or systems in which microscopic electric and elastic fields in the particles may be solved by using self-consistent scheme approximation. Other cases such as systems allowable for pair interaction approximation or the nearest interaction approximation may also exist and Eqs.(2.10.66)–(2.10.68) can also be simplified. It is seen that reasonable approximations may not only simplify the many-body interaction problem, but may also make it possible to obtain necessary statistical information in many cases.

2.10.4 Effective Properties of Composite with Random Microstructure

The problem of interest in this section is to see how the proposed statistical continuum multipole approach can be used to find explicitly the effective properties of the composite and the statistical anisotropy and shape effects of the microstructures on the overall properties of the composites with randomly distributed inhomogeneities. For simplicity, we shall study elastic dielectric composites in which the electroelastic coupling effect is ignored. The problem of finding the effective permittivity and the effective elastic properties of the composite is thus fully separated. We may now consider the following microscopic constitutive relations:

$$\boldsymbol{D} = \left(\varepsilon^e + \sum_{\alpha=1}^{M} \Delta\varepsilon^e\gamma^\alpha\right) \cdot \boldsymbol{E} \tag{2.10.78}$$

and

$$\boldsymbol{t} = \left(\boldsymbol{C} + \sum_{\alpha=1}^{M} \Delta\boldsymbol{C}\gamma^\alpha\right) \cdot \boldsymbol{e} \tag{2.10.79}$$

where γ^α is the indicative function of the αth inhomogeneous particle. The ensemble average of relations (2.10.78) and (2.10.79) then gives

$$\langle \boldsymbol{D} \rangle = \varepsilon^e \cdot \langle \boldsymbol{E} \rangle + \Delta\varepsilon^e \cdot \left\langle \sum_{\alpha=1}^{M} \gamma^\alpha \boldsymbol{E} \right\rangle \tag{2.10.80}$$

and

$$\langle \boldsymbol{t} \rangle = \boldsymbol{C} \cdot \langle \boldsymbol{e} \rangle + \Delta\boldsymbol{C} \cdot \left\langle \sum_{\alpha=1}^{M} \gamma^\alpha \boldsymbol{e} \right\rangle \tag{2.10.81}$$

Due to the linearity of the problems, the microscopic electric and elastic strain fields in the αth inclusion (α = 1, 2,..., M) may be expressed by

$$E_i^{(\alpha)}(\xi; \Omega^1,\ldots,\Omega^M) = T_{ij}^{\alpha}(\xi; \Omega^1,\ldots,\Omega^M)E_j^o, \qquad (\xi \in V_I) \tag{2.10.82}$$

and

$$e_{ij}^{(\alpha)}(\xi; \Omega^1,\ldots,\Omega^M) = A_{ijkl}^{\alpha}(\xi; \Omega^1,\ldots,\Omega^M)e_{kl}^o, \qquad (\xi \in V_I) \tag{2.10.83}$$

where tensors $\boldsymbol{T}^{\alpha}$ and $\boldsymbol{A}^{\alpha}$, in general, are unknown beforehand and have to be determined by solving the equations of microscopic electric and elastic strain fields for the given applied (external) electric field $\boldsymbol{E}^{o}$ and the elastic strain field $\boldsymbol{e}^{o}$.

With the use of the electric and elastic multipole modeling, the microscopic electric and elastic fields in the composite may be written, respectively, as

$$E_i(\boldsymbol{x}; \Omega^1,\ldots,\Omega^M) = F_{ij}(\boldsymbol{x}; \Omega^1,\ldots,\Omega^M)E_j^o \tag{2.10.84}$$

$$e_{ij}(\boldsymbol{x}; \Omega^1,\ldots,\Omega^M) = B_{ijkl}(\boldsymbol{x}; \Omega^1,\ldots,\Omega^M)e_{kl}^o \tag{2.10.85}$$

in which

$$F_{ij} = \delta_{ij} + \sum_{\beta=1}^{M} R_{ij}^{\beta} \tag{2.10.86}$$

and

$$B_{ijkl} = \delta_{ik}\delta_{jl} + \sum_{\beta=1}^{M} Y_{ijkl}^{\beta} \tag{2.10.87}$$

where

$$R_{ij}^{\beta} = \sum_{k=1}^{\infty} \frac{(-1)^{k-1}}{(k-1)!} G^{e}_{,is_1\ldots s_k}(\boldsymbol{x}-\boldsymbol{x}^{\beta}) \int_{V_I} \Delta\varepsilon^{e}_{ps_1} T^{\beta}_{pj} y^{\beta}_{s_2}\ldots y^{\beta}_{s_k} d\xi \tag{2.10.88}$$

and

$$Y_{ijkl}^{\beta} = \sum_{k=1}^{\infty} \frac{(-1)^{k-1}}{(k-1)!} g_{ijns_1\ldots s_k}(\boldsymbol{x}-\boldsymbol{x}^{\beta}) \int_{V_I} \Delta C_{pqns_1} A^{\beta}_{pqkl} y^{\beta}_{s_2}\ldots y^{\beta}_{s_k} d\xi \tag{2.10.89}$$

with

$$g_{ijns_1\ldots s_k}(\boldsymbol{x}-\boldsymbol{x}^{\beta}) = \frac{1}{2}[G_{in,js_1\ldots s_k}(\boldsymbol{x}-\boldsymbol{x}^{\beta}) + G_{jn,is_1\ldots s_k}(\boldsymbol{x}-\boldsymbol{x}^{\beta})] \tag{2.10.90}$$

The effective properties of the composite can thus be determined by comparing its definition

$$\langle \boldsymbol{D} \rangle = \varepsilon^{eff} \cdot \langle \boldsymbol{E} \rangle \tag{2.10.91}$$

and

$$\langle \boldsymbol{t} \rangle = \boldsymbol{C}^{eff} \cdot \langle \boldsymbol{e} \rangle \tag{2.10.92}$$

with Eqs.(2.10.80) and (2.10.81), which leads to the following result:

$$\varepsilon_{ij}^{eff} = \varepsilon_{ij}^{e} + \Delta\varepsilon_{ip}^{e} \langle \sum_{\alpha=1}^{M} \gamma^{\alpha} T_{pk}^{\alpha} \rangle \langle F_{kj} \rangle^{-1} \tag{2.10.93}$$

and

$$C_{ijkl}^{eff} = C_{ijkl} + \Delta C_{ijpq} \langle \sum_{\alpha=1}^{M} \gamma^{\alpha} A_{pqmn}^{\alpha} \rangle \langle B_{mnkl} \rangle^{-1} \tag{2.10.94}$$

Obviously to perform the ensemble average, the details of the statistical information on the microstructure of the composite have to be given. In addition, to complete the calculation, one has to find the microscopic electric and elastic fields in the particles, which seems to be impossible in practice because of the complicated many-body interaction problems as well as their statistical character. Some physically reasonable assumptions are, therefore, needed to simplify the problem and make it solvable. For instance, for small concentration suspensions, a first-order approximation can be made by neglecting interaction between particles, where classical solutions of a single inhomogeneous particle embedded in an infinite medium may be used [see Eshelby (1961), Mura (1982) and Böttcher (1973)]. Higher-order approximations accounting for some interaction effects have been also proposed as, for instance, the nearest interaction model of using a solution of a pair of interactive particles [see Jeffrey (1973, 1974), McCoy and Beran (1976)] and the classical self-consistent scheme [see Hershey (1954) and Hashin (1968)]. For weakly inhomogeneous composites, theories of bounds and classical perturbation method have been used quite successfully [see Hashin and Shtrikman (1962), Beran (1968), Hori (1973) and Willis (1977)].

In this section, we shall use the statistical continuum multipole approach to study the effective properties of an elastic and dielectric composite with M statistically distributed identical inhomogeneous elastic and dielectric ellipsoidal particles with the elastic moduli $\boldsymbol{C}^*$ and the permittivity ε^* embedded in a homogeneous matrix with the elastic moduli $\boldsymbol{C}$ and the permittivity ε^e. To see how the statistical continuum multipole approach works, we consider a dilute suspension system, that is, the effect of interaction between particles may be neglected. Under the assumption of the statistical homogeneity but not necessarily statistical isotropic on the geometric arrangement of the inhomogeneous particles, the probability density function (see Section 2.9.2) can be introduced as

$$\rho(\Omega') = M\rho^*(\Gamma')/V_R \tag{2.10.95}$$

in which

$$\rho^*(\Gamma') = \sum_{\alpha=1}^{M} \frac{1}{M} W^*_\alpha(\Gamma') \tag{2.10.96}$$

where

$$W^*_\alpha = \int_\Gamma \ldots \int_\Gamma w(\Gamma^1,\ldots,\Gamma^M) d\Gamma^1 \ldots d\Gamma^{\alpha-1} d\Gamma^{\alpha+1} \ldots d\Gamma^M \tag{2.10.97}$$

with the normalization condition

$$\int_\Gamma \ldots \int_\Gamma w(\Gamma^1,\ldots,\Gamma^M) d\Gamma^1 \ldots d\Gamma^M = 1 \tag{2.10.98}$$

In the dilute approximation, the microscopic electric and elastic strain fields in the αth ellipsoidal inclusion (particle) may be found, as a single inhomogeneous ellipsoid with a certain orientation in an infinite matrix subjected to the external electric field $\boldsymbol{E}^o$ and the external strain field $\boldsymbol{e}^o$, to be

$$E_m^{(\alpha)} = Q_{mi}(\Gamma^\alpha) Q_{nj}(\Gamma^\alpha) T_{ij} E_n^o \tag{2.10.99}$$

and

$$e_{mn}^{(\alpha)} = Q_{mi}(\Gamma^\alpha) Q_{nj}(\Gamma^\alpha) Q_{pk}(\Gamma^\alpha) Q_{ql}(\Gamma^\alpha) A_{ijkl} e_{pq}^o \tag{2.10.100}$$

where T_{ij} and A_{ijkl} are the transformation tensors that characterize, respectively, the electric and elastic field in an ellipsoidal particle with its axes coincide with a chosen coordinate system (see Fig. 2.7) in which $\theta = 0$, $\psi = 0$ and $\omega = 0$. They are both known constant tensors [see Stratton (1941) and Mura (1982)].

The electric and elastic multipoles defined by Eqs.(2.10.61)–(2.10.63) can thus be found by inserting Eqs.(2.10.99) and (2.10.100). For instance, the electric dipole is found as

$$P_m^{e(\alpha)} = V_I \Delta\varepsilon_{mk}^e Q_{ki}(\Gamma^\alpha) Q_{nj}(\Gamma^\alpha) T_{ij} E_n^o \tag{2.10.101}$$

where we have set $\boldsymbol{P}^{o(\alpha)} = 0$ for inhomogeneous particles without the spontaneous polarization, and the elastic monopole is found as

$$P_{st}^{(\alpha)} = -V_I \Delta C_{stmn} Q_{mi}(\Gamma^\alpha) Q_{nj}(\Gamma^\alpha) Q_{pk}(\Gamma^\alpha) Q_{ql}(\Gamma^\alpha) A_{ijkl} e_{pq}^o \tag{2.10.102}$$

where only elastic contribution is considered.

An interesting result in the case of statistical homogeneity is found that since the statistical continuum electric and elastic multipoles defined in Eqs.(2.10.75)–(2.10.77) are independent of **x**' in the dilute approximation, Eq.(2.10.64) can be reduced to

$$\langle E_m \rangle = E_m^o - \frac{1}{\varepsilon^e} \bar{P}_i^e L_{im}^e \tag{2.10.103}$$

where $\bar{\boldsymbol{P}}^e$ is the statistical continuum electric dipole moment given by

$$\bar{P}_k^e = f \Delta \varepsilon^e \left[\int_\Gamma \rho^*(\Gamma') Q_{ki}(\Gamma') Q_{nj}(\Gamma') d\Gamma' \right] T_{ij} E_n^o \tag{2.10.104}$$

with $f = MV_I/V_R$ being the volume fraction of inhomogeneous particles in the composite and $\boldsymbol{L}^e$ the electric depolarizing tensor given by

$$L_{im}^e = -\varepsilon^e \int_{V_R} G_{,im}^e(\boldsymbol{x} - \boldsymbol{x}') dx' = \frac{1}{3} \delta_{im} \tag{2.10.105}$$

for $\boldsymbol{x}$ inside the spherical volume V_R. Here, the matrix and the particle are both assumed to be isotropic for simplicity.

Also, from Eq.(2.10.65), one can find the ensemble average of the elastic strain fields in the composite by

$$\langle e_{ij} \rangle = e_{ij}^o + L_{ijkl} \bar{P}_{lk} \tag{2.10.106}$$

where $\bar{P}_{lk}$ is the statistical continuum elastic monopole moment given by

$$\bar{P}_{lk} = -f \Delta C_{lkmn} \left[\int_\Gamma \rho^*(\Gamma') Q_{mi}(\Gamma') Q_{nj}(\Gamma') Q_{ps}(\Gamma') Q_{qt}(\Gamma') d\Gamma' \right] A_{ijst} e_{pq}^o \tag{2.10.107}$$

and L_{ijkl} is the elastic depolarization tensor defined by

$$L_{ijkl} = -\frac{1}{2} \int_{S_R} (G_{il,j'} + G_{jl,i'}) n_k' dS' \tag{2.10.108}$$

which can be given explicitly from Eq.(2.10.25) for isotropic elastic matrix media.

Now, comparing Eq.(2.10.103) with Eq.(2.10.84), we may find

$$\langle F_{mk} \rangle = \delta_{mk} - \frac{\Delta \varepsilon^e f}{3 \varepsilon^e} T_{ij} \left[\int_\Gamma \rho^*(\Gamma') Q_{mi}(\Gamma') Q_{kj}(\Gamma') d\Gamma' \right] \tag{2.10.109}$$

And, comparing Eq.(2.10.106) with Eq.(2.10.85), we may get

$$\langle B_{ijpq}\rangle = \delta_{ip}\delta_{jq}$$

$$-fL_{ijkl}\Delta C_{lkmn}\left[\int_{\Gamma}\rho^*(\Gamma')Q_{mi}(\Gamma')Q_{nj}(\Gamma')Q_{ps}(\Gamma')Q_{qt}(\Gamma')d\Gamma'\right]A_{ijst} \quad (2.10.110)$$

Furthermore, by noting Eqs.(2.10.82) and (2.10.99), we may find

$$\left\langle \sum_{\alpha=1}^{M}\gamma^{\alpha}T_{kl}^{\alpha}\right\rangle = f\int_{\Gamma}\rho^*(\Gamma')T_{ij}Q_{ki}(\Gamma')Q_{lj}(\Gamma')d\Gamma' \quad (2.10.111)$$

And, by noting Eqs.(2.10.83) and (2.10.100), we get

$$\left\langle \sum_{\alpha=1}^{M}\gamma^{\alpha}A_{mnpq}^{\alpha}\right\rangle = f\int_{\Gamma}\rho^*(\Gamma')A_{ijkl}Q_{mi}(\Gamma')Q_{nj}(\Gamma')Q_{pk}(\Gamma')Q_{ql}(\Gamma')d\Gamma'$$

$$(2.10.112)$$

Thus, we have obtained, in general, the analytical expressions, Eqs.(2.10.93), (2.10.94), and Eqs.(2.10.109)–(2.10.112) for the determination of the effective permittivity and of the effective elastic moduli of the statistically anisotropic dielectric and elastic composite in the dilute suspension approximation provided that the probability density function $\rho^*(\Gamma')$ on the statistical distribution of orientations of the particles is given.

For simplicity as well as for the similarity of the dielectric and elastic problems, we shall now consider a concrete example in which we would like to find the effective permittivity of a (rigid) dielectric composite with M statistically distributed identical inhomogeneous rigid ellipsoidal particles. We assume that the dielectric composite contains M ellipsoidal inhomogeneous particles which are statistically distributed in a uniform matrix. The orientations of the particles is assumed to be characterized by the Gaussian distribution:

$$\rho^*(\Gamma') = C\exp(-\sigma^2|\Gamma'-\Gamma^o|^2) \quad (2.10.113)$$

where σ is a statistical parameter determined and controllable by the method of manufacture of the composites, and C is the constant determined by the following normalization condition:

$$\int_{\Gamma}\rho^*(\Gamma')d\Gamma' = 1 \quad (2.10.114)$$

According to Stratton (1941), we may write explicitly

$$Q_{ki}(\Gamma')Q_{lj}(\Gamma')T_{ij} = \sum_{i=1}^{3} \lambda_i Q_{ki}(\Gamma')Q_{li}(\Gamma') \tag{2.10.115}$$

where

$$\lambda_i = \left(1 + abc\frac{\Delta\varepsilon^e}{2\varepsilon^e}B_i\right)^{-1} \tag{2.10.116}$$

in which

$$B_1 = \int_0^\infty \frac{1}{(s+a^2)^{3/2}\sqrt{(s+b^2)(s+c^2)}}ds \tag{2.10.117}$$

and

$$B_2 = \int_0^\infty \frac{1}{(s+b^2)^{3/2}\sqrt{(s+a^2)(s+c^2)}}ds \tag{2.10.118}$$

and

$$B_3 = \int_0^\infty \frac{1}{(s+c^2)^{3/2}\sqrt{(s+a^2)(s+b^2)}}ds \tag{2.10.119}$$

where a, b, and c are the axes of the ellipsoid.

It can then be seen from Eqs.(2.10.93), (2.10.109), and (2.10.111) that the shape effect of the inhomogeneous particles on the overall electric permittivity of the composite is taken into account in the parameters λ_i ($i = 1, 2, 3$) while the effect of statistical orientations of the particles is accounted by the orientation tensor $\mathbf{Q}(\Gamma')$ as well as the probability density function $\rho^*(\Gamma')$. In the case of spherical inhomogeneous particles, the problem is much simplified and we have

$$\lambda_1 = \lambda_2 = \lambda_3 = \frac{3\varepsilon^e}{3\varepsilon^e + \Delta\varepsilon^e} \tag{2.10.120}$$

and

$$\langle F_{kj} \rangle = \frac{3\varepsilon^e + (1-f)\Delta\varepsilon^e}{3\varepsilon^e + \Delta\varepsilon^e}\delta_{kj} \tag{2.10.121}$$

and

$$\langle \sum_{\alpha=1}^{M} \gamma^\alpha T_{kj}^\alpha \rangle = \frac{3\varepsilon^e f}{3\varepsilon^e + \Delta\varepsilon^e}\delta_{kj} \tag{2.10.122}$$

The effective permittivity of the composite with the dilute spherical inhomogeneous particles can thus be obtained as

$$\varepsilon^{eff} = \frac{2\varepsilon^e + \varepsilon^* + 2f\Delta\varepsilon^e}{2\varepsilon^e + \varepsilon^* - f\Delta\varepsilon^e}\varepsilon^e \tag{2.10.123}$$

which is in accordance with the Rayleigh mixture formula [Rayleigh (1892)] obtainable by using a volume average method. The equivalence of the ensemble average and the volume average is then proved in this case. It should be, however, noticed that real physical systems satisfy the ergodic hypothesis very rarely [Kröner (1986)]. The result also shows that the derivation of the Rayleigh mixture formula for a composite with statistically homogeneous distributed dilute spherical particles is independent of the assumption of statistical isotropy for the composite provided that one ignores the interaction among particles.

To see explicitly the shape effect of inhomogeneous particles on the overall effective properties of the dielectric composite, let us consider the ellipsoidal particles (with $a = b$). If one assumes that all directions of the orientations of particles are equally probable, the probability of the particle orientation within the range $(\theta', \theta'+d\theta')$ and $(\psi', \psi'+d\psi')$ then reads

$$\rho^*(\Gamma')d\Gamma' = \frac{1}{4\pi}\sin\theta' d\theta' d\psi' \tag{2.10.124}$$

Using Eqs.(2.10.58), (2.10.93), (2.10.109), and (2.10.111), the effective permittivity of the dielectric composite can thus be derived as

$$\varepsilon^{eff} = \frac{9\varepsilon^e + 2f\Delta\varepsilon^e(2\lambda_1 + \lambda_3)}{9\varepsilon^e - f\Delta\varepsilon^e(2\lambda_1 + \lambda_3)}\varepsilon^e \tag{2.10.125}$$

in which the shape effect parameters λ_1 and λ_2 are given by Eqs.(2.10.116)–(2.10.119) with $a = b$. For $c >> a$, one has the rod or needle type of particle. For $c << a$ and close to zero, one has the disc type of particles. For $c = a$, one gets the spherical particles.

It is shown that for fully random orientations of the ellipsoidal particles, the ensemble average behavior of the dielectric composite displays an isotropic property. If now the orientations of the particles ($a = b$) have a preferred direction, say in the direction of x_3-axis ($\theta^o = 0$), and if its statistical distribution is represented by the Gaussian distribution (2.10.113), we can then write

$$\rho^*(\Gamma')d\Gamma' = C\exp(-\sigma^2\theta'^2)\sin\theta' d\theta' d\psi' \tag{2.10.126}$$

where σ is the statistical parameter characterizing the standard deviation of θ' and C is a constant determined by Eq.(2.10.114), which can be written as

$$C = \left[2\pi\int_0^\pi \exp(-\sigma^2\theta'^2)\sin\theta' d\theta'\right]^{-1} \tag{2.10.127}$$

After some calculations, we can get

$$[\varepsilon_{ij}^{eff}] = \varepsilon^e \frac{3\varepsilon^e + 2f\Delta\varepsilon^e A_1}{3\varepsilon^e - 2f\Delta\varepsilon^e A_1} \begin{bmatrix} 1 & 0 & 0 \\ 0 & 1 & 0 \\ 0 & 0 & \eta \end{bmatrix} \tag{2.10.128}$$

where the dimensionless parameter η is defined by

$$\eta = \frac{(3\varepsilon^e + 2f\Delta\varepsilon^e A_2)(3\varepsilon^e - 2f\Delta\varepsilon^e A_1)}{(3\varepsilon^e + 2f\Delta\varepsilon^e A_1)(3\varepsilon^e - 2f\Delta\varepsilon^e A_2)} \tag{2.10.129}$$

with the constants A_1 and A_2 being given, respectively, by

$$A_1 = \pi \int_0^\pi C\exp(-\sigma^2\theta^2)[\lambda_1(1 + \cos^2\theta) + \lambda_3\sin^2\theta]\sin\theta d\theta \tag{2.10.130}$$

and

$$A_2 = 2\pi \int_0^\pi C\exp(-\sigma^2\theta^2)[\lambda_1\sin^2\theta + \lambda_3\cos^2\theta]\sin\theta d\theta \tag{2.10.131}$$

It is shown that the overall effective permittivity of the composite can be anisotropic provided that the orientations of the micro ellipsoidal particles have statistically a preferred direction. In the considered (transversely isotropic) case, the macroscopic effective permittivities are found to have two independent constants and they are given in Eq.(2.10.128).

2.10.5 Overall Behavior of Elastic Dielectric Composites

This section is concerned with the study of the overall behavior of the elastic dielectric composite in which some effects of interaction between electric field and elastic field are taken into account. The constitutive relations characterizing the overall behavior of the elastic dielectric composite will be derived with the use of the statistical continuum multipole approach, in which no preassumptions on the macroscopic constitutive relations of the composite are needed. This result thus makes the statistical continuum multipole approach superior than classical effective medium theory where the effective constitutive relations of composites has to be pre-assumed. To make a distinction, we shall call the derived governing equations for the overall behavior of the elastic dielectric composites the overall macroscopic constitutive relations and call their coefficients the overall macroscopic material coefficients rather than the effective constitutive relations and the effective material coefficients defined in effective medium theory.

Consider an elastic dielectric composite with M randomly distributed elastic dielectric inhomogeneous particles. The material properties of the matrix and the particles are supposed to be the same as those described in Section 2.10.1. With the use of the result given in Section 2.10.1, we may find that the ensemble average electric polarization and elastic stress fields can be expressed in the following form:

$$\langle P_i \rangle = \varepsilon_o \chi \langle E_i \rangle + \varepsilon_o \Delta\chi \langle \sum_{\alpha=1}^{M} \gamma^\alpha E_i \rangle \tag{2.10.132}$$

and

$$\langle t_{ij} \rangle = \lambda \langle e_{kk} \rangle \delta_{ij} + 2G \langle e_{ij} \rangle + \langle \sum_{\alpha=1}^{M} \gamma^\alpha (\Delta\lambda e_{kk} \delta_{ij} + 2\Delta G e_{ij} - P_j E_i) \rangle \tag{2.10.133}$$

To perform the ensemble average, the detailed solution on the microscopic fields has to be obtained. For simplicity, we consider an elastic dielectric composite with dilute suspension of spherical inhomogeneous particles and make the assumptions of the statistical homogeneity and the statistical isotropy for the composite. To the first-order approximation, the interaction between particles is ignored for the dilute system, and the influence of the small change of particle orientations caused by small elastic deformation on the overall electric field is neglected. In such a case, after some calculations, we can obtain

$$\langle \boldsymbol{P} \rangle = \varepsilon_o \chi^{mac} \langle \boldsymbol{E} \rangle \tag{2.10.134}$$

where the overall macroscopic dielectric susceptibility can be written as

$$\chi^{mac} = \chi + \frac{3\varepsilon^e f \Delta\chi}{3\varepsilon^e + (1-f)\Delta\varepsilon^e} \tag{2.10.135}$$

which is in accordance with classical Rayleigh mixture formula for the effective dielectric susceptibility of the composite with dilute spherical particles in its rigid-body approximation.

The ensemble average strain field may be written as

$$\langle e_{ij} \rangle = e^o_{ij} + L_{ijkl} \bar{P}_{kl} \tag{2.10.136}$$

where the statistical continuum elastic monopole moment $\bar{\boldsymbol{P}}$ may be found by Eqs.(2.10.63), (2.10.74), and (2.10.77) as

$$\bar{P}_{ii} = -\frac{(3K+4G)f}{3K+4G+3\Delta K}\left\{3\Delta K e^o_{ii} - \frac{(3\varepsilon^e)^2 \varepsilon_o \chi^*}{[3\varepsilon^e(1-f)\Delta\varepsilon^e]^2} \langle E_i \rangle \langle E_i \rangle\right.$$

$$+\frac{3\varepsilon^e\Delta\varepsilon^e-\varepsilon_o\chi^*}{(3\varepsilon^e+\Delta\varepsilon^e)^2}(p^o)^2\Bigg\} \qquad (2.10.137)$$

where sum is made over the suffix i, and for $i \neq j$

$$\bar{P}_{ij} = -\frac{5G(3K+4G)f}{5G(3K+4G)+\Delta G(6K+12G)}$$

$$\times\left\{2\Delta G e^o_{ij}-\frac{(3\varepsilon^e)^2\varepsilon_o\chi^*}{[3\varepsilon^e(1-f)\Delta\varepsilon^e]^2}\langle E_i\rangle\langle E_j\rangle\right\} \qquad (2.10.138)$$

By noting Eqs.(2.10.25) and (2.10.103), we finally arrive at the result that the overall macroscopic stress in the elastic dielectric composite with dilute spherical particles in the absence of permanent electric polarizations may be written as

$$\langle t_{ij}\rangle = \left(K^{mac}-\frac{2}{3}G^{mac}\right)\langle e_{kk}\rangle\delta_{ij}+2G^{mac}\langle e_{ij}\rangle$$

$$-\left[\frac{\xi-\tau}{3}\langle P_k\rangle\langle E_k\rangle\delta_{ij}+\tau\langle P_j\rangle\langle E_i\rangle\right] \qquad (2.10.139)$$

where the overall macroscopic elastic moduli read

$$K^{mac} = K+\frac{(3K+4G)f\Delta K}{3K+4G+3(1-f)\Delta K} \qquad (2.10.140)$$

$$G^{mac} = G+\frac{5G(3K+4G)f\Delta G}{5G(3K+4G)+6\Delta G(1-f)(K+2G)} \qquad (2.10.141)$$

which are in accordance with classical results for the effective elastic moduli of elastic composite with dilute spherical particles in the case of no electric forces (see, for instance, Christensen (1979)).

By taking into account the electroelastic coupling effect, it is found from Eq.(2.10.139) that there are two new dimensionless macroscopic electroelastic parameters responsible the overall behavior of the elastic dielectric material composites defined by

$$\xi = \frac{f\chi^*}{[1+3\Delta K(1-f)/(3K+4G)][1+\Delta\varepsilon^e(1-f)/(3\varepsilon^e)]^2\chi^{mac}} \qquad (2.10.142)$$

and

$$\tau = \frac{f\chi^*}{[1 + 6\Delta G(1-f)(K+2G)/(5G(3K+4G))][1 + \Delta\varepsilon^e(1-f)/(3\varepsilon^e)]^2 \chi^{mac}} \tag{2.10.143}$$

where f denotes the volume fraction of the particles.

The obtained Eqs.(2.10.134) and (2.10.139) may be used approximately as the overall macroscopic constitutive relations of the composite, provided that applied electric and mechanical loads do not have significant variations within any representative volume which is macroscopically small compared with the total volume of the composite, but is microscopically large enough to contain many particles. However, for treating composites with loads having significant variations within the length scale of several particle distance, such as cracked composites, the effect of high order multipoles has to be, in general, taken into account. In addition, if the particles in the composites have spontaneous electrical polarizations $p^o\boldsymbol{n}^\alpha$ ($\alpha = 1, 2, \ldots, M$), a stress term $\beta_o(p^o)^2\delta_{ij}$ with

$$\beta_o = \frac{(3\varepsilon^e + \Delta\varepsilon^e - \varepsilon_o\chi^*)f}{3[1 + 3\Delta K(1-f)/(3K+4G)](3\varepsilon^e + \Delta\varepsilon^e)^2} \tag{2.10.144}$$

is found to be present in Eq.(2.10.139). This stress term exists even in the absence of external electric and mechanical loads, and, therefore, may be considered as the internal stress in the composite. The presence of the internal stress is physically understandable since the internal electric field caused by all permanent electric dipole moments of the particles will generate electric forces and torques on the particles, which must be balanced by mechanical forces in equilibrium state. The presented example shows the advantage of the statistical continuum multipole model in comparison with the classical effective medium theory where the formal effective macroscopic constitutive relations for the composite have to be assumed beforehand, which, however, as shown, may be unknown for complex electromagnetic deformable material composites. Similar results may also be obtained for certain elastic magnetic material composites due to the analogy between the electric polarization and the magnetization.

2.11 SOME EXAMPLES OF BOUNDARY–VALUE PROBLEMS

The aim of this section is to introduce some approaches often used in continuum mechanics for analyzing boundary-value problems, and to illustrate those concepts introduced above on material properties of solids. Since these approaches and concepts might not be familiar to electrical engineers, we shall start with the study of two simple elastic problems, and then a photothermoelastic problem as illustrative examples.

2.11.1 Circular Cylindrical Tube Subjected to Pressure

This is a boundary-value problem, in which an infinite elastic circular cylindrical tube is subjected to an internal pressure p_o and an external pressure p_1, as shown in Fig. 2.8. To determine the displacement and stress fields in the tube, we have the following boundary conditions in a cylindrical coordinate system (r, φ, z) with the z taken along the axis of the cylinder:

$$t_{rr} = -p_o \quad \text{at } r = a, \quad \text{and} \quad t_{rr} = -p_1 \quad \text{at } r = b \tag{2.11.1}$$

The material solid of the tube is supposed to be linear elastic and isotropic. Thus we have the following constitutive equation:

$$t_{ij} = \lambda \varepsilon_{kk} \delta_{ij} + 2G\varepsilon_{ij} \tag{2.11.2}$$

where λ and G are the Lamé constants.

For small deformation as assumed here, the strain tensor ε_{ij} can be expressed by

$$\varepsilon_{ij} = \frac{1}{2}(u_{i,j} + u_{j,i}) \tag{2.11.3}$$

where $\boldsymbol{u}$ is the displacement vector. Thus, we may write the mechanical motion equation for the linear elastic solid in the form:

$$(\lambda + 2G)\nabla(\nabla \cdot \boldsymbol{u}) - G\nabla \times (\nabla \times \boldsymbol{u}) + \rho\left(\boldsymbol{f} - \frac{\partial^2 \boldsymbol{u}}{\partial t^2}\right) = 0 \tag{2.11.4}$$

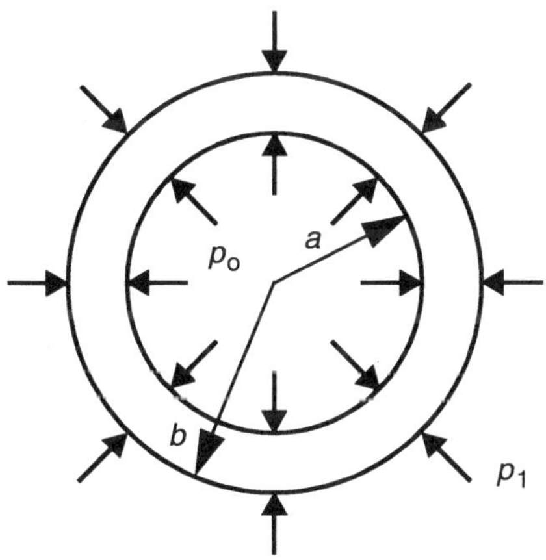

Figure 2.8 Infinite circular cylindrical tube subject to internal and external pressure.

where ρ is the mass density, and f the body force. In the absence of body forces, and in static cases, which we are now considering in this example, Eq.(2.11.4) becomes

$$(\lambda + 2G)\nabla(\nabla \cdot \boldsymbol{u}) - G\nabla \times (\nabla \times \boldsymbol{u}) = 0 \tag{2.11.5}$$

which is the mechanical equilibrium equation.

In our problem, since any point on a cylindrical surface r = constant will not move in either the φ or z direction due to the symmetry of the problem, we have the displacement components, $u_\varphi = u_z = 0$, and the displacement component $u_r = u_r(r)$. Thus, in the cylindrical coordinate system, the mechanical equilibrium equation (2.11.5) reduces to a single equation:

$$\frac{d}{dr}\left[\frac{1}{r}\frac{d}{dr}(ru_r)\right] = 0 \tag{2.11.6}$$

By Eqs.(2.11.2) and (2.11.3), the stress components can be related to the displacement component u_r in this case by

$$t_{rr} = \frac{\lambda}{r}\frac{d(ru_r)}{dr} + 2G\frac{du_r}{dr} \tag{2.11.7}$$

$$t_{\varphi\varphi} = \frac{\lambda}{r}\frac{d(ru_r)}{dr} + 2G\frac{u_r}{r} \tag{2.11.8}$$

$$t_{zz} = \frac{\lambda}{r}\frac{d(ru_r)}{dr} \tag{2.11.9}$$

and $t_{rz} = t_{\varphi z} = t_{r\varphi} = 0$. Thus, by Eq.(2.11.7), the boundary conditions in Eq.(2.11.1) can be expressed by

$$\left[\frac{\lambda}{r}\frac{d(ru_r)}{dr} + 2G\frac{du_r}{dr}\right]_{r=a} = -p_o \tag{2.11.10}$$

$$\left[\frac{\lambda}{r}\frac{d(ru_r)}{dr} + 2G\frac{du_r}{dr}\right]_{r=b} = -p_1 \tag{2.11.11}$$

The general solution of Eq.(2.11.6) is

$$u_r = C_1 r + \frac{C_2}{r} \tag{2.11.12}$$

where C_1 and C_2 are two constants that can be determined by Eqs.(2.11.10) and (2.11.11). The result is

$$C_1 = \frac{a^2 p_o - b^2 p_1}{2(\lambda + G)(b^2 - a^2)} \tag{2.11.13}$$

$$C_2 = \frac{a^2 b^2 (p_o - p_1)}{2G(b^2 - a^2)} \tag{2.11.14}$$

The stress fields in the cylindrical tube can then be found by using Eqs.(2.11.7)–(2.11.9). The results are

$$t_{rr} = p_o \frac{1 - b^2/r^2}{b^2/a^2 - 1} - p_1 \frac{1 - a^2/r^2}{1 - a^2/b^2} \tag{2.11.15}$$

$$t_{\varphi\varphi} = p_o \frac{1 + b^2/r^2}{b^2/a^2 - 1} - p_1 \frac{1 + a^2/r^2}{1 - a^2/b^2} \tag{2.11.16}$$

$$t_{zz} = \frac{\lambda}{\lambda + G} \frac{a^2 p_o - b^2 p_1}{b^2 - a^2} \tag{2.11.17}$$

and $t_{rz} = t_{\varphi z} = t_{r\varphi} = 0$.

A special case of $p_1 = 0$ might be of interest, where we have

$$t_{rr} = p_o \frac{1 - b^2/r^2}{b^2/a^2 - 1} \tag{2.11.18}$$

$$t_{\varphi\varphi} = p_o \frac{1 + b^2/r^2}{b^2/a^2 - 1} \tag{2.11.19}$$

It is shown that if $p_1 = 0$ and $p_o > 0$, t_{rr} is always a compressive stress, and $t_{\varphi\varphi}$ is always a tensile stress, the maximum value of which occurs at the inner radius and is always numerically greater than the internal pressure. For high-pressure vessels, proper mechanical design is always required to ensure the mechanical strength of the vessels.

As we may know, in superconducting energy storage systems, the energy stored may be estimated by the value of VB^2/μ_o, where B is the magnitude of magnetic induction field, and V is the volume of the superconducting magnet. Obviously, to have a large energy storage system, we require a large volume V and/or a large magnetic field B. On the other hand, the high magnetic field may cause a high pressure p_m acting on the superconducting magnet, as we shall discuss late in the next chapter. Quantitatively, the "magnetic" pressure, p_m =

$B^2/(2\mu_0)$ has a value of 10^7 N/m^2 for $B = 5$ tesla. This value can be compared with the yield stresses of aluminum alloys, copper alloys, and ferritic steels, which are in the range of 6×10^7–2×10^9 N/m^2 at 4.2 K. It can, therefore, be understood that sufficiently large magnetic field may destroy the superconducting magnet. Indeed, the available mechanical strength of large superconducting energy storage systems has become one of the limiting factors that have to be considered in the design of such systems.

2.11.2 The Rayleigh Surface Wave

Next, we shall give an example of analyzing elastic dynamic boundary-value problem. We shall demonstrate that a surface elastic wave, called the Rayleigh surface wave, may occur on the free surface of a homogeneous, isotropic, semi-infinite elastic solid. This is an important type of wave because the largest disturbances caused by an earthquake recorded on a distant seismogram are usually those of the Rayleigh surface waves.

In this example, we consider an elastic half-space $y \geq 0$, as shown in Fig. 2.9. The surface $y = 0$ is stress-free. Thus we have the following boundary conditions:

$$t_{yx} = t_{yy} = t_{yz} = 0 \qquad \text{on } y = 0 \tag{2.11.20}$$

The equation of motion for the problem in the absence of body forces is

$$c_L^2 \nabla(\nabla \cdot \boldsymbol{u}) - c_T^2 \nabla \times (\nabla \times \boldsymbol{u}) = \frac{\partial^2 \boldsymbol{u}}{\partial t^2} \tag{2.11.21}$$

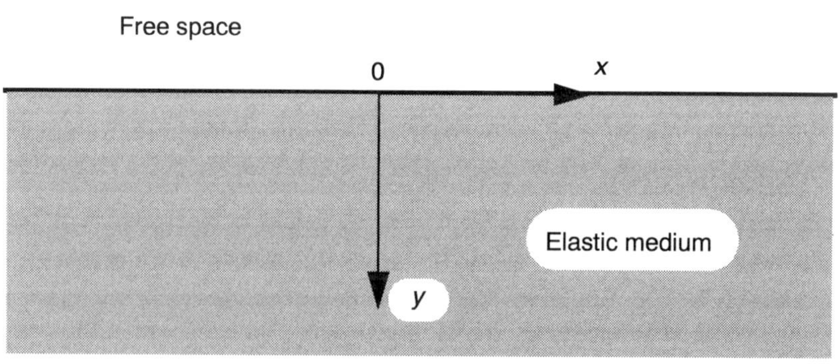

Figure 2.9 An elastic half-space.

where c_L is given by

$$c_L = \sqrt{\frac{\lambda + 2G}{\rho}} \qquad (2.11.22)$$

which is the phase velocity of longitudinal elastic waves, and c_T is given by

$$c_L = \sqrt{\frac{G}{\rho}} \qquad (2.11.23)$$

which is the phase velocity of transverse elastic waves. These two phase velocities, c_L and c_T can be understood easily by studying plane elastic waves in an infinitely large elastic medium [see, e.g., Fung (1965)].

We now consider a possible solution for the displacements represented by the real part of the following expressions:

$$u_x = Ae^{-\alpha y}\exp[ik(x - \mathrm{v}t)] \qquad (2.11.24)$$

$$u_y = Be^{-\alpha y}\exp[ik(x - \mathrm{v}t)] \qquad (2.11.25)$$

$$u_z = 0 \qquad (2.11.26)$$

where A and B are complex constants. The coefficient α is supposed to be real and positive so that the amplitude of the wave decreases exponentially with increasing y, and tends to zero as $y \to \infty$, which fulfills the criterion for surface waves. Substituting Eqs.(2.11.24)–(2.11.26) into Eq.(2.11.21), we may obtain

$$[c_T^2\alpha^2 + (\mathrm{v}^2 - c_L^2)k^2]A - i(c_L^2 - c_T^2)\alpha k B = 0 \qquad (2.11.27)$$

$$-i(c_L^2 - c_T^2)\alpha k A + [c_L^2\alpha^2 + (\mathrm{v}^2 - c_T^2)k^2]B = 0 \qquad (2.11.28)$$

The condition for the existence of a non-trivial solution for this set of equations is the vanishing of the determinant of the coefficients, which may be written in the form:

$$[c_L^2\alpha^2 - (c_L^2 - \mathrm{v}^2)k^2][c_T^2\alpha^2 - (c_T^2 - \mathrm{v}^2)k^2] = 0 \qquad (2.11.29)$$

This gives the following roots for α:

$$\alpha' = k\sqrt{1 - \frac{\mathrm{v}^2}{c_L^2}}, \qquad \alpha'' = k\sqrt{1 - \frac{\mathrm{v}^2}{c_T^2}} \qquad (2.11.30)$$

The requirement of α is real implies that $v < c_T < c_L$. Corresponding to $\alpha'$and α'', respectively, the ratio B/A can be solved from Eqs.(2.11.27) and (2.11.28). Hence, a general solution, satisfying the equation of motion, may be written as

$$u_x = A'e^{-\alpha' y}\exp[ik(x - vt)] + A''e^{-\alpha'' y}\exp[ik(x - vt)] \tag{2.11.31}$$

$$u_y = -\frac{\alpha'}{ik}A'e^{-\alpha' y}\exp[ik(x - vt)] + \frac{ik}{\alpha''}A''e^{-\alpha'' y}\exp[ik(x - vt)] \tag{2.11.32}$$

$$u_z = 0 \tag{2.11.33}$$

To satisfy the boundary conditions (2.11.20), we use the elastic constitutive equation (2.11.2), the strain-displacement relation (2.11.3), and this general solution for the displacements. After some manipulations, in order to have a non-zero solution, we may obtain the following characteristic equation for v:

$$\frac{v^2}{c_T^2}\left[\frac{v^6}{c_T^6} - 8\frac{v^4}{c_T^4} + v^2\left(\frac{24}{c_T^2} - \frac{16}{c_L^2}\right) - 16\left(1 - \frac{c_T^2}{c_L^2}\right)\right] = 0 \tag{2.11.34}$$

which is called the *Rayleigh equation*. The equation for the ratio of A''/A' is

$$\frac{A''}{A'} = -\frac{2\sqrt{(1 - v^2/c_L^2)(1 - v^2/c_T^2)}}{2 - v^2/c_T^2} \tag{2.11.35}$$

Equation (2.11.34) has always a non-zero root v in the range $(0, c_T)$. For instance, for an incompressible solid $c_L \to \infty$, Eq.(2.11.34) becomes

$$\frac{v^6}{c_T^6} - 8\frac{v^4}{c_T^4} + 24\frac{v^2}{c_T^2} - 16 = 0 \tag{2.11.36}$$

This cubic equation in v^2 has a real root at $v^2 = 0.91275c_T^2$, corresponding to the Rayleigh surface wave with the phase velocity v of about $0.95538c_T$.

2.11.3 Thermal Stress-Induced Index Profile Distortion in an Optical Fiber

It is known from Section 2.8.3 that the presence of thermal stresses in an optical fiber may affect optical properties of the optical fiber through photothermoelastic effects. In this section, we shall first present a method of analyzing thermal stresses in optical fibers, which may be caused by manufacture processes or by varying environmental temperature. We shall then show how the optical index profile can be distorted by the thermal stresses in the optical fiber.

We shall assume that the optical fiber considered has the geometric shape of a long cylinder with outer radius b, located in a temperature field, being symmetric about the axis of the optical fiber. We shall also assume that the temperature field is independent of the axial coordinate z and can be obtained by solving an uncoupled heat conduction equation as the first approximation. The ends of the fiber may be subjected to a total force F along the axis of the optical fiber.

By noting the axial symmetry of the problem, we may write the strain field in a cylindrical coordinate system (r, φ, z) by

$$e_{rr} = \frac{\partial u_r}{\partial r}, \qquad e_{\varphi\varphi} = \frac{u_r}{r}, \qquad e_{zr} = e_{\varphi z} = e_{r\varphi} = 0 \tag{2.11.37}$$

with e_{zz} being supposed to be constant in space coordinates, and u_r the radial displacement. Besides, we have the following mechanical equilibrium equation:

$$\frac{\partial}{\partial r}(rt_{rr}) - t_{\varphi\varphi} = 0 \tag{2.11.38}$$

in the absence of body forces.

The thermoelastic constitutive equations for the optical fiber in the cylindrical coordinate system can be expressed by

$$e_{rr} = \alpha\theta + \frac{1}{E}[t_{rr} - \nu(t_{zz} + t_{\varphi\varphi})] \tag{2.11.39}$$

$$e_{\varphi\varphi} = \alpha\theta + \frac{1}{E}[t_{\varphi\varphi} - \nu(t_{zz} + t_{rr})] \tag{2.11.40}$$

$$e_{zz} = \alpha\theta + \frac{1}{E}[t_{zz} - \nu(t_{rr} + t_{\varphi\varphi})] \tag{2.11.41}$$

where E is Young´s modulus, and ν Poisson´s ratio. α is the thermal coefficient of linear expansion, which is related to the thermal modulus β by $\alpha = \beta(1-2\nu)/E$.

Introducing now the quantity:

$$P = t_{rr} + t_{\varphi\varphi} + t_{zz} \tag{2.11.42}$$

we may find from Eqs.(2.11.38)–(2.11.41)

$$t_{rr} = \frac{1}{r^2}\int_0^r f(r)rdr \qquad \left(\to \frac{1}{2}f(0) \text{ as } r \to 0 \right) \tag{2.11.43}$$

with the integrand f defined by

$$f = \frac{P - E(e_{zz} - \alpha\theta)}{1 + \nu} \tag{2.11.44}$$

From Eq.(2.11.38), we can then obtain

$$t_{\varphi\varphi} = f(r) - \frac{1}{r^2}\int_0^r f(r)rdr \tag{2.11.45}$$

and

$$t_{zz} = \nu f(r) + E(e_{zz} - \alpha\theta) \tag{2.11.46}$$

from Eqs.(2.11.41), (2.11.43), and (2.11.45).

Furthermore, by noting the strain-displacement relations (2.11.37) and the constitutive equations (2.11.39)–(2.11.41), we may arrive at the following a compatibility equation to be satisfied

$$\frac{\partial}{\partial r}\left[\frac{(1-\nu)P}{E}\right] - t_{rr}\frac{\partial}{\partial r}\left(\frac{1+\nu}{E}\right) = \frac{\partial}{\partial r}(e_{zz} - 2\alpha\theta) \tag{2.11.47}$$

For glass fibers with constant elastic moduli, the second term on the left-hand side of Eq.(2.11.47) vanishes, and the equation can be readily integrated. For optical fibers of step-index profile with material properties independent of the radius r in the cladding and in the core of the optical fiber, an exact general solution of Eq.(2.11.47) can be found by

$$P = \frac{E_1}{1-\nu_1}(c_1 - 2\alpha_1\theta + e_{zz}) \qquad r < a \tag{2.11.48}$$

$$P = \frac{E_2}{1-\nu_2}(c_2 - 2\alpha_2\theta + e_{zz}) \qquad a < r < b \tag{2.11.49}$$

with two undetermined quantities c_1 and c_2, which are not functions of the radius r. Here, E_1, ν_1, and α_1 are the elastic and thermal expansion constants of the core glass. E_2, ν_2, and α_2 are the elastic and thermal expansion constants of the cladding glass.

By Eq.(2.11.44), we can then find

$$f = \frac{E_1}{1-\nu_1^2}[c_1 - (1+\nu_1)\alpha_1\theta + \nu_1 e_{zz}] \qquad r < a \tag{2.11.50}$$

$$f = \frac{E_2}{1-\nu_2^2}[c_2 - (1+\nu_2)\alpha_2\theta + \nu_2 e_{zz}] \qquad a < r < b \tag{2.11.51}$$

The quantities c_1 and c_2 can be determined by the boundary condition of that the radial stress t_{rr} must be zero at the outer radius b, and by the interface conditions of that the radial stress t_{rr} and the radial displacement u_r must be continuous across the interface at $r = a$ (the core radius), that is,

$$\frac{a^2 E_1}{2(1-\nu_1^2)}c_1 + \frac{(b^2-a^2)E_2}{2(1-\nu_2^2)}c_2 = \int_0^b \frac{E[(1+\nu)\alpha\theta - \nu e_{zz}]}{1-\nu^2} r\,dr \tag{2.11.52}$$

and

$$c_1\left[1 + \frac{(1+\nu_2)E_1 - (1+\nu_1)E_2}{2(1-\nu_1^2)E_2}\right] - c_2 = \frac{(1+\nu_2)E_1 - (1+\nu_1)E_2}{a^2(1-\nu_1^2)E_2} \times \int_0^a [(1+\nu_1)\alpha_1\theta - \nu_1 e_{zz}] r\,dr \tag{2.11.53}$$

from which we can obtain

$$c_1 = \frac{1}{d_o}\left\{\frac{(b^2-a^2)[(1+\nu_2)E_1 - (1+\nu_1)E_2]}{2a^2(1-\nu_1^2)(1-\nu_2^2)}\int_0^a [(1+\nu_1)\alpha_1\theta - \nu_1 e_{zz}] r\,dr + \int_0^b \frac{E[(1+\nu)\alpha\theta - \nu e_{zz}]}{1-\nu^2} r\,dr\right\} \tag{2.11.54}$$

and

$$c_2 = c_1\left[1 + \frac{(1+\nu_2)E_1 - (1+\nu_1)E_2}{2(1-\nu_1^2)E_2}\right] - \frac{(1+\nu_2)E_1 - (1+\nu_1)E_2}{a^2(1-\nu_1^2)E_2}\int_0^a [(1+\nu_1)\alpha_1\theta - \nu_1 e_{zz}] r\,dr \tag{2.11.55}$$

with the parameter d_o given by

$$d_o = \frac{2a^2(1-\nu_2^2)E_1 + (b^2 - a^2)[(1+\nu_1)(1-2\nu_1)E_2 + (1+\nu_2)E_1]}{4(1-\nu_1^2)(1-\nu_2^2)} \tag{2.11.56}$$

For an optical fiber with continuous index profiles, the elastic moduli and the thermal expansion coefficient of the fiber vary with the radius r. If the variation is small and no jumps in these coefficients, a successive procedure may be used to solve Eq.(2.11.47) [Scherer (1979)]. At the first order approximation by ignoring the second term on the left-hand side of Eq.(2.11.47), we may find

$$P = \frac{E}{1-\nu}(c_o - 2\alpha\theta + e_{zz}) \qquad r < b \tag{2.11.57}$$

and from Eq.(2.11.44)

$$f = \frac{E}{1-\nu^2}[c_o - (1+\nu)\alpha\theta + \nu e_{zz}] \qquad r < b \tag{2.11.58}$$

for optical fibers with continuously distributed index profiles, such as graded-index optical fibers. The undetermined quantity c_0 can be found by using the free-stress boundary condition at the outer radius b, as

$$c_o = \frac{\int_0^b \{E[(1+\nu)\alpha\theta - \nu e_{zz}]/(1-\nu^2)\} r\,dr}{\int_0^b [E/(1-\nu^2)] r\,dr} \tag{2.11.59}$$

We may notice that no restriction has been made here for the thermal expansion coefficient α, which can have relatively large variation with the radius r.

To determine the axial strain component e_{zz}, we may use St. Venant's principle [Fung (1965)] and have the following condition for the resultant force

$$\int_0^b 2\pi r t_{zz} dr = F \tag{2.11.60}$$

Thus, by Eqs.(2.11.47), (2.11.50), (2.11.51), and (2.11.58), we have from Eq.(2.11.60)

$$\int_0^b \frac{c_o E\nu}{1-\nu^2} r\,dr - \int_0^b \frac{E\alpha\theta}{1-\nu} r\,dr + \int_0^b \frac{E e_{zz}}{1-\nu^2} r\,dr = \frac{F}{2\pi} \tag{2.11.61}$$

for optical fibers with continuously distributed index profile, or

$$\frac{a^2 E_1 \nu_1}{2(1-\nu_1^2)} c_1 + \frac{(b^2 - a^2) E_2 \nu_2}{2(1-\nu_2^2)} c_2 - \int_0^b \frac{E\alpha\theta}{1-\nu} r dr + \int_0^b \frac{E e_{zz}}{1-\nu^2} r dr = \frac{F}{2\pi} \tag{2.11.62}$$

for optical fibers with step index profile.

By inserting c_o from Eq.(2.11.59), or c_1 and c_2 from Eqs.(2.11.54) and (2.11.55), we can obtain the equation for the determination of e_{zz}. For instance, for optical fibers with continuously distributed index profile, we may find Eqs.(2.11.59) and (2.11.61)

$$e_{zz} = \frac{\int_0^b [E\alpha\theta/(1-\nu)] r dr}{\int_0^b [E/(1-\nu)] r dr} \tag{2.11.63}$$

provided that the ends of the fiber are free from external force ($F = 0$).

Some special cases of the optical glass fiber with ends free of forces ($F = 0$) may be of interest. If the elastic and thermal moduli of the glass fiber are all constant and the temperature is uniform in the cross section of the fiber, we have then

$$e_{zz} = \alpha\theta, \qquad f = 0, \qquad c_o = \alpha\theta \tag{2.11.64}$$

from Eqs.(2.11.63), (2.11.58), and (2.11.59), which implies $t_{rr} = t_{\varphi\varphi} = t_{zz} = 0$. The result shows that no thermoelastic stresses present in the glass fiber having homogeneous elastic and thermal moduli in a uniform temperature field, as we may expect.

If, however, the fiber with homogeneous thermoelastic properties is placed in a non-uniform temperature field, we have

$$e_{zz} = \frac{2\alpha}{b^2} \int_0^b \theta r dr \tag{2.11.65}$$

$$c_o - \frac{2\alpha}{b^2} \int_0^b \theta r dr \tag{2.11.66}$$

$$f = \frac{E\alpha}{1-\nu} \left(\frac{2}{b^2} \int_0^b \theta r dr - \theta \right) \tag{2.11.67}$$

from which the thermoelastic stresses in the glass fiber can be obtained by

$$t_{rr} = \frac{E\alpha}{1-\nu}\left(\frac{1}{b^2}\int_0^b \theta r dr - \frac{1}{r^2}\int_0^r \theta r' dr'\right) \tag{2.11.68}$$

$$t_{\varphi\varphi} = \frac{E\alpha}{1-\nu}\left(\frac{1}{b^2}\int_0^b \theta r dr + \frac{1}{r^2}\int_0^r \theta r' dr' - \theta\right) \tag{2.11.69}$$

$$t_{zz} = \frac{E\alpha}{1-\nu}\left(\frac{2}{b^2}\int_0^b \theta r dr - \theta\right) = t_{rr} + t_{\varphi\varphi} \tag{2.11.70}$$

This result is in accordance with the result given by Timoshenko and Goodier (1970).

For step-index optical fibers in a uniform temperature field, we may find from Eqs.(2.11.54)–(2.11.56)

$$c_1 = \frac{1}{d_o}(A_1 - Q_1 e_{zz}) \tag{2.11.71}$$

$$c_2 = \frac{1}{d_o}(A_2 + Q_2 e_{zz}) \tag{2.11.72}$$

where the parameters: A_1, Q_1, A_2 and Q_2 are given, respectively, by

$$A_1 = \frac{\left\{(b^2 - a^2)\left[1 - \frac{(1+\nu_1)E_2}{(1+\nu_2)E_1} + \frac{2(1-\nu_1)E_2\alpha_2}{E_1\alpha_1}\right] + 2a^2(1-\nu_2)\right\}E_1\alpha_1\theta}{4(1-\nu_1)(1-\nu_2)} \tag{2.11.73}$$

$$Q_1 = \frac{\left\{(b^2 - a^2)\left[1 + \nu_2 - \frac{(1+\nu_1)E_2}{E_1} + \frac{2(1-\nu_1^2)E_2\nu_2}{E_1\nu_1}\right] + 2a^2(1-\nu_2^2)\right\}E_1\nu_1}{4(1-\nu_1^2)(1-\nu_2^2)} \tag{2.11.74}$$

$$A_2 = \frac{\left\{A_1\left[\frac{(1-2\nu_1)E_2}{E_1} + \frac{1+\nu_2}{1+\nu_1}\right] - d_o\alpha_1\theta\left[1 + \nu_2 - \frac{(1+\nu_1)E_2}{E_1}\right]\right\}E_1(1+\nu_1)}{2(1-\nu_1^2)E_2} \tag{2.11.75}$$

$$Q_2 = \frac{d_o \nu_1[(1+\nu_2)E_1 - (1+\nu_1)E_2] - Q_1[(1+\nu_1)(1-2\nu_1)E_2 + (1+\nu_2)E_1]}{2(1-\nu_1^2)E_2} \tag{2.11.76}$$

The axial strain component e_{zz} can be found from Eq.(2.11.62) for $F = 0$ and from Eqs.(2.11.71) and (2.11.73) by

$$e_{zz} = \frac{a^2(1-\nu_2^2)E_1[(1+\nu_1)\alpha_1\theta d_o - A_1\nu_1]}{a^2(1-\nu_2^2)E_1(d_o - Q_1\nu_1) + (b^2-a^2)(1-\nu_1^2)(d_o + Q_2\nu_2)E_2}$$
$$+ \frac{(b^2-a^2)(1-\nu_1^2)[(1+\nu_2)\alpha_2\theta d_o - A_2\nu_2]E_2}{a^2(1-\nu_2^2)E_1(d_o - Q_1\nu_1) + (b^2-a^2)(1-\nu_1^2)(d_o + Q_2\nu_2)E_2} \tag{2.11.77}$$

It is shown that thermal stresses may exist in an inhomogeneous optical fiber.

Numerically, we may consider an example of a step-index optical fiber, which has the following properties: E_1 = 67 Gpa, $\nu_1 = 0.2$, $\alpha_1 = 32\times10^{-7}/^{o}\mathrm{C}$ for the core of the optical fiber, and E_2 = 71 Gpa, $\nu_2 = 0.134$, $\alpha_2 = 4.1\times10^{-7}/^{o}\mathrm{C}$ for the cladding of the optical fiber. The geometry of the optical fiber is characterized by its core radius a = 4.3 μm and its outer radius of the cladding b = 62.5 μm. The optical fiber with $F = 0$ is supposed to be cooled from its reference temperature T_o = 520 °C to the room temperature of T = 20 °C. Thus, we have $\theta = T - T_o$ = –500 °C. Thus, we may find that there may exist thermal residual stresses in the step-index optical fiber. In particular, for the numerical example, it is found numerically that the axial stress t_{zz} is compressive (t_{zz} = –0.56 Mpa) in the cladding, but tensile (t_{zz} = 117 Mpa) in the core of the optical fiber. Such a compressive stress in the cladding is sometime useful to strengthen the optical fiber by reducing the driving force of opening of the surface cracks of the glass fiber [Krohn and Cooper (1969)]. The thermal stress components t_{rr} and $t_{\varphi\varphi}$ are found to be equal constant (= 62.3 Mpa) and tensile in the core of the optical fiber, but they vary with the radius r in the cladding of the fiber [Zhou (1990)]. The radial stress component t_{rr} is also tensile in the cladding of the fiber, while the circumferential stress component $t_{\varphi\varphi}$ is compressive in the cladding of the fiber.

It should be noticed that the calculation given in the above numerical example does not account for the effect of possible temperature-dependence of elastic and thermal properties of the optical glass fiber. Since a large temperature interval of several hundred degrees is usually involved in the analysis of thermal residual stresses, where the reference temperature T_o may be taken to be, for instance, at the strain point temperature (about 500 °C for borosilicates), we may have to take into account of the effect of temperature-dependence of thermoelastic properties of the glass fiber [Zhou (1990)].

It has been shown that thermal stresses may exist in an inhomogeneous optical fiber, such as an step-index optical fiber, even in a uniform temperature field. The presence of thermal stresses in an optical fiber may cause the index-profile distortion of the optical fiber due to the photothermoelastic effect. For triaxial stresses in an optical fiber, we may find from Eqs.(2.8.31)–(2.8.33) that the index change of the optical fiber is given by

$$\Delta n_r = \pi^*\theta + B_1 t_{rr} + B_2(t_{\varphi\varphi} + t_{zz}) \tag{2.11.78}$$

$$\Delta n_\varphi = \pi^*\theta + B_1 t_{\varphi\varphi} + B_2(t_{rr} + t_{zz}) \tag{2.11.79}$$

$$\Delta n_z = \pi^*\theta + B_1 t_{zz} + B_2(t_{\varphi\varphi} + t_{rr}) \tag{2.11.80}$$

As a numerical example, we may use the result of thermal stresses obtained in above section for the step-index optical fiber with an unstressed step-index profile: $n_1 = 1.4679$ for the core, and $n_2 = 1.46335$ for the cladding. We take the stress-optical coefficients measured by Primak and Post (1959) for fused silica: $B_1 = -6.5\times10^{-7}$/Mpa, and $B_2 = -4.2\times10^{-6}$/Mpa. The negative sign of the stress-optical coefficients implies that the refractive index decreases when the stress is tensile, and increases when the stress is compressive. The numerical results are shown illustratively in Fig. 2.10 [Zhou (1990)]. Here, we have ignored the thermal-optical term ($\pi^*\theta$). In some cases, we may, however, notice the significance of the thermal-optical effect in, for instance, the study of thermally induced optical phase effects in optical fibers [Hughes and Priest (1980)].

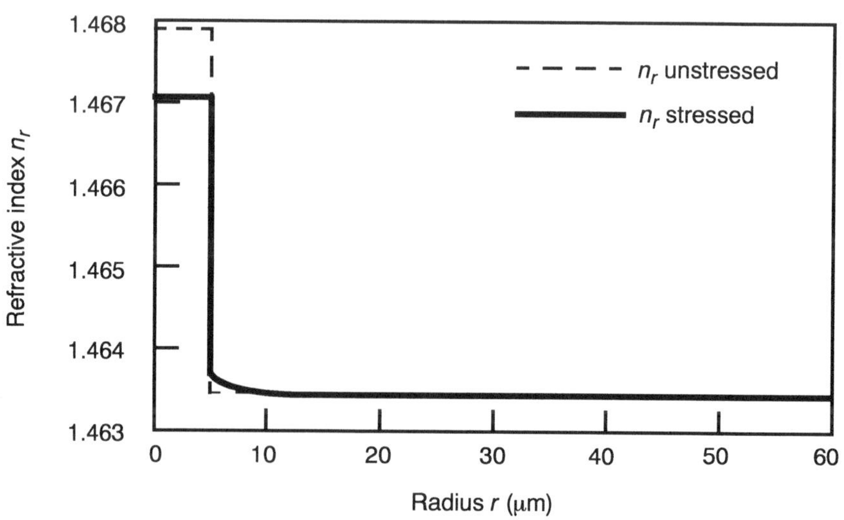

Figure 2.10 Index profile distortion due to stresses in a step-index optical fiber.

3

Electrodynamics of Superconductors in Weak Fields

The discovery of superconductivity and its recent advances in high-temperature superconducting materials have led to the rapid development and advancement of superconductivity technology. Although the development of quantum mechanics in the 1920s led to the understanding of the normal process of electrical conduction in metals and the BCS microscopic mechanism in 1957 for conventional superconductors, our understanding of the microscopic mechanism of recently discovered high-temperature oxide superconductors remains uncompleted so far. In order to develop the new technology utilizing superconducting materials, a complete appreciation of the macroscopic electromagnetic behaviors of superconductors is needed before one can undertake the most basic engineering calculations for any concrete problem of applications. This situation encouraged the development of phenomenological theories, which had established that most superconducting phenomena can be derived from a small number of empirical postulates. The purpose of this chapter is therefore to introduce some basic phenomena of superconductivity, the macroscopic electrodynamic theory for superconductors in weak fields, and the well-known BCS microscopic mechanism on superconductivity. Some examples of practical interest to engineering applications are also discussed.

3.1 BASIC PHENOMENA OF SUPERCONDUCTIVITY

3.1.1 Zero Resistance and Transition Temperature

It is known that the electric current in a conductor is carried by conduction electrons which are free to move through the material. These conduction electrons have a wave-like nature, and an electron traveling through a material can be represented by a plane wave progressing in the same direction. In a perfect crystal, each atom has a definite equilibrium location in regular array. A property of the plane wave is that it can pass through a perfectly periodic structure without being scattered into other directions. Hence an electron is able to pass through a perfect crystal without any loss of momentum in its original direction. In other words, if in a perfect crystal we start a current flowing, the current will experience no resistance. However, any fault in the periodicity of the crystal will scatter the electron wave and introduce some resistance. Actual crystals are never perfect. Defects such as missing atoms, atoms out of place, irregularities in the spacing of rows of atoms and the presence of impurities may exist in the crystal. Besides, at temperatures above absolute zero, the atoms are vibrating and will be displaced by various amounts from their equilibrium positions, interrupting the perfect periodicity of a crystal. Both the thermal vibrations and any impurities or imperfections scatter the moving conduction electrons and give rise to electrical resistance.

At present, electrical conductors, even the very best, resist to some extent the flow of charge through them at ordinary room temperature. When the temperature of a metal or alloy is lowered, its electrical resistivity decreases generally since the thermal vibrations of the atoms decrease and the conduction electrons are less frequently scattered. However, even at absolute zero, any real specimen of metal containing imperfections has a certain residual resistivity. The more impure the metal, the larger will be its residual resistivity. Such a qualitative behavior of electrical conductors was considered to be obvious for all metals or alloys until the discovery of superconductivity by the Dutch physicist, Heike Kamerlingh Onnes in 1911.

Onnes found that, down to 4.2 K, the resistance of a mercury sample decreased with temperature as other metals did. At the temperature of about 4.2 K, though, its resistance fell sharply to as close to zero as his instruments could measure (see Fig. 3.1). The mercury has practically a zero dc-resistance value at temperatures below 4.2 K. This new phenomenon of zero-resistance at low

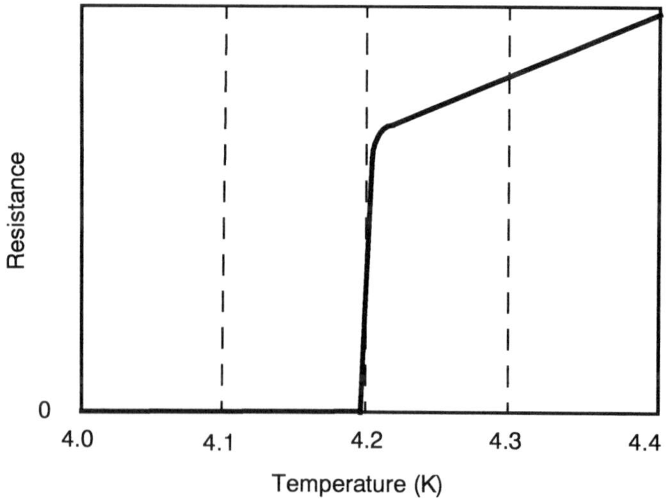

Figure 3.1 Resistance of a mercury sample at low temperature.

temperature was soon found in many other metals. In particular, the phenomenon of the zero dc-resistance may occur even if the metal is so impure that it would otherwise have had a large residual resistivity. H. K. Onnes received the Nobel prize in physics in 1913 for his work in "researches on the properties of matter at low temperatures."

An important characteristic of the loss of dc-resistance observed is the sharpness of the transition. Careful experiments by de Hass and Voogd (1931) suggested that in "ideal' conditions the transition from the normal resistance state of the testing material to its superconducting state would be practically discontinuous. The temperature at which superconductivity first occurs in a material is thus termed the superconducting transition (or critical) temperature of the material and is denoted by T_c. The transition temperatures of some materials are, for instance, 1.2 K for Aluminum, 7.2 K for Lead, 9.2 K for Niobium, 18.1 K for Nb_3Sn, and 23.2 K for Nb_3Ge.

The highest transition temperature of superconductors did not over 23.2 K found in the compound Nb_3Ge until 1986 when two scientists, J. G. Bednorz and K. A. Muller (1986) reported possible superconductivity in a mixture of La and Ba copper oxides at temperature of 30 K. Following their discovery, which led them to receive the 1987 Nobel prize in Physics, the high-T_c superconductors have, so far, been found in various ceramic oxides having the transition temperatures of, for instance, 95 K for $YBa_2Cu_3O_{7-\delta}$, 125 K for $Tl_2Ba_2Ca_2Cu_3O_{10}$, and 133 K for $HgBa_2Ca_2Cu_3O_{8+\delta}$. The importance of these discoveries is based on the fact that the superconductivity in such ceramics can be maintained in cryostats with relatively cheap liquid nitrogen (boiling point 77.4 K) rather than liquid helium so that large-scale applications of the superconductivity become economically viable.

Practical measurement of the transition temperature T_c of a superconductor is usually made with the use of the following three basic methods. The first one is to measure the change in resistance of the material. For pure samples containing few metallurgical defects, the transition is sharp, with a typical width less than 0.01 K, and changes little with current level. Inhomogeneous samples, however, have broad transitions. In this case, T_c is usually defined as the point where the resistivity of the testing material decreases to half its normal value. The primary problem with the resistance method is the current-dependent shift of T_c for the inhomogeneous samples. Therefore, the resistance method is best suited to homogeneous samples. The second method is to measure the change in magnetic permeability of the material, which is based on the fact of the large decrease in magnetic permeability of the material as it passes into the superconducting state. The advantage of this method is that it requires no direct electrical contacts with the sample, and consequently, it can be used for the samples which are small, irregular or in powder form. However, this method has its disadvantage of that the highest T_c component of an inhomogeneous sample can shield lower T_c components, thus dominating the measurement. The third method utilizes the fact that the specific heat of a superconductor shows a discontinuity at the transition temperature T_c.

Since the discovery of superconductivity, one has once wondered that whether in the superconducting state the resistance has indeed become zero or whether it has merely fallen to a very small value. Of course, it can never be proven directly by experiment that the resistance is in fact zero; the resistance of any specimen may always be just less than the sensitivity of our apparatus that allows us to detect. Indirectly, experiments of measuring the decay of currents flowing in superconducting wire loops have been made, in which currents have been set up in superconducting wire loops and the resulting magnetic fields monitored, sometimes for years. No decrease in such currents has ever been found: superconductors do have no resistance at all!

It is tempting to suppose that the superconducting electrons (superelectrons in short) in a superconductor behave like electrons in vacuum. The electrons in the beam of a cathode ray tube, for example, are resistanceless in the sense that they flow without undergoing any collisions. There is, however, a significant difference between the two cases. It is possible to maintain a potential drop along an electron beam while the current remains at a constant value. This is because, though the current must be the same all along the path of the beam, the electron density need not remain constant. Hence the electrons accelerate from the cathode towards the anode and the electron density is relatively high near the cathode and decreases as the anode is approached. However, the product of the electron density and electron velocity, that is, the current, remains constant along the beam. The fact that the electrons are able to accelerate allows us to maintain the electric field. In a superconductor, however, conditions are different. The superconductor must remain everywhere electrically neutral and, since the positive ions are fixed in the crystal, the electron density cannot vary through the material. Hence for a constant current to be maintained through the superconductor, the velocity of all the electrons along the current path must be the same. The electrons, therefore, do not accelerate and an electric field cannot exist in the superconductor carrying the constant current.

The fact that there is no voltage drop along a superconductor when a current is passed through it, and no power is generated by the passage of the current is only strictly true for direct current of constant value. If the current is changing, an electric field is developed and some power is dissipated. It has been found that the behavior of zero dc-resistance of a superconductor at temperatures below its critical temperature is modified significantly at very high frequencies of alternating current up to 100 MHz, as shown by London (1940). When the frequency went to the infra-red frequencies of the order 10^{13} Hz, the resistance of the superconductors would be the same and independent of temperature both in the normal and superconducting state.

This behavior of superconductors may be explained by a two-fluid model, which was suggested by thermodynamic arguments based on the results of specific heat and similar measurements on superconductors. In the two-fluid model, a superconductor below its transition temperature appears to be permeated by two electron fluids, one of normal electrons, which can be scattered and so experience resistance, and one of superelectrons, which can pass through the superconductor without resistance. The relative electron density in the two fluids depends on the temperature. At absolute zero, all conduction electrons behave like superelectrons, but, if the temperature is raised, a few begin to behave as normal electrons, and on further heating the proportion of normal electrons increases. Eventually, at the transition temperature, all the electrons have become normal electrons and the material loses its superconductive properties.

According to the two-fluid model, in a superconductor, the current can be carried generally by both the normal electrons and superelectrons. However, in the special case of a constant dc current, all the current is carried by the superelectrons. This is because if the current is to remain constant, there must be no electric field in the superconductor, otherwise the superelectrons would be accelerated continuously in this field and the current would increase indefinitely. If there is no field there is nothing to drive the normal electrons and so there is no normal current. In an alternating field in which current is changing with time, an electric field must be present to accelerate the electrons due to the *inertia* of electrons. Because there now is an electric field present, some of the current will be carried by the normal electrons, causing power loss by the resistance resulting from their being scattered in the material. Since the inertia of an electron is very small and so, unless we go to extremely high frequencies, only a very low electric field presents and a tiny fraction of the current is carried by the normal electrons. Thus, the corresponding power loss is minute. However, if the frequency of an applied field is sufficiently high, a superconductor responds in the same way as a normal conductor. The behavior of a superconductor at optical frequencies is therefore no different from that of a normal metal, and there is, for instance, no change in the visual appearance of a superconductor as it is cooled below its transition temperature.

3.1.2 Critical Magnetic Field

The discovery of zero dc-resistance of superconductors in 1911 had led to interests in producing intense magnetic fields with the aid of superconducting coils without iron cores. However, it was soon found by H. K. Onnes in 1913 that the observed superconducting behavior was quenched in the presence of small current densities and quite ordinary levels of magnetic fields. Onnes thus found another important properties of superconductors, namely, that when the superconductor is placed in a sufficiently strong magnetic field, superconductivity can be destroyed. The superconductivity, however, reappears when the applied magnetic field is removed. The minimum magnetic field required to destroy the superconductivity depends on the shape and orientation of the specimen at a given temperature. If the specimen is in the shape of a long cylinder and its axis is placed parallel to the applied magnetic field, the transition is sharp and the minimum magnetic field required to destroy the superconductivity is called the critical (magnetic) field and denoted by B_c.

Within only small deviations, the temperature dependence of B_c was found to be well represented by a parabolic relation, as shown in Fig. 3.2,

$$B_c(T) = B_{co}\left(1 - \left(\frac{T}{T_c}\right)^2\right) \tag{3.1.1}$$

where B_{co} denotes the critical field at zero temperature and T_c the superconducting transition temperature, which are properties of superconductors. For most superconducting elements, B_{co} is of the order of 10^{-2} tesla. Table 3.1 lists some of well-known superconducting elements and their characteristic B_{co} and T_c. The practical determination of the critical field is usually from the measurements of the superconducting magnetization curves, and the critical field is simply the field at which the magnetization becomes zero.

The experimental evidence of that the current density in a superconducting wire could only be increased to a certain threshold value until a voltage occurred had led the introduction of the concept of critical current density. According to a hypothesis put forward by Silsbee in 1916, the threshold values of the current and the magnetic field are simply related: The magnetic field produced by the critical current J_c at the surface of the superconductor equals the critical magnetic field, which gives $J_c(T) = B_c(T)/(\mu_o\lambda(T))$. λ is the field penetration depth to be discussed later.

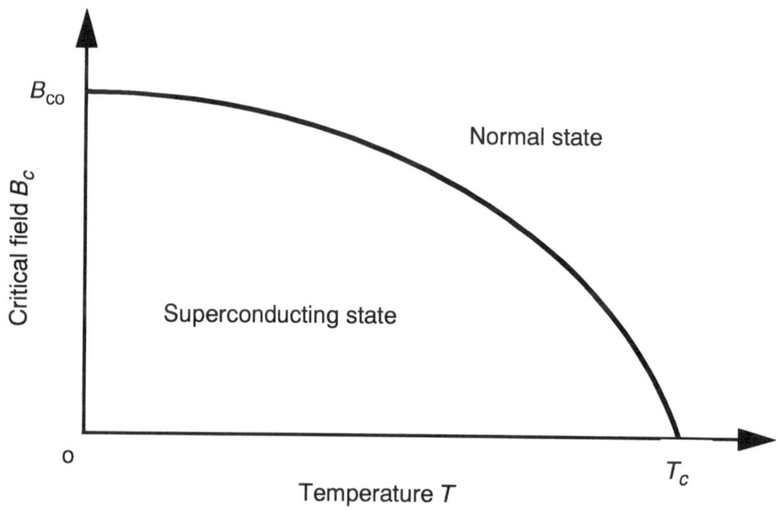

Figure 3.2 The critical magnetic field as a function of temperature.

Table 3.1 Some Superconducting Elements

Element	T_c (K)	B_{co} (Tesla)
Aluminum	1.2	0.99×10^{-2}
Gallium	1.1	0.52×10^{-2}
Indium	3.4	2.76×10^{-2}
Lead	7.2	8.04×10^{-2}
Mercury α	4.2	4.15×10^{-2}
Mercury β	4.0	3.39×10^{-2}
Niobium	9.3	0.2
Rhenium	1.7	2.01×10^{-2}
Tantalum	4.5	8.30×10^{-2}
Thalium	2.4	1.76×10^{-2}
Tin	3.7	3.02×10^{-2}
Vanadium	5.4	0.141
Zinc	0.9	0.53×10^{-2}
Techneitium	7.8	0.141
Zirconium	0.6	0.47×10^{-2}

3.1.3 The Meissner Effect

For many years after the discovery of superconductivity, it was tacitly assumed that the electromagnetic behaviors of a superconductor could be deduced from its infinite conductivity (i.e., zero resistivity). According to classical theory of electrodynamics, the infinite conductivity implies that the electric field $\boldsymbol{E}$ is zero inside a superconductor according to Ohm's law. It follows that $\nabla \times \boldsymbol{E} = 0$ and hence by one of Maxwell's equations (1.4.18) that $\partial \boldsymbol{B}/\partial t = 0$, which means that the magnetic induction $\boldsymbol{B}$ inside is constant in time, and would be, therefore, dependent of the past history of the state of the superconductor. For instance, the final state of a perfect conductor depends on whether the material is cooled first and then the magnetic field $\boldsymbol{B}_a$ is applied, or vice versa. This property of superconductor and its consequences were taken for granted until 1933 when Meissner and Ochsenfeld (1933) measured the field surrounding a superconductor and concluded that the magnetic induction field $\boldsymbol{B}$ inside a macroscopic specimen of pure superconductor always be zero (strictly speaking, the experiment showed that it was very small compared with its value in the normal state), independent of initial conditions (See Fig. 3.3). Here, the macroscopic specimen means that its size is large enough so that the size effect is not important since, otherwise, the magnetic behavior would be seriously modified if the specimen size becomes comparable to the penetration depth ($\sim 10^{-7}$ m), which will be discussed later.

The absence of any magnetic flux in a pure superconductor independent of the initial conditions is an additional fundamental property of the superconductor since it cannot be deduced from the perfect conductivity. This means that the superconductor behaves actually not simply like a perfect conductor (which implies only $\partial \boldsymbol{B}/\partial t = 0$) but also like a perfect diamagnetic ($\boldsymbol{B} = 0$). This fundamental property of the superconductor is now called the Meissner effect. In the thermodynamic sense, the Meissner effect indicates that the superconductive state in a given external magnetic field is a single stable state to which the laws of thermodynamics apply, and the magnetically induced transition between the superconducting state and the normal state of a superconductor is, in principle, a reversible phase transition.

The fascinating phenomena of superconductivity and their potential applications had attracted not only many experimentalists for finding new superconductors, but also attracted many theoreticians in order to understand the physical phenomena of superconductivity and to describe the electromagnetic and thermodynamic behaviors of superconductors. Extensive theoretical studies have been made along the two main lines, the microscopic theories were developed to study the possible mechanism of superconductivity and its properties, while the macroscopic theories were introduced to describe

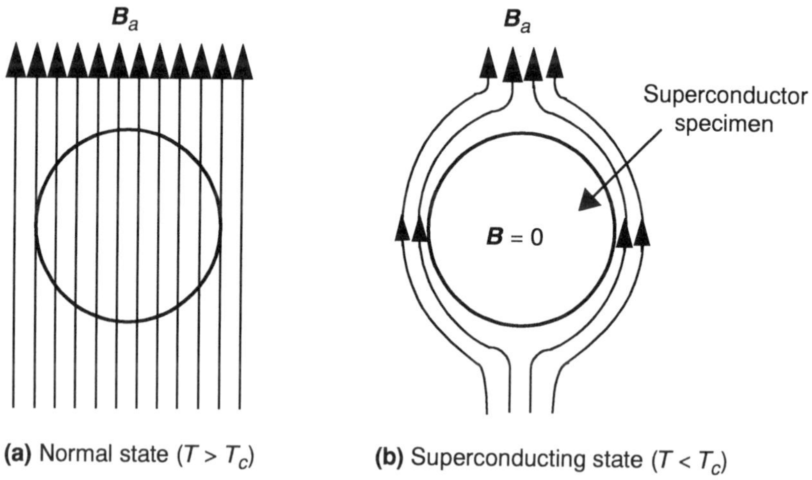

Figure 3.3 The Meissner effect of a superconductor.

phenomenologically the macroscopic behaviors of superconductors. We shall first start with the introduction of the well-known macroscopic theory for superconductors developed by the brothers F. and H. London in 1935. The microscopic theory of superconductivity, developed by Bardeen, Cooper and Schrieffer in 1957, will be discussed briefly in the end of this chapter.

3.2 THE LONDON THEORY OF SUPERCONDUCTORS

3.2.1 Free Electron Model

Before the discovery of the Meissner effect in superconductors, Becker et al. (1933) analyzed the electrodynamic behaviors of superconductors as if they were perfect conductors with the use of a simple free electron model, in which the electrons accelerate without any resistance under the exertion of an electric field $\boldsymbol{E}$. Becker, Heller, and Sauter argued that if the electrons encountered no resistance, an applied electric field would accelerate them steadily according to Newton's law:

$$m^* \frac{d\mathbf{v}_s}{dt} = e^* \boldsymbol{E} \tag{3.2.1}$$

where m^* and e^* are, respectively, the effective mass and charge of the superelectrons in the phenomenological theory. (It was found later by the BCS theory that the superelectrons are formed in the Cooper pairs in superconductors so that we have $e^* = 2e$, where e denotes the charge of one electron, being negative, and $m^* = 2m_e$ with m_e being the mass of one electron.)

If there are n_s numbers of superelectrons per unit volume moving with the mean local velocity $\mathbf{v}_s$, we may introduce a supercurrent density $\boldsymbol{J}_s$ by

$$\boldsymbol{J}_s = n_s e^* \mathbf{v}_s = -n_s |e^*| \mathbf{v}_s \tag{3.2.2}$$

where the minus sign indicates that the supercurrent flows in a direction opposite to the direction of the motion of the superelectrons according to the definition. Substituting this equation into Eq.(3.2.1) and taking a linear approximation, we may obtain the following equation:

$$\mu_o \lambda_L^2 \frac{\partial \boldsymbol{J}_s}{\partial t} = \boldsymbol{E} \tag{3.2.3}$$

where λ_L is a parameter of length dimension defined by

$$\lambda_L = \sqrt{\frac{m^*}{\mu_o n_s e^{*2}}} \tag{3.2.4}$$

By taking the curl of both sides of Eq.(3.2.3) and using Maxwell's equations (1.4.18) for the superconductor, we may then write inside a superconductor

$$\mu_o \lambda_L^2 \frac{\partial}{\partial t}(\nabla \times \boldsymbol{J}_s) = -\frac{\partial \boldsymbol{B}}{\partial t} \tag{3.2.5}$$

Since, for non-magnetic superconductors, any flux density in the superconducting body must be due to the current, we can obtain

$$\lambda_L^2 \frac{\partial}{\partial t}(\nabla \times (\nabla \times \boldsymbol{B})) = -\frac{\partial \boldsymbol{B}}{\partial t} \tag{3.2.6}$$

from Eqs.(3.2.5) and (1.4.16) with the neglection of displacement current at the magneto-quasistatic approximation. Equation (3.2.6) may also be written, by noting the divergence of $\boldsymbol{B}$ being zero, in the form:

$$\lambda_L^2 \nabla^2 \left(\frac{\partial \boldsymbol{B}}{\partial t}\right) = \frac{\partial \boldsymbol{B}}{\partial t} \tag{3.2.7}$$

which implies that the magnetic field would fall off exponentially inside a superconductor to a "trapped" field $\boldsymbol{B}_0$, depending on the initial state of the superconductor. This prediction is clearly contradicted to the observed Meissner´s effect which states that the internal magnetic induction field in an ideal superconductor is always zero, independent of the initial states. Thus, the brothers H. and F. London (1935) suggested that since the macroscopic theory of a perfect conductor makes correct predictions about superconductors for the special case $\boldsymbol{B}_0 = 0$, it might be reasonable to suppose that the magnetic behavior of a superconductor may be correctly described according to the Meissner effect if Eq.(3.2.7) applied not only to $\partial \boldsymbol{B}/\partial t$ but to $\boldsymbol{B}$ itself, that is,

$$\lambda_L^2 \nabla^2 \boldsymbol{B} = \boldsymbol{B} \tag{3.2.8}$$

Equivalently, we may also write

$$\boldsymbol{B} = -\mu_o \lambda_L^2 \nabla \times \boldsymbol{J}_s \tag{3.2.9}$$

It may be noticed that, in the London assumption, $\boldsymbol{B}_0 = 0$ has been introduced arbitrarily to agree with experiment, but we shall see later that the exclusion of other possible values of $\boldsymbol{B}_0$ is in accordance with the idea of the superconductor as the macroscopic quantum mechanism.

The set of equations (3.2.3) and (3.2.9) describes thus the electromagnetic behaviors of the superconductor according to the London theory. Equation (3.2.3) is now called the first London equation, describing the resistanceless property of a superconductor, and Eq.(3.2.9) is called the second London equation, characterizing the Meissner effect. It can be seen from Eq.(3.2.3) that there is no electric field in a superconductor unless the current is changing with time. The only material parameter λ_L introduced in the model is called the London penetration depth, which is assumed to be constant in space in the London theory. In particular, λ_L is assumed to be independent of the strength of the applied magnetic field and also of the dimensions of the specimen though its value may depend on temperatures.

By introducing a magnetic vector potential $\boldsymbol{A}$ defined by $\boldsymbol{B} = \nabla\times\boldsymbol{A}$ and by choosing a proper gauge, we may also write from Eq.(3.2.9)

$$\boldsymbol{J}_s = -\frac{1}{\mu_o\lambda_L^2}\boldsymbol{A} \tag{3.2.10}$$

which gives a special form of the second London equation being somewhat reminiscent of Ohm's law. It should, however, be noticed that the relation (3.2.10) is valid only for problems involving simply connected superconductor due to the specially chosen gauge (the London gauge: $\nabla\cdot\boldsymbol{A} = 0$ and $\boldsymbol{A}\cdot\boldsymbol{n} = 0$ on the boundary surface of the superconductor with $\boldsymbol{n}$ being the unit normal vector).

An important question may arise. Whether can either of the London equations (3.2.3) or (3.2.9) be regarded as more fundamental than the other? Since these two equations describe the zero dc resistance and the Meissner effect, respectively, it is evidently important to decide whether it is at all possible to say that one of these effects is a consequence of the other or whether they must be regarded as entirely separate phenomena requiring separate explanations in a fundamental theory. A good discussion about this question was given by Shoenberg (1952).

We have already seen that the first London (acceleration) equation (3.2.3) leads not to Eq.(3.2.9) but only to its time derivative (3.2.5). The omission of the integration constant which gives Eq.(3.2.9) is thus a separate assumption. If, instead, we start from Eq.(3.2.9) and use, in addition, the Maxwell equation $\nabla\times\boldsymbol{E} = -\partial\boldsymbol{B}/\partial t$, we may find

$$\nabla\times\left(\boldsymbol{E} - \mu_o\lambda_L^2\frac{\partial\boldsymbol{J}_s}{\partial t}\right) = 0$$

which is, of course, weaker than Eq.(3.2.3) and shows only that

$$\boldsymbol{E} - \mu_o\lambda_L^2\frac{\partial\boldsymbol{J}_s}{\partial t} = \nabla\phi$$

where ϕ is a scalar. For a simply connected superconductor, there must be no component of $\nabla\phi$ in the direction of $\boldsymbol{J}_s$, for otherwise, there would be energy dissipation at the expense of the static magnetic field, which would violate conservation of energy. In this case then, if we ignore any component of $\nabla\phi$

perpendicular to $\boldsymbol{J}_s$, we can obtain the first London equation (3.2.3) from the second London equation (3.2.9). Just, here, however, the first London equation (3.2.3) is of no interest since Eq.(3.2.9) is by itself completely adequate to describe the behavior of a simply connected body with no currents led in from outside. It is for a multiply-connected body such as a ring, or for a current fed in from outside, that it is important to prove Eq.(3.2.3), and this turns out to be impossible without further hypotheses. For a ring, for instance, there is nothing to indicate that a non-vanishing value of $\nabla\phi$ is inadmissible since ϕ may be a many-valued potential function, though such a value would, of course, mean that the current in the ring would die away and the flux locked in the ring would disappear. The fact that this does not happen must, therefore, be regarded as an experimental fact and cannot be deduced mathematically from Eq.(3.2.9).

From the above discussion, there is a suggestion that Eq.(3.2.9) is more fundamental than Eq.(3.2.3) since, for a simply-connected body, Eq.(3.2.3) can be deduced from Eq.(3.2.9) though the reverse is not possible. F. London has, however, shown that Eq.(3.2.9) can be expressed in a rather more significant way, as shown here by Eq.(3.2.10) for the special case of a simply-connected superconductor not fed by a current from outside. This result is in striking contrast to that for a normal conductor where it can be shown that we must have $\boldsymbol{J} = 0$. Thus, a superconductor is distinguished from a normal conductor in that something prevents the mean momentum $\boldsymbol{p}$ from assuming the local value of $e^*\boldsymbol{A}$. The mechanism for this prevention is essentially non-classical. In the following section, we shall, therefore, introduce a macroscopic quantum wave model for superconductors.

3.2.2 Macroscopic Quantum Wave Model

In the 1940s, Fritz London had realized that superconductivity is an inherently quantum mechanical phenomenon that manifests itself on macroscopic scales, and the equation (3.2.9) could also be derived from a more fundamental idea if the superelectron fluid could be treated as a quantum mechanical entity. In this section, we shall introduce the macroscopic quantum wave model for superconductors. We shall show that the model not only encompasses the classical London equations, but also has its importance consequence on the macroscopic quantum behavior of superconductors.

In the macroscopic quantum wave model, a fundamental assumption adopted is that there exists a macroscopic (ensemble average) wave function, $\psi(\boldsymbol{x}, t)$, with a well-defined amplitude and phase angle that can be used to describe the behavior of entire ensemble of superelectrons in the superconductor. Explicitly, the macroscopic wave function is expressed by

$$\psi = |\psi| e^{i\theta} \tag{3.2.11}$$

where θ is a real scalar function representing the phase of the wave function, and $|\psi|$ denotes the amplitude of the wave function, related to the density n_s of superelectrons by

$$|\psi|^2 = n_s \tag{3.2.12}$$

It is assumed here that $|\psi|^2$ is independent of position for superconductors in weak fields. Problems of possible spatial variation of distribution of superelectrons will be studied later.

Since for a particle of charge e^* and mass m^* with velocity $\mathbf{v}_s$ in the presence of a magnetic field, the classical canonical momentum is

$$\boldsymbol{p} = m^*\mathbf{v}_s + e^*\boldsymbol{A} \tag{3.2.13}$$

we may write the canonical momentum density as

$$n_s\boldsymbol{p} = n_s(m^*\mathbf{v}_s + e^*\boldsymbol{A}) \tag{3.2.14}$$

for a distribution of superelectrons with the density n_s, all having the same momentum,

Using the quantum-mechanical description in which $n_s\boldsymbol{p}$ is the expectation value of the canonical-momentum operator $-i\hbar\nabla$, operating on the macroscopic wave function (3.2.11), we can get the following relation:

$$\hbar\nabla\theta = m^*\mathbf{v}_s + e^*\boldsymbol{A} \tag{3.2.15}$$

which may also be written, by using Eq.(3.2.2), as

$$\frac{\hbar}{e^*}\nabla\theta = \mu_o\lambda_L^2\boldsymbol{J}_s + \boldsymbol{A} \tag{3.2.16}$$

where $\hbar$ (= 1.054×10^{-34} joule-second) is called the reduced Plank's constant.

We may recall that, in classical electrodynamics, the magnetic vector potential $\boldsymbol{A}$ as well as the electric scalar potential ϕ are not unique quantities, but are only defined in terms of their derivatives relating to the physical quantities of electric and magnetic fields, as shown in Section 1.4.5. Since the gauge transformation must leave all measurable quantities unchanged, the gauge transformation by Eqs.(1.4.34) and (1.4.35) will result in a necessary change of the local phase of the wave function (3.2.11) by

$$\theta \to \theta' = \theta + \frac{e^*}{\hbar}\Lambda \tag{3.2.17}$$

in order to leave the velocity $\mathbf{v}_s$ (a measurable quantity) unchanged.

To show the model is consistent with the classical London model, let us first take the curl of both sides of Eq.(3.2.16), which results immediately in the second London equation by noting $\boldsymbol{B} = \nabla \times \boldsymbol{A}$. We then take the partial derivative with respect to time of Eq.(3.2.16). We find

$$\frac{\hbar}{e^*}\nabla\left(\frac{\partial\theta}{\partial t}\right) = \mu_o\lambda_L^2\frac{\partial \boldsymbol{J}_s}{\partial t} + \frac{\partial \boldsymbol{A}}{\partial t} \tag{3.2.18}$$

Since the macroscopic wave function ψ is supposed to obey a Schrödinger-like equation for the ensemble of superelectrons in an electromagnetic field, we may find the following relation between the electric potential ϕ and the phase θ of the macroscopic wave function by

$$\frac{\hbar}{e^*}\frac{\partial\theta}{\partial t} = -\phi \tag{3.2.19}$$

provided that the density n_s is a constant and the effect of magnetic field is negligible, which is consistent with the assumptions used in the London theory at the linear approximation. Thus, by noting that the electric field $\boldsymbol{E}$ can be given by

$$\boldsymbol{E} = -\nabla\phi - \frac{\partial \boldsymbol{A}}{\partial t} \tag{3.2.20}$$

we recover, therefore, the first London equation (3.2.3) according to Eq.(3.2.18) which is derived from the macroscopic quantum model.

We shall now show an important consequence of the macroscopic quantum wave model, which is not embedded in the classical particle model for free electrons. Let us consider a multiply connected superconductor in a weak magnetic field, and let L be a closed loop located entirely within the superconductor and embracing one of these holes, as shown in Fig. 3.4. Thus, any open surface S bounded by L will be partially in a superconducting and partially in a non-superconducting region. Integrating Eq.(3.2.16) over the closed loop L, and applying Stokes´ theorem, we obtain

$$\frac{\hbar}{e^*}\oint_L \nabla\theta \cdot d\boldsymbol{L} = \Phi_s + \oint_L \mu_o\lambda_L^2\boldsymbol{J}_s \cdot d\boldsymbol{L} \tag{3.2.21}$$

where Φ_s is the magnetic flux defined by

$$\Phi_s = \int_S \boldsymbol{B} \cdot d\boldsymbol{S} \tag{3.2.22}$$

The physical requirement that the wave function ψ is a single-valued function at each point indicates that the term on the left-hand side of Eq.(3.2.21) can only

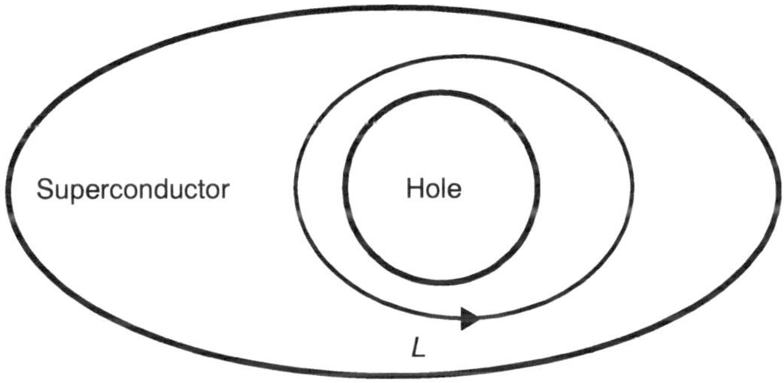

Figure 3.4 A multiply connected superconductor with one hole.

be of the form:

$$\frac{\hbar}{e^*} \int_L \nabla\theta \cdot d\boldsymbol{L} = n\Phi_o \qquad (n = 0, \pm 1, \pm 2, \ldots) \tag{3.2.23}$$

where Φ_0 is the flux quantum (or fluxon) defined by

$$\Phi_o = \frac{2\pi\hbar}{|e^*|} = 2.07 \times 10^{-15} \text{ (Wb)} \tag{3.2.24}$$

In Eq.(3.2.23), $n = 0$ corresponds to a simply connected superconductor.

In many cases, for instance, when the cross-section of the superconductor is large in comparison with the penetration depth, we may choose a loop with $\boldsymbol{J}_s$=0 on L so that the second term on the right-hand side of Eq.(3.2.21) vanishes, and we thus have

$$\Phi_s = n\Phi_o \qquad (n = 0, \pm 1, \pm 2, \ldots) \tag{3.2.25}$$

which shows the fact that the magnetic flux enclosed by the loop is quantized, that is, any magnetic flux contained within a superconductor should only exist as multiples of the flux quantum Φ_0. In general, $\boldsymbol{J}_s$ is generated to just produce the nearest integral number of fluxons [see Byers and Chang (1961)]. The predicted value of the fluxon is extremely small. Experimentally the value of the fluxon has been measured by Deaver and Fairbank (1961) and Doll and Nabauer (1961). The fact that the value of the fluxon is found to be, as predicted, Planck's constant divided by twice the electronic charge, is strong evidence that the supercurrent is correctly considered to be carried by pairs of electrons. The measurements also indicate the existence of the predicted quantum-mechanical effect on a large scale.

By noting the terms on the right-hand side of Eq.(3.2.21), we may introduce a quantity, called the fluxoid (London (1950)), defined by

$$\Phi_L = \int_S \boldsymbol{B} \cdot d\boldsymbol{S} + \oint_L \mu_o \lambda_L^2 \boldsymbol{J}_s \cdot d\boldsymbol{L} \tag{3.2.26}$$

Thus, within the approximation of the London theory, we may find

$$\frac{\partial \Phi_L}{\partial t} = \frac{\partial}{\partial t} \int_S \boldsymbol{B} \cdot d\boldsymbol{S} + \oint_L \boldsymbol{E} \cdot d\boldsymbol{L} = \int_S \left(\frac{\partial \boldsymbol{B}}{\partial t} + \nabla \times \boldsymbol{E} \right) \cdot d\boldsymbol{S} = 0 \tag{3.2.27}$$

by using Maxwell's equation (1.4.18), which is valid on the whole surface S. Equation (3.2.27) indicates that the fluxoid Φ_L is a constant in time. It can be seen that the fluxoid does not depend on the shape of the curve L as long as it embraces the same hole but once. This result comes from the fact that the fluxoid within any closed curve not surrounding a non-superconducting region is zero. Consequently, we have shown that the fluxoid is not a property of the curve L but rather one of the hole in question for the superconductor at rest.

3.2.3 Classical Two-Fluid Model

From the first London equation (3.2.3), we may notice that, for alternating fields, an electric field must be present to accelerate the electrons which have a small inertial mass and so the supercurrent does not rise instantaneously but only at the rate at which the electrons accelerate in the electric field, that is, the supercurrent will lag behind the field because of the inertia of the superelectrons. Since there now is an electric field present, some of the current will be carried by the normal electrons which always exist at finite temperatures according to the two-fluid model proposed by Gorter and Casimir (1934). Although the original intention of the two-fluid model is to study thermodynamic properties of superconductors, the combination of this model with London's theory has turned out to be of much wider application.

Noting the fact that the transition of the normal state to the superconducting state of a superconductor is not sudden, as the first-order transition from liquid to solid is, but more gradual, being a second-order transition without latent heat in the absence of magnetic field, it is assumed in the two-fluid model that the electrons in the superconductor may occupy either of two sets of states, the superconducting state in which electrons are paired and resistanceless, and the normal state in which they behave like normal conduction electrons. The electron assembly as a whole consists of these two inter-penetrating electron fluids having quite different properties. Thus, in the classical two-fluid model, it is postulated that conduction electrons in a superconductor below its critical temperature T_c can be divided into two distinct groups. A fraction of the conduction electrons is condensed into a superconducting aggregate while the remainder is in the normal state. The total electron density N is the sum of the densities of the superelectrons N_s (= $2n_s$) and the normal electrons N_n. Gorter and Casimir found that the best agreement with the thermal properties of superconductors could be obtained if the fraction was chosen in the form:

$$\frac{N_s}{N} = 1 - \left(\frac{T}{T_c}\right)^4, \qquad \frac{N_n}{N} = \left(\frac{T}{T_c}\right)^4 \tag{3.2.28}$$

which shows that the density of superelectrons rises rapidly (while the density of normal electrons decreases rapidly) when the temperature T falls below T_c.

In non-stationary processes, the presence of normal conduction electrons requires a modification of London's electrodynamic equations. In a simple model, we may assume that the dynamics of the conduction electron fluid in the presence of an electric field may be described by the following set of equations:

$$m^* \frac{d\mathbf{v}_s}{dt} = e^* \boldsymbol{E} \tag{3.2.29}$$

$$m_e \frac{d\mathbf{v}_n}{dt} + \frac{m_e}{\tau_n} \mathbf{v}_n = e\boldsymbol{E} \tag{3.2.30}$$

where $\mathbf{v}_s$ and $\mathbf{v}_n$ are the average velocities of the superelectrons and of the normal electrons, respectively. τ_n is the momentum relaxation time accounting for the effect of collisions of normal electrons.

By introducing the superconducting current density and the normal conduction current density, respectively, by

$$\boldsymbol{J}_s = N_s e \mathbf{v}_s \tag{3.2.31}$$

$$\boldsymbol{J}_n = N_n e \mathbf{v}_n \tag{3.2.32}$$

we may find that Eqs.(3.2.29) and (3.2.30) may be expressed in terms of the current densities as

$$\frac{\partial \boldsymbol{J}_s}{\partial t} = \frac{1}{\mu_o \lambda_L^2} \boldsymbol{E} \tag{3.2.33}$$

$$\tau_n \frac{\partial \boldsymbol{J}_n}{\partial t} + \boldsymbol{J}_n = \frac{N_n}{N} \sigma_N \boldsymbol{E} \tag{3.2.34}$$

ignoring nonlinear effects. Here, λ_L is the London penetration depth defined in Eq.(3.2.4), and $\sigma_N = N\tau_n e^2/m_e$ is the normal-state conductivity of the superconductor. If we introduce the normal conductivity $\sigma_n(T) = \sigma_N N_n(T)/N$, and the London penetration depth $\lambda_L(T)$ at temperature $T \le T_c$, we have

$$\lambda_L(T) = \frac{\lambda_o}{\sqrt{1-(T/T_c)^4}}, \qquad \sigma_n(T) = \sigma_N\left(\frac{T}{T_c}\right)^4 \tag{3.2.35}$$

with λ_0 denoting the value of λ_L at the absolute zero temperature. The relation shows that at lower temperatures, the penetration depth predicted by the classical two-fluid model is nearly independent of temperature while, at temperatures above about 0.8 of the transition temperature, the penetration depth increases rapidly, and approaches infinity as the temperature approaches the critical temperature T_c. In Table 3.2, material parameters for some superconductors are presented, where ξ_0 denotes the coherence length of the superconductor [Turneaure et al. (1991)], which will be discussed later.

According to the two-fluid model, in time-harmonic field ($\sim e^{i\omega t}$), the total current density $\boldsymbol{J}$ may be expressed in the following form:

$$\boldsymbol{J} = \boldsymbol{J}_s + \boldsymbol{J}_n = \sigma \boldsymbol{E} \tag{3.2.36}$$

where σ is an effective complex conductivity, defined by

$$\sigma(T) = \sigma_n(T) - \frac{i}{\omega\mu_o\lambda_L^2(T)} \tag{3.2.37}$$

in the case of $(\omega\tau_n)^2 << 1$.

Table 3.2 Material Parameters for Some Superconductors

Superconductor	λ_0 (nm)	ξ_0 (nm)	T_c (K)
Al	16	1500	1.2
In	25	400	3.3
Sn	28	300	3.7
Pb	28	110	7.2
Nb	32	39	8.95–9.3
Nb_3Sn	50	6	18
$YBa_2Cu_3O_x$	140	1.5	90

3.2.4 The London Theory for Superconductors in AC Fields

As we may know, London´s equations (3.2.3) and (3.2.9) do not replace Maxwell's equations, which, of course, still apply to all currents and the fields they produce. The London equations are additional conditions obeyed by the supercurrent. For a superconductor in ac fields, the total current density $\boldsymbol{J}$ is, in general, the sum of a normal current density $\boldsymbol{J}_n$ and a supercurrent density $\boldsymbol{J}_s$ according to the two-fluid model. In a steady state, when fields and currents are not changing with time, the only current is the supercurrent, and we need only employ the London equations (3.2.3) and (3.2.9). For alternating current (ac) fields, an electric field must however be present to accelerate the electrons which have a small inertial mass and so the supercurrent does not rise instantaneously but only at the rate at which the electrons accelerate in the electric field, that is, the supercurrent will lag behind the field because of the inertia of the superelectrons. Because there is an electric field present, some of the current will be carried by the normal electrons. However, the inertia of an electron is very small and so, unless we go to extremely high frequencies, only a tiny fraction of the current is carried by the normal electrons. Thus a correspondingly minute dissipation of power exists in the superconductor in ac fields. If the frequency of an applied field is sufficiently high, however, a superconducting metal responds in the same way as a normal metal. The behavior of a superconductor at optical frequencies is, therefore, no different from that of a normal metal.

To discuss quantitatively the time-varying cases, we may transform the set of electrodynamic equations for the superconductors into a convenient form to be analyzed. We shall suppose for the moment that the superconductor is isotropic and homogeneous, and has the normal conductivity σ_n, the permittivity ε, and the permeability μ_0. By means of Eqs.(3.2.9), (3.2.33), (3.2.36), and (1.4.18), we may write

$$\mu_o \lambda_L^2 \frac{\partial \boldsymbol{J}}{\partial t} = \boldsymbol{E} + \sigma_n \mu_o \lambda_L^2 \frac{\partial \boldsymbol{E}}{\partial t} \tag{3.2.38}$$

and

$$-\mu_o \lambda_L^2 \nabla \times \boldsymbol{J} = \boldsymbol{B} + \sigma_n \mu_o \lambda_L^2 \frac{\partial \boldsymbol{B}}{\partial t} \tag{3.2.39}$$

From these two equations and Maxwell's equations (1.4.16)–(1.4.19), we may obtain the following four equations for $\boldsymbol{B}$, $\boldsymbol{E}$, $\boldsymbol{J}$, and ρ_e:

$$\nabla \times (\nabla \times \boldsymbol{E}) + \frac{1}{\lambda_L^2}\boldsymbol{E} + \sigma_n \mu_o \frac{\partial \boldsymbol{E}}{\partial t} + \varepsilon \mu_o \frac{\partial^2 \boldsymbol{E}}{\partial t^2} = 0 \tag{3.2.40}$$

$$\nabla \times (\nabla \times \boldsymbol{B}) + \frac{1}{\lambda_L^2}\boldsymbol{B} + \sigma_n \mu_o \frac{\partial \boldsymbol{B}}{\partial t} + \varepsilon \mu_o \frac{\partial^2 \boldsymbol{B}}{\partial t^2} = 0 \tag{3.2.41}$$

$$\nabla \times (\nabla \times \boldsymbol{J}) + \frac{1}{\lambda_L^2}\boldsymbol{J} + \sigma_n \mu_o \frac{\partial \boldsymbol{J}}{\partial t} + \varepsilon \mu_o \frac{\partial^2 \boldsymbol{J}}{\partial t^2} = 0 \tag{3.2.42}$$

$$\frac{1}{\lambda_L^2}\rho_e + \sigma_n \mu_o \frac{\partial \rho_e}{\partial t} + \varepsilon \mu_o \frac{\partial^2 \rho_e}{\partial t^2} = 0 \tag{3.2.43}$$

Here, the equation for the total free charge density ρ_e is an ordinary differential equation in time t alone with constant coefficients at a given temperature, and can thus be solved immediately in full generality by

$$\rho_e = A_1 e^{-\gamma_1 t} + A_2 e^{-\gamma_2 t} \tag{3.2.44}$$

where A_1 and A_2 are integration constants independent of time but arbitrary functions of space, and γ_1 and γ_2 are given by

$$\gamma_1 = \frac{\sigma_n}{2\varepsilon}\left(1 + \sqrt{1 - \frac{4\varepsilon}{\mu_o \sigma_n^2 \lambda_L^2}}\right) \approx \frac{\sigma_n}{2\varepsilon} \tag{3.2.45}$$

$$\gamma_2 = \frac{\sigma_n}{2\varepsilon}\left(1 - \sqrt{1 - \frac{4\varepsilon}{\mu_o \sigma_n^2 \lambda_L^2}}\right) \approx \frac{1}{\mu_o \sigma_n \lambda_L^2} \tag{3.2.46}$$

We may introduce a relaxation time τ by the slower of the two exponentials

$$\tau = \frac{1}{\gamma_2} \approx \mu_o \sigma_n \lambda_L^2 \tag{3.2.47}$$

which is typically less than one picosecond. Thus, any free charges which might occur in the superconductor would disappear within this extremely short time. A quasistatic approximation may, therefore, be introduced to treat superconducting phenomena of frequencies less than a few GHz, depending on specific properties of the superconductor in question. At the quasistatic approximation, Eqs.(3.2.40)–(3.2.42) may be reduced to be

$$\nabla^2 \boldsymbol{E} = \frac{1}{\lambda_L^2}\boldsymbol{E} \tag{3.2.48}$$

$$\nabla^2 \boldsymbol{B} = \frac{1}{\lambda_L^2}\boldsymbol{B} \tag{3.2.49}$$

$$\nabla^2 \boldsymbol{J} = \frac{1}{\lambda_L^2}\boldsymbol{J} \tag{3.2.50}$$

together with the quasistatic form of the continuity equation of charges

$$\nabla \cdot \boldsymbol{J} = 0 \tag{3.2.51}$$

where $\boldsymbol{J}$ is the total current density being the sum of the supercurrent density $\boldsymbol{J}_s$ and the normal current density $\boldsymbol{J}_n$ which satisfies the classical Ohm's law. Here, we have also used the fact that the divergence of electric field in the superconductor is zero, that is, $\nabla \cdot \boldsymbol{E} = 0$ at the quasistatic approximation.

For superconductors in ac fields of higher frequencies, at which the displacement current is still negligible, we may have the following set of field equations:

$$\nabla^2 \boldsymbol{E} = \frac{1}{\lambda_L^2}\boldsymbol{E} + \sigma_n \mu_o \frac{\partial \boldsymbol{E}}{\partial t} \tag{3.2.52}$$

$$\nabla^2 \boldsymbol{B} = \frac{1}{\lambda_L^2}\boldsymbol{B} + \sigma_n \mu_o \frac{\partial \boldsymbol{B}}{\partial t} \tag{3.2.53}$$

$$\nabla^2 \boldsymbol{J} = \frac{1}{\lambda_L^2}\boldsymbol{J} + \sigma_n \mu_o \frac{\partial \boldsymbol{J}}{\partial t} \tag{3.2.54}$$

with $\nabla \cdot \boldsymbol{E} = 0$.

To complete the theoretical model, the interface conditions at the material interface between a superconductor and a normal conductor (or a dielectric, including free space) may be given by

$$\boldsymbol{n} \times [\boldsymbol{E}] = 0 , \qquad \boldsymbol{n} \cdot [\boldsymbol{D}] = \alpha_f \tag{3.2.55}$$

$$\boldsymbol{n} \times [\boldsymbol{B}] = 0 , \qquad \boldsymbol{n} \cdot [\boldsymbol{B}] = 0 \tag{3.2.56}$$

$$\boldsymbol{n} \cdot [\boldsymbol{J}] = 0 \tag{3.2.57}$$

Furthermore, at the interface of two different superconductors, we have the condition:

$$\boldsymbol{n} \times [\lambda_L^2 \boldsymbol{J}] = 0 \tag{3.2.58}$$

where $[\ F\ ] = F^+ - F^-$ denotes the jump of the quantity F across the interface S with $\boldsymbol{n}$ being the unit normal vector always drawn from S^- to S^+.

It might be worth mentioning of that there exist really two ways of describing the electromagnetic behaviors of superconductors in literature. To avoid any possible confusion, some discussion is necessary. In the first way, as we have done in the presentation of the London theory, the two field quantities $\boldsymbol{E}$ and $\boldsymbol{B}$ are sufficient in the study of electrodynamics of the superconductors since, in the London theory, we only deal with non-magnetic superconductors [Note that magnetic superconducting materials have already been found and formulated in the work of, e.g., Zhou (1991a).] The Meissner effect is due to the induced superconducting currents, which are, in effect, screening currents serving to cancel the magnetic flux deep inside the superconductor.

In the second way, we regard the superconductor as perfect diamagnetic materials (if the penetration of the magnetic field is neglected) with, however, a modified current density expression. In this description, two more field quantities $\boldsymbol{H}$ and $\boldsymbol{M}$ are introduced. More specifically, the dc field equation for the superconductor writes in the second description:

$$\nabla \times \boldsymbol{H} = \boldsymbol{J} , \qquad \nabla \times \boldsymbol{B} = 0 \tag{3.2.59}$$

with $\boldsymbol{B} = \mu_0(\boldsymbol{H} + \boldsymbol{M})$ and $\boldsymbol{E} = 0$ in the superconductor, and with the interface conditions:

$$\boldsymbol{n} \times [\boldsymbol{H}] = \boldsymbol{K}_f , \qquad \boldsymbol{n} \cdot [\boldsymbol{B}] = 0 \tag{3.2.60}$$

where $\boldsymbol{K}_f$ and $\boldsymbol{J}$ denote, respectively, the external surface current density and the volume current density, such as the applied external current density and the persistent current density which may exist in multiply connected superconductors. The magnetization $\boldsymbol{M} = -\boldsymbol{H}$ is used to describe the perfect diamagnetic effect (Meissner effect) in the superconductor if we may neglect the penetration of the magnetic field into its very thin layer near the surface of the superconductor. In particular, we may easily find that the magnetic intensity field $\boldsymbol{H}$ inside a long superconducting cylinder with the negligible demagnetizing effect is equal to the external magnetic intensity field $\boldsymbol{H}^e = \boldsymbol{B}^e/\mu_0$, applied in the direction parallel to the axis of the cylinder.

It can be seen that the difference between the two ways of description arises from what is meant by the symbol $\boldsymbol{J}$ for the current density in their relevant field equations. In the first way, the current density $\boldsymbol{J}$ stands for all types of currents, the externally applied currents as well as the screening currents, while, in the second way, it stands only for the externally applied currents (including the persistent currents such as the circulating current in a superconducting ring). We may find later that the second description is particularly convenient to be used to study, for instance, thermodynamic properties of the superconductors. Finally, it should be noticed that the current form of the London theory of superconductors is only valid for studying superconductors in weak magnetic fields so that the Meissner effect and the assumption of the independence of the penetration depth on applied magnetic fields apply.

3.2.5 Energy Theorem and Uniqueness Theorem

The energy theorem for superconductors may follow from Maxwell's equations in the usual way as shown in Section 1.5. The energy integral (1.5.50), which, for a homogeneous isotropic non-magnetic superconductor, may be written into its local form:

$$\nabla \cdot \boldsymbol{S} + \frac{\partial}{\partial t}\left(\frac{1}{2}\left(\varepsilon E^2 + \frac{1}{\mu_o}B^2\right)\right) = -\boldsymbol{J} \cdot \boldsymbol{E} \tag{3.2.61}$$

where $\boldsymbol{S}$ is the Poynting vector, and $\boldsymbol{J}\cdot\boldsymbol{E}$ is the work done by the field on the moving electric charges. Unlike normal conductors, a superconductor, however, does not waste this work completely by transformation into irreversible Joule heat. This can be seen by writing

$$\boldsymbol{J} \cdot \boldsymbol{E} = \boldsymbol{J}_s \cdot \boldsymbol{E} + \boldsymbol{J}_n \cdot \boldsymbol{E} = \frac{\partial}{\partial t}\left(\frac{1}{2}\mu_o \lambda_L^2 J_s^2\right) + \boldsymbol{J}_n \cdot \boldsymbol{E} \tag{3.2.62}$$

with the aid of the first London equation (3.2.3) for the supercurrent $\boldsymbol{J}_S$.

Thus, we may express the energy equation (3.2.61) in the following form:

$$\nabla \cdot (\boldsymbol{E} \times \boldsymbol{H}) + \frac{\partial}{\partial t}\left(\frac{1}{2}\left(\varepsilon E^2 + \frac{1}{\mu_o}B^2\right) + \frac{1}{2}\mu_o \lambda_L^2 J_s^2\right) = -\boldsymbol{J}_n \cdot \boldsymbol{E} \tag{3.2.63}$$

in which the term $\partial(\mu_0(\lambda_L J_s)^2/2)/\partial t$, is evidently positive when the supercurrent is generated, negative when it is switched off. Hence, the energy $\mu_0(\lambda_L J_s)^2/2$, which is the kinetic-energy density of the supercurrent, represents reversible work which can be entirely recovered. In fact, the normal current also has kinetic energy, which, however, is usually neglected since Ohm's law does not count the inertia of the normal conduction electrons. The term on the right-hand side of Eq.(3.2.63) is always negative or zero. This term is the energy dissipation by the Joule heat due to the normal current. For stationary conditions, where there is no normal current, it is exactly zero, and for quasistationary conditions, it is extremely small. Hence, there is almost no energy dissipation in a superconductor except in the case of very rapidly alternating fields.

It is shown by Eq.(3.2.63) that the electromagnetic energy in a superconductor with volume V_S may be expressed by

$$U_s = \int_{V_S} \left(\frac{1}{2\mu_o}B^2 + \frac{1}{2}\mu_o \lambda_L^2 J^2\right) dV \tag{3.2.64}$$

for stationary cases where we have $\boldsymbol{E} = 0$, $\boldsymbol{J}_n = 0$ and $\boldsymbol{J} = \boldsymbol{J}_S$. The first term on the right-hand side of Eq.(3.2.64) is the magnetic field energy, and the second term denotes the kinetic energy of superelectrons. Consider now a superconductor embedded in an insulating medium at the stationary condition, as shown in Fig. 3.5. We may write the total energy U by

$$U = \int_{V_I} \frac{1}{2\mu_o} B^2 dV + \int_{V_S} \left(\frac{1}{2\mu_o} B^2 + \frac{1}{2}\mu_o \lambda_L^2 J^2 \right) dV \tag{3.2.65}$$

where V_I denotes the insulating medium. By using the magnetic vector potential $\boldsymbol{A}$ ($\boldsymbol{B} = \nabla \times \boldsymbol{A}$), we may write

$$U = \int_{V_I} \frac{1}{2\mu_o} \nabla \cdot (\boldsymbol{A} \times \boldsymbol{B}) dV + \int_{V_S} \left(\frac{1}{2\mu_o} \nabla \cdot (\boldsymbol{A} \times \boldsymbol{B}) + \frac{1}{2}\boldsymbol{J} \cdot (\boldsymbol{A} + \mu_o \lambda_L^2 \boldsymbol{J}) \right) dV \tag{3.2.66}$$

since the current $\boldsymbol{J} = 0$ in the insulator.

Applying Gauss' theorem, and noting that the continuity of the normal component of $\boldsymbol{A} \times \boldsymbol{B}$ across the interfaces between the superconductor and the insulator, we may write

$$U = \int_{S_o} \frac{1}{2\mu_o} (\boldsymbol{A} \times \boldsymbol{B}) \cdot d\boldsymbol{S} + \int_{V_S} \frac{\hbar}{2e^*} \boldsymbol{J} \cdot \nabla \theta dV \tag{3.2.67}$$

where we have made use of Eq.(3.2.16) with θ being the superpotential according to London (1950). Since the volume bounded by the surfaces S, S_1^+, and S_1^- is a simply connected region, we may apply Gauss' theorem to the volume integral in Eq.(3.2.67), which results in the following expression:

$$U = \int_{S_o} \frac{1}{2\mu_o} (\boldsymbol{A} \times \boldsymbol{B}) \cdot d\boldsymbol{S} + \frac{1}{2}\Phi_1 I_1 \tag{3.2.68}$$

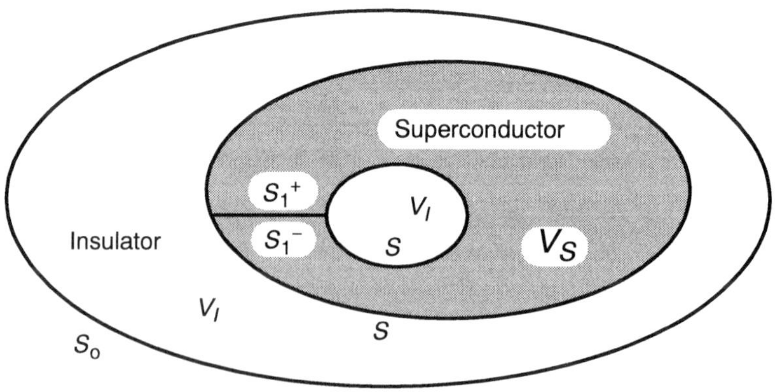

Figure 3.5 Scheme of a superconductor embedded in an insulator.

where Φ_1 denotes the fluxoid of the hole, as discussed in Section 3.2.2, and I_1 is the total current through the surface S_1, that is,

$$I_1 = \int_{S_1} \boldsymbol{J} \cdot d\boldsymbol{S} \tag{3.2.69}$$

The surface integral in Eq.(3.2.68) disappears if we let the surface S_0 move to infinity, provided that the field $\boldsymbol{B}$ disappears at great distance more rapidly than $1/R^2$, as it should be if all sources of the field are assumed to be located within a finite range. Thus we obtain the total energy for the system of the superconductor with one hole: $U = \Phi_1 I_1/2$. This result indicates that there cannot exist the magnetic field as well as the supercurrent inside a simply connected superconductor at the stationary condition without external field sources. This can be seen from Eq.(3.2.65) that the integral over a quantity which is everywhere non-negative can evidently disappear only if $\boldsymbol{B}$=0 and $\boldsymbol{J}$=0 everywhere.

With the use of the above result, we may now give a proof of the uniqueness theorem for the London theory of superconductors by following the same classical procedure [London (1950)]. We assume that we have two solutions of the same differential equations, say,

$$\boldsymbol{B}',\ \boldsymbol{A}',\ \boldsymbol{J}',\ \Phi_1',\ I_1'$$

and

$$\boldsymbol{B}'',\ \boldsymbol{A}'',\ \boldsymbol{J}'',\ \Phi_1'',\ I_1''$$

which need not disappear at infinity. We suppose, however, that at the surface S_0, which encloses the whole system and may be assumed to be located at infinity, the magnetic fields are the same. Then the field

$$\boldsymbol{B} = \boldsymbol{B}' - \boldsymbol{B}''$$

$$\boldsymbol{A} = \boldsymbol{A}' - \boldsymbol{A}''$$

$$\boldsymbol{J} = \boldsymbol{J}' - \boldsymbol{J}''$$

with $\Phi_1 = \Phi_1' - \Phi_1''$ and $I_1 = I_1' - I_1''$ is also a solution of the same differential equations since all are linear differential equations. However, this new field fulfills the boundary condition $\boldsymbol{B} = 0$ on S_o; therefore, the surface integral in Eq.(3.2.68) disappears and we have simply

$$\int_{V_I} \frac{1}{2\mu_o} B^2 dV + \int_{V_S} \left(\frac{1}{2\mu_o} B^2 + \frac{1}{2}\mu_o \lambda_L^2 J^2\right) dV = \frac{1}{2}\Phi_1 I_1 = 0$$

provided that either the current $I_1 = 0$, or the fluxoid $\Phi_1 = 0$ (i.e., either the current through the surface S_1, or the fluxoid of the hole is given). This integral over a quantity which is everywhere non-negative can evidently disappear only if $\boldsymbol{B} = 0$ and $\boldsymbol{J} = 0$ everywhere; this means that the two solutions $(\boldsymbol{B}', \boldsymbol{J}')$ and $(\boldsymbol{B}'', \boldsymbol{J}'')$, assumed above, must be identical.

In summary, we can say that the fields and currents inside a superconductor are uniquely determined by the magnetic field at the surface S_o (or at infinity) if, in addition, either the current I_1 or the fluxoid Φ_1 is given. In particular, for the simply connected superconductor where the fluxoid Φ is zero for any closed curve, the field and current are entirely determined by the field at great distance. The above discussion may easily be generalized for superconductors with a number of holes [see London (1950)].

The uniqueness property of London´s theory of superconductors touch upon a very essential point. As we know, the "perfect-conductor" theory implied an infinity of states belonging to the same external conditions. Only after Meissner´s discovery did it become clear that in fact not as many states are possible within a superconductor as would be expected for a perfect conductor. In fact, according to London´s theory, we have shown that the magnetic field and current in a superconductor is uniquely determined by the applied field.

In time-harmonic cases, the complex form of Poynting's theorem [see Eq.(1.5.62)] for a superconductor of volume V can be expressed in the following form:

$$\int_V \frac{i\omega}{2}(\boldsymbol{B}\cdot\boldsymbol{H}^* + \mu_o\lambda_L^2\boldsymbol{J}_s\cdot\boldsymbol{J}_s^* - \boldsymbol{E}\cdot\boldsymbol{D}^*)dV + \int_V \frac{1}{2}\boldsymbol{E}\cdot\boldsymbol{J}_n^* dV = -\int_{\partial V}\frac{1}{2}\boldsymbol{S}_c\cdot\boldsymbol{n}dS \tag{3.2.70}$$

using the first London equation (3.2.3). Taking the imaginary parts of Eq.(3.2.70), we may get

$$-\int_{\partial V}\frac{1}{2}Im\{\boldsymbol{E}\times\boldsymbol{H}^*\}\cdot\boldsymbol{n}dS = \int_V 2\omega(w_m + w_k - w_e)dV \tag{3.2.71}$$

in which w_m and w_e are known to be respectively the time-average of the stored energy densities of the magnetic field and of the electric field in V, as discussed in Section 1.5, while w_k is a new term for the superconductor, given by

$$w_k = \frac{1}{4}\mu_o\lambda_L^2\boldsymbol{J}_s\cdot\boldsymbol{J}_s^* \tag{3.2.72}$$

which characterizes the time-averaged kinetic energy of superelectrons moving inside the superconductor. The kinetic energy term has its important consequence in the analysis of superconducting electronic circuits, as we shall see later.

3.2.6 Electromagnetic Pressure

Many of us may have seen the fascinating demonstration of superconductivity with a small magnet floating on a superconductor as long as the superconductor remains its superconducting state. Obviously, there exists some kind of force acting on the magnet to balance the earth gravitational force. Equally, there is a force acting on the superconductor according to Newton's laws. Interestingly, the magnet is found to be not in directly contact with the superconductor due to the presence of the electromagnetic force. To calculate the total electromagnetic force acting on the superconductor at the magnto-quasistatic approximation, we may make use of Eq.(2.6.2) for the electromagnetic stress tensor. Thus the total electromagnetic force on a superconductor can be calculated by

$$\boldsymbol{F}^{em} = \int_S \frac{1}{\mu_o}\left((\boldsymbol{B}\cdot\boldsymbol{n})\boldsymbol{B} - \frac{1}{2}(\boldsymbol{B}\cdot\boldsymbol{B})\boldsymbol{n}\right) dS \tag{3.2.73}$$

where $\boldsymbol{n}$ is the unit normal vector on the surface S of the superconductor.

When the superconductor is at the Meissner state in a weak magnetic field, we may consider the fact that the supercurrent density is only distributed within a very thin layer (with the penetrating depth λ_L) near the boundary surface S of the superconductor. Thus we may introduce the concept of surface current by making a limiting process of letting $\lambda_L \to 0$ and denoting the surface current density by $\boldsymbol{K}_f = \lim \boldsymbol{J}\lambda_L$. Then, Eq.(3.2.73) may be written as

$$\boldsymbol{F}^{em} = -\int_S \frac{1}{2\mu_o}|\boldsymbol{B}^+|^2 \boldsymbol{n}\, dS = -\frac{\mu_o}{2}\int_S |\boldsymbol{K}_f|^2 \boldsymbol{n}\, dS \tag{3.2.74}$$

by noting that $\boldsymbol{B}^+$ is the magnetic induction field at the outer side of the boundary of the superconductor, and that the normal component of $\boldsymbol{B}^+$ vanishes at the boundary due to the perfect diamagnetism ($\boldsymbol{B} = 0$ in the superconductor). Here, we have used the following magnetic interface condition:

$$\boldsymbol{n}\times[\boldsymbol{B}] = \mu_o \boldsymbol{K}_f\,, \qquad \boldsymbol{n}\cdot[\boldsymbol{B}] = 0 \tag{3.2.75}$$

which gives the expression $\boldsymbol{B}^+ = \mu_0 \boldsymbol{K}_f \times \boldsymbol{n}$ on S. Similarly, we can find the following interface relation at the interface S between a normal conductor and a superconductor with perfect diamagnetism, having the free surface current $\boldsymbol{K}_f$,

$$\boldsymbol{n}\cdot[\boldsymbol{t}^{em}] = -\frac{\mu_o}{2}|\boldsymbol{K}_f|^2\boldsymbol{n} \quad \text{on } S \tag{3.2.76}$$

where $\boldsymbol{n}$ is a unit normal vector of the interface S, drawn from the superconductor to the normal conductor.

It is shown by Eq.(3.2.74) that the superconductor is subjected to an electromagnetic pressure with the magnitude of $\mu_0|\boldsymbol{K}_f|^2/2$ (or $|\boldsymbol{B}^+|^2/2\mu_0$) on its boundary surface. The electromagnetic pressure acting on the superconductor is different from a normal magnetic pressure due to the jump in the magnetization of a magnetic body. The effect of electromagnetic pressure on superconductors makes possible superconducting suspensions, bearings, and motors by means of properly shaped magnetic fields [Newhouse (1964)]. Typical examples of the effect of superconducting pressure are, for instance, the early experiment made by Arkadiev (1945, 1947) which showed that a permanent magnet can float in equilibrium above a superconducting cup, and the experiment by Simon (1953) who showed the ability of suitably shaped magnetic fields to support a superconducting sphere.

3.3 SOME BOUNDARY–VALUE PROBLEMS IN LONDON'S THEORY

3.3.1 Superconducting Half-Space in a Static Magnetic Field

To illustrate the use of the London theory, we shall first present here a simple analytical solution for a superconducting half-space in a uniform static magnetic field $\boldsymbol{B}^e$, applied in the y-direction at its surface $x = 0$, as shown in Fig. 3.6. The applied static magnetic field is supposed to be not strong so that the superconductor remains at the Meissner state. To find the distribution of the magnetic field inside the superconducting half space, we may use Eq.(3.2.49), which reads here

$$\frac{\partial^2 B_y}{\partial x^2} = \frac{1}{\lambda_L^2} B_y \tag{3.3.1}$$

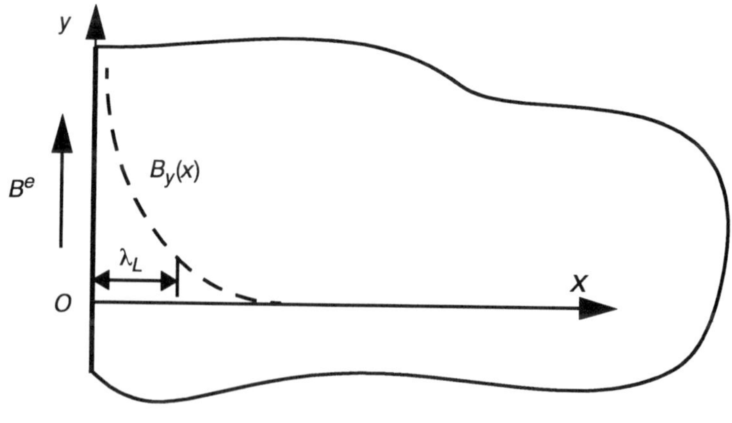

Figure 3.6 Magnetic field applied parallel to a superconducting half-space.

This equation has obviously the solution

$$B_y(x) = B^e exp(-x/\lambda_L) \tag{3.3.2}$$

which shows that the magnetic field in the superconducting half space decays exponentially with the increase of the distance x. The measure of the penetration of the external field into the superconductor is by the London penetration depth λ_L. Since the penetration depth is usually on the order of 10^{-7} m so that the exponential decay of the magnetic field is extremely rapid. This nearly complete flux exclusion within the superconductor is known as the Meissner effect. The induced superconducting current density can also be found by

$$J_z = \frac{1}{\mu_o}\frac{\partial B_y}{\partial x} = -\frac{B^e}{\mu_o \lambda_L} exp(-x/\lambda_L) \tag{3.3.3}$$

which shows that the superconducting screen current flows within also a thin layer near the surface of the superconductor.

3.3.2 Superconducting Cylinder Carrying DC Current

The second example is to find the superconducting current density distribution in an infinite long cylinder with a radius R and carrying a total amount of dc current I flowing along its axis, as shown in Fig. 3.7. In this case, we may use Eq.(3.2.50), which in a cylindrical coordinate system writes

$$\frac{d^2 J_z}{dr^2} + \frac{1}{r}\frac{dJ_z}{dr} - \frac{1}{\lambda_L^2} J_z = 0 \tag{3.3.4}$$

A general solution of this equation can be found as

$$J_z(r) = C I_o(r/\lambda_L) \tag{3.3.5}$$

where C is a constant determined by the condition of the total current

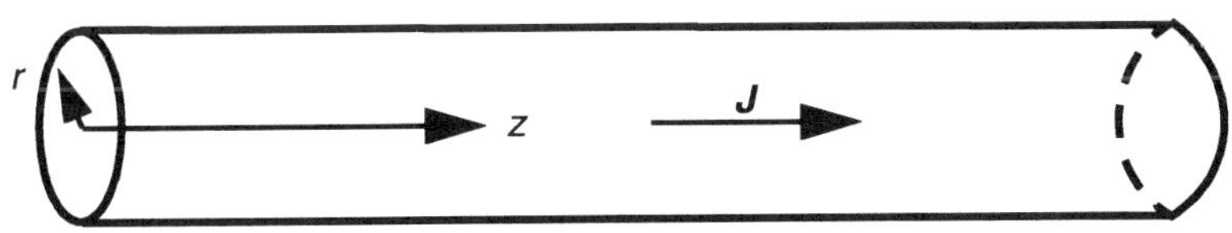

Figure 3.7 Current passed through an infinite long cylindrical superconductor.

$$I = 2\pi\int_0^R J_z(r) r dr \tag{3.3.6}$$

The result gives

$$J_z(r) = \frac{I}{2\pi R\lambda_L}\frac{I_o(r/\lambda_L)}{I_1(R/\lambda_L)} \quad \text{for } 0 \le r \le R \tag{3.3.7}$$

where I_o and I_1 are the modified Bessel functions of the first kind of the zeroth order and of the first order, respectively.

Two limiting cases may be of interest, which are

$$J_z(r) = \frac{I}{2\pi R\lambda_L}\sqrt{\frac{R}{r}} exp\left(-\frac{R-r}{\lambda_L}\right) \quad \text{for } \lambda_L \ll r \le R \tag{3.3.8}$$

and

$$J_z(r) = \frac{I}{\pi R^2} \quad \text{for } r \le R \ll \lambda_L \tag{3.3.9}$$

which shows a uniform current density distribution in the case of $R << \lambda_L$.

The maximum current density in Eq.(3.3.8) is found to be $J_{zmax} = I/(2\pi R\lambda_L)$ at $r = R$. The magnetic field in the superconductor may be found from the second London equation (3.2.9). In particular, at the surface of the superconducting cylinder, the magnetic field may be found by Ampere´s circuital law:

$$B_\theta(R) = \frac{\mu_o I}{2\pi R} \tag{3.3.10}$$

Since there exists a critical magnetic field for the superconductor, we can expect that there is a corresponding critical current density, which is given here by $J_{zc} = B_c/(\mu_o\lambda_L)$. This example confirms, therefore, the Silsbee hypothesis discussed in Section 3.1.2.

3.3.3 Superconducting Coaxial Line

The third example which we shall consider is to study the electromagnetic behavior of a superconducting coaxial line, as shown in Fig. 3.8. We assume that the coaxial line is infinitely long and has the circular symmetry with respect to its axis along *oz*-direction. There are currents of amount of I flowing in both the inner superconductor and the outer superconductor with opposite direction. Due to the symmetry of the problem, the current density vector $\boldsymbol{J}$ in both the inner and outer superconductors has only a *z*-component, $J_z(r)$, which depends only on the radial coordinate r in the cylindrical coordinate system. If only the superconducting current is considered, Eq.(3.3.4) is the field equation for the determination of the superconducting current density in both the inner and outer superconductors. For the coaxial line problem, we may obtain the following general solution:

$$J_z(r) = C\mathrm{I}_o(r/\lambda_L) \qquad \text{for } r \le a \tag{3.3.11}$$

and

$$J_z(r) = A\mathrm{I}_o(r/\lambda_L) + D\mathrm{K}_o(r/\lambda_L) \quad \text{for } b \le r \le c \tag{3.3.12}$$

where the coefficients A and D are determined by the boundary conditions, while C can be determined by the condition of total current I inside the inner superconductor [see Eq.(3.3.6)]:

$$C = \frac{I}{2\pi a \lambda_L \mathrm{I}_1(a/\lambda_L)} \tag{3.3.13}$$

where I_1 is the modified Bessel function of the first kind of the first order.

By the second London equation (3.2.9), we may find the magnetic field $\boldsymbol{B} = (0, B_\theta, 0)$ with

$$B_\theta = \mu_o \lambda_L^2 \frac{\partial J_z}{\partial r} \tag{3.3.14}$$

in the superconductors. Using Eqs.(3.3.11) and (3.3.12), we then obtain

$$B_\theta = \frac{I\mu_o \mathrm{I}_1(r/\lambda_L)}{2\pi a \mathrm{I}_1(a/\lambda_L)} \qquad (r \le a) \tag{3.3.15}$$

and

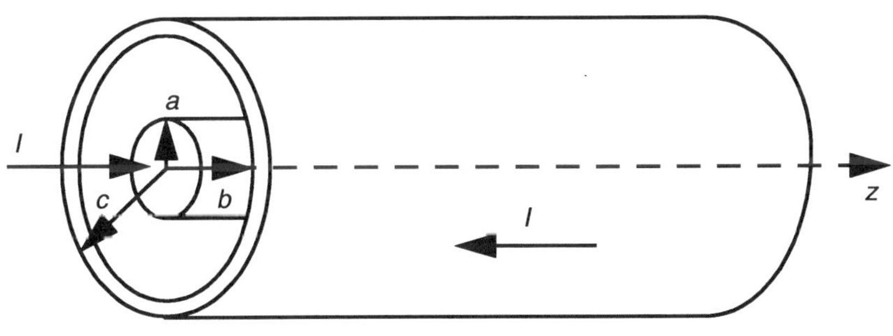

Figure 3.8 Scheme of a coaxial line.

$$B_\theta(r) = \mu_o\lambda_L A \mathrm{I}_1(r/\lambda_L) - \mu_o\lambda_L D K_1(r/\lambda_L) \qquad (b \le r \le c) \tag{3.3.16}$$

The magnetic field component B_θ in the free space between the inner and outer superconductors can be found by using Ampere´s circuital law:

$$B_\theta(r) = \frac{\mu_o I}{2\pi r} \qquad (a < r < b) \tag{3.3.17}$$

By the continuity condition of the magnetic field component B_θ at $r = b$ and $B_\theta = 0$ at $r = c$, we get

$$A = \frac{I K_1(c/\lambda_L)}{2\pi b\lambda_L[K_1(c/\lambda_L)\mathrm{I}_1(b/\lambda_L) - K_1(b/\lambda_L)\mathrm{I}_1(c/\lambda_L)]} \tag{3.3.18}$$

and

$$D = \frac{I\ \mathrm{I}_1(c/\lambda_L)}{2\pi b\lambda_L[K_1(c/\lambda_L)\mathrm{I}_1(b/\lambda_L) - K_1(b/\lambda_L)\mathrm{I}_1(c/\lambda_L)]} \tag{3.3.19}$$

The derived magnetic field B_θ and superconducting current density J_z can be used to calculate, for instance, the line inductance L of the superconducting coaxial line by $L = L_{ext} + L_{int}$, where L_{ext} denotes the external inductance per unit length of the coaxial line given by

$$L_{ext} = \frac{2\pi}{\mu_o I^2}\int_a^b |B_\theta|^2 r dr = \frac{\mu_o}{2\pi}\ln\left(\frac{b}{a}\right) \qquad (\mathrm{H/m}) \tag{3.3.20}$$

and L_{int} is the internal inductance per unit length given by

$$L_{int} = \frac{2\pi}{\mu_o I^2}\left(\int_0^a |B_\theta|^2 r dr + \int_b^c |B_\theta|^2 r dr\right) + \frac{2\pi\mu_o}{I^2}\left(\int_0^a \lambda_L^2 |J_z|^2 r dr + \int_b^c \lambda_L^2 |J_z|^2 r dr\right) \tag{3.3.21}$$

in which the second term on the right-hand side of Eq.(3.3.21) denotes the kinetic inductance per unit length of the superconducting coaxial line.

In a special case where we let λ_L be very large ($\to \infty$) so that the current distribution inside the conductors can be considered as uniform (corresponding to the normal state of the coaxial line in dc or low-frequency cases), we may find, by noting $\mathrm{I}_1(x) \to x/2$ and $K_1(x) \to 1/x$ for $x \to 0$, the magnetic field distribution B_θ by

$$B_\theta = \frac{\mu_o I r}{2\pi a^2} \qquad (r \le a) \tag{3.3.22}$$

and

$$B_\theta(r) = \frac{\mu_o I}{2\pi(c^2 - b^2)}\left(\frac{c^2}{r} - r\right) \qquad (b \le r \le c) \tag{3.3.23}$$

The magnetic field in the free space between the inner and outer conductors is the same as that given by Eq.(3.3.17). These results are in consistent with those derived for a normal conducting coaxial line in the case where the current distribution inside the conductors can be considered as uniform [see Ramo et al. (1984)]. In such a case, the internal inductance per unit length of the normal coaxial line can be found by

$$L_{int} = \frac{\mu_o}{8\pi} + \frac{\mu_o}{2\pi}\left[\frac{c^4 \ln(c/b)}{(c^2 - b^2)^2} + \frac{b^2 - 3c^2}{4(c^2 - b^2)}\right] \qquad (\mathrm{H/m}) \tag{3.3.24}$$

Thus, for frequencies low enough to assume uniform current distribution in the non-magnetic conductors, the total inductance per unit length of the coaxial line is the sum of Eq.(3.3.20) and Eq.(3.3.24).

3.3.4 Superconducting Cylinder in a Static Magnetic Field

In this section, we shall study the field distribution in a superconducting cylinder of infinite length and of radius R in a uniform static magnetic field B^e parallel to its axis (oz). To solve this problem, we may use the field equation (3.2.49), from which we can obtain

$$B_z(r) = \frac{B^e \mathrm{I}_o(r/\lambda_L)}{\mathrm{I}_o(R/\lambda_L)} \qquad \text{for } 0 \le r \le R \tag{3.3.25}$$

The average induced magnetization in the superconducting cylinder may then be expressed by

$$M_z = \frac{2B^e}{\mu_o R^2}\int_0^R \left(\frac{\mathrm{I}_o(r/\lambda_L)}{\mathrm{I}_o(R/\lambda_L)} - 1\right) r\,dr = -\frac{B^e}{\mu_o}\left(1 - \frac{2\lambda_L}{R}\frac{\mathrm{I}_1(R/\lambda_L)}{\mathrm{I}_o(R/\lambda_L)}\right) \tag{3.3.26}$$

which, in the limiting cases of $R >> \lambda_L$ and $R << \lambda_L$, may be reduced to

$$M_z = -\frac{B^e}{\mu_o}\left(1 - \frac{2\lambda_L}{R}\right) \qquad \text{for } R \gg \lambda_L \tag{3.3.27}$$

and

$$M_z(r) = -\frac{B^e R^2}{8\mu_o \lambda_L^2} \qquad \text{for } R \ll \lambda_L \tag{3.3.28}$$

Equation (3.3.27) shows that the superconducting cylinder presents almost perfect diamagnetic effect provided that its radius R is much larger than its penetration depth λ_L. In general, we have shown that a superconductor in an applied magnetic field does not show a perfect diamagnetic effect since screening currents cannot be confined entirely to the surface. The penetration of the magnetic flux into the thin surface layer of the superconductor is characterized by the penetration depth λ_L. Though this layer is so thin, it plays an important part in determining the behavior of the superconductor.

3.3.5 Magnetic Field Shielding by Superconductor

Magnetic field shielding is of importance since essentially field-free regions are often necessary or desirable for experimental purposes or for the reliable working of some electronic devices. In the following example, we shall give an analysis on the field distribution around a superconducting shell of spherical shape with an outer radius R and an inner radius a, embedded in a uniform external static magnetic field $\boldsymbol{B}^e$, which is supposed to be along the z-axis, as shown in Fig. 3.9. According to the London theory, the field equations for the superconducting region can be expressed by

$$\nabla^2 \boldsymbol{J} = \frac{1}{\lambda_L^2}\boldsymbol{J} \qquad (a < r < R) \tag{3.3.29}$$

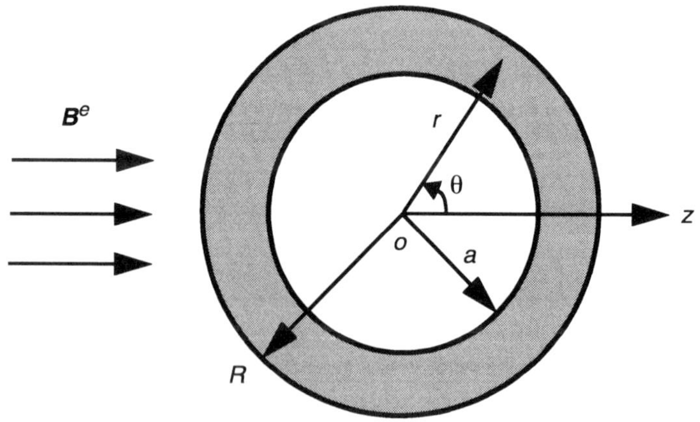

Figure 3.9 A superconducting shell is embedded in an external magnetic field.

and

$$\boldsymbol{B} = -\mu_o\lambda_L^2\nabla\times\boldsymbol{J} \qquad (a<r<R) \tag{3.3.30}$$

By noting the symmetry of the system and of the external field assumed, we can choose a spherical coordinate system (r, θ, φ) so that we may write $\boldsymbol{J} = (0, 0, J_\varphi(r, \theta))$ according to the symmetry of the problem. Furthermore, we may let $J_\varphi = f(r)\sin\theta$ and substitute it into the Eq.(3.3.29) for J_φ, which leads to an ordinary differential equation for $f(r)$:

$$\frac{d^2f}{dr^2} + \frac{2}{r}\frac{df}{dr} - \left(\frac{2}{r^2} + \frac{1}{\lambda_L^2}\right)f = 0 \qquad (a<r<R) \tag{3.3.31}$$

Equation (3.3.31) has a general solution of the form:

$$f(r) = \frac{A}{r^2}\left(\sinh\left(\frac{r}{\lambda_L}\right) - \frac{r}{\lambda_L}\cosh\left(\frac{r}{\lambda_L}\right)\right) + \frac{C}{r^2}\left(\cosh\left(\frac{r}{\lambda_L}\right) - \frac{r}{\lambda_L}\sinh\left(\frac{r}{\lambda_L}\right)\right) \tag{3.3.32}$$

where A and C are two integration constants, and functions $\cosh(x)$ and $\sinh(x)$ are defined, respectively, by

$$\cosh(x) = \frac{e^x + e^{-x}}{2}, \qquad \sinh(x) = \frac{e^x - e^{-x}}{2} \tag{3.3.33}$$

The magnetic field in the superconducting shell may then be found from Eq.(3.3.30).

The field equations in the free space are simply the static magnetic equations

$$\nabla\times\boldsymbol{B} = 0, \qquad \nabla\cdot\boldsymbol{B} = 0 \qquad (\text{for } R<r<\infty \text{ or } r<a) \tag{3.3.34}$$

The symmetry of the problem suggests also trying the solution from a superposition of the uniform applied magnetic field $\boldsymbol{B}^e$ and the field of a magnetic dipole at the origin with dipole moment **m**, oriented in the direction of the applied field, that is,

$$B_r = \left(B^e + \frac{2\mu_o \mathrm{m}}{4\pi r^3}\right)\cos\theta \tag{3.3.35}$$

$$B_\theta = \left(-B^e + \frac{\mu_o \mathrm{m}}{4\pi r^3}\right)\sin\theta \tag{3.3.36}$$

$$B_\varphi = 0 \tag{3.3.37}$$

for $R < r < +\infty$. Furthermore, the magnetic field in the cavity ($r < a$) may be tried by the form

$$B_r = B^c \cos\theta , \quad B_\theta = -B^c \sin\theta , \quad B_\varphi = 0 \quad \text{for } r < a \tag{3.3.38}$$

Here, the magnitude of the magnetic field B^c in the cavity and the magnetic dipole moment m can be determined together with the integration constants A and C by the interface conditions that the magnetic field $\boldsymbol{B}$ is continuous across the surfaces at $r = a$ and $r = R$ [see Zhou (1991a)]. It can be shown that inside the cavity there is a uniform magnetic field parallel to $\boldsymbol{B}^e$. In the special case of $a = 0$, we may find

$$\mathrm{m} = -\frac{2\pi R^3 B^e}{\mu_o}\left(1 + \frac{3\lambda_L^2}{R^2} - \frac{3\lambda_L}{R}\coth\left(\frac{R}{\lambda_L}\right)\right) \tag{3.3.39}$$

which recovers the result for a solid superconducting sphere of radius R in a uniform external magnetic field $\boldsymbol{B}^e$ [London (1950)]. In particular, if the radius R of the superconducting sphere is much larger than its penetration depth λ_L, we have approximately

$$\mathrm{m} \approx -\frac{2\pi B^e}{\mu_o}(R - \lambda_L)^3 \tag{3.3.40}$$

Furthermore, we may define an induced average magnetization in the superconducting sphere by

$$M = \frac{3\mathrm{m}}{4\pi R^3} = -\frac{3B^e(R - \lambda_L)^3}{2\mu_o R^3} \approx -\frac{3B^e}{2\mu_o} \tag{3.3.41}$$

which shows that the superconducting sphere presents the perfect diamagnetic effect if we ignore the penetration of the field in the case of $R >> \lambda_L$.

Next, we may consider the case of that the radii a and R are both much larger than the penetration depth of the superconducting shell, that is, $a >> \lambda_L$ and $R >> \lambda_L$, which is usually the case of practical interest. The magnetic field inside the cavity can then be found approximately by

$$B^c \approx \frac{R\lambda_L B^e}{a^2} exp\left(-\frac{R-a}{\lambda_L}\right) , \quad (a, R \gg \lambda_L) \tag{3.3.42}$$

which shows the perfect magnetic shielding if $\lambda_L \to 0$. A comparison can be made between the magnetic field shielding by using the superconductor and the magnetic field shielding by using the permeable material with high relative permeability μ_r. In the case of using the permeable material with high relative permeability μ_r, the magnetic field inside the cavity is given by [Jackson (1975)]

$$B^c \approx \frac{9B^e}{2\mu_r(1 - a^3/R^3)}, \qquad (\mu_r \gg 1) \tag{3.3.43}$$

Quantitatively, by taking the numerical values of $a = 0.1$ m, $R = 0.11$ m ($R - a = 10$ mm) for the shield shell made of magnetically permeable material with $\mu_r = 10^6$ (e.g., supermalloy), we find from Eq.(3.3.43) that $B^c/B^e \sim 1.8\times10^{-5}$. However, if we take the numerical values of $a = 0.1$ m, $R = 0.1001$ m ($R - a = 0.1$ mm) for the shield shell (though which is quite thin) made of the superconductor having the London penetration depth $\lambda_L \sim 10^{-7}$ m, we can find from Eq.(3.3.42) that $B^c/B^e \sim 10^{-439}$, which is extremely small. Thus we have shown that a shield made of superconducting materials can cause a much more great reduction in the field inside it than the one made of high-permeability materials even with the highest permeability available. In practice, we may use a composite of a normal (for instance, dielectric) matrix covered with a thin superconducting film at its surface as the magnetic shielding material, where the normal matrix material is mainly used to stand for necessary mechanical loadings and other technical requirements.

3.3.6 Superconducting Half-Space in a Time-Harmonic Magnetic Field

We shall now consider an ac electrodynamic problem of superconductors in which the effect due to the presence of normal conduction electrons will be taken into account by using the classical two-fluid model and London's theory. The geometry of the superconductor is taken to be the same as the example shown in Section 3.3.1 (a half-space of superconductor) with, however, now a time-harmonic uniform magnetic field $B^e e^{i\omega t}$, applied in the y-direction, as shown in Fig. 3.6.

By using Eq.(3.2.53), we may find

$$B_y(x, t) = B^e exp(i\omega t - kx) \tag{3.3.44}$$

with the parameter k given by

$$k = \frac{1}{\lambda_L}\sqrt{1 + i2\left(\frac{\lambda_L}{\delta_n}\right)^2} \tag{3.3.45}$$

where δ_n denotes the normal skin-depth, defined by

$$\delta_n = \sqrt{\frac{2}{\omega\mu_o\sigma_n}} \tag{3.3.46}$$

Since the normal conductivity σ_n is temperature-dependent by Eq.(3.2.35) according to the two-fluid model, the normal skin-depth δ_n is also temperature-dependent.

Introducing the complex surface impedance defined by

$$Z_s = \frac{E_z(x=0)}{\int_0^\infty J_z dx} \quad (\Omega) \tag{3.3.47}$$

we may obtain

$$Z_s = R_s + iX_s = \frac{i\omega\mu_o\lambda_L\sqrt{1 - i2(\lambda_L/\delta_n)^2}}{\sqrt{1 + 4(\lambda_L/\delta_n)^4}} \tag{3.3.48}$$

in which R_s is called the surface resistance given by

$$R_s = \frac{\omega\mu_o\lambda_L}{\sqrt{2}}\sqrt{\frac{\sqrt{1 + 4(\lambda_L/\delta_n)^4} - 1}{1 + 4(\lambda_L/\delta_n)^4}} \tag{3.3.49}$$

and X_s called the surface inductive reactance given by

$$X_s = \frac{\omega\mu_o\lambda_L}{\sqrt{2}}\sqrt{\frac{\sqrt{1 + 4(\lambda_L/\delta_n)^4} + 1}{1 + 4(\lambda_L/\delta_n)^4}} \tag{3.3.50}$$

Here, two special cases may be of interest. The first one is that the superconductor is in the normal state. In such a case, the surface resistance and the surface inductive reactance are equal, and they are given by

$$R_{sn} = X_{sn} = \sqrt{\frac{\omega\mu_o}{2\sigma_n}} \tag{3.3.51}$$

The second case is that the superconductor is in the superconducting state, where the normal conduction current is supposed to be much smaller than the supercurrent in the superconductor ($\lambda_L/\delta_n \ll 1$). Quantitatively, the conductivity of copper at 77 K is about $\sigma_n = 5\times10^8/\Omega\text{m}$. This gives the normal skin depth $\delta_n = 5.6\ \mu\text{m}$ at the frequency $\omega = 10^{12}$ rad/s. The superconducting penetration depth λ_L of high-T_c superconductors is, however, mostly around a few hundreds nanometers, provide that the gap frequency of the superconductor is larger than the operating frequency and the operating temperature is not too close to the critical temperature of the superconductor. In such a case, we may get

$$R_s = \frac{\omega\mu_o\lambda_L^3}{\delta_n^2} = \frac{\sigma_n\mu_o^2\omega^2\lambda_L^3}{2} \tag{3.3.52}$$

and

$$X_s = \omega\mu_o\lambda_L \tag{3.3.53}$$

which shows that the surface resistance R_s, predicted by this model for the superconductor, increases as the square of the frequency, whereas for normal conductors, the surface resistance R_{sn} increase only as the square root of the frequency, as shown by Eq.(3.3.51).

Since both the normal conductivity σ_n and the London penetration depth λ_L are temperature-dependent by Eq.(3.2.35) according to the two-fluid model, the surface resistance R_s and the surface inductive reactance are also temperature-dependent. It has been recognized that the frequency and temperature dependence of the surface impedance of superconductors predicted by the two-fluid approach are close to observed experimental behaviors. As a result many circuit and waveguide calculations are based on the two-fluid approach. However, at very high (microwave) frequencies, appreciable deviations may occur. Nevertheless, the simple two-fluid approach is often used in engineering calculations.

3.3.7 Reflection and Transmission of Normally Incident Plane Wave

In this section, we shall study the problem on the behavior of an electromagnetic plane wave being normally incident on a plane of a superconductor, as shown in Fig. 3.10. We consider a semi-infinite superconductor characterized by λ_L, σ_n, and μ_o, occupying the region for $z \geq 0$, and a lossless dielectric medium with ε (> 0) and μ_o, occupying the region for $z < 0$. A uniform plane wave is propagating in the $+z$ direction within the incident dielectric medium, having been generated by sources infinitely far from the boundary at $z = 0$. If the electric and magnetic fields immediately to the left of the boundary at $z = 0^-$ are non-zero, then the fields to the right ($z = 0^+$) are also non-zero because the tangential components of $\boldsymbol{E}$ and $\boldsymbol{H}$ (also $\boldsymbol{B}$ here) must be continuous across the boundary having no surface current. Thus a transmitted wave may generally exist in the region of $z > 0$. Usually, a reflected wave is generated that travels back toward the incident sources so that the total fields may satisfy all of the boundary conditions (see also Section 1.10.3).

Suppose that the superposition of an incident [x-polarized, i.e, $\boldsymbol{E} = (E_x, 0, 0)$] plane wave and a reflected wave in the incident dielectric medium can be expressed by

$$E_x = E_I e^{i(\omega t - \beta_I z)} + E_R e^{i(\omega t + \beta_I z)} \qquad (z < 0) \tag{3.3.54}$$

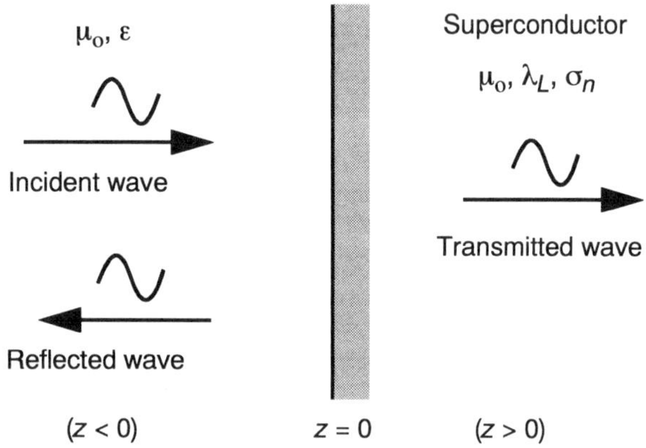

Figure 3.10 Incident, reflected, and transmitted waves at a planar boundary.

and

$$B_y = \frac{i}{\omega}\frac{\partial E_x}{\partial z} = \frac{\beta_I}{\omega}[E_I e^{i(\omega t - \beta_I z)} - E_R e^{i(\omega t + \beta_I z)}] \qquad (z < 0) \tag{3.3.55}$$

where E_I is the complex amplitude of the incident wave, E_R is the complex amplitude of the reflected wave, and $\beta_I = \omega(\varepsilon\mu_o)^{1/2}$ is the phase constant (or the wave number) of the plane wave in the dielectric medium.

The transmitted wave in the superconductor ($z > 0$) is expressed by

$$E_x = E_T e^{i(\omega t - \beta_T z)} \qquad (z > 0) \tag{3.3.56}$$

and

$$B_y = \frac{\beta_T}{\omega} E_T e^{i(\omega t - \beta_T z)} \qquad (z > 0) \tag{3.3.57}$$

where E_T is the complex amplitude of the transmitted wave. The phase constant β_T, being generally a complex value, can be found from Eq.(3.2.52) as

$$\beta_T = \frac{1}{\sqrt{2}\lambda_L}\left[\sqrt{\sqrt{1 + 4\left(\frac{\lambda_L}{\delta_n}\right)^4} - 1} - i\sqrt{\sqrt{1 + 4\left(\frac{\lambda_L}{\delta_n}\right)^4} + 1}\right] \tag{3.3.58}$$

where δ_n is the normal skin-depth defined by Eq.(3.3.46).

Because the tangential components of the electric field $\boldsymbol{E}$ must be continuous at the boundary ($z = 0$) between the incident medium and the superconductor, we have

$$E_I + E_R = E_T \tag{3.3.59}$$

Besides, since the tangential magnetic field is continuous at $z = 0$, we have

$$\beta_I(E_I - E_R) = \beta_T E_T \tag{3.3.60}$$

Thus we may find the following two equations:

$$\Gamma = \frac{E_R}{E_I} = \frac{\eta - 1}{\eta + 1} \tag{3.3.61}$$

$$\Xi = \frac{E_T}{E_I} = \frac{2\eta}{\eta + 1} = 1 + \Gamma \tag{3.3.62}$$

with η being the normalized wave impedance, defined by

$$\eta = \frac{\beta_I}{\beta_T} \tag{3.3.63}$$

which is in general complex. Here, we have introduced the complex reflection coefficient Γ, and the complex transmission coefficient Ξ by Eq.(3.3.61) and Eq.(3.3.62), respectively.

The time-averaged Poynting vector may be calculated for the x-polarized plane wave in each of the two regions, which are, respectively,

$$\bar{S_z} = \frac{1}{2}Re(E_x H_y{}^*) = \frac{\beta_I |E_I|^2}{2\mu_o\omega}(1 - |\Gamma|^2) \qquad (z < 0) \tag{3.3.64}$$

and

$$\bar{S_z} = \frac{1}{2}Re(E_x H_y{}^*) = \frac{\beta_{Tr}\Xi^2 |E_I|^2}{2\mu_o\omega} e^{\beta_{Ti} z} \qquad (z > 0) \tag{3.3.65}$$

where $\beta_{Tr} = Re(\beta_T)$ and $\beta_{Ti} = Im(\beta_T)$ are, respectively, the real part and the imaginary part of the phase constant β_T. We may notice that the time-averaged power flowing in the incident lossless dielectric medium is independent of coordinate z, while the time-averaged power flowing in the superconductor is dependent of coordinate z.

Some special cases may be of interest. If the loss due to normal electrons in the superconductor is negligible, we have $\beta_{Tr} = 0$ and $\beta_{Ti} = -1/\lambda_L$ from Eq.(3.3.58), which implies that the time-averaged power is zero in the region of superconductor, as shown by Eq.(3.3.65). In such a case, we find that the magnitude of the reflection coefficient is unity ($|\Gamma| = 1$), which implies that no power flows across the planar interface, and all of the power is reflected.

If the superconductor is in a normal state, we have then $\beta_{Tr} = 1/\delta_n$ and $\beta_{Ti} = -1/\delta_n$, which implies that there is power transmitted into the normal conductor.

However, it can be seen from Eq.(3.3.65) that nearly all of the power that flows into the conductor is dissipated within a distance δ_n. We may express the power dissipated per unit area (W/m^2) P_d in the normal conductor as

$$P_d = \frac{1}{2}Re(E_x H_y^*)\Big|_{z=0} = \frac{|H_y(z=0)|^2}{4}\sqrt{\frac{2\omega\mu_o}{\sigma_n}} \tag{3.3.66}$$

In general, for superconductors in the superconducting state, taking into account the normal conduction loss, we may find

$$P_d = \frac{1}{2}Re(E_x H_y^*)\Big|_{z=0} = \frac{\mu_o \omega \beta_{Tr} |H_y(z=0)|^2}{2|\beta_T|^2} \tag{3.3.67}$$

in which the phase constant β_T is given by Eq.(3.3.58).

3.4 ELECTRODYNAMIC BEHAVIORS OF SUPERCONDUCTORS AT HIGH FREQUENCIES

3.4.1 Full-Wave Solution of a Superconducting Planar Waveguide

Recent advances in both broadband communications and supercomputer technologies have led to the development of ultrafast, very high frequency electronics. Electrical pulses containing frequency components above 1 THz is now possible to be generated by optoelectronic techniques. The demand for a mean to transmit waveforms of such very short signal pulses from one location to another while maintaining fidelity of the signals is thus increasing due also to the recent progress in millimeter-wave integrated analog and digital circuits. Studies of broad-bandwidth waveform propagation in room-temperature normal conducting transmission lines have revealed that signal distortion in these lines can be quite serious due largely to a frequency-dependent propagation velocity and a frequency-dependent attenuation in the normal conducting transmission lines. Thus superconducting transmission lines have been proposed and designed to transmit very short (picosecond) signal pulses among ultrafast electronic devices. An early systematic study of superconducting transmission lines can be found in the work of Matick (1969).

In order to study high-frequency behaviors of superconductors, in this section, we shall derive a full-wave solution for a superconducting planar waveguide, as shown in Fig. 3.11, based on the Maxwell field theory. We shall solve the problem for TM modes, in which electromagnetic fields are described by the transverse magnetic field $\boldsymbol{B} = (0, B_y, 0)$ and the electric field $\boldsymbol{E} = (E_x, 0, E_z)$. The superconductor in the waveguide is assumed to be isotropic.

By the London theory for isotropic superconductors, the magnetic field component B_y can be determined by the following equation:

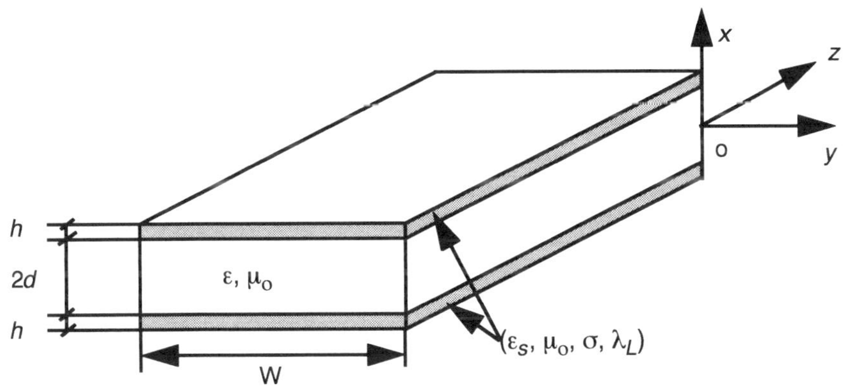

Figure 3.11 A superconducting planar waveguide of finite thickness.

$$\nabla^2 B_y + \left(\omega^2\mu_o\varepsilon_s - \frac{1}{\lambda_L^2} - i\omega\mu_o\sigma_n\right)B_y = 0 \tag{3.4.1}$$

where σ_n is the normal conductivity, λ_L the London penetration depth, and ε_s the permittivity, which are in general functions of temperature T and radian frequency ω. Here, we consider only non-magnetic superconductors so that the magnetic permeability is nearly equal to μ_0. We suppose that the wave solution has the form of $B_y = B_y(x)\exp(i(\omega t - \beta z))$ with $B_y(x)$ satisfies the equation:

$$\frac{\partial^2 B_y}{\partial x^2} + (k^2 - \beta^2)B_y = 0 \tag{3.4.2}$$

where β is as yet an unknown quantity. For a medium having losses, β is generally a complex quantity, which may be expressed by

$$\beta = \beta_o - i\alpha \tag{3.4.3}$$

with β_0 being the real propagation constant, and α the attenuation coefficient. The propagation constant β_0 defines the wavelength and phase velocity of the wave propagating in the waveguide at frequency ω. The parameter k in Eq.(3.4.2) is defined by

$$k^2 = \omega^2\mu_o\varepsilon_s - \frac{1}{\lambda_L^2} - i\omega\mu_o\sigma_n \qquad (d < |x| < d + h) \tag{3.4.4}$$

$$k^2 = \omega^2\mu_o\varepsilon \qquad (0 < |x| < d) \tag{3.4.5}$$

$$k^2 = \omega^2\mu_o\varepsilon_o \qquad (d+h<|x|<\infty) \tag{3.4.6}$$

where ε_o is the permittivity in vacuum, and ε (= $\varepsilon'-i\varepsilon''$) the permittivity of the dielectric medium in the waveguide.

The electric field $\boldsymbol{E}$ can be found to be related to the magnetic field $\boldsymbol{B}$ by Maxwell´s equations

$$\boldsymbol{E} = -\frac{i\omega}{k^2}\nabla\times\boldsymbol{B} \tag{3.4.7}$$

To find the solution for even TM modes, we may note the fact that the magnetic field $\boldsymbol{B}$ external to the waveguide vanishes since the field from the currents in the $+z$ and $-z$ direction will cancel at all external points (but will add at internal points) provided that $W/2d$ is supposed to be sufficiently large. In fact, it has been shown quantitatively that the fringe field effect is very small for the ratio $W/2d \geq 15$ and for $h/2d \leq 50$ [Chang (1979)]. In such a case, the solution for even TM modes can be found in the following form:

$$B_y = A\frac{\cos(\kappa d)}{e^{\gamma h}-e^{-\gamma h}}[e^{-\gamma(|x|-d-h)}-e^{\gamma(|x|-d-h)}] \qquad (d<|x|<d+h) \tag{3.4.8}$$

$$B_y = A\cos(\kappa x) \qquad (|x|<d) \tag{3.4.9}$$

for the magnetic field component B_y. The two components of the electric field $\boldsymbol{E}$ can then be found from Eq.(3.4.7). In particular, by using the interface condition for the continuity of the electric field component E_z at $x = \pm\, d$, we obtain the following eigenvalue equation:

$$tg(\kappa d) = \frac{(\omega/c)^2\varepsilon_r(1-i\varepsilon''/\varepsilon')}{(\omega/c)^2\varepsilon_{sr}-\lambda_L^{-2}-i(\omega/c)^2\sigma/(\omega\varepsilon_o)} \times \frac{\gamma}{\kappa} \times \coth(\gamma h) \tag{3.4.10}$$

where c is the phase velocity of light in vacuum, and ε_r (= $\varepsilon'/\varepsilon_0$) and ε_{sr} (= $\varepsilon_s/\varepsilon_0$) are the relative permittivity of the dielectric medium and of the superconductor, respectively. The parameters γ and κ are in general two complex quantities, given, respectively, by the following equations:

$$\gamma^2 = \beta^2-\omega^2\mu_o\varepsilon_s+\frac{1}{\lambda_L^2}+i\omega\mu_o\sigma \tag{3.4.11}$$

and

$$\kappa^2 = \omega^2\mu_o\varepsilon-\beta^2 \tag{3.4.12}$$

Equation (3.4.10) determines, in principle, the value of the propagation constant

β_o as well as the attenuation coefficient α with the aid of Eqs.(3.4.11) and (3.4.12). However, since the equation is a transcendental containing complex quantities, it is not easy to solve in the general case. In some special cases of practical interest, approximate solutions may, however, be obtained.

In the case of superconducting waveguide, the displacement current in the superconductors may be ignored since it is usually small as compared to the supercurrent at microwave frequencies. In addition, we know that β^2 should be of the same order as $\omega^2\mu_o\varepsilon$ for guide modes. Thus we may write

$$\gamma^2 = \frac{1}{\lambda_L^2} + i\omega\mu_o\sigma_n \tag{3.4.13}$$

The surface impedance of the superconductor can then be found by

$$Z_s = \frac{i\omega\mu_o\lambda_L}{\sqrt{1 + i2(\lambda_L/\delta_n)^2}}\coth\left(\frac{h}{\lambda_L}\sqrt{1 + i2(\lambda_L/\delta_n)^2}\right) \tag{3.4.14}$$

in which δ_n is the normal skin depth. Obviously, the parameter γ is complex, and can be expressed as $\gamma = \gamma_o + i\gamma_I$. Equation (3.4.14) determines the surface impedance of the superconductor of thickness h including both the normal skin effect and the Meissner effect, expressed by London´s equations.

If the London penetration depth λ_L is much smaller than the normal skin depth δ_n, Eq.(3.4.14) may be reduced to

$$Z_s = i\omega\mu_o\lambda_L\coth\left(\frac{h}{\lambda_L}\right) \tag{3.4.15}$$

which shows that the surface impedance of the superconductor is reactive and, therefore, lossless in the superconductor. In the normal state, we have then

$$Z_s = \frac{\gamma}{\sigma}\coth(h\gamma) \tag{3.4.16}$$

with

$$\gamma = \frac{1}{\delta_n}(1 + i) \tag{3.4.17}$$

which is a well-known result for the normal conductor if the effect of relaxation time is ignored. It should be noted that the above formulas represents the impedance of one conductor only.

We now study further the propagation properties of the superconducting waveguide. For the even TM modes considered, we can expect β^2 to be near the value of $\omega^2\mu_o\varepsilon$, found for the ideal case. Thus we have $tg(\kappa d) \approx \kappa d$. With this approximation, we can then determine the propagation constant β_o and the attenuation coefficient α with the aid of Eq.(3.4.12). In particular, if the displacement current is negligible and the condition of $\min\{\lambda_L/\delta_n, 2\lambda_L h/\delta_n^2\} << 1$ is satisfied, we can derive the following formulas:

$$\beta_o = \frac{\omega\sqrt{\varepsilon_r}}{c}\sqrt{a} \times \left[1 + \frac{1}{8}\left(\frac{\varepsilon''}{\varepsilon'} + \frac{b}{a}\right)^2\right] \tag{3.4.18}$$

for the determination of the propagation constant, and

$$\alpha = \frac{\omega\sqrt{\varepsilon_r}}{2c}\sqrt{a} \times \left|\frac{\varepsilon''}{\varepsilon'} - \frac{b}{a}\right| \tag{3.4.19}$$

for the determination of the attenuation coefficient. Here, the parameters a and b are given, respectively, by

$$a = 1 + \frac{\lambda_L}{d}\coth\left(\frac{h}{\lambda_L}\right) + \frac{h}{d}\left(\frac{\lambda_L}{\delta_n}\right)^4\left(\sinh\left(\frac{h}{\lambda_L}\right)\right)^{-2} \tag{3.4.20}$$

$$b = -\frac{\lambda_L^3}{d\delta_n^2}\left[\coth\left(\frac{h}{\lambda_L}\right) + \frac{h}{\lambda_L}\left(\sinh\left(\frac{h}{\lambda_L}\right)\right)^{-2}\right] \tag{3.4.21}$$

On the other hand, if the thickness h of the superconductor is sufficiently large, we may find that the propagation constant β_o and the attenuation coefficient α can be expressed by

$$\beta_o = \frac{\omega}{\sqrt{2}c}\sqrt{\varepsilon_r}\left\{\left(a_1 + b_1\frac{\varepsilon''}{\varepsilon'}\right) + \sqrt{(a_1^2 + b_1^2)\left(1 + \left(\frac{\varepsilon''}{\varepsilon'}\right)^2\right)}\right\}^{1/2} \tag{3.4.22}$$

and

$$\alpha = \frac{\omega}{\sqrt{2}c}\sqrt{\varepsilon_r}\left\{\sqrt{(a_1^2 + b_1^2)\left(1 + \left(\frac{\varepsilon''}{\varepsilon'}\right)^2\right)} - \left(a_1 + b_1\frac{\varepsilon''}{\varepsilon'}\right)\right\}^{1/2} \tag{3.4.23}$$

where parameter a_1 and b_1 read, respectively,

$$a_1 = 1 + \frac{\lambda_L^2\gamma_o}{d(1 + 4(\lambda_L/\delta_n)^4)}\left(1 + \frac{2\gamma_I}{\gamma_o}\left(\frac{\lambda_L}{\delta_n}\right)^2\right) \tag{3.4.24}$$

and

$$b_1 = \frac{\lambda_L^2\gamma_o}{d(1 + 4(\lambda_L/\delta_n)^4)}\left(\frac{\gamma_I}{\gamma_o} - 2\left(\frac{\lambda_L}{\delta_n}\right)^2\right) \tag{3.4.25}$$

in which γ_o and γ_I are given, respectively, by

$$\gamma_o = \frac{1}{\sqrt{2}\lambda_L}\sqrt{1+\sqrt{1+4(\lambda_L/\delta_n)^4}} \tag{3.4.26}$$

$$\gamma_I = \frac{1}{\sqrt{2}\lambda_L}\sqrt{\sqrt{1+4(\lambda_L/\delta_n)^4}-1} \tag{3.4.27}$$

These formulas can be further simplified if $\lambda_L \ll \delta_n$. In such a case, we may write

$$\beta_o = \frac{\omega\sqrt{\varepsilon_r}}{c}\sqrt{1+\frac{\lambda_L}{d}} \times \left(1+\frac{1}{8}\left(\frac{-\varepsilon''}{\varepsilon'}+\frac{\lambda_L^2}{\delta_n^2(1+d/\lambda_L)}\right)^2\right) \tag{3.4.28}$$

and

$$\alpha = \frac{\omega\sqrt{\varepsilon_r}}{2c}\sqrt{1+\frac{\lambda_L}{d}} \times \left|\frac{\varepsilon''}{\varepsilon'}+\frac{\lambda_L^2}{\delta_n^2(1+d/\lambda_L)}\right| \tag{3.4.29}$$

provided that $|\varepsilon''/\varepsilon'| \ll 1$. It can be seen that the dielectric loss in the dielectric medium of the waveguide has relatively little effect (second-order effect) on the propagation speed. It has, however, the effect (first-order effect) on the attenuation of the wave propagation in the superconducting waveguide.

It is noted that, in the derivation of formulas (3.4.28)–(3.4.29), no special assumption has been made for the value of the ratio λ_L/d. However, the condition $\lambda_L \ll \delta_n$ should be fulfilled. If the dielectric loss in the dielectric core of the waveguide is negligible for certain low loss dielectric materials, and $\lambda_L^2/[\delta_n^2(1+d/\lambda_L)] \ll 1$, we may further get the following approximate formula for the propagation constant:

$$\beta_o = \frac{\omega\sqrt{\varepsilon_r}}{c}\sqrt{1+\frac{\lambda_L}{d}} \tag{3.4.30}$$

and the formula for the attenuation coefficient:

$$\alpha = \frac{\omega\sqrt{\varepsilon_r}}{c}\sqrt{1+\frac{\lambda_L}{d}} \times \frac{\lambda_L^2}{2\delta^2(1+d/\lambda_L)} \tag{3.4.31}$$

Equations (3.4.30) and (3.4.31) are in agreement with the result given by Matick (1969) for superconducting waveguides with the neglection of both the normal conduction loss and the dielectric loss. It can be shown that, to have lower loss and higher propagation speed for high-speed circuits, we should design the superconducting waveguide such that the ratio λ_L/d is kept smaller. On the other hand, if delay lines are desired, we could slow down the propagation speed by increasing the ratio λ_L/d. This can be done by choosing smaller d and/or thinner h superconducting films.

In addition, for the normal state, we can find from Eqs.(3.4.22)–(3.4.25)

$$\beta_o = \frac{\omega\sqrt{\varepsilon_r}}{c}\left(1 + \frac{\delta_n}{4d}\right)\left[1 + \frac{1}{8}\left(\frac{-\varepsilon''}{\varepsilon'} + \frac{\delta_n}{2d}\right)^2\right] \tag{3.4.32}$$

and

$$\alpha = \frac{\omega\sqrt{\varepsilon_r}}{2c}\left(1 + \frac{\delta_n}{4d}\right)\left|\frac{\varepsilon''}{\varepsilon'} + \frac{\delta_n}{2d}\right| \tag{3.4.33}$$

for the case $\delta_n/d << 1$ and $|\varepsilon''/\varepsilon'| << 1$. Here, the effect of relaxation time is ignored. When the dielectric loss is negligible, these two formulas (3.4.32) and (3.4.33) recover the well-known result for a normal conducting waveguide in the absence of the dielectric loss [Matick (1969)].

So far, we have derived some formulas for calculating the propagation constant β_0, the attenuation coefficient α, and the surface impedance Z_S of the superconducting planar waveguide. In order to study further the dispersion and attenuation distortion behaviors of the superconducting planar waveguide, relations characterizing the frequency-dependent behaviors of material properties, such as the normal conductivity σ_n, the permittivity ε, and the penetration depth λ_L have to be determined for specific materials under consideration. In the following section, we shall first introduce a modified two-fluid model to describe phenomenologically the frequency-dependent behaviors of the conductivity and the penetration depth in superconductors.

3.4.2 Modified Two-Fluid Model

It has been known that, in a non-stationary process, the presence of normal conduction electrons requires the modification of London´s electrodynamic equations. This has led to the early proposal of a two-fluid model, as shown in Section 3.2.3. The classical two-fluid model has been used widely and successfully to describe behaviors of superconductors at frequencies well below the energy gap frequency. In particular, it has been shown in Section 3.3.6 that the surface resistance of superconductors has a frequency-dependence of ω^2 according to the classical two-fluid model, while the surface resistance for normal conductors is proportional to $\omega^{1/2}$. Appreciable deviations between the predicted data using the classical two-fluid model and experimental data at higher frequencies were, however, observed [Pippard (1954), Glover and Tinkham (1957)], and had been studied systematically by Pippard in early 1950s, which resulted in the proposal of nonlocal modification of electrodynamics for superconductors. Pippard´s nonlocal theory is useful to describe superconductors with relatively large coherence length, which will be discussed in Section 3.4.5. It has been found that recently discovered high-T_c oxide superconductors have very small coherence lengths and their surface resistance measurements do not agree always with the frequency-dependence ω^2 behavior predicted by the classical two-fluid model in the GHz frequency range. A variety of modified forms of the two-fluid model have, therefore, been proposed. Inthis section, we introduce a modified two-fluid model in which both the temperature and frequency dependence of the superelectron density and of the normal electron density may be taken into account within the phenomenological framework [Zhou (1994a)].

According to the BCS theory, because of the energy gap, energy can be absorbed only by already excited states (quasi-particles) for frequencies smaller than a gap frequency, defined by

$$\omega_g = \frac{2\Delta}{\hbar} \tag{3.4.34}$$

at absolute zero. Here, Δ denotes the gap energy. As an example, for a BCS type of superconductor of $T_c = 90$ K, the gap-frequency ω_g is about 4×10^{13} rad/s at absolute zero, well above THz frequency. The density of superconducting electrons is supposed to be frequency-independent up to the gap frequency. Thus, according to this model, at absolute zero, the conduction electrons of a superconductor will not absorb energy for photon frequencies smaller than the gap frequency. However, it has been observed experimentally that there exists a certain type of residual rf resistance even at $T = 0$ for both low T_c metallic and high T_c oxide superconductors. According to the BCS theory, the residual rf resistance seems to be not related to the condensed electrons. Thus it has been claimed to be due to surface imperfection and/or granular structures of the superconducting samples.

At finite temperature, quasi-particles can be excited thermally, which cause the energy absorption even when photon frequency is smaller than the gap frequency. In general, the gap frequency is temperature-dependent. Obviously, the densities of both the superconducting electrons and the normal conduction electrons are temperature-dependent. Experimentally, it has been found that the fraction of superelectrons varies approximately as $1-\Theta^4$, where Θ $(= T/T_c)$ denotes the reduced temperature. Thus the classical two-fluid model has been proposed, as discussed in Section 3.2.3. The possible effect of frequency was, however, not considered in the classical two-fluid model. In the modified two-fluid model, we assume that, according to quantum–mechanical arguments, there exists the possibility of the breaking of Cooper pairs of superconducting electrons even when photon frequency is smaller than the gap frequency. This probability should increase quickly as the photon frequency ω approaching to the gap frequency ω_g, and reach to one when the photon frequency ω is larger than the gap frequency ω_g. Indeed, it was noticed by Glover and Tinkham (1957) that the photon frequency ω at which the superconductor behaves as a normal conductor can be several times its gap frequency ω_g. This implies the fact that the transition of superconducting state to the normal state is not sharp at the gap frequency. Intuitively, the proposed frequency-dependent behavior of the superconducting electron density may be shown in Fig. 3.12.

Phenomenalogically, we may assume that the density of superconducting electrons and the density of normal conduction electrons can be described, respectively, by the following relations:

$$\frac{n_s(\Theta, \omega)}{n} = (1 - \Theta^4)g(\Theta, \omega) \tag{3.4.35}$$

and

$$\frac{n_n(\Theta, \omega)}{n} = 1 - (1 - \Theta^4)g(\Theta, \omega) \tag{3.4.36}$$

where n denotes the total number of electrons per unit volume. The choice of the function g is somewhat arbitrary in the phenomenological model. By noting the probability of Cooper pair occupancy in the BCS theory, the function g is given here in the following form:

$$g(\Theta, \omega) = 1 - \frac{1}{\sqrt{1 + \Gamma\left(\frac{\omega_g(\Theta)}{\omega}\right)^\eta}} \tag{3.4.37}$$

where Γ and η are two phenomenalogical parameters, relevant to the superconductive material under consideration. The gap frequency ω_g is generally a function of temperature, which may be estimated by the BCS theory for gap parameter $\Delta(\Theta)$. Approximately, the variation of the gap frequency $\omega_g(\Theta)$ with temperature may well be evaluated within 2% by the following equation:

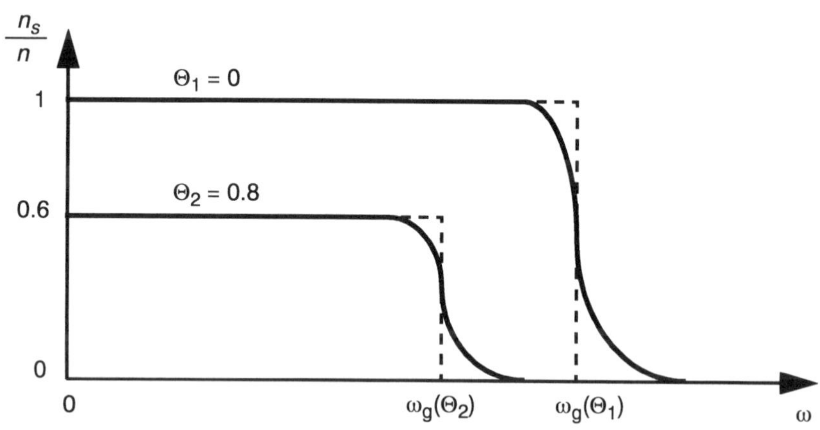

Figure 3.12 Scheme of frequency-dependence of superconducting electron density.

$$\frac{\omega_g(\Theta)}{\omega_g(0)} = \sqrt{\cos\left(\frac{\pi}{2}\Theta^2\right)} \tag{3.4.38}$$

We may note that the function $g(\Theta,\omega)$ given by Eq.(3.4.37) is quite similar to the function characterizing the probability of pair occupancy in the BCS theory. Since the choice of the function g is based on phenomenological consideration, the validity of this model has to be checked properly with experiments. Though there exists the possibility that one might be able to derive it from fundamental microscopic theory, this should not prevent us from applying this model, just like many other phenomenological models in practical engineering applications.

Based on this modified two-fluid model, we may write the total current density in the superconductor, using the expression of complex conductivity, by

$$\boldsymbol{J} = \sigma\boldsymbol{E} = (\sigma_1 - i\sigma_2)\boldsymbol{E} \tag{3.4.39}$$

where

$$\frac{\sigma_1}{\sigma_{DC}} = \frac{1-(1-\Theta^4)g(\Theta,\omega)}{1+4(\lambda_o/\delta_o)^4} \tag{3.4.40}$$

with $\sigma_{DC} = n\tau_n e^2/m_e$ denoting the dc-conductivity of the superconductor at the normal state, which is generally a function of temperature. Note that the effect of the momentum relaxation time τ_n has also been taken into account here. In Eq.(3.4.40), $\lambda_o = (m_e/\mu_o n e^2)^{1/2}$ is called here the dc-penetration depth of the superconductor at absolute zero. δ_o is an introduced quantity defined by

$$\delta_o = \sqrt{\frac{2}{\mu_o\omega\sigma_{DC}}} \tag{3.4.41}$$

The quantity σ_2 in Eq.(3.4.39) is given by

$$\frac{\sigma_2}{\sigma_{DC}} = \frac{(1-\Theta^4)g(\Theta,\omega)+4(\lambda_o/\delta_o)^4}{\mu_o\omega\sigma_{DC}\lambda_o^2(1+4(\lambda_o/\delta_o)^4)} \tag{3.4.42}$$

where the effect of displacement current in the superconductor has been ignored.

Furthermore, by identifying $\sigma_2 = (\omega\mu_o\lambda_L{}^2)^{-1}$ according to Eq.(3.2.37), we may introduce an rf penetration depth $\lambda_L(\Theta,\omega)$, given by

$$\lambda_L(\Theta,\omega) = \frac{\lambda_o\sqrt{1+4(\lambda_o/\delta_o)^4}}{\sqrt{(1-\Theta^4)g(\Theta,\omega)+4(\lambda_o/\delta_o)^4}} \tag{3.4.43}$$

which is now both temperature-dependent and frequency-dependent. It is noticed that, even at the normal state ($\Theta = 1$), the rf penetration depth does not go to infinite, but is given by

$$\lambda_{LN} = \frac{\delta_o^2}{2\lambda_o}\sqrt{1 + 4(\lambda_o/\delta_o)^4} \tag{3.4.44}$$

Indeed, the frequency-dependence of the penetration depth of some high-T_c superconductors at extremely high frequencies over 500 GHz has been shown recently by some experiments [Nuss et al. (1991)].

It can be shown that Eqs.(3.4.40) and (3.4.43) may recover their well-known forms given by Eq.(3.2.35) if we ignore the effect of the relaxation time, and the frequency-dependence of the density of superconducting electrons, which are fairly good approximation at relatively low frequencies. However, these effects may become important at high frequencies, especially when the operating frequency is close to the gap-frequency of the superconductor.

3.4.3 Surface Resistance of a Superconductor at High Frequency

Based on the modified two-fluid model, we shall now study quantitatively the surface resistance of superconductors at high frequencies, especially close to the gap frequency of the superconductor. At very high frequencies, both effects of the displacement current and the relaxation time may be thought to be of importance. A simple estimate shows that, at frequency around 1 THz, the effect of displacement current will be of interest only when $\varepsilon_s/\varepsilon_0$ is on the order of 10^9. Indeed, Ginsburg and Landau (1950) considered that a superconductor might have such an enormous dielectric constant, $\varepsilon_s/\varepsilon_0$, varying with temperature, but of the order of magnitude of 10^9 in order to account for the magnitude of the resistance at microwave frequencies. However, according to Pippard (1954), it does not seems a helpful suggestion if it cannot also explain the anomalous frequency variation, and more seriously, it would contradict the well-known Kramer-Kronig relations (see Section 1.4.6) if the relaxation effect is ruled out. Recent study of high-T_c superconductors has also indicated the possibility of the existence of negative dielectric function ε_s for the superconductors. Interested readers about this topic may refer to the work of, for instance, Dolgov et al. (1987) and Mei and Liang (1991).

In order to estimate the possible effect of the momentum relaxation time τ_n, let us consider some cases where the effect of the momentum relaxation time is of importance. On one hand, by Eq.(3.4.42), we can show that the effect of the momentum relaxation time may contribute to the penetration depth λ_L provided that the value of $\tau_n\sigma_n/\varepsilon_0$ is on the order of 10^9 at THz frequency. This corresponds to the value of τ_n on the order of 100–1000 ps if the superconductor has the value of σ_n around 10^5–10^6 S/m and λ_0 about 100 nm, which are typical for high-T_c oxide superconductors. On the other hand, the conductivity by Eq.(3.4.40) will be influenced by the effect of the momentum relaxation time if τ_n is on the order of 1 ps. Thus it is possible that, for some superconductors in certain frequency range, the effect of the momentum relaxation time will influence mainly the conductivity but not the penetration depth. Indeed, this is a desired situation since it will improve the dispersion properties of the superconducting waveguide. We have, however, to note that, at very high frequencies over the gap frequency of superconductors, the effect of the momentum relaxation time may influence significantly the electrodynamic behaviors of the superconductor. Though, at such high frequencies, all superconducting electron pairs are broken, the surface resistance may reduce due to the effect of the momentum relaxation time.

Now, let us check the possible anomalous skin effect. We make an estimate on the skin depth δ_n and the mean free path of electrons l_e in some high-T_c superconductors with $\sigma_n = 10^6$ S/m and $\xi = 1$ nm, and $\lambda_0 = 100$ nm at frequency $\omega=10^{12}$ rad/s. We may find that δ_n is about 1 μm, while, for $\tau_n = 0.1$ ps, l_e is about 0.01–0.1 μm for the mean (Fermi) velocity of electrons around 10^5–10^6 m/s. This shows that the anomalous skin effect is negligible in this case. Thus we may expect that, in the high-T_c oxide superconductors, the effect of the momentum relaxation time may dominate the anomalous skin effect at certain frequency range due to the lower charge-carrier density, lower conductivity, and smaller Fermi velocity (very short coherence length) in high-T_c oxide superconductors, as compared to low-T_c metallic superconductors. Furthermore, the nonlocal anomalous effect concerning superconducting current is also negligible since the coherence length ξ, about 1 nm for high-T_c oxide superconductors, is much smaller than the penetration depth λ_0, around 100 nm.

In the case where the local electrodynamics of the superconductors is valid, the normalized surface resistance of the superconductor may be obtained by

$$\frac{R_s}{R_{sn}} = \frac{2(\lambda_L/\delta_o)^2}{1 + 4(\lambda_L/\delta_n)^4} \times \frac{2\gamma_o(\lambda_L/\delta_n)^2 - \gamma_I}{\gamma_{oN} - 2\gamma_{IN}(\lambda_o/\delta_o)^2} \tag{3.4.45}$$

in which the parameters γ_o and γ_I are given by Eqs.(3.4.26) and (3.4.27), and δ_0 is given by Eq.(3.4.41). The rf penetration depth λ_L is given by Eq.(3.4.43) in the modified two-fluid model. The parameters γ_{oN} and γ_{IN} are given, respectively, by

$$\gamma_{oN} = \frac{1}{\delta_o\sqrt{1 + 4(\lambda_o/\delta_o)^4}}\sqrt{2(\lambda_o/\delta_o)^2 + \sqrt{1 + 4(\lambda_o/\delta_o)^4}} \tag{3.4.46}$$

$$\gamma_{IN} = \frac{1}{\delta_o\sqrt{1 + 4(\lambda_o/\delta_o)^4}}\sqrt{\sqrt{1 + 4(\lambda_o/\delta_o)^4} - 2(\lambda_o/\delta_o)^2} \tag{3.4.47}$$

In the classical two-fluid model, in which the effect of relaxation time is ignored and the function g is taken to be identical one, the normalized surface resistance is then given by

$$\frac{R_s}{R_{sn}} = 2\left(\frac{\lambda_L}{\delta_o}\right)^2 \times \frac{2\gamma_o\delta_o(\lambda_L/\delta_n)^2 - \gamma_I\delta_o}{1 + 4(\lambda_L/\delta_n)^4} \tag{3.4.48}$$

in which the parameters γ_o and γ_I are still given by Eqs.(3.4.26) and (3.4.27), while the normal skin depth δ_n and the penetration depth λ_L are given, respectively, by Eqs.(3.3.46) and (3.2.35) in the classical two-fluid model. A special case may be of interest. If the photon frequency ω is not very high so that $\lambda_L << \delta$, we may then get approximately

$$\frac{R_s}{R_{sn}} = \frac{\mu_o^{3/2}\sigma_n^{3/2}\lambda_o^3\omega^{3/2}t^4}{\sqrt{2}(1-t^4)^{3/2}} \tag{3.4.49}$$

with the neglection of the effect of the momentum relaxation time. This equation shows the well-known frequency-dependent behavior ($\omega^{3/2}$) of the normalized surface resistance of superconductors predicted by the classical two-fluid model. This prediction has been found experimentally to be correct at relatively lower frequency as compared to the gap frequency of the superconductor.

By the modified two-fluid model, we may now study the behaviors of the surface resistance of a superconductor even at extremely high frequencies close to the gap frequency ω_g of the superconductor. As a numerical example, Fig. 3.13 shows quantitatively the normalized surface resistance predicted by the

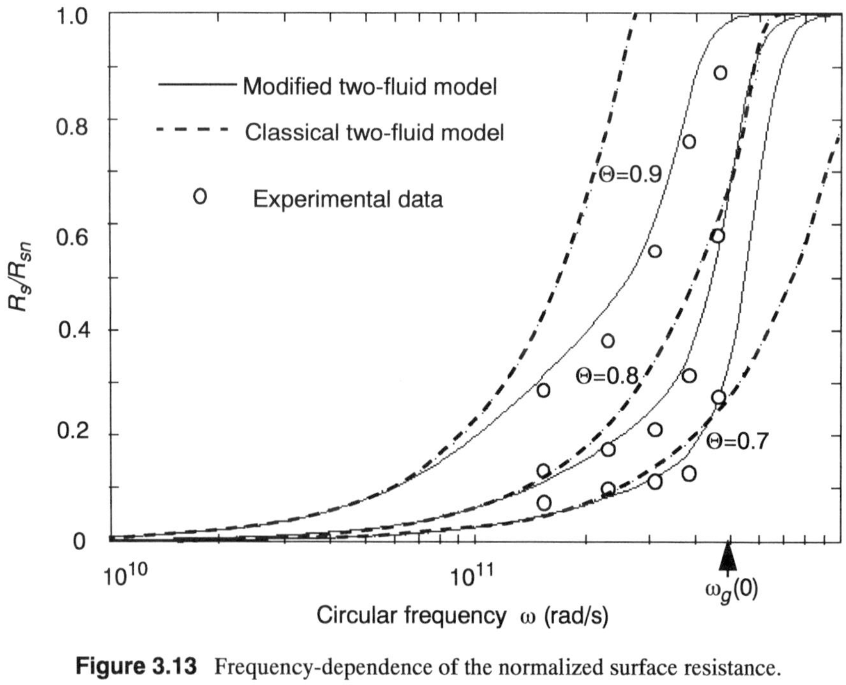

Figure 3.13 Frequency-dependence of the normalized surface resistance.

modified two-fluid model as compared to that given by the classical two-fluid model. In the calculation, parameters taken for the superconducting aluminium are $T_c = 1.178$ K; $\omega_g(0) = 4.93\times10^{11}$ rad/s; $\sigma_{DC} = 6.8\times10^8/\Omega\cdot$m; $\lambda_o = 50$ nm; $\Gamma = 100$, $\eta = 12$. It is shown that the modified two-fluid model gives more reasonable description of the high-frequency behavior of the surface resistance of the superconductor than the classical two-fluid model does, in comparison with the experimental data [Biondi and Garfunkel (1959)]. In particular, it is shown that the classical two-fluid model fails at high frequencies close to the gap frequency of the superconductor, while the modified two-fluid model can give a reasonable estimate of the surface resistance of the superconductor at high frequencies. We also find that, at lower frequencies (less than 10 GHz, here), the two models coincide as we may expect. This example shows that, although the aluminum is the type of superconductor fulfilling the conditions for the extreme anomalous limit, the modified two-fluid model may still be used to describe effectively the surface resistive behavior of the superconductor.

Furthermore, shown in Fig. 3.14 is the numerical results from the modified two-fluid model as compared with the experimental data for the aluminium at both frequencies below and above the energy gap frequency ω_g [Biondi and Garfunkel (1959)]. It is shown again that the modified two-fluid model gives reasonable prediction of the behavior of surface resistance at both frequencies either slight lower or slight higher than the gap frequency of the superconductor. Note that such a prediction cannot be made by the classical two-fluid model.

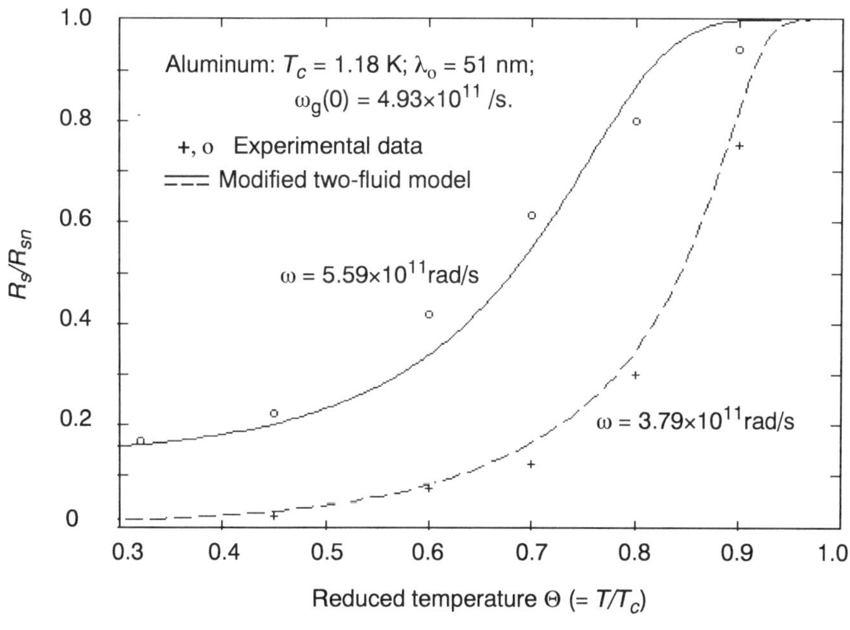

Figure 3.14 Temperature-dependence of normalized surface resistance.

3.4.4 Dispersion and Attenuation Distortion

Dispersion of waves in a waveguide may cause the spread of a signal pulse propagating in the waveguide. The pulse spreading reduces the bandwidth and limits information capacity of the waveguide. In addition, the quality of speech and data signals comprising a complex mixture of pure, single-frequency components may be degraded after passing through the waveguide due to different attenuation of the signals at different frequencies. The effect of the tonal degradation of the received signals is known as the "attenuation distortion" or "frequency attenuation distortion". The study of the dispersion and attenuation distortion properties of superconducting waveguides is therefore of practical interest. In this section, we shall simulate numerically the propagation behaviors of the superconducting planar waveguide based on the analytical solution given in Section 3.4.1 and with the use of the modified two-fluid model introduced in Section 3.4.2. In what follows, we shall assume that the superconductor considered is of the BCS type and has the following basic properties: $\lambda_o = 100$ nm, $T_c = 90$ K ($\omega_g = 4.16\times10^{13}$ rad/s at absolute zero), and $\sigma_n = 10^6$ S/m. These values are not changed in the following calculations. Other properties may vary, but they will be indicated in relevant places.

By Eq.(3.4.18), we may find that the phase velocity v of the superconducting planar waveguide with superconductors of finite thickness h can be determined by

$$\mathrm{v} = \frac{\omega}{\beta_o} = \frac{c}{\sqrt{a\varepsilon_r}} \times \left[1 + \frac{1}{8}\left(\frac{\varepsilon''}{\varepsilon'} + \frac{b}{a}\right)^2\right]^{-1} \tag{3.4.50}$$

in which the parameters a and b are given, respectively, by Eqs.(3.4.20) and (3.4.21). An estimation on the condition of $\min\{\lambda_L/\delta_n, 2\lambda_L h/\delta_n^2\}$ being less than one shows that Eq.(3.4.50) is approximately applicable for the superconducting planar waveguide at the excitation frequency ω up to 10^{13} rad/s in 77 K or lower temperature (with h on the order of 1 μm).

As a numerical example, shown in Fig. 3.15 is the dispersion behavior of the superconducting planar waveguide predicted by the modified two-fluid model. Different values of h of the superconducting planar waveguide are studied, as shown in the figure. It is shown that the superconducting planar waveguide is essentially non-dispersive at a frequency ω up to about 10% of the gap frequency ω_g. It is also shown that the phase velocity v decreases with decreasing thickness h of the superconductor. Besides, we may also show that the reduction of the thickness ($2d$) of the dielectric layer of the waveguide reduces also the phase velocity v. Such a size effect may, in fact, be used in the design of superconducting delay lines.

In the above calculation, although the effect of the momentum relaxation time is included ($\tau_n = 10$ ps in the example), it can be shown that the neglection of the effect of the momentum relaxation time has little influence on the dispersion property of the superconducting planar waveguide in the example.

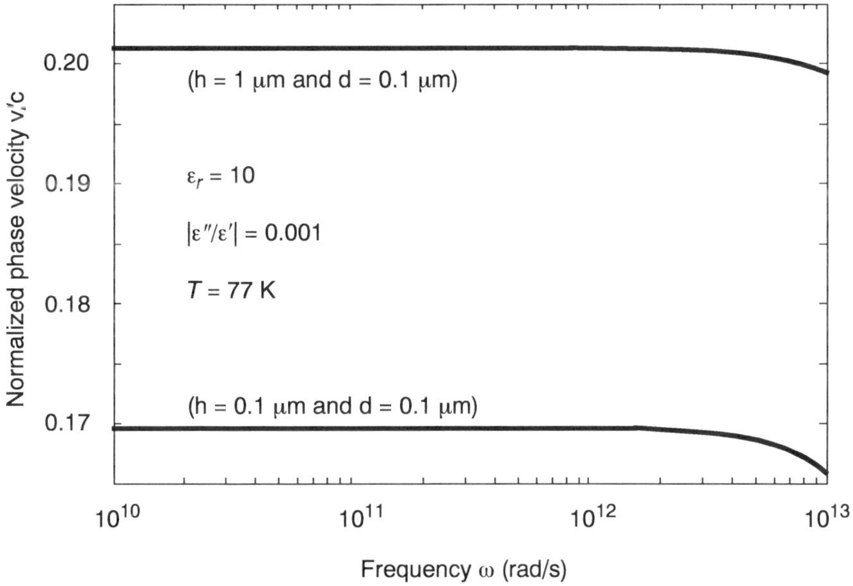

Figure 3.15 Dispersion behavior of the superconducting planar waveguide.

Also, it can be shown that the dielectric loss has little influence on the dispersion of the superconducting waveguide, but changing slightly the phase velocity.

In Fig. 3.16, by using the same model for the same type of superconductor, we show the comparison of the dispersion behavior of the superconducting planar waveguide with that of the normal conducting planar waveguide. In the calculation, both the effect of the momentum relaxation time and the effect of thickness h have been ignored by assuming that h is sufficiently large. It is shown that the normal conducting waveguide is much more dispersive than the superconducting waveguide. This is true even for very good conductors, such as copper which has the conductivity of $\sigma_n \approx 5\times10^8$ S/m at 77 K.

Let us now study the attenuation property of the superconducting planar waveguide as well as its comparison with that of the normal conducting planar waveguide. If the thickness h of the superconductor is large ($h >> 1$ μm), we may ignore the effect of the thickness h in the calculation. Thus, by Eq.(3.4.23), the numerical results on the attenuation coefficient α of the superconducting waveguide and the normal conducting waveguide are shown in Fig. 3.17. It is shown that the superconducting planar waveguide has much lower attenuation than the normal conducting planar waveguide at frequencies up to 1 THz. The effect of the dielectric loss in either the superconducting waveguide or the normal conducting waveguide are also shown in the figure. It can be seen that the dielectric loss in the superconducting waveguide is of particular significance at high frequencies.

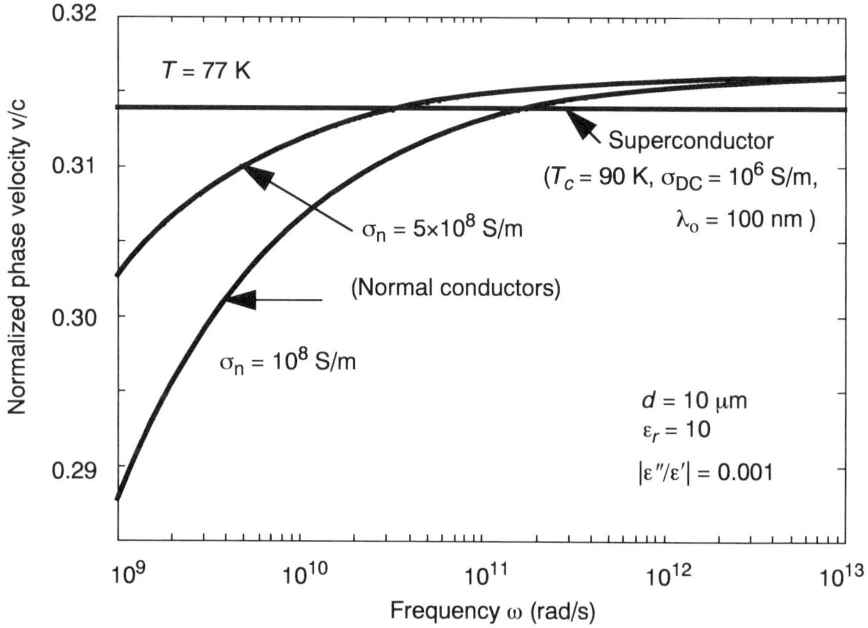

Figure 3.16 Dispersion behaviors of superconducting and normal planar waveguides.

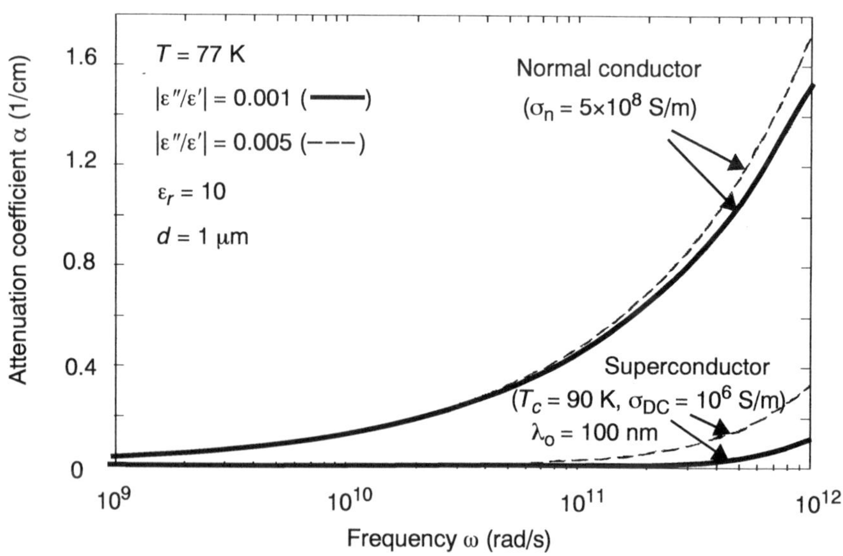

Figure 3.17 Comparison of the frequency-dependent behavior of the attenuation coefficient α between superconducting waveguide and normal waveguide.

The frequency-dependent behavior of the attenuation coefficient α also shows that the attenuation distortion in the normal conducting waveguides will be generally larger than that in the superconducting waveguide, provided that the excitation frequency ω is not too close to the gap frequency ω_g of the superconductor.

Finally, in Figs. 3.18 and 3.19, we study the size effect of the superconducting planar waveguide on the attenuation coefficient α. It is shown in Fig. 3.18 that the attenuation of the superconducting planar waveguide increases with decreasing the thickness of the dielectric layer of the waveguide. It is also shown that the attenuation coefficient α of the superconducting waveguide is nearly independent of the thickness h of the superconductor as long as h is larger than the penetration depth λ_L of the superconductor.

Shown in Fig. 3.19 is the size effect of the dielectric layer ($2d$) on the attenuation coefficient α. It is shown that the attenuation of the normal conducting waveguide increases much more rapidly than that of the superconducting waveguide as the thickness of the dielectric layer of the waveguide decreases. Thus we can expect that delay lines made of superconducting waveguides will have much lower attenuation than those made of normal conducting waveguide if they all use the effect of the dependence of phase velocity v on the thickness of the dielectric layer of the waveguide by reducing the thickness ($2d$).

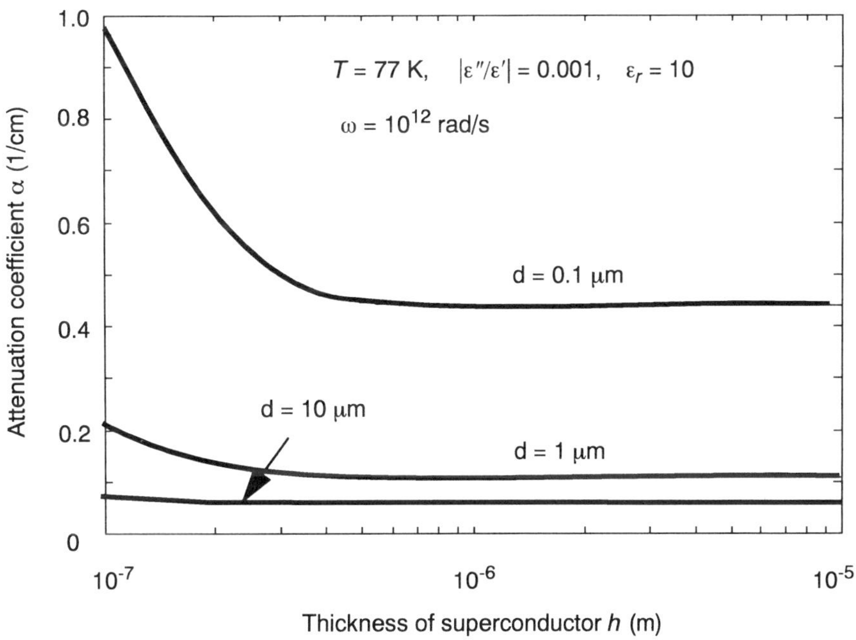

Figure 3.18 Size effect of the superconducting waveguide on the attenuation coefficient α.

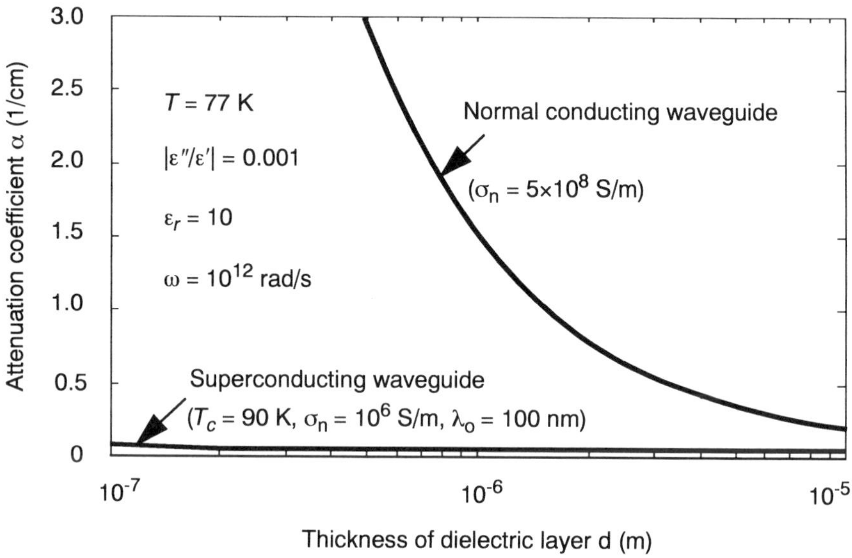

Figure 3.19 Size effect of the dielectric layer ($2d$) on the attenuation coefficient α.

3.4.5 Pippard´s Coherence Length and Nonlocal Relation

It is well known that, in normal conductors, when frequency is increased to a point where wavelength is comparable to the mean free path of an electron, one sees deviations from the usual skin effect, that is, the anomalous skin effect. In this situation, the electrons which contribute to the current at a particular point have momentum acquired from the field at other locations and earlier times. Therefore, the current at a point is determined not only by the field at that same location (as is the case for lower frequencies), so one must attempt to account for nonlocal effects. In superconductors, analogous situation may exist. Since the superconducting current is comprised of pairs of electrons, we are concerned about the extent of nonlocal effects due to the pair size, that is, the range over which a pair can remain coherent. London´s theory is a local theory in the sense that the current densities are related to the electromagnetic potentials at the same point in space, shown, for instance, by Eq.(3.2.10). It was found by Pippard that the London theory had to be modified for certain superconductors in order to explain some experimental observations from his series of experiments for the measurements of the penetration depth of various types of superconductors, and its dependence on applied magnetic fields [Pippard (1950a)] and on impurities in superconductors [Pippard (1950b)] as well as its anisotropy [Pippard (1953)]. In particular, Pippard found that the penetration depth was noticeably dependent upon the impurity content, which could not be explained by the local theory of London since the density of superelectron and its effective mass could only be weak functions of the impurity concentration. Thus, guided by the nonlocal expression relating current density and electric field in normal conductors [Reuter and Sondheimer (1948) and Chambers (1952)] for explaining anomalous skin effect:

$$J_n = \frac{3\sigma_n}{4\pi l_e} \int_V \frac{\boldsymbol{r} \cdot \boldsymbol{E}(\boldsymbol{x}', t)\boldsymbol{r}}{r^4} exp\left(-\frac{r}{l_e}\right) dV' \tag{3.4.51}$$

with $\boldsymbol{r} = \boldsymbol{x} - \boldsymbol{x}'$ and $r = |\boldsymbol{x} - \boldsymbol{x}'|$, which is reduced to Ohm's law $\boldsymbol{J}_n = \sigma_n \boldsymbol{E}$ for fields varying slowly over the electron mean free path l_e, Pippard proposed that the local $\boldsymbol{J}_S$–$\boldsymbol{A}$ relation (3.2.10) should be replaced for certain superconductors by a nonlocal relation of the form:

$$J_s = \frac{-3}{4\pi\mu_o\lambda_L^2\xi_o} \int_V \frac{\boldsymbol{r} \cdot \boldsymbol{A}(\boldsymbol{x}', t)\boldsymbol{r}}{r^4} exp\left(-\frac{r}{\xi}\right) dV' \tag{3.4.52}$$

where the integral is to be taken over the whole volume V of the superconductor. In the nonlocal relation (3.4.52), Pippard introduced two new parameters: ξ_0 and ξ. The parameter ξ_0 is called the intrinsic coherence length, which is independent of impurities, and can be given empirically by

$$\xi_o = \gamma \frac{\hbar v_F}{k_B T_c} \tag{3.4.53}$$

with the empirical constant γ. It was found that $\gamma = 0.18$ by the BCS microscopic theory. Here, v_F is the electron velocity at the Fermi surface and k_B the Boltzmann constant. The numerical values of ξ_0 are found roughly on the order of 10^{-7}–10^{-6} m for most pure metals (see Table 3.2). The parameter ξ is called the effective coherence length which may be given by an empirical relation:

$$\frac{1}{\xi} = \frac{1}{\xi_o} + \frac{1}{\alpha l_e} \tag{3.4.54}$$

with l_e being the electron mean free path and α being a constant on the order of unity.

The introduction of the concept of the coherence length in the nonlocal relation (3.4.52) characterizes the fact that spatial variation of the density of superelectrons cannot occur over arbitrarily small distances and is only possible within a certain distance ξ. Furthermore, the nonlocal model shows that the superconducting current density at a certain point depends on an average of the magnetic vector potential $\boldsymbol{A}$ over a volume of radius about ξ around the point of interest. To study quantitatively the variation of the penetration depth of superconductors with, for instance, the change of impurities in the superconductor, we may define, independent of any particular penetration law, an effective penetration depth by

$$\lambda = \frac{1}{B(0)}\int_0^\infty B(x)dx \tag{3.4.55}$$

where $B(0)$ is the magnetic induction at the surface, $B(x)$ the magnetic field at a distance x inside the superconductor, where x is measured from the surface of the superconductor. (It can be shown that Eq.(3.4.55) clearly yields the right value of λ when $B(x)$ is exponential as shown, e.g., by the example in Section 3.3.1.)

We may consider two extreme cases where Pippard's nonlocal relation (3.4.52) can be reduced to a simpler local form. In the first extreme case of $\lambda >> \xi$, which implies the slow variation of the magnetic vector potential A over the region $r < \xi$, the nonlocal relation (3.4.52) can be reduced to the following local form:

$$\boldsymbol{J}_s = -\frac{\xi}{\xi_o \mu_o \lambda_L^2}\boldsymbol{A} \tag{3.4.56}$$

Comparison with the London relation (3.2.10) shows that the penetration depth λ for this case is

$$\lambda = \lambda_L\sqrt{\frac{\xi_o}{\xi}} \tag{3.4.57}$$

which implies that λ will be increased if ξ is reduced by impurities or geometry in accordance to the experimental observation [Pippard (1953)]. For very impure specimens or for thin film, ξ may be taken equal to the mean free path (limited by impurity and/or boundary scattering). It can be seen that the London theory is rigorous only in the case of $\lambda >> \xi$. Noting the temperature-dependence of the penetration depth, all superconductors become local when their operating temperature is very near T_c since the penetration depth increases rapidly with increasing temperature according to the two-fluid model.

In the other extreme case of $\lambda << \xi_o$, which holds for most pure bulk metallic superconductors at temperatures not too close to T_c, Faber and Pippard (1955) have shown that

$$\lambda = \left(\frac{\sqrt{3}}{2\pi}\xi_o\lambda_L^2\right)^{1/3} \tag{3.4.58}$$

in which the intrinsic coherence length ξ_o are, for instance, $\xi_o = 2.1\times10^{-7}$ m for tin and 12.3×10^{-7} m for Al. Thus, when nonlocality is important, the actual penetration depth λ will exceed the London penetration depth λ_L by a modest factor of order $(\xi_0/\lambda_L)^{1/3}$.

It is the contribution of Pippard to the physics of superconductivity that the concepts of his nonlocality and coherence play important roles in determining the properties of the superconductors to be discussed further in Chapter 4.

3.4.6 The Mattis-Bardeen Theory

Mattis and Bardeen (1958) introduced a theory to describe the anomalous skin effect in normal and superconducting metals based on the BCS theory of superconductivity. In this theory, Mattis and Bardeen developed a mathematical relation between current density and an ac applied field, which is similar to Pippard's expression (3.4.52). By the Mattis-Bardeen theory, Miller had calculated the surface impedance of tin and aluminum by considering the electron mean free path $l_e \to \infty$. Later, the Mattis-Bardeen formula was simplified by Tuneaure (1967) in a form that is tractable for numerical calculation. In general, the Mattis-Bardeen formula can be expressed in terms of the Fourier components of the overall current density $\boldsymbol{J}(q)$ and vector potential $\boldsymbol{A}(q)$ as

$$\boldsymbol{J}(q) = -K(q)\boldsymbol{A}(q) \tag{3.4.59}$$

where the kernel function $K(q)$ is given by [see, e.g., Pöpel (1989)]

$$K(q) = \frac{-3}{4\pi\hbar v_F \lambda_L^2(0)} \int_0^{\infty} \int_{-1}^{1} (1-u^2) I(\omega, R, T) e^{iquR} e^{-R/l_e} du dR \tag{3.4.60}$$

in which

$$\begin{aligned} I(\omega, R, T) = {} & -i\pi \int_{\Delta-\hbar\omega}^{\Delta} [1-2f(E+\hbar\omega)][g(E)\cos(\alpha\Delta_2) - i\sin(\alpha\Delta_2)]e^{i\alpha\Delta_1} dE \\ & - i\pi\int_{\Delta}^{\infty}[1-2f(E+\hbar\omega)][g(E)\cos(\alpha\Delta_2) - i\sin(\alpha\Delta_2)]e^{i\alpha\Delta_1} dE \\ & + i\pi\int_{\Delta}^{\infty}[1-2f(E)][g(E)\cos(\alpha\Delta_1) + i\sin(\alpha\Delta_1)]e^{-i\alpha\Delta_2} dE \end{aligned} \tag{3.4.61}$$

where $\Delta_1 = (E^2 - \Delta^2)^{1/2}$, $\Delta_2 = ((E^2 + \hbar\omega)^2 - \Delta)^{1/2}$, $\alpha = R/(\hbar v_F)$, and

$$g(E) = \frac{1}{\Delta_1\Delta_2}(E^2 + \Delta^2 + \hbar\omega E) \tag{3.4.62}$$

$$f(E) = \frac{1}{1 + \exp(E/(k_B T))} \tag{3.4.63}$$

Here, k_B is the Boltzmann constant, $\hbar$ the reduced Plank constant, v_F the Fermi velocity, and $\lambda_L(0)$ is the London penetration depth at absolute zero.

The integration with respect to u in Eq.(3.4.60) can be easily done as

$$\int_{-1}^{1} (1-u^2)e^{iquR}du = \frac{4}{(qR)^2}\left(\frac{1}{qR}\sin(qR) - \cos(qR)\right) \tag{3.4.64}$$

Thus, with the substitution $x = qR$, the kernel $K(q)$ becomes

$$K(q) = \frac{-3}{\pi\hbar v_F \lambda_L^2(0) q}\int_0^{\infty}\left(\frac{\sin(x)}{x^3} - \frac{\cos(x)}{x^2}\right) I(\omega, x/q, T) e^{-x/(ql_e)} dx \tag{3.4.65}$$

Having found the Mattis-Bardeen kernel $K(q)$, we may calculate the surface impedance using an equation of Reuter and Sondheimer (1948) for diffuse scattering of the electrons at the surface of the conductor:

$$Z_s = i\omega\mu_o\pi\left[\int_0^{\infty} \ln\left(1 + \frac{K(q)}{q^2}\right) dq\right]^{-1} \tag{3.4.66}$$

A simple approximation to the surface impedance is the extreme anomalous limit approximation, in which the penetration of field is small compared with the coherence distance. In this approximation, we may set $\alpha = R/(\hbar v_F) = 0$ [Mattis and Bardeen (1958)]. It is then convenient to introduce a complex conductivity $\sigma = \sigma_1 - i\sigma_2$ for the superconducting state. We then have

$$\frac{\sigma_1}{\sigma_n} = \frac{2}{\hbar\omega}\int_{\Delta}^{\infty} [f(E) - f(E+\hbar\omega)]g(E)dE + \frac{1}{\hbar\omega}\int_{\Delta-\hbar\omega}^{-\Delta} [1 - 2f(E+\hbar\omega)]g(E)dE \tag{3.4.67}$$

and

$$\frac{\sigma_2}{\sigma_n} = \frac{1}{\hbar\omega}\int_{\Delta-\hbar\omega,\,-\Delta}^{\Delta} [1 - 2f(E+\hbar\omega)]g(E)dE \tag{3.4.68}$$

where the second term of Eq.(3.4.67) is zero unless $\hbar\omega > 2\Delta$ in which case the lower limit of the integral in Eq.(3.4.68) is $-\Delta$ instead of $\Delta - \hbar\omega$. Here, σ_n is the normal-state conductivity, and $g(E)$ is given by Eq.(3.4.62), which is so-called coherence factor in the BCS theory.

The set of Eqs.(3.4.67) and (3.4.68) determine the temperature and frequency dependence of the dynamic conductivity of a clean superconductor, which in most cases corresponds to type I superconductor ($\lambda < \xi$). The first term in Eq.(3.4.67) is the contribution of thermally excited normal quasiparticles which exist only at $T \neq 0$, whereas the second term is the contribution resulting from breaking of the Cooper pairs by an external field when the frequency $\omega > 2\Delta/\hbar$. At $T = 0$, there is a threshold 2Δ for excitations in a superconductor. Thus, for frequencies $\omega < 2\Delta/\hbar$, there should be no intrinsic loss according to the model. At finite but low temperatures ($< 0.5T_c$), the resulting surface resistance is given approximately by [see, e.g., Duzer and Turner (1981)]

$$R_s = \frac{C\omega^{3/2}}{T}\exp\left[-\frac{\Delta(T)}{k_B T}\right] \tag{3.4.69}$$

where C is some material-dependent constant, and $\Delta(T)$ is the energy gap. It is shown that the surface resistance by Eq.(3.4.69) decreases exponentially at low temperatures whereas the London theory with the classical two-fluid model predicts a slower power-law decrease.

The Mattis-Bardeen theory has been used by a number of researchers to investigate high-frequency behaviors of superconductors for both conventional low-T_c metallic superconductors and high-T_c oxide superconductors [Miller (1960), Kautz (1978), Nuss and Goossen (1989), Pöpel (1989), Turneaure et al. (1991), and Zhou et al. (1992)]. Readers who are interested in the subject may consult those original works.

Due to its simplicity, engineers often make use of the classical two-fluid model introduced in Section 3.2.3 in their design of superconducting devices. Studies have shown that qualitatively the London theory with the classical two-fluid model may provide a quite satisfactory agreement with experimental data for many superconductors, especially for dirty superconductors, where the extreme anomalous limit approximation is not valid. One may find that quantitatively the comparison of temperature and frequency dependence of the surface impedance predicted by the classical two-fluid model with experimental data may show some differences, but even the Mattis-Bardeen theory is unable to achieve reliable estimates of losses in many practical situations. Thus, for not too low temperatures and frequencies on the order of 10 GHz, the local electrodynamics based on the London theory together with the classical two-fluid model can well be used for practical estimates of losses in superconducting resonators and transmission lines, especially if the appropriate penetration depth is used.

3.5 THE LONDON ELECTRODYNAMIC MODEL FOR ANISOTROPIC SUPERCONDUCTORS

3.5.1 The London Equations for Anisotropic Superconductors

There are a number of superconductive materials having anisotropic properties. Some of recently discovered high-temperature oxide superconductors, such as Tl- and Bi-based high-T_c superconductors, are strongly anisotropic. Although there are no complete material property data available for these new high temperature oxide superconductors yet, a good survey of material properties of these superconductors may be found in a recent textbook by Poole et al. (1995). Some typical data reported are, for instance, the resistivity of $YBa_2Cu_3O_{7-\delta}$ is around two orders of magnitude greater along the c-axis than parallel to the ab-plan of the superconductor, that is, $\rho_c/\rho_{ab} \sim 100$ with ρ_{ab} being about 200 $\mu\Omega\cdot$cm at 100 K, and the resistivity of $Bi_2Sr_{2.2}Cu_2O_8$ is around five orders of magnitude greater along the c-axis than parallel to the ab-plan of the superconductor, that is, $\rho_c/\rho_{ab} \sim 10^5$ with ρ_{ab} being about 55 $\mu\Omega\cdot$cm at 100 K. Besides, for the field penetration depths of $YBa_2Cu_3O_{7-\delta}$ (T_c = 92 K), one has $\lambda_c/\lambda_{ab} \sim 6$ with λ_{ab} being about 90 nm. For $HgBa_2Ca_2Cu_3O_{8+\delta}$ (T_c = 133 K), one has $\lambda_c/\lambda_{ab} \sim 27$ with λ_{ab} being about 130 nm.

To model electrodynamic properties of anisotropic superconductors in weak fields, the simplest way is to express the anisotropy in terms of a phenomenological effective-mass tensor m_{kl} in the framework of the London theory. Extending the free-electron model to include anisotropic effects, we may write the first London equation in its anisotropic form:

$$\mu_o \frac{\partial}{\partial t}(\Lambda \cdot \boldsymbol{J}^{(s)}) = \boldsymbol{E} \tag{3.5.1}$$

where Λ_{kl} is a coefficient tensor, related to the effective mass tensor m_{kl} by

$$\Lambda_{kl} = \frac{1}{n_s e^{*2}} m_{kl} \tag{3.5.2}$$

which is supposed to be symmetric and positive definite. For isotropic superconductors, we have simply $\Lambda_{kl} = \lambda_L{}^2 \delta_{kl}$ with λ_L being the London penetration depth.

Similarly, the second London equation for the anisotropic superconductor can be written as

$$\mu_o \nabla \times (\Lambda \cdot \boldsymbol{J}^{(s)}) = -\boldsymbol{B} \tag{3.5.3}$$

To include the effect of normal conducting electrons, we may generalize the classical two-fluid model for anisotropic superconductors so that the total current density $\boldsymbol{J}$ can be expressed by

$$J = J^{(s)} + J^{(n)} \tag{3.5.4}$$

where $J^{(n)}$ is the normal conduction current density obeying Ohm´s law:

$$J_k^{(n)} = \sigma_{kl} E_l \tag{3.5.5}$$

in which σ_{kl} is the normal conductivity tensor, related to the reduced temperature $\Theta = T/T_c$ in the classical two-fluid model by

$$\sigma_{kl} = \sigma_{kl}^{DC} \Theta^4 \tag{3.5.6}$$

where σ_{kl}^{DC} is the normal dc-conductivity tensor of the anisotropic superconductor.

Furthermore, in the classical two-fluid model, we have

$$\Lambda_{kl} = \frac{1}{1 - \Theta^4} \Lambda_{kl}^{o} \tag{3.5.7}$$

where $\Lambda^{o}{}_{kl}$ is the coefficient tensor characterizing the dc-penetration depths of the anisotropic superconductor at temperature of absolute zero. For an isotropic superconductor, we have simply $\Lambda_{kl} = \lambda_L{}^2 \delta_{kl}$ with $\lambda_L = \lambda_o/(1-\Theta^4)$. Thus $\Lambda^{o}{}_{kl} = \lambda_o{}^2 \delta_{kl}$ with λ_o being the dc-penetration depth of the isotropic superconductor at temperature of absolute zero.

Thus, for homogeneous non-magnetic anisotropic superconductors in weak fields, we may derive the following field equation:

$$e_{kpq} e_{mst} \Lambda_{qm} \frac{\partial^2 B_t}{\partial x_p \partial x_s} - \mu_o e_{kpq} \Lambda_{qm} \sigma_{mn} \frac{\partial E_n}{\partial x_p} = -B_k \tag{3.5.8}$$

with the use of Eqs.(3.5.3)–(3.5.5) and the Maxwell equation of Ampere´s circuital law. Here, e_{kpq} is the permutation symbol. Equation (3.5.8) indicates that the electric field components may not be eliminated simply by using Maxwell's equations for the anisotropic superconductor. This implies that we have to, in general, solve the coupled set of partial differential equations for both electric and magnetic fields simultaneously for the anisotropic superconductor. The field equation derived for the determination of the magnetic field $\boldsymbol{B}$ should be supplemented with appropriate boundary conditions for practical problems.

Illustratively, let us study a simple example of an anisotropic superconductor which is supposed to occupy a half infinite space, as shown in Fig. 3.20. We choose a coordinate system with its axes (*x*, *y*, *z*) in the principal axes (*a*, *b*, *c*) of the superconductor. In such a coordinate system, the coefficient tensors Λ_{kl} and σ_{kl} can be expressed in the matrix form:

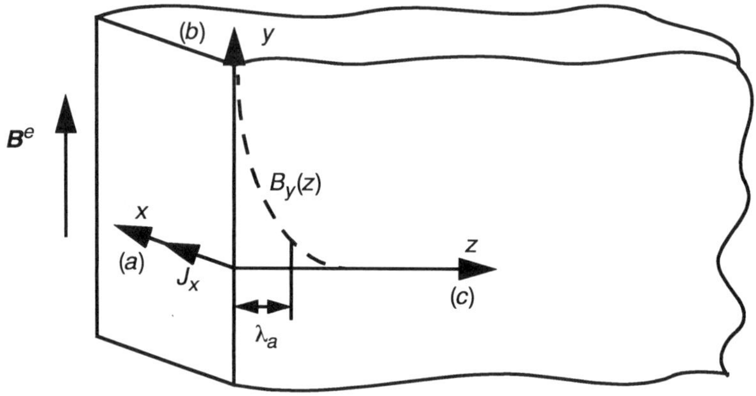

Figure 3.20 Scheme of an anisotropic superconductor occupying a half-space.

$$[\Lambda_{kl}] = \begin{bmatrix} \lambda_a^2 & 0 & 0 \\ 0 & \lambda_b^2 & 0 \\ 0 & 0 & \lambda_c^2 \end{bmatrix} \quad \text{and} \quad [\sigma_{kl}] = \begin{bmatrix} \sigma_a & 0 & 0 \\ 0 & \sigma_b & 0 \\ 0 & 0 & \sigma_c \end{bmatrix} \tag{3.5.9}$$

In particular, if a static magnetic field $\boldsymbol{B}^e$ is applied along the y-axis to the superconductor, we may find that the magnetic field inside the superconductor can be expressed as $\boldsymbol{B} = (0, B_y, 0)$, where the magnetic field component B_y is determined by the following equation:

$$\lambda_a^2 \frac{\partial^2 B_y}{\partial z^2} = B_y \tag{3.5.10}$$

which indicates that the magnetic field in the superconductor decays exponentially with the increase of z (> 0) and that the measure of the field penetration is by λ_a.

3.5.2 Field Solution of an Anisotropic Superconducting Planar Waveguide

In this section, we shall study the field solution of an anisotropic superconducting planar waveguide (strip) of finite thickness, as shown in Fig. 3.21, in which the waveguide is made from a dielectric layer of $2d$ thickness, covered by two identical anisotropic superconducting strips of thickness h and width W. We shall assume that the surfaces of the two superconductors are

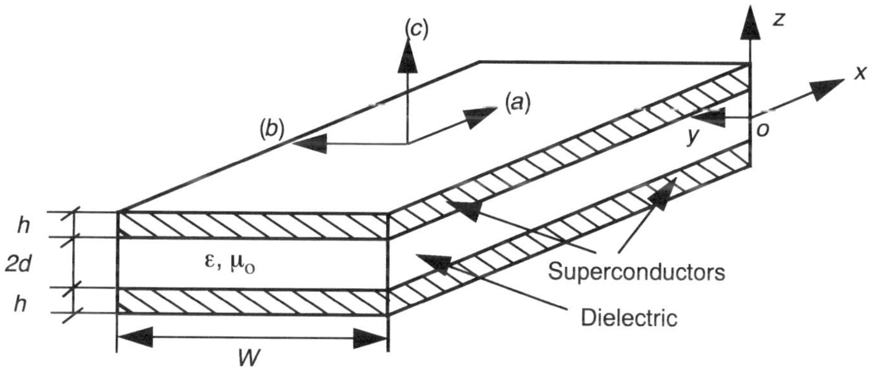

Figure 3.21 Scheme of an anisotropic superconducting planar waveguide.

parallel to the *a-b* plane of the anisotropic superconducting planar waveguide. We choose a reference frame aligned with the principal axes of the superconductors such that the *z*-axis is along the *c*-axis, which is perpendicular to the surfaces of the superconducting strips. The *x*-axis and *y*-axis are assumed to be, respectively, along the *a*-axis and *b*-axis directions of the superconductor. For simplicity, we shall ignore the fringing effect at the edges of the waveguide, assuming that the thickness of the dielectric layer is much less than the width of the waveguide (i.e., $2d << W$).

Since we are studying a problem which may involve extremely high frequencies, we shall use the full-wave analysis. We look for a possible field solution having the magnetic field components, $\boldsymbol{B} = (0, B_y, 0)$, and the electric field components, $\boldsymbol{E} = (E_x, 0, E_z)$, which describes a type of electromagnetic (TM) wave propagating in the superconducting strip waveguide. We shall assume that no external magnetic fields are applied to the waveguide and the self-generated magnetic field by currents flowing in the waveguide is small so that the superconductor is in the Meissner state. Thus no magnetic flux-flow effects will be considered here. We also assume that the wave is propagating and attenuated in the positive *x*-direction ($\sim e^{i(\omega t - \beta x)}$) so that we may write $\beta = \beta_R - i\alpha$ with β_R (> 0) being the propagation constant and α (> 0) being the attenuation coefficient.

In such a case, Eq.(3.5.8) becomes:

$$\left(\lambda_c^2 \frac{\partial^2}{\partial x^2} + \lambda_a^2 \frac{\partial^2}{\partial z^2}\right) B_y - \mu_o \left(\lambda_c^2 \sigma_c \frac{\partial E_z}{\partial x} - \lambda_a^2 \sigma_a \frac{\partial E_x}{\partial z}\right) = B_y \tag{3.5.11}$$

in the anisotropic superconductors, and

$$\frac{\partial^2 B_y}{\partial z^2} + \kappa^2 B_y = 0 \tag{3.5.12}$$

with $\kappa^2 = \mu_o\varepsilon\omega^2 - \beta^2$ in the dielectric ($\varepsilon = \varepsilon_o\varepsilon_r(1 - i\tan\delta)$) or in free-space ($\varepsilon = \varepsilon_o$). Here, μ_o and ε_o are, respectively, the permeability and the permittivity of free-space. ε_r is the relative permittivity of the dielectric material, and $\tan\delta$ is the loss tangent characterizing the dielectric loss of the dielectric material, which is supposed here to be isotropic. λ_a and λ_c are the London penetration depths with corresponding currents along the a-axis and the c-axis, respectively. σ_a and σ_c are the normal conductivities along the a-axis and the c-axis, respectively. Here, a common set of the principal axes has been assumed to exist for those material coefficient tensors involved for the anisotropic superconductors.

By noting the geometric symmetry of the waveguide studied, we find that the possible field solution may be given in the following form:

$$B_y = Ae^{-\gamma(|z| - d - h)} + Be^{\gamma(|z| - d - h)} \qquad (d < |z| < d + h) \tag{3.5.13}$$

$$B_y = C\cos(\kappa z) \qquad (|z| < d) \tag{3.5.14}$$

$$B_y = De^{-i\kappa_o(|z| - d - h)} \qquad (d + h < |z|) \tag{3.5.15}$$

where κ_o is the root of $\kappa^2 = \mu_o\varepsilon_o\omega^2 - \beta^2$ with negative imaginary part.

By Ampere´s circuital law and the first London equation (which, in our case, may be written as $J_x^{(s)} = E_x/(i\omega\mu_o\lambda_a^2)$ and $J_z^{(s)} = E_z/(i\omega\mu_o\lambda_c^2)$ together with the two-fluid model of $\boldsymbol{J} = \boldsymbol{J}^{(s)} + \boldsymbol{J}^{(n)}$ with $\boldsymbol{J}^{(s)}$ being the supercurrent density and $\boldsymbol{J}^{(n)}$ the normal conduction current density, we may find that the electric field components, E_x and E_z in the superconductor can be related to the magnetic field component B_y by

$$E_x = \frac{-i\omega\lambda_a^2}{1 + i\omega\mu_o\lambda_a^2\sigma_a}\frac{\partial B_y}{\partial z} \tag{3.5.16}$$

and

$$E_z = \frac{i\omega\lambda_c^2}{1 + i\omega\mu_o\lambda_c^2\sigma_c}\frac{\partial B_y}{\partial x} \tag{3.5.17}$$

The presence of the electric field component E_z indicates that there is a current component flowing along the c-axis direction, which is the case even for isotropic superconducting planar waveguides [Meyers (1961)].

By substituting of Eqs.(3.5.13), (3.5.16), and (3.5.17) into Eq.(3.5.11), we may find that the parameter γ, which is generally a complex quantity ($\gamma = \gamma_R + i\gamma_I$), can be given by

$$\gamma^2 = \frac{1}{\lambda_a^2} + i\omega\sigma_a\mu_o + \frac{\beta^2\lambda_c^2}{\lambda_a^2}\left(\frac{1 + i\omega\mu_o\lambda_a^2\sigma_a}{1 + i\omega\mu_o\lambda_c^2\sigma_c}\right) \tag{3.5.18}$$

Now, with the use of the interface conditions, namely the continuity of tangential components of the fields B_y and E_x at $|z| = d$ and $|z| = d+h$, we may obtain the following eigenvalue equation:

$$-\frac{\kappa q}{\varepsilon}\tan(\kappa d) = \frac{\sinh(\gamma h) + (i\kappa_o q/\varepsilon_o)\cosh(\gamma h)}{\cosh(\gamma h) + (i\kappa_o q/\varepsilon_o)\sinh(\gamma h)} \tag{3.5.19}$$

with the parameter q defined by $q = (1+i\omega\mu_o\sigma_a\lambda_a^2)/(\mu_o\omega^2\gamma\lambda_a^2)$. For $h\to\infty$, we may find that Eq.(3.5.19) recovers the result given by Zhou (1994c). Equation (3.5.19) is a transcendental equation, the analytical solution of which can only be found approximately.

If we now consider situations where we have $W \gg d$, $h/\lambda_a \geq 0.1$, $\lambda_a \leq 1$ μm, $\varepsilon_r \geq 10$, and $\omega \leq 10^{12}$ rad/s, which are usually the case in most practical applications of superconducting waveguides, we may assume that the magnetic field outside the waveguide (i.e., $|z| > d+h$) is zero. In such cases, we may find simply the following field solutions:

$$B_y = \frac{\mu_o I}{2W\sinh(\gamma h)}[e^{-\gamma(|z|-d-h)} - e^{\gamma(|z|-d-h)}] \qquad (d < |z| < d+h) \tag{3.5.20}$$

$$B_y = \frac{\mu_o I}{W\cos(\kappa d)}\cos(\kappa z) \qquad (|z| < d) \tag{3.5.21}$$

where I is the total current flowing along the superconducting strip of thickness h and width W. Furthermore, the eigenvalue equation (3.5.19) becomes

$$\tan(\kappa d) = -\frac{\varepsilon\mu_o\omega^2\gamma\lambda_a^2}{\kappa(1+i\omega\mu_o\sigma_a\lambda_a^2)}\coth(\gamma h) \tag{3.5.22}$$

It may be seen that, if we choose λ_a as the characteristic London penetration depth in the problem, the derived set of Eqs.(3.5.20)–(3.5.22) has the same form as for an isotropic superconducting planar waveguide, except that the parameter γ is now given by Eq.(3.5.18) which accounts for the effect of anisotropy of the superconductor.

To see quantitatively the effect of anisotropy, let us now make some rough estimates as follows. We compare the third term with the second term on the right-hand side of Eq.(3.5.18). By noting $\beta^2 \sim \omega^2\mu_o\varepsilon$ for guided wave mode, we may find the following ratio:

$$r_a = \left|\left(\frac{\varepsilon\omega}{\sigma_a}\right)\left(\frac{\lambda_c}{\lambda_a}\right)^2\left(\frac{1+i\omega\mu_o\lambda_a^2\sigma_a}{1+i\omega\mu_o\lambda_c^2\sigma_c}\right)\right| \tag{3.5.23}$$

Suppose that $|\varepsilon/\varepsilon_o| = 25$, which is a typical value for $LaAlO_3$ as the dielectric, and $\sigma_a = 10^5$ S/m, which is a typical value for some high-T_c oxide superconductors.

We may find that, if $\lambda_c/\lambda_a > 60$ as for some strongly anisotropic superconductors, the ratio r_a could be larger than 1 for $\omega > 1.3\times10^{11}$ rad/s (20 GHz). This implies that the third term on the right-hand side of Eq.(3.5.18) may not be negligible as compared with the effect of the normal conduction current for some strongly anisotropic superconductors at high frequencies. However, for isotropic superconductors, the third term may well be ignored even for the waveguide operating at microwave frequencies on the order of terahertz. Indeed, this has been the classical approximation used, so far, in almost all analyses for superconducting waveguides. Clearly, we see that if the third term on the right-hand side of Eq.(3.5.18) is ignored, the well-known result for an isotropic superconducting planar waveguide [Matick (1969)] is recovered, as it should be.

Now, by Eq.(3.5.22) and noting the fact of $\beta^2 \sim \omega^2\mu_0\varepsilon$ for guided wave mode, we may find the complex propagation parameter β at the first-order approximation by

$$\beta^2 = \frac{\varepsilon_r(\omega/c)^2(1-i\tan\delta)[1+\coth(\gamma h)/(\gamma d)]}{1-\varepsilon_r(\omega/c)^2(1-i\tan\delta)\lambda_c^2\coth(\gamma h)/[\gamma d(1+i\omega\mu_o\sigma_c\lambda_c^2)]} \tag{3.5.24}$$

Here, γ is given by Eq.(3.5.18), in which the complex propagation parameter β^2 is replaced approximately by $\beta_0{}^2$ as

$$\beta_o^2 = \frac{\omega^2}{c^2}\varepsilon_r(1-i\tan\delta)\left[1+\frac{\lambda_a\coth\left(h\sqrt{1/\lambda_a^2+i\omega\mu_o\sigma_a}\right)}{d\sqrt{1+i\omega\mu_o\sigma_a\lambda_a^2}}\right] \tag{3.5.25}$$

where $c = 1/(\mu_0\varepsilon_0)^{1/2}$ is the speed of light in free-space. It may be noticed that β_0 from Eq.(3.5.25) has actually the explicit form of the complex propagation parameter for isotropic superconducting planar waveguide with the London penetration depth λ_a and the normal conductivity σ_a [Matick (1969)].

3.5.3 Effect of Material Anisotropy on Kinetic Inductance

In this section, we shall derive an analytical formula for the calculation of kinetic inductance L_k per unit length of the anisotropic superconducting planar waveguide with the aid of the field solution obtained in the above section. The line kinetic inductance L_k may be determined here by

$$L_k = \frac{\mu_o}{|I|^2}\int_S (\lambda_a^2|J_x^{(s)}|^2 + \lambda_c^2|J_z^{(s)}|^2)dS \tag{3.5.26}$$

where the surface S of the integration includes both the cross-sections of superconducting strips on the *y-z* plane, shown in Fig. 3.21. It may be noticed that Eq.(3.5.26) includes a term characterizing the part of kinetic energy due to the transverse supercurrent component $J_z^{(s)}$. Though such a kinetic energy term appears also in isotropic superconductors, it is usually ignored, since the effect is small compared with the part of kinetic energy due to the main transport (longitudinal) supercurrent component $J_x^{(s)}$, as already noticed by Swihart (1961). However, for anisotropic superconductors, this term may not be simply ignored since the penetration depth λ_c can be much larger than λ_a in some strongly anisotropic superconductors.

By noting that the superconducting current density components may be given here by

$$J_x^{(s)} = \frac{-1}{\mu_o(1 + i\omega\mu_o\lambda_a^2\sigma_a)}\frac{\partial B_y}{\partial z} \tag{3.5.27}$$

and

$$J_z^{(s)} = \frac{1}{\mu_o(1 + i\omega\mu_o\lambda_c^2\sigma_c)}\frac{\partial B_y}{\partial x} \tag{3.5.28}$$

with the obtained field solution from Eqs.(3.5.20) and (3.5.21), we may find from Eq.(3.5.26) the kinetic inductance L_k per unit length of the waveguide by

$$L_k = \frac{\mu_o}{W[\cosh(2\gamma_R h) - \cos(2\gamma_I h)]}\left\{\frac{\lambda_a^2(\gamma_R^2 + \gamma_I^2)}{1 + \omega^2\mu_o^2\sigma_a^2\lambda_a^4}\left[\frac{\sinh(2\gamma_R h)}{\gamma_R} + \frac{\sin(2\gamma_I h)}{\gamma_I}\right]\right.$$
$$\left. + \frac{\lambda_c^2(\beta_R^2 + \alpha^2)}{1 + \omega^2\mu_o^2\sigma_c^2\lambda_c^4}\left[\frac{\sinh(2\gamma_R h)}{\gamma_R} - \frac{\sin(2\gamma_I h)}{\gamma_I}\right]\right\} \tag{3.5.29}$$

in which the parameters γ_R, γ_I, β_R, and α can be determined, respectively, from Eqs.(3.5.18), (3.5.24), and (3.5.25). The derived analytical formulae (3.5.29) can be used to calculate the kinetic inductance L_k per unit length of the anisotropic superconducting planar waveguide, in which both the effect of the normal conduction current and the effect of frequency are taken into account.

Some special cases may be of interest. For isotropic superconductors, by noting the fact that $|\beta|^2$ is usually much smaller than $|\gamma|^2$, we may derive from Eq.(3.5.29):

$$L_k = \frac{\mu_o/W}{[\cosh(2\gamma_R h) - \cos(2\gamma_I h)]}\left\{\frac{\lambda_L^2(\gamma_R^2 + \gamma_I^2)}{1 + \omega^2\mu_o^2\sigma_N^2\lambda_L^4}\left[\frac{\sinh(2\gamma_R h)}{\gamma_R} + \frac{\sin(2\gamma_I h)}{\gamma_I}\right]\right\} \tag{3.5.30}$$

in which

$$\gamma_R = \frac{1}{\sqrt{2}\lambda_L}\sqrt{1 + \sqrt{1 + (\lambda_L^2\omega\mu_o\sigma_N)^2}} \tag{3.5.31}$$

$$\gamma_I = \frac{1}{\sqrt{2}\lambda_L}\sqrt{\sqrt{1 + (\lambda_L^2\omega\mu_o\sigma_N)^2} - 1} \tag{3.5.32}$$

where λ_L and σ_N are, respectively, the London penetration depth and the normal conductivity of the isotropic superconductor at a given temperature. Furthermore, if the effect of the normal conduction current is negligible so that $\gamma_I \to 0$ and $\gamma_R \to 1/\lambda_L$, we may find from Eq.(3.5.30)

$$L_k = \frac{\mu_o}{W}\left[\lambda_L \coth\left(\frac{h}{\lambda_L}\right) + \frac{h}{[\sinh(h/\lambda_L)]^2}\right] \tag{3.5.33}$$

by using the identity relations: $\cosh(2x) - 1 = \sinh(2x)\tanh(x)$ and $\sinh(2x) = 2\sinh(x)\cosh(x)$. Thus the well-known result [Duzer and Turner (1981)] for the kinetic inductance L_k per unit length of an isotropic superconducting planar waveguide of finite thickness is shown to be recovered here as a special case of the general formulae (3.5.29) for anisotropic superconducting planar waveguides of finite thickness.

3.5.4 Effect of Material Anisotropy on Wave dispersion and Attenuation

To study quantitatively the effect of material anisotropy of the superconductor on wave dispersion and attenuation behaviour in the anisotropic superconducting planar waveguide, let us assume that the superconductor layers are sufficiently thick so that the geometric effect of thickness h is negligible for simplicity. In such a case, Eq.(3.5.24) for the propagation parameter β becomes

$$\beta^2 = \left(\frac{\omega}{c}\right)^2 \frac{\varepsilon_r(1 - i\tan\delta)[1 + 1/(\gamma d)]}{1 - \omega^2\lambda_c^2\varepsilon_r(1 - i\tan\delta)/[c^2\gamma d(1 + i\omega\mu_o\sigma_c\lambda_c^2)]} \tag{3.5.34}$$

in which the parameter γ is given by

$$\gamma = \left[\frac{1}{\lambda_a^2} + i\omega\mu_o\sigma_a + \beta_o^2\left(\frac{\lambda_c}{\lambda_a}\right)^2 \frac{1 + i\omega\mu_o\sigma_a\lambda_a^2}{1 + i\omega\mu_o\sigma_c\lambda_c^2}\right]^{1/2} \tag{3.5.35}$$

where

$$\beta_o^2 = \frac{\omega^2}{c^2}\varepsilon_r(1 - i\tan\delta)\left(1 + \frac{\lambda_a}{d\sqrt{1 + i\omega\mu_o\sigma_a\lambda_a^2}}\right) \tag{3.5.36}$$

As we may know, recent vortex-lattice decoration experiments have shown that in $YBa_2Cu_3O_{7-\delta}$, the penetration depths, $\lambda_a = \Lambda_1^{1/2}$, $\lambda_b = \Lambda_2^{1/2}$, and $\lambda_c = \Lambda_3^{1/2}$ are approximately in the ratios $\lambda_a:\lambda_b:\lambda_c = 1.2:1:5.5$. This implies that the penetration depths λ_a and λ_b are much smaller than λ_c because the screening currents along the CuO_2 planes (in the a- or b-direction) flow much more easily than in the c-direction. λ_b is somewhat smaller than λ_a because current parallel to the CuO_2 planes flows more easily along the CuO chains (in the b-direction) than perpendicular to these chains (in the *a*-direction). For Bi- and Tl- based HTSC, they exhibit even stronger anisotropy, having, for instance, the anisotropic ratio λ_c/λ_a that could be larger than 60. In addition, the conductivities σ_a, σ_b in the *a-b* plane of some high-temperature oxide superconductors can be several orders of magnitude larger than the conductivity σ_c in the direction of the *c*-axis due to the layered structure of these superconductors.

In order to make the result be of general interest, the following numerical calculation do not limited to any specific superconductive material, rather a wide range of possible values of material anisotropic properties of the superconductor are examined in the numerical calculation. Some chosen material parameters are: $\varepsilon_r = 25$ and $\tan\delta = 10^{-5}$, which are typical values for $LaAlO_3$ as the dielectric material; and $\lambda_{ao} = 150$ nm and $\sigma_{ao} = 10^5$ S/m, which are typical values for some high-T_c superconductors. At any temperature T below the critical temperature T_c, we have $\lambda_a(\Theta) = \lambda_{ao}/(1-\Theta^4)^{1/2}$ and $\sigma_a(\Theta) = \sigma_{ao}\Theta^4$ by the classical two-fluid model. The material parameters λ_c and σ_c will be specified in relevant figures from the numerical calculation.

Shown in Figs. 3.22 and 3.23 are some numerical results on the effects of anisotropy of superconductors to the dispersion and attenuation of wave propagation in the superconducting waveguide. It is shown in Fig. 3.22 that the phase velocity ($v = \omega/\beta_0$) of the wave propagation in the waveguide is essentially not changed in a certain range of the values of the anisotropic ratios: λ_c/λ_a and σ_c/σ_a, while it may be reduced for some strongly anisotropic superconductors though such a reduction may be small as shown in the figure. Figure 3.23 shows a somewhat interesting result, which reveals that a certain combination of the values of the anisotropic ratios: λ_c/λ_a and σ_c/σ_a of the superconductor may result in a much higher attenuation than others. It is not at all the case that the waveguide made of the strongest anisotropic superconductor will necessarily have the highest attenuation. Thus the result shown could be useful in choosing proper superconductors to construct a low-loss superconducting waveguide, which is supposed to be operated at very high frequencies and at temperatures close to T_c, if these superconductors are intrinsically anisotropic, such as high-T_c oxide superconductors.

Some other numerical results from Eqs.(3.5.34)–(3.5.36) on the effect of anisotropy of the superconductor have shown by Zhou (1994c) that, at frequencies less than 10 GHz, the effect of the anisotropy of the superconductor on the characteristics of the superconducting waveguide is negligible, which is in accordance with the result of Lee et al. (1992). However, at higher frequencies on the order of, for instance, 100 GHz, it is found that the anisotropy of the superconductor may affect the attenuation behavior of wave propagation in the superconducting waveguide, made of some strongly anisotropic superconductors. The effect of the anisotropy of the superconductor on the wave dispersion is, however, moderate, and it increases with increasing temperature. Both the anisotropic penetration depths and the anisotropic conductivities of the superconductor are found to influence the characteristics of the superconducting waveguide.

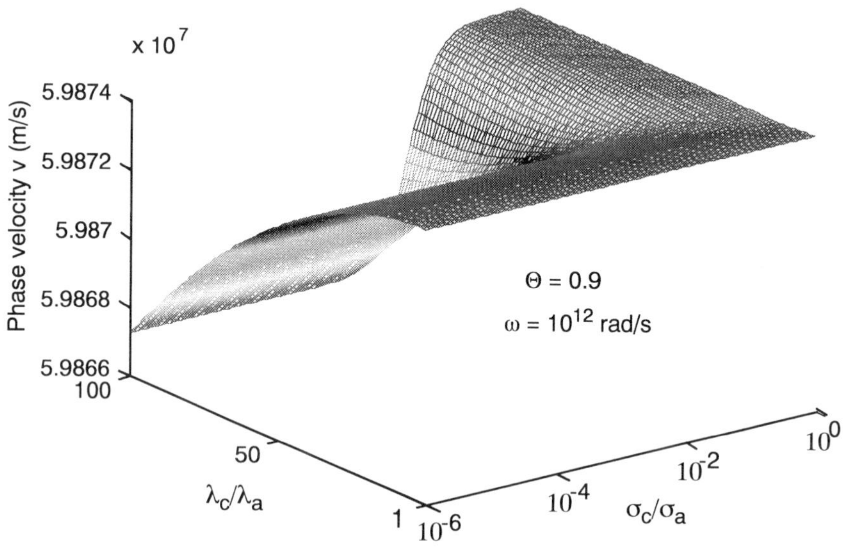

Figure 3.22 The effect of anisotropy of the superconductor on the wave dispersion behaviour of the planar superconducting waveguide at different temperature.

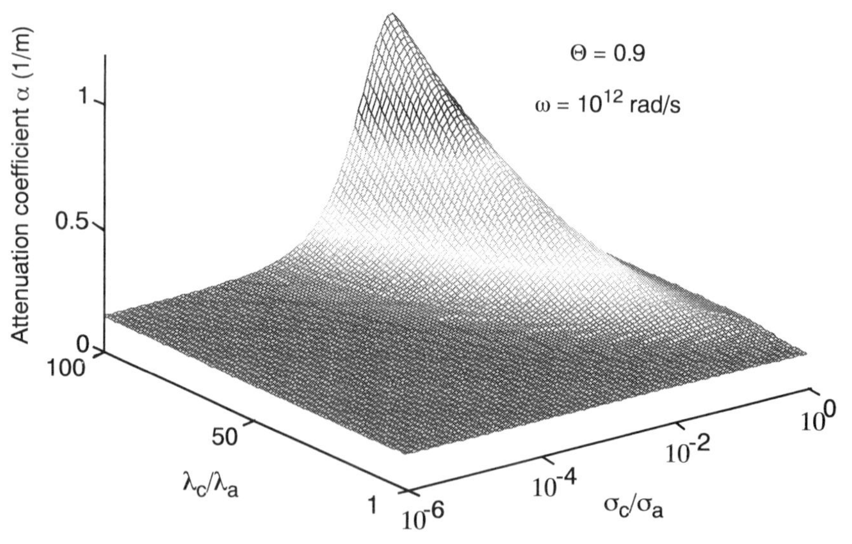

Figure 3.23 The effect of anisotropy of the superconductor on the wave attenuation in the planar superconducting waveguide at different frequencies.

3.6 MICROSCOPIC MECHANISM OF SUPERCONDUCTIVITY

3.6.1 Isotope Effect

Since the discovery of superconductivity by H. Kamerlingh Onnes in 1911, it took many decades for scientists to understand this fascinating phenomenon. A crucial step toward the emergence of the BCS microscopic theory on superconductivity is the isotope effect discovered by Reynolds et al. (1950) and Maxwell (1950), which shows that the superconducting transition temperature T_c and the critical field H_c depend on the mass of ions forming the crystal lattice of the material in the form:

$$T_c \propto \mathrm{M}^{-1/2} \qquad \text{and} \qquad H_c \propto \mathrm{M}^{-1/2} \tag{3.6.1}$$

This isotope effect implies that superconductivity is not a purely electronic phenomenon since the mass of the ions manifests itself only when lattice vibrations are taken into account. On the basis of this fact, Fröhlich (1950) and Bardeen (1950, 1951) demonstrated independently that the electrons, while residing in the crystal lattice, are capable of attracting one another. The physical origin of such an attraction may be seen by considering that if one puts an electron into a polarizable lattice, the positively charged lattice deforms to lower the electrostatic energy by crowding positive charge nearer the electron's position. If now there is a second electron to be introduced, the concentration of positive charge around the first electron makes that region also favorable to the second electron, which leads to an effective attractive force between the two electrons in addition to their direct mutual repulsive Coulomb interaction.

After the discovery of the attraction between electrons, it seems to become immediately possible to work out the theory of superconductivity. However, it took again several years for the formation of the concept of pairing of electrons. In 1956, it was Cooper who first studied theoretically what happens when two electrons with kinetic energies E_1 and E_2 are added to a metal at absolute zero. He concluded that, in the presence of even very weak attractive interaction, the Fermi sea of single electrons is unstable and any small perturbation that moves two electrons above E_F, where scattering is possible, will lower the system energy by pairing [Cooper (1956)]. Because of the importance of the pairing concept in the theory of superconductivity, we shall present here some analysis on the formation of the bound Cooper pair of electrons.

3.6.2 The Cooper Pair of Electrons

According to Cooper, when two electrons with kinetic energies E_1 and E_2 are added to a metal at absolute zero, both energies E_1 and E_2 of the electrons must be above E_F in order not to violate Pauli's exclusion principle since all the eigenstates with kinetic energies up to E_F (Fermi energy) are occupied at absolute zero. The lowest values of E_1 and E_2, which are above E_1, lie within an energy about $\hbar\omega_c$ of the Fermi energy, where ω_c denotes the cutoff frequency. Consider now the scattering between the electron states resulting from interaction between the electrons. In the electron-phonon scattering mechanism, the electron occupying state $\boldsymbol{k}_1$ emits a phonon (i.e., vibrates the lattice) which is absorbed by the electron occupying the state $\boldsymbol{k}_2$. The total momentum $\boldsymbol{K} = \boldsymbol{k}_1 + \boldsymbol{k}_2$ in the final state is the same as in the initial state (i.e., the total momentum is conserved in the scattering event). It is, however, not necessary to require conservation of energy due to Heisenberg's uncertainty principle. The largest number of allowed scattering processes, yielding the maximum lowering of the energy as we shall see later, is obtained by pairing electrons with equal and opposite momenta, that is, $\boldsymbol{K} = 0$ for a translationally invariant system carrying no current.

The wave function of such a pair of electrons with opposite momenta and with a weak interaction can thus be expressed, according to a perturbation theory [see, e.g., Schiff (1968)], by

$$\Psi(\boldsymbol{x}_1, \boldsymbol{x}_2) = \sum_{\boldsymbol{k}} a(\boldsymbol{k}) \exp(i\hbar\boldsymbol{k} \cdot \boldsymbol{r}) \tag{3.6.2}$$

with $\boldsymbol{r} = \boldsymbol{x}_1 - \boldsymbol{x}_2$ and $\boldsymbol{k} = (\boldsymbol{k}_1 - \boldsymbol{k}_2)/2$, where $a(\boldsymbol{k})$ is the probability amplitude of finding an electron in the state $\hbar\boldsymbol{k}$ and another electron in the state $-\hbar\boldsymbol{k}$.

The Schrödinger equation for the two electrons with the interaction potential $V(\boldsymbol{r})$ reads

$$-\frac{\hbar^2}{2m_e}(\nabla_1^2 + \nabla_2^2)\Psi + V(\boldsymbol{r})\Psi = (E + 2E_F)\Psi \tag{3.6.3}$$

where the energy E of the pair is measured from the state where the two electrons are at the Fermi level. The equation for $a(\boldsymbol{k})$ can be found to be

$$\frac{\hbar^2}{m_e}k^2 a(\boldsymbol{k}) + \sum_{\boldsymbol{k}'} a(\boldsymbol{k}')V_{\boldsymbol{k}\boldsymbol{k}'} = (E + 2E_F)a(\boldsymbol{k}) \tag{3.6.4a}$$

with

$$a(\boldsymbol{k}) = 0 \quad \text{for } k < k_F \tag{3.6.4b}$$

where $V_{\boldsymbol{k}\boldsymbol{k}'}$ is the scattering matrix element defined by

$$V_{\boldsymbol{k}\boldsymbol{k}'} = \langle \hbar\boldsymbol{k}'|V|\hbar\boldsymbol{k}\rangle = \frac{1}{\Omega}\int_{\Omega} V(\boldsymbol{r})\exp[i\hbar((\boldsymbol{k} - \boldsymbol{k}') \cdot \boldsymbol{r})]d\Omega \tag{3.6.4c}$$

which characterizes the strength of the potential for scattering a pair of electrons with momenta $(\boldsymbol{k}', -\boldsymbol{k}')$ to momenta $(\boldsymbol{k}, -\boldsymbol{k})$. Here, Ω denotes the volume of the system and V the interaction potential. The condition (3.6.4b) expresses the fact that states $k < k_F$ are already occupied (Pauli´s exclusion principle).

Cooper pointed out that bound states exist for $E < 0$ provided that V is attractive, whatever the magnitude of this interaction. To see this, one may consider a simple model in which $V_{\boldsymbol{kk}'}$ is assumed to be a constant, that is,

$$V_{kk'} = -V \quad \text{for } E_F < \frac{\hbar^2 k^2}{2m_e} < E_F + \hbar\omega_c \quad \text{and} \quad E_F < \frac{\hbar^2 k'^2}{2m_e} < E_F + \hbar\omega_c \tag{3.6.5}$$

with $V > 0$ and $V_{\boldsymbol{kk}'} = 0$ everywhere else. Thus Eq.(3.6.4a) becomes

$$\left(\frac{\hbar^2 k^2}{m_e} - E - 2E_F\right) a(\boldsymbol{k}) = \left[\sum_{\boldsymbol{k}'} a(\boldsymbol{k}')\right] V \tag{3.6.6}$$

where the summation over $\boldsymbol{k}'$ is restricted to the band $\hbar\omega_c$ above E_F.

Dividing each term of Eq.(3.6.6) by $(\hbar^2 k^2/m_e - E - 2E_F)$ and then summing over $\boldsymbol{k}$ yields

$$\frac{1}{V} = \sum_{\boldsymbol{k}} \frac{1}{\hbar^2 k^2/m_e - E - 2E_F} \tag{3.6.7}$$

The discrete summation over $\boldsymbol{k}$ may be replaced by an integral over the energy with the range bounded by $E_F < \hbar^2 k^2/(2m_e) < E_F + \hbar\omega_c$, or $0 < \varepsilon(k) < \hbar\omega_c$ by introducing a new definition of the energy

$$\varepsilon(k) = \frac{\hbar^2 k^2}{2m_e} - E_F \tag{3.6.8}$$

so that the integral reads simply

$$\frac{1}{V} = \int_0^{\hbar\omega_c} \frac{N(\varepsilon)}{2\varepsilon - E} d\varepsilon \tag{3.6.9}$$

in which $N(\varepsilon)$ denotes the density of state (per unit energy and unit volume) by

$$N(\varepsilon) = \frac{4\pi k^2}{(2\pi)^3} \frac{dk}{d\varepsilon} \tag{3.6.10}$$

If $\hbar\omega_c \ll E_F$, the density of the states $N(\varepsilon)$ is nearly constant and equal to its value at the Fermi surface $N(0)$ over the range of integration so that Eq.(3.6.9) may be integrated to give

$$E = \frac{2\hbar\omega_c}{1 - \exp(2/(VN(0)))} \tag{3.6.11}$$

For weak interaction having $VN(0) << 1$, we have approximately

$$E = -2\hbar\omega_c \exp\left(-\frac{2}{VN(0)}\right) \tag{3.6.12}$$

which states that an allowed energy state exists with $E < 0$.

The calculation thus shows that, in the presence of even very weak attractive interaction, the Fermi sea of single electrons is unstable and any small perturbation that moves two electrons above E_F, where scattering is possible, will lower the system energy by pairing.

3.6.3 The BCS Theory of Superconductivity

With the concept of pairing of electrons and the electron-phonon interaction mechanism, which provides the necessary attractive interaction strong enough to overcome the repulsive screened Coulomb interaction between electrons, Bardeen, Cooper, and Schrieffer (1957) were able to give the first microscopic theory of superconductivity, now known as the BCS theory.

In the BCS theory, it is assumed that the ground state can be expressed wholly in terms of paired electrons having the same total momentum, which are in a singlet state such that if the state $\boldsymbol{k}\uparrow$ is occupied, so is $-\boldsymbol{k}\downarrow$; and similarly if $\boldsymbol{k}\uparrow$ is vacant, $-\boldsymbol{k}\downarrow$ is also vacant. The total reduced Hamiltonian treated by BCS for electron pairs with zero net momentum is taken to be of the following form:

$$H_{red} = 2\sum_{k > k_F} \varepsilon_k b_k{}^* b_k + 2\sum_{k < k_F} |\varepsilon_k| b_k b_k{}^* + \sum_{kk'} V_{kk'} b_{k'}{}^* b_k \tag{3.6.13}$$

where $\varepsilon_k = \hbar^2 k^2/(2m_e) - E_F$ is the kinetic energy of a Bloch state measured relative to the Fermi level. The first two terms on the right-hand side of Eq.(3.6.13) denote the kinetic energy of the electron pairs relative to the normal ground state and the last term is the potential energy relative to the normal state, the reduced potential energy. The factors of 2 appear in the expression since there are two electrons in the pair, that is, there are two possible spins for each $\boldsymbol{k}$. The summation puts a spin-up electron in each $\boldsymbol{k}$ state and a spin-down electron in the corresponding $-\boldsymbol{k}$ state with the result that each state for $k < k_F$ contains both spin-up electron and spin-down electrons.

The operators b_k and $b_k{}^*$ used in the formalism of second quantization denote, respectively, the pair operators defined by

$$b_k = c_{-k\downarrow} c_{k\uparrow} \quad \text{and} \quad b_k{}^* = c^*{}_{k\uparrow} c^*{}_{-k\downarrow} \tag{3.6.14}$$

where $c^*{}_{k\uparrow}$ is the single-electron creation operator, which places an electron in state $\boldsymbol{k}$ with spin up, and $c_{k\uparrow}$ is the single-electron annihilation operator, which causes the elimination of an electron [details about operator algebra is referred to in the work of, e.g., Schiff (1968)].

To find the distribution of pair occupancy in the superconducting state, BCS introduced a ground state function with the use of a Hartree-like approximation in which the probabilities that a specific configuration of pairs occurs in the wave function is given by a product of occupancy probabilities for the individual pair states, that is,

$$|\Psi\rangle = \prod_{k}[u_k + v_k b_k^*]|0\rangle \qquad (3.6.15)$$

where $|0>$ denotes the vacuum state with no electrons present, v_k^2 is the probability of pair occupancy, and $u_k^2 = 1 - v_k^2$ (for all $\boldsymbol{k}$) is the probability of pair vacancy, so that the wave function (3.2.15) is correctly normalized to unity.

The probability that the pair ($\boldsymbol{k}\uparrow$, $-\boldsymbol{k}\downarrow$) is occupied reads

$$\langle\Psi|c^*_{k\uparrow}c_{k\uparrow}|\Psi\rangle = v_k^2 \qquad (3.6.16)$$

and the probability that both pairs $\boldsymbol{k}$ and $\boldsymbol{k}'$ are occupied is

$$\langle\Psi|c^*_{k\uparrow}c_{k\uparrow}c^*_{k'\uparrow}c_{k'\uparrow}|\Psi\rangle = v_k^2 v_{k'}^2 \qquad (3.6.17)$$

and the probability that $\boldsymbol{k}$ is occupied and $\boldsymbol{k}'$ unoccupied is

$$\langle\Psi|c^*_{k\uparrow}c_{k\uparrow}(1 - c^*_{k'\uparrow}c_{k'\uparrow})|\Psi\rangle = v_k^2 u_{k'}^2 \qquad (3.6.18)$$

It is assumed here that only those many body states in which the two members of a pair with the pair state of spin-antiparallel ($\boldsymbol{k}\uparrow$, $-\boldsymbol{k}\downarrow$) are either both occupied or both unoccupied are chosen. It is also assumed that the probability amplitude v_k^2, for the occupation of a particular pair of state ($\boldsymbol{k}\uparrow$, $-\boldsymbol{k}\downarrow$), does not depend on the occupation of the other states. Essentially, the basic approximation of the BCS theory of superconductivity rests in their assumption that it is the two-body correlations that are responsible for the qualitative features of superconductivity and that of the two-body correlations there is a very strong preference for singlet zero momentum pairs (so strong that one can get an adequate description of superconductivity by treating these correlations alone [Blatt (1964)]). The ground wave function (3.6.15) chosen by BCS has been found to be very satisfied from the empirical point of view since the BCS theory is excellently agreement with experiment. However, there are some questions remained yet to be studied from the theoretical point of view, especially for the new high-T_c superconductors.

The energy of the superconducting ground state relative to the normal ground state is the expectation value of the reduced Hamiltonian

$$W = \langle\Psi|H_{red}|\Psi\rangle \tag{3.6.19}$$

Substituting (3.6.13) and (3.6.15) into (3.6.19), we obtain

$$W = 2\sum_{k>k_F} \varepsilon_k v_k^2 + 2\sum_{k<k_F} |\varepsilon_k| u_k^2 + \sum_{kk'} V_{kk'} u_k v_k u_{k'} v_{k'} \tag{3.6.20}$$

In the equilibrium state, the energy W is minimized with respect to v_k^2 and with the restriction condition $u_k^2 = 1 - v_k^2$. The resulting probability of occupancy reads

$$v_k^2 = \frac{1}{2}\left(1 - \frac{\varepsilon_k}{\sqrt{\varepsilon_k^2 + \Delta_k^2}}\right) \tag{3.6.21}$$

in which Δ_k is called the gap parameter defined by

$$\Delta_k = -\sum_{k'} V_{kk'} u_{k'} v_{k'} \tag{3.6.22}$$

Substitution of Eq.(3.6.21) into Eq.(3.6.22) gives

$$\Delta_k = -\sum_{k'} V_{kk'} \frac{\Delta_{k'}}{2E_{k'}} \tag{3.6.23}$$

with $E_{k'} = (\varepsilon_{k'}^2 + \Delta_{k'}^2)^{1/2}$. Equation (3.6.23) is a nonlinear integral equation since $\Delta_{k'}$ is present in the denominator in $E_{k'}$. This integral equation has obviously the trivial solution $\Delta_k = 0$ for all $\boldsymbol{k}$ leading to normal state. The condition for superconductivity is thus whether the integral equation possess a non-trivial solution leading to a lower energy.

For general interactions this equation has to be solved numerically. In the simple BCS model, an assumption is made that the matrix element $V_{kk'}$ can be replaced by a constant average matrix element

$$V_{kk'} = \begin{cases} -V & \text{for } |\varepsilon_k| \text{ and } |\varepsilon_{k'}| < \hbar\omega_c \\ 0 & \text{otherwise} \end{cases} \tag{3.6.24}$$

with $V > 0$ (meaning the attractive interaction between electrons in a pair). Here, the cutoff frequency ω_c is taken to be the Debye phonon frequency ω_D in the electron-phonon mechanism.

For this simple interaction, Δ_k is zero if $|\varepsilon_k| > \hbar\omega_c$ and it is a constant Δ if $|\varepsilon_k| < \hbar\omega_c$ and it is determined from the following equation

$$\frac{1}{V} = \sum_{k'} \frac{1}{2\sqrt{\varepsilon_{k'}^2 + \Delta^2}} \tag{3.6.25}$$

which has a solution only if V is positive. Replacing the summation over $\boldsymbol{k}'$ by an integral over the corresponding energy range and by noting the fact that the density of states is nearly constant close to the Fermi surface, so N(ε) ≈ N(0) over the range $|\varepsilon_{\boldsymbol{k}}| < \hbar\omega_c$, Eq.(3.6.25) becomes

$$\frac{1}{VN(0)} = \int_0^{\hbar\omega_c} \frac{1}{\sqrt{\varepsilon^2 + \Delta^2}} d\varepsilon \tag{3.6.26}$$

Solving for Δ, we obtain

$$\Delta = \frac{\hbar\omega_c}{\sinh(1/(N(0)V))} \tag{3.6.27}$$

If $N(0)V << 1$ for the weak coupling between electrons and phonons as in most of elemental materials, Eq.(3.6.27) is reduced to be

$$\Delta = 2\hbar\omega_c \exp\left(-\frac{1}{N(0)V}\right) \tag{3.6.28}$$

which has a typical value of about 1 meV.

The ground state energy relative to the normal ground state at the absolute zero can then be found from Eqs.(3.6.20), (3.6.21), (3.6.24), and (3.6.28)

$$W_s - W_n = -\frac{1}{2}N(0)\Delta^2\left[1 - \exp\left(-\frac{2}{N(0)V}\right)\right] \approx -\frac{1}{2}N(0)\Delta^2 \tag{3.6.29}$$

where the last form holds in the weak coupling limit.

By equating Eq.(3.6.29) with the thermodynamic condensation energy at $T = 0$, we find

$$\frac{\mu_o H_o^2}{2} = (g_n - g_s)_{T=0} = \frac{1}{2}N(0)\Delta^2 \tag{3.6.30}$$

It is shown that, although the BCS theory of the ground state has all electrons paired, only those in a narrow energy range of order Δ participate in the condensation. Those farther below the Fermi level are described mathematically as pairs with no loss of veracity but they are too far from the surface to be scattered by the electron–phonon interaction, and hence do not participate in the reduction of the system energy. This shows that the electron pairs are not true Bose particles otherwise the ground state would be formed by placing all the Bosons in the lowest state. The effect of Pauli's exclusion principle acting on the individual electrons forming the pair is shown from the fact that no $\boldsymbol{k}\uparrow$ and $-\boldsymbol{k}\downarrow$ state may be occupied by more than one pair at a time, which then makes the probability $v_{\boldsymbol{k}}^2$ of the occupancy of the pair state ($\boldsymbol{k}\uparrow$, $-\boldsymbol{k}\downarrow$) is less than 1 as expressed by Eq.(3.6.21).

To study the physical properties of the system, we have to treat the excitation states of the system which deviates from its ground state due to electrons being excited out of the ground state by thermal lattice vibrations or by incident photons. To find the excited states of the BCS reduced Hamiltonian, we consider adding an electron to the system in the state $\boldsymbol{k}\uparrow$ with its mate $-\boldsymbol{k}\downarrow$ being empty. The only effect of this process is to block the pair state ($\boldsymbol{k}\uparrow$, $-\boldsymbol{k}\downarrow$) from participating in the pairing interaction due to Pauli's exclusion principle. The state of the system, with an excitation in state $\boldsymbol{k}_1$ and all other states occupied by pairs, reads

$$|\Psi_{k_1}\rangle = c^*{}_{k_1\uparrow}\prod_{k\neq k_1}[u_k + v_k b_k^*]|0\rangle \tag{3.6.31}$$

where the creation operator $c^*{}_{k1\uparrow}$ gives sure occupancy in state $\boldsymbol{k}_1\uparrow$, but the state $-\boldsymbol{k}_1\downarrow$ remains empty. The kinetic energy of state (3.6.31) relative to the ground state is reduced by the absence of the pair occupying ($\boldsymbol{k}_1\uparrow$, $-\boldsymbol{k}_1\downarrow$) with probability $v_{\boldsymbol{k}}^2$ and increased by the sure occupancy of $\boldsymbol{k}_1\uparrow$ by the excitation, and the reduced potential energy is changed by eliminating the pair scattering from and into ($\boldsymbol{k}_1\uparrow$, $-\boldsymbol{k}_1\downarrow$). The total change of energy from the ground state to the excited state (3.6.31) is

$$W_{k_1\uparrow} - W_o = \varepsilon_{k_1}(1 - 2v_{k_1}^2) + 2u_{k_1}v_{k_1}\Delta_{k_1} = E_{k_1} \tag{3.6.32}$$

which shows that the parameter E_{k1} is just the energy required to create an excitation (a quasi-particle) in state $\boldsymbol{k}_1\uparrow$.

A plot of E_k as a function of k is given in Fig. 3.24. It is shown that the excitation energy has a minimum value Δ, which means that excitations cannot be created with an arbitrarily small amount of energy as in the case of the normal metal. To break a pair, two excitations must be created and, thus, the minimum energy required is at least 2Δ, Δ for removing an electron from one state and Δ for placing it in another state. The occurrence of thermally excited particles at finite temperature will result in increase of energy and entropy. At any given temperature, the number of excitations at thermal equilibrium is determined by minimizing the free energy of the system.

In the BCS theory, the free energy of the system at finite temperature is taken to be of the following form [Bardeen et al. (1957)]:

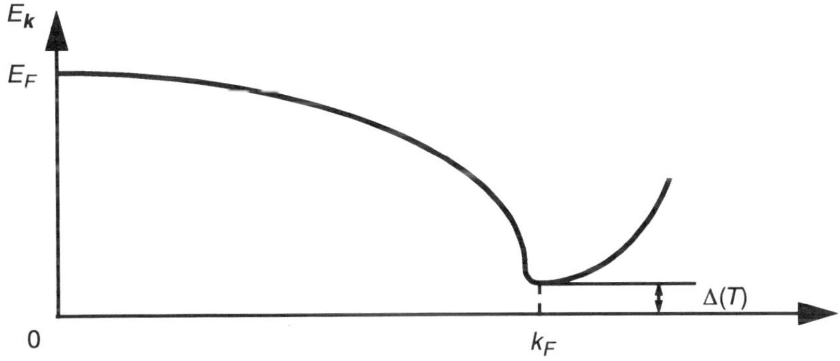

Figure 3.24 Temperature-dependent gap parameter Δ in the excitation spectrum.

$$F = 2\sum_k |\varepsilon_k|[f_k + (1-2f_k)v_k^2(|\varepsilon_k|)]$$
$$+ \sum_{kk'} V_{kk'}u_k v_k u_{k'} v_{k'}(1-2f_k)(1-2f_{k'}) - TS \tag{3.6.33}$$

where the entropy S is

$$S = -2k_B \sum_k [f_k \ln(f_k) + (1-f_k)\ln(1-f_k)] \tag{3.6.34}$$

By minimizing the free energy of the system with respect to the probability of pair occupancy v_k^2, we can find the same expression (3.6.21) for v_k^2 but with the new expression for the gap parameter

$$\Delta = -V\left\{\sum_{k'} u_{k'} v_{k'}(1-2f_{k'})\right\} \tag{3.6.35}$$

where we have used the assumption given by Eq.(3.6.24).

By minimizing F with respect to the probability of excitation occupancy f_k, we find the expression for the probability of excitation occupancy f_k by

$$f_k = \frac{1}{1 + \exp(E_k/(k_B T))} \tag{3.6.36}$$

which is the usual Fermi-Dirac distribution function. Thus the excitations act completely like a set of independent fermions whose energies are given by the dispersion law $E_k = (\varepsilon_k^2 + \Delta^2)^{1/2}$.

By substituting Eqs.(3.6.21) and (3.6.36) into Eq.(3.6.35), a self-consistent condition for the determination of the gap parameter Δ_k is found

$$\frac{1}{V} = \sum_{k'} \frac{1}{2E_{k'}} \left[1 - \frac{2}{1 + \exp(E_{k'}/(k_B T))} \right] \tag{3.6.37}$$

which may, further, be expressed in the following integral form, by assuming a constant density of state $N(0)$,

$$\frac{1}{N(0)V} = \int_0^{\hbar\omega_c} \tanh\left(\frac{\sqrt{\varepsilon^2 + \Delta^2}}{2k_B T} \right) \frac{1}{\sqrt{\varepsilon^2 + \Delta^2}} d\varepsilon \tag{3.6.38}$$

This equation is an implicit relation for the determination of the temperature-dependent gap parameter $\Delta(T)$. As temperature T increases from zero, Δ decreases and vanishes at the transition temperature T_c. Thus T_c is given by

$$\frac{1}{N(0)V} = \int_0^{\hbar\omega_c} \frac{1}{\varepsilon} \tanh\left(\frac{\varepsilon}{2k_B T_c} \right) d\varepsilon \tag{3.6.39}$$

In the weak-coupling limit ($N(0)V << 1$), this gives

$$k_B T_c = 1.14 \hbar\omega_c \exp\left(-\frac{1}{N(0)V} \right) \tag{3.6.40}$$

Since $N(0)V$ is expected to be independent of isotopic mass M, we see that the transition temperature T_c is proportional to $\hbar\omega_c$, which is consistent with the isotope effect since $\omega_c \propto M^{1/2}$ in the electron–phonon interaction mechanism.

By comparing Eq.(3.6.40) with Eq.(3.6.28), we find the following relation for the energy gap at $T = 0$

$$2\Delta(0) = 3.52 k_B T_c \tag{3.6.41}$$

where the actual values of the constant, 3.52, is found to be in reasonably good agreement with experiment for weak-coupling superconductors. In general, it lies within about 30% of 3.52.

In general, the temperature-dependent energy gap $\Delta(T)$ may be calculated numerically from Eq.(3.6.38). Near the transition temperature T_c, the gap parameter can be found from the following approximate relation:

$$\Delta(T) \approx 3.2 k_B T_c \sqrt{1 - \frac{T}{T_c}} \tag{3.6.42}$$

which has the form consistent with Landau's theory of second-order phase transition.

So far, nothing has been said about the electrodynamic properties of superconductors, such as the disappearance of resistance and the Meissner effect. Both the ground state and the excited states which we have discussed have a perfectly isotropic distribution of electrons in momentum space, so that there are as many electrons traveling one way as the other, and no macroscopic net current flows. It is, however, possible to visualize a situation in which each Cooper pair, instead of having zero total momentum, has a resultant momentum $\hbar\boldsymbol{K}$, which is the same for all pairs. In this case the pair wave function is then of the form

$$\Psi_K = \Psi \exp(i\hbar\boldsymbol{K}\cdot\boldsymbol{R}) \tag{3.6.43}$$

with $\boldsymbol{K} = \boldsymbol{k}_1 + \boldsymbol{k}_2$ and $\boldsymbol{R} = (\mathbf{x}_1 + \mathbf{x}_2)/2$ being the position of the centre of mass of the pair. In this picture the current is carried by pairs of electrons which have a total momentum $\boldsymbol{K}$. When a current is carried by an ordinary conductor, such as a normal metal or a semiconductor, resistance is inevitably present because the current carriers (either electrons or holes) can be scattered with a change in momentum so that their free acceleration in the direction of the electric field is hindered. This scattering may be due to impurity atoms, lattice defects, or thermal vibrations.

In the case of a superconductor, the electrons which make up a Cooper pair are constantly scattering each other, but since the total momentum remains constant in such a process there is no change in the current flowing. The only scattering process which can reduce the current flow is one in which the total momentum of a pair in the direction of the current changes, and this can only happen if the pair is broken up. This de-pairing requires a minimum amount of energy 2Δ so the scattering can only happen if this energy can be supplied from somewhere. Thus, up to a certain critical value, there is a supercurrent, which flows without resistance. Macroscopically, we have shown that the electrodynamic behaviors of the superconductors may be described by an ensemble-average wave function in the form of $\psi = |\psi|\exp(i\theta)$, where θ is the phase of the electron pairs. More detailed discussion about this point may be found in the work of Blatt (1964).

After the BCS microscopic theory appeared, alternative and more general formulations were developed, among which an important and useful method of Green´s functions and Feynman diagrams was introduced to the theory of superconductivity by Gor´kov (1958). Other topics, such as the microscopic theory of superconducting tunnel junctions and the possible micromechanisms other than the electron-phonon mechanism responsible for the superconductivity were also studied intensively. In particular, researches for revealing the micromechanism and developing theories for the newly discovered high–T_c oxide superconductors are still in progress. Readers who are interested in those specific topics may refer to the works of, for instance, Josephson (1962, 1969), Barone and Paterno (1982), Khurana (1987), Lundqvist et al. (1987), and Poole et al. (1995).

4

Electrodynamics of Superconductors in Strong Fields

At the beginning of the 1960s, it became universally recognized that there exists a whole new class of superconductors, called the type II superconductors. The type II superconductors may show the Meissner effect in weak magnetic fields, but when the field is increased to a value larger than a lower critical field B_{c1}, the Meissner effect breaks down and the magnetic flux begins to penetrate into the interior of the superconductor. However, the superconductivity is not destroyed. The type II superconductor in the mixed state may continue to show essentially no dc resistance up to a much higher upper critical field B_{c2}, where bulk superconductivity finally disappears. Recently discovered high-temperature oxide superconductors are examples of such superconductors, having the upper critical magnetic field on the order of hundreds of tesla. Since the discovery of type II superconductors, which can maintain their superconducting state in very strong magnetic fields, large-scale applications of superconductivity have become possible. The dream of Onnes on large superconducting magnets has already been realized after almost fifty years since his discovery of superconductivity. This chapter presents theoretical models for the study of electromagnetic properties as well as thermomagnetoelectric effects in type II superconductors in the mixed state. Some electrodynamic methods will also be introduced to analyze electromagnetic behaviors of these superconductors.

4.1 THERMODYNAMICS OF PHASE TRANSITIONS IN SUPERCONDUCTORS

4.1.1 Thermodynamic Functions for Superconductors

We start with the study of some thermodynamic behaviors and properties of superconductors. The idea of applying thermodynamics to the transition between the superconducting and normal states was originally suggested and developed by Keesom (1924), Rutgers (1934), and Gorter and Casimir (1934).

To study the thermodynamics of the superconducting transition in various fields, the simplest way of obtaining relevant thermodynamic functions is to treat the superconductive material in the superconducting state as a magnetic substance, described at the end of Section 3.2.4, and in the normal state as a non-magnetic substance. (Here, we do not consider magnetic superconductors [Zhou (1991a)].)

For a magnetic body placed in a magnetic field whose (current) sources are fixed, the change of the Gibbs free energy of the body corresponding to an infinitesimal change in the field can be found as [Landau et al. (1984)]

$$\delta G = -S\delta T - \int_V \boldsymbol{M} \cdot \delta \boldsymbol{B}^e dV \tag{4.1.1}$$

where V is the volume of the magnetic body. The last term on the right-hand side of Eq.(4.1.1) represents the differential amount of work done on the body when the external field is increased by an amount of $\delta \boldsymbol{B}^e$ for fixed (current) sources, which are independent of the field that they produce. The magnetization vector $\boldsymbol{M}$ in Eq.(4.1.1) can be generally a nonlinear functional of the magnetic field. Here, we have ignored mechanical deformation effects by assuming a rigid material body under constant pressure. The mechanical effects on the superconductivity will, however, be studied later.

In the case of a constant temperature, Eq.(4.1.1) becomes

$$\delta G = -\int_V \boldsymbol{M} \cdot \delta \boldsymbol{B}^e dV = -\delta \int_V \left(\mu_o \int_0^{H^e} \boldsymbol{M} \cdot d\boldsymbol{H}^e \right) dV \tag{4.1.2}$$

Thus we derive that any substance, which in an applied field $\boldsymbol{H}^e$ ($= \boldsymbol{B}^e/\mu_o$) acquires a magnetization $\boldsymbol{M}$, changes its Gibbs free energy per unit volume, denoted here by g, at a given temperature by an amount

$$\delta g = -\delta \left(\mu_o \int_0^{H^e} \boldsymbol{M} \cdot d\boldsymbol{H}^e \right) \tag{4.1.3}$$

This result shows that, in the case of the field producing a positive magnetization, that is, the magnetization in the same direction as the magnetic field (for isotropic magnetic materials), the Gibbs free energy is lowered.

For a magnetic body of the shape of an ellipsoid of rotation with an axis of symmetry being parallel to the direction of the applied (uniform) field $\boldsymbol{H}^e$, then the internal magnetization $\boldsymbol{M}$ and the field $\boldsymbol{H}$ in the body are everywhere uniform and can be expressed by

$$\boldsymbol{H} = \boldsymbol{H}^e - N_d \boldsymbol{M} \tag{4.1.4}$$

where N_d is the demagnetizing factor along the axis of the ellipsoid parallel to

$\boldsymbol{H}^e$ and is a function of its ellipticity only. Some numerical values of the demagnetizing factor are, for instance, $N_d = 1/3$ for a sphere, $N_d = 1/2$ for a cylinder with its axis normal to the direction of the applied field, and $N_d = 0$ for a cylinder with its axis parallel to the direction of the applied field or for a flat infinitely thin plate parallel to the direction of the applied field.

For a superconducting specimen, the application of a magnetic field produces a negative magnetization which, if penetration of the field is neglected, exactly cancels the flux due to the applied field, so that $\boldsymbol{M} = -\boldsymbol{H}$ in the superconducting body. In particular, for a superconducting body of the shape of an ellipsoid of rotation with an axis of symmetry being parallel to the direction of the applied field $\boldsymbol{H}^e$, we can find by Eq.(4.1.4)

$$\boldsymbol{M} = -\frac{\boldsymbol{H}^e}{1-N_d} \tag{4.1.5}$$

and, therefore,

$$g_s(T, H^e) = g_s(T, 0) + \frac{\mu_o}{2}\frac{(H^e)^2}{(1-N_d)} \tag{4.1.6}$$

where $g_s(T, 0)$ denotes the Gibbs free energy per unit volume of the superconducting body in the absence of the applied magnetic field and at the superconducting state. Equation (4.1.6) shows that, if we apply a magnetic field to a superconductor, its Gibbs free energy increases due to the Meissner effect. In particular, for the superconducting specimen of a long cylinder with its axis parallel to the applied field, the Gibbs free energy per unit volume is then increased to the value

$$g_s(T, H^e) = g_s(T, 0) + \frac{\mu_o}{2}(H^e)^2 \tag{4.1.7}$$

with negligible demagnetization effect from the ends of the specimen.

The specimen in the normal state is however virtually non-magnetic and acquires negligible magnetization in an applied magnetic field. Consequently, the application of a magnetic field does not change the Gibbs free energy of the normal state though it raises that of the superconducting state. If the field strength is increased enough, the Gibbs free energy of the superconducting state will be raised above that of the normal state, and, in this case, the specimen will not remain superconducting but will become normal. There is, therefore, a maximum magnetic field strength that can be applied to a superconductor if it is to remain in the superconducting state. From Eq.(4.1.7), we may find that this critical magnetic (intensity) field strength is given by

$$H_c(T) = \sqrt{\frac{2}{\mu_o}(g_n(T) - g_s(T, 0))} \tag{4.1.8}$$

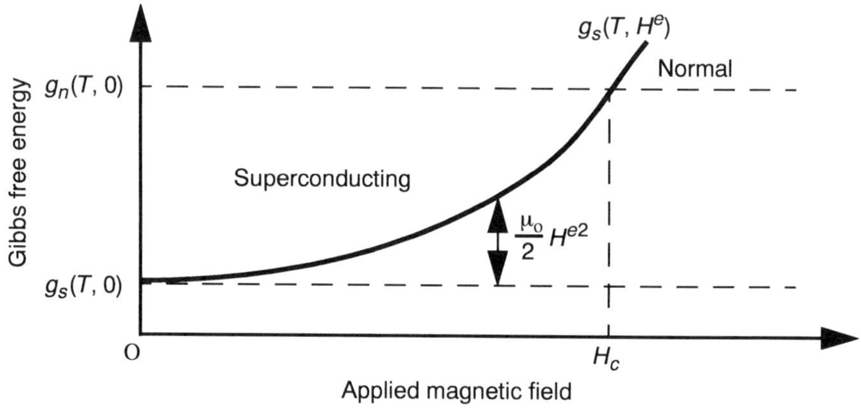

Figure 4.1 Effect of applied magnetic field on Gibbs free energy.

where $g_n(T)$ is the Gibbs free energy density of the specimen in its normal state. This critical magnetic field strength can be measured by, for instance, applying a magnetic field parallel to a wire or a long rod of superconductor and observing the strength at which resistance appears. From Eq.(4.1.8), we can see that the critical magnetic field is temperature-dependent and that it falls from some value at low temperatures to zero at the superconducting transition temperature T_c, as shown illustratively in Fig. 4.1. Experimentally, it has been found that the variation of the critical magnetic field can be conveniently described with a good approximation by a parabolic curve described by Eq.(3.1.1) with $B_c(T) = \mu_0 H_c(T)$.

The existence of a critical magnetic field of the superconductor has been made use of in a controlled switch, called the cryotron, which is composed of a straight wire called the gate and wire wound in a long single-layer coil called the control. The straight wire is insetted in the wire coil and both wires are in superconducting state at low temperatures. The superconducting materials are chosen such that the critical magnetic field of the coil wire is higher than that of the straight wire. The current through the gate can then be controlled by a small current in the control using the fact that the gate may be driven to normal by the magnetic field generated by the control current in the coil while the control coil remains resistanceless. Such a device is analogous to a relay. Large cryotons may be used to control the currents in superconducting magnet circuits.

4.1.2 First and Second-Order Phase Transitions in Superconductors

From Eqs.(4.1.7) and (4.1.8), we have therefore, in an applied magnetic field of strength H^e, a difference in the Gibbs free energy density between the normal and the superconducting states given by

$$g_n(T) - g_s(T, H^e) = \frac{\mu_o}{2}(H_c^2 - H^{e2}) \tag{4.1.9}$$

If the applied magnetic field strength are kept constant but the temperature T is varied by an amount δT, there will be also a change of the Gibbs free energy by

$$\delta G = -S\delta T \tag{4.1.10}$$

with

$$S = -\left(\frac{\partial G}{\partial T}\right)_{H^e} \tag{4.1.11}$$

The entropy per unit volume can thus be written

$$s = -\left(\frac{\partial g}{\partial T}\right)_{H^e} \tag{4.1.12}$$

Substituting Eq.(4.1.9) into this expression, we get for the superconductor (ignoring any flux penetration)

$$s_n - s_s = -\mu_o H_c \frac{\partial H_c}{\partial T} \tag{4.1.13}$$

which shows that the entropy of the superconducting state is less than that of the normal state, that is, the superconducting state has a higher degree of order than the normal state since the critical magnetic field always decreases with increase of temperature, so $\partial H_c/\partial T$ is always negative. The critical field H_c falls to zero as the temperature is raised towards T_c. Therefore, according to Eq.(4.1.13), the entropy difference between the normal and superconducting states vanishes at this temperature. Furthermore, by the third law of thermodynamics, s_n must also equal s_s at $T = 0$, which means that $\partial H_c/\partial T$ must be zero at 0 K since the critical field H_c is not zero. This is in accordance with the experimental observation that, for all superconductors, the slope of the H_c versus T curve appears to become zero as the temperature approaches 0 K.

In the presence of an applied magnetic field, there is a latent heat when a superconductive specimen undergoes the superconducting-normal transition. The latent heat L_h for the transition between two phases α and β is given by $L_h = vT(s_\alpha - s_\beta)$, so from Eq.(4.1.13) we have

$$L_h = -vT\mu_o H_c \frac{\partial H_c}{\partial T} \tag{4.1.14}$$

where v is the volume per unit mass. This latent heat arises because at temperatures between T_c and 0 K the entropy of the normal state is greater than that of the superconducting state, so heat must be supplied if the transition is to take place at constant temperature. Thus, in the presence of an applied magnetic field, the superconducting-normal transition is of the first-order, that is, although g is continuous, $\partial g/\partial T$ is not. It can be seen that, in the absence of any magnetic field, the transition occurs at the transition temperature T_c and $H_c = 0$, but if there is a magnetic field the transition occurs at some lower temperature T where $H_c > 0$.

In the absence of the applied magnetic field, since, at the transition temperature T_c, $s_n = s_s$, we have, for the superconducting-normal transition at T_c,

$$\left(\frac{\partial g}{\partial T}\right)_n = \left(\frac{\partial g}{\partial T}\right)_s \tag{4.1.15}$$

which shows that it is a phase transition of the second order (that is, not only g is continuous but also $\partial g/\partial T$ is continuous at the transition).

4.1.3 Discontinuity of Specific Heat at Transition Temperature

The second-order phase transition has two important characteristics: At the transition, there is no latent heat, and there is a jump in the specific heat. The first characteristic follows immediately from the fact that $\Delta Q = T\Delta S$ and we have seen that, at the transition temperature, there is no change in entropy and, therefore, no latent heat. The second condition follows from the fact that the specific heat of a material is given by

$$C_v = vT\frac{\partial s}{\partial T} \tag{4.1.16}$$

and the difference in the specific heats of a superconductor in the normal state and in the superconducting state can be obtained from Eq.(4.1.12):

$$C_{vs} - C_{vn} = vT\mu_o H_c \frac{\partial^2 H_c}{\partial T^2} + vT\mu_o\left(\frac{\partial H_c}{\partial T}\right)^2 \tag{4.1.17}$$

In particular, at the transition temperature T_c, we have $H_c = 0$ and so have for the transition in the absence of an applied magnetic field

$$(C_{vs} - C_{vn})_{T_c} = vT_c\mu_o\left(\frac{\partial H_c}{\partial T}\right)^2_{T_c} \tag{4.1.18}$$

which is known as Rutgers´ formula. It predicts the value of the discontinuity in the specific heat of a superconductor at the transition temperature (see Fig. 4.2). It may be noticed that $\partial H_c/\partial T$ is considered here to be a property of the material whose value does not depend on whether or not a field is actually present.

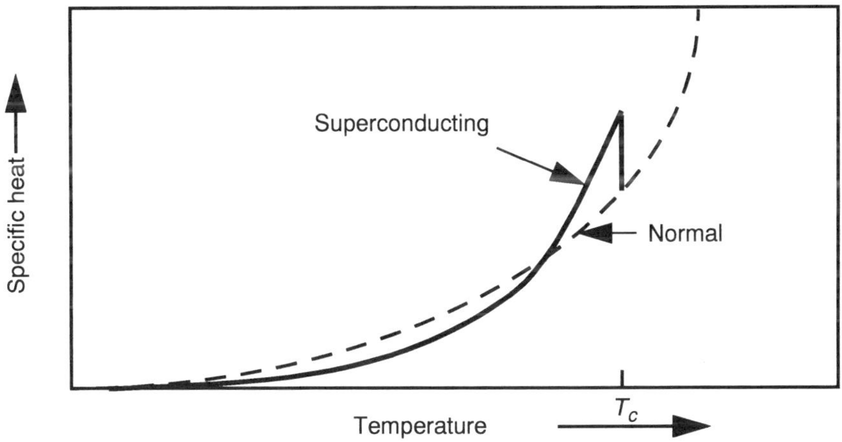

Figure 4.2 Specific heat of a superconductor in normal and superconducting states.

It is an interesting suggestion of Kok (1934) that, if the specific heat of the metal in the superconducting state is assumed to vary as T^3 and the specific heat of it in the normal state is assumed to vary as $aT+bT^3$, then it is possible to predict the parabolic form (3.1.1) of describing the temperature-dependent behavior of the critical field $H_c(T)$, which is in fair agreement with experiment.

4.2 THE GINZBURG-LANDAU THEORY OF SUPERCONDUCTORS

4.2.1 Complex Order Parameter

It has been shown in Chapter 3 that the fundamental assumptions used in London´s theory are the spatial invariance of the superconducting electron density n_s and the independence of n_s on the magnetic field. These assumptions are only justified for certain cases where the magnetic field is weak and where no inhomogeneities are present. In many circumstances of physical or technological interest, however, these restrictions are not appropriate. In particular, we have seen from Pippard's work that the electromagnetic properties of superconductors in the superconducting state could only be understood if there existed a certain degree of long-range order, extending typically over the coherence distance. To account for the possible spatial variation of the superconducting electron density, Ginzburg and Landau (1950) proposed an extension of the London theory with the aid of Landau's general theory of second-order phase transitions (1937), in which Landau introduced an important concept of the order parameter.

It is assumed in the Ginzburg-Landau (G-L) theory that the behavior of the superconducting electrons may be described by a complex order parameter, being an "effective wave function" $\psi=|\psi|e^{i\theta}$ with the square of its amplitude $|\psi|$ equal to the superelectron density n_s, that is, $|\psi|^2 = n_s$. The order parameter ψ goes to zero at the superconducting-normal transition. Furthermore, it is assumed that the free energy of the superconducting state differs from that of the normal state by an amount which can be expanded in powers of ψ with the coefficients of the expansion are regular functions of the temperature in the absence of the magnetic field (where one ignores possible effects of mechanical deformation). Thus, at T close to T_c where ψ is small, they wrote

$$f_s - f_n = \alpha|\psi|^2 + \frac{1}{2}\beta|\psi|^4 + \gamma|\nabla\psi|^2 + \dots \tag{4.2.1}$$

where f_s and f_n are the free energies per unit volume in the superconducting state and in the normal state, respectively. α, β and γ are phenomenological coefficients. The gradient term of ψ in Eq.(4.2.1) accounts for the non-uniform of ψ in space. It is argued that, if ψ varies slowly in space, it should be sufficient to keep only the leading term in $|\nabla\psi|^2$.

To be in accordance with the Landau general theory of second-order phase transition, only even powers may appear in the expansion of (4.2.1) due to the fact that the stability of the system at the transition point (at which $\psi = 0$) requires f_s to attain a minimum for $\psi = 0$. In addition, for the minimum in f_s to occur at finite values of $|\psi|^2$, we must have $\beta > 0$. Otherwise, if β were negative, there would be no limit to the decrease of the free energy with increasing $|\psi|^2$ where the expansion is clearly invalid. Furthermore, since higher-order terms have been omitted, Eq.(4.2.1) is clearly only valid in the limit of small ψ, which implies temperatures close to the transition temperature T_c. To avoid inessential complications, we shall deal with here only isotropic superconductors. Superconductors of anisotropic behaviors will however be studied later.

In a homogeneous superconductor with no external field, the order parameter ψ is independent of the coordinates in space. In such a case, the gradient term $|\nabla\psi|^2$ disappears and the equilibrium value of $|\psi|^2$ is determined by the condition for the free energy density f_s to be a minimum. For $\alpha > 0$ the minimum occurs at $|\psi|^2 = 0$ corresponding to the normal state and to the case $T > T_c$. On the other hand, for $\alpha < 0$, the minimum occurs at

$$|\psi|^2 = |\psi_o|^2 \equiv -\frac{\alpha}{\beta} \tag{4.2.2}$$

corresponding to the superconducting state where $T < T_c$. Thus, by Eqs.(4.2.1) and (4.2.2), we get the free energy density difference

$$f_s - f_n = -\frac{\alpha^2}{2\beta} \tag{4.2.3}$$

which shows that the free energy of the superconducting state is lower than that of the normal state. By noting that α must change its sign at $T = T_c$, we may write the expansion near the transition temperature T_c

$$\alpha(T) = \alpha_o(T - T_c) \quad \text{and} \quad \beta(T) \approx \beta(T_c) \tag{4.2.4}$$

with the constant $\alpha_0 > 0$. Thus, by Eq.(4.2.2), we have

$$|\psi_o|^2 = \frac{\alpha_o}{\beta(T_c)}(T_c - T) \tag{4.2.5}$$

Equation (4.2.3) may then be written in the form:

$$f_s - f_n = -\frac{\alpha_o^2}{2\beta(T_c)}(T_c - T)^2 \tag{4.2.6}$$

from which, we can find

$$\frac{\partial(f_s - f_n)}{\partial T} = \frac{\alpha^2}{2\beta}(T_c - T) \tag{4.2.7}$$

which goes to zero for $T \to T_c$, indicating a phase transition at least of second order. The discontinuity in the specific heat at the transition point can then be found by

$$C_{vs} - C_{vn} = -vT_c\frac{\partial^2(f_s - f_n)}{\partial T^2} = vT_c\frac{\alpha_o^2}{\beta(T_c)} \tag{4.2.8}$$

We may also define, in analogy to the definition of the critical magnetic field by Eq.(4.1.8), a thermodynamic critical field B_c (= $\mu_0 H_c$) for all types of superconductors by means of $f_n - f_s = B_c^2/(2\mu_0)$ at zero field. Thus, from Eqs.(4.2.1) and (4.2.2), we may obtain

$$B_c(T) = \sqrt{\frac{\mu_o\alpha^2(T)}{\beta(T)}} \tag{4.2.9}$$

which can also be written by using Eq.(4.2.4) as

$$B_c(T) = T_c\sqrt{\frac{\mu_o\alpha_o^2}{\beta(T_c)}}\left(1 - \frac{T}{T_c}\right) \tag{4.2.10}$$

near the transition temperature T_c ($T < T_c$).

4.2.2 The Ginzburg-Landau Equations

In inhomogeneous superconductors, to take into account the non-local character [Pippard (1950, 1953)] of the superconducting state, that is, of the fact that the values of $\psi(\boldsymbol{x})$ at any point is described by integral equations which involve the values of $\psi(\boldsymbol{x})$ at all neighboring points, the term of lowest order $|\nabla\psi(\boldsymbol{x})|^2$ accounting for the spatial variation in $\psi(\boldsymbol{x})$ has to be added to the free energy. If also magnetic field exists, Ginzburg and Landau proposed that a gauge invariant free energy density for the superconductor can be expanded in the following form:

$$f_s - f_n = \alpha|\psi|^2 + \frac{1}{2}\beta|\psi|^4 + \frac{1}{2m^*}|(-i\hbar\nabla - e^*\boldsymbol{A})\psi|^2 + \frac{1}{2\mu_o}B^2 \qquad (4.2.11)$$

where the third term on the right-hand side represents presumably the kinetic energy of the superconducting electrons. This may be seen by identifying $|\psi|^2 = n_s$ with n_s being the number of the Cooper pairs of superconducting electrons per unit volume. The kinetic-energy density is then $m^* n_s \mathrm{v}_s^2/2$, where m^* may approximately be identified as the mass of the Cooper pairs (total mass of the two electrons) according to the experimental evidence by Zimmermann and Mercereau (1965). The effect of a magnetic field is then introduced by resorting to a theorem in classical mechanics which states that the effect of the Lorentz force ($e^*\mathbf{v}_s \times \boldsymbol{B}$) on the motion of a charged particle in a magnetic field $\boldsymbol{B}$ ($= \nabla \times \boldsymbol{A}$) may be completely accounted for by replacing the momentum $\boldsymbol{p}$, wherever it occurs in the expression for the kinetic energy by $\boldsymbol{p} - e^*\boldsymbol{A}$. Finally, to make the transition to quantum mechanics description, $\boldsymbol{p}$ is replaced by the operator $-i\hbar\nabla$ in the expression of the kinetic energy. It is shown that the phenomenological coefficient γ in Eq.(4.2.1) can be identified here by $\gamma = \hbar^2/(2m^*)$.

The thermodynamic equilibrium of the system characterized by the free energy $F = \int f dV$ as a functional of the three independent functions ψ, ψ^* and $\boldsymbol{A}$ requires the following set of differential equations to be satisfied, which are now called the Ginzburg-Landau (G-L) equations:

$$\alpha\psi + \beta|\psi|^2\psi + \frac{1}{2m^*}(-i\hbar\nabla - e^*\boldsymbol{A})^2\psi = 0 \quad \text{in } V \qquad (4.2.12)$$

$$\boldsymbol{J} = \frac{e^*\hbar}{2im^*}(\psi^*\nabla\psi - \psi\nabla\psi^*) - \frac{e^{*2}}{m^*}|\psi|^2\boldsymbol{A} \quad \text{in } V \qquad (4.2.13)$$

where V denotes the volume of the superconductor. We have written $\boldsymbol{J}$ for $\boldsymbol{J}_s$ since in the thermodynamic equilibrium there is no normal currents. By noting $\psi = |\psi|\exp(i\theta)$, Eq.(4.2.13) for the supercurrent density may also be written

$$J = \frac{e^*\hbar|\psi|^2}{m^*}\left(\nabla\theta - \frac{e^*}{\hbar}A\right) \tag{4.2.14}$$

which shows that the gradient of the phase of the wave function ψ determines the observable quantity, the supercurrent density. In the presence of an external magnetic field, Eq.(4.2.14) is shown to be of gauge-invariance as it should be.

To complete the theory, proper boundary conditions have to be specified. For the field, we have the continuity of the magnetic induction field $\boldsymbol{B}$ across the interface for non-magnetic superconductors. To the G-L equations, we have the following boundary condition:

$$\left(\nabla\psi - \frac{ie^*}{\hbar}A\psi\right)\cdot n = \lambda_b\psi \tag{4.2.15}$$

where λ_b is a real constant, being zero for a superconductor-insulator (S-I) interface, and non-zero for a superconductor-normal metal (S-N) interface [de Gennes (1964) and Deutscher and de Gennes (1969)]. It is easy to verify that, for real λ_b, the boundary condition (4.2.15) is consistent with the fact that the normal component of the superconducting current by Eq.(4.2.13) is zero across the S-N boundary since no nondissipative current can flow in a normal metal. However, the mutual effect of the electrons at the two sides of the interface between the superconductor and the normal metal can be extended much farther than the interatomic distance. The quantity $1/\lambda_b$ must be of the order of the coherence length ξ, but its exact value can only be determined from a microscopic theory of superconductivity. Some detailed discussions based on the BCS microscopic theory for the coefficient λ_b may be found in the work of, for instance, Zaitsev (1965, 1966) and Deutscher and de Gennes (1969).

Physically, the boundary condition (4.2.15) implies that a thin layer of normal region adjacent to a superconductor becomes superconducting itself. This effect is called the proximity effect. Due to the proximity effect, a sufficiently thin film of superconductor deposited on a bulk normal metal may cease to be superconductive, while a thin film of normal metal deposited on a bulk superconductor may become superconducting. These phenomena have been detected experimentally [Meissner (1958, 1959, 1960) and Werthamer (1963)]. The possibility of proximity-induced superconductivity in a macroscopic region of a normal metal is important for the fabrication of SNS Josephson junctions [Likharev (1979)], as well as for the design of composite current-carrying superconducting cables [Carr (1983)].

4.2.3 Critical Fluctuation and Validity of the Ginzburg-Landau Theory

It is known that physical quantities which describe a macroscopic body in equilibrium are, almost always, very nearly equal to their mean values. Nevertheless, deviations from the mean values, though probably small, do occur, which characterize the fluctuation of the physical quantities. In the discussion of a phase transition of the second kind, the fluctuation of the order parameter, which is the wave function ψ for the superconducting-normal phase transition, may increase rapidly when the transition point is approached due to the anomalous behavior of the thermodynamic functions of the body at the actual transition point [Lifshitz and Pitaevskii (1980)]. Thus, to have small fluctuation of the order parameter, one has to be restricted to the situations outside of the fluctuation region near the transition point. On the other hand, the validity of the expansion (4.2.1) requires the order parameter to be small, which implies here the condition $|T_c - T| << T_c$ has to be satisfied. Therefore, for the G-L theory to be valid, the temperature of the superconductor to be considered should be within a certain range, away from the fluctuation region near the transition temperature so that both the order parameter and its fluctuation can be kept small. A quantitative estimate of the fluctuation region for the superconducting-normal phase transition was made by Ginzburg (1961), who showed that the fluctuation regions for common superconductors were extremely narrow so that the observed superconducting-normal phase transition was practically of a "normal" character (i.e., there are no anomalies of the specific heat, etc.). From this it follows that the fluctuation is not important in the thermodynamics of bulk superconductors. Nevertheless, situations are possible where the role of fluctuation increases noticeably for some observed effects, say paraconductivity, in small objects as thin films and filaments [Aslamazov and Larkin (1968) and Glover (1971)]. Recently discovered high-T_c oxide superconductors may have relatively large fluctuation region due to their small coherence lengths (on the order of 10 Å), which implies some modifications of the G-L theory might be required.

One may further ask what the range of the validity of the G-L theory is on the low-temperature side. One condition is known to be $|1 - T/T_c| << 1$. The other is due to the limitation of G-L theory being a local theory. To study the limitation of locality, we may introduce the following two characteristic lengths ξ and λ from the G-L equations (4.2.12) and (4.2.13). The first characteristic length ξ can be introduced by writing the G-L equation (4.2.12) into the following form:

$$\xi^2 \nabla^2 \psi' + \psi' - \psi' |\psi'|^2 = 0 \tag{4.2.16}$$

where we have introduced a new non-dimensional function $\psi' = \psi/|\psi_o|$ with $|\psi_o|$ being the equilibrium solution given by Eq.(4.2.2). The quantity ξ is defined by

$$\xi(T) = \frac{\hbar}{\sqrt{2m^* |\alpha(T)|}} \tag{4.2.17}$$

which clearly measures the range of variation of ψ' (i.e., which is the smallest distance over which the order parameter ψ can be of large fractional changes). Therefore, it is also called the coherence length at temperature T (or called the

Ginzburg-Landau coherence length) in accordance to the concept of Pippard's coherence length. However, it should be noticed that this coherence length $\xi(T)$ has a different significance from those of the (temperature-independent) Pippard coherence lengths, the intrinsic coherence length ξ_0 and the mean-free-path-dependent coherence length $\xi(l_e)$ given by Eq.(3.4.54).

The second characteristic length λ comes into play if we introduce electromagnetic effects. Consider a superconductor in a weak field, to the first order in $\boldsymbol{B}$, $|\psi|^2$ can be replaced by $|\psi_0|^2$, and the G-L equation (4.2.13) may be written

$$\boldsymbol{J} = \frac{e^*\hbar}{2im^*}(\psi^*\nabla\psi - \psi\nabla\psi^*) - \frac{e^{*2}}{m^*}|\psi_o|^2\boldsymbol{A} \quad \text{in } V \tag{4.2.18}$$

Taking the curl of $\boldsymbol{J}$, we then obtain

$$\nabla\times\boldsymbol{J} = -\frac{e^{*2}}{m^*}|\psi_o|^2\boldsymbol{B} \quad \text{in } V \tag{4.2.19}$$

which is equivalent to the London equation (3.1.9) with the penetration depth defined by

$$\lambda(T) = \sqrt{\frac{m^*}{\mu_o e^{*2}|\psi_o|^2}} = \sqrt{\frac{m^*\beta}{\mu_o e^{*2}|\alpha(T)|}} \tag{4.2.20}$$

This temperature-dependent penetration depth characterizes the range of variation of electromagnetic induction field $\boldsymbol{B}$.

It is shown that, by defining an effective density of the superelectron pairs $n_{seff} = |\psi_0|^2$, the weak field solutions of the G-L theory are the same of that of the London theory when the gradient term of ψ is neglected. This is known to be able to give accurate results for local electrodynamics of the superconductors. For some purposes of practical calculations, we may also assume it to be useful as an approximation in the nonlocal cases by taking

$$|\psi_o(T)|^2 = \frac{m^*}{\mu_o e^{*2}\lambda^2(T)} \tag{4.2.21}$$

with $\lambda(T)$ being the appropriate penetration depth, subject to direct measurement.

In order for the G-L theory to be valid, both these lengths, ξ and λ have to be large in comparison with the intrinsic coherence length ξ_0 so that all quantities vary sufficiently slowly in space. Such a condition can be, in general, satisfied near the transition point since, by noting Eqs.(4.2.4), (4.2.17), and (4.2.20), both lengths, ξ and λ increase in proportion to $|T - T_c|^{-1/2}$ as the transition temperature is approached. Although the G-L theory is strictly speaking only valid at temperatures close to the transition temperature, in practice, their numerical predictions do not seem to be in error by more than 50% at any temperature [Goodman (1966)].

4.2.4 The Ginzburg-Landau Parameter κ

Due to the divergence of both two characteristic lengths $\xi(T)$ and $\lambda(T)$ as $T \rightarrow T_c$, it is thus of interest to consider their ratio

$$\kappa = \frac{\lambda(T)}{\xi(T)} = \frac{m^*}{\hbar|e^*|}\sqrt{\frac{2\beta}{\mu_o}} \qquad (4.2.22)$$

which is called the Ginzburg-Landau parameter of the superconductor. This parameter is independent of temperature within the framework of the G-L theory, and has its importance of characterizing superconducting materials. By Eqs.(4.2.9) and (4.2.20), this parameter can also be related to observable quantities, the critical field B_c and the penetration depth λ by

$$\kappa = \frac{\sqrt{2}|e^*|}{\hbar}\lambda^2 B_c \qquad (4.2.23)$$

which shows that the G-L parameter κ may be determined by measuring the penetration depth λ in low field and the thermodynamic critical field B_c. In practice, there are, however, other better ways of obtaining the values of κ [see Saint-James et al. (1969) and Kuper (1968)].

Further studies of the microscopic foundation of the G-L theory by Gor´kov (1959a,b) based on the BCS microscopic theory has helped to improve our understanding of the G-L equations and their limitations. It was found by Gor´kov that the G-L parameter κ may be expressed as

$$\kappa = \kappa_p = 0.96\frac{\lambda_L(0)}{\xi_o} \quad \text{for pure superconductors } (\xi_o \ll l_e) \qquad (4.2.24)$$

and

$$\kappa = \kappa_d = 0.72\frac{\lambda_L(0)}{l_e} \quad \text{for dirty superconductors } (\xi_o \gg l_e) \qquad (4.2.25)$$

where $\lambda_L(0)$ is the London penetration depth at the temperature of absolute zero, ξ_0 is Pippard´s intrinsic coherence length introduced in (3.4.53), and l_e is the electron mean free path. Equation (4.2.25) shows that the κ of a superconductor is increased if the electron mean free path is shortened by a high impurity concentration, which also implies that the electric resistance of the superconductor in the normal state is increased. The general dependence of κ upon the normal state resistivity ρ_n has been given by Goodman (1962)

$$\kappa = \kappa_p + 2.4 \times 10^6 \rho_n \sqrt{\gamma_e} \tag{4.2.26}$$

where γ_e is the electronic specific heat constant, and ρ_n is the normal state resistivity. With the use of Eq.(4.2.26), good agreement with experiment results has been reported by Livingston (1963). In addition, near T_c in the pure and dirty limit, the microscopic theory gives

$$\xi = \xi_p(T) = 0.74\xi_o \sqrt{\frac{T_c}{T_c - T}} \quad \text{(pure)} \tag{4.2.27}$$

and

$$\xi = \xi_d(T) = 0.85\sqrt{\xi_o l_e} \sqrt{\frac{T_c}{T_c - T}} \quad \text{(dirty)} \tag{4.2.28}$$

and

$$\lambda = \lambda_p(T) = \frac{\lambda_L(0)}{\sqrt{2}} \sqrt{\frac{T_c}{T_c - T}} \quad \text{(pure)} \tag{4.2.29}$$

and

$$\xi = \lambda_d(T) = 0.615\lambda_L(0) \sqrt{\frac{\xi_o T_c}{l_e(T_c - T)}} \quad \text{(dirty)} \tag{4.2.30}$$

Further generalization of the G-L theory to all temperatures has been made for dirty superconducting alloys in high magnetic fields based on the BCS microscopic theory [see, e.g., Maki (1964) and de Gennes (1966)], where we can find that the quantity κ is temperature-dependent. In the following section, we shall see that superconductors can be classified according to their particular values of the G-L parameter κ.

Table 4.1 shows numerical data for the coherence length ξ, the penetration depth λ and the Ginzburg-Landau parameter κ of some superconductors.

Table 4.1 Data for ξ, λ and κ of Some Superconductors

Material	T_c (K)	ξ (nm)	λ (nm)	κ (= λ/ξ)
Nb	9.25	39	50	1.28
Nb-Ti	9.5	4	300	75
Nb_3Ge ($A15$)	23.2	3	90	30
$YBa_2Cu_3O_7$	89	1.8	170	95

4.2.5 Upper Critical Magnetic Field B_{c2}

It has been known that some pure superconductors undergoes a first-order transition into the normal state at a critical magnetic field B_c (= $\mu_0 H_c$), and the density of superelectrons has almost a constant value up to B_c and then drops abruptly to zero. In this section, with the use of G-L theory, we shall study superconductors which present a second-order phase transition at a certain critical field and that the order parameter ψ approaches zero continuously when the critical magnetic field is reached. Let us now consider an infinite superconductive medium, and take the z-axis along the uniform applied magnetic field $\boldsymbol{B}^e$, that is, $\boldsymbol{B}^e = (0, 0, B^e)$. When the applied field is sufficiently large the medium is in the complete normal state with $|\psi| = 0$. If the applied field is gradually reduced, we may ask what value this field must be reduced in order that the nucleation of superconducting regions start to occur, that is, that Eq.(4.2.12) may just have solutions other than $\psi = 0$. Since during the first appearance of superconducting regions, ψ is small, we can linearize the G-L equation for ψ by dropping the nonlinear term $\beta|\psi|^2\psi$. In addition, the condition of small ψ implies that field penetration is virtually complete so that the magnetic field in the medium may be written approximately as $\boldsymbol{B} = \boldsymbol{B}^e$. Thus the vector potential $\boldsymbol{A}$ appropriate to the field $\boldsymbol{B}$ may be chosen as

$$\boldsymbol{A} = (0, \mu_o H^e x, 0) \tag{4.2.31}$$

The linearized form of G-L equation (4.2.12) now writes

$$-\frac{\hbar^2}{2m^*}\left(\frac{\partial^2\psi}{\partial x^2} + \frac{\partial^2\psi}{\partial z^2}\right) + \frac{1}{2m^*}\left(-i\hbar\frac{\partial}{\partial y} - e^*\mu_o H^e x\right)^2\psi = |\alpha|\psi \tag{4.2.32}$$

which has the form of the Schrödinger equation for a particle of mass m^*, charge e^* and zero spin moving in a uniform magnetic field directed along the z-axis [Landau and Lifshitz (1977)], where $|\alpha|$ plays the part of the eigenvalue. Thus we may seek ψ in the following form:

$$\psi(\boldsymbol{x}) = u(x)exp(ik_y y)exp(ik_z z) \tag{4.2.33}$$

where $u(x)$ is yet to be determined.

By substituting Eq.(4.2.33) into Eq.(4.2.32), we find

$$-\frac{\hbar^2}{2m^*}\frac{\partial^2 u}{\partial x^2} + \frac{(e^*\mu_o H^e)^2}{2m^*}(x - x_o)^2 u = \left(|\alpha| - \frac{\hbar^2 k_z^2}{2m^*}\right)u \tag{4.2.34}$$

which has the form of the Schrödinger equation for a particle moving in a harmonic potential well centered at

$$x_o = \frac{\hbar k_y}{e^* \mu_o H^e} \tag{4.2.35}$$

The eigenvalues of Eq.(4.2.34) are thus given by

$$|\alpha| = \left(n + \frac{1}{2}\right)\hbar\omega_c + \frac{\hbar^2 k_z^2}{2m^*} \tag{4.2.36}$$

where ω_c is the cyclotron resonance frequency defined by

$$\omega_c = \frac{\mu_o |e^*| H^e}{m^*} \tag{4.2.37}$$

Equation (4.2.36) can also be written in the form:

$$H^e = \frac{2m^*}{(2n+1)|e^*|\hbar\mu_o}\left(|\alpha| - \frac{\hbar^2 k_z^2}{2m^*}\right) \tag{4.2.38}$$

Obviously the field of interest is the highest field for which superconductivity begins to occur. This corresponds to $n = 0$ and $k_z = 0$ in Eq.(4.2.38) and the field denoted by B_{c2} ($= \mu_0 H_{c2}$) is

$$B_{c2} = \frac{2m^*|\alpha|}{|e^*|\hbar} = \frac{\Phi_o}{2\pi\xi^2} \tag{4.2.39}$$

where Φ_0 is the flux quantum given by Eq.(3.2.24), and ξ is the G-L coherence length given by Eq.(4.2.17). The corresponding eigenfunction is the ground-state wave function of the harmonic oscillator given by

$$u(x) = exp\left(-\frac{(x - x_o)^2}{2\xi^2}\right) \tag{4.2.40}$$

By noting Eqs.(4.2.9) and (4.2.22), we can rewrite Eq.(4.2.39) as

$$B_{c2} = \kappa\sqrt{2}B_c \tag{4.2.41}$$

which is valid whatever the value of κ. Shown in Table 4.2 are numerical data on the upper critical magnetic field B_{c2} for some type II superconductors [Brechna (1973), Hein (1974), Osamura (1994), and Poole et al. (1995)].

The result obtained above is for an infinite medium, where the influence of the specimen surfaces is ignored. For superconductors of finite sizes, it has been shown [Saint-James and de Gennes (1963) and Saint-James et al. (1969)] that there exists a superconducting sheath with the thickness being of the order of ξ

Table 4.2 B_{c2} for Some Type II Superconductors

Material	T_c (K)	$B_{c2}(0)$ (T)
Nb	9.25	0.3
$Nb_{37}Ti_{63}$	9.08	15
Nb_3Sn	18	23
Nb_3Al	18.6	31
Nb_3Ge	23	38
V_3Ga	15	23
V_3Si	16	20
$PbMo_6S_8$	15.3	60
YBaCuO	92	120–200
TlBaCaCuO	125	60
HgBaCaCuO	131	190

close to the surface of the body, in which $\psi \neq 0$ for magnetic fields up to $B_{c3}(\varphi)$ where φ is the angle between the direction of the field and the normal to the surface. For $\varphi = \pi/2$, that is, for the surface parallel to the applied field, $B_{c3}(\varphi)$ has its maximum value of

$$B_{c3} = 1.695 B_{c2} \tag{4.2.42}$$

which is larger than B_{c2}. The presence of this superconducting sheath has been verified experimentally [Bon Mardion et al. (1964), Cardona and Rosenblum (1964), and Tomasch and Joseph (1964)]. However, such a sheath does not exist if the surface of the superconductor is in contact with a normal metal rather than an insulator because of the different boundary condition on the normal derivative of ψ.

4.2.6 Classification of Superconductors According to the G-L Theory

From Eq.(4.2.41), we can see that, if $\kappa > 1/\sqrt{2}$, there exist magnetic fields greater than B_c but less than B_{c2} in which the normal state is unstable with respect to the establishment of a certain degree of superconducting order ($\psi \neq 0$)

[Ginzburg and Landau (1950)]. We may, therefore, distinguish between two types of magnetic behaviors of a superconductor in a decreasing field. If $\kappa<1/\sqrt{2}$ (type I behavior), the normal state is the only stable state for $B > B_c$. When we decrease the field we first meet the value B_c, at which the complete Meissner effect takes place in the type I superconductors. If $\kappa > 1/\sqrt{2}$ (type II behavior), a new kind of superconducting state (characterized by the partial exclusion of a magnetic field) is stable for $B_c < B < B_{c2}$. The value $\kappa = 1/\sqrt{2}$ is thus the point that separates the two kinds of superconductors.

It may be shown that the surface energy associated with the interphase boundary between a normal and a superconducting region is negative for type II superconductors, and is positive for type I superconductors [see, e.g., Saint-James et al. (1969)]. A simple explanation of the origin of the surface energy may be given here according to Rose-Innes and Rhoderick (1969). At the boundary of superconducting and normal phases, there is not a sudden change from fully normal behavior to fully superconducting behavior. The magnetic flux density penetrates a distance characterized by λ into the superconducting region, and in the superconducting region, the number of superelectrons per unit volume increases slowly over a distance characterized by ξ (see Fig. 4.3). Thus, at the boundary, the degree of order (i.e., the number of superelectrons n_s) rises gradually over a distance determined by the coherence length ξ, so the decrease in free energy due to the increasing order of the electrons takes place over the same distance. On the other hand, the free energy rises over a distance of about the penetration depth λ due to the positive "magnetic" contribution. In general, ξ and λ are not the same so that the two contributions do not cancel near the boundary. If the coherence length is longer than the penetration depth, the total free energy density is increased close to the boundary, that is, there is a positive

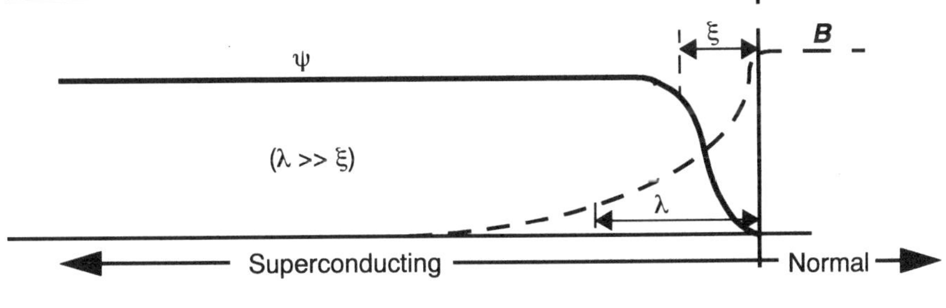

Figure 4.3 Schematic diagram of variation of B and ψ in a domain wall.

surface energy. If the coherence length is shorter than the penetration depth, the total free energy density is decreased close to the boundary, that is, there is a negative surface energy. This means that, for type I materials ($\kappa < 1/\sqrt{2}$), the existence of an interphase boundary leads to a raising of the Gibbs free energy over that of a uniformly superconducting sample. Thus it is energetically unfavorable for such interphase boundaries to exist in a sample of type I material with zero demagnetization factor.

For samples of other shapes with non-zero demagnetization factor, such as a sphere, the energy involved in the distortion of the applied field makes a subdivision into the superconducting and the normal phases energetically favorable and, thus, results in the presence of a so-called intermediate state in the superconducting sample. Such an intermediate state may be understood from the following consideration: Consider a sphere of type I superconductor in a uniform applied magnetic field $\boldsymbol{B}^e$. When the applied field $\boldsymbol{B}^e$ is raised gradually, the magnetic field strength at the equator of the sphere will at first reach to the critical value B_c for $B^e = 2B_c/3$. We might expect that the sphere would then be driven into a normal state. However, if it were so, the magnetic intensity field inside the sphere would be equal to $B^e = 2B_c/3$, which is less than B_c. We should then have the impossible situation of a completely normal body in a field smaller than the critical value B_c. Thus a possible situation is that the magnetic flux may penetrate into the superconducting sphere such that the interior region of the sphere is divided into a certain structure of superconducting and normal domains. For the two phases to exist in equilibrium, the magnetic field at the interphase boundary must be tangential and equal to the critical magnetic field B_c. The positive free energy contribution of the interphase surface will determine the equilibrium configuration of superconducting and normal domains. Such a configuration is known as the intermediate state, which has a relatively coarse structure and has been confirmed experimentally [see Shoenberg (1952) and Bodmer et al. (1972) for more discussions].

For type II materials ($\kappa > 1/\sqrt{2}$), the presence of interface boundaries in the superconducting sample even having a zero demagnetization factor are energetically favorable. Thus, when the type II superconductor is placed in a magnetic field above a certain value, the superconductor will be in a mixed state, that is, the superconducting medium splits into some fine-scale mixture of superconducting and normal regions whose boundaries lie parallel to the applied field. It can be seen that the intermediate state in type I superconductors is fundamentally different from the mixed state in type II superconductors. In addition, the structure of the intermediate state is relatively coarse and the gross features may be made visible to the naked eye. The structure of the mixed state is, however, on a much finer scale with a periodicity generally less than 10^{-7} m, which was first theoretically predicted by Abrikosov (1957) and was later verified experimentally by Essmann and Träuble (1967).

4.2.7 The Abrikosov Mixed State in Type II Superconductors

In this section, we shall present the analysis of Abrikosov (1957) on the vortex structures in the mixed state of type II superconductors in a field near the upper critical magnetic intensity field H_{c2} (= B_{c2}/μ_o). In the analysis, only the regime with applied fields slightly less than H_{c2} is treated, where the solution ψ of the complete G-L equations must have strong similarity to a certain solution ψ_L of the linearized G-L equations. This may be seen by considering a sample of type II superconductive material with negligible demagnetization effect placed in an external magnetic field $\boldsymbol{B}^e$ (= $\mu_o \boldsymbol{H}^e$). If we let $\boldsymbol{B}^e$ decrease gradually, the nucleation of the superconducting phase in the interior of the sample will begin when the magnetic intensity field H^e in the sample becomes equal to the upper critical field H_{c2}. Thus, for the field H^e slightly less than H_{c2}, the order parameter $|\psi|$ will be small and the form of ψ may be obtained from the linearized G-L equations. However, if the field H^e decreases appreciably below H_{c2}, the order parameter $|\psi|$ will become larger, and then, the complete nonlinear G-L equations have to be used.

In the case of the applied field slightly less than H_{c2}, as the first-order approximation, we have the linearized G-L equation for the determination of ψ_L, which reads

$$\alpha\psi_L + \frac{1}{2m^*}(-i\hbar\nabla - e^*\boldsymbol{A}_o)^2\psi_L = 0 \tag{4.2.43}$$

with $\boldsymbol{A}_o = (0, \mu_o H_{c2}x, 0)$ so that

$$\nabla \times \boldsymbol{A}_o = (0, 0, \mu_o H_{c2}) \tag{4.2.44}$$

where we have assumed that the applied field is along the z-direction.

From Section 4.2.5, we have known that the solutions of Eq.(4.2.43) are of the form:

$$\psi_k = e^{iky} exp\left(-\frac{(x-x_o)^2}{2\xi^2}\right) \tag{4.2.45}$$

with

$$x_o = \frac{\hbar k}{e^*\mu_o H_{c2}} \tag{4.2.46}$$

where k is an arbitrary parameter. The solutions describe a Gaussian band of superconductivity of width $\xi(T)$ extending perpendicular to the x-axis at the location $x = x_o(k)$. A general solution ψ_L must be a linear combination of the ψ_k. Following the original calculation of Abrikosov, we consider the solution being periodic both in x- and y-direction. The periodicity in y-direction is achieved by setting

$$k = k_n = nq \tag{4.2.47}$$

yielding the period

$$\Delta y = \frac{2\pi}{q} \tag{4.2.48}$$

Thus the general solution for ψ_L can be expressed by

$$\psi_L = \sum_n C_n e^{inqy} exp\left(-\frac{(x-x_n)^2}{2\xi^2}\right) \tag{4.2.49}$$

with

$$x_n = \frac{\hbar nq}{e^* \mu_o H_{c2}} \tag{4.2.50}$$

The periodicity in x-direction can be established if the coefficients C_n are periodic functions of n, such that $C_{n+\nu} = C_n$, where ν is some integer. The particular choice of ν determines the type of periodic lattice structure (ν = 1: square lattice; ν = 2: triangular lattice). From Eqs.(4.2.48) and (4.2.50), we note that the periodicity in x-direction is

$$\Delta x = \frac{\Phi_o}{2\pi \mu_o H_{c2}} \times \frac{2\pi}{\Delta y} \tag{4.2.51}$$

yielding

$$\Delta x \Delta y \mu_o H_{c2} = \Phi_o \tag{4.2.52}$$

which means that each unit cell of the periodic array contains one flux quantum Φ_o.

From the form ψ_L in Eq.(4.2.49), some conclusions independent of the choices for C_n and q may be drawn. The first conclusion concerns the supercurrent $\boldsymbol{J}_L$ associated with ψ_L, which is the solution of Eq.(4.2.43) corresponding to the lowest eigenvalue $|\alpha| = \hbar\omega_c/2$. It is found that the lines of current flow $\boldsymbol{J}_L$ coincide with the lines of constant $|\psi_L|$ [see, e.g., De Gennes (1966) and Huebener (1979)]. In particular, we may find that the local magnetic field $\boldsymbol{B}_s$ associated with the supercurrent by $\nabla \times \boldsymbol{B}_s = \mu_o \boldsymbol{J}_L$ is given by

$$\boldsymbol{B}_s = -\frac{\mu_o \pi \hbar^2}{\Phi_o m^*} |\psi_L|^2 z^o \tag{4.2.53}$$

where z^o is the unit vector along the direction of z-axis. Thus Eq.(4.2.53) concludes that the lines of constant $|\psi_L|$ coincide with the lines of the constant local magnetic field $\boldsymbol{B}_s$.

Let us now study the correction to the first-order solution ψ_L from the perturbation of the field which is slightly less than the critical field H_{c2}. We consider the normalization of ψ_L, which is rather important due to the nonlinearity of the complete G-L equations. The normalization will determine the strength of the supercurrent, and therefore, the macroscopic magnetic flux density and the mean free energy density. We assume that the free energy F remains stationary if ψ_L is replaced by the function $(1+\varepsilon)\psi_L$, where ε is a small quantity independent of the spatial coordinate $\boldsymbol{x}$. To the first order in ε, the variation in the free energy is

$$\delta F = 2\varepsilon\int\left\{\alpha|\psi_L|^2 + \beta|\psi_L|^4 + \frac{1}{2m^*}|(-i\hbar\nabla - e^*\boldsymbol{A})\psi_L|^2\right\}dV \tag{4.2.54}$$

By defining the following form of volume average integrals

$$\langle|\psi_L|^2\rangle = \frac{1}{V}\int_V |\psi_L|^2 dV \tag{4.2.55}$$

with V being the macroscopic volume, the normalization condition $\delta F = 0$ yields

$$\alpha\langle|\psi_L|^2\rangle + \beta\langle|\psi_L|^4\rangle + \frac{1}{2m^*}\langle|(-i\hbar\nabla - e^*\boldsymbol{A})\psi_L|^2\rangle = 0 \tag{4.2.56}$$

By setting

$$\boldsymbol{A} = \boldsymbol{A}_o + \boldsymbol{A}_1 \tag{4.2.57}$$

where $\boldsymbol{A}_o$ [defined by Eq.(4.2.44)] is the vector potential which exists in the presence of the magnetic field $B_{c2} = \mu_o H_{c2}$. $\boldsymbol{A}_1$ represents the correction that arises from the facts that the applied field is slightly less than $\mu_o H_{c2}$ and that the supercurrent also contributes to the field. Noting that ψ_L must satisfy Eq.(4.2.43) and keeping only terms up to first order in $\boldsymbol{A}_1$, we obtain

$$\beta\langle|\psi_L|^4\rangle - \langle\boldsymbol{A}_1\cdot\boldsymbol{J}_L\rangle = 0 \tag{4.2.58}$$

with

$$\boldsymbol{J}_L = \frac{e^*\hbar}{2im^*}(\psi_L^*\nabla\psi_L - \psi_L\nabla\psi_L^*) - \frac{e^{*2}}{m^*}|\psi_L|^2\boldsymbol{A}_o \tag{4.2.59}$$

We note that $\boldsymbol{J}_L$ is the current associated with the unperturbed solution. Integrating the second term in (4.2.58) in parts and setting $\nabla\times\boldsymbol{A}_1 = \boldsymbol{B}_1$ and $\nabla\times\boldsymbol{B}_s = \mu_o\boldsymbol{J}_L$, we find

$$\beta\langle|\psi_L|^4\rangle - \frac{1}{\mu_o}\langle \boldsymbol{B}_1 \cdot \boldsymbol{B}_s\rangle = 0 \tag{4.2.60}$$

Due to the symmetry of the problem, $\boldsymbol{B}_1$ and $\boldsymbol{B}_s$ are everywhere parallel to the applied field $\boldsymbol{B}^e$ in the z-direction, we can write

$$B_1 = B^e - \mu_o H_{c2} + B_s \tag{4.2.61}$$

Inserting (4.2.53) and (4.2.61) into (4.2.60), we get

$$\beta\langle|\psi_L|^4\rangle - \frac{\pi\hbar^2}{\Phi_o m^*}\left\langle|\psi_L|^2\left(B^e - \mu_o H_{c2} - \frac{\mu_o \pi\hbar^2}{\Phi_o m^*}|\psi_L|^2\right)\right\rangle = 0 \tag{4.2.62}$$

which can be further written, by noting Eq.(4.2.22) for κ, as

$$(2\kappa^2 - 1)\frac{\mu_o \pi\hbar^2}{\Phi_o m^*}\langle|\psi_L|^4\rangle - (\mu_o H_{c2} - B^e)\langle|\psi_L|^2\rangle = 0 \tag{4.2.63}$$

Setting again $|\psi_L| = |\psi_o|\psi'_L$ and using Eqs.(4.2.17), (4.2.20), (4.2.22), and (4.2.39), we can get

$$\left(1 - \frac{1}{2\kappa^2}\right)\langle\psi'^{\,4}_L\rangle - \left(1 - \frac{B^e}{B_{c2}}\right)\langle\psi'^{\,2}_L\rangle = 0 \tag{4.2.64}$$

This equation represents a rather general result, which is independent of the detailed behavior of the function ψ_L, that is, independent of the type of the periodic lattice configuration of ψ_L.

For a particular lattice type, as determined by selecting the wave number q and the periodicity of the coefficients C_n, we can calculate the quantity

$$\beta_A = \frac{\langle\psi'^{\,4}_L\rangle}{\langle\psi'^{\,2}_L\rangle^2} \tag{4.2.65}$$

which is a function of the geometry of the vortex arrangement only, and is larger than 1 (Schwartz inequality). The ratio β_A takes the value unit if ψ is spatially constant and becomes increasingly large for functions which more and more peaked locally. It is only the quantity β_A, which must be determined numerically for obtaining $<\psi'^{\,2}_L>$ and $<\psi'^{\,4}_L>$ from Eqs.(4.2.64) and (4.2.65). The macroscopic magnetic flux density $<B>$ and the free energy density can then be calculated immediately. By using (4.2.53), the macroscopic magnetic flux density $<B>$ may be expressed as

$$\langle B\rangle = B^e + \langle B_s\rangle = B^e - \frac{\mu_o \pi \hbar^2}{\Phi_o m^*}\langle|\psi_L|^2\rangle \tag{4.2.66}$$

Using Eqs.(4.2.20) and (4.2.23), we find

$$\langle B\rangle = B^e - \frac{\mu_o H_c}{\sqrt{2}\kappa}\langle\psi'^2_L\rangle \tag{4.2.67}$$

Eliminating $<\psi'^2_L>$ by using Eqs.(4.2.64) and (4.2.65), we finally find

$$\langle B\rangle = B^e - \frac{\mu_o(H_{c2} - H^e)}{(2\kappa^2 - 1)\beta_A} \tag{4.2.68}$$

and equivalently the macroscopic magnetization $\boldsymbol{M}$

$$M = \frac{1}{\mu_o}\langle B\rangle - H^e = -\frac{H_{c2} - H^e}{(2\kappa^2 - 1)\beta_A} \tag{4.2.69}$$

The magnetization is shown to vanishes at $H^e = H_{c2}$ and the transition is of the second order. The slope of the magnetization is finite and is given by

$$\frac{dM}{dH^e} = \frac{1}{(2\kappa^2 - 1)\beta_A} \tag{4.2.70}$$

which becomes very large when κ approaches the value $1/\sqrt{2}$ from above, and diverges at this value. It is such a discontinuous rise in M, which characterizes type I superconductors as shown in Fig. 4.4.

Using the identity for the mean Gibbs free energy density g

$$\left(\frac{\partial g}{\partial H}\right)_T = -\mu_o M \tag{4.2.71}$$

we may calculate the free energy density g by integrating down from the normal state at H_{c2}, where $g_s(T, H_{c2}) = g_n(T, H_{c2})$, and obtain

$$g_s(T, H^e) - g_n(T, H_{c2}) = -\frac{\mu_o}{2}\frac{(H_{c2} - H^e)^2}{(2\kappa^2 - 1)\beta_A} \tag{4.2.72}$$

which applies to the regime $H^e < H_{c2}$ and $\kappa > 1/\sqrt{2}$. Equation (4.2.72) indicates that the configuration with the smallest value of β_A is thermodynamically most stable. Numerical calculations show that the square lattice and the triangular

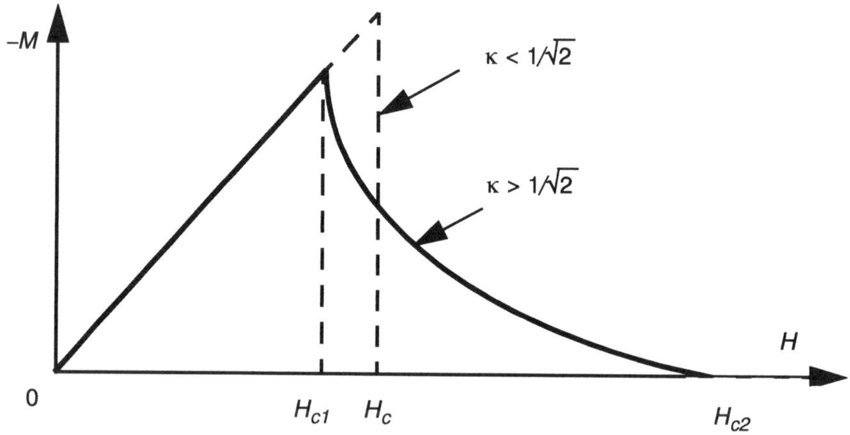

Figure 4.4 Variation of the magnetization M versus the field H in a superconductor.

lattice yield the values $\beta_A = 1.18$ and $\beta_A = 1.16$, respectively [Kleiner et al. (1964)]. This means that a triangular flux arrangement will be lowest in energy and, therefore, will be the most stable form, at least, in isotropic superconductors. The theoretical result has well been supported experimentally. The experimental observation of such triangular vortex lattice in a Nb disk in a perpendicular magnetic field was made by using a high-resolution Bitter method [see Essmann and Träuble (1967)]. Figure 4.5 shows illustratively the mixed state of a type II superconductor in an applied magnetic field.

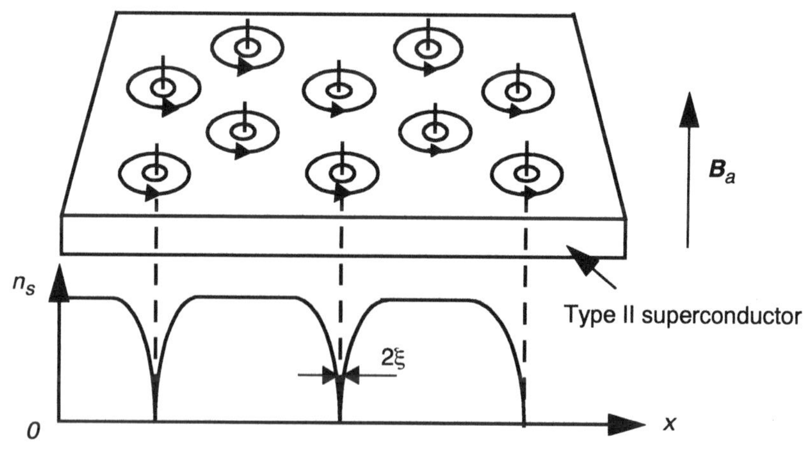

Figure 4.5 The mixed state showing encircling supercurrent vortices and variation with position of concentration of superelectrons.

4.2.8 A Solution of the Ginzburg-Landau Equations for Thin Film

Superconducting films with its advantage of being of high switching speed, compared to that of bulk superconductors limited by eddy currents, has led to their use in superconductive electronic devices. Because of possible device applications and theoretical interest, they have been studied intensively. In the following two sections, we shall present some basic properties of the thin films by using the Ginzburg-Landau theory. To begin with, let us first introduce the following non-dimensional quantities:

$$x' = \frac{x}{\lambda}, \qquad \psi' = \frac{\psi}{|\psi_o|}, \qquad A' = \frac{A}{\sqrt{2}\mu_o H_c \lambda} \tag{4.2.73}$$

where λ, $|\psi_o|^2$, and $\mu_0 H_c$ (= B_c) are, respectively, given by Eq.(4.2.20), (4.2.2), and (4.2.9). Thus the G-L equation (4.2.12) may be written in the following non-dimensional form:

$$|\psi'|^2\psi' - \psi' + \left(-\frac{i}{\kappa}\nabla' - A'\right)^2 \psi' = 0 \quad \text{in } V \tag{4.2.74}$$

and the insulate-superconducting boundary condition (4.2.15) is

$$\boldsymbol{n}\cdot\left(-\frac{i}{\kappa}\nabla' - A'\right)\psi' = 0 \tag{4.2.75}$$

where the parameter κ is defined by Eq.(4.2.22).

Suppose now the film with the thickness d occupies the region $-d/2 < x < d/2$ and the applied field is along the z-axis, that is, $B^e = (0, 0, B^e)$, we may choose $\boldsymbol{A} = (0, A(x), 0)$ so that the magnetic field in the film may be expressed as $\boldsymbol{B} = (0, 0, dA/dx)$. Thus Eqs.(4.2.74) and (4.2.75) become

$$\frac{d^2\psi'}{dx'^2} = \kappa^2\psi'(A'^2 - 1 + |\psi'|^2) \quad \text{for } -\frac{d}{2\lambda} < x' < \frac{d}{2\lambda} \tag{4.2.76}$$

and

$$\frac{d\psi'}{dx'} = 0 \quad \text{for } x' = \pm\frac{d}{2\lambda} \tag{4.2.77}$$

Since $\kappa << 1$ for many pure (type I) superconductors, we shall solve the problem in this case. At the zeroth approximation, we may neglect the term on the right-hand side of Eq.(4.2.76), and find

$$\psi' = \psi_o' = const \tag{4.2.78}$$

Substituting this, as yet unknown, constant into Eq.(4.2.13) and using Maxwell´s equation of Ampere´s law, which reads now

$$\frac{d^2A'}{dx'^2} = |\psi_o'|^2 A' \qquad \text{for } -\frac{d}{2\lambda} < x' < \frac{d}{2\lambda} \tag{4.2.79}$$

together with the boundary condition $B(\pm d/2) = B^e$, we can obtain

$$B(x) = B^e \frac{\cosh(x|\psi_o'|/\lambda)}{\cosh(d|\psi_o'|/(2\lambda))} \tag{4.2.80}$$

$$A(x) = \frac{B^e \lambda}{|\psi_o'|} \frac{\sinh(x|\psi_o'|/\lambda)}{\cosh(d|\psi_o'|/(2\lambda))} \tag{4.2.81}$$

Consider the next approximation by substituting $\psi' = \psi_o' + \psi_1'$ into Eq.(4.2.76), we get

$$\frac{d^2\psi_1'}{dx'^2} = \kappa^2 \psi_o'(A'^2 - 1 + |\psi_o'|^2) \tag{4.2.82}$$

Noting the boundary condition (4.2.77), we find

$$\int_{-d/(2\lambda)}^{d/(2\lambda)} (A'^2 - 1 + |\psi_o'|^2)dx' = 0 \tag{4.2.83}$$

which then gives

$$\left(\frac{B^e}{\sqrt{2}\mu_o H_c}\right)^2 = \frac{2|\psi_o'|^2(1 - |\psi_o'|^2)\left(\cosh\left(\frac{d|\psi_o'|}{2\lambda}\right)\right)^2}{\frac{\lambda}{d|\psi_o'|}\sinh\left(\frac{d|\psi_o'|}{\lambda}\right) - 1} \tag{4.2.84}$$

This relation is valid at any B^e and gives an implicit relation for $|\psi_o'|$ in terms of B^e. It is shown that, though the order parameter $|\psi_o'|$ is treated as independent of space coordinate, it depends on the applied magnetic field.

From Eq.(4.2.80), we can calculate the mean magnetization of the film by

$$M = \frac{1}{\mu_o}\left(\frac{1}{d}\int_{-d/2}^{d/2} B\,dx - B^e\right) = -\frac{B^e}{\mu_o}\left(1 - \frac{2\lambda}{d|\psi_o'|}\tanh\left(\frac{d|\psi_o'|}{2\lambda}\right)\right) \tag{4.2.85}$$

which shows that, for $|\psi_o'|d/\lambda >> 1$, we have approximately $\mathrm{M} = -H^e = -B^e/\mu_o$, corresponding to the Meissner effect. If, however, $|\psi_o'|d/\lambda << 1$, then, from Eqs.(4.2.84) and (4.2.85), we find

$$|\psi_o'|^2 \approx 1 - \frac{d^2}{12\lambda^2}\left(\frac{B^e}{\sqrt{2}\mu_o H_c}\right)^2 \tag{4.2.86}$$

and

$$M \approx -\frac{d^2 B^e |\psi_o'|^2}{12\mu_o\lambda^2} \approx -\frac{d^2 B^e}{12\mu_o\lambda^2}\left(1 - \frac{d^2}{12\lambda^2}\left(\frac{B^e}{\sqrt{2}\mu_o H_c}\right)^2\right) \tag{4.2.87}$$

4.2.9 Critical Field for Thin Film

The results in above section show that, for sufficiently thin film, the order parameter $|\psi_0'|$ and, therefore, the magnetization goes smoothly to zero with the increase of the magnetic field. Thus the corresponding superconducting-normal phase transition is of the second-order, even for the film of type I superconductors. The corresponding critical magnetic field denoted by B_{cf} for the thin film may be found from Eq.(4.2.86) by setting $|\psi_0'|^2 = 0$, that is,

$$B_{cf} = \frac{\lambda\sqrt{24}}{d}B_c \tag{4.2.88}$$

which shows that, for very thin films, there is a considerable increase in the critical magnetic field B_{cf} over the bulk value B_c for the same material.

We may now calculate the Gibbs free energy, obtained by adding $-\boldsymbol{B}^e\cdot\boldsymbol{B}/\mu_0 + (B^e)^2/(2\mu_0)$ to Eq.(4.2.11). With ψ being constant, we have

$$\frac{G_s - G_n}{V} = \frac{1}{d}\int_{-d/2}^{d/2}\left\{\alpha|\psi|^2 + \frac{\beta}{2}|\psi|^4 + \frac{e^{*2}}{2m^*}A^2|\psi|^2 + \frac{1}{2\mu_o}(B - B^e)^2\right\}dx \tag{4.2.89}$$

where V is the volume of the thin film. Substituting Eqs.(4.2.80) and (4.2.81) into Eq.(4.2.89) and using Eq.(4.2.73), we can find

$$\Delta g' = \frac{G_s - G_n}{V\mu_o H_c^2/2} = |\psi_o'|^4 - 2|\psi_o'|^2 + \left(\frac{B^e}{\mu_o H_c}\right)^2\left(1 - \frac{2\lambda}{d|\psi_o'|}\tanh\left(\frac{d|\psi_o'|}{2\lambda}\right)\right) \tag{4.2.90}$$

In a field equal to the critical magnetic field B_{cf}, we must have $\Delta g' = 0$, which means

$$\left(\frac{B_{cf}}{B_c}\right)^2 = \frac{|\psi_o'|^2(2 - |\psi_o'|^2)}{1 - \frac{2\lambda}{d|\psi_o'|}\tanh\left(\frac{d|\psi_o'|}{2\lambda}\right)} \tag{4.2.91}$$

Now consider the limiting case where the film thickness is large ($d >> \lambda$). We can expect a first-order phase transition for the film of type I superconductor, which implies that $|\psi_o'|$ is close to unity at the transition point. Thus we may let $|\psi_o'| = 1-\varepsilon'$, where $\varepsilon' << 1$. We can find approximately ε' from Eq.(4.2.84), and then, by Eq.(4.2.91), find the critical magnetic field for the film with $d >> \lambda$ by

$$B_{cf} = \left(1 + \frac{\lambda}{d}\right)B_c \tag{4.2.92}$$

It can be expected that at a certain thickness d_c of the film, the first-order phase transition changes to a second-order phase transition. Indeed, a representative form of the variation of $|\psi_c/\psi_o|$ with respective to d/λ may be shown illustratively in Fig. 4.6, where the value of d_c is found to be $\lambda\sqrt{5}$ [Douglass (1962) and Abrikosov (1988)]. Since λ is temperature-dependent, even relatively large samples may undergo a second-order transition at temperatures sufficiently close to the transition temperature T_c.

In the above analysis, we have calculated the critical field of thin films of type I superconductors with $\kappa << 1$, which enabled us to assume ψ = constant throughout the film. The obtained result may also be generalized to thin films of type II superconductors with their thickness d less than the coherence length ξ since for which the order parameter ψ must not vary throughout the film thickness. Practically, experiments with films of thicknesses down to about 10 nm (in comparison with ξ being of the order of 1 μm in some pure crystals) are entirely feasible. Thus, for such thin films of type II superconductors with $\xi << \lambda$, we can use the formulas for $d<<\lambda$. In particular, the critical field of the transition to the normal state is given by Eq.(4.2.88).

We have, so far, been discussing the case of a field parallel to the surface of

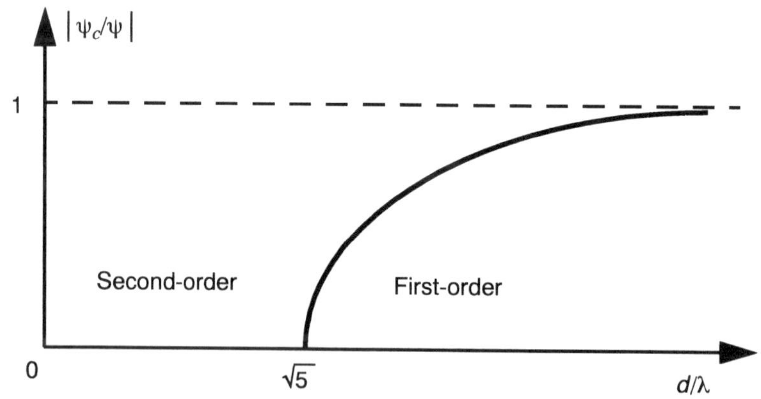

Figure 4.6 Critical value of order parameter at the superconducting-normal transition as a function of film thickness.

the film. The actual behavior of a thin film in a magnetic field depends markedly upon the direction of that field. For fields normal to the film, the flux line picture of Abrikosov is likely to be of importance even for films of type I superconductors. The field penetrates the film by way of individual flux lines, spaced some distance apart. Eventually, an "upper critical field" is reached at which the flux lines touch and the film makes the transition to the normal state. According to Tinkham (1963), the perpendicular critical field $B_{c\perp}$ for sufficiently thin films ($d < d_c$) of type I superconductors may be written as

$$B_{c\perp} = \sqrt{2}\kappa(d)B_c \tag{4.2.93}$$

where the G-L parameter $\kappa(d)$ depends on the film thickness d and can be expressed approximately into the following form:

$$\kappa(d) = 0.715\lambda_L(0)\left(\frac{1}{l_{eb}} + \frac{C}{d}\right) \tag{4.2.94}$$

in which l_{eb} denotes the electron mean free path of the bulk material, $\lambda_L(0)$ is the London penetration depth at absolute zero, and C is a constant. It can be seen that, for very thin films, $B_{c\perp}$ is proportional to $1/d$.

In the case of thick films ($d > d_c$) which behaves like bulk slabs, we may write

$$B_{c\perp} = \left(1 - \sqrt{\frac{C'\delta}{d}}\right)B_c \tag{4.2.95}$$

where δ is an interphase surface-energy parameter and C' is a constant being of order 1 [Cody and Miller (1968)]. The critical thickness d_c may be estimated by equating the two transition fields as

$$d_c \approx \frac{C'\delta}{(1 - 2\kappa^2)^2} \tag{4.2.96}$$

which is usually of the order of $10^{-6} \sim 10^{-7}$ m. Experimentally, the transition from type I to type II behavior in superconducting films in a perpendicular magnetic field has been studied by, for instance, Guyon et al. (1964), Miller and Cody (1968), Kunze et al. (1974), and Dolan (1974).

4.2.10 Critical Current in Thin Film

We shall now apply the G-L equations to calculate the critical current in superconducting thin films at which superconductivity breaks down. At first, we consider a thin film $d << \xi$ and $d << \lambda$ so that $|\psi|$ and $\boldsymbol{J}_s$ may be supposed to be constant and uniform over the sample cross-section of the thin film. We can set $\psi = |\psi|e^{i\theta(x)}$ with $|\psi|$ being independent of $\boldsymbol{x}$. Equation (4.2.14) for the supercurrent yields

$$\boldsymbol{J}_s = \frac{e^*}{m^*}(\hbar\nabla\theta - e^*\boldsymbol{A})|\psi|^2 = e^*|\psi|^2\mathbf{v}_s \tag{4.2.97}$$

where $\mathbf{v}_s$ denotes the mean velocity of superelectrons. The free energy density in Eq.(4.2.11) may thus be expressed as

$$f_s = f_n + \left(\alpha + \frac{\beta}{2}|\psi|^2 + \frac{1}{2}m^*\mathrm{v}_s^2\right)|\psi|^2 + \frac{1}{2\mu_o}B^2 \tag{4.2.98}$$

Minimizing f_s with respect to $|\psi|$, we find

$$\alpha + \frac{\beta}{2}|\psi|^2 + \frac{1}{2}m^*\mathrm{v}_s^2 = 0 \tag{4.2.99}$$

and, therefore,

$$\mathrm{v}_s^2 = \frac{2\alpha}{m^*}(|\psi'|^2 - 1) \tag{4.2.100}$$

and get the magnitude of the supercurrent:

$$J_s = |e^*||\psi|^2\mathrm{v}_s = |e^*||\psi_o|^2\sqrt{\frac{2|\alpha|}{m^*}}|\psi'|^2\sqrt{1-|\psi'|^2} \tag{4.2.101}$$

It is shown that J_s equals zero for $|\psi'|^2 = 1$, and J_s reaches a maximum for $|\psi'|^2 = 2/3$ corresponding to the critical current density

$$J_c = \frac{2|e^*||\psi_o|^2}{3\sqrt{3}}\sqrt{\frac{2|\alpha|}{m^*}} \tag{4.2.102}$$

which may be transformed, with the aid of Eqs.(4.2.9), (4.2.17), and (4.2.20), into the following expression containing more readily available quantities as

$$J_c = \frac{2\sqrt{2}\,B_c(T)}{3\sqrt{3}\mu_o\lambda(T)} \tag{4.2.103}$$

It is shown that near the transition temperature T_c, J_c varies as $(T_c-T)^{3/2}$ since B_c varies as (T_c-T) and λ varies as $(T_c-T)^{-1/2}$. The $(T_c-T)^{3/2}$ behavior of the critical current density near T_c has often been observed experimentally [Huebener (1979)]. For current densities larger than J_c, there do not exist solutions with $|\psi'| \neq 0$. As a consequence, for $J > J_c$ the superconductor becomes normal, and $|\psi'|$ changes abruptly from $(2/3)^{1/2}$ to zero. The value of the critical current density given by Eq.(4.2.103) can be quite large. For typical values of $B_c = 0.05$ tesla and $\lambda = 10^{-7}$ m, we find that J_c is about 2×10^{11} A/m^2.

Next, we consider thick films of type I superconductors ($d >> \lambda$), the critical current density J_c for such thick films may be found simply by using Silsbee´s criterion for bulk superconductors, which states that a superconductor loses its zero resistance when, at any point on the surface, the total magnetic field strength due to transport current and applied magnetic field exceeds the critical field strength of the superconductor. Thus, by using Eq.(4.2.80) and Maxwell´s equation $\nabla\times\boldsymbol{B} = \mu_o\boldsymbol{J}$, we can find

$$J_c = \frac{B_c|\psi_o'|}{\mu_o\lambda}\tanh\left(\frac{d|\psi_o'|}{2\lambda}\right) \approx \frac{B_c(T)}{\mu_o\lambda(T)} \tag{4.2.104}$$

for sufficiently large d ($>> \lambda$). The comparison of Eq.(4.2.103) and Eq.(4.2.104) shows that the critical current density for the very thin film is reduced in contrast with the fact that its critical magnetic field is increased. It may be worth mentioning of that Silsbee´s criterion is not valid for type II superconductors in high magnetic fields. The critical currents of type II superconductors in high magnetic fields are, in fact, almost completely controlled by the imperfections in the materials, which we shall discussed in later sections.

4.3 VORTEX DYNAMICS AND THE LONDON-BEAN MODEL

4.3.1 The London Model of Vortex Line and Lower Critical Field B_{c1}

It has been shown that type II superconductors have solutions of the G-L equations with $|\psi| > 0$ until the field B_{c2} is reached. In particular, for sufficiently large applied magnetic fields, the Abrikosov solution indicates that there is a regular array of magnetic flux (vortex) lines in the superconductor when it is in the mixed state. In this section, we shall study problems on vortex lines in type II superconductors in fields slightly higher than a critical field B_{c1}, called the lower critical field, defined by the lowest field at which the first vortex line appears in the superconductor. In such a case, only a few vortex lines occur and, thus, the separation of these vortex lines may be much greater than the penetration depth λ so that it is possible to treat them individually as isolated vortex lines by neglecting the interaction among these vortex lines at the first approximation. Thus we may introduce the London model for the study of these vortex lines. In this model, it is assumed that the diameter ($\sim 2\xi$) of the normal vortex core is very small compared with the penetration depth λ (i.e., $\kappa >> 1$). With this restriction, the London model represents a good approximation at all temperatures (not too close to T_c) and for magnetic fields in the range of $B_{c1} < B << B_{c2}$, where the interaction between vortex lines is not too strong.

At first, let us study an isolated vortex line. For $\kappa >> 1$, the local magnetic field $\boldsymbol{b}$ may be described by the following modified London equation:

$$\mu_o\lambda^2\nabla \times \boldsymbol{J} + \boldsymbol{b} = \Phi_o\delta(\boldsymbol{r} - \boldsymbol{r}_o)z^o \tag{4.3.1}$$

where Φ_0 is the flux quantum of the vortex line, z^0 is the unit vector along the axis of the vortex line, and $\delta(\boldsymbol{r} - \boldsymbol{r}_0)$ is the two-dimensional delta function with $\boldsymbol{r}-\boldsymbol{r}_0$ being the radial distance vector from the center of the vortex line located at $\boldsymbol{r}_0$. With the use of Maxwell's equations $\nabla\times\boldsymbol{b} = \mu_0\boldsymbol{J}$ and $\nabla\cdot\boldsymbol{b} = 0$, Eq.(4.3.1) becomes

$$\lambda^2\nabla^2\boldsymbol{b} - \boldsymbol{b} = -\Phi_o\delta(\boldsymbol{r} - \boldsymbol{r}_o)z^o \tag{4.3.2}$$

The solution for this equation in the circular cylindrical coordinates (r, φ, z) with its origin on the vortex line ($\boldsymbol{r}_0 = 0$) may be found as

$$\boldsymbol{b}(r) = \frac{\Phi_o}{2\pi\lambda^2}K_o\left(\frac{r}{\lambda}\right)z^o \tag{4.3.3}$$

and the circulating supercurrent reads

$$\boldsymbol{J}(r) = \frac{\Phi_o}{2\pi\mu_o\lambda^3}K_1\left(\frac{r}{\lambda}\right)\varphi^o \tag{4.3.4}$$

where K_0 and K_1 are the Hankel functions of imaginary argument of zero order and of first order, respectively [Morse and Feshbach (1953)]. Here, φ^0 denotes the unit vector along the direction of coordinate φ. With the use of the asymptotic approximations for K_0, we have

$$b(r) \approx \frac{\Phi_o}{2\pi\lambda^2}ln\left(\frac{\lambda}{r}\right), \qquad \text{for } \xi \le r \ll \lambda \tag{4.3.5}$$

and

$$b(r) \approx \frac{\Phi_o}{2\pi\lambda^2}\sqrt{\frac{\pi\lambda}{2r}}exp\left(-\frac{r}{\lambda}\right), \qquad \text{for } r \gg \lambda \tag{4.3.6}$$

In the London model of the vortex line, it is assumed that $|\psi|$ is constant everywhere except at $r = 0$, where there is a singularity. To eliminate the artificial singularity, we have used Eq.(4.3.5) by cutting off the solution at $r = \xi$.

The energy U_v per unit length of the vortex line can be calculated from

$$U_v = \int_{S-S_o} \frac{1}{2\mu_o}(\boldsymbol{b}^2 + \lambda^2(\nabla \times \boldsymbol{b})^2)dS \tag{4.3.7}$$

where the plane surface S is perpendicular to the vortex line, and S_0 denotes the small cross surface of the core of the vortex ($0 \le r < \xi$), which is excluded from the integration. After some manipulations, we may find

$$U_v \approx \frac{\Phi_o^2}{4\pi\mu_o\lambda^2} ln(\kappa), \qquad \text{for } \kappa \gg 1 \tag{4.3.8}$$

which shows that the energy of a vortex line depends logarithmically upon the G-L parameter κ, and quadratically upon the flux quantum Φ_o per vortex line for high-κ materials.

From the energy U_v of one vortex line, we can calculate the lower critical field B_{c1} by noting the fact that the Gibbs free energies must be equal at the common boundary between the Meissner state (or phase) with no vortex lines and the mixed state with a few vortex lines at B_{c1}, that is,

$$nU_v = n\Phi_o \frac{B_{c1}}{\mu_o} \tag{4.3.9}$$

where n is the number of vortex lines in the superconductor assumed to be a long and thin sample parallel to the magnetic field so that the internal field is uniform and equal to the applied field. From Eqs.(4.3.9) and (4.3.8), we can find

$$B_{c1} = \mu_o U_v / \Phi_o = \frac{\Phi_o}{4\pi\lambda^2} ln(\kappa), \qquad \text{for } \kappa \gg 1 \tag{4.3.10}$$

This equation may also be expressed in terms of the thermodynamic field B_c by using Eq.(4.2.23)

$$B_{c1} = \frac{B_c}{\sqrt{2}\kappa} ln(\kappa), \qquad \text{for } \kappa \gg 1 \tag{4.3.11}$$

Equation (4.3.11) corresponds to the relation (4.2.41) between B_{c2} and B_c. Thus, apart from the $ln(\kappa)$ term, $B_c/B_{c1} = B_{c2}/B_c = \kappa\sqrt{2}$, so that B_c is approximately the geometric mean of B_{c1} and B_{c2}.

Shown in Fig. 4.7 is a typical B–T phase diagram for a type II superconductor of an ideal cylindrical shape. It is shown that, for weak fields $B < B_{c1}$, there is the Meissner phase with complete flux expulsion. For $B_{c1} < B < B_{c2}$, there is the mixed state of the superconductor with incomplete flux penetration. Complete flux penetration takes place at the field B_{c2}, called the upper critical field. The curve $B_{c2}(T)$ on the phase diagram is the line of the second-order phase transition between superconducting and normal states. This second-order phase transition is in contrast to the first-order phase transition of a type I superconductor placed in a magnetic field.

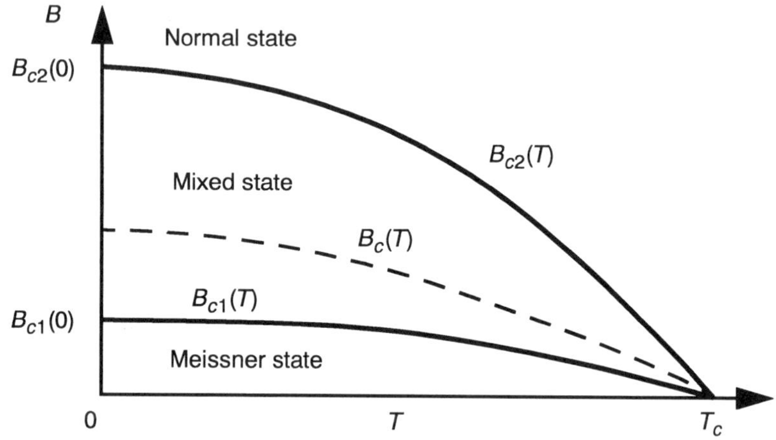

Figure 4.7 The phase diagram of a typical type II superconductor.

4.3.2 Interaction Between Two Vortex Lines

Let us now study the interaction between two vortex lines in a type II superconductor based on the London model. Since the London model is linear, we may easily find, by superposition, the total field of two vortex lines, centers of which are located at $\boldsymbol{r}_1$ and $\boldsymbol{r}_2$, respectively. Suppose that the two vortex lines are parallel to each other, the magnetic field is by superposition

$$\boldsymbol{b}(\boldsymbol{r}) = \boldsymbol{b}_1(\boldsymbol{r}) + \boldsymbol{b}_2(\boldsymbol{r}) = \frac{\Phi_o}{2\pi\lambda^2}\left(K_o\left(\frac{|\boldsymbol{r}-\boldsymbol{r}_1|}{\lambda}\right) + K_o\left(\frac{|\boldsymbol{r}-\boldsymbol{r}_2|}{\lambda}\right)\right)z^o \tag{4.3.12}$$

The total energy for the two flux lines can be written

$$U = \int_{S-S_1-S_2} \frac{1}{2\mu_o}(\boldsymbol{b}^2 + \lambda^2(\nabla\times\boldsymbol{b})^2)dS \tag{4.3.13}$$

where $\boldsymbol{b}$ is substituted from Eq.(4.3.12). After some manipulations [De Gennes (1966)], we may find that the total energy is composed of two individual vortex line energies, and an interaction energy U_{12} given by

$$U_{12} = \frac{\Phi_o^2}{2\pi\mu_o\lambda^2}K_o\left(\frac{|\boldsymbol{r}_1-\boldsymbol{r}_2|}{\lambda}\right) \tag{4.3.14}$$

The force due to the interaction can be obtained by taking derivative of U_{12}, which gives, for instance,

$$f_2 = J_1(r_2) \times \Phi_o z^o \tag{4.3.15}$$

by noting $f_{2x} = -\partial U_{12}/\partial x$ and $f_{2y} = -\partial U_{12}/\partial y$, and using the Maxwell equation: $\nabla \times \boldsymbol{b} = \mu_o \boldsymbol{J}$. This is the force per unit length acting on vortex line 2 (located at $\boldsymbol{r}_2$), arising from the supercurrent of vortex line 1. Similarly, we may find the interaction force on vortex line 2. It can be shown that the interaction force is repulsive if the two vortex lines are parallel in the same direction. The force is attractive if the two vortex lines are parallel in opposite direction, which may be shown illustratively in Fig. 4.8.

If we generalize the result to an arbitrary array of vortices, we have then

$$f = J_s \times \Phi_o z^o \tag{4.3.16}$$

where z^o is the unit vector along the local vortex line. $\boldsymbol{J}_s$ represents now the total supercurrent density due to all other vortices, including any net transport current at the location of the core of the vortex line in question. It can be seen that a vortex array can be in static equilibrium only if the total force on each vortex is zero. This can be accomplished if each vortex line is surrounded by a symmetrical array, as in the square array or triangular array discussed above, in the absence of transport current. The square array turns out to have only unstable equilibrium, so that small displacement tends to grow, while the triangular array is stable since it has the lowest energy. In the presence of transport current, vortices even in the triangular array may move unless they are pinned in place by pinning centers, such as inhomogeneities in the medium.

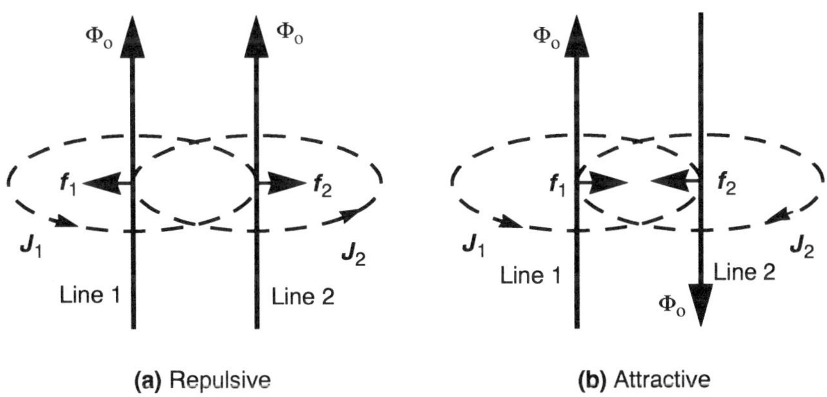

Figure 4.8 Interaction forces between two vortex lines.

4.3.3 Flux Flow, Pinning, and Critical State

It has been shown that in a high magnetic field greater than the lower critical field B_{c1}, but smaller than the upper critical field B_{c2}, the type II superconductor is in a mixed state with a distribution of vortex lines. For an ideal (soft) type II superconductor carrying no bulk transport currents, the vortex lines may interact and arrange themselves in equilibrium with no net force exerted on any vortex line. If, however, there exists a bulk transport current $\boldsymbol{J}^{\text{ext}}$ flowing in the superconductor, the vortex lines may move under the action of the Lorentz force on the vortex lines, given by

$$f_L = \boldsymbol{J}^{\text{ext}} \times \Phi_o z^o \tag{4.3.17}$$

Multiplying by the number n of vortex lines per unit cross-sectional area, we may obtain the Lorentz body force

$$\boldsymbol{F}_L = \boldsymbol{J}^{\text{ext}} \times \boldsymbol{B}^{(\text{v})} \tag{4.3.18}$$

where $\boldsymbol{B}^{(\text{v})} = n\Phi_0 z^0$ denotes the mean magnetic induction field.

Suppose that the motion of a vortex line is opposed by a viscous friction force defined by

$$f_{\text{v}} = -\eta \mathbf{v}_L \tag{4.3.19}$$

where $\mathbf{v}_L$ denotes the mean velocity of vortex motion and η the viscosity coefficient, which may, in general, be a coefficient tensor for anisotropic superconductors. By the force balance equation $f_{\text{v}} + f_L = 0$, we may find that $\mathbf{v}_L$ is given by

$$\mathbf{v}_L = \frac{\Phi_o}{\eta} \boldsymbol{J}^{ext} \times z^o \tag{4.3.20}$$

which shows that the vortex line move perpendicular to z^0 and $\boldsymbol{J}^{\text{ext}}$. The motion of the vortex lines in the mean magnetic induction field $\boldsymbol{B}^{(\text{v})}$ can lead to the generation of a vortex electric field, defined by

$$\boldsymbol{E}^{(\text{v})} = -\mathbf{v}_L \times \boldsymbol{B}^{(\text{v})} \tag{4.3.21}$$

In the case of the vortex magnetic field $\boldsymbol{B}^{(\text{v})}$ being locally perpendicular to the transport current density vector $\boldsymbol{J}^{\text{ext}}$, we have

$$\boldsymbol{E}^{(\text{v})} = \frac{\Phi_o B^{(\text{v})}}{\eta} \boldsymbol{J}^{ext} \tag{4.3.22}$$

The generation of the vortex electric field along the transport current $\boldsymbol{J}^{\text{ext}}$

implies the appearance of electric resistance, which can be defined by

$$\rho_v = \frac{\partial E^{(v)}}{\partial J^{ext}} = \frac{\Phi_o B^{(v)}}{\eta} \tag{4.3.23}$$

For a homogeneous superconductor, when it becomes normal at the critical field B_{c2}, the flow resistivity ρ_v is equal to the normal-state resistivity ρ_n of the material. Thus we may find

$$\eta = \frac{\Phi_o B_{c2}}{\rho_n} \tag{4.3.24}$$

By this equation, we may write Eq.(4.3.23) in the form:

$$\rho_v = \rho_n \frac{B^{(v)}}{B_{c2}} \tag{4.3.25}$$

which shows that, for a given strength of applied magnetic field, the flow resistivity ρ_v is proportional to the normal-state resistivity ρ_n of the material. Furthermore, the flow resistivity increases with increasing strength of the applied magnetic field, approaching the normal-state resistivity as the applied field strength approaches B_{c2}. The given simple expressions (4.3.24) and (4.3.25) are valid by order of magnitude at low temperature $T << T_c$ and small magnetic field $B^{(v)} << B_{c2}$, since the true viscosity coefficient η depends on many mechanisms and not only on the energy dissipation in the normal cores of the vortices. Some theoretical efforts to evaluate η in terms of material parameters may be found in the work of Bardeen and Stephen (1965).

It has been shown that an ideal type II superconductor, which is free of lattice defects, in the mixed state is unable to carry transport currents without losses since the Lorentz force acting on the vortex structure will move the vortex lines which will then generate electric fields in the superconductor, resulting in dissipation. In order to transport larger resistanceless currents in type II superconductors, the motion of vortex lines has to be impeded since when the vortex lines are pinned, there is no electromotive force induced and no current flows in the normal cores of the vortex lines, and the transport current flows around the cores without energy loss. Thus, for engineering applications, most of type II superconductors contain imperfections, such as dislocations, grain boundaries, precipitates, inhomogeneities, and the like, which may create various pinning centers to impede the motion of vortex lines [Campbell and Evetts (1972) and Ullmaicr (1975)]. The local changes of the superconductor from these defects result in position-dependence of the free energy of the vortex lines. The difference between the free energy of a vortex line in the pinning center and in the surrounding medium characterizes the strength of the pinning center. In such superconductors, called often the hard superconductors, driving forces on vortex lines from transport currents or temperature gradients are thus balanced by the pinning forces at the pinning centers. Microscopically, not every individual vortex core is directly pinned to the material, but the interaction between the vortex lines is sufficient to give the vortex lattice a certain rigidity, so that, if only a few vortex core are pinned, the whole pattern may still be immobilized. What matters, therefore, is the average pinning force per vortex core. When the deriving force density exceeds a critical value of the mean pinning force density, the vortex lines in the superconductors will move and result in the dissipation in the superconductor.

To provide a phenomenological description of the behavior of hard superconductors, the concept of the "critical state" was introduced by London (1962), Bean (1962), and Kim et al. (1962), which reduced the variables to a single material-sensitive property, the mean pinning force density $\boldsymbol{P}_{\mathrm{v}}$. In the equilibrium state, we can write the local force balance equation in the mixed state

$$\boldsymbol{J}^{\mathrm{ext}} \times \boldsymbol{B}^{(\mathrm{v})} + \boldsymbol{P}_{\mathrm{v}} = 0 \tag{4.3.26}$$

The mean pinning force density $\boldsymbol{P}_{\mathrm{v}}$ can take any value up to a certain maximum, leading to a maximum current density, or flux density gradient, given by the following critical-state equation:

$$\boldsymbol{B}^{(\mathrm{v})} \times \boldsymbol{J}_c = \boldsymbol{P}_{\mathrm{v}max} \tag{4.3.27}$$

which may be further written, if $\boldsymbol{B}^{(\mathrm{v})}$ is perpendicular to $\boldsymbol{J}^{\mathrm{ext}}$,

$$J_c = \frac{P_{\mathrm{v}max}}{B^{(\mathrm{v})}} \tag{4.3.28}$$

This maximum current density $\boldsymbol{J}_c$ is called the critical current density for the bulk superconductor since it determines the maximum transport current density that a hard superconductor can carry without losses. For superconducting magnets, very high values of J_c are required in high fields. It has been reported that a niobium-titanium filament has J_c as high as 3.7×10^5 A/cm^2 at 5 T [Chengren and Larbalestier (1987)]. The critical current densities for high-T_c oxide superconductors reported are, for instance, $J_c = 10^4$ A/cm^2 at 77 K and at 0.1 T for $YBa_2Cu_3O_{7-\delta}$ monocrystals [Crabtree et al. (1987)], $J_c = 5\times10^6$ A/cm^2 at 77 K in the absence of external applied magnetic field, and $J_c = 4\times10^4$ A/cm^2 at 5.5 T for $YBa_2Cu_3O_{7-\delta}$ epitaxial thin films [Roas et al. (1990)], where the applied magnetic field is perpendicular to the *ab*-plane.

Some efforts have been made to calculate theoretically the mean pinning force. The calculation is usually divided into two steps: calculating the force of interaction of the vortex lattice with a separate defect, and then, averaging the force over different randomly distributed pinning centers. However, it is often convenient and sufficient to assume that the maximum pinning force density

$\boldsymbol{P}_{vmax}$ is a phenomenological function of the mean magnetic field at a given temperature. For instance, in the Bean model [Bean (1962)], P_{vmax} is supposed to be proportional to $B^{(v)}$, while in the Kim model [Kim et al. (1962)], P_{vmax} is assumed to be independent of $B^{(v)}$. Generally, to obtain high critical currents in practical type II superconductors, we may produce the maximum pinning forces in the material by introducing proper metallurgical defects, such as point defects as voids and small second-phase particles, line defects as dislocations, surface defects as grain boundaries, and volume defects as large precipitates. Optimum pinning occurs when the defect size is comparable to the vortex core size that is about ξ in size. If the defect size is smaller than the vortex core size, the core can span several defects at once, averaging out their effect and drastically reducing the pinning force. Therefore, the maximum pinning force is reduced rapidly when the magnetic field is applied to be approaching to the critical value B_{c2} of the material since, in such a case, the vortex core size becomes effectively very large.

If the transport current density in the superconducting material is slightly larger than its critical value J_c, the flux vortices begin to move in the hard type II superconductors. The mean flux flow velocity $\mathbf{v}_L$ may be obtained by

$$\mathbf{v}_L = \frac{\Phi_o}{\eta}(\boldsymbol{J}^{\text{ext}} - \boldsymbol{J}_c) \times \boldsymbol{z}^o \tag{4.3.29}$$

from the balance of three forces: the driving force (Lorentz force), the maximum pinning force, and the viscous friction force. The vortex electric field is then given from Eq.(4.3.22) by

$$\boldsymbol{E}^{(\text{v})} = \frac{\Phi_o B^{(\text{v})}}{\eta}(\boldsymbol{J}^{\text{ext}} - \boldsymbol{J}_c) \tag{4.3.30}$$

in the case of $\boldsymbol{B}^{(\text{v})}$ being perpendicular to $\boldsymbol{J}^{\text{ext}}$. It can be seen that the flow resistivity ρ_v of the hard type II superconductor is still given by Eq.(4.3.23). This means that the (differential) flow resistivity ρ_v, shown in the simple analysis, depends only on the parameters of ideal type II superconductors and is not influenced by pinning effects, which is qualitatively in accordance with experimental observations [Kim et al. (1965) and Kim and Stephen (1969)].

4.3.4 Flux Creep and the Anderson-Kim Model

According to the critical state model discussed above, as long as the driving force (Lorentz body force) is smaller or just equal to the maximum pinning force density, the vortex arrangement remains stationary and no heat is generated in the superconductor since no energy is dissipated from the motion of magnetic vortex lines. This critical state model does provide a sufficiently accurate description of the behavior of a hard superconductor for most practical purposes. However, very sensitive flux measurements reveal a very slow motion of flux at current densities even slightly lower than the critical current density. Such a phenomenon, called the flux creep attributed to thermally activated jumps of so-called flux bundles, was first observed by Kim et al. (1962) and Anderson (1962).

The amount of heat generated by the flux creep is usually sufficiently small not to be of practical importance, but because of the very small specific heat and thermal conductivity of the superconductors, the heat may produce relatively large local change in temperature. Since the mean pinning force density decreases rapidly as the temperature rises, the driving force may become larger than the maximum mean pinning force, resulting in the flow of the vortex lines and, thus, causes power dissipation in the superconductor. When the power dissipation exceeds a certain limit, a thermomagnetic instability appears and it may result in catastrophic consequences if the magnetic energy stored in, for instance, a superconducting magnet is suddenly converted into thermal energy.

To study the phenomenon of flux creep, a theoretical model (Anderson-Kim model) was developed to describe dynamic effects arising from the thermally activated motion of the fluxoids past the pinning barriers. In this model, the concept of "flux bundles" was introduced. Because of relatively long-range interaction among flux lines, local perturbation of the line density are very unfavourable energetically. Thus the arrangement can be irregular only on a scale greater than λ. Locally, the density must be uniform, and any local variation must be only slight increase in the local density spread out over a region of radius $> \lambda$. This is the idea behind the concept of "flux bundles"—that, in fact, while it is probably the individual flux line's internal structure (of size ξ) which is caught by a pinning center, that flux line individually cannot jump over the barrier alone, because it would get badly out of equilibrium with the local density in its neighborhood; but rather a whole bundle of flux lines, of radius $\sim \lambda$, must move simultaneously. Therefore, of course, it is the force on the total bundle which acts against the pinning barrier. The term "flux bundle" describes the cluster of neighboring fluxoids, coupled by their mutual interactions, which is supposed to act collectively in each thermally activated event.

According to the Anderson-Kim theory of flux creep, the rate at which "flux bundles" jump over the pinning barriers is given by the usual Arrhenius expression:

$$\nu = \nu_o exp\left(-\frac{U_o}{k_B T}\right) \tag{4.3.31}$$

where ν_0 is some characteristic frequency of flux-line vibration, generally unknown in detail, but assumed to lie in the range 10^5 to 10^{11} sec^{-1}, and U_o is the activation free energy, or barrier energy, which must be overcome to allow flux motion. U_o may be related to the critical current density J_c by $U_o = J_c B^{(v)} V_b d$ with V_b being the volume of the flux bundle and d the distance traveled by the flux line bundle in one thermally activated jump. Anderson and Kim suggested that U_o should scale as $H_c{}^2\xi^3/2\mu_o$, where H_c is the thermodynamic critical field and ξ is the superconducting coherence length. For conventional type II superconductors, this usually comes out to several electron-volts in agreement with earlier experiments. For YBCO, we may take $\xi^3 = \xi_{ab}{}^2\xi_c$ and obtain $U_o \approx 0.15$ eV at low temperatures [Yeshurun and Malozemoff (1988)]. By a heuristic scaling argument, Yeshurun and Malozemoff argue that U_o should have a form:

$$U_o = \frac{\beta H_c^2 \xi \Phi_o}{\mu_o B^{(v)}} \tag{4.3.32}$$

where H_c (= B_c/μ_0) is the thermodynamic critical field, ξ the coherence length, Φ_0 the flux quantum, $B^{(v)}$ the flux density in the material, and the parameter β (presumed ~ 1) is introduced to absorb all numerical factors. Equation (4.3.32) is considered to be an empirical formula. According to Tinkham (1988), Eq.(4.3.32) can be transformed to

$$\frac{U_o}{k_B T} = A(1-\Theta)^{3/2}/B^{(v)} \tag{4.3.33}$$

with $A = 4CJ_{co}/T_c$. Here, J_{co} is the value of J_c at $B^{(v)} = 0$ and without reduction by thermal fluctuations. C is a numerical factor. $\Theta = T/T_c$ is the reduced temperature. For YBCO, with $T_c = 92$ K, the factor $4C/T_c \approx 3.5\beta$, where β is expected to be of order unity. Although the temperature-dependence behavior in expression (4.3.33) builds in the two-fluid empirical approximations near T_c. The result is actually accurate to better than ±4% all the way from $\Theta = 1$ down to $\Theta = 1/2$. To permit extension to lower temperatures, Malozemoff et al. (1989) have used $1-\Theta^2$ rather than $1-\Theta$ in the expression (4.3.33) to permit extension to lower temperatures.

In the absence of any flux-density gradient, or driving forces to move the flux lines, jumps are as likely to occur in one direction as the other, and no net creep velocity exists. Now, if we introduce a driving force of density f, the force on a flux bundle of cross-section A and of length L is fV_b ($V_b = AL$ is the volume of the flux bundle), so the work done in moving it a distance d (a hopping distance) is $\Delta U = fV_b d$. This will lead to a net jump rate in the direction of the force f of

$$\Delta\nu = \nu_o e^{-U_o/k_B T}(e^{\Delta U/k_B T} - e^{-\Delta U/k_B T}) \tag{4.3.34}$$

This amounts to a net creep velocity of

$$\mathrm{v} = 2V_o \sinh\left(\frac{fV_b d}{k_B T}\right) \tag{4.3.35}$$

with $V_0 = d\nu_0 \exp(-U_o/k_BT)$. At relatively low temperatures, V_o is very small, and the net creep velocity will be unobservably small unless the driving term in Eq.(4.3.35) is huge. In this case, we will then have simply

$$\text{v} = V_o exp\left(\frac{fV_b d}{k_B T}\right) \tag{4.3.36}$$

which corresponds to the situation for $(J_c - J)/J_c << 1$ if f is the Lorentz body force.

On the other hand, in the small-J limit ($fV_b d/k_BT << 1$), we may write

$$\text{v} = \frac{2\nu_o f V_b d}{k_B T} exp\left(-\frac{U_o}{k_B T}\right) \tag{4.3.37}$$

which is considered to be in the thermally activated flux-flow (TAFF) regime. In high-temperature superconductors, U_o is comparable to k_BT and the small current regime is observable. The basic difference between the TAFF model and the flux-creep model is the larger thermal effects in the former case. TAFF occurs only in superconductors with an intrinsically low pinning barrier and at temperatures high enough to overcome this barrier. Experimentally, when a very small ac magnetic field δB is superimposed on a large dc field B_o, such that B is almost uniform, that is, $B = B_o + \delta B$ with $\delta B << B_o$, the current density is very small over the entire superconductor and one is in the TAFF regime.

Consider a case in which the vortex magnetic field $\mathbf{B}^{(\text{v})}$ is along z-axis direction, and the local velocity $\mathbf{v}$ of the moving vortices is along x-axis direction, we have then the vortex electric field $E_y = \text{v}B^{(\text{v})}$, according to Eq.(4.3.22), which gives

$$E_y = 2B^{(\text{v})} d\nu_o exp\left(-\frac{U_o}{k_B T}\right) \sinh\left(\frac{fV_b d}{k_B T}\right) \tag{4.3.38}$$

by noting Eq.(4.3.35).

For small currents ($J << J_c$), we may find that the resistivity ρ, defined by $\rho = E_y/J_y$,

$$\rho = \frac{2(B^{(\text{v})})^2 \nu_o V_b d^2}{k_B T} exp\left(-\frac{U_o}{k_B T}\right) \tag{4.3.39}$$

where $E(J)$ is linear. For large currents ($J \approx J_c$), where the effect of flux creep becomes pronounced, we may arrive at

$$E = B^{(\text{v})} \nu_o d exp\left(-\frac{U_o(1 - J/J_{co})}{k_B T}\right) \tag{4.3.40}$$

and

$$\rho = \frac{B^{(v)}v_o d}{J} exp\left(-\frac{U_o(1-J/J_{co})}{k_B T}\right) \tag{4.3.41}$$

which indicates that $E(J) - \rho J$ increases exponentially with J. The critical current determined in experiments depends on the lowest voltage which can be measured. If this is E_c, then, we have

$$J_c = J_{co}\left[1 - \frac{k_B T}{U_o} ln\left(\frac{B^{(v)}v_o d}{E_c}\right)\right] \tag{4.3.42}$$

where J_{co} is the critical current in absence of thermal activation. For typical parameters [Campbell and Evetts (1972)], the logarithm is about 30, but since $k_B T_c/U_o$ is of order 10^{-3} in conventional low-temperature superconductors, the thermal activation term is a negligible 3% correction. By contrast, for YBCO, $k_B T_c/U_o$ is of order 0.05, so that the thermal activation can be of order unity. Thus, through Eq.(4.3.42), critical-current measurements and even magnetic measurements of $H_{c2}(T)$ may be strongly affected by thermally activated flux creep. In particular, the creep term in Eq.(4.3.42) could contribute to the large temperature dependence of the measured critical current.

By substitution of Eq.(4.3.33) into Eq.(4.3.42), we may find the condition for zero J_c:

$$1 - \Theta = \left[\frac{B^{(v)}}{A} ln\left(\frac{B^{(v)}v_o d}{E_c}\right)\right]^{2/3} \tag{4.3.43}$$

This result offers a simple explanation for the phenomenon of "irreversibility line" or "quasi de Almeida-Thouless line" found recently in high-temperature oxide superconductors, such as YBCO. This line separates the region near T_c in the (H,T) plane in which the sample shows a unique reversible magnetization $M(H,T)$, from the region in which $M(H,T)$ depends on the previous path in the (H,T) plane [see Yeshurun and Malozemoff (1988), and Yeh (1991)].

By the continuity equation of the flux line, we have

$$\frac{\partial B^{(v)}}{\partial t} = -\frac{\partial}{\partial x}(vB^{(v)}) \tag{4.3.44}$$

When the driving force is the Lorentz force with the transport current along y-axis, we have

$$f = -\frac{1}{2\mu_o}\frac{\partial}{\partial x}(B^{(v)2}) \tag{4.3.45}$$

Here, we have ignored the difference between the vortex magnetic field and the total magnetic field. Taking the time derivative of Eq.(4.3.45), and interchanging the order of time and space differentiation, we may obtain by using Eq.(4.3.44)

$$\frac{\partial f}{\partial t} = \frac{\partial}{\partial x}\left[\frac{B^{(v)}}{\mu_o}\frac{\partial}{\partial x}(vB^{(v)})\right] \approx \frac{B^{(v)2}}{\mu_o}\frac{\partial^2 v}{\partial x^2} \tag{4.3.46}$$

Furthermore, by noting Eq.(4.3.35) and integrating Eq.(4.3.46), we may find

$$f = f_c - C_1 ln(t) \tag{4.3.47}$$

near the critical state. Here, f_c is the critical pinning force density, and C_1 is the parameter defined by $C_1 = k_B T/(V_b d)$. Applying this expression to the case of flux trapped in a hollow superconducting cylinder with wall thickness h, where $f \approx B^{(v)2}/(2h\mu_o)$, we have

$$B^{(v)} \approx B_c\left(1 - \frac{C_1}{2f_c} ln(t)\right) \tag{4.3.48}$$

where B_c is approximately the value of $B^{(v)}$ giving the critical-state condition. It should be noted that the infinity at $t = 0$ is inaccessible because of the choice of origin of time. Also, this simple form will hold only for moderately small fractional changes in $B^{(v)}$; thus, no significance should be attached to the fact that Eq.(4.3.47) changes sign after extremely long times. This logarithmic time dependence has been well verified experimentally [Kim et al. (1963)].

4.3.5 The London-Bean Model for Hard Superconductors

It has been shown in above sections that a type II superconductor may exhibit the Meissner effect in weak fields. With the increase of the applied field, the Meissner effect breaks down and magnetic flux begins to penetrate into the interior of the superconductor, but the superconductor continues to show essentially no dc resistance up to a much higher "upper" critical field B_{c2}, where bulk superconductivity finally disappears. The macroscopic average magnetization density M_{vor} from the vortex structure formed in the superconductor becomes a small fraction of the magnetic field B for applied field well above B_{c1} (see Fig. 4.4) and may, therefore, be neglected compared with the H contribution for most cases of interest. In time-varying fields, ac losses may, however, appear at all frequencies for practical superconducting materials. Though these losses are often small compared with the loss in a copper wire at room temperature, they may generate heat that occurs at low temperature and requires a large amount of refrigerator power for its removal in order to avoid the increase of temperature of the superconductor, which may lead to the catastrophic destruction of superconductivity.

To analyze the ac losses in superconductors, we shall first introduce here a simple model (often called the London-Bean model) for the study of ac losses in hard (type II) superconductors in high fields. The London-Bean model makes use of the following assumptions [Bean (1962)]:

1. The magnetic flux structure in type II superconductors in the mixed state is capable of sustaining lossless macroscopic current up to a critical current density J_c;
2. The J_c is independent of magnetic field, which is equivalent to the assumption that the applied fields are much less than the critical fields of the vortices;
3. The interstices of the flux mesh are fitted with a soft superconductor of the critical field B_c;
4. The magnetic field is shielded by the soft superconductor to a field B_c; then further shielding is accomplished by currents flowing in the flux mesh. These currents flow to the full amount J_c for a depth necessary to reduce the field to B_c.

Essentially, these assumptions are made that there exists a limiting macroscopic superconducting current density J_c that a hard superconductor can carry; and further, that any electromotive force, however small, will induce this full current to flow locally. We may draw the J–E curve illustratively in Fig. 4.9. Mathematically, we may introduce the J–E relation:

$$\boldsymbol{J} = J_c \frac{\boldsymbol{E}}{E} \quad \text{for } \boldsymbol{E} \neq 0 \tag{4.3.49}$$

with $\boldsymbol{J} = 0$ or $|\boldsymbol{J}| = J_c$ for $\boldsymbol{E} = 0$ depending upon the magnetic history. If $\boldsymbol{E}$ is zero due to decay from some previous nonzero value, $|\boldsymbol{J}| = J_c$ and its direction is given by $\boldsymbol{E}/E$ as E approaches zero. It can been seen that, in this model, the magnitude of $\boldsymbol{J}$ can take only its critical value or zero, and the critical value exists everywhere in a material that has been fully penetrated. The critical current density J_c is supposed to be independent of magnetic field in the London-Bean model. In addition, the vortex structure and the Meissner effect

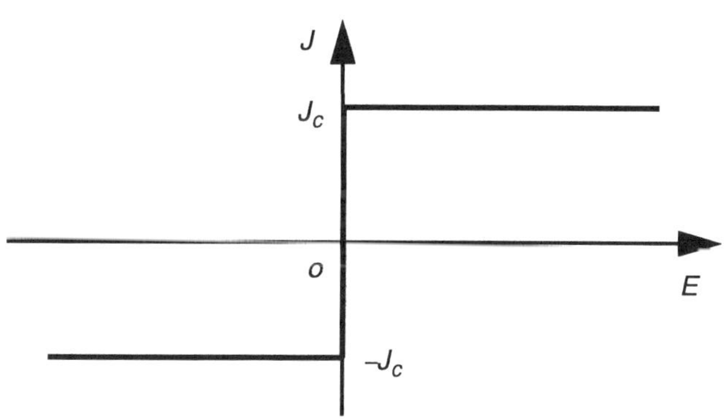

Figure 4.9 A plot of J–E relation in the London-Bean model.

have been ignored in the model so that the magnetic field in the superconductor is only the sum of the applied field and the field from bulk transport and shielding currents. Obviously the model will be particularly good for treating the case of a small ac field superimposed upon a dc bias such that the total magnetic field is close to the upper critical field B_{c2}. Furthermore, the magneto-quasistatic approximation is used in the model.

4.3.6 Magnetization of High-Field Superconductor

It is known that the magnetic properties of hard (type II) superconductors differ from that of type I superconductors. In particular, the hard superconductors often exhibit hysteresis. The magnetization of hard superconductors may be studied by the London-Bean model. Illustratively, let us first consider an infinite superconducting plate of thickness $2W$ exposed to an external magnetic field B_a parallel to the surface, as shown in Fig. 4.10.

As the field is applied, screening currents will be induced near the surface of the superconducting plate. In the London-Bean model, if the applied field B_a is less than B_c, the shielding is complete if W is much greater than the penetration depth λ_L of the superconductor. This gives

$$B_z = 0; \qquad 0 \le |x| \le W, \qquad 0 \le B_a \le B_c \tag{4.3.50}$$

As the field is further increased, the region near the surface of the superconducting plate becomes normal and currents are induced to flow inside the plate. For the field is an amount $B_a - B_c$ above the bulk critical field, then by the circuital form of Ampere's law, we may find

$$B_z = B_a - B^*\left(1 + \frac{x}{W}\right) \qquad \text{for } -W \le x \le -x_o \tag{4.3.51}$$

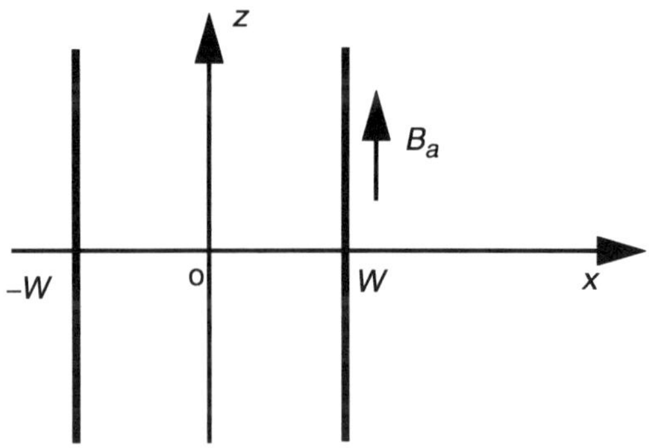

Figure 4.10 A superconducting plate in a magnetic field.

$$B_z = B_a - B^*\left(1 - \frac{x}{W}\right) \qquad \text{for } x_o \le x \le W \tag{4.3.52}$$

and $B_z(x) = 0$ for $|x| < x_o$ when the applied field B_a is smaller than $B_c + B^*$, as shown in Fig. 4.11. Here, we have introduced a field value of $B^* = \mu_o J_c W$ so that when the applied field B_a is equal to $B_c + B^*$, we have $x_o = 0$. Here, x_o is defined to be the distance at which $B_z(|x| = x_o) = B_c$. Thus we may find

$$x_o = W\left(1 - \frac{B_a - B_c}{B^*}\right) \tag{4.3.53}$$

When all the bulk properties have been destroyed for $B_a > B_c + B^*$, we have then

$$B_z = B_a - B^*\left(1 + \frac{x}{W}\right) \qquad \text{for } -W \le x \le 0 \tag{4.3.54}$$

$$B_z = B_a - B^*\left(1 - \frac{x}{W}\right) \qquad \text{for } 0 \le x \le W \tag{4.3.55}$$

The average magnetization of the specimen can be obtained by

$$\langle M \rangle = \frac{1}{\mu_o}(\langle B \rangle - B_a) \tag{4.3.56}$$

where B_a is the externally applied magnetic field, and $\langle B \rangle$ is the average internal magnetic field, given here by

$$\langle B \rangle = \frac{1}{2W}\int_{-W}^{W} B_z(x)dx \tag{4.3.57}$$

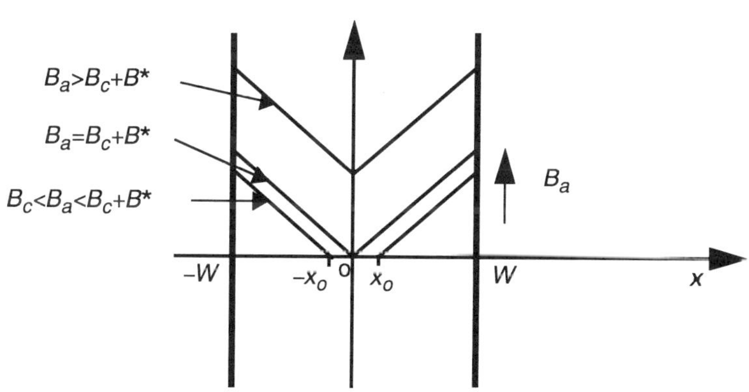

Figure 4.11 A plot of increasing local fields in the superconducting plate.

Thus the average magnetization of the specimen at the three different field regions can be obtained, respectively, by

$$\langle M \rangle = -\frac{B_a}{\mu_o}; \qquad 0 \le B_a \le B_c \tag{4.3.58}$$

$$\langle M \rangle = -\frac{B_a}{\mu_o} + \frac{B_a^2 - B_c^2}{2\mu_o B^*}; \qquad B_c \le B_a \le B_c + B^* \tag{4.3.59}$$

$$\langle M \rangle = -\frac{B^*}{2\mu_o}; \qquad B_c + B^* \le B_a \tag{4.3.60}$$

We may also consider the situation in which the applied field B_a is reduced to be $B_a - b_o$ in the case of, for instance, $B_a > B^*$. For some strongly type II superconductors having very small lower critical field, we may often ignored the effect of B_c by simply taking $B_c = 0$ in the analysis for some high-field applications. For simplicity, we shall ignore here the effect of B_c. As the field is reduced, the surface of the plate feels an *emf* oppositely directed to the one felt as the field was increasing—hence the surface currents reverse. Thus we may find that the magnetic field B_z in the plate becomes

$$B_z = B_a - b_o + B^*\left(1 + \frac{x}{W}\right) \qquad \text{for } -W \le x \le -x_1 \tag{4.3.61}$$

$$B_z = B_a - b_o + B^*\left(1 - \frac{x}{W}\right) \qquad \text{for } x_1 \le x \le W \tag{4.3.62}$$

where x_1 is the distance at which we have the maximum trapped flux $B_m = B_a - b_o/2$, as shown in Fig. 4.12. Thus x_1 may be found by

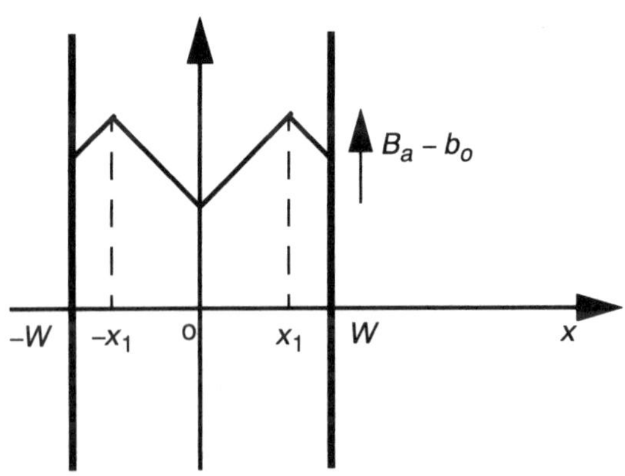

Figure 4.12 A plot of decreasing local fields in the superconducting plate.

$$x_1 = W\left(1 - \frac{b_o}{2B^*}\right) \tag{4.3.63}$$

When the applied magnetic field drops below zero and a negative applied field forms a critical state in the opposite direction. Determination of internal magnetic fields in the field-lowering process is similar to the way discussed above.

Similarly, we may also consider a superconducting cylinder of radius R. We assume that a magnetic field B_a is applied parallel to the axis of the cylinder, being along the z-axis direction. If the applied field is less than the bulk critical field B_c, the shielding is complete if the radius R is much greater than the London penetration depth λ_L. This gives

$$B_z = 0; \qquad 0 \le r \le R, \qquad 0 \le B_a \le B_c \tag{4.3.64}$$

As the field is further increased, the region near the surface of the superconducting plate becomes normal and currents are induced to flow inside the cylinder. For the field is an amount B_a–B_c above the bulk critical field, then by the circuital form of Ampere's law, we have (noting $J_\theta = -J_c$ in the critical state)

$$\frac{\partial B_z}{\partial r} = \mu_o J_c \tag{4.3.65}$$

from which we may get

$$B_a - B_c = \mu_o J_c (R - r_o) \tag{4.3.66}$$

where r_o is the radius at which $B_z(r_o) = B_c$. For $r < r_o$, we have $B_z(r) = 0$. We can define a field value of $B^* = \mu_o J_c R$, so that, when the applied field B_a is equal to $B_c + B^*$, we have $r_o = 0$. Thus we have

$$B_z = 0; \qquad 0 \le r \le R\left(1 - \frac{B_a - B_c}{B^*}\right) \tag{4.3.67}$$

$$B_z = B_a - B^*\left(1 - \frac{r}{R}\right); \qquad R\left(1 - \frac{B_a - B_c}{B^*}\right) \le r \le R \tag{4.3.68}$$

for $B_c < B_a < B_c + B^*$.

When all the bulk properties have been destroyed for $B_a > B_c + B^*$, we have then

$$B_z = B_a - B^*\left(1 - \frac{r}{R}\right); \qquad 0 \le r \le R \tag{4.3.69}$$

The average magnetization of the superconducting cylinder at the three different field regions can be calculated, respectively, by

$$\langle M \rangle = -\frac{B_a}{\mu_o}; \qquad 0 \le B_a \le B_c \tag{4.3.70}$$

$$\langle M \rangle = -\frac{B_a}{\mu_o} + \frac{B_a^2 - B_c^2}{\mu_o B^*} + \frac{B_c^2(3B_a - 2B_c) - B_a^3}{3\mu_o B^{*2}}; \qquad B_c \le B_a \le B_c + B^* \tag{4.3.71}$$

$$\langle M \rangle = -\frac{B^*}{3\mu_o}; \qquad B_c + B^* \le B_a \tag{4.3.72}$$

It may be noticed that a striking result of the calculation by using the London-Bean model is that the average magnetization of the hard superconductor depends on the macroscopic dimensions of the samples. Some experiments were performed to confirm this result [see Bean (1962)].

4.3.7 AC Loss in a Thin Superconducting Plate

In this section, the ac loss in a thin superconducting plate in an applied field B^e, which is in the plane of the plate along the positive direction of the y-axis as shown in Fig. 4.13, will be analyzed with the use of the London-Bean model.

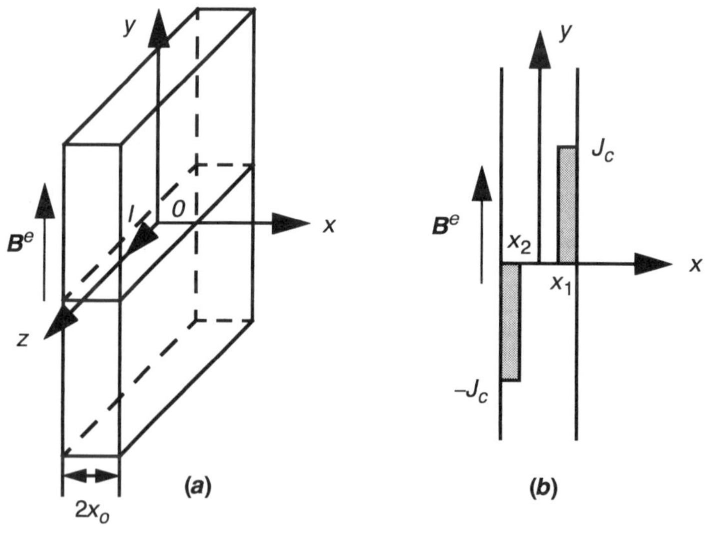

Figure 4.13 Configuration of a superconducting plate with the magnetic field applied along the y-axis and transport current flowing along the z-axis.

It is assumed that the plate may have a net transport current per unit width of the plate along the z-axis, given by

$$I = \int_{-x_o}^{x_o} J_z \, dx \tag{4.3.73}$$

In such a case, the magnetic fields on the plate surfaces are

$$B_y(x_o) = B^e + \frac{1}{2}\mu_o I \tag{4.3.74}$$

$$B_y(-x_o) = B^e - \frac{1}{2}\mu_o I \tag{4.3.75}$$

Consider that, for a small magnetic field and small transport currents, the flux penetration and also the shielding current penetration are only appreciable within the penetration depth λ. However, with the increase of the magnetic field and the transport current, the total current density from the transport current and the induced shielding current may reach the value of the critical current density J_c. If we increase further the field, the London-Bean model assumes that the total current density remains constant and the flux penetrates deeper into the material, where more shielding currents are induced. In such a case (as shown in Fig. 4.13(*b*) in which we have assumed that the applied magnetic field is dominant), the magnetic field in the region of $-J_c$ may be found as

$$B_y(x) = B^e - \frac{1}{2}\mu_o I - \mu_o J_c(x + x_o), \qquad (-x_o < x < x_2) \tag{4.3.76}$$

and, in the region of J_c, reads

$$B_y(x) = B^e + \frac{1}{2}\mu_o I + \mu_o J_c(x - x_o), \qquad (x_1 < x < x_o) \tag{4.3.77}$$

which show that $B_y(x)$ is a linear function of the coordinate x in the London-Bean model.

The quantities x_1 and x_2, characterizing the field penetration, are determined by the conditions that the magnetic field is zero at $x = x_1$ and at $x = x_2$ in the case of virgin state, that is,

$$0 = B^e - \frac{1}{2}\mu_o I - \mu_o J_c(x_2 + x_o) \tag{4.3.78}$$

and

$$0 = B^e + \frac{1}{2}\mu_o I + \mu_o J_c(x_1 - x_o) \tag{4.3.79}$$

which give

$$x_1 = x_o - \frac{1}{\mu_o J_c}\left(B^e + \frac{1}{2}\mu_o I\right) \tag{4.3.80}$$

and

$$x_2 = -x_o + \frac{1}{\mu_o J_c}\left(B^e - \frac{1}{2}\mu_o I\right) \tag{4.3.81}$$

In particular, the penetration field B^p at which full penetration occurs, may be determined by the condition $x_1 = x_2$

$$B^p = x_o \mu_o J_c \tag{4.3.82}$$

for the case described by Eqs.(4.3.80) and (4.3.81), where the applied magnetic field is dominant. In general, the penetration field B^p will be dependent on the previous history of the superconductor.

For time-varying fields, the electric field in the superconductor may be determined from Maxwell's equations as

$$E_z(x) = \left(\frac{\partial B^e}{\partial t} - \frac{1}{2}\mu_o \frac{\partial I}{\partial t}\right)(x - x_2), \qquad (-x_o < x < x_2) \tag{4.3.83}$$

and

$$E_z(x) = \left(\frac{\partial B^e}{\partial t} + \frac{1}{2}\mu_o \frac{\partial I}{\partial t}\right)(x - x_1), \qquad (x_1 < x < x_o) \tag{4.3.84}$$

where we have used the conditions that E_z must vanish at the inner moving boundaries x_1 and x_2. Equations (4.3.83) and (4.3.84) determine generally the sign of the current density. The instantaneous power loss may now be calculated by

$$P(t) = \int \boldsymbol{J} \cdot \boldsymbol{E} dV \tag{4.3.85}$$

where the integral is over the volume of the superconductor.

With the use of Eq.(4.3.49), the loss per unit total volume for the considered case may be given by

$$\frac{P}{V} = \frac{1}{2x_o}\int_{-x_o}^{x_o} \boldsymbol{J} \cdot \boldsymbol{E} dx = \frac{J_c}{2x_o}\int_{-x_o}^{x_o} |E_z| dx \tag{4.3.86}$$

which holds quite generally for the London-Bean model.

In the case of E_z given by Eqs.(4.3.83) and (4.3.84), we have

$$\frac{P}{V} = \frac{J_c}{2x_o}\left(\left|\frac{\partial B^e}{\partial t} - \frac{1}{2}\mu_o\frac{\partial I}{\partial t}\right|\int_{-x_o}^{x_2}|x - x_2|dx + \left|\frac{\partial B^e}{\partial t} + \frac{1}{2}\mu_o\frac{\partial I}{\partial t}\right|\int_{x_1}^{x_o}|x - x_1|dx\right)$$

$$= \frac{J_c}{4x_o}\left(\left|\frac{\partial B^e}{\partial t} - \frac{1}{2}\mu_o\frac{\partial I}{\partial t}\right|(x_o + x_2)^2 + \left|\frac{\partial B^e}{\partial t} + \frac{1}{2}\mu_o\frac{\partial I}{\partial t}\right|(x_o - x_1)^2\right) \qquad (4.3.87)$$

For the particular simple case of the virgin state, where the fields are initially equal to zero, then, the use of Eqs.(4.3.80) and (4.3.81) leads to

$$\frac{P}{V} = \frac{1}{4x_o\mu_o^2J_c}\left[\left|\frac{\partial B^e}{\partial t} - \frac{1}{2}\mu_o\frac{\partial I}{\partial t}\right|\left(B^e - \frac{1}{2}\mu_oI\right)^2 + \left|\frac{\partial B^e}{\partial t} + \frac{1}{2}\mu_o\frac{\partial I}{\partial t}\right|\left(B^e + \frac{1}{2}\mu_oI\right)^2\right]$$

$$= \frac{1}{6\mu_o^2I_c}\left(\left|\frac{\partial}{\partial t}\left(B^e - \frac{1}{2}\mu_oI\right)^3\right| + \left|\frac{\partial}{\partial t}\left(B^e + \frac{1}{2}\mu_oI\right)^3\right|\right) \qquad (4.3.88)$$

where $I_c = 2x_oJ_c$ denotes the critical current per unit width.

For the case of a cyclic state with partial penetration, the positions of the moving boundaries x_1 and x_2 may be found as

$$x_1 = x_o - \frac{1}{2\mu_oJ_c}\left|B^e(t) + \frac{1}{2}\mu_oI(t) - B^e(t_1) - \frac{1}{2}\mu_oI(t_1)\right| \qquad (4.3.89)$$

$$x_2 = -x_o + \frac{1}{2\mu_oJ_c}\left|B^e(t) - \frac{1}{2}\mu_oI(t) - B^e(t_2) + \frac{1}{2}\mu_oI(t_2)\right| \qquad (4.3.90)$$

where t_1 is the time when $x_1 = x_o$ and t_2 the time when $x_2 = -x_o$. It is shown that the speed of the moving boundary for the cyclic state is one-half that for the virgin state, assuming the same rate of change of the field at the surface.

Substitution of Eqs.(4.3.89) and (4.3.90) into Eq.(4.3.87) gives

$$\frac{P}{V} = \frac{1}{8\mu_o^2I_c}\left[\left|\frac{\partial B^e}{\partial t} - \frac{1}{2}\mu_o\frac{\partial I}{\partial t}\right|\left(B^e - \frac{1}{2}\mu_oI - B^e(t_2) + \frac{1}{2}\mu_oI(t_2)\right)^2\right.$$
$$\left. + \left|\frac{\partial B^e}{\partial t} + \frac{1}{2}\mu_o\frac{\partial I}{\partial t}\right|\left(B^e + \frac{1}{2}\mu_oI - B^e(t_1) - \frac{1}{2}\mu_oI(t_1)\right)^2\right] \qquad (4.3.91)$$

or

$$\frac{P}{V} = \frac{1}{24\mu_o^2I_c}\left\{\left|\frac{\partial}{\partial t}\left[B^e - \frac{1}{2}\mu_oI - B^e(t_2) + \frac{1}{2}\mu_oI(t_2)\right]^3\right|\right.$$
$$\left. + \left|\frac{\partial}{\partial t}\left[B^e + \frac{1}{2}\mu_oI - B^e(t_1) - \frac{1}{2}\mu_oI(t_1)\right]^3\right|\right\} \qquad (4.3.92)$$

In particular, consider a sample in a cyclic state where the applied field, in the plane of the plate, is cycled from $B^e = -B^m$ to $B^e = B^m$ and back to $-B^m$ without reentrant loops, that is, the sign of $\partial B^e/\partial t$ changes only at the ends of the half-cycles. If B^m is insufficient to cause full penetration, the instantaneous power loss may be calculated by Eq.(4.3.92) with $I = 0$ and $t_1 = t_2$ in the London-Bean approximation

$$\frac{P}{V} = \frac{1}{12\mu_o^2 I_c}\left|\frac{\partial}{\partial t}[B^e(t) - B^e(t_1)]^3\right| \tag{4.3.93}$$

where in the first half-cycle $B^e(t_1) = -B^m$ and for the second half-cycle $B^e(t_1) = B^m$. Thus the loss per cycle can be found by

$$\int \frac{P}{V}dt = 2\int_{-B^m}^{B^m} \frac{P}{V}dt = \frac{1}{6\mu_o^2 I_c}\int_{-B^m}^{B^m} \frac{\partial}{\partial t}[B^e(t) + B^m]^3 dt = \frac{4(B^m)^3}{3\mu_o^2 I_c} \tag{4.3.94}$$

which shows that the loss is characterized by the appearance of the cubic terms in the field. For further details about the calculations of some other types of ac loss problems for superconductors in time-varying fields as well as some improvements for the ac loss calculation, readers may refer to the works of Carr (1983) and Wilson (1983).

4.3.8 Magnetothermal Instability of Hard Superconductors

The discovery of hard superconductors has led to the large-scale applications of superconductivity in the construction of, for instance, superconducting power cables and high-field magnets. However, as we have learned, the penetration of magnetic flux into a hard superconductor is essentially a dissipative process in which heat is generated. The generation of heat may cause a temperature rise which decreases the pinning strength and allows further flux penetration. Thus the penetration of flux is a regenerative process which may become catastrophic if certain conditions prevail. Besides, the dissipation of energy associated with flux creep in hard superconductors may also lead to disastrous consequences if it leads to a thermal runaway in which the material rapidly heats up and the entire energy stored in the magnet is suddenly converted into thermal energy. To prevent this, the system must be thermally stable; that is, if some region of the material acquires a temperature increment δT, stability requires that δT decay to zero, not continue to grow. To find the condition for thermal stability, let us first consider the case of flux creep.

For the case of flux creep, we may derive the resistive power dissipated per unit volume in the material by

$$P = \mathrm{v}f \tag{4.3.95}$$

where v is the velocity of the flux creep and f is the driving force density. By Eq.(4.3.36) for the flux-creep velocity v, we may obtain

$$\frac{T\partial P}{P\partial T} = \frac{U_o}{k_B T} - \frac{fV_b d}{k_B T} - \frac{1}{k_B}\frac{\partial U_o}{\partial T} \tag{4.3.96}$$

For a typical value of 100 [Tinkham (1975)] for the above expression, we may find that a small increase in temperature will lead to a very much larger increase in heating power, and instability will result unless cooling is very efficient.

To study this balance, we consider the heat-flow equation:

$$C\frac{\partial T}{\partial t} = \kappa\nabla^2 T + P \tag{4.3.97}$$

where C is the specific heat per unit volume, κ the thermal conductivity, and P the power input per unit volume. In steady state, $T(\boldsymbol{r})$ is the solution of $\kappa\nabla^2 T + P = 0$. Suppose a small temperature fluctuation δT occurs. Whether it grows or decays is determined by the equation:

$$C\frac{\partial \delta T}{\partial t} = \kappa\nabla^2 \delta T + \frac{\partial P}{\partial T}\delta T \tag{4.3.98}$$

For a fluctuation localized in a volume of radius ~ r, the condition for thermal stability can be estimated as [Anderson and Kim (1964) and Tinkham (1975)]

$$\frac{\Delta T}{T} < 10^{-2} \tag{4.3.99}$$

which indicates that a very tiny rise in temperature may presage a complete thermal breakdown. Thus, to avoid thermal runaway, magnets should always be operated under conditions such that the steady-state dissipation due to flux creep is so small that the resulting temperature rise is less than ~ 1 percent. In practice, superconducting materials are usually properly stabilized so that they can tolerate some individual flux jumps, which are common feature of mixed-state behavior for type II superconductors with currents near critical levels where the flux creep rate is enhanced. This may be achieved by using composite superconductive materials, such as composite superconducting cables with many thin superconducting filaments embedded in a copper matrix which has high thermal and electrical conductivity. The high conductivity matrix assists in removing heat from the superconductor and slows down the propagation of magnetic disturbances. In addition, to reduce the eddy current losses due to the coupling between superconducting filaments via the normal conducting matrix, most modern superconducting materials have twisted filaments [Carr (1983)].

In a multifilamentary composite superconductor, the magnetothermal instability can occur at two levels. First, at the local level, the magnetic flux jumping occurs in one or simultaneously in several superconducting filaments. Second, at the global level, the magnetothermal instability develops in the entire cross-section of a multifilamentary composite superconductor at once. In this case, the magnetic flux jumping arises as a result of the coupling between the temperature and the electromagnetic field perturbations in the superconducting filaments and those in the normal-metal matrix. In what follows, we shall introduce an example of studying the magnetothermal instability in current-carrying multifilamentary composite superconductors by following the work of Mints (1998).

We consider the problem of magnetothermal instability arising in a multifilamentary composite superconductor in the global level. We assume that the number of the superconducting filaments in the cross-section of the composite is large. Thus temperature and electromagnetic fields vary little on a space scale of the order of the interfilament spacing and the diameter of the superconducting filaments. This enable us to treat the composite superconductor as a homogeneous medium with some effective material parameters [Carr (1983)].

Let us consider a homogeneous superconducting wire carrying an increasing transport current $I(t)$. Suppose that initially there are no superconducting currents in the wire. In this case, the transport current progressively penetrates the wire from the outside in and the region with $J \neq 0$ has the form of a ring, as shown in Fig. 4.14. Here, the outer radius of the wire is denoted by a, and the inner radius of the ring, a_i, depends on the current. If the London-Bean critical state model is used, we may find the value of a_i by

$$a_i = a\sqrt{1 - \frac{I}{I_c}} \tag{4.3.100}$$

where I_c is defined by $I_c = \pi a^2 J_c$ with $J_c(T)$ being the critical current density of the wire.

Due to the time-varying transport current $I(t)$, there exists an induced (background) electric field in the region $a_i < r < a$, which can be calculated by

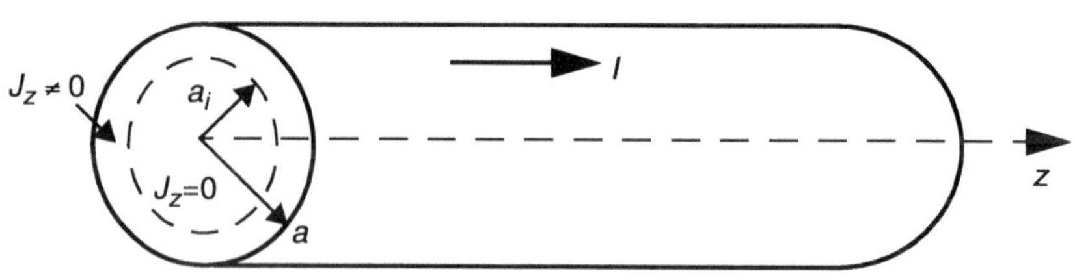

Figure 4.14 Current distribution in a homogeneous superconducting wire.

the Maxwell equations in a cylindrical coordinate system (r, θ, z):

$$\frac{\partial(rB_\theta)}{\partial r} = \mu_o r J_c \qquad (a_i < r < a) \tag{4.3.101}$$

$$\frac{\partial E_z}{\partial r} = \frac{\partial B_\theta}{\partial t} \qquad (a_i < r < a) \tag{4.3.102}$$

with the boundary conditions:

$$B_\theta = 0 \quad \text{for } r = a_i \quad \text{and} \quad E_z = 0 \quad \text{for } r = a_i \tag{4.3.103}$$

where B_θ is the θ component of the magnetic field induced by the current flowing in the saturated region. From this set of equations (4.3.101)–(4.3.103), we may obtain

$$E_z = \frac{\mu_o}{2\pi}\frac{dI}{dt}\ln\left(\frac{r}{a_i}\right) \qquad (a_i < r < a) \tag{4.3.104}$$

which shows that the time-varying transport current $I(t)$ results in a longitudinal background electric field that is proportional to the current ramp rate, that is, $E_z \propto \mu_0 dI/dt$. By Eq.(4.3.104), we may find the following expression:

$$\langle E_z \rangle = \frac{1}{\pi a^2}\int_0^{2\pi}\int_0^{a} E_z r dr d\theta = -\frac{\mu_o}{4\pi}\frac{dI}{dt}\left[\frac{I}{I_c} + \ln\left(1 - \frac{I}{I_c}\right)\right] \tag{4.3.105}$$

for the value of E_z averaged over the cross-sectional area of the wire.

Let us now suppose that a temperature perturbation $\delta T > 0$ arises in the superconducting wire. This temperature increase δT causes a decrease in the superconducting current. To keep the superconducting state stable, that is, to keep the total current at the same level, an electric field perturbation δE_z arises. The additional electric field δE_z causes an additional heat release $\delta \Upsilon$. The superconducting state is stable if the additional heat release $\delta \Upsilon$ can be removed to the coolant by the additional heat flux δQ, resulting from the temperature perturbation δT. Thus the superconducting-state stability criterion for the wire has the form:

$$\delta \Upsilon = \int_A J_z \delta E_z dA < \delta Q = \int_P h \delta T dP \tag{4.3.106}$$

where J_z is the current density, δE_z the longitudinal (parallel to the superconducting wire) electric field perturbation, h the heat transfer coefficient to the coolant, P the cooling perimeter of the wire, and A is the cross-sectional area of the wire.

To find a relation between δT and δE_z, we let δJ_- denote the decrease of the current density resulting from the temperature perturbation δT, and δJ_+ the increase of the current density resulting from the electric field perturbation δE_z. If the superconducting state is stable, then the total current density stays constant, that is,

$$\delta J_z = \delta J_- + \delta J_+ = 0 \tag{4.3.107}$$

In the superconducting state, the current density J_z is close to $J_c(T)$. Thus the decrease of the current density due to the temperature perturbation is given approximately by

$$\delta J_- = -\left|\frac{\partial J_c}{\partial T}\right|\delta T \tag{4.3.108}$$

The increase of the current density due to the electric field perturbation is given by

$$\delta J_+ = \frac{\partial J_z}{\partial E_z}\delta E_z \tag{4.3.109}$$

In most cases of practical interest when the superconductor is in the flux-creep region for low frequency applications, the total current density J_z may take the following form:

$$J_z = J_c + J_1 \ln\left(\frac{E_z}{E_c}\right) \tag{4.3.110}$$

in which J_1 and E_c are two phenomenological parameters. The second term on the right-hand side of this equation characterizes effectively the resistive component of the total current density, caused by the flux motion. Generally, we have $J_1 << J_c$ in most case of practical interest. In particular, for commercial Nb-Ti-based multifilamentary composite superconductors, the dependence of the ratio J_1/J_c on temperature and magnetic field is assumed to be weak, $J_c \sim 10^9$ A/m^2 and $J_1 \sim 3\times10^7$ A/m^2, that is, $J_1/J_c \sim 0.03 << 1$.

Thus, by using Eqs.(4.3.108)–(4.3.110), we may find from Eq.(4.3.107) the relation:

$$\delta E_z = \frac{E_z}{J_1}\left|\frac{\partial J_c}{\partial T}\right|\delta T \tag{4.3.111}$$

If we assume that the physical properties of the superconducting wire are uniform, we may find that the superconducting-state stability criterion (4.3.106) can be written in the following form [Mints and Rakhmanov (1982, 1988)]:

$$\langle E_z \rangle < E_{max} = \frac{Ph}{A}\frac{J_1}{J_c}\left|\frac{\partial J_c}{\partial T}\right|^{-1} \tag{4.3.112}$$

which implies that the longitudinal background electric field E_z averaged over the cross-sectional area of the wire has to be less than a certain critical value E_{max}, in order to let the superconducting wire be stable at the global level of magnetothermal stability.

Substituting now Eq.(4.3.105) into Eq.(4.3.112), we may obtain the following equation:

$$-\frac{\mu_o}{4\pi}\frac{dI}{dt}\left[\frac{I}{I_c} + \ln\left(1 - \frac{I}{I_c}\right)\right]\Bigg|_{I = I_{max}} = \frac{Ph}{\pi a^2}\frac{J_1}{J_c}\left|\frac{\partial J_c}{\partial T}\right|^{-1} \tag{4.3.113}$$

for the determination of the maximum superconducting current I_{max}. Introducing a characteristic current ramp rate defined by

$$\frac{dI_q}{dt} = \frac{4Ph}{\mu_o a^2}\frac{J_1}{J_c}\left|\frac{\partial J_c}{\partial T}\right|^{-1} \tag{4.3.114}$$

Eq.(4.3.113) can be rewritten in the form:

$$\left[\frac{I_{max}}{I_c} + \ln\left(1 - \frac{I_{max}}{I_c}\right)\right] + \frac{dI_q/dt}{dI/dt} = 0 \tag{4.3.115}$$

It can be shown from this equation that the value of the maximum superconducting current I_{max} tends to the critical current I_c for low current ramp rates, that is, for $dI/dt < dI_q/dt$. A considerable reduction of the current-carrying capacity occurs if the current ramp rate is high, that is, for $dI/dt >> dI_q/dt$. In this range of current ramp rate, the value of the maximum superconducting current I_{max} is much less than the critical current I_c. Some more detailed discussion on this subject may be found in the work of Mints (1998).

4.4 ELECTRODYNAMIC MODEL FOR TYPE II SUPERCONDUCTORS AT HIGH FREQUENCIES

4.4.1 Vortex Electric Field in Type II Superconductors in AC Fields

It has been shown that type II superconductors may exhibit the Meissner effect in small magnetic fields, but when the field is increased to a value larger than a lower critical field B_{c1}, the Meissner effect breaks down and the magnetic flux begins to penetrate into the interior of the superconductor. However, the superconductivity is not destroyed. The type II superconductor in the mixed state may continue to show essentially no dc resistance up to a much higher upper critical field B_{c2}, where bulk superconductivity finally disappears though surface superconductivity may persist up to a still higher critical field. Such a phenomenon has been explained physically by introducing the pinning mechanism in which strong pinning forces may exist in hard type II superconductors. These pinning forces in hard superconductors are so strong that magnetic flux-lines can be pinned at the pinning centers, where the driving forces are balanced by the pinning forces up to a certain maximum value. In the flux-pinned state, flux-lines are not moved in the superconductor in the dc case, and therefore, cause no dc losses.

In ac fields, losses are, however, found to exist in the type II superconductor in the mixed state. The magnetic flux-lines driven by time-varying forces may, for instance, vibrate locally around their pinning centers since these flux-lines are not attached rigidly to the pinning centers. When the driving force on a flux-line increases to a value larger than the maximum pinning force, the flux-line may move away from its pinning center and be driven into a flux-flow state, causing losses even in the dc case. When flux-lines move in the superconductor, there appears a macroscopic electric field:

$$\boldsymbol{E}^{(\mathrm{v})} = \boldsymbol{B}^{(\mathrm{v})} \times \mathbf{v}^{(\mathrm{f})} \tag{4.4.1}$$

where $\boldsymbol{B}^{(\mathrm{v})}$ is the local vortex magnetic field vector and $\mathbf{v}^{(\mathrm{f})}$ is the local velocity of the flux-lines.

Clearly, there are two distinct flux-motion states: one is the localized vibration of flux-lines around their pinning centers in ac fields when the driving forces are smaller than the maximum pinning force, and the other one is the global motion of flux-lines driven away from their pinning centers when the driving forces are larger than the maximum pinning force.

In the London-Bean (critical-state) model discussed in the above section, the effect of localized vibration of flux-lines in ac fields is ignored by assuming that the flux-lines are rigidly pinned to their pinning centers so that one has $\boldsymbol{E}^{(\mathrm{v})} = 0$ in the superconductor below a critical state, while the effect of global motion of flux-lines may induce a finite macroscopic vortex electric field in the superconductor above the critical state. This critical state is reached at $\boldsymbol{J} = \boldsymbol{J}_c$ with $\boldsymbol{J}$ being the local current density and $\boldsymbol{J}_c$ the critical current density of the superconductor, which characterizes effectively the pinning effect in hard superconductors. The transition from the flux-pinned state with $\boldsymbol{E}^{(\mathrm{v})} = 0$ to the flux-flow state with $\boldsymbol{E}^{(\mathrm{v})} \neq 0$ is considered implicitly to be instantaneous in the London-Bean model. Such an approximation made in the London-Bean model may be reasonable, and have, indeed, been verified by many experiments in situations where relatively low frequency phenomena are involved.

For high-frequency electromagnetic phenomena in type II superconductors in the mixed state, we may, however, wonder whether the London-Bean model is applicable since a finite change in velocity of flux-lines, which results in the finite change in the vortex electric field, cannot happen instantaneously due to the inertia of flux-lines in the superconductors. Besides, the localized vibration of flux-lines in ac fields may induce a macroscopic vortex electric field in the case of $\boldsymbol{J} < \boldsymbol{J}_c$, which may not be negligible in the superconductor in ac fields of extremely high frequencies, such as microwave frequencies.

To model high-frequency electrodynamic behavior of type II superconductors in the mixed state, we first introduce here a phenomenological relation between the local mean velocity $\mathbf{v}^{(f)}$ of the flux-lines and the Lorentz force $\boldsymbol{J} \times \boldsymbol{B}^{(v)}$ acting on the flux-lines:

$$\mathrm{v}_k^{(f)} = (B^{(v)})^{-2} R_{kl} e_{lpq} J_p B_q^{(v)} \tag{4.4.2}$$

where e_{ijk} denotes the permutation symbol, and $B^{(v)}$ is the magnitude of the vortex magnetic field $\boldsymbol{B}^{(v)}$. R_{kl} is generally a phenomenological coefficient tensor, characterizing effectively the dynamic mobility of the flux-lines, which can be temperature- and field-dependent. Here, we assume that the phenomenological relation (4.4.2) is valid for the type II superconductor in the mixed state, which can be either below or above the critical state. In general, the coefficient tensor R_{kl} could be determined experimentally in the phenomenological model.

To see more clearly the role of the coefficient tensor R_{kl}, let us study the vortex electric field $\boldsymbol{E}^{(v)}$ in the superconductor due to the motion of flux-lines. In the continuum model, the vortex electric field $\boldsymbol{E}^{(v)}$ induced by the motion of flux-lines may be expressed in the following form:

$$E_k^{(v)} = \rho_{kl}^{(v)} J_l \tag{4.4.3}$$

in which $\rho_{kl}^{(v)}$ is the flux-flow resistivity tensor, defined by

$$\rho_{kl}^{(v)} = e_{kmn} e_{lpq} R_{mp} n_q n_n \tag{4.4.4}$$

where $\boldsymbol{n}$ is the unit direction vector of the vortex magnetic field $\boldsymbol{B}^{(v)}$ (that is,, $\boldsymbol{n} = \boldsymbol{B}^{(v)}/B^{(v)}$ with $B^{(v)}$ being the magnitude of the vortex magnetic field).

For isotropic type II superconductors in the mixed state, we have simply $R_{kl} = \rho^{(v)}\delta_{kl}$, and

$$\rho_{kl}^{(v)} = \rho^{(v)}(\delta_{kl} - n_k n_l) \tag{4.4.5}$$

where δ_{kl} is the Kronecker delta. Thus the vortex electric field $\boldsymbol{E}^{(v)}$ in an isotropic type II superconductor can be expressed by

$$\boldsymbol{E}^{(v)} = \rho^{(v)}[\boldsymbol{J} - (\boldsymbol{J} \cdot \boldsymbol{n})\boldsymbol{n}] \tag{4.4.6}$$

This result indicates that the local vortex electric field $\boldsymbol{E}^{(v)}$ vanishes when the local current density vector $\boldsymbol{J}$ is parallel to the local vortex magnetic field $\boldsymbol{B}^{(v)}$ in the model for isotropic type II superconductors in the mixed state. For $\boldsymbol{J}$ being perpendicular to $\boldsymbol{B}^{(v)}$, we have then $\boldsymbol{E}^{(v)} = \rho^{(v)}\boldsymbol{J}$, which shows that the phenomenological coefficient $\rho^{(v)}$ characterizes effectively the flux-flow resistivity of the isotropic type II superconductor in the mixed state.

4.4.2 Possible Effect of Inertia of Flux Lines

To study possible effects of inertia of flux lines [Zhou (1995b)], let us consider an isotropic type II superconductor and ignore possible thermomagnetic effects for simplicity. We investigate first a simple one-dimensional situation in which a single straight flux line moves in a viscous medium under the exertion of the Lorentz force $f_d = J\Phi_o$ per unit length of the flux line, just after the break-up of an elastic spring that connects the flux-line to its pinning center, as shown in Fig. 4.15. In the figure, J is the driving current density and Φ_o is the flux-quantum of the flux line. In such a case, we may write the motion equation for the flux-line in the following form:

$$m_o \frac{dv}{dt} + \eta_o v = f_d \tag{4.4.7}$$

Now, assuming that the flux line has a velocity v_o at the moment that the elastic spring breaks at $J = J_c$, we may introduce an initial condition: $v = v_o$ at $t = 0$ for the differential equation (4.4.7). We suppose that the driving force f_d increases with respect to the time t in the form of $f_d = \Phi_o J = \Phi_o J_c(1+\omega t)$ for $t \geq 0$. Furthermore, we may assume that the initial velocity v_o is smaller than the value: $\Phi_o J_c/\eta_o$, which would be the velocity of the flux line given by Eq.(4.4.7) in the case of ignoring the effect of inertia of the flux line at $J = J_c$. Thus we may solve Eq.(4.4.7) with the given initial condition, and find the following solution:

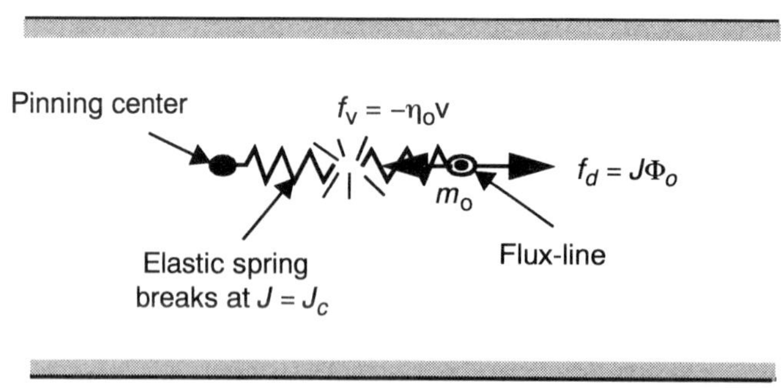

Figure 4.15 A scheme of a moving flux-line just after the break-up of an "elastic spring" connecting the flux line and a pinning center in a type II superconductor.

$$\mathrm{v}(t) = \mathrm{v}^*(t) + (\mathrm{v}_o - \mathrm{v}_o{}^*)exp\left(-\frac{t}{\tau_o}\right), \qquad (t \geq 0) \tag{4.4.8}$$

in which $\mathrm{v}^*(t) - A+Ct$ with the constants A and C given, respectively, by $A = \Phi_o J_c(1-\omega\tau_o)/\eta_o$ and $C = \Phi_o J_c\omega/\eta_o$. $\mathrm{v}_o{}^*$ is given by $\mathrm{v}_o{}^* = \mathrm{v}^*(0) = A$, and τ_o is a time constant defined by $\tau_o = m_o/\eta_o$.

By noting the given form of the driving force f_d, we may write $t=(J/J_c-1)/\omega$. Thus the solution (4.4.8) may also be written in the following form:

$$\mathrm{v}(J) = \mathrm{v}^*(J) + (\mathrm{v}_o - \mathrm{v}_o{}^*)exp\left(\frac{1-J/J_c}{\omega\tau_o}\right), \qquad (J \geq J_c) \tag{4.4.9}$$

It can be seen from Eq.(4.4.9) that the velocity v of the flux line is continuous at the initial state ($J = J_c$), that is, at the moment of break-up of the elastic spring, and it increases continuously and rapidly from its initial value v_o to the value of $\mathrm{v}^*(J)$, at which the flux line moves steadily under the exertion of the driving force for $J > J_c$, provided that the flux line has not yet been trapped by other pinning centers. The time scale for such a transition is shown here to be characterized by the time constant τ_o, called the vortex relaxation time.

The situation discussed above may occur in type II superconductors with, for instance, a dilute distribution of pinning centers. This does not however mean that the distribution of flux lines has to be dilute in the superconductor. Due to interactions among flux lines, it may well be that flux lines move through the superconductor in a form of flux "bundles". In such a case, flux lines may be pinned collectively by individual pinning centers. For these superconductors in ac fields of extremely high frequencies, the time period of the excitation field may be comparable with the time needed for the flux line (or flux bundles) moving from one pinning center to another. In such a case, we have to consider carefully the local vortex dynamics of the flux lines (or flux bundles). When the local pinning force at individual pinning centers dominates the flux-pinning and flux-flow process, we may get the following expression for the flux-flow velocity $\mathbf{v}^{(\mathrm{f})}$ in time-harmonic fields of frequency ω:

$$\mathbf{v}^{(\mathrm{f})} = \frac{\boldsymbol{J} \times \boldsymbol{B}^{(\mathrm{v})}}{\eta(1 + i(\omega\tau_{\mathrm{f}} - \omega_o/\omega))}, \qquad (|\boldsymbol{J}| < J_c) \tag{4.4.10}$$

$$\mathbf{v}^{(\mathrm{f})} = \frac{\boldsymbol{J} \times \boldsymbol{B}^{(\mathrm{v})}}{\eta(1 + i\omega\tau_{\mathrm{f}})} + (\mathbf{v}_c - \mathbf{v}_c{}^*)exp\left(\frac{1-|\boldsymbol{J}|/J_c}{\omega\tau_{\mathrm{f}}}\right), \qquad (|\boldsymbol{J}| \geq J_c) \tag{4.4.11}$$

by analyzing the vortex dynamics of flux lines in the isotropic type II superconductor in the mixed state and accounting for the effect of inertia of flux lines. Here, we have introduced the quantity: $\mathbf{v}_c = \boldsymbol{J}_c \times \boldsymbol{B}^{(\mathrm{v})}/\eta[1+ i(\omega\tau_{\mathrm{f}}-\omega_o/\omega)]$, which is the flux-flow velocity at $\boldsymbol{J} = \boldsymbol{J}_c$ in the presence of the elastic spring force, and the quantity: $\mathbf{v}_c{}^* = \boldsymbol{J}_c \times \boldsymbol{B}^{(\mathrm{v})}/\eta(1+i\omega\tau_{\mathrm{f}})$, which is the flux-flow velocity at $\boldsymbol{J} = J_c$ in the absence of the elastic spring force. Here, τ_f is the vortex relaxation time, defined by the ratio of the effective mass density m_f of flux-lines to the flux-flow viscosity η (i.e., $\tau_f = m_f/\eta$). ω_0 is the depinning frequency defined by $\omega_0 = k/\eta$ with k being the elastic spring constant. $\boldsymbol{B}^{(v)}$ is the vortex magnetic field, and J_c is the critical current density of the type II superconductor.

Thus we may write the complex ac flux-flow resistivity $\rho^{(v)}$ in the following form:

$$\rho^{(v)} = \frac{\rho_v(1 - i(\omega\tau_f - \omega_o/\omega))}{1 + (\omega\tau_f - \omega_o/\omega)^2}, \qquad (|\boldsymbol{J}| < J_c) \tag{4.4.12}$$

$$\rho^{(v)} = \frac{\rho_v(1 - i\omega\tau_f)}{1 + (\omega\tau_f)^2}\left[1 - \frac{(\omega_o/\omega)^2 - \omega_o\tau_f - i\omega_o/\omega}{1 + (\omega\tau_f - \omega_o/\omega)^2} \times \frac{J_c}{|\boldsymbol{J}|} \times exp\left(\frac{1 - |\boldsymbol{J}|/J_c}{\omega\tau_f}\right)\right]$$
$$(|\boldsymbol{J}| \geq J_c) \tag{4.4.13}$$

It can be seen that, by noting the continuity of the complex ac flux-flow resistivity $\rho^{(v)}$ from Eqs.(4.4.12) and (4.4.13), the vortex electric field $\boldsymbol{E}^{(v)}$ given by equation (4.4.6) is continuous at the critical state ($|\boldsymbol{J}| = J_c$) in our model, in contrast to the result of discontinuous vortex electric field at the critical state in the London-Bean model, as shown in Fig. 4.9. The model introduced here is supposed to be valid for the type II superconductor in the mixed state both below the critical state and above the critical state.

We may notice that, in this theoretical model [Zhou (1995b)], besides the critical current density J_c, there are three independent material parameters, ρ_v, ω_0 and τ_f to characterize electrodynamic properties of an isotropic type II superconductor in the mixed state in ac fields. ρ_v is the dc flux-flow resistivity, and ω_0 is the depinning frequency, which separates the low-frequency regime where pinning forces dominate, and the high-frequency regime where viscous forces dominate. τ_f is the vortex relaxation time, characterizing effectively the transition time between the flux-pinned state and the flux-flow state in the type II superconductor in the mixed state. Theoretically, the model is valid when the period of ac fields is longer than the vortex relaxation time τ_f, which was reported to be typically less than one picosecond for conventional (low-T_c) superconductors [Suhl (1965)]. Thus we consider here only cases with $\omega\tau_f < 1$. In general, the material parameters: ρ_v, ω_0 and τ_f can be determined experimentally. Theoretical calculations of these parameters are possible, however, and can be of interest in understanding the basic physics of type II superconductors in the mixed state, especially for high-temperature oxide superconductors. It may be noted that the essential idea presented in this section may be generalized to study anisotropic type II superconductors in the mixed state, details of which will, however, not be discussed here. Instead, we shall show quantitatively how the effect of inertial of flux-lines may influence the $E^{(v)}$–J behavior of an isotropic type II superconductor in the mixed state, and make a comparison of this theoretical model with the London-Bean critical-state model.

For simplicity, let us consider a case in which the current density vector $\boldsymbol{J}$ is perpendicular to the vortex magnetic field $\boldsymbol{B}^{(v)}$ in an isotropic type II superconductor in the mixed state. We plot some numerical results from Eqs.(4.4.6), (4.4.12), and (4.4.13) on the $E^{(v)}$–J behavior of the superconductor in Fig. 4.16, where $E^{(v)}$ denotes the magnitude of the vortex electric field ($E^{(v)} = |\boldsymbol{E}^{(v)}|$). Relevant material parameters are given in the figure. It is shown in Fig. 4.16 that the vortex electric field $E^{(v)}$, predicted by this new model, is continuous at the critical state ($\boldsymbol{J} = \boldsymbol{J}_c$), while the London-Bean model indicates a discontinuous change in the vortex electric field from its zero value jumping to a finite value. In addition, it is shown in Fig. 4.16 that the vortex electric field $E^{(v)}$ is nearly zero for $J < J_c$ at relatively low frequencies (less than 1 GHz), and it almost jumps to its finite value for $J > J_c$. Such an $E^{(v)}$–J behavior is consistent with the result from the London-Bean model with the difference that the values of the vortex electric field $E^{(v)}$ near the critical state is now well-defined mathematically in the new model.

At higher frequencies (~ 10 GHz), we find however that the vortex electric field $E^{(v)}$ is not negligible even for the superconductor with flux lines at the flux-pinned states ($J < J_c$) due to the effect of localized vibration of the flux lines around their pinning centers. The actual value of the vortex electric field depends very much on the depinning frequency ω_o of the superconductor, as shown in the figure. Besides, we may find from Fig. 4.16 that there exists a transition region near the critical state, in which the vortex electric field $E^{(v)}$

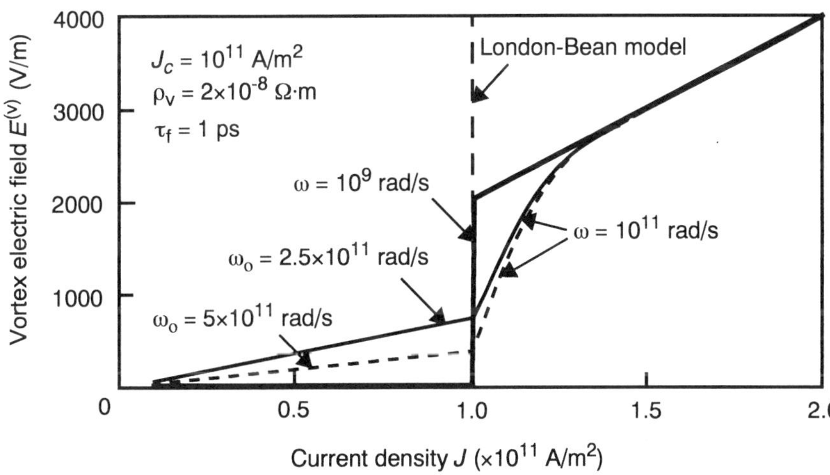

Figure 4.16 $E^{(v)}$–J behavior of a type II superconductor nearby the critical state.

increases rapidly from its value at $J = J_c$ to its value at the entry of a "fully developed" flux-flow state for $J >> J_c$, characterized essentially by the dc flux-flow resistivity ρ_v.

To study the width of the transition region and its relation with the vortex relaxation time τ_f, we also plot some numerical results showing the effect of the vortex relaxation time τ_f on the $E^{(v)}$–J behavior of the superconductor above the critical state ($J > J_c$) in Fig. 4.17. It is shown that the smaller the vortex relaxation time τ_f is, the narrower the transition region is. The same is true for the excitation frequency ω, that is, the lower the frequency ω is, the narrower the transition region is. Such an $E^{(v)}$–J behavior of the type II superconductor at very high frequencies shown by the new model deviates obviously from that predicted by the London-Bean model, especially for some type II superconductors with a relatively large vortex relaxation time τ_f.

It is can be seen that the $E^{(v)}$–J behavior near the critical state ($J = J_c$) is highly nonlinear. The model introduced here is expected to be able to analyze nonlinear electrodynamic behaviors of type II superconductors at high frequencies, and it can also recover the results predicted by the London-Bean critical-state model at low frequencies. Furthermore, it is expected that this new model will be more easily treated in numerical calculations than the London-Bean model since the new model has better mathematical behavior at the critical state. So far, it seems that there are few experimental studies on nonlinear electrodynamic behaviors of type II superconductors in the mixed state at extremely high frequencies. Further efforts are, therefore, required.

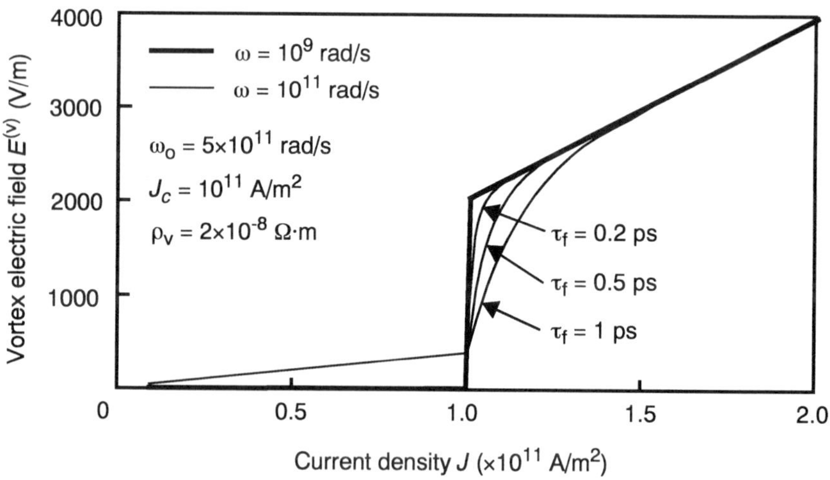

Figure 4.17 Effect of vortex relaxation time τ_f on $E^{(v)}$–J behavior of the superconductor for $J \geq J_c$.

4.4.3 *E–J* Relation for Anisotropic Type II Superconductors in the Mixed State

In this section, we shall study the *E–J* relation of anisotropic type II superconductors in the mixed state. We start from the first generalized London equation for anisotropic type II superconductors in the mixed state, from which we may write the total electric field $\boldsymbol{E}$ in the superconductor by

$$\boldsymbol{E} = \boldsymbol{E}^{(\mathrm{v})} + \mu_o \Lambda \cdot \frac{\partial \boldsymbol{J}^{(s)}}{\partial t} \tag{4.4.14}$$

where μ_0 is the permeability of free space, $\boldsymbol{J}^{(s)}$ is the local supercurrent density vector, and Λ_{kl} is a coefficient tensor characterizing effectively the penetration depths of the anisotropic superconductor. $\boldsymbol{E}^{(\mathrm{v})}$ is the vortex electric field. Taking into account of the effect of the normal conduction current density $\boldsymbol{J}^{(n)}$, we may write the total current density vector $\boldsymbol{J}$ by

$$\boldsymbol{J} = \boldsymbol{J}^{(s)} + \boldsymbol{J}^{(n)} \tag{4.4.15}$$

with Ohm's law for the normal conduction current component:

$$J_k^{(n)} = \sigma_{km}^{(n)} E_m \tag{4.4.16}$$

in a two-fluid model. Here, $\sigma_{km}^{(n)}$ is the normal conductivity tensor of the superconductor.

In such a case, by using Eqs.(4.4.3), (4.4.14)–(4.4.16), we may find the following ***E–J*** relation:

$$E_k + \mu_o \Lambda_{kl} \sigma_{lm}^{(n)} \frac{\partial E_m}{\partial t} = \rho_{kl}^{(\mathrm{v})} J_l + \mu_o \Lambda_{kl} \frac{\partial J_l}{\partial t} \tag{4.4.17}$$

for anisotropic type II superconductors in the mixed state, where $\rho_{kl}^{(\mathrm{v})}$ denotes the flux-flow resistivity tensor, defined in Eq.(4.4.4). For isotropic type II superconductors, Eq.(4.4.17) becomes

$$\boldsymbol{E} + \mu_o \lambda_L^2 \sigma_n \frac{\partial \boldsymbol{E}}{\partial t} = \rho^{(\mathrm{v})} [\boldsymbol{J} - (\boldsymbol{J} \cdot \boldsymbol{n}) \boldsymbol{n}] + \mu_o \lambda_L^2 \frac{\partial \boldsymbol{J}}{\partial t} \tag{4.4.18}$$

where λ_L is the London penetration depth, and σ_n the normal conductivity of the superconductor. In general, Eq.(4.4.17) [or (4.4.18)] is highly nonlinear since the flux-flow resistivity coefficients $\rho_{kl}^{(\mathrm{v})}$ are strongly field-dependent, especially nearby the critical state of the superconductor, as we have shown in the last section. Besides, the penetration depths of type II superconductors in the mixed state may also be field-dependent. It can be seen from Eq.(4.4.17) that both the incomplete Meissner effect ($\Lambda \neq 0$) and the normal conduction fluid effect ($\sigma^{(n)} \neq 0$) are taken into account here in the theoretical model, which accounts for the vortex dynamic effect in addition.

In some cases, linearization of these equations is however possible. An example is to investigate propagation behaviors of microwaves with small ac field amplitudes in hard type II superconductors in strong dc magnetic fields. In such a case where linearization is possible, we may get the following effective constitutive equation from Eq.(4.4.17) for anisotropic type II superconductors in the mixed state in time-harmonic fields:

$$E_m = \rho_{mn}^{(eff)} J_n, \qquad (|\boldsymbol{J}| < J_c) \tag{4.4.19}$$

which has a form similar to Ohm's law. Here, $\rho_{mn}^{(eff)}$ may be called the effective ac resistivity tensor, defined by

$$\rho_{mn}^{(eff)} = [\delta_{km} + i\omega\mu_o\Lambda_{kl}\sigma_{lm}^{(n)}]^{-1}[i\omega\mu_o\Lambda_{kn} + \rho_{kn}^{(v)}] \tag{4.4.20}$$

If the effect of normal conduction current is negligible in some cases, Eq.(4.4.20) becomes simply

$$\rho_{kl}^{(eff)} = i\omega\mu_o\Lambda_{kl} + \rho_{kl}^{(v)} \tag{4.4.21}$$

For isotropic superconductors, we have, then, $\Lambda_{kl} = \lambda_L{}^2\delta_{kl}$ and

$$\rho_{kl}^{(v)} = \frac{\rho_v(1 + i\omega_o/\omega)}{1 + (\omega_o/\omega)^2}(\delta_{kl} - n_k n_l) \tag{4.4.22}$$

where ρ_v is the dc flux-flow resistivity, and ω_o is the depinning frequency of the superconductor.

Furthermore, for anisotropic type II superconductors in the mixed state, we may introduce a critical current density vector $\boldsymbol{J}_c$, defined by $\boldsymbol{J}_c = J_c\boldsymbol{J}/|\boldsymbol{J}|$ with J_c being the magnitude of the critical current density vector $\boldsymbol{J}_c$, which depends on the direction of the local current density vector $\boldsymbol{J}$ and is generally a function of temperature and magnetic fields in the superconductor. A simple phenomenological relation describing the magnitude of the critical current density J_c along any direction of $\boldsymbol{j}$ ($= \boldsymbol{J}/|\boldsymbol{J}|$) may be supposed to be of the following form:

$$\frac{1}{J_c^2} = \frac{j_x^2}{J_{c1}^2} + \frac{j_y^2}{J_{c2}^2} + \frac{j_z^2}{J_{c3}^2} \tag{4.4.23}$$

in which J_{c1}, J_{c2}, J_{c3} denote, respectively, the critical current density components along the principal axes (x, y, z) of the anisotropic superconductor, which may be, respectively, the a-axis, b-axis, and c-axis of high-T_c oxide superconductors. In general, these critical current density components may be determined experimentally in the phenomenological model.

For high-T_c oxide superconductors, it has been observed experimentally that superconducting current flows more easily in the Cu-O (ab-) planes of these superconductors than perpendicular to these planes. The critical current for current flow in the ab-planes is usually much greater than that for current flow perpendicular to these planes, that is, parallel to the c direction. Some experimental data and discussion on the critical current in these superconductors may be found in the work of, for instance, Poole et al. (1995).

4.4.4 Nonlinear Field Equations and Linearization

To analyze electrodynamic behaviors of anisotropic type II superconductors in the mixed state, a set of electrodynamic field equations may be formulated as follows. Generally, we have the following Maxwell's equations:

$$\nabla \times \boldsymbol{E} = -\frac{\partial \boldsymbol{B}}{\partial t} \tag{4.4.24}$$

$$\nabla \times \boldsymbol{H} = \boldsymbol{J} \tag{4.4.25}$$

in which the effect of displacement current in the superconductor is ignored at the magneto-quasistatic approximation. To characterize electrodynamic behaviors of anisotropic type II superconductors in the mixed state, besides the first generalized London equation (4.4.14), we may write the second generalized London equation:

$$\boldsymbol{B} = \boldsymbol{B}^{(\mathrm{v})} - \mu_o \nabla \times (\Lambda \cdot \boldsymbol{J}^{(s)}) \tag{4.4.26}$$

where $\boldsymbol{B}$ is the magnetic induction field in the superconductor. In addition, we have the following vortex continuity equation [Goodman (1966)]:

$$\nabla \times \boldsymbol{E}^{(\mathrm{v})} = -\frac{\partial \boldsymbol{B}^{(\mathrm{v})}}{\partial t} \tag{4.4.27}$$

Rewriting Eq.(4.4.26) as

$$\boldsymbol{B}^{(\mathrm{v})} = \boldsymbol{B} + \nabla \times [\Lambda \cdot (\nabla \times \boldsymbol{B})] - \mu_o \nabla \times [\Lambda \cdot (\sigma^{(n)} \cdot \boldsymbol{E})] \tag{4.4.28}$$

gives the local variation of the vortex magnetic field. Equation (4.4.28) shows how the vortex and total magnetic fields differ due to the incomplete Meissner effect and the presence of the normal conduction fluid. For isotropic homogeneous type II superconductors, Eq.(4.4.28) becomes

$$\boldsymbol{B}^{(v)} = \boldsymbol{B} - \lambda_L^2 \nabla^2 \boldsymbol{B} + \mu_o \lambda_L^2 \sigma_n \frac{\partial \boldsymbol{B}}{\partial t} \tag{4.4.29}$$

if the B dependence of the penetration depth λ_L can be ignored. In such a case, by taking the time derivative of Eq.(4.4.29) and using Eqs.(4.4.27) and (4.4.6), we may get a single nonlinear vector partial differential equation for the total magnetic field $\boldsymbol{B}(\mathbf{x}, t)$ in an isotropic type II superconductor in the mixed state:

$$\frac{\partial \boldsymbol{B}}{\partial t} - \lambda_L^2 \nabla^2 \left(\frac{\partial \boldsymbol{B}}{\partial t} \right) + \mu_o \lambda_L^2 \sigma_n \frac{\partial^2 \boldsymbol{B}}{\partial t^2} = -\nabla \times \left\{ \frac{\rho^{(v)}}{\mu_o} [\nabla \times \boldsymbol{B} - ((\nabla \times \boldsymbol{B}) \cdot \boldsymbol{n})\boldsymbol{n}] \right\} \tag{4.4.30}$$

where the flux-flow resistivity $\rho^{(v)}$ is generally a nonlinear function of the local fields. An equivalent form of this equation has also been derived by Coffey (1992) in the case of ignoring the critical-state effect, and has been used to study the generation of some nth-order harmonics in an isotropic type II superconductor in the mixed state.

For anisotropic type II superconductors in the mixed state, we may, however, find a coupled set of nonlinear vector partial differential equations for the total magnetic field $\boldsymbol{B}(\mathbf{x}, t)$ and the total electric field $\boldsymbol{E}(\mathbf{x}, t)$:

$$\frac{\partial \boldsymbol{B}}{\partial t} + \nabla \times \left[\Lambda \cdot \left(\nabla \times \frac{\partial \boldsymbol{B}}{\partial t} \right) \right] = -\frac{1}{\mu_o} \nabla \times [\rho^{(v)} \cdot (\nabla \times \boldsymbol{B})] + \mu_o \nabla \times \left[\Lambda \cdot \left(\sigma^{(n)} \cdot \frac{\partial \boldsymbol{E}}{\partial t} \right) \right] \tag{4.4.31}$$

$$\boldsymbol{E} + \mu_o \Lambda \cdot \left(\sigma^{(n)} \cdot \frac{\partial \boldsymbol{E}}{\partial t} \right) = \frac{1}{\mu_o} [\rho^{(v)} \cdot (\nabla \times \boldsymbol{B})] + \Lambda \cdot \left(\nabla \times \frac{\partial \boldsymbol{B}}{\partial t} \right) \tag{4.4.32}$$

in which the flux-flow resistivity tensor $\rho^{(v)}$ is defined generally in Eq.(4.4.4). In general, solutions of nonlinear partial differential equations together with proper boundary conditions are much involved and numerical approaches are usually adopted. Efforts in both experiments and numerical analyses are, therefore, required to study nonlinear electrodynamic behaviors of type II superconductors and their possible applications.

In some cases, linearization of the field equations is possible. For instance, we may consider a situation in which a type II superconductor is subject to a large dc magnetic field $\boldsymbol{B}_a$ and a small superimposed ac (time-harmonic) field $\boldsymbol{b}_a$ ($|\boldsymbol{b}_a| << |\boldsymbol{B}_a|$). In such a case, there exist a static magnetic field $\boldsymbol{B}_o$, averaged from the static vortex-magnetic field, and an ac field $\boldsymbol{b}$ in the superconductor. Here, we assume for simplicity that B_o ($>> B_{c1}$) is uniform in the superconductor, and $\boldsymbol{b}$ is so small ($|\boldsymbol{b}| << |\boldsymbol{B}_o|$) that we have everywhere $|\boldsymbol{J}| < J_c$ in the type II superconductor in the mixed state. Thus, at the first order approximation, we may assume that all material parameters with B-dependence are only functions of the uniform static magnetic field $\boldsymbol{B}_o$ in the superconductor. In such a case, by noting the total magnetic field $\boldsymbol{B}(\mathbf{x}, t) = \boldsymbol{B}_o + \boldsymbol{b}(\mathbf{x}, t)$, we may derive the following linearized field equation for anisotropic type II superconductors:

$$\boldsymbol{b} + \nabla \times [\Lambda \cdot (\nabla \times \boldsymbol{b})] = \frac{i}{\omega\mu_o}\nabla \times [\rho^{(v)} \cdot (\nabla \times \boldsymbol{b})] + \mu_o \nabla \times \{\Lambda \cdot [\sigma^{(n)} \cdot (\rho^{(eff)} \cdot (\nabla \times \boldsymbol{b}))]\} \quad (4.4.33)$$

where we have made use of Eq.(4.4.19). In a principal coordinate system, this equation can be much simplified. The resulting linearized field equation has recently been used to analyze some properties of wave propagation in a planar superconducting waveguide, made of anisotropic superconducting materials [Zhou (1994c)].

For isotropic type II superconductors, the linearized form of the field equation (4.4.30) becomes

$$\tilde{\lambda}^2 \nabla^2 \boldsymbol{b} = \boldsymbol{b} + \frac{i\delta_{VC}^2/2}{1 + 2i\lambda_L^2/\delta_n^2}\nabla \times [(\nabla \times \boldsymbol{b}) \cdot \boldsymbol{n}]\boldsymbol{n} \quad (4.4.34)$$

for determining the ac field $\boldsymbol{b}$ in the isotropic superconductor in the mixed state for $|\boldsymbol{J}| < J_c$. Here, $\boldsymbol{n}$ denotes the unit direction vector of $\boldsymbol{B}_o$, that is, $\boldsymbol{n} = \boldsymbol{B}_o/B_o$, and $\tilde{\lambda}$ is the complex penetration depth defined by

$$\tilde{\lambda}^2 = \frac{\lambda_L^2 - i\delta_{VC}^2/2}{1 + 2i\lambda_L^2/\delta_n^2} \quad (4.4.35)$$

where $\delta_n = (2/\mu_o\omega\sigma_n)^{1/2}$ is the normal skin depth, and $\delta_{VC} = (2/\mu_o\omega\sigma_v)^{1/2}$ is the complex skin depth with $\sigma_v = \sigma_{v1} - i\sigma_{v2}$ being the complex flux-flow conductivity. μ_o is the permeability in free space, and ω is the radian frequency, Here, we may consider the B-field dependence of the London penetration depth λ_L by using, for instance, a known relation: $\lambda_L^2(B_o) = \lambda_L^2(0)/(1 - B_o/B_{c2})$ with B_{c2} being the upper critical field of the superconductor. In the next section, we shall show an example of applying the theoretical model [Zhou (1995a)].

4.5 MICROWAVE PROPERTIES OF PLANAR SUPERCONDUCTING WAVEGUIDE IN DC MAGNETIC FIELD

4.5.1 Field Equation and Wave Solution

In this section, we shall formulate the problem of electromagnetic wave propagating in a planar superconducting waveguide in the presence of an applied static and uniform magnetic field $\boldsymbol{B}_o=(B_o, 0, 0)$, as shown in Fig. 4.18. In this case, we may express the total magnetic field by $\boldsymbol{B}(\boldsymbol{x}, t) = \boldsymbol{B}_o(\boldsymbol{x}) + \boldsymbol{b}(\boldsymbol{x}, t)$ with $\boldsymbol{b}$ being the time-varying magnetic field due to the electromagnetic wave propagation in the waveguide. We assume $|\boldsymbol{b}| << |\boldsymbol{B}_o|$. The superconducting film constructing the waveguide is supposed to be an isotropic type-II superconductor, which is in the mixed state under the exertion of the dc magnetic field $\boldsymbol{B}_{\rm o}$ ($B_{c1} << B_{\rm o} << B_{c2}$). We also assume that the continuum approximation of the vortex magnetic field in the superconductor can be expressed by $\boldsymbol{B}_{\rm o}$ at the first-order approximation. Thus the linearized wave equation (4.4.34) for the isotropic type II superconductor in the mixed state can be applied.

We now consider a wave problem for TM modes, which are described by a transverse magnetic field $\boldsymbol{b} = (0, b_y, 0)$ and an electric field $\boldsymbol{E} = (E_x, 0, E_z)$. Thus, by Eq.(4.4.34), the magnetic induction field component in the superconductor, $b_y = b_y(x)\exp(i(\omega t-\beta z))$ with $b_y(x)$ satisfies the following equation:

$$\frac{\partial^2 b_y}{\partial x^2} - \frac{1}{1 - i\delta_{VC}^2/(2\lambda_L^2)}\left(\frac{1}{\lambda_L^2} + i\omega\mu_o\sigma_n + \beta^2\right)b_y = 0 \tag{4.5.1}$$

where β is as yet an unknown quantity, which is, in general, a complex quantity that may be expressed by

$$\beta = \beta_o - i\alpha \tag{4.5.2}$$

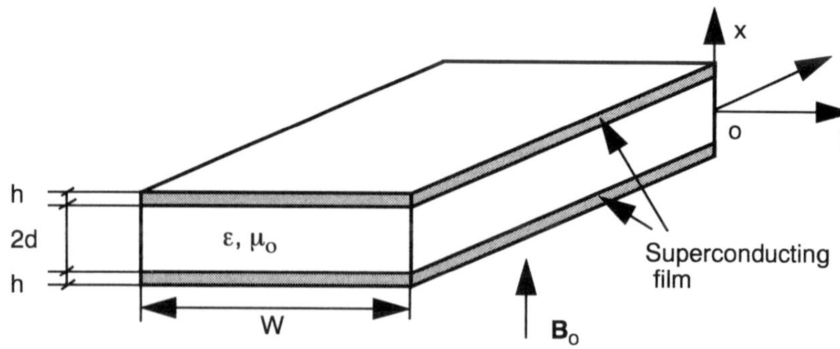

Figure 4.18 A planar superconducting waveguide of finite thickness.

In the expression, β_o is the propagation constant and α the attenuation coefficient. It has been noticed that the second term on the right-hand side of Eq.(4.4.34) is not zero in the case considered. This term shows the effect of the current parallel to the vortex magnetic field along the x-axis direction.

Ignoring the fringe field effect in the case of $W/2d >> 1$, we may find the following solution:

$$b_y = A\frac{\cos\kappa d}{e^{\gamma h} - e^{-\gamma h}}[e^{-\gamma(|x| - d - h)} - e^{\gamma(|x| - d - h)}] \qquad (d < |x| < d + h) \tag{4.5.3}$$

$$b_y = A\cos(\kappa x) \qquad (|x| < d) \tag{4.5.4}$$

in which the parameters γ and κ are given, respectively, by the following equations:

$$\gamma^2 = \frac{1}{1 - i\delta_{VC}^2/(2\lambda_L^2)}\left(\frac{1}{\lambda_L^2} + i\omega\mu_o\sigma_n + \beta^2\right) \tag{4.5.5}$$

and

$$\kappa^2 = \omega^2\mu_o\varepsilon - \beta^2 \tag{4.5.6}$$

where ε $(= \varepsilon' - i\varepsilon'')$ is the permittivity of dielectric medium in the waveguide.

The electric field components, E_x and E_z, can be obtained from Maxwell´s equations. In the case considered, we may notice that the electric field induced by the motion of magnetic flux lines is perpendicular to the direction of the vortex magnetic field. Thus the classical first London equation describing the relation between the superconducting current and the electric field along the direction of the vortex magnetic field is not modified at the approximation considered, where possible flux bending and flux cutting effects are ignored.

By using the interface condition for the continuity of the electric field component E_z at $x = \pm d$, we may obtain the following equation:

$$tg(\kappa d) = \frac{\frac{\omega^2}{c^2}\varepsilon_r\left(1 - i\frac{\varepsilon''}{\varepsilon'}\right)\left(1 - i\frac{\delta_{VC}^2}{2\lambda_L^2}\right)}{-\frac{1}{\lambda_L^2} - i\frac{\omega^2}{c^2}\frac{\sigma_n}{\omega\varepsilon_o}} \times \frac{\gamma}{\kappa} \times \coth(\gamma h) \tag{4.5.7}$$

where $\varepsilon_r = \varepsilon'/\varepsilon_o$ is the relative permittivity of dielectric medium, ε_o is the permittivity of free space, and c is the spped of light in free space. Equation (4.5.7) determines, in principle, the value of the propagation constant β_o and the attenuation coefficient α with the aid of Eqs.(4.5.5) and (4.5.6).

4.5.2 Surface Impedance of Type II Superconductor in the Mixed State

In this section, we shall study the surface impedance of a type II superconductor of finite thickness in the mixed state. By using the solution given in the above section, we may find the surface impedance of the type II superconductor in the mixed state by

$$Z_s = \frac{i\omega\mu_o[1 - i\delta_{VC}^2/(2\lambda_L^2)]}{i\omega\mu_o\sigma_n + 1/\lambda_L^2} \times \gamma\coth(\gamma h) \tag{4.5.8}$$

which is, in general, a complex quantity. We may write $Z_S = R_s + iX_s$, where R_s is the surface resistance, and X_s is the surface reactance. By noting the fact that β^2 is of the same order as $\omega^2\mu_0\varepsilon$ for guide modes and is negligible in the quantity γ from Eq.(4.5.5), as compared with conduction current effect, we may obtain, after some manipulations, the following equations:

$$R_s = \frac{\omega\mu_o[\gamma_1\sinh(2h\gamma_o) + \gamma_o\sin(2h\gamma_1)]}{(\gamma_o^2 + \gamma_1^2)[\cosh(2h\gamma_o) - \cos(2h\gamma_1)]} \tag{4.5.9}$$

for the surface resistance, and

$$X_s = \frac{\omega\mu_o[\gamma_o\sinh(2h\gamma_o) - \gamma_1\sin(2h\gamma_1)]}{(\gamma_o^2 + \gamma_1^2)[\cosh(2h\gamma_o) - \cos(2h\gamma_1)]} \tag{4.5.10}$$

for the surface reactance. Here, we have introduced the complex quantity $\gamma = \gamma_0 + i\gamma_1$ with γ_0 and γ_1 being given, respectively, by

$$\gamma_o = \frac{1}{\sqrt{2}}\sqrt{f_1 + \sqrt{f_1^2 + f_2^2}} \tag{4.5.11}$$

$$\gamma_1 = \frac{1}{\sqrt{2}}\sqrt{\sqrt{f_1^2 + f_2^2} - f_1} \tag{4.5.12}$$

in which

$$f_1 = \frac{\omega\mu_o[\omega\mu_o\sigma_{v1}(\omega\mu_o\sigma_{v2}\sigma_n + \sigma_{v1}/\lambda_L^2)]}{(\omega\mu_o\sigma_{v1})^2 + (\omega\mu_o\sigma_{v2} + 1/\lambda_L^2)^2} - \frac{\omega\mu_o(\omega\mu_o\sigma_{v1}\sigma_n - \sigma_{v2}/\lambda_L^2)(\omega\mu_o\sigma_{v2} + 1/\lambda_L^2)}{(\omega\mu_o\sigma_{v1})^2 + (\omega\mu_o\sigma_{v2} + 1/\lambda_L^2)^2} \tag{4.5.13}$$

$$f_2 = \frac{\omega\mu_o[\omega\mu_o\sigma_{v1}(\omega\mu_o\sigma_{v1}\sigma_n - \sigma_{v2}/\lambda_L^2)]}{(\omega\mu_o\sigma_{v1})^2 + (\omega\mu_o\sigma_{v2} + 1/\lambda_L^2)^2} + \frac{\omega\mu_o(\omega\mu_o\sigma_{v2}\sigma_n + \sigma_{v1}/\lambda_L^2)(\omega\mu_o\sigma_{v2} + 1/\lambda_L^2)}{(\omega\mu_o\sigma_{v1})^2 + (\omega\mu_o\sigma_{v2} + 1/\lambda_L^2)^2} \tag{4.5.14}$$

Equations (4.5.9)–(4.5.14) describe analytically the surface resistance R_s and the surface reactance X_s of a type II superconductor of finite thickness, including the normal skin effect, the London penetration-depth effect, and the vortex dynamic effect. Some well-known results may be recovered for the given formulas. For instance, by letting $\lambda_L \to \infty$, we find $f_1 = 0$ and $f_2 = \omega\mu_o\sigma_n$, which recovers the result for a normal conducting waveguide, as it should be. If the superconductor is in the Meissner state, we let $\delta_{VC} \to 0$, which gives $f_1 = 1/\lambda_L{}^2$ and $f_2 = \omega\mu_o\sigma_n$. The result is then in accordance with that for a superconducting waveguide obtained by for instance Matick (1969). In particular, if the thickness h of the superconductor is very large, we may get the result of Clem and Coffey (1992) by simply letting $h \to \infty$.

So far, we have derived the analytical formula for calculating the surface impedance Z_s for the planar superconducting waveguide of finite thickness in the presence of an applied dc magnetic field $\boldsymbol{B}_o$. To complete the formulation, explicit relations characterizing the field-dependence, frequency-dependence, and temperature-dependence of the material properties, such as the conductivity, the permittivity, and the penetration depth, have to be specified for each material under consideration.

For the complex flux-flow conductivity, $\sigma_v = \sigma_{v1} - i\sigma_{v2}$, we may write

$$\sigma_{v1} = \frac{\sigma_n(T)B_{c2}(T)}{B_o} \quad \text{and} \quad \sigma_{v2} = \frac{2\pi J_c(T, B_o)}{\omega\sqrt{B_o\Phi_o}} \tag{4.5.15}$$

if the Bardeen-Stephen model (1965) and the Gittleman-Rosenblum model (1966) are adopted. Here, σ_n is the normal state conductivity, B_{c2} the upper critical magnetic field, J_c the critical current density, and Φ_o the flux quantum ($\Phi_o = 2.07\times10^{-15}$ Wb). The ratio of the real part and the imaginary part of the complex flux-flow conductivity gives $\sigma_{v2}/\sigma_{v1} = \omega_o/\omega$, which defines a quantity ω_o, called the depinning frequency. The depinning frequency ω_o can be found from Eq.(4.5.15) as

$$\omega_o(T, B_o) = \frac{2\pi J_c(T, B_o)\sqrt{B_o}}{\sigma_n(T)B_{c2}(T)\sqrt{\Phi_o}} \tag{4.5.16}$$

which shows that the depinning frequency ω_o is generally dependent of temperature and magnetic field. Numerically, if we take some values of B_{c2} = 110 T, $J_c = 10^{11}$ A/m^2, $\sigma_n = 5\times10^5$ S/m, and B_o = 1 T, we have $\omega_o \approx 2.5\times10^{11}$ rad/s.

Intuitively, the effect of the depinning frequency ω_o can be seen qualitatively from Fig. 4.19, which plots the numerical simulation for the normalized surface resistance $R_s(\omega)/R_s(\infty)$ with respect to the normalized frequency ω/ω_o. It is shown that, when the excitation frequency ω raised approaches to the depnning frequency ω_o, the surface resistance increases rapidly. Such a behavior was indeed observed experimentally in the work of Gittleman and Rosenblum (1966).

It is noticed that the formulas (4.5.15) and (4.5.16) are valid for describing the effect of vortex dynamics in relatively low-field and low-temperature region where flux creep can be neglected. In the high-field and high-temperature region where flux creep plays an essential role, Eqs.(4.5.15) and (4.5.16) may be modified to include the effect of flux creep. If the Clem-Coffey model (1992) is used, we may write the complex flux-flow conductivity σ_f:

$$\sigma_f = \frac{\sigma_n B_{c2}}{B_o} \times \frac{\zeta + (\omega/\omega_f)^2 - i(1-\zeta)(\omega/\omega_f)}{\zeta^2 + (\omega/\omega_f)^2} \tag{4.5.17}$$

where ζ is the flux-creep factor given by $\zeta = 1/[I_o(\nu)]^2$, and ω_f is a characteristic frequency ($1/\omega_f$ is the characteristic relaxation time), which can be related to the depinning frequency ω_o by

$$\frac{\omega_f}{\omega_o} = \frac{I_1(\nu)I_o(\nu)}{I_o^2(\nu) - 1} \tag{4.5.18}$$

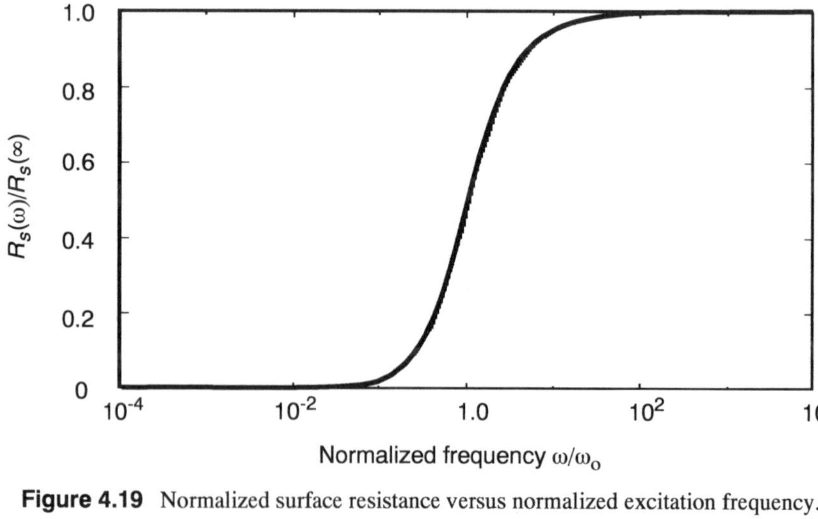

Figure 4.19 Normalized surface resistance versus normalized excitation frequency.

Here, $I_n(\nu)$ is the modified Bessel function of the first kind of order n, and $\nu(B_o,T)=U_o(B_o,T)/2k_BT$ is the normalized height of the energy barrier for thermal activation of a vortex out of a pinning site. A model for $U_o(B_o,T)$ has been used [Clem and Coffey (1992)], which gives: $U_o(B_o,T) = U(1-T/T_{c2})^{3/2}/B_o$ with U = 0.15 eV·T [T_{c2} is defined as the temperature at which $B_o = B_{c2}(T)$]. It can be seen that when the pinning energy is sufficiently high (or the temperature is sufficiently low) so that the normalized barrier height obeys $\nu(B_o,T) >> 1$, the flux-creep factor ζ becomes exponentially small, $\zeta << 1$. In this case, the flux-creep effect can be ignored, and the complex flux-flow conductivity σ_f and the characteristic frequency ω_f become $\sigma_f = \sigma_v$, and $\omega_f = \omega_o$, as we may expect. Some numerical results on the surface impedance of type II superconductors in the mixed state has also been reported in the work of Clem and Coffey (1992), where the effect of flux creep is taken into account though the effect of thickness of the superconductor is not studied.

4.5.3 DC Magnetic Field Effect on Wave Attenuation and Dispersion

It is known that dispersion of waves in a waveguide may cause the spread of a signal pulse propagating in the waveguide. The pulse spreading reduces the bandwidth and limits information capacity of the waveguide. In addition, the quality of speech and data signals comprising a complex mixture of pure, single-frequency components may be degraded after passing through the waveguide due to different attenuation of the signals at different frequencies. The effect of the tonal degradation of the received signals is known as the "attenuation distortion" or "frequency attenuation distortion". The study of the dispersion and attenuation distortion properties of superconducting waveguides is, therefore, of much practical interest. In this section, we shall study the effect of dc magnetic field on attenuation and dispersion of electromagnetic wave propagating in the planar superconducting waveguide of finite thickness. In principle, we may find the propagation constant β_o and the attenuation coefficient α from Eq.(4.5.7) with the aid of Eqs.(4.5.5) and (4.5.6). However, since Eq.(4.5.7) is transcendental containing complex quantities, it is not easy to solve analytically in the general case. For some special cases of practical interest, approximate solutions may be obtained. By noting that β^2 is near the value, $\omega^2\mu_o\varepsilon$, found for the ideal case, we may derive at the first-order approximation the following expressions for the propagation constant β_o and for the attenuation coefficient α of the planar superconducting waveguide:

$$\beta_o = \frac{1}{\sqrt{2}}\sqrt{a_1 + \sqrt{a_1^2 + a_2^2}} \tag{4.5.19}$$

$$\alpha = \frac{1}{\sqrt{2}}\sqrt{-a_1 + \sqrt{a_1^2 + a_2^2}} \tag{4.5.20}$$

in which

$$a_1 = \frac{\omega\varepsilon'}{d}\left(-R_s\frac{\varepsilon''}{\varepsilon'} + \omega\mu_o d + X_s\right) \tag{4.5.21}$$

$$a_2 = \frac{\omega\varepsilon'}{d}\left[R_s + (\omega\mu_o d + X_s)\frac{\varepsilon''}{\varepsilon'}\right] \tag{4.5.22}$$

where R_s and X_s are the surface resistance by Eq.(4.5.9) and surface reactance by Eq.(4.5.10).

Shown in Fig. 4.20 and Fig. 4.21 are some numerical results of the effect of the dc magnetic-field B_0 on the wave attenuation and dispersion properties of the planar waveguide. In the numerical calculation (including those given in the next section), some commonly used models are taken:

$$J_c(T) = \frac{\kappa_{po}(1-\Theta^2)^2}{2\pi\sqrt{B_o\Phi_o}} \tag{4.5.23}$$

$$B_{c2}(T) = \frac{(1-\Theta^2)B_{c2}(0)}{(1+\Theta^2)} \tag{4.5.24}$$

$$\sigma_n(T) = \sigma_N(T)\Theta^4\left\{1-(1-\Theta^4)\left[1-\frac{B_o}{B_{c2}(T)}\right]\right\} \tag{4.5.25}$$

with $\sigma_N(T) = \sigma_0/(1+\sigma_0\Gamma T)$; and the field-dependent penetration depth:

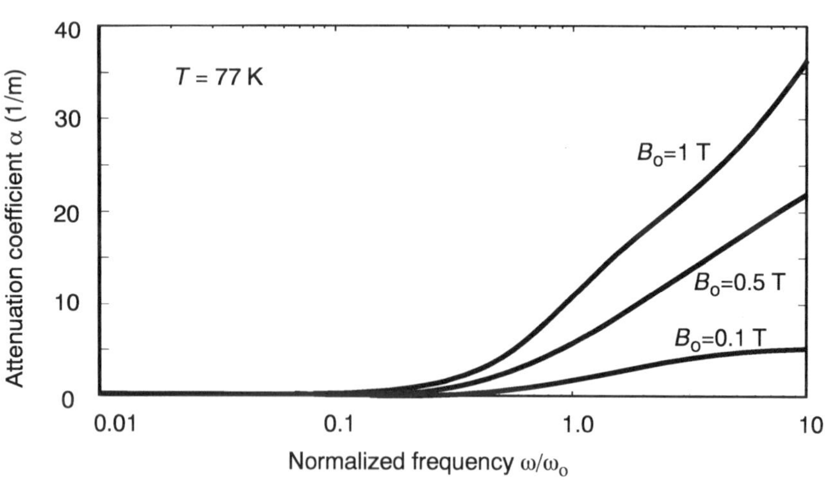

Figure 4.20 Attenuation coefficient α versus ω/ω_0 at different dc magnetic fields.

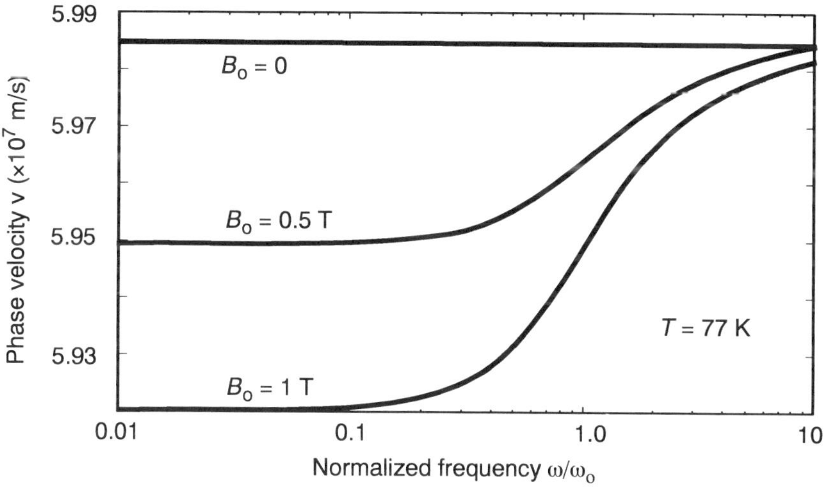

Figure 4.21 Phase velocity v versus ω/ω_o at different dc magnetic fields.

$$\lambda_L = \frac{\lambda_o}{\sqrt{(1-\Theta^4)[1-B_o/B_{c2}(T)]}} \tag{4.5.26}$$

Here, Θ is the reduced temperature ($\Theta = T/T_c$). In particular, in the numerical calculations (including those given in the next section), we take the following data: $\kappa_{po} = 2.1\times10^4$ N/m^2, $B_{c2}(0) = 110$ T, $\sigma_o = 5\times10^5$ S/m, $\Gamma = 1.1\times10^{-8}$ Ω·m/K, $\lambda_o = 140$ nm, and $T_c = 91$ K. (These material parameters are on the order of those for $YBa_2Cu_3O_{7-x}$, but the particular values are chosen to be illustrative rather than to describe any particular material sample.) In addition, dielectric material parameters are taken to be $\varepsilon'/\varepsilon_o = 25$ and $\varepsilon''/\varepsilon' = 10^{-5}$, which are typical values for $LaAlO_3$. The thickness 2d of dielectric medium is taken to be 2d = 0.1 mm, and the thickness h of the superconducting film is taken to be h = 1 μm, unless we give specific indication.

It is shown by Fig. 4.20 and Fig. 4.21 that both the attenuation and the dispersion of microwaves propagating in the superconducting planar waveguide exhibit clear increases for wave frequencies near the depinning frequency, especially at high dc magnetic field. In Fig. 4.21, it is also shown that the phase velocity is reduced in the presence of dc magnetic field in the frequency range considered. Thus, in order to reduce attenuation and dispersion of some extremely high-frequency electromagnetic waves propagating in the superconducting waveguide, it would be very desirable to increase the depinning frequency ω_o, which could be achieved by, for instance, increasing the critical current density of the superconductor, as shown by Eq.(4.5.16).

4.5.4 Magnetic Field Dependence of Quality Factor *Q* and Resonant Frequency

Many applications of microwave devices, such as resonators and filters, require high quality factor Q. The very low surface impedance of superconductors, as compared with normal conductors, provides a novel way of achieving very high Q for a variety of superconducting microwave devices. In this section, we shall analyze the effect of dc magnetic field on the quality factor Q and resonant frequency of superconducting resonators made of planar superconducting waveguide of finite length. We shall study the quality factor Q by

$$Q = \frac{\beta_o}{2\alpha} \tag{4.5.27}$$

Here, we have ignored radiation loss, but have included both conduction loss and dielectric loss in the superconducting planar waveguide resonator. Thus, by Eqs.(4.5.19) and (4.5.20), the quality factor Q can be calculated. Some numerical results are shown in Fig. 4.22 and Fig. 4.23.

Shown in Fig. 4.22 is the dependence of Q on dc magnetic field B_0 at a number of temperatures. It is shown that Q decreases with increasing dc magnetic field and with increasing temperature. Shown in Fig. 4.23 is the dependence of Q on the thickness h of superconducting films of the waveguide. It is shown that Q increases with increasing the film thickness h up to a few microns, and then becomes independent of h for further increase of the film thickness h. The increase of Q with h is not negligible as we may see. It is, therefore, of practical interest to choose properly the film thickness h in designing superconducting waveguide resonators. In addition, a simple

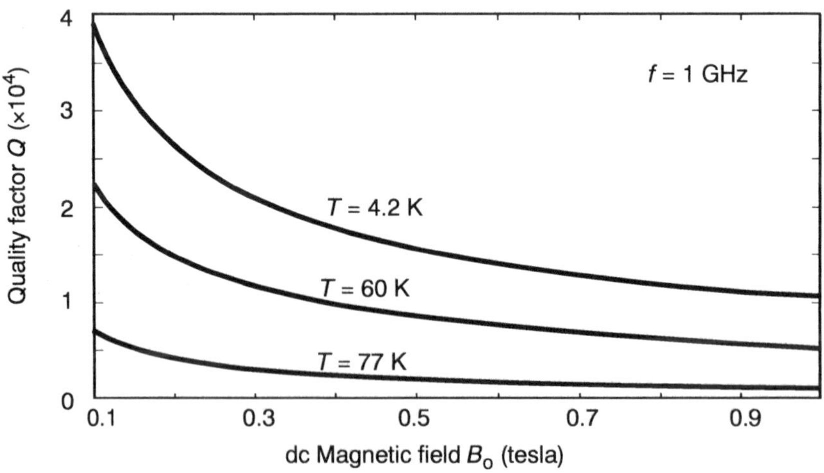

Figure 4.22 Dependence of Q on dc magnetic field at different temperatures.

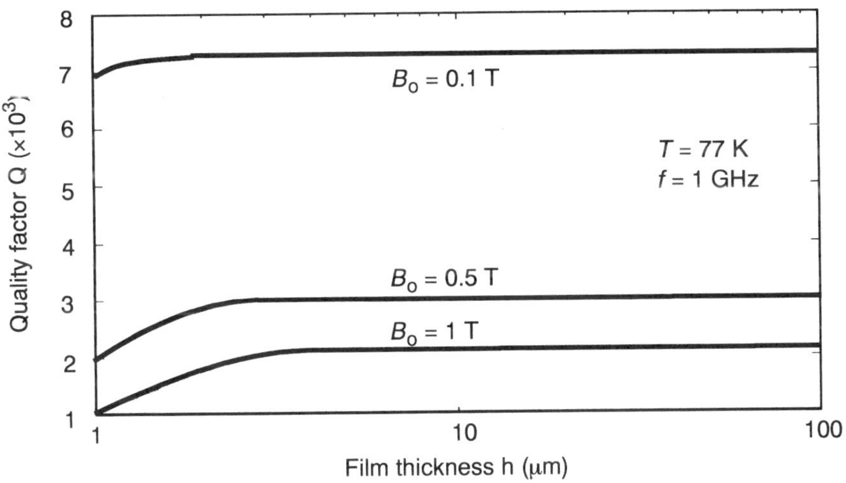

Figure 4.23 Dependence of Q on film thickness h at different dc magnetic fields.

calculation shows that the quality factor Q for a planar normal conducting waveguide resonator is only around 150 for copper films with $\sigma_n = 5\times10^8$ S/m at 77 K. This value of Q is much lower than that for the corresponding superconducting waveguide resonator at 77 K, even in the presence of dc magnetic field of $B_0 = 1$ tesla in the example considered.

Let us now study the problem on the change of resonant frequency of the planar superconducting waveguide in the presence of dc magnetic field B_0. With the aid of Eq.(4.5.19), we may find the relative change of resonant frequency by

$$-\frac{\Delta f}{f_o} = 1 - \frac{\beta_o(B_o = 0)}{\beta_o(B_o)} \tag{4.5.28}$$

where f_0 denotes the resonant frequency in the absence of dc magnetic field ($B_0 = 0$), and $\Delta f = f - f_0$ is the change of the resonant frequency in the presence of the dc magnetic field B_0.

Shown in Fig. 4.24 is the numerical result on the dependence of the relative change of resonant frequency on the dc magnetic field. A nearly linear dependence of the relative change of resonant frequency on the applied dc magnetic field is found. Figure 4.24 also shows that the relative change of resonant frequency increases with increasing temperature.

We may notice that the result shown here is different from those given by, for instance, Lam et al. (1992) on the dependence of the relative change of resonant frequency on rf magnetic fields in a superconducting stripline resonator, where a nonlinear dependence is observed for some relatively high-

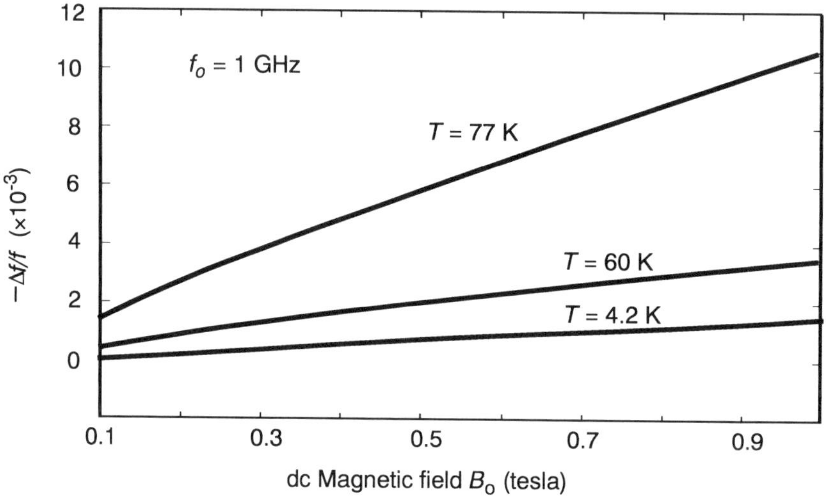

Figure 4.24 Dependence of relative change of resonant frequency on dc magnetic field at different temperatures.

power microwave applications. This is due to the fact that the rf magnetic field considered there is generated by the rf current flowing in the stripline. This results generally in a nonlinear problem, in which we may also find that the surface resistance R_S shows a nonlinear-dependence on the rf magnetic field [Oates et al. (1992)]. In the case of the dc magnetic field, which is generated by external sources, a linearization of the problem is possible if the rf magnetic field induced by microwaves in the waveguide is much smaller than the dc magnetic field, as in the case considered here.

4.6 THERMOMAGNETOELECTRIC EFFECTS IN TYPE II SUPERCONDUCTORS IN THE MIXED STATE

4.6.1 Formulation of Thermomagnetoelectric Effects in Type II Superconductors

Thermomagnetoelectric effects are of both practical and theoretical interest in many respects. In type II superconductors in the mixed state, the flux-flow Hall effect and thermomagnetic effects have been observed since the 1960s. Some theoretical studies on the thermomagnetoelectric effects have also been made during the past 30 years. However, most of these studies deal with isotropic superconductors [Solomon and Otter (1967), Caroli and Maki (1967), Kim and Stephen (1969)]. Recently, the strongly anisotropic properties shown in new high-temperature oxide superconductors have stimulated researches on the modeling of anisotropic effects in type II superconductors. Although the Ginzburg-Landau equations with a phenomenological mass tensor may be used to describe the anisotropic type II superconductor in the mixed state near B_{c2}, the Ginzburg-Landau equations cannot be solved in closed form due to their nonlinearity for the intermediate field $B_{c1} << B << B_{c2}$. Such an intermediate field region is, however, of particular interest for recently discovered high-temperature oxide superconductors due to their small lower critical field and large upper critical field, as shown in Table 4.3, in comparison with the conventional low-T_c metallic Nb superconductor. To study the behavior of type II superconductors in the intermediate field region, the London approximation has been used in modeling type II superconductors in the mixed state, where the vortex dynamic effect is taken into account with the aid of a self-consistent approach, as we have shown in above sections.

In this section, we shall introduce a theoretical model for studying anisotropic type II superconductors in the mixed state, in which the anisotropic effect, the vortex dynamic effect, and the thermomagnetoelectric effects can all be taken into account self-consistently [Zhou (1995c)]. Essentially, the theoretical model proposed here is phenomenological and is based on London's theory with the modification of the presence of the vortex magnetic field.

It is known that when magnetic field penetrates into the interior of a type II superconductor, a vortex structure is formed about each quantized unit of flux. At the microscopic level, a flux line consists of superelectrons moving with a certain density and velocity distribution around the centre of the flux line. This means that forces on a flux line will actually be experienced by the electrons and can, therefore, be caused by electric and/or magnetic fields. Under conditions of non-isothermal flux motion, a force due to temperature gradient may also act on the flux line. For dynamic driving forces, flux lines may move around their equilibrium sites due to pinning forces acting on the flux-lattice

Table 4.3 Basic Values of Some Type II Superconductors

Material Properties	YBaCuO	TlBaCaCuO	Nb
Transition temperature T_c	up to 95 K	up to 130 K	9.2 K
Coherence length ξ_{ab}	1.5 nm	3.0 nm	39 nm
Coherence length ξ_c	0.2 nm	0.1 nm	—
Penetration depth λ_{ab}	150 nm	200 nm	50 nm
Lower critical field B_{c1}	10–100 mT	10 mT	0.13 T
Upper critical field B_{c2}	100–200 T	60 T	0.3 T

which has certain rigidity and elasticity. At the macroscopic level, we are dealing with phenomena that can be described in terms of average macroscopic fields and currents. In a continuum model, we may write the following equation of motion for the flux-lattice continuum in an anisotropic type II superconductor in the mixed state:

$$m_{kl}^{(v)}\frac{\partial^2 w_l}{\partial t^2} = -\eta_{kl}\frac{\partial w_l}{\partial t} - K_{kl}w_l + \varepsilon_{kpq}J_pB_q^{(v)} - S_{kl}\frac{\partial T}{\partial x_l} + Q_vE_k^{(v)} \tag{4.6.1}$$

where $m_{kl}^{(v)}$ is a coefficient tensor characterizing effectively the anisotropic mass density associated with the flux-lattice continuum, which may be assumed to be a symmetric and positive-definite tensor. $\boldsymbol{w}$ is the vortex displacement vector, measured from an equilibrium pinning site in the medium. η_{kl} is the flux-flow viscosity tensor for the anisotropic superconductor. Phenomenologically, the flux-flow viscosity tensor η_{kl} is supposed to satisfy the Onsager reciprocity theorem [De Groot and Mazur (1962)]. The vortex magnetic field $\boldsymbol{B}^{(v)}$ and the current density vector $\boldsymbol{J}$ are both averaged quantities over microscopic fields and current distributions in the superconductor. In Eq.(4.6.1), we have adopted a simplified model [Gittleman and Rosenblum (1966), Coffey and Clem (1991)], in which, for the case of local vibration of flux lines without global flux-flow, the effects of the pinning force and the deformation of the flux-lattice are modeled by a simple restoring force of the form, $-K_{kl}w_l$ for the anisotropic type II superconductor, where $\boldsymbol{K}$ is the spring coefficient tensor, which may phenomenologically be supposed to be a symmetric and positive-definite tensor.

In Eq.(4.6.1), besides the well-known Lorentz-type of magnetic driving force acting on flux lines, which is associated with the electric current and represented by the third term on the right-hand side of this equation, there are two additional driving forces, the thermal driving force and the electric driving force, shown, respectively, by the fourth term and the fifth term on its right-hand side. The thermal driving force represents an effective force on flux lines due to a temperature gradient when the superconductor is in a non-isothermal state. The presence of the thermal driving force on flux lines because of a temperature gradient may be explained qualitatively by considering that the degree of disorder in the vortex system is larger at higher temperatures and the entropy can be increased by flux lines moving toward the colder part of the superconductor [Ullmaier (1975)].

Thus this thermal force on flux lines will tend to move the flux lines from hotter to colder regions. Here, S_{kl} may be expressed as $S_{kl} = S_fD_{kl}$ by introducing a quantity S_f, characterizing effectively the mean entropy of transport per unit volume associated with the displacement of the vortex line. Should S_f be a function of the vortex magnetic field, we assume the relation: $S_f(\boldsymbol{B}^{(v)}) = S_f(-\boldsymbol{B}^{(v)})$ is satisfied. The normalized coefficient tensor D_{kl} is supposed to be symmetric and positive definite. Thus D_{kl} is diagonal with $D_1D_2D_3 = 1$ if the reference frame chosen is aligned with the principal axes.

For isotropic superconductors, we have simply $D_{kl} = \delta_{kl}$ with δ_{kl} being the Kronecker delta. Here, we have noticed the comment given by Hake (1968) for isotropic superconductors in the mixed state, which stated that no thermal force on the flux line can exist perpendicular to the temperature gradient because such a force would permit extraction of work with no heat transported in violation of the second law. Independent of possible arguments on whether the statement together with its assumptions used is still valid for anisotropic superconductors in the mixed state, the assumption that we have made on D_{kl}, being a symmetric and positive-definite tensor, excludes the existence of the thermal force perpendicular to the temperature gradient in the principal coordinate system. However, we have allowed the possibility of that the thermal force could, generally, be different along different principal axes of the anisotropic superconductor. Should the thermal force be orientationally independent from temperature gradient, D_{kl} is simply the Kronecker delta.

The electric driving force $Q_v \boldsymbol{E}^{(v)}$ on flux lines introduced by the fifth term on the right-hand side of Eq.(4.6.1) is closely related to the so-called Magnus force discussed often in the literature of physics [see, e.g., Bardeen and Stephen (1965), Wang and Ting (1992)]. Here, we adopt a phenomenological approach and assume that Q_v is a phenomenological parameter, which may be called the effective charge density for the flux lines. This phenomenological parameter Q_v can be positive or negative. If Q_v is negative, we may say that the effective charge of the flux lines is electron-like, while if Q_v is positive, the effective charge of the flux lines is hole-like. Should Q_v be a function of the vortex magnetic field, we assume that the relation: $Q_v(\boldsymbol{B}^{(v)}) = Q_v(-\boldsymbol{B}^{(v)})$ is satisfied. Here, $\boldsymbol{E}^{(v)} = \boldsymbol{B}^{(v)} \times (\partial \boldsymbol{w}/\partial t)$ is the vortex electric field induced by the motion of the flux lines. For simplicity, we have ignored the possible effect of flux creep in the formulation.

In many cases, physical phenomena of superconductors can be analyzed with the aid of knowledge gained from studying the behavior of the superconductor in time-harmonic fields ($\sim e^{i\omega t}$). In time-harmonic fields, we may obtain from Eq.(4.6.1) the following relation:

$$w_k = \frac{1}{i\omega(B^{(v)})^2} R_{kl}\left(e_{lpq} J_p B_q^{(v)} - S_{lp}\frac{\partial T}{\partial x_p}\right) \tag{4.6.2}$$

in which ω is the radian frequency, e_{ijk} is the permutation symbol, and R_{kl} is defined by

$$R_{kl} = \left\{\frac{1}{(B^{(v)})^2}[\eta_{kl} - i(K_{kl} - \omega^2 m_{kl}^{(v)})/\omega] - \frac{Q_v}{B^{(v)}} e_{kpl} n_p\right\}^{-1} \tag{4.6.3}$$

where $\boldsymbol{n}$ is the unit direction vector of the vortex magnetic field ($\boldsymbol{n} = \boldsymbol{B}^{(v)}/B^{(v)}$), and $B^{(v)}$ is the magnitude of the vortex magnetic field. Here, a tensor $(C_{kl})^{-1}$ is defined as the inverse form of the tensor C_{kl} such that $C_{mk}(C_{kl})^{-1} = \delta_{ml}$. By

Eq.(4.6.2), the flux-flow velocity **v** can be given by

$$v_k = \frac{\partial w_k}{\partial t} = \frac{1}{(B^{(v)})^2} R_{kl}\left(e_{lpq} J_p B_q^{(v)} - S_{lp}\frac{\partial T}{\partial x_p}\right) \tag{4.6.4}$$

in time-harmonic fields.We may now introduce a flux-flow resistivity tensor $\rho_{kl}^{(v)}$ and a flux-flow thermoelectric coefficient tensor $\Pi_{kl}^{(v)}$ for anisotropic type II superconductors in the mixed state, defined in the following expression:

$$E_k^{(v)} = \rho_{kl}^{(v)} J_l + \Pi_{kl}^{(v)} \frac{\partial T}{\partial x_l} \tag{4.6.5}$$

By Eq.(4.6.4) and $\boldsymbol{E}^{(v)} = \boldsymbol{B}^{(v)} \times \mathbf{v}$, we may find the following expressions:

$$\rho_{kl}^{(v)} = e_{kmn} e_{lpq} R_{mp} n_q n_n \tag{4.6.6}$$

and

$$\Pi_{kl}^{(v)} = \frac{S_f}{B^{(v)}} e_{kmn} R_{mp} D_{pl} n_n \tag{4.6.7}$$

It can be shown that the flux-flow resistivity tensor $\rho_{kl}^{(v)}$ satisfy the Onsager reciprocity theorem in the presence of the vortex magnetic field, that is,

$$\rho_{kl}^{(v)}(\boldsymbol{B}^{(v)}) = \rho_{lk}^{(v)}(-\boldsymbol{B}^{(v)}) \tag{4.6.8}$$

which is in accordance with our earlier assumptions about η_{kl}, K_{kl}, and $m_{kl}^{(v)}$, which are all supposed to satisfy the Onsager reciprocity theorem. In general, $\rho_{kl}^{(v)}$ can be separated into a symmetric part and an antisymmetric part.

It is well known that the antisymmetric part of a resistivity tensor characterizes the Hall effect. The result thus implies that the electric driving force term ($Q_v \neq 0$) is not the only cause that may induce the flux-flow Hall effect in the anisotropic type II superconductor in the mixed state. For isotropic superconductors, the introduction of the electric driving force on flux-lines may be reasonable and sufficient to account for the flux-flow Hall effect. However, for anisotropic superconductors, the anisotropic flux-flow Hall effect observed recently in some high-temperature oxide superconductors indicates the possible existence of additional mechanisms, relevant to, most probably, the field-dependent anisotropic viscous flux-flow in the anisotropic type II superconductor in the mixed state. Obviously, the anisotropic flux-flow Hall effect could be modeled with the aid of this phenomenological model since the antisymmetric part of the flux-flow resistivity tensor has, in general, three independent components. Further studies based on microscopic theories and proper experiments are, however, needed to clarify the real physical mechanism that causes the flux-flow Hall effect. In the phenomenological model, we can only give some rather preliminary arguments.

To model electrodynamics of anisotropic type II superconductor in the mixed state, we shall use the Maxwell equations (4.4.24) and (4.4.25), and the second modified London equation:

$$\mu_o e_{kpq} \frac{\partial}{\partial x_p}(\Lambda_{qm} J_m) = B_k^{(\mathrm{v})} - B_k \tag{4.6.9}$$

and the first modified London equation:

$$\mu_o \frac{\partial}{\partial t}(\Lambda_{km} J_m) = E_k - E_k^{(\mathrm{v})} \tag{4.6.10}$$

together with the vortex continuity equation (4.4.27). Here, μ_o is the permeability of free space, $\boldsymbol{J}$ is the supercurrent density, $\boldsymbol{B}$ is the magnetic induction field, and $\boldsymbol{B}^{(\mathrm{v})}$ is the local vortex magnetic field which is, in general, not equal to the magnetic induction field $\boldsymbol{B}$. The coefficient tensor Λ_{kl} in Eq.(4.6.9) characterizes effectively the penetration depths of the anisotropic superconductor. In order to show more clearly the effect of flux-motion on the thermomagnetoelectric effects in type II superconductors in the mixed state, we ignore here the possible effect of normal conduction current. In practice, this approximation may be valid if we are limited to cases where the temperature of a superconductor is not close to the critical temperature and the frequencies involved are not close to the gap frequency of the superconductor.

Furthermore, to describe the heat conduction phenomenon, we have

$$\rho_M C_{\mathrm{v}} \frac{\partial T}{\partial t} = -\nabla \cdot \boldsymbol{q} + Q_h \tag{4.6.11}$$

which is the heat conduction equation. Here, ρ_M denotes the mass density of the superconductor, C_{v} is the specific heat per unit mass, $\boldsymbol{q}$ is the heat current density, and Q_h is the intensity of heat source per unit volume, which may be due to Joule's heat or other heat sources. The heat current density $\boldsymbol{q}$ may be separated into two parts here. One part of the heat current density is due to the flux motion, $q_k^{(\mathrm{f})} = TS_{kl}\mathrm{v}_l^{(\mathrm{f})}$ since the entropy density within the normal core of a flux line is larger than that in the surrounding superconducting phase. The another part of the heat current density, $q_k^{(T)} = -\kappa_{kl}^{(o)}\partial T/\partial x_l$, which is associated with temperature gradient. Here, $\kappa_{kl}^{(o)}$ is the thermal conductivity tensor of the anisotropic superconductor in the absence of flux-motion. The thermal conductivity tensor should satisfy the Onsager reciprocity theorem in the presence of magnetic fields. With the aid of above equations, we could, in principle, derive a complete set of self-consistent field equations, which is generally nonlinear. In what follows, we shall introduce a linearized version of this theory when nonlinear effects can be ignored.

4.6.2 Linearized Model for Thermomagnetoelectric Effects

In this section, we shall consider a homogeneous type II superconductor, which is initially in an applied dc magnetic field $\boldsymbol{B}_0$ and a constant temperature T_0 such that the superconductor is in the mixed state. The magnitude of the applied magnetic field B_o is supposed to be higher than the lower critical field B_{c1}, but much less than the upper critical field B_{c2} so that the London approximation can be used. We shall study a situation in which the superconductor is then subject to small time-varying electromagnetic and thermal loadings. We would like to investigate the thermomagnetoelectric response of the superconductor in the mixed state.

To start with, we write the total magnetic induction field by $\boldsymbol{B} = \boldsymbol{B}_o + \boldsymbol{b}$ and the total temperature $T = T_0 + \theta$ in the superconductor. Here, $\boldsymbol{b}$ and θ are, respectively, the small perturbed time-varying magnetic induction field ($|\boldsymbol{b}| << |\boldsymbol{B}_0|$) and the small perturbed time-varying temperature field ($|\theta| << T_0$) to be determined. With these considerations in mind, we may obtain from the vortex continuity equation (4.4.27) the following relation at the linear approximation:

$$\boldsymbol{B}^{(\text{v})} = \boldsymbol{B}_o - \nabla \times (\boldsymbol{B}_o \times \boldsymbol{w}) \tag{4.6.12}$$

$$\boldsymbol{E}^{(\text{v})} = -\frac{\partial \boldsymbol{w}}{\partial t} \times \boldsymbol{B}_o \tag{4.6.13}$$

which shows that the vortex magnetic field $\boldsymbol{B}^{(\text{v})}$ in the superconductor deviates from the applied magnetic field $\boldsymbol{B}_0$. Such a deviation is usually small if B_o is not too close to B_{c1}. Thus, as the first-order approximation, we may assume that all material properties of the superconductor involved are only functions of the applied dc magnetic field $\boldsymbol{B}_0$ and the constant temperature T_0, which are supposed to be uniform in the superconductor.

By Eqs.(4.6.5) and (4.6.10), we may find the following relation at the linear approximation:

$$E_k = \rho_{kl} J_l + \Pi_{kl}^{(\text{v})} \frac{\partial \theta}{\partial x_l} \tag{4.6.14}$$

where ρ_{kl} is the effective complex resistivity tensor, defined by

$$\rho_{kl} = i\omega\mu_o \Lambda_{kl} + \rho_{kl}^{(\text{v})} \tag{4.6.15}$$

and $\Pi_{kl}^{(\text{v})}$ is the flux-flow thermoelectric coefficient tensor, defined by Eq.(4.6.7) with $B^{(\text{v})}$ being replaced here by B_0 at the linear approximation.

Furthermore, by noting $q_k^{(\text{f})} = TS_{kl} v_l$, we may obtain

$$q_k = T_o \Gamma_{kl}^{(v)} J_l - \kappa_{kl} \frac{\partial \theta}{\partial x_l} \tag{4.6.16}$$

where κ_{kl} is the effective complex thermal conductivity tensor, defined by

$$\kappa_{kl} = \frac{T_o S_f^2}{B_o^2} D_{km} R_{mn} D_{nl} + \kappa_{kl}^{(o)} \tag{4.6.17}$$

and $\Gamma_{kl}^{(v)}$ is defined by

$$\Gamma_{kl}^{(v)} = -\frac{S_f}{B_o} e_{lmn} R_{pm} D_{kp} n_n \tag{4.6.18}$$

where R_{pm} is defined by Eq.(4.6.3) with $B^{(v)}$ being replaced here by B_o at the linear approximation. It can be shown that these coefficient tensors satisfy the proper parity relations: $\Gamma_{kl}^{(v)}(\boldsymbol{B}_o) = \Pi_{lk}^{(v)}(-\boldsymbol{B}_o)$, $\rho_{kl}(\boldsymbol{B}_o) = \rho_{lk}(-\boldsymbol{B}_o)$, and $\kappa_{kl}(\boldsymbol{B}_o) = \kappa_{lk}(-\boldsymbol{B}_o)$, which may be expected physically for the derived phenomenological constitutive equations (4.6.14) and (4.6.16) to describe effectively the properties of the anisotropic type II superconductor in the mixed state in time-harmonic fields. The obtained set of effective constitutive equations (4.6.14) and (4.6.16) are quite general, which include the anisotropic effect, the vortex dynamic effect, the flux-flow Hall effect, and the flux-flow galvanomagnetic and thermomagnetic effects in anisotropic type II superconductors in the mixed state.

To show illustratively the relevance of the thermomagnetoelectric effects to the properties of flux-lattice continuum, we shall now consider an isotropic type II superconductor in the mixed state as a special example. We let the applied dc magnetic field $\boldsymbol{B}_o$ be along the x_3-axis of a chosen rectangular Cartesian coordinate system (x_1, x_2, x_3). We may find from Eq.(4.6.3)

$$[R_{kl}] = \begin{bmatrix} \dfrac{\sigma_v}{\sigma_v^2 + Q_v^2/B_o^2} & \dfrac{-Q_v}{B_o(\sigma_v^2 + Q_v^2/B_o^2)} & 0 \\ \dfrac{Q_v}{B_o(\sigma_v^2 + Q_v^2/B_o^2)} & \dfrac{\sigma_v}{\sigma_v^2 + Q_v^2/B_o^2} & 0 \\ 0 & 0 & \sigma_v^{-1} \end{bmatrix} \tag{4.6.19}$$

where Q_v is the effective charge density of the flux-lattice continuum, and $\sigma_v = \sigma_{v1} - i\sigma_{v2}$ is the complex flux-flow conductivity, defined here by

$$\sigma_{v1} = \frac{\eta}{B_o^2} \quad \text{and} \quad \sigma_{v2} = \frac{K - m^{(v)}\omega^2}{\omega B_o^2} \tag{4.6.20}$$

Now, by Eqs.(4.6.6) and (4.6.19), we may obtain from Eq.(4.6.15)

$$[\rho_{kl}] = \begin{bmatrix} i\omega\mu_o\lambda_L^2 + \dfrac{\sigma_v}{\sigma_v^2 + Q_v^2/B_o^2} & \dfrac{-Q_v}{B_o(\sigma_v^2 + Q_v^2/B_o^2)} & 0 \\ \dfrac{Q_v}{B_o(\sigma_v^2 + Q_v^2/B_o^2)} & i\omega\mu_o\lambda_L^2 + \dfrac{\sigma_v}{\sigma_v^2 + Q_v^2/B_o^2} & 0 \\ 0 & 0 & i\omega\mu_o\lambda_L^2 \end{bmatrix} \tag{4.6.21}$$

The presence of $\rho_{12} = -\rho_{21}$ due to the flux-motion indicates the flux-flow Hall effect. Thus the effective complex resistivity tensor derived here includes both the effect of flux-flow conductivity and the flux-flow Hall effect in the isotropic type II superconductor in the mixed state.

In addition, we may find from Eqs.(4.6.7) and (4.6.19) the following expression for the flux-flow thermoelectric coefficient tensor:

$$[\Pi_{kl}^{(v)}] = \begin{bmatrix} \dfrac{S_f Q_v}{B_o^2\sigma_v^2 + Q_v^2} & \dfrac{S_f \sigma_v B_o}{B_o^2\sigma_v^2 + Q_v^2} & 0 \\ -\dfrac{S_f \sigma_v B_o}{B_o^2\sigma_v^2 + Q_v^2} & \dfrac{S_f Q_v}{B_o^2\sigma_v^2 + Q_v^2} & 0 \\ 0 & 0 & 0 \end{bmatrix} \tag{4.6.22}$$

for the isotropic type II superconductor in the mixed state. The result on the presence of the coefficients, $\Pi_{11}^{(v)}$, $\Pi_{22}^{(v)}$, $\Pi_{12}^{(v)}$, and $\Pi_{21}^{(v)}$, indicates the presence of thermomagnetic effects.

Furthermore, by Eqs.(4.6.17) and (4.6.19), we may find

$$[\kappa_{kl}] = \begin{bmatrix} \kappa_{11}^{(o)} + \dfrac{T_o\sigma_v S_f^2}{B_o^2\sigma_v^2 + Q_v^2} & \kappa_{12}^{(o)} - \dfrac{T_o Q_v S_f^2}{B_o(B_o^2\sigma_v^2 + Q_v^2)} & 0 \\ \kappa_{21}^{(o)} + \dfrac{T_o Q_v S_f^2}{B_o(B_o^2\sigma_v^2 + Q_v^2)} & \kappa_{22}^{(o)} + \dfrac{T_o\sigma_v S_f^2}{B_o^2\sigma_v^2 + Q_v^2} & 0 \\ 0 & 0 & \kappa_{33}^{(o)} + \dfrac{T_o S_f^2}{B_o^2\sigma_v} \end{bmatrix} \tag{4.6.23}$$

with $\kappa_{12}^{(o)} = -\kappa_{21}^{(o)}$, and $\kappa_{11}^{(o)} = \kappa_{22}^{(o)} = \kappa_{33}^{(o)}$ for the isotropic type II superconductor in the mixed state. Equation (4.6.23) shows that the flux-motion contributes not only to the heat conductivity, but also to the flux-flow Righi-

Leduc effect in the isotropic type II superconductor in the mixed state, which will be discussed in the following section.

4.6.3 Flux-Flow Hall Effect and Flux-Flow Righi-Leduc Effect

We first study the flux-flow Hall effect in type II superconductors in the mixed state. To illustrate the Hall effect, let us show a typical experimental arrangement for investigating the Hall effect in Fig. 4.25, where a magnetic field B_o is applied in the z direction perpendicular to a slab and a driving current density J_y is flowing in the y direction. The Hall effect implies the existence of an electric field component E_x in the x direction (positive or negative). The Hall coefficient R_H is defined generally as the ratio:

$$R_H = \frac{E_x}{J_y B_o} \tag{4.6.24}$$

For an isotropic type II superconductor in the mixed state, by Eqs.(4.6.14) and (4.6.21), we may find the flux-flow Hall coefficient:

$$R_H = \frac{-Q_v}{B_o^2 \sigma_v^2 + Q_v^2} \tag{4.6.25}$$

at isothermal condition. It is shown that the phenomenological parameter Q_v characterize effectively the flux-flow Hall effect in the isotropic type II superconductor in the mixed state. Furthermore, we may find the flux-flow Hall angle Θ_H by

$$\tan(\Theta_H) = \frac{|E_x|}{|E_y|} = \frac{|Q_v|}{B_o \sigma_v} \tag{4.6.26}$$

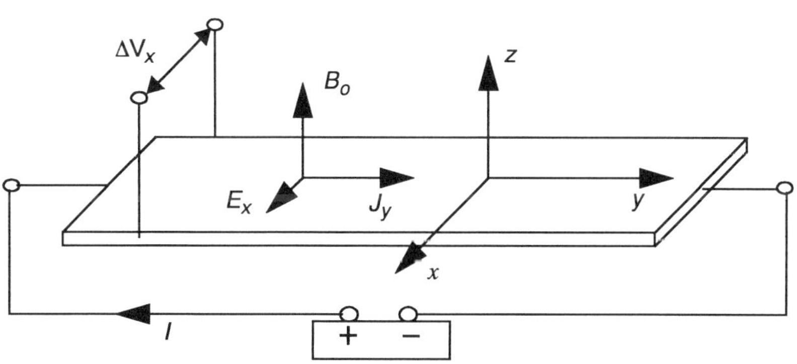

Figure 4.25 An experimental arrangement for the Hall effect measurement.

It can be seen from Eq.(4.6.25) that the flux-flow Hall coefficient R_H is positive if the effective charge density Q_v is negative, or R_H is negative if Q_v is positive. Indeed, experimentally, it has been observed that the flux-flow Hall coefficient R_H can be either positive or negative, depending on the strength of the applied magnetic field, temperature as well as materials. Some experimental results on the flux-flow Hall effect may be found in the work of, for instance, Hagen et al. (1990) and Freimuth et al. (1991) for some high-T_c oxide superconductors.

If we ignore the flux-flow Hall effect by simply letting $Q_v = 0$, we may get the following result:

$$[\rho_{kl}] = \begin{bmatrix} i\omega\mu_o\lambda_L^2 + \sigma_v^{-1} & 0 & 0 \\ 0 & i\omega\mu_o\lambda_L^2 + \sigma_v^{-1} & 0 \\ 0 & 0 & i\omega\mu_o\lambda_L^2 \end{bmatrix} \tag{4.6.27}$$

Furthermore, if the flux-flow effects are ignored, we may then arrive at the well-known result: $\rho_{kl} = i\omega\mu_o\lambda_L{}^2\delta_{kl}$ for isotropic superconductors in the Meissner state, as we expect.

We now study the flux-flow Righi-Leduc effect in type II superconductors in the mixed state. The Righi-Leduc effect is the exact thermal analogue of the Hall effect. The Righi-Leduc effect refers to the rise of a transverse temperature gradient $\partial T/\partial y$ along the y axis due to a heat current q_x along the x axis in the presence of magnetic field in the z direction. For an isotropic type II superconductor in the mixed state, the flux-flow Righi-Leduc coefficient R_L, defined by

$$R_L = \frac{\partial T/\partial y}{B_o q_x} \tag{4.6.28}$$

can be obtained from Eqs.(4.6.16) and (4.6.23) as

$$R_L = \frac{B_o^2\sigma_v^2 + Q_v^2}{Q_v T_o S_f^2} \tag{4.6.29}$$

when electric currents are absent in the superconductor.

It is shown by Eqs.(4.6.25) and (4.6.29) that the flux-flow Righi-Leduc coefficient R_L can be related to the flux-flow Hall coefficient R_H by

$$R_L = \frac{-1}{R_H T_o S_f^2} \tag{4.6.30}$$

which indicates that R_L has an opposite sign as R_H if R_L is defined according to Eq.(4.6.29).

4.6.4 Flux-Flow Seebeck Effect and Flux-Flow Nernst Effect

Next, we shall study the flux-flow Seebeck effect and the flux-flow Nernst effect in type II superconductors in the mixed state. The Seebeck (thermal power) effect is a phenomenon that an electric field may be developed in the direction of a temperature gradient along a sample which has no electric current flowing through it. For an isotropic type II superconductor in the mixed state, the flux-flow Seebeck coefficient S_B, defined by

$$E_x = S_B \frac{\partial T}{\partial x} \tag{4.6.31}$$

can be obtained from Eqs.(4.6.14) and (4.6.22) as

$$S_B = \frac{S_f Q_v}{B_o^2 \sigma_v^2 + Q_v^2} \tag{4.6.32}$$

From Eqs.(4.6.25) and (4.6.32), we may notice that the flux-flow Seebeck coefficient S_B can be related to the flux-flow Hall effect R_H. In particular, we may find that $S_B = -S_f R_H$, which indicates that S_B has an opposite sign as R_H if S_B is defined according to Eq.(4.6.32).

Besides, to describe the temperature-dependence of the flux-flow Seebeck coefficient for type II superconductors in the mixed state, a useful model using the Caroli and Maki theory (1967) or a counter-flow model [Huebener et al. (1990)] may be given here by

$$S_B(T) = \frac{\rho^{(v)}(T)}{\rho_n(T)} S_{Bn}(T) \tag{4.6.33}$$

where $\rho^{(v)}$ is the flux-flow resistivity, ρ_n and S_{Bn} are the normal-state values of the resistivity and of the Seebeck coefficient, respectively. Some experimental measurements of the flux-flow Seebeck effect for high-T_c oxide superconductors in the mixed state have been reported in the work of, for instance, Ri et al. (1991).

The Nernst effect refers to the generation of a transverse electric field due to a longitudinal thermal gradient in a perpendicular magnetic field. It is perhaps the most frequently studied thermomagnetic effect for type II superconductors in the mixed state. Since the flux-flow Nernst effect is very small for superconductors in the normal state, the appearance of the Nernst signal for type II superconductors in the mixed state is a rather striking feature which contrast sharply with the small signal measured in the normal state. In conventional low-T_c metallic type II superconductors, the flux-flow Nernst effect was observed first by Otter and Solomon (1966). Reports on measurements of the Nernst effect in high-T_c oxide superconductors may be found in the work of, for instance, Galffy et al. (1990) and Ri et al. (1994). According to our model, we may find from Eqs.(4.6.14) and (4.6.22) the flux-flow Nernst coefficient Q_N by its definition:

$$Q_N = -\frac{E_y}{B_o(\partial T/\partial x)} = \frac{\rho^{(v)}S_f}{B_o^2} \tag{4.6.34}$$

where $\rho^{(v)}$ is the flux-flow resistivity given by

$$\rho^{(v)} = \frac{\sigma_v}{\sigma_v^2 + Q_v^2/B_o^2} \tag{4.6.35}$$

according to Eq.(4.6.21). The flux-flow Nernst coefficient from Eq.(4.6.34) may also be expressed in a form often used in literature of physics:

$$Q_N = \frac{\rho^{(v)}S_\Phi}{\Phi_o B_o} \tag{4.6.36}$$

where S_Φ is the transport entropy per unit length of flux line, and Φ_o is the flux-quantum.

4.6.5 Plane Thermomagnetic Wave in Superconductors in the Mixed State

In this section, we shall first look for the field equations for describing temperature and electromagnetic fields in type II superconductors in the mixed state at the linear approximation, and then, give illustratively an analytical example for the problem of plane thermomagnetic wave propagation in type II superconductors in a dc magnetic field.

By Eq.(4.6.9), we may derive, with the aid of Eqs.(4.6.5) and (4.6.14), the following linearized field equation:

$$e_{kpq}e_{mst}\Lambda_{qm}\frac{\partial^2 b_t}{\partial x_p \partial x_s} = -b_k - \frac{1}{i\omega\mu_o}e_{kmn}e_{pst}\rho_{np}^{(v)}\frac{\partial^2 b_t}{\partial x_m \partial x_s} - \frac{1}{i\omega}e_{kmn}\Pi_{ns}^{(v)}\frac{\partial^2 \theta}{\partial x_m \partial x_s} \tag{4.6.37}$$

Furthermore, by Eq.(4.6.11), we may get, with the aid of Eq.(4.6.16), the following linearized heat conduction equation:

$$\kappa_{kl}\frac{\partial^2 \theta}{\partial x_k \partial x_l} - i\omega\rho_M C_v\theta = \mu_o^{-1}T_o\Gamma_{kl}^{(v)}e_{lmn}\frac{\partial^2 b_n}{\partial x_m \partial x_k} \tag{4.6.38}$$

where the heat source is ignored. It is shown that the derived set of linearized field equations (4.6.37) and (4.6.38) for the determination of the perturbed time-varying temperature θ and magnetic induction field $\boldsymbol{b}$ are, in general, coupled. This set of field equations may be used to analyze thermomagnetoelectric behavior of anisotropic type II superconductors in the mixed state in time-harmonic fields.

As a special case, let us consider an isotropic superconductor in the mixed state. We shall ignore the flux-flow Hall effect and the flux-flow Righi-Leduc effect for simplicity. Thus Eqs.(4.6.37) and (4.6.38) can be reduced to be of the following form:

$$\left(\lambda_L^2 - \frac{i}{\mu_o \omega \sigma_v}\right)\nabla^2 \boldsymbol{b} = \boldsymbol{b} + \frac{i}{\mu_o \omega \sigma_v}\nabla \times [(\nabla \times \boldsymbol{b}) \cdot \boldsymbol{n}]\boldsymbol{n} + \frac{iS_f}{\omega B_o \sigma_v}\nabla \times (\boldsymbol{n} \times \nabla\theta) \tag{4.6.39}$$

and

$$\left(\kappa^{(o)} + \frac{T_o S_f^2}{B_o^2 \sigma_v}\right)\nabla^2\theta - i\omega\rho_M C_v \theta = -\frac{T_o S_f}{\mu_o B_o \sigma_v}[\boldsymbol{n} \cdot (\nabla^2 \boldsymbol{b})] \tag{4.6.40}$$

where $\boldsymbol{n}$ is the unit direction vector of $\boldsymbol{B}_o$, defined by $\boldsymbol{n} = \boldsymbol{B}_o/B_o$. σ_v is the complex flux-flow conductivity. It is shown that the set of Eqs.(4.6.39) and (4.6.40) is still coupled due to the thermomagnetic effect, induced by the flux-motion in the superconductor. The field equations which we have obtained above should, of course, be supplemented with appropriate boundary conditions for practical problems, as we have discussed in earlier chapters.

Illustratively, let us now consider an infinitely large superconductor in a dc magnetic field $\boldsymbol{B}_o$, in which there may exist a propagating plane thermomagnetic wave of the form:

$$b_z = b_o exp[i(\omega t - \beta x)] \tag{4.6.41}$$

$$\theta = \theta_o exp[i(\omega t - \beta x)] \tag{4.6.42}$$

where b_o and θ_0 are certain constants relevant to the wave amplitude. Other field components are $b_x = b_y = 0$. ω is the radian frequency of the wave, and β is the propagation constant which is, in general, a complex quantity. Here, we shall choose a rectangular Cartesian coordinate system (x,y,z), in which $\boldsymbol{B}_0$ is supposed to be along the direction of the z-axis. Thus Eqs.(4.6.41) and (4.6.42) describe a type of plane thermomagnetic wave propagating along the x-axis direction. Substitution of Eqs.(4.6.41) and (4.6.42) into Eqs.(4.6.39) and (4.6.40) gives the following algebraic equations:

$$-\frac{iS_f\beta^2}{\omega B_o \sigma_v}\theta_o + \left[1 + \beta^2\left(\lambda_L^2 - \frac{i}{\mu_o \omega \sigma_v}\right)\right]b_o = 0 \tag{4.6.43}$$

$$\left[\beta^2\left(\kappa^{(o)} + \frac{T_o S_f^2}{\sigma_v B_o^2}\right) + i\omega\rho_M C_v\right]\theta_o + \frac{T_o S_f \beta^2}{\sigma_v B_o \mu_o} b_o = 0 \qquad (4.6.44)$$

for the determination of b_o and θ_0. The condition for the existence of a non-trivial solution of Eqs.(4.6.43) and (4.6.44) gives the following dispersion equation:

$$\frac{iT_o S_f^2 \beta^4}{\mu_o \omega B_o^2 \sigma_v^2} + \left[1 + \beta^2\left(\lambda_L^2 - \frac{i}{\mu_o \omega \sigma_v}\right)\right]\left[\beta^2\left(\kappa^{(o)} + \frac{T_o S_f^2}{\sigma_v B_o^2}\right) + i\omega\rho_M C_v\right] = 0 \qquad (4.6.45)$$

from which the propagation constant β may be determined. It is shown that the effect of vortex dynamics may cause the presence of the plane thermomagnetic wave in the type II superconductor in the mixed state, depending on the material properties of the superconductor considered.

If the Bardeen-Stephen model (1965) and the Gittleman-Rosenblum model (1966) are adopted, the complex flux-flow conductivity σ_v may be expressed by $\sigma_v = \sigma_{v1} - i\sigma_{v2}$ with $\sigma_{v1} = \sigma_N B_{c2}/B_0$ and $\sigma_{v2} = 2\pi J_c/[\omega(B_0\Phi_0)^{1/2}]$. Here, σ_N is the normal state conductivity, B_{c2} is the upper critical magnetic field, J_c is the critical current density, and Φ_0 is the flux quantum ($\Phi_0 = 2.07\times10^{-15}$ Wb). In addition, the transport entropy density S_f may be given by $S_f = -B_0 \partial H_{c1}/\partial T$, close to B_{c1} [Huebener and Seher (1969)]. Thus material parameters involved in Eq.(4.6.46) have been related to some experimentally measurable quantities. As a numerical example, let us consider a case with $B_o = 0.15$ T, $T_o = 4.2$ K, and $\omega = 10^6$ rad/s. We take the following numerical values for material parameters of the superconductor: $T_c = 9.2$ K; $\sigma_N = 10^8$ S/m; $S_f/B_o = 1.5\times10^4$ A/m·K; $B_{c2} = 0.25$ T; $\kappa^{(o)} = 10$ W/m·K; $C_v = 40$ J/kg·K; $\rho_M = 9\times10^3$ kg/m^3; $\lambda_L = 40$ nm; $J_c = 3\times10^7$ A/m^2. The material parameters given here are only illustrative though they could be typical for niobium at the temperature and magnetic field considered.

By Eq.(4.6.45), we may find, after some manipulations, that the propagation constant β is numerically close to the complex value given by $\beta \approx 0.707(1-i)(\omega\rho_M C_v/\kappa^{(o)})^{1/2}$ in this example. Obviously, the wave is dissipative. The phase velocity of propagation of the coupled thermomagnetic wave is shown to be slightly lower than that of uncouple thermal wave if we simply ignore the first-term (the coupling term) on the left-hand side of Eq.(4.6.45). This example shows that the thermomagnetic coupling effect due to the flux-motion may cause the presence of plane thermomagnetic wave propagating in type II superconductors in the mixed state.

5

Electrodynamics of Josephson Junctions and Circuits

The concept of quantum-mechanical tunneling of a particle through a potential barrier was known already in the 1920s. The discovery of tunneling of electron pairs across a superconductor–insulator–superconductor (SIS) junction was made by Josephson in 1962 and experimentally confirmed first by Anderson and Rowell in 1963. The phenomenon of that the electron-pair tunneling may produce a dissipationless current which may flow at zero bias voltage opened not only a new chapter in solid-state physics, but also provided a wide variety of stimulating applications. In this chapter, we introduce the phenomenology of the Josephson effect, formulate the electrodynamics of the Josephson junctions, and outline some circuits of practical interest. Some electrodynamic analyses are given to superconducting quantum interference devices (SQUIDs). A brief discussion about the possible use of SQUIDs and about the physical limits of various switching technologies in comparison with the superconducting electronic technology is also provided at the end of this chapter.

5.1 MACROSCOPIC QUANTUM STATE AND JOSEPHSON EFFECTS

5.1.1 Macroscopic Quantum State and DC Josephson Effect

Tunneling is a quantum mechanical process arising from the wave nature of the electron. It is manifest by the transport of electrons through spaces that are forbidden by classical physics either because of a potential barrier or an energy gap. We have shown in Section 3.2.2 that superconductivity is a macroscopic quantum phenomenon. The central idea of the macroscopic quantum state is represented by assigning a macroscopic number of electrons to a single "wave" function $\psi = |\psi|e^{i\theta}$. These electrons are assumed to somehow have condensed into a single state due to the fact in the BCS theory that the superelectrons are considered to be formed in the Cooper pairs which behave like the Bose particles. This condensation results in a macroscopic density of the Cooper-pair particles $|\psi|^2$ sharing the same quantum phase θ for all the pairs. Both $|\psi|^2$ and θ

may be functions of space and time. In what follows, we shall present the theoretical basis of electrodynamics for the Josephson junction, which is composed of two superconductors separated from each other by an insulator (a semiconductor or a normal metal for weak links), as shown illustratively in Fig. 5.1.

We start with the use of the concept of macroscopic quantum state [Feynman (1965) and Mercereau (1969)] for a superconductor in which all electron pairs may be considered to have condensed into the same quantum state described by $\psi = |\psi|e^{i\theta}$. All superelectron pairs have also the same energy U. Since all the electrons must do exactly the same thing, the time rate for the macroscopic state ψ must be the same as for a single pair, that is, $i\hbar\partial\psi/\partial t = U\psi$. To analyze a junction between two superconductors, we write ψ_1 being the common wave function of all the electrons on one side, and ψ_2 being the corresponding function on the other side. Thus, in the absence of applied magnetic field, we have the following relations:

$$i\hbar\frac{\partial\psi_1}{\partial t} = U_1\psi_1 + K\psi_2 \tag{5.1.1}$$

$$i\hbar\frac{\partial\psi_2}{\partial t} = U_2\psi_2 + K\psi_1 \tag{5.1.2}$$

where the constant K is a characteristic of the junction. If K is zero when the insulator barrier is very thick, there is no coupling via tunneling between the two pieces of superconductors. The two sides act like separate superconductors with different values of the phase θ (i.e., θ_1 and θ_2 are independent from each other). In the case of $K \neq 0$ when the insulator barrier is very thin, there exists coupling between the two pieces of superconductors. As a special case, if the

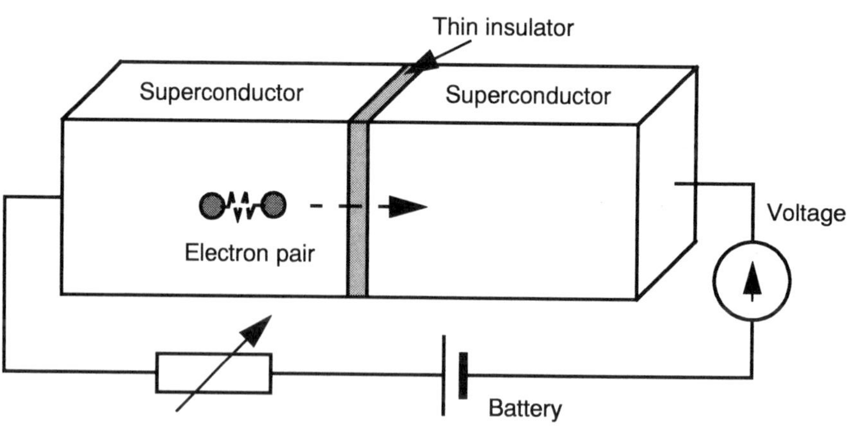

Figure 5.1 Scheme of a Josephson junction.

thickness of the insulator barrier goes to zero, these two superconductors will become a single superconductor. Thus, in the case of $K \neq 0$, by substituting $\psi = |\psi| e^{i\theta}$ into above equations and equating real and imaginary parts, we can derive

$$\frac{\partial |\psi_1|^2}{\partial t} = \frac{2K}{\hbar} |\psi_1||\psi_2| \sin\varphi' \tag{5.1.3}$$

$$\frac{\partial |\psi_2|^2}{\partial t} = -\frac{2K}{\hbar} |\psi_1||\psi_2| \sin\varphi' \tag{5.1.4}$$

and

$$\frac{\partial \theta_1}{\partial t} = -\frac{K|\psi_2|}{\hbar|\psi_1|} \cos\varphi' - \frac{U_1}{\hbar} \tag{5.1.5}$$

$$\frac{\partial \theta_2}{\partial t} = -\frac{K|\psi_1|}{\hbar|\psi_2|} \cos\varphi' - \frac{U_2}{\hbar} \tag{5.1.6}$$

where $\varphi' = \theta_2 - \theta_1$. Equations (5.1.3) and (5.1.4) show that $\partial|\psi_1|^2/\partial t = -\partial|\psi_2|^2/\partial t$ which means that one side loses charge at the same rate as the other side accumulates it. But whatever charge is lost it will be immediately replenished by the active element (voltage or current source) in the circuit. Thus we may write for the current density being normal to the barrier flowing from side 1 to side 2

$$J = J_o \sin\varphi' \tag{5.1.7}$$

where J_o denotes the maximum current density which can flow across the junction and is defined here by $J_o = 2K|\psi_o|^2/\hbar$ with $|\psi_o|^2 = |\psi_1|^2 = |\psi_2|^2$. This equation tells us that, if $\varphi' \neq 0$ (there is no reason why φ' should be equal to zero), then a finite current flows across the insulator without causing any voltage drop. So, in fact, the insulator behaves as a superconductor. The relationship (5.1.7) between the supercurrent density and the phase is known as the *Josephson current–phase* relation. We may notice that for the Josephson junction, the supercurrent density is related to the phase difference, $\varphi' = \theta_2 - \theta_1$, between the two sides of the junction, while the supercurrent density in a single superconductor is related to the phase gradient $\nabla\theta$, as we have learned.

It can be seen that, since the magnitude of $\sin\varphi'$ varies between zero and one, the magnitude of supercurrent density J varies from zero to J_o, which may be called the critical current density of the Josephson junction. Based on a microscopic theory under BCS approximation, the magnitude of J_o for an insulating barrier was derived by Ambegaokar and Baratoff (1963)

$$J_o(T) = \frac{\pi\Delta(T)}{|e^*|R_n} \tanh\left(\frac{\Delta(T)}{2k_B T}\right) \tag{5.1.8}$$

for identical superconductors on the two sides of the junction. Here, R_n is the resistance per unit area of the junction in the normal state, T the absolute temperature, k_B the Boltzmann constant, and Δ the energy gap. In particular, at absolute zero, we have

$$J_o(0) = \frac{\pi\Delta(0)}{|e^*|R_n} \tag{5.1.9}$$

where $\Delta(0)$ is the energy gap at the temperature of aboslute zero. Near the critical temperature T_c, we have

$$J_o(T) = \frac{\pi\Delta^2(T)}{2|e^*|R_n k_B T_c} \tag{5.1.10}$$

The other pair of equations (5.1.5) and (5.1.6) tells us

$$\frac{\partial\varphi'}{\partial t} = \frac{e^* V}{\hbar} \tag{5.1.11}$$

which can also be integrated at constant temperature as

$$\varphi'(t) = \varphi'_o + \frac{e^*}{\hbar}\int V(t)dt \tag{5.1.12}$$

where $V = (U_1 - U_2)/e^*$ denotes the voltage across the junction. Equation (5.1.11) is known as the *Josephson voltage–phase relation.*

It can be seen from Eqs.(5.1.7) and (5.1.12) that, if we have zero voltage across the junction, we can get a current! With no voltage the current density can be any amount between $+J_0$ and $-J_0$, depending on the value of φ'_0. This is the essence of the *dc Josephson effect* [Josephson (1962)]. Figure 5.2 shows schematically the current-voltage behavior of a Josephson junction. It is shown that, as the current I is increased from zero at which the phase is $\varphi' = 0$, no voltage appears across the junction for currents less than the critical current I_0, at which the phase becomes $\varphi' = \pi/2$. When the critical current is exceeded, there is a jump from zero voltage to a finite value of voltage V_g, called the gap voltage, as shown by the normal tunnel current curve. The first observation of such a "strange" behavior was made by Anderson and Rowell (1963). An experimental evaluation of the voltage across the junction in the dc Josephson regime was given by Smith (1965).

For real junctions at non-zero temperature, the I–V characteristic is somewhat rounded. The temperature dependence of the gap voltage near T_c can be approximately given by $V_g(T) = 1.74V_g(0)(1 - T/T_c)^{1/2}$, which is useful over quite a wide temperature range [Rogalla (1998)].

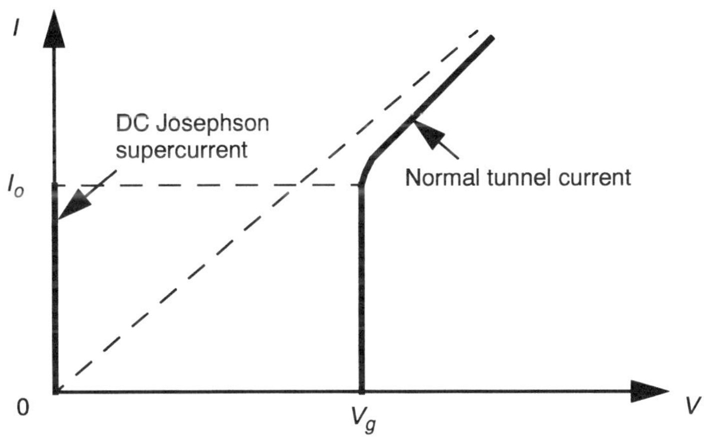

Figure 5.2 *I–V* characteristic of a Josephson Junction.

5.1.2 Electrodynamic Equations for the Josephson Junctions

In this section, we shall derive a set of electrodynamic equations for the study of the Josephson junctions in the presence of applied magnetic fields. In such a case, a gauge invariance condition has to be imposed on Eqs.(5.1.7) and (5.1.11). The gauge invariance means that the physics of any situation is unchanged if the magnetic vector potential $\boldsymbol{A} \rightarrow \boldsymbol{A}+\nabla\chi$, the electric scalar potential $\phi \rightarrow \phi-\partial\chi/\partial t$, and $\theta \rightarrow \theta+(e^*/\hbar)\chi$, where χ is an arbitrary scalar quantity. Therefore, Eqs.(5.1.7) and (5.1.11) may now be written in a gauge invariant form:

$$J = J_o \sin\left(\varphi' + \frac{e^*}{\hbar}\int_2^1 \boldsymbol{A}\cdot d\boldsymbol{L}\right) \tag{5.1.13}$$

and

$$\frac{\partial}{\partial t}\left(\varphi' + \frac{e^*}{\hbar}\int_2^1 \boldsymbol{A}\cdot d\boldsymbol{L}\right) = \frac{e^* V}{\hbar} \tag{5.1.14}$$

Equations (5.1.13) and (5.1.14) present the two basic equations of the general theory of the Josephson junction. By noting that

$$\frac{e^* V}{\hbar} = -\frac{e^*}{\hbar}\int_2^1 \boldsymbol{E}\cdot d\boldsymbol{L} \tag{5.1.15}$$

at constant temperature, and $\boldsymbol{E} = -\nabla\phi - \partial\boldsymbol{A}/\partial t$, we may obtain

$$\frac{\partial \varphi'}{\partial t} = \frac{e^*}{\hbar}\int_2^1 \nabla\phi \cdot d\boldsymbol{L} \tag{5.1.16}$$

To derive an equation for describing the spatial variation of the phase difference, we may consider Fig. 5.3, where we let P_1, P_2 and Q_1, Q_2 be two pairs of points, the members of each pair being adjacent to each other but on opposite sides of the barrier.

From the supercurrent–phase relation (4.2.14), we may obtain

$$\varphi'(P) - \varphi'(Q) = \int_{\Gamma_1+\Gamma_2} \frac{e^*}{\hbar}(\boldsymbol{A} + \mu_o\lambda_L^2\boldsymbol{J}_s)\cdot d\boldsymbol{L} \tag{5.1.17}$$

By choosing a suitable integration contour on which $\boldsymbol{J}_s$ is either zero or is made to be small (by considering that the dominant part of the current flows parallel to the barrier surface, shielding the field from the interior of the bulk superconductors), the second term on the right-hand side of Eq.(5.1.17) becomes negligibly small in comparison with the first term, and thus we can get, by adding the line integrals from P_2 to P_1 and Q_1 to Q_2 to both sides and replacing the line integral of $\boldsymbol{A}$ around the closed loop by the surface integral of $\boldsymbol{B}$,

$$\varphi(P) - \varphi(Q) = \frac{e^*}{\hbar}\Phi_s(\Gamma) \tag{5.1.18}$$

where $\Phi_s(\Gamma)$ is the magnetic flux enclosed by the contour Γ (= Γ_1+Γ_2+P_2P_1+Q_1Q_2) and φ is the gauge-invariant generalization of the relative pair phase defined by

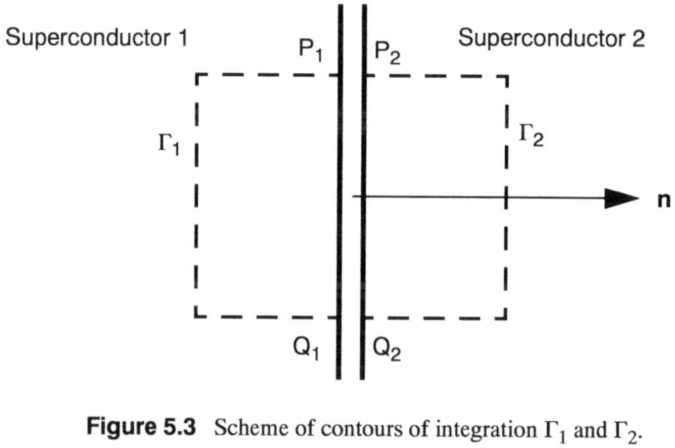

Figure 5.3 Scheme of contours of integration Γ_1 and Γ_2.

$$\varphi(P) = \varphi'(P) + \frac{e^*}{\hbar}\int_{P_2}^{P_1} \boldsymbol{A}\cdot d\boldsymbol{L} \tag{5.1.19}$$

where $\varphi'(P) = \theta(P_2) - \theta(P_1)$.

If P and Q are close to each other, we may write

$$\varphi(P) - \varphi(Q) = \nabla\varphi \cdot \overrightarrow{QP} \tag{5.1.20a}$$

$$\Phi_s(\Gamma) = \boldsymbol{B}\cdot(\delta\boldsymbol{n}\times\overrightarrow{QP}) = (\boldsymbol{B}\times\delta\boldsymbol{n})\cdot\overrightarrow{QP} \tag{5.1.20b}$$

from which we arrive at

$$\nabla\varphi = \frac{e^*\delta}{\hbar}\boldsymbol{B}\times\boldsymbol{n} \tag{5.1.21}$$

with $\delta = d+\lambda_{L1}+\lambda_{L2}$. Here, d is the thickness of the barrier, λ_{L1} and λ_{L2} are the effective London penetration depths of the two superconductors forming the junction. ***B*** represents the actual magnetic field in the plane of the junction including both externally applied magnetic field and the field induced by the currents flowing in the junction. ***n*** is the unit normal vector of the junction surface and directed from superconductor 1 to superconductor 2.

Equations (5.1.13), (5.1.14), and (5.1.21) form a set of constitutive relations for the Josephson junction composed of thick superconductors. These constitutive equations relate the current density to the phase and phase to the electromagnetic fields in the junction. The electrodynamic equations of Josephson junctions are then obtained by combining Maxwell's equations with the constitutive current–phase–field relations.

In the case of a junction that lies in the *x*-*y* plane with ***n*** along the *z*-axis direction, we have from Maxwell's equations

$$\frac{\partial B_y}{\partial x} - \frac{\partial B_x}{\partial y} = \mu_o J_z + \mu_o c_s \frac{\partial V}{\partial t} \tag{5.1.22}$$

where $c_s\partial V/\partial t$ is the displacement current. Expressing B_y and B_x from Eq.(5.1.21), V from Eq.(5.1.14), J_z from Eq.(5.1.13) and substituting them into Eq.(5.1.22), we may obtain a single differential equation in φ:

$$\frac{\partial^2\varphi}{\partial x^2} + \frac{\partial^2\varphi}{\partial y^2} - \frac{1}{v^2}\frac{\partial^2\varphi}{\partial t^2} = \frac{1}{\lambda_J^2}\sin\varphi \tag{5.1.23}$$

where v is the phase velocity given by [Swihart (1961)]

$$\text{v} = \frac{1}{\sqrt{\mu_o c_s \delta}} = c\sqrt{\frac{d}{\varepsilon_r \delta}} \tag{5.1.24}$$

with $\varepsilon_r = \varepsilon/\varepsilon_0$ being the effective relative dielectric constant, c the velocity of light in vacuum, and λ_J is the Josephson penetration depth defined by

$$\lambda_J = \sqrt{\frac{\hbar}{|e^*|\delta\mu_o J_o}} \tag{5.1.25}$$

which is of the order of a millimeter for typical junctions.

If we further include the current due to the tunnelling of normal electrons, we may write approximately [Solymar (1972)]

$$J_z = J_o \sin\varphi + g_o V \tag{5.1.26}$$

with $1/g_o$ characterizing the resistance of the junction, and Eq.(5.1.23) becomes

$$\frac{\partial^2 \varphi}{\partial x^2} + \frac{\partial^2 \varphi}{\partial y^2} - \frac{1}{\text{v}^2}\frac{\partial^2 \varphi}{\partial t^2} - \frac{\beta}{\text{v}^2}\frac{\partial \varphi}{\partial t} = \frac{1}{\lambda_J^2}\sin\varphi \tag{5.1.27}$$

where $\beta = g_o/c_S$. This differential equation together with proper boundary conditions has no general analytical solution [see Barone et al. (1982)]. It is, however, possible to obtain solutions in some special cases [see, e.g., Solymar (1972), and Barone and Paterno (1982)]. In what follows, some illustrative examples will be given for explaining the basic phenomena of Josephson junctions and their uses in superconducting quantum interference devices (SQUIDs). Possible applications of the SQUIDs are discussed briefly.

5.1.3 AC Josephson Effect

It has been shown from the basic Josephson relations (5.1.7) and (5.1.11) that a finite supercurrent of maximum value J_o can flow with zero voltage drop (the dc Josephson effect). We may now consider the case that, if we put on a dc voltage V_o, the argument of the sine function in Eq.(5.1.7) becomes then $\varphi'_0+(e^*/\hbar)V_o t$, and we find

$$J = J_o \sin\left(\varphi'_o + \frac{e^* V_o}{\hbar}t\right) \tag{5.1.28}$$

which means that, if a constant voltage is maintained at the junction, an alternating current will flow across the junction. The frequency of the alternating current is $\omega_J = |e^* V_o/\hbar|$ ($\omega_J/2\pi V_o \approx$ 484 GHz/mV quantitatively), which is called the *Josephson frequency.* Such an effect is known as the *ac Josephson effect.* Physically, the ac Josephson effect may be understood as follows. When the total current through the junction is larger than the critical current J_o, part of the total current must be carried by normal electrons. These normal electrons pass through the junction, producing a non-zero voltage V_o across the junction. This non-zero voltage across the junction results in a time-dependent phase difference $\varphi'(t)$ between the two sides of the junction, which produces a time-dependent supercurrent through the junction according to the Josephson current–phase relation (5.1.7).

Practically the ac Josephson effect can be used in the construction of detectors of high-frequency electromagnetic radiation. To illustrate this point, let us consider a junction across which a constant dc voltage V_0 is applied, resulting in the alternating supercurrent of the ac Josephson effect. Suppose now that high-frequency (e.g., microwave) electromagnetic radiation of frequency ω_a is incident on the junction, inducing an ac voltage, $V_1\cos(\omega_a t)$, across the junction. Thus, by setting the total voltage V across the junction as

$$V = V_o + V_1\cos(\omega_a t) \tag{5.1.29}$$

we can obtain from Eq.(5.1.12) in the absence of applied magnetic fields

$$\varphi'(t) = \varphi'_o + \omega_f t + \frac{V_1 e^*}{\omega_a \hbar}\sin(\omega_a t) \tag{5.1.30}$$

with $\omega_f = e^*V_o/\hbar$. In the case of $V_1 << V_0$, using the approximation: $\sin(x+\Delta x) \approx \sin(x)+\Delta x\cos(x)$, we can obtain

$$J = J_o\left[\sin(\varphi'_o + \omega_J t) + \frac{e^* V_1}{\omega_a \hbar}\sin(\omega_a t)\cos(\omega_J t + \varphi'_o)\right] \tag{5.1.31}$$

The first term is zero on the average, but the second term is not if $\omega_a = \omega_f$. This means that there should be a dc current component if the ac voltage has just the Josephson frequency [Shapiro (1963)]. In general, by using the Fourier-Bessel series identity:

$$e^{ib\sin(x)} = \sum_{k=-\infty}^{+\infty} |J_k(b)| e^{ikx} \tag{5.1.32}$$

where J_k is the kth order Bessel function of the first kind, which satisfies the following relation:

$$J_{-k}(b) = (-1)^k J_k(b) \tag{5.1.33}$$

we may find a useful mathematical relation [Orlando and Delin (1991)]:

$$\sin[a + b\sin(x)] = \sum_{k=-\infty}^{+\infty} (-1)^k J_k(b)\sin(a - kx) \tag{5.1.34}$$

Thus we may express the Josephson current density by

$$\begin{aligned} J &= J_o \sin\left[\varphi'_o + \omega_f t + \frac{V_1 e^*}{\omega_a \hbar}\sin(\omega_a t)\right] \\ &= J_o \sum_{k=-\infty}^{\infty} (-1)^k J_k\left(\frac{V_1 e^*}{\omega_a \hbar}\right) \sin[(\omega_f - k\omega_a)t + \varphi'_o] \end{aligned} \tag{5.1.35}$$

This equation is a sum of components at infinitely many frequencies, $\omega_f - k\omega_a$ ($k = \pm 1, \pm 2, \ldots$). We may see that the current is phase modulated. If ω_f is equal to an integer multiple $k\omega_a$ ($k = \pm 1, \pm 2, \ldots$) of the microwave frequency of the ac voltage, the Josephson current density contains a time-independent (dc) current density component whose magnitude is

$$J_{DC}^{(k)} = \left| J_o J_k\left(\frac{e^* V_1}{\omega_a \hbar}\right) \sin\varphi'_o \right| \tag{5.1.36}$$

The condition $\omega_f = k\omega_a$ is corresponding to the condition:

$$V_n = \frac{k\hbar\omega_a}{e^*} \qquad (k = \pm 1, \pm 2, \ldots) \tag{5.1.37}$$

These zero-frequency supercurrents correspond to the appearance of the dc Josephson effect at a non-zero voltage V_o across the junction when the junction is irradiated with electromagnetic radiation of frequency ω_a. These results predict a series of spikes of dc supercurrent superimposed on the dc current–voltage characteristic of the junction. The voltage separation between adjacent spikes is $\hbar\omega_a / |e^*|$, which is determined by the radiation frequency ω_a. The first observation of such dc supercurrent steps (called the *Shapiro steps*) was reported by Shapiro (1963), as shown illustratively in Fig. 5.4.

From Eq.(5.1.36), we may notice that the amplitude of step k of the dc supercurrent, induced by the radiation, is proportional to the Bessel function of order k, $J_k(x)$, and its argument $x = e^* V_1/(\hbar\omega_a)$. If we consider the step for $k = 0$, which corresponds to zero voltage by Eq.(5.1.37), we have

$$J_{DC}^{(0)} = \left| J_o J_0\left(\frac{e^* V_1}{\omega_a \hbar}\right) \sin\varphi'_o \right| \tag{5.1.38}$$

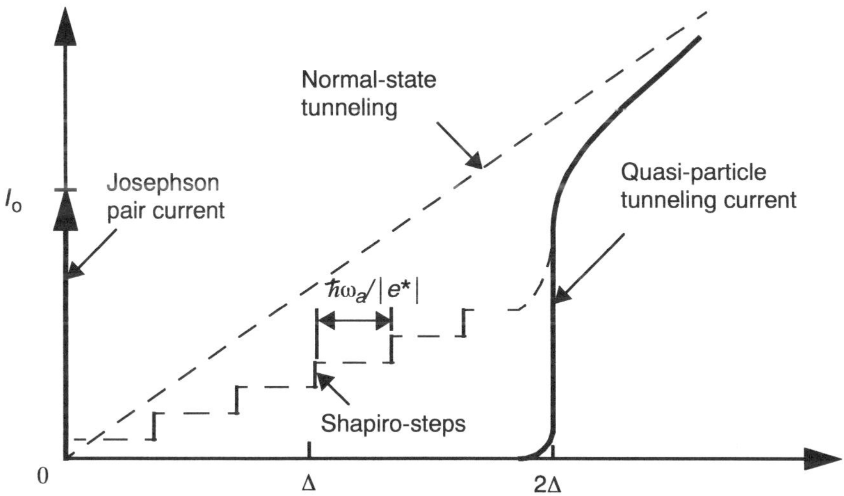

Figure 5.4 Voltage–current characteristic of Josephson junction to microwave signals.

Since the Bessel function of order zero, $J_0(x)$ decreases monotonically with increasing x, for x less than approximately 2.4. Thus the amplitude of $J_{DC}^{(0)}$ of the $k = 0$ step responds to incident electromagnetic radiation by decreasing with increasing V_1 (at low values of V_1), which implies that $J_{DC}^{(0)}$ decreases with increasing microwave power incident on the junction. Furthermore, since Eq.(5.1.37) is satisfied for any frequency ω_a of incident electromagnetic radiation at the case of $k = 0$, $J_{DC}^{(0)}$, which is sometimes also called the critical current of the junction, will respond to broadband radiation, containing many frequencies.

Based on the ac Josephson effect, many different types of Josephson junctions have been used to construct electromagnetic radiation detectors in a number of way, depending on their response to incident electromagnetic radiation. For more detailed discussion on the ac Josephson effect and their related devices, the reader may refer to the work of, for instance, Clark (1974), and Barone and Paterno (1982).

5.2 SUPERCONDUCTING QUANTUM INTERFERENCE DEVICES

5.2.1 Single Josephson Junction in Magnetic Fields

In this section, we shall study the influence of applied magnetic fields on the behavior of the Josephson junctions. Let us first look at the case where there is a constant (both temporally and spatially) magnetic field, $\boldsymbol{B} = (B_x, B_y, 0)$ in the plane of the junction lying in the x-y plane. In such a case, Eq.(5.1.21) may be integrated, with the approximation of ignoring the self-magnetic field generated by the supercurrent, to give

$$\varphi = \frac{e^*\delta}{\hbar}(B_y x - B_x y) + \alpha \tag{5.2.1}$$

where α is an integration constant. The total current across the junction can now be expressed by

$$I_z = \int J_o \sin\varphi dS = Im\int J_o \exp\left\{i\left[\frac{e^*\delta}{\hbar}(B_y x - B_x y) + \alpha\right]\right\}dS \tag{5.2.2}$$

where *Im* denotes the imaginary part and the integration is taken over the area of the junction. The maximum supercurrent through the insulator is given by the maximum value of this expression with respect to changes in α, which is

$$I_o = \left|\int J_o \exp\left\{i\left[\frac{e^*\delta}{\hbar}(B_y x - B_x y)\right]\right\}dS\right| \tag{5.2.3}$$

This integral may be calculated if the spatial variation of J_o is known. For most cases of interest J_o may be taken as a constant, Eq.(5.2.3) is then reduced to be

$$I_o = J_o a_x a_y \left|\frac{\sin(u_1 a_x)}{u_1 a_x}\right|\left|\frac{\sin(u_2 a_y)}{u_2 a_y}\right| \tag{5.2.4}$$

with

$$u_1 = \frac{\pi e^*\delta}{\hbar}B_y, \qquad u_2 = \frac{\pi e^*\delta}{\hbar}B_x \tag{5.2.5}$$

in the case of a rectangular junction of dimensions a_x and a_y. The quantity $J_o a_x a_y$ is the maximum supercurrent in the absence of a magnetic field.

If the applied magnetic field is parallel with one of the edges of the junction (say $B_x = 0$), we have

$$I_o = J_o a_x a_y \left|\frac{\sin(\pi\Phi_J/\Phi_o)}{\pi\Phi_J/\Phi_o}\right| \tag{5.2.6}$$

where Φ_J denotes the magnetic flux enclosed, that is, $\Phi_J = \delta a_x B$. Thus, whenever the magnetic flux is an integral multiple of the flux quantum Φ_o, no supercurrent will flow across the junction. This effect was first observed by Rowell (1963). Figure 5.5 shows quantitatively the dependence of the normalized maximum supercurrent on the normalized magnetic flux, which has been well confirmed experimentally by Langenberg et al. (1966).

If we now consider the case of no externally applied magnetic field, we should use, by including the self-magnetic field effect, Eq.(5.1.27) which, in the time-independent case and by assuming a one-dimensional geometry, becomes

$$\frac{\partial^2 \varphi}{\partial x^2} = \frac{1}{\lambda_J^2} \sin\varphi \tag{5.2.7}$$

For small φ (valid for small magnetic fields [Josephson (1969)]), Eq.(5.2.7) may be written as

$$\frac{\partial^2 \varphi}{\partial x^2} = \frac{1}{\lambda_J^2} \varphi \tag{5.2.8}$$

which has the solution of the form $\varphi \sim \exp(-x/\lambda_J)$, in analogy to the Meissner effect. It can be seen that, if the Josephson penetration depth λ_J is small, the current (also the magnetic field) is confined to the edge of the junction and the maximum supercurrent is drastically reduced. If, however, λ_J is large in comparison with the dimensions of the junction, φ can be regarded a constant

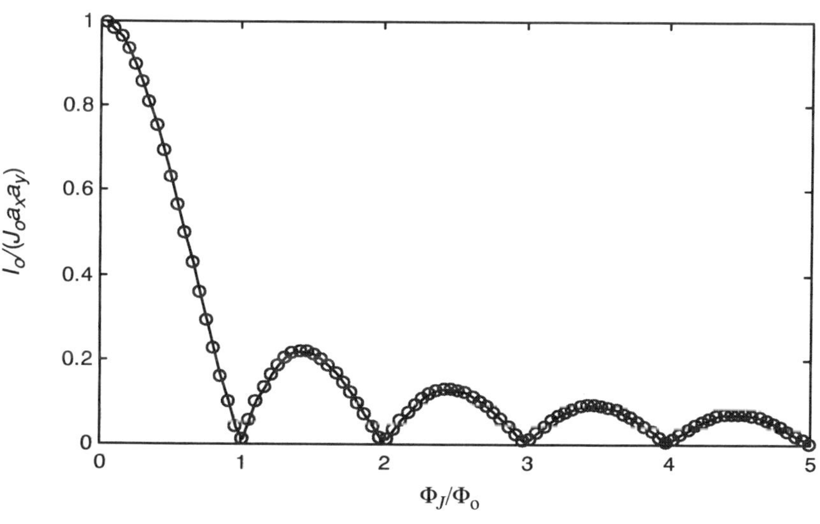

Figure 5.5 The normalized maximum supercurrent as a function of magnetic flux.

and the total current can be calculated by simply multiplying the current density by the area of the junction. It should be noticed that the exclusion of the magnetic field from the junction (in analogy with type II superconductors) will be destroyed by sufficiently large fields and quantized flux lines will enter the barrier. Such cases have to be studied by solving Eq.(5.2.7) without assuming that φ is small [see Solymar (1972)].

5.2.2 Double Josephson Junctions in Magnetic Fields

It has been shown in Eq.(5.2.6) that, for a single Josephson junction, I_o cannot be made to depend very sensitively on applied magnetic field since the effective junction width cannot be increased above the Josephson penetration depth λ_J. The situation is, however, quite different for double junctions. Let us now consider a combination of two identical Josephson junctions in parallel, as shown in Fig. 5.6. For simplicity, we shall ignore the effect of self-inductance in the following discussion.

The total current through the double junction is equal to the sum of the supercurrents through the individual contacts and is a function of φ_a and Φ_s (the externally applied magnetic flux enclosed by the loop)

$$I(\Phi_s, \varphi_a) = I_o(\sin\varphi_a + \sin\varphi_b) = I_o\left[\sin\varphi_a + \sin\left(\varphi_a + \frac{e^*}{\hbar}\Phi_s\right)\right]$$

$$= 2I_o \cos\left(\frac{e^*}{2\hbar}\Phi_s\right) \sin\left(\varphi_a + \frac{e^*}{2\hbar}\Phi_s\right) \tag{5.2.9}$$

The maximum current that can flow through the junctions in the superconducting loop without any voltage appearing across the junctions may

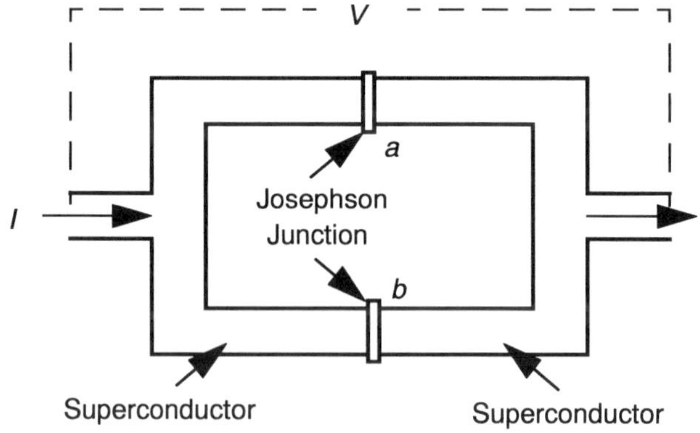

Figure 5.6 A double Josephson junction configuration.

be obtained by maximizing the above expression with respect to φ_a at a given value of Φ_s. The result reads

$$I_{max}(\Phi_s) = 2I_o\left|\cos\left(\frac{e^*}{2\hbar}\Phi_s\right)\right| = 2I_o\left|\cos\left(\frac{\pi\Phi_s}{\Phi_o}\right)\right| \tag{5.2.10}$$

which defines a critical current (current at which a voltage first appears across the junction) for the double junction in the superconducting loop. One of the significant features of this I_{max} vs. Φ_s dependence is that $I_{max} = 0$ when $\Phi_s = (n+1/2)\Phi_0$ and $I_{max} = 2I_o$ when $\Phi_s = n\Phi_0$. In practice, it is hard to observe experimentally the result $I_{max} = 0$, which might partly be due to an asymmetry of the junctions and to the self-induced flux [see, e.g., de Bruyn Ouboter (1977)].

Equation (5.2.10) shows that the critical current, I_{max} of the device is a periodic function of Φ_s, the flux through the loop, with the period being equal to one flux quantum Φ_0, in contrast to the case of the single junction critical current by Eq.(5.2.6) which is periodic in Φ_J, the flux through the junction. Since the double junction loop area can be made arbitrarily large, the critical current of the junction pair becomes a very sensitive measure of applied field.

We may now consider the situation when the current exceeds the maximum value possible with zero voltage. Note that Eq.(5.2.9) is still valid for the dc supercurrent flowing in the circuit only φ_a is now related to the voltage. Adding the normal current term V/R_n, we may write the total current with the use of Eq.(5.1.16)

$$I = 2I_o\cos\left(\frac{e^*}{2\hbar}\Phi_s\right)\sin\left(\varphi_a + \frac{e^*}{2\hbar}\Phi_s\right) + \frac{\hbar}{e^*R_n}\frac{\partial\varphi_a}{\partial t} \tag{5.2.11}$$

Following McCumber (1968), we may introduce the dimensionless variables

$$\tau = \frac{2e^*R_nI_o\cos\left(\frac{e^*}{2\hbar}\Phi_s\right)}{\hbar}t, \qquad \zeta = \frac{I}{2I_o\cos\left(\frac{e^*}{2\hbar}\Phi_s\right)} \tag{5.2.12}$$

$$\eta(\tau) = \frac{V}{2R_nI_o\cos\left(\frac{e^*}{2\hbar}\Phi_s\right)} = \frac{df}{d\tau} \tag{5.2.13}$$

yielding

$$\zeta = \frac{df}{d\tau} + \sin f \tag{5.2.14}$$

with $f = \varphi_a + e^*\Phi_s/(2\hbar)$ by noting that Φ_s is independent of time.

In the case of $|\zeta| \geq 1$, Eq.(5.2.14) may be written in the form

$$\frac{df}{\zeta - \sin f} = d\tau \tag{5.2.15}$$

which may be integrated to give

$$f = 2\tan^{-1}\left[\frac{1}{\zeta} + \sqrt{1 - \frac{1}{\zeta^2}}\tan\left(\frac{\tau}{2}\sqrt{\zeta^2 - 1}\right)\right] \tag{5.2.16}$$

We can then get

$$\eta(\tau) = \frac{df}{d\tau} = \frac{\zeta^2 - 1}{\zeta + \sin[\tau\sqrt{\zeta^2 - 1} + \tan^{-1}(1/\sqrt{\zeta^2 - 1})]} \tag{5.2.17}$$

The average value of $\eta(\tau)$ may be obtained by integrating Eq.(5.2.17) over a period, yielding

$$\overline{\eta(\tau)} = \sqrt{\zeta^2 - 1} \tag{5.2.18}$$

Thus the time-average (dc) voltage V is found

$$\overline{V(\tau)} = R_n\sqrt{I^2 - I_{max}^2} \tag{5.2.19}$$

where I_{max} is given by Eq.(5.2.10).

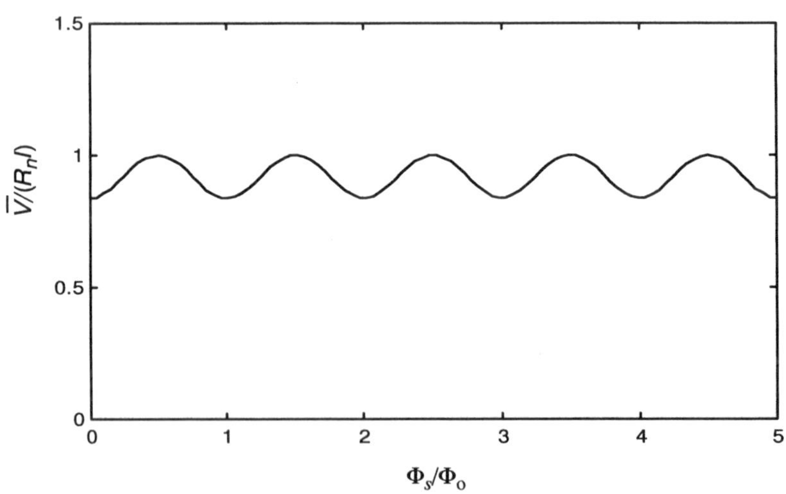

Figure 5.7 The dependence of dc voltage $V/(R_nI)$ on the magnetic flux Φ_s/Φ_0.

It follows clearly from Eq.(5.2.19) that the dc voltage as a function of magnetic field is in anti-phase with I_{max}. Whenever I_{max} is a minimum, V is maximum and vice versa. It is also shown that the dc voltage oscillates as a function of the external applied magnetic flux with the same period Φ_0 in the case that the applied constant current I exceeds the critical value I_{max} (see Fig. 5.7). The experimental evidence of such phenomenon was observed by Zimmerman and Silver (1966).

5.2.3 SQUIDs and Their Applications

In the above section, we have shown that the dependence of the maximum supercurrent or the dc voltage on the applied magnetic field may be used for the measurement of small magnetic fields and their small changes. Such a device is called the dc *superconducting quantum interference device* (SQUID), composed of two Josephson junctions mounted on a superconducting ring. Since both I_{max} and the dc voltage across the double junctions are periodic functions of the magnetic field, we can gain information about the change in magnetic field by measuring the change in voltage [Zimmmerman and Silver (1966)]. The resolution obtained by such a device with a 1 sec time constant was about 10^{-15} tesla [Mercereau (1970)]. The device may also be used to measure larger magnetic fields by counting the number of flux quanta moving in or out of the ring [Forgacs and Warnick (1967)]. Details about some technical problems for such devices are referred to in the work of, for instance, Clarke (1983) and Solymar (1972).

Another type of device is the rf (radio-frequency) SQUID. An rf SQUID consists basically of a single Josephson junction mounted on a superconducting ring of inductance L (typically 10^{-9} H), shunted by a resistance and a capacitance for eliminating the hysteresis of the current-voltage character [Ziemmerman et al. (1970), Mercereau (1970), and Clarke (1989)]. For radio frequency operation of the device, the ring is inductively coupled to a coil of an LC-resonant circuit that is excited by a sinusoidal current I_{rf} at its resonant frequency (typically 20 or 30 MHz). The rf voltage, V_T developed across the tank circuit is amplified and detected with a diode (for example). If one plots V_T (vertically) versus I_{rf} (horizontally), one obtains a series of "steps" and "risers". On the steps, V_T is nearly independent of I_{rf}. If the external magnetic flux Φ_s is slowly changed, the voltage at which the steps appear oscillate as a periodic function of Φ_s with period Φ_0. The measurement of magnetic field may then proceed in the same manner as for the dc SQUID. Detailed discussions about rf SQUIDs may be found elsewhere [Clarke (1993)].

Shown in Fig. 5.8 is some historical perspective on SQUIDs.

The extreme sensitivity of SQUIDs to magnetic field signals of the order of 10^{-15} $\mathrm{THz}^{-1/2}$, and their extreme energy sensitivity of the order of 10^{-32} JHz^{-1}, have led to many applications of practical interest in using SQUIDs. Listed below are some known applications of SQUIDs:

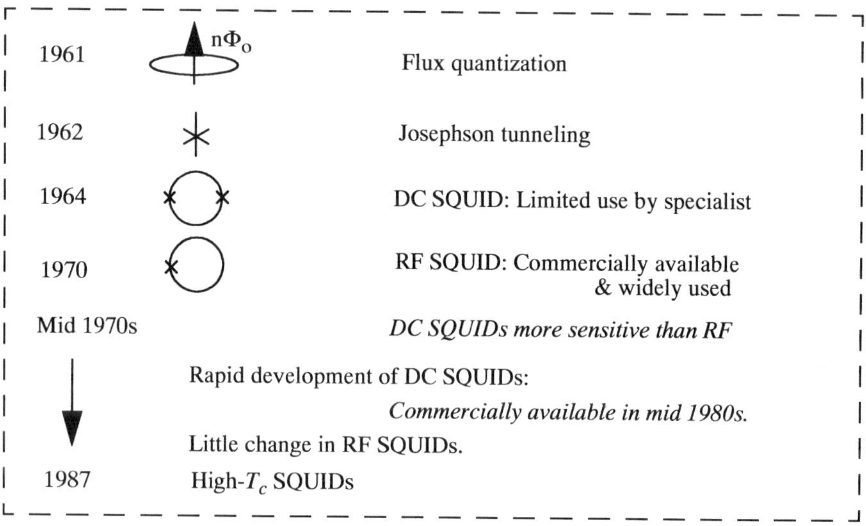

Figure 5.8 Some historical perspective on SQUIDs.

1. SQUIDs to biomedical applications. The measurements of the magnetic signals from various parts of human body can yield new information about the organs that generate electric currents, not available to surface electrodes, and also about organs which contain foreign ferromagnetic particles. Signals measured from the heart are called magnetocardiograms (MCG) [Cohen (1975)], those from neural activity within the brain are called magnetoencephalograms (MEG) [Cohen (1968) and Hari and Ilmoniemi (1986)], those from muscle action are called magnetomyogram (MMG) [Cohen and Givler (1972)], and those due to eye movements are called magnetooculograms (MOG). Magnetic signals have also been detected from blood flow, injured tissues, fetuses in utero, and the magnetite dust in stomach and lung [Robinson (1981)].

Shown in Fig. 5.9 are some biomagnetic signal amplitudes relative to the earth's field and common urban electromagnetic noise levels. it is seen that all these biomagnetic signals are within the sensitivity of SQUIDs.

The first observation of the magnetocardiogram was reported by Baule and McFee (1963), where a peak field of about 50 pT was observed. This achievement marks the birth of experimental biomagnetism, which is the area of studying the weak magnetic fields associated with biological activity. The interest in finding theoretical models for presenting data was then motivated by desire to decrease the effort in taking data and to establish a format that conveys a more intuitive appreciation for the actual heart action. One operationally simple approach is to assume that the observed fields originate from a magnetic dipole, called the "magnetic heart vector," which is located at the center of the

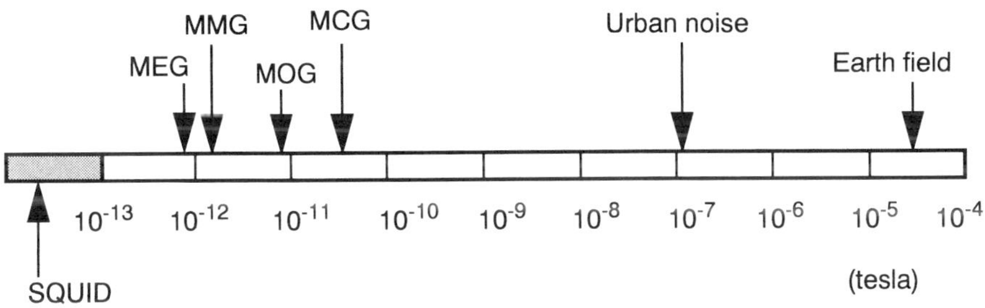

Figure 5.9 Some biomagnetic signal amplitudes relative to the earth´s field and urban noise level.

heart and whose magnitude and direction vary through the cardiac cycle [Wynn et al. (1975)]. More sophisticated models based on current dipole and higher magnetic multipoles were also proposed [Horacek (1973) and Grynszpan and Geselowitz (1973)].

The observation of a neuromagnetic field produced by neural activity was made later by Cohen (1968), where the strongest MEG signals are typically about 2 pT. Using a SQUID magnetometer, Cohen and Givler (1972) had also found that human skeletal muscles at the elbow produce both dc and ac magnetic fields when contracted. The fields are found to be quite large near the skin, having an amplitude of about 20 pT.

Some recent advances in biomagnetism may be found, for instance, in the Proceedings of the international conference on biomagnetism edited by Atsumi et al. (1988).

2. SQUIDs to geophysics for studying the geological history of the earth, for determining the properties of the earth deep below the surface and for measuring the magnetic properties of rocks under various temperatures and pressures [Goubau (1980)]. For instance, seawater motion, seismic stresses and electric conductivity of the earth crust may be studied by measuring the fluctuating magnetic gradients generated, respectively, by sea water oscillation across the earth´s magnetic field [Podney (1975)], by the piezomagnetic effects [Stacey (1964)], and by the distortion of magnetic fields of electric currents flowing in the ionosphere [Cagniard (1953)]. Information from the measurement of magnetic fields caused by piezomagnetism of rocks may also be useful in earthquake prediction since there is evidence [Clarke (1977)] that the magnetic field along a fault line changes over a period of a few days prior to an earthquake.

3. SQUIDs to nuclear gyros which may be used in navigation system. The principle of the operation of a nuclear gyroscope is as follows: For a species of nuclei with a magnetic moment $\boldsymbol{m}$ in a magnetic field $\boldsymbol{B}$, the nuclear moments precess about the direction of the field at the Larmor frequency ω. If the frame

of reference containing the nuclear free precession detection system were fixed in an inertial frame, the precession frequency measured by an observer in that frame would be the Larmor frequency. If, however, the whole apparatus experiences a rotation about an axis parallel to the axis of the magnetic field, then the measured precessional frequency will be shifted by the rotation frequency. With the use of the exceptional sensitivity of SQUID in place of conventional NMR free precession detection methods, the nuclear sample can be made much smaller or operated in lower ambient field [Adams et al. (1980)].

4. SQUIDs to submarine communications at extremely low frequencies (from 30 to 3000 Hz), which may permit submarines to operate much deeper than that of using conventional communication systems, with which receiving antennas have to be placed within a few meters of the ocean surface for maximum signal-to-noise ratio (SNR) [Davis and Nisenoff (1977)].

5. SQUIDs to radiation detections at millimeter and submillimeter frequencies, which may be used, for instance, as receivers and detectors of millimeter and far infrared radiation from plasma diagnostics to analyze electron and ion temperature and to measure plasma drift velocities and the density of the plasma [Pedersen (1980) and Ulrich and Tutter (1980)].

6. SQUIDs to opto-magnetic investigation for studying basic interactions in solids, which have the advantage of being capable of measuring the magnetic interactions in crystal-compounds without destroying the sensitive electronic structure by an external field of the measuring apparatus [Heidrich and Mataew (1980)].

7. SQUIDs to nondestructive evaluation (NDE) of materials have essentially two main categories of applications: (1) the detection of magnetic anomalies associated with materials, such as ferromagnetic materials; and (2) the detection of magnetic field anomalies associated with electric currents, which could be induced by inhomogeneities or cracks inside the materials.

Other applications of SQUIDs are, for instance, the tracking of magnetic objects, such as a moving magnetic dipole [Wynn *et al.* (1975)], the use as computer elements [Wolf (1977)] and the use as high resolution voltmeters [Clarke (1966)].

5.3 THE JOSEPHSON LOGIC CIRCUITS AND QUANTUM ELECTRONIC DEVICES

5.3.1 Circuit Model of the Josephson Junction

It has been shown in the above sections that electrodynamic behaviors of the Josephson junction depend generally on its space-geometry as well as the physical properties of the materials forming the junction. In many practical applications, the modeling of the Josephson junction can be much simplified if the geometric size of the junction is small enough so that it could be treated practically as a lumped element. Such a simplification is common in practical circuit design. The situation is similar to our modeling of lumped elements, such as resistors, capacitors, and inductors in conventional electronic circuits.

As a lumped circuit element, the current–voltage (*I–V*) behavior of the Josephson junction can often be modeled by the so-called RSJ model (Resistively Shunted Junction model) introduced by McCumber (1968) and Stewart (1968). Mathematically, the RSJ model can be expressed by the following basic equations:

$$I = I_c \sin\varphi + \frac{1}{R_n}V + C\frac{dV}{dt} + I_F \tag{5.3.1}$$

$$\frac{d\varphi}{dt} = \frac{e^*}{\hbar}V \tag{5.3.2}$$

in which I_c is the critical current of the junction, and φ the Josephson phase. The first term on the right-hand side of Eq.(5.3.1) is the supercurrent, V/R_n is the normal current through the resistor R_n, CdV/dt is the displacement current through the capacitor C, and I_F denotes the possible fluctuation current, characterizing, for instance, thermal noise. Equation (5.3.1) shows that the total current I flowing through the Josephson junction is the sum of these current components. This set of equations enables one to calculate the current $I(t)$ provided that the voltage $V(t)$ is known and vice versa. Intuitively, the equivalent circuit model can be shown in Fig. 5.10.

It has been shown that the *I–V* characteristic of the Josephson junction is generally nonlinear, and only in some special cases, the set of nonlinear dynamic equations (5.3.1) and (5.3.2) can be solved analytically. Besides, to avoid some discrepancy between the RSJ model and the real Josephson tunnel junctions, a nonlinear-resistive (RSJN) model was also introduced. Detailed discussion about the dynamics of Josephson junctions and circuits may be found in the excellent book by Likharev (1986).

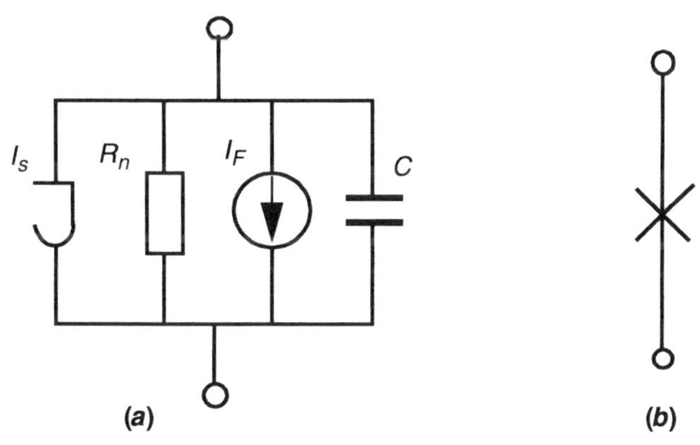

Figure 5.10 (*a*) Equivalent circuit; and (*b*) simplified notation of the Josephson junction.

5.3.2 Voltage-State Logic Circuits

It has been known since the 1960s that the Josephson junctions with their novel *I–V* characteristics may be used to construct computer logic circuits. There are two large categories of the Josephson logic circuits, classified according to whether they are voltage transfer devices or flux transfer devices. In voltage transfer devices, the information is presented by the voltage across the Josephson junction, while in flux transfer devices, information is coded by single flux quanta in superconducting quantum interferometers.

For voltage-state logic circuits, there exist essentially two types of circuits: (1) the overdrive type of circuits, and (2) the inductive type of circuits based on Nb-technology. Shown in Fig. 5.11 is the over-drive type of logic circuit. For this type of logic circuits, switching takes place when the input currents, I_{in}, overdrive the maximum zero-voltage current I_c. If the circuit were an OR gate, either I_1 or I_2 would be sufficient to produce an input greater than I_c. If it were an AND gate both I_1 and I_2 would be required.

The current passes through the junction until it reaches I_c. After switching, the junction becomes a high resistance and the current is diverted to the load R_L, which would typically be a matched transmission line with characteristic impedance of R_L. The current transferred to the load is the difference between I_{in} and the small current remaining in the junction. In this type of logic circuits also called latching logic circuits, a bias dc current slightly less than the critical current I_c of the junction is usually applied. An arriving signal current drives the total current beyond I_c, and induces its switching to its resistive state with $V \neq 0$. The reset of the resistive state of the junction to the superconducting state of the junction with $V = 0$ cannot, however, be achieved by merely turning the input signal off. One has to reset the gate by switching off the bias current.

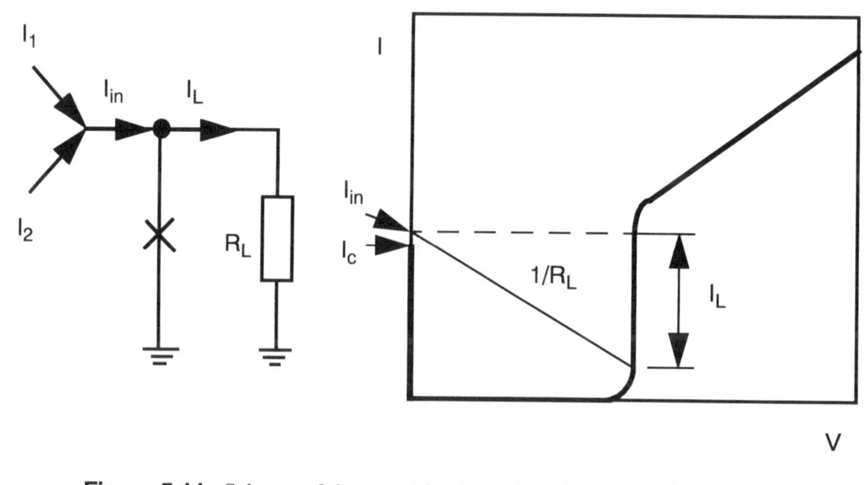

Figure 5.11 Scheme of the overdrive type of the Josephson logic circuit.

Thus, in the latching logics, periodic reset of all gates is achieved by using rf rather than dc current supply of all the gates. This operation mode has been found to have some drawbacks, such as limited clock frequencies, and poor isolation of the logic gates.

5.3.3 Single Flux Quantum Logic Circuit

To overcome the drawbacks of the voltage-state logic using hysteretic Josephson junctions, the logic family utilizing forms of single quantum flux transfer for logic signals has been proposed. This type of logic circuit uses nonhysteretic junctions, which has, in addition, the advantage of being able to use nonhysteretic high-T_c junctions. Essentially, there exist two types of single quantum flux logic circuits, (1) the rapid single flux quantum (RSFQ) logic circuits, and (2) the quantum flux parametron (QFP) logic circuits. Here, we introduce only RSFQ logic circuits [Likharev et al. (1985) and Hosoya et al. (1991)]. In the RSFQ logic circuit, binary information is presented not by the dc voltage (as in all semiconductor transistor logics, as well as in superconductor latching logics), but by very short (picosecond) voltage pulses $V(t)$ of a quantized area:

$$\int V(t)dt = \Phi_o = 2.07 \ \ mV \times ps \tag{5.3.3}$$

In operation, the binary unity may be encoded, for instance, by presence of the single-flux-quantum (SFQ) pulse during a period limited by two similar clock pulses, while binary zero, by absence of the signal pulse during this period. The essential idea is that these SFQ pulses can be quite naturally generated, reproduced, amplified, memorized, and processed by elementary circuits comprising overdamped Josephson junctions. Figure 5.12(*a*) shows the dc I–V curve of the overdamped Josephson junction with a bias current I_b. It is shown that, in contrast to the underdamped case as shown in Fig. 5.11, the curve is single-valued. It implies that, after a current pulse $I_{in}(t)$, the junction is self-reset again to its original superconducting state.

In Fig. 5.12(*b*), one shows a circuit diagram of the RS flip-flop, which contains the essential features of all RSFQ circuity. In the circuit, a dc bias current is applied to J_1, J_2, and each Josephson junction in the Josephson transmission line at 80% of their critical currents. The circuit has two modes of operation. In the first mode, a current pulse enters I_{set}, causing junction J_1 to enter the voltage state long enough to trap a single flux quantum in the loop (J_1, L_1, J_2). In this case, junction J_1 has undergone a phase change of 2π radians, and a persistent current flows through L_1 in a clockwise direction, causing a bias current to flow through J_2. A current pulse from I_{reset} is then sufficient to exceed the critical current J_{c2}, causing a fluxon to be generated and to flow in the Josephson transmission line toward the input of the next logic gate. Note

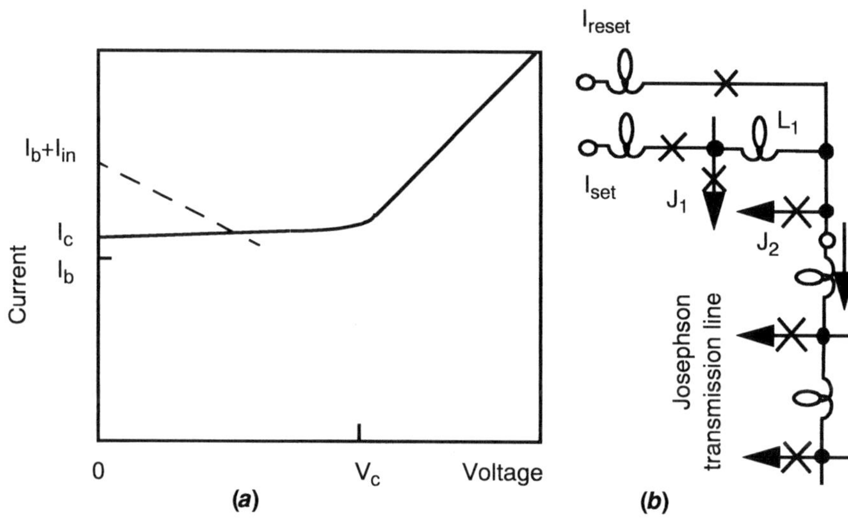

Figure 5.12 (*a*) I–V curve for an overdamped Josephson junction; and (*b*) RS flip-flop connected to a Josephson transmission line.

that this event resets the interferometer J_1, L_1, J_2 to its initial state.

In the second mode of operation, no set pulse is generated, and thus no bias current flows through J_2. When the reset current pulse I_{reset} is applied, no pulse is generated at the output. This example shows the basic principles of the RSFQ logic: first, the transmission of information using fluxons, and second, the storage and recall of information in the interferometer loops via fluxons entering and leaving these loops. Test RSFQ circuits have been made and they have demonstrated their workability at clock frequencies in excess of 100 GHz with quite decent parameter margins, despite of a relatively primitive 5-μm (all-Nb) technology used for their fabrication [Kaplunenko et al. (1989)]. Some numerical simulations show that transfer to the 1-μm technology can increase the speed beyond the 300-GHz level [Likharev and Semenov (1991)]. The RSFQ logic gates have also very low power consumption. Their $P\tau$ products are limited only by unavoidable thermal fluctuations, which is about 1000 k_BT. For T = 4 K, this is about 0.055 aJ. The minimum feature size of the RSFQ logic family, like all other Josephson junction circuits, is limited by the magnetic field penetration effects at the order of $\lambda(T) \sim 0.1$ μm, which is quite similar to the limits typical for semiconductor transistors. This factor limits integration scales of both technologies by a maximum figure of order 10^7 gates per cm^2.

5.3.4 Superconducting Transistor

Shown in Fig. 5.13 is a type of superconducting field-effect transistor proposed by Clark et al. (1980). This is a gated proximity effect Josephson weak link device. The gate controls the carrier concentration in the link/channel, thereby changing the normal coherence length ξ_n and critical current I_c of the junction. The superconducting path from source to drain is controlled by an electric potential applied to the gate electrode.

Experimentally, gate controlled supercurrents were observed for the first time in silicon-coupled, lead–lead junctions by Nishino et al. (1985). Application of gate voltage up to 200 mV changes the critical current considerably. Samples that carry no supercurrent at all can be made to carry supercurrent above 75 mV. Since the control mechanism in the superconducting FET is identical to that in conventional semiconducting FETs, with the same gate voltage swing required in both cases to produce a given change in channel carrier concentration, switching the device on and off will require the same CV^2 energy per cycle in both cases, as long as the same change in carrier density is required. Also, making the channel superconducting does not change carrier transit time, which represents a limit to the speed of the FET device.

The role of superconductivity in a superconducting FET is to change the device characteristics, allowing a zero on-state resistance. Voltage gain of the superconducting FET may be achieved, not by superconductivity, but by controlling channel resistance as in an ordinary semiconducting FET [Kleinsasser (1989)].

There is also another type of superconducting transistor, called the superconducting-base transistor. In conventional semiconductor transistors, the usual approach for extracting higher current gain and higher cut-off frequency is to make the base layer as thin as possible. A thin base layer, however, inevitably causes high base circuit resistance which, in turn, limits the

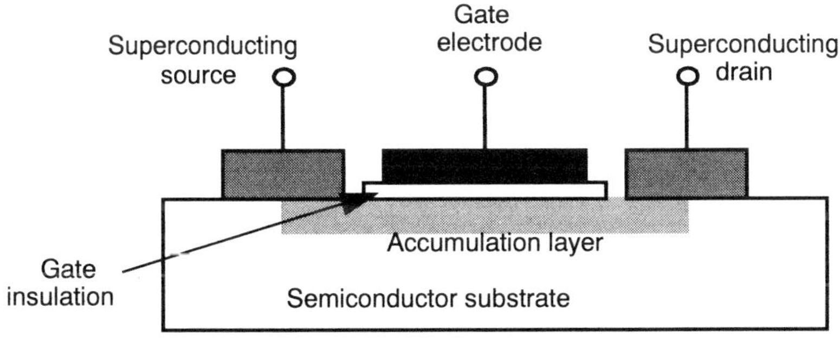

Figure 5.13 Cross-section of a superconducting field effect device, in which a superconducting path from source to drain is controlled by a potential applied to the gate.

maximum oscillation frequency. To resolve this problem, superconducting-base transistors have been proposed in the literature [Frank et al. (1985)]. Figure 5.14 shows schematically the type of superconducting-base transistor. This structure resembles that of the quiteron, but the emitter and collector need not be superconducting and there is a relatively thick semiconductor layer between base and collector, rather than a tunnel barrier. The chosen semiconductor isolation is to be in ideal ohmic contact with the base and collector electrodes. Quasi-particles (essentially unpaired electrons) are injected into the base film by tunnelling. The two "fluids" in the base are the pairs (majority carriers) and quasi-particles (minority carriers) in the superconductor, in close analogy to a bipolar transistor, but with zero base resistance. Initial estimates have shown that superconducting-base transistors should be a fast device with response in 10 ps range or better. Using high-T_c superconductors, the cut-off frequency and the maximum oscillation frequency for a superconducting-base transistor are estimated both to be 1 THz, and switching energy is estimated to be 10^{-20} J/gate [Tazoh (1991)].

The hybrid superconductor-semiconductor structures and devices may have a promising future, perhaps even more promising since the discovery of high-T_c superconductors. Many semiconductors are known to achieve improved performance at the boiling point of liquid nitrogen (77 K). The high-T_c superconductors may also provide one with a mean to connect the Josephson devices operated at 4.2 K to semiconductor devices cooled to 77 K. This also widens the applicability of the Josephson devices. At present, because of the new ceramics, superconductors can, in principle, become a generic part of any semiconductor circuit operating at these temperatures.

Recent exploration of new physical ideas that may allow larger integration scales has resulted in the prediction and experimental observation of a new

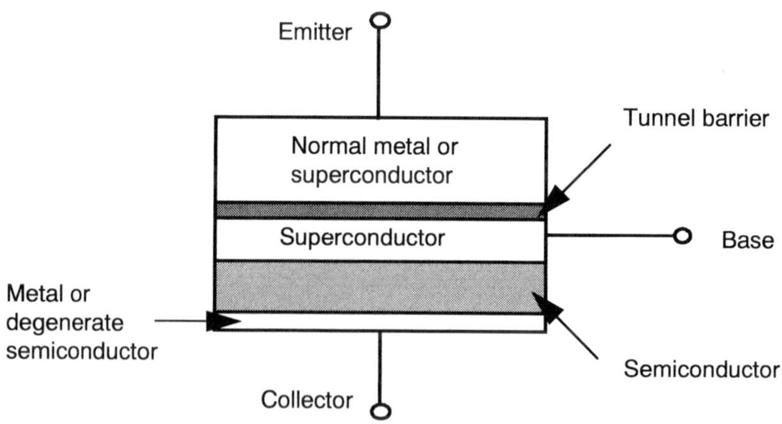

Figure 5.14 A scheme of a superconducting-base transistor.

effect, the correlated single-electron transfer in ultrasmall tunnel junctions. This new effect has created a physical basis for novel electronic devices including integrated circuits of extremely large densities up to 10^{10} gates per cm^2. In analogy with the RSFQ device, where the digital bits are presented by single quanta of magnetic flux, by the single electron tunnelling device, the single digital bits may be coded by single electrons rather than by dc voltage/current levels as in transistor circuits. Several versions of the single electron logic (SEL) devices have been proposed in literature.

5.3.5 The Coulomb Blockade and Single-Electron Tunneling

The Coulomb blockade, known in the early 1960s, is a phenomenon of suppression of tunneling of an electron through a tunnel junction if the charge Q at the junction is less than $+e/2$ and greater than $-e/2$ since tunneling in any direction would increase the energy of the system. However, if the charge Q at the junction is greater than $+e/2$ (or less than $-e/2$), an electron can tunnel through the junction in a particular direction, subtracting e from Q. If now the junction is connected to a source of constant current, a phenomenon of single-electron tunneling (SET) oscillations occurs. Simply, the single-electron oscillation may be explained as follows [Likharev and Claeson (1992)]: Suppose that the surface charge Q at the junction is zero initially, then the system is within the Coulomb blockade limits, and tunneling is suppressed. The current flowing from the source through wires will start to change the charge Q continuously. If the charge reaches and slightly exceeds, say, $+e/2$, tunneling becomes possible. One electron will then cross the junction, making its charge slightly greater than $-e/2$. Hence, the system is within the Coulomb blockade range again, and tunneling is not possible. The current continues to add positive charge to the junction at a constant rate, and Q grows until it exceeds $+e/2$ again. The repetition of this process produces the single-electron tunneling oscillations: the voltage changes periodically with a frequency equal to the current divided by the fundamental unit of charge, e. The SET phenomenon has been used to construct single-electron transistors, which are two-tunnel-junction devices that share a middle electrode. The current through the device can be controlled by changing the charge of the middle electrode just as the current through usual semiconductor transistors. Details about the SET devices is referred to in the work of, for instance, Averin and Likharev (1991).

Recently, there is an interesting report on a nanometre-scale mechanical electrometer [Cleland and Roukes (1998)]. By monitoring either the charge-induced shift in the rest point of a torsional (mechanical) resonator, or the change in the resonance frequency due to the charge modulation of the torsional spring coefficient, the sensitivity of charge detection may ultimately reach 10^{-6} $e/Hz^{1/2}$, comparable with charge detection capabilities of cryogenic single-electron transistors. The advantage of the electrometer is its ability of operating at higher temperature and response over larger bandwidth than SET.

5.3.6 Quantum-Effect Devices

Quantum-effect devices have mesoscopic structures that are large on the atomic scale but small enough that quantum mechanical coherence of the electron waves dominates the device physics. Here, there are three characteristic lengths of importance; the electron wavelength, λ_e, the electron mean free path, L_e, and the phase coherence length L_ϕ. The electron wavelength λ_e is typically in the range 10–100 nm, which is about hundred times shorter than those of visible light. Hence, the dimension of electron wave devices will have to be exceedingly small so that wavelike phenomena, such as interference, might be observable. The electron mean free path L_e is the distance an electron wave propagates before it scatters in a phase deterministic way. Devices shorter than L_e are ballistic, that is, without any scattering. At low temperatures (< 1 K), L_e can be a few micrometers, depending on materials. The phase coherence length L_ϕ is the distance an electron wave propagates before it loses phase information. The phase coherence length L_ϕ increases with increasing temperature. For instance, at low temperature (< 1 K), L_ϕ can be larger than 10 μm in a GaAs/AlGaAs heterostructure, while at room temperature $L_\phi \cong 10$ nm [Timp and Howard (1991)].

Since the wave nature of the electron is fundamental to the performance of quantum-effect devices, one or more device dimensions have to be comparable to the electron wavelength λ_e, which is typically much smaller than the phase coherence length L_ϕ. At present, it is possible to fabricate individual quantum-effect device with controllable dimensions on the order of several tens of nanometers with the most advanced nanofabrication technology. Nowadays, quantum wires and/or quantum dots can be routinely made in a number of research laboratories. Here, the quantum wire is a narrow conductive conduit with a few tens of nanometers across. An electron in the quantum wire may move freely along its length, but the electron cannot slide off to the side.

Experiments in which two quantum wires are merged have demonstrated the effects of interference, showing electrons to be traveling matter waves. However, unlike an electromagnetic wave in optical or microwave waveguide, an electron wave is sensitive to an applied electric or magnetic field because it possesses a charge. Indeed, this has resulted in the three-terminal quantum interference devices, which explores the Aharonov-Bohm effect, in which a magnetic field affects the phase of the electron wave. Quantum dots have the cross section of about a few tens of nanometers in diameter. The number of electrons per dot is in the range from 10 to 20. An array of quantum dots is interesting from the point of view of possible applications in memory and computation. Since dots contain only a few electrons, information could, in principle, be encoded via discrete quantum parameters, such as spin, charge, and so on. Also, since these dots are very small, the density of information can be extremely high, approaching to 10 billion per square centimeter. Arrays of quantum dots interacting in a predictable way, combined with cellular automata architecture may someday find their applications hopefully.

5.4 SOME PHYSICAL LIMITS IN SWITCHING TECHNOLOGIES

5.4.1 Basic Physical Limits

Superconducting switching technology has been studied for about 40 years. Today, there is revival of interest in this technology due to recently discovered high-T_c superconductors and the rapid development in information technology as well as in computer industry. In order to assess the technology, we need, however, to understand the following issues: (a) What are the realistic possibilities for speed, power dissipation, and size for superconductive switching devices? (b) What advantages do superconductive switches offer as compared to conventional and future semiconductor electronic and optical switches? These issues have been studied earlier by, for instance, Keyes (1975, 1993). In this section, we shall further investigate these questions, by taking into account recent research results on new superconducting electronics and mesoscopic electronic devices, and by considering interconnects and packaging issues from a system point of view.

According to established knowledge in physics, there are fundamental limits on possible speed, power dissipation, and size of any elementary switching device. These limits are discussed as follows. Shown in Fig. 5.15 are some basic physical limits on semiconductor, optical, and superconducting switching devices. Though this figure should be updated from time to time with the progress in semiconductor industry, it may reveal some physical limits that are fundamental.

Minimum thermal energy limit: According to thermodynamic argument [Landauer (1961)], a single yes–no logic operation in information process requires the thermodynamical minimum of energy on the order of magnitude k_BT. Here, k_B is Boltzman's constant (k_B = 1.3807×10^{-23} J/K), and T is the absolute temperature. At room temperature (T = 300 K), the minimum amount of thermal energy required is about 4×10^{-3} aJ (0.025 eV), and at low temperature T = 4 K is about 5.52×10^{-5} aJ (0.00034 eV).

Quantum limit: According to quantum mechanical uncertainty principle, a physical measurement in digital information process that is performed in a time interval Δt must involve an energy: $\Delta E \geq h/(4\pi\Delta t)$, where h (= 6.63×10^{-34} J·s) is Planck's constant. Thus the quantity: $P_o = h/(4\pi\tau^2)$ may be regarded as a quantum mechanical lower bound on the power dissipation of a continuously operating logic device, where τ denotes the switching time. This limit is plotted in Fig. 5.15, labeled as quantum limit.

Photon-number fluctuation limit: According to statistical arguments, a certain number of photons is needed in order to prevent random transitions in the state of an optical bistable device, caused by background thermal energy. Thus, though the minimum energy needed for switching is in principle one photon, a larger number of photons must be used to guarantee that the switching action almost always occurs whenever desired. A minimum of 100 photons may

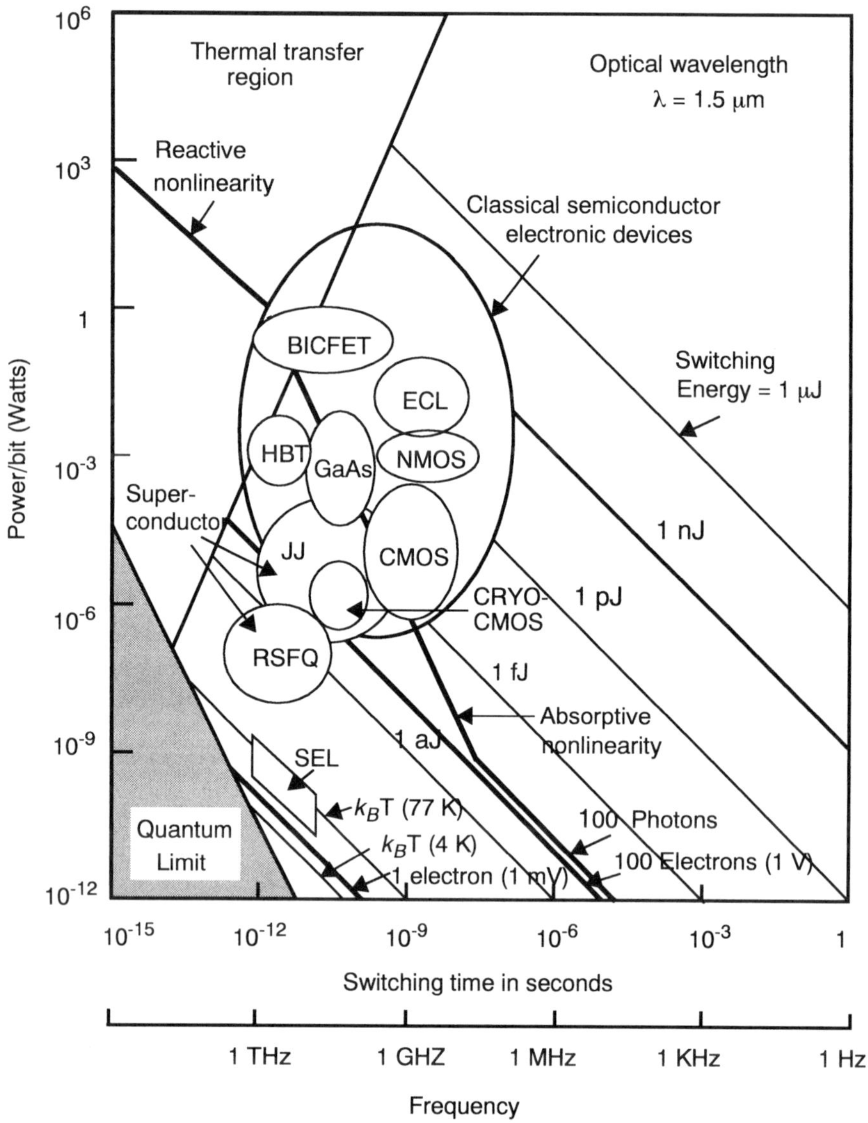

Figure 5.15 Operation ranges of some switching elements based on different technologies.

be used as a reference to define a switching state. For light of wavelength λ = 1.5 μm, the 100-photon energy reads $E = 100hc/\lambda = 13.3$ aJ, which is much greater than the thermal unit of energy k_BT at room temperature. Here, c is the velocity of light in free space.

Electron-number fluctuation limit: The limit of 100-quanta switching energy applies also to electronic switching devices. Since the scale of operating voltages in semiconductor devices and the Josephson devices are typically of order 1.5 V and of order a few mV, respectively, they yield the limits of 100-electron switching energy of about 25 aJ and 0.1 aJ, respectively.

Thermal-transfer limit: In large switching systems or computers in which a large number of logic elements are closely packed together, the heat dissipated by the logic circuits must be removed properly in order to avoid the destruction of the system due to large heat accumulation. The maximum rate at which the dissipated power must be removed sets a limit, making the combination of very short switching times and very high switching energies untenable. The thermal-transfer limit on systems using switching elements each of which dissipates power P reads: $Q_h(c_1\tau)^2 > m^2P$, where c_1 is the velocity of signal propagation, and m the number of the average inter-element spacing that signals must travel between logical operations [Keyes (1970)]. Q_h denotes some maximum rate of heat transfer. Assuming a practical value of Q_h = 100 W/cm^2, which is appropriate for liquid-cooled elements and a maximum acceptable temperature rise of 20 ^{o}C, we may plot this thermal-transfer limit in Fig. 5.15 for c_1 = 2×10^{10} cm/s and m = 1. It is noted that the thermal-transfer limit is less restrictive if the device is not operated at continuous mode, so that the energy of one switching operation has more than a bit time to be dissipated.

Nonlinearity limit: Optical switching utilizes in essence one optical pulse to control a second optical pulse by virtue of their interaction with a common atomic system. The primary limitation on the all-optical switching is a result of the weakness of the nonlinear effects in currently available materials, which makes the required switching energy rather large. A device operating by absorbing light that saturates an atomic transition and changes the optical properties of the material needs to operate to the right of the "absorptive nonlinearity", shown in Fig. 5.15, which is determined under the assumption of a large dipole moment for an atom (the product of electron charge and Bohr radius): $P > 1.2\times10^{-24}\times(\lambda/\tau)^2$ (W), where λ is the wavelength of light and τ is the switching time [Smith (1982)]. It should, however, be noted that in recent MQW devices and with the use of Fabry-Perot resonators, the effective dipole moment can become larger, shifting the line of absorptive nonlinearity to the left [Gibbs (1985)]. For a material exhibiting an optical Kerr effect, we may find a "refractive nonlinearity" limit given by: $P > n_o\lambda^3/(2n_2c\tau)$, where n_o is the (linear) refractive index of the material, and n_2 is the optical Kerr coefficient. The line labeled as "reactive nonlinearity" in Fig. 5.15 is calculated using n_2 = 6×10^{-12} cm^2/W for polydiacetylene PTS.

From Fig. 5.15, it can be seen that 1 ps switching time has been approached even in semiconductor (silicon) devices (heterojunction bipolar transistor (HBT) and bipolar inversion channel field effect transistor (BICFET)), which, however, require considerable higher switching power. Though CMOS devices, such as CRYO-CMOS, have lower power dissipation, their speed is somewhat limited currently to about 10–100 ps, even for the devices with gate length of 0.1 μm.

Optical switching elements consume, in general, much more energy than semiconductor or Josephson switching elements and, therefore, have much higher power dissipation, due to weak nonlinear optical effects, as shown in Fig. 5.15. In addition, the minimum size of an optical switching element is limited by the confinement and interaction lengths associated with wavelengths of the light, which implies that integrated optical logic circuits look rather large in comparison to electronic circuits. Besides, fabrication of integrated optical logic circuits is more difficult and expensive than the well-developed manufacturing processes for electronic circuits since optical circuits occupying a large substrate are more likely to be ruined by small defects. Optical logic devices have, therefore, not been serious contenders for general-purpose logic.

Another message from Fig. 5.15 is that the Josephson switching devices can operate with both very short (picosecond) switching time and very low power (below one microwatt). The Josephson logic devices emerge from the discovery of the Josephson junction, which can change its state very fast from a superconductive state to a resistive state by application of a magnetic field or injection of a current to the junction. These two states provide a natural mechanism for the manipulation of digital information.

Also, included in Fig. 5.15 is a newly conceived and experimentally confirmed type of device, the "correlated single-electron transfer" device [Fulton and Dolan (1987) and Likharev (1988)], which is based on quantum effects in ultrasmall tunnel junctions. Such a single-electron logic (SEL) device has been experimentally tested, which shows even lower power dissipation close to the quantum limit. However, so far, it has been known that the very small capacitance of the tunnel junctions needed to observe the single-electron effect (Coulomb blockade) are difficult to fabricate and are easily overwhelmed by stray capacitance. The energies involved are so small that they are dominated by thermal energy unless the temperature is so low (i.e., at 4 K for tunnel junctions with sizes on the order of 10 nm) that $k_B\mathrm{T} << (e^2/C_J)$, where e is the electron charge, and C_J is the capacitance of a tunnel junction. Random background charges, noticed recently, may also easily disturb I–V characteristics of the device. In addition, the energy levels are broadened and the Coulomb blockade effect is obscured unless the residence times of electrons on the node are long. The impedances involved are much larger than $h/e^2 = 25$ kΩ. Only small currents can be controlled because of the small amounts of charge involved and the high impedances. As a result, communication with other devices is very restricted; there is no ability to drive lengthy lines or to provide large fan-out [Keyes (1993)]. The potential for its application to large electronic switching systems seems to be rather limited unless novel circuit architecture could be developed.

Recently, with the progress in manufacturing exceedingly small structures out of semiconductors and metals, research efforts have been conducted to explore electronic devices based on quantum principles. New techniques for manipulating atoms, such as molecular-beam epitaxy and scanning-tunneling microscopy, now come close to the ideal of selecting one atom at a time and setting it in place. The so-called quantum wires or quantum dots have been fabricated in laboratories and have been used to construct quantum interference devices with dimensions smaller than 100 nm. However, because random thermal energy of atoms at room temperature is great enough to disturb the quantum states of electrons in motion, interference experiments with quantum wires, for instance, require, so far, temperatures too low to be feasible for general use. Generally speaking, operation of quantum-effect devices at 77 K will require confinement to dimensions in the range 10–50 nm [Ismail et al. (1991)]. With field-effect-controlled devices, the confining potential well would be several times smaller than minimum lithographic features. Furthermore, it will likely be necessary to control line-width variations to less than about 10% of the minimum line-widths of these devices in VLICs so that quantum levels are predictable and repeatable. At present, the foreseeable nanofabrication technology based on x-ray nanolithography has the line-width limit of about 10 nm. Clearly, for commercial uses of these quantum devices, one may have to await the breakthrough in technology needed to form quantum wires and relevant devices many times smaller than it is possible to form today. Recent research activities in various growth techniques [Oshinowo et al. (1994)] for making nanostructures are very interesting and promising.

Besides nanofabrication techniques, we have to also face to the challenge of novel device concepts and interconnection issues in the quantum world as well as its interface to the classical world which we live. At present, massive efforts in basic research are required to understand and control quantum effects in artificial nanostructures before we could reach clear-cut ideas for developing electronic systems based on quantum effects. On the other hand, as we may know, superconductivity is a macroscopic quantum phenomenon. Superconducting quantum interference devices (SQUIDs) have already been commercialized. In this promising field, there is also a steam of new ideas concerning superconductive electronic devices and circuits that have not yet been explored, and requiring further research efforts.

5.4.2 Logic-Level Voltage Limit

Like optical logic devices, electronic logic devices make also use of nonlinearity. To produce the necessary nonlinear response to electrical signals in semiconductors, a change in energy per electron of many k_BT is needed to override the effect of random thermal motion of electrons. This implies the existence of a lower limit on the voltage that is applied to change the energy of electrons in different regions of, say, a p–n junction. Obviously, the minimum voltage is on the order of $k_B T/|e|$, which is about 25 mV at 300 K. In practice, voltage required is usually much larger compared to $k_B T/|e|$ due to many factors, such as the factor that the properties of devices are not perfectly reproducible and controllable by processing technologies. From the voltage requirement, one may also see that a large number of electrons is usually needed to process information with present technology, since the number of electrons on a capacitance is $n=CV/|e|$. By noting the voltage requirement $V >> k_B T/|e|$, we have then $n >> Ck_B T/e^2$. Thus the value of C that corresponds to one electron is less than 5 aF at room temperature, which is much less than the capacitance of manufacturable devices and interconnections and even of stray capacitances. The voltage limit discussed above plays an important role in power dissipation and in the device limits. Clearly, reducing temperature is one way of reducing voltage required.

For the Josephson tunneling devices, there exists also a requirement analogous to the voltage requirement ($V >> k_B T/|e|$), which states that the gate current I has to be much larger than $k_B T/\Phi_o$ [Keyes et al. (1970)]. Here, $\Phi_o = h/(2|e|)$ is the flux quantum. This amount of flux must be produced by a current I flowing through an inductance L in the performance of a switching operation. The value of the inductance that permits the minimum energy dissipation, approaching to $k_B T$, can be estimated to be of order: $L_{min} \sim \Phi_o^2/k_B T$, which is about 4 nH at 77 K. The inductance of order of 1 nH can be easily fabricated, permitting operation of superconducting circuits with a single quantum of flux, in contrast to ordinary electronics, for which many quanta of charge, electrons, are needed with manufacturable capacitance.

5.4.3 Physical Limits on Interconnect and Packaging

Although fast switching speed is important, switching speed of individual logic device alone cannot produce the fastest computers or communications switching systems. Issues like wiring and system integration are also very important. In computers or communications switching systems of future, limiting speeds may not be set by the speed of the logic devices, but instead by the quality and length of the wires that connect the logic devices in the system. We have observed the progression of microelectronics driven by the scaling of integrated circuit (IC) features to smaller dimensions. The forces driving this tremendous miniaturization are cost, reliability, and power. Huge decreases in cost and in power dissipation and large increases in reliability have made it possible to construct systems with enormous numbers of electronic components. The scaling rules for reduced feature dimensions have been applied not only to active devices, but also, to avoid increasing the fraction of IC area consumed by interconnections, to the cross section of interconnections. In a large system, logic circuits must operate, transmit, and receive signals. The close packing of devices, wires, and chips allows elements to influence their neighbors, causing thermal and EMC problems. In any case, the system must function correctly in spite of these hostile aspects of the environment. To identify potential technologies and design properly the integrated circuits, one has to consider carefully some physical limits on the interconnection and packaging of ICs.

Wiring and resistance limit: Probably, the hardest part of making an IC chip is making the wires. The amount of wire, measured in the right way, grows faster than the number of activc dcviccs and circuits. With thc increasing density of active components by miniaturization, the density of wire channels, which may be described by the total length per unit area of the channels in which a wire can be placed, has to increase in order to provide necessary communication channels among these components. Extrapolation of the growth of the density of wire channels indicates that one may soon reach the wire density of order of 10^4/cm [Keyes (1993)]. Due to the limited area of chips, the increasing wire density reduces the width of wire. Besides, since the thickness h of wire is decreased with its width *W* due to the limitations of manufacturing processes, the cross-sectional area of wire decreases nearly as the square of their width. This will cause an increase in the resistance of wire by miniaturization.

Shown in Fig. 5.16 is the dependence of dc wire resistance on wire density for copper wires at room temperature. Here, two-layer wiring with an equal width *W* of separation of wire channels, having a width-to-thickness ratio of *W*/h = 3, is assumed. It is seen that the dc wire resistance increases rapidly when the wire density exceeds 10^4/cm. The high value of resistance would be a source of difficulty. In addition, the finer the wire, the smaller the defect that can cause an open or short circuit; and the larger the area, the greater the probability that it contains a defect.

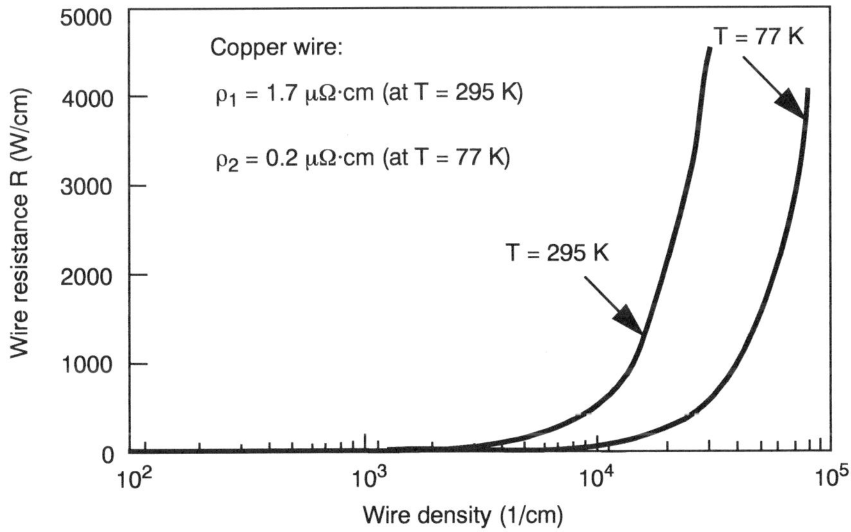

Figure 5.16 Dependence of wire resistance on wire density.

The increasing wire resistance has obviously a fundamental origin by miniaturization. With normal conductors, the wire resistance increases with the inverse of the scaling factor. This is intolerable for high-speed circuits, not only because there is an undesirable voltage drop on the wires which cuts into the circuit margins but the unchanged RC product of the lines makes it difficult to utilize the increased device speed. Also, the higher wire resistance is the higher power dissipation and the higher Johnson noise, which is of prime importance in memory sense lines where very low level signals are encountered. Furthermore, the increasing current density makes the wiring vulnerable to ruptures caused by the effect of electromigration, which is a phenomenon of failure of wires due to the diffusive motion of atoms in the wires caused by electric forces. Nowadays, electromigration may place stringent limits on reliability and performance of integrated circuits as the miniaturization continues.

One way of overriding the resistance limit is to use several layers of wires in a chip. At present four layers of wire are formed on some chips, but over 30 could be formed in multichip packaging substrates. The solution of using multilayer wiring is however expensive because it requires a large number of processing steps which also lower the fabrication yield. It is, therefore, difficult to accept the idea of an unlimited number of wire layers, and it is hard to escape the reality of a resistance limit.

On the other hand, superconducting wire has a considerable advantage because the dc wire resistance is simply zero. As a consequence, the zero dc electric field may prevent electromigration. For high-frequency signals approaching 100 GHz, superconducting interconnects have very low attenuation and dispersion. We may pass picosecond waveforms for distances well exceeding typical chip size with a low crosstalk. Ultrafast digital signals can thus be passed along the chips with a propagation speed approaching that of light. This fact makes that superconductivity may first appear in supercomputers or ultrafast switches in the form of simple wiring. In addition, the superconductive wiring of a given dimension can carry a thousand times more current than the same-sized copper wire. Thus, for fixed current level, superconductive wire can be made much smaller, liberating precious chip real estate for other purposes. Also, expensive and low-yield multilayer wiring techniques are not necessary in a superconducting environment.

The superconductive wiring has some limitations, too. When the wire dimensions become comparable with the London penetration depth ($\sim 0.1\ \mu$m), then the inductances no longer scale properly. Furthermore, the wire cross-section has to stay large enough for the critical current of the wires to remain appreciably above the operating currents. When one of these limits is reached, then multi-layer wiring will have to be introduced. However, the number of layers in a superconducting environment should still be considerably smaller than in a normal-conducting one.

Power dissipation limit in IC packaging: One of the important limitations on the performance of IC chips is the ability to remove heat from them.

Convection cooling in air allows removal of about 1 W of heat per cm^2 of chip area. Forced air cooling may remove heat of about 5 W/cm^2. Thermal conduction cooling, whereby a metallic piston is in contact with the chip and conducts the heat away to a water-cooled plate allows roughly removal of 10 W/cm^2 of heat. Direct liquid cooling of a chip by immersion in a suitable fluorocarbon liquid, which carries the heat away by nucleate boiling, with about 20 W/cm^2 is not much better. Using microchannel liquid cooling technology, 160 W/cm^2 of heat may be removed from the IC chip with temperature increasing of order 10 degree. Though there are some extrapolations on future cooling technologies of removing heat of several thousand watts per square centimeter, it has recently been argued that the realistic limit for the cooling technology is about a few hundred watts per square centimeter.

Illustratively, we may consider a semiconductor chip, predicted recently to have a capacity of 16 Gbits/chip operating at 1 GHz with the chip size of 10 cm^2 and feature size of 0.1 μm in the beginning of the next century. An estimation shows that the required power dissipation for such a chip would reach 800 W/cm^2, assuming each bit dissipates 0.5 μW without even taking into account of the power dissipation from interconnects. Of course, in many circuits, only a small fraction of devices operate at any given time cycle, so the limits set by power removal may not be reached. However, the performance of some special circuits can indeed be limited by the heat dissipation tolerances of present packaging technology. Besides, the supply of large amount of electric power to such chips will become increasingly difficult. The fact that the pursuit of miniaturization causes power dissipation to become an increasingly difficult aspect of technology is, therefore, very fundamental.

Recently, there have been some studies on the comparison of power dissipations between optical and electrical interconnects [Feldman (1988) and Miller (1989)]. It was argued that although optics cannot be justified on logic-device switching energies, optical devices and interconnects may offer lower communication energies than electrical devices and interconnects. Since much of the energy in electronic chips is for communication between logic devices, these studies analyzed quantitatively some energy dissipations in both optical and electrical interconnects, and concluded that optical interconnects may be advantageous even for intrachip communication in large area VLSI circuits or wafer scale integrated circuits, especially when high data rates are required. This conclusion was thought to be not changed even for perfectly lossless (e.g., superconducting) electrical interconnects [Miller (1989)]. We should, however, notice that in superconducting logic circuits, the operating voltages are on the order of 1 mV rather than 1 V. Since the minimum energy required for an electrical interconnect is $C_T V^2$ with C_T being the total capacitance of the interconnect including input and output capacitances and V the logic-level voltage, we may easily show that superconducting interconnects in Josephson logic circuits consume much less energy (with a factor of about 10^{-5}) than normal electric interconnects in semiconductor logic circuits, where the logic-level voltage is on the order of 1 V. If we now compare the result with that of optical interconnects, we may find that not only superconducting logic devices consume much less energy than optical logic devices do, but also the superconducting interconnects may consume less energy than optical interconnects, including optical sources and detectors, for intrachip communication even at foreseeable high data rates. Besides, superconducting interconnects have advantages of wider dynamic range and virtually unlimited number of taps, as compared to optical interconnects.

Light-speed limit: Although power dissipation sets a practical limitation on IC chips, there is a fundamental limitation on the finite propagation speed of signals on the chip. Here, we have to consider the issue of risetime effects for logic gates driving signal lines and other gates. In discussing risetime effects, we often differentiate between a line that is electrically "long" and one that is electrically "short". A long line is defined as a line for which the risetime is less than twice the signal propagation time of the line. The signal risetime for an electrically short line is given by the RC time constant, and the total delay is proportional to the square of the line length. This is the time required to charge the line to the new signal voltage level. The total delay thus is dominated by the charging time, and the time of light-flight is negligible. As the line is made longer, propagation effects dominate until the transmission line regime is entered. Thus the total delay for a short line should slowly increase as a function of line length until it merges into the light-speed (transmission line) case. As the conceived miniaturization proceeds and chip sizes become larger and larger, the interconnects may become physically longer, relative to their cross-sections. Thus microelectronic system interconnections will be no longer well behaved, short interconnections, but instead, less well-behaved, longer interconnections. Therefore, in future high-speed IC chips, with clock frequencies of the order of probably hundreds gigahertz, the effect of the finite propagation speed of light has to be carefully examined.

For a synchronous system (chip) controlled by a single master clock of frequency f_c, there exists obviously a fundamental limit on the propagation speed of signals on the chip. The maximum propagation speed attainable is the speed of light. This places a fundamental upper limit on the length of links that can be used without encountering synchronization errors. Since the master clock must communicate with gates all over the chip, this implies an upper limit on the linear dimensions of a chip. Quantitatively, the limitation can be stated as: $A < (c/f_c)^2$, where A is the area of the chip and c the speed of light [Bate (1985)]. In Fig. 5.17, we show the light-speed limit on the maximum clock frequency attainable in a synchronous system for a given chip area.

It is shown that only asynchronous or self-timed chips can have areas or clock frequencies greater than this limit. Here, we see that the global synchronization of an IC at clock frequency over 100 GHz is impractical due to the light-speed limit. Thus expected high-speed ICs at clock frequencies of order of 100 GHz can only be asynchronous or self-timed. Self-timed systems require, however, more hardware than the synchronous systems because of the extra logic required for signaling and communication. They are also harder to design. A recent study [Likharev and Semenov (1991)] has indicated that the natural intrinsic memory of superconducting RSFQ logic cells may open the way to very hardware-saving digital devices with bit-level conveying. The introduction of local self-timing will, therefore, allow natural design of ultrafast superconducting digital processor units at ultrahigh clock frequencies beyond 100 GHz.

As a further consequence of the light-speed limit, one may imagine that future ultrafast switches or supercomputers face a natural size limit. Consider the fact that light can only travel a distance of about 0.3 mm within a time interval of one picosecond. If two logical devices having a switching speed of one picosecond are connected by a wire of the length 10 mm, then, even with the light speed, the fastest attainable speed in the universe, it will require, at least, 66 picoseconds for them to act cooperatively due to the signal exchange between these two logical devices. Because of this, a future core unit of an ultrafast switch or a supercomputer that breaks the nanosecond barrier will have to be no larger than a "teacup" in order to avoid being slowed down by the signals traveling within it.

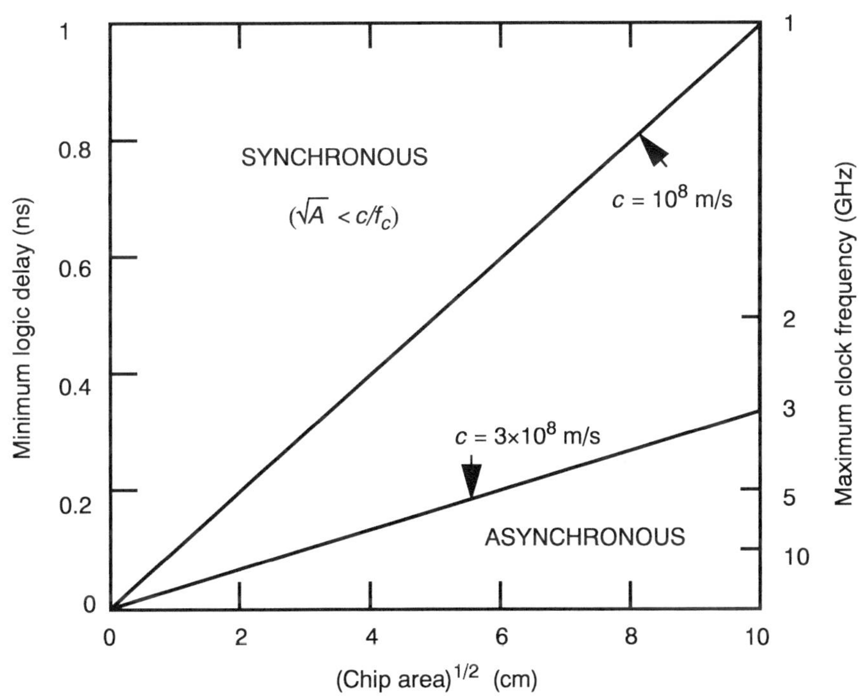

Figure 5.17 Light-speed limit on a synchronous IC for a given chip area.

6

Electromagnetic Analysis of Transmission Line and Waveguide

Interchange of electric and magnetic energy is known to give rise not only to the propagation of electromagnetic waves in free space, but also to waves that are guided by conducting or dielectric boundaries. The guided waves are of great importance in guiding electromagnetic energy from a source to a distant device or system. One of the simplest and very important types of waveguides is the two-conductor transmission line. In this chapter, the theory of transmission lines will first be introduced based on a distributed-circuit model. A relationship between the circuit theory and electromagnetic field theory is then established for the transmission lines. By the field theory based on Maxwell´s equations, we shall demonstrate how the distributed circuit elements, such as the line resistance, line inductance, line capacitance, and line conductance of the transmission lines of complex geometry and material composites can be analyzed with the use of the finite element method at the quasi-TEM approximation. To illustrate the limit of classical quasi-TEM approximation, a full-wave analysis is carried out to investigate properties of anisotropic superconducting transmission lines which may operate at extremely high frequencies. Electromagnetic behaviors of coupled transmission lines and directional couplers are studied. Full-wave analysis for waveguide with conducting boundary is introduced. Electromagnetic properties of microwave resonators, such as discrete element resonant circuits, transmission line resonators, and cavity resonators are also discussed.

6.1 TRANSMISSION LINE THEORY

6.1.1 Formulation of Transmission Lines

A transmission line consists of two or more parallel conductors, which can be either normal conductors or superconductors. A typical example is the two-wire line, as shown in Fig. 6.1. Other examples are, for instance, the coaxial line and various types of striplines or microstrip lines. In the ideal case, when the

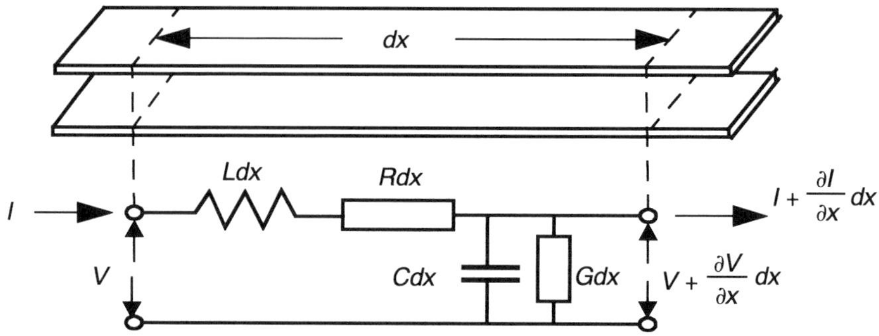

Figure 6.1 Equivalent circuit of a differential length of transmission line.

conductors can be considered to be perfect conducting, the basic mode of the line is the TEM mode. Practical lines are always not perfect, even for superconducting lines, and this results in a perturbation or change from a TEM mode to a mode that has a small axial component of electric field. Many practical lines have such small losses that we may analyze their behavior by considering them as ideal lines, and then make a simple perturbation calculation to obtain the effect of losses. Although the TEM transmission line wave is but one special case of guided wave propagation, it is so important practically that it is usually treated as ordinary transmission line theory quite early in the training of electrical engineers. In this section, we shall introduce a set of the basic equations for describing propagation behavior of the TEM transmission line wave. Since a unique voltage and current wave may be associated with the electric and magnetic fields of a TEM mode on a transmission line [see Collin (1966)], we may describe the transmission line as a distributed-circuit network..

Shown in Fig. 6.1 is the equivalent circuit of a section of transmission line of differential length dx. If the voltage and current at the input are $V(x, t)$, $I(x, t)$, and at the output are $V+(\partial V/\partial x)\, dx$, $I+(\partial I/\partial x)dx$, we have then, according to the Kirchhoff law,

$$V - \left(V + \frac{\partial V}{\partial x} dx\right) = IRdx + Ldx\frac{\partial I}{\partial t} \tag{6.1.1}$$

which gives

$$\frac{\partial V}{\partial x} = -RI - L\frac{\partial I}{\partial t} \tag{6.1.2}$$

Similarly, we find

$$\frac{\partial I}{\partial x} = -GV - C\frac{\partial V}{\partial t} \tag{6.1.3}$$

Equations (6.1.2) and (6.1.3) constitute the set of basic equations for transmission lines. Equation (6.1.2) states that the potential difference between the input and output is equal to the potential drop across the line resistance R (Ω/m) and the line inductance L (H/m). Equation (6.1.3) states that the output current is less than the input current by an amount equal to the shunt current flowing through the line capacitance C (C/m) and line conductance G (1/Ω-m).

With time-harmonic excitation ($\sim \exp(i\omega t)$), the set of transmission line equations (6.1.2) and (6.1.3) becomes

$$\frac{\partial V(x,\omega)}{\partial x} = -(R + i\omega L)I(x,\omega) \tag{6.1.4}$$

$$\frac{\partial I(x,\omega)}{\partial x} = -(G + i\omega C)V(x,\omega) \tag{6.1.5}$$

In general, the line parameters R, C, G, and L are functions of the frequency ω (rad/s). Here, the direction of wave propagation is chosen along the x-axis direction. From this set of transmission line equations, we may obtain solutions of the general form:

$$V = V_A e^{-\gamma x} + V_B e^{\gamma x} \tag{6.1.6}$$

where γ is the complex propagation constant, given by

$$\gamma^2 = (R + i\omega L)(G + i\omega C) \tag{6.1.7}$$

By writing $\gamma = \alpha + i\beta$ with $\alpha \geq 0$ and $\beta > 0$, we may get from Eq.(6.1.6)

$$\alpha = Re\{\sqrt{(R + i\omega L)(G + i\omega C)}\} \tag{6.1.8}$$

$$\beta = Im\{\sqrt{(R + i\omega L)(G + i\omega C)}\} \tag{6.1.9}$$

where α is the attenuation constant and β the real propagation (phase) constant.

It is shown from Eq.(6.1.4) that the current I is given according to the general solution by

$$I = \frac{-\gamma}{R + i\omega L}(-V_A e^{-\gamma x} + V_B e^{\gamma x}) \tag{6.1.10}$$

Introducing the impedance Z_o of a transmission line, defined by the ratio of voltage to current at a given point along the line, we have

$$Z_o = \frac{V}{I} = \frac{R + i\omega L}{-\gamma} \times \frac{V_A e^{-\gamma x} + V_B e^{\gamma x}}{-V_A e^{-\gamma x} + V_B e^{\gamma x}} \tag{6.1.11}$$

If we assume that the line is infinitely long, then it is necessary that the constant V_B is equal to zero for a wave propagating along the positive x-direction of the line; otherwise, as x becomes large, the terms with positive exponent become very large, an undesirable and unrealistic condition. Thus, by setting $V_B = 0$, we find

$$Z_o = \sqrt{\frac{R + i\omega L}{G + i\omega C}} \tag{6.1.12}$$

which is usually called the characteristic impedance of the line.

Obviously, if the transmission line is lossless such that $R = G = 0$, then the characteristic impedance becomes simply $Z_o = (L/C)^{1/2}$, which is a pure resistance, independent of frequency so long as L and C are independent of frequency. For low-loss lines with $R << \omega L$ and $G << \omega C$, we may find approximately the phase constant by

$$\beta = \omega\sqrt{LC} \tag{6.1.13}$$

and the attenuation constant

$$\alpha = \frac{1}{2}\sqrt{LC}\left(\frac{R}{L} + \frac{G}{C}\right) \tag{6.1.14}$$

The physical meaning of the phase constant β and the attenuation constant α can be understood by considering the voltages V_o and V_l at two points, $x = 0$ and $x = l$, on the line. Their ratio gives then

$$\frac{V_l}{V_o} = e^{-\alpha l}e^{-i\beta l} \tag{6.1.15}$$

It is obvious that, for a lossless line with $R = G = 0$, we have $\alpha = 0$, which means that a line with no losses has no attenuation as we may expect. Here, the phase shift for the given length l of the line is thus determined by Eq.(6.1.15) for $\alpha = 0$. The phase shift per unit length along the line is a measure of the velocity of propagation of a wave along the line:

$$\mathrm{v} = \frac{\omega}{\beta} = \frac{1}{\sqrt{LC}} \tag{6.1.16}$$

where ω is the radian frequency of the wave.

6.1.2 Incident and Reflected Waves Along a Transmission Line

In this section, we study the properties of a transmission line terminated in an arbitrary load impedance Z_l, as shown in Fig. 6.2. For this problem, we have the following boundary conditions:

$$V(0) = V_s - Z_s I(0) \tag{6.1.17}$$

$$V(l) = Z_l I(l) \tag{6.1.18}$$

where V_s and Z_s denote the source voltage and the source impedance, respectively. By the general solution from Eq.(6.1.6), we may find that the two constants V_A and V_B can be determined by

$$V_A + V_B = V_s - \frac{Z_s}{Z_o}(V_A - V_B) \tag{6.1.19}$$

and

$$Z_l = Z_o \frac{V_A e^{-\gamma l} + V_B e^{\gamma l}}{V_A e^{-\gamma l} - V_B e^{\gamma l}} \tag{6.1.20}$$

It can be seen from the general solution (6.1.6) that V_A is the amplitude of the positively traveling component, and V_B is the amplitude of the negatively traveling component of the voltage wave.

We may introduce a voltage reflection coefficient Γ_V at the load ($x = l$), defined by

$$\Gamma_V = \frac{V_B e^{\gamma l}}{V_A e^{-\gamma l}} = \frac{V_B}{V_A} e^{2\gamma l} \tag{6.1.21}$$

From Eq.(6.1.20), we may find

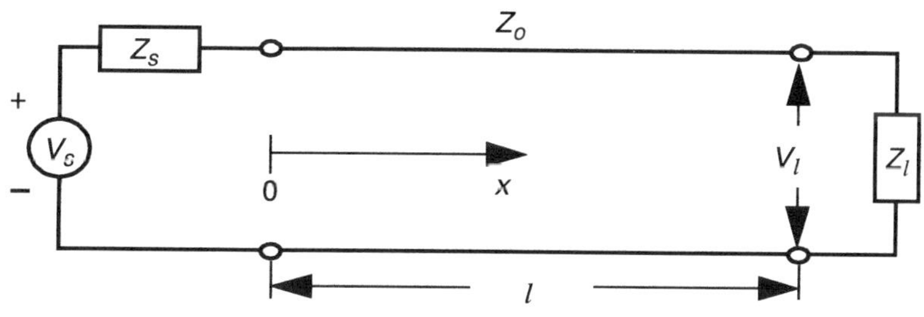

Figure 6.2 A transmission line terminated in a load impedance Z_l.

$$\Gamma_V = \frac{Z_l - Z_o}{Z_l + Z_o} \tag{6.1.22}$$

Similarly, we may introduce a current reflection coefficient Γ_I at the load, given by

$$\Gamma_I = \frac{I_B e^{\gamma l}}{I_A e^{-\gamma l}} = \frac{Z_o - Z_l}{Z_l + Z_o} \tag{6.1.23}$$

with the use of Eq.(6.1.10). It is shown that the current reflection coefficient is just the negative of the voltage reflection coefficient, that is, $\Gamma_I = -\Gamma_V$.

The reflection coefficients characterize the amount of voltage (or current) which is reflected back, away from the load. Noting that the amount voltage transmitted to the load is the sum of the incident wave plus reflected wave, we may introduce a voltage transmission coefficient T_V, defined by

$$T_V = \frac{V_A e^{-\gamma l} + V_B e^{\gamma l}}{V_A e^{-\gamma l}} = 1 + \Gamma_V \tag{6.1.24}$$

which is the ratio of the voltage at the load to the incident voltage. Similarly, a current transmission coefficient T_I can be defined and given by

$$T_I = 1 + \Gamma_I \tag{6.1.25}$$

It is apparent from Eqs.(6.1.22) and (6.1.23) that, when $Z_l = Z_o$, both reflection coefficients are zero, which implies that there can be no reflected traveling wave on an impedance matched line. In this case, all the power in the incident wave is transmitted to the load and none of it is reflected back toward the generator. By the power delivered to the load given by

$$P_l = \frac{1}{2} Re(V_l I_l^*) \tag{6.1.26}$$

we may find $P_l = |V_l|^2/(2Z_0)$ for the case of a lossless line without reflection. Here, the superscript * denotes the complex conjugation.

If the load impedance Z_l is not equal to the characteristic impedance of the line Z_o, the load is mismatched to the line and a reflected wave is produced. In such a case, the power delivered to the load is given by

$$P_l = \frac{1}{2} Re(V_l I_l^*) = \frac{1}{2Z_o} |V_A|^2 (1 - |\Gamma_V|^2) \tag{6.1.27}$$

where $|V_A|$ is the magnitude of the incident voltage. This result indicates that the power delivered to the load is the incident power minus that reflected from the load.

In the absence of reflection and losses, the magnitude of the voltage along the line is a constant equal to $|V_A|$. When a reflected wave also exists, the incident and reflected waves interfere to produce a standing-wave pattern along the line. The voltage at any given point along the line is then given by

$$V = V_A e^{-\gamma x}[1 + \Gamma_V e^{-2\gamma(l-x)}] \tag{6.1.28}$$

For a lossless line, the voltage has a magnitude by

$$V = |V_A|\left|1 + \Gamma_V e^{-2i\beta(l-x)}\right| \tag{6.1.29}$$

If we let $\Gamma_V = |\Gamma_V| \exp(i\theta)$, we have then

$$V = |V_A|\left\{(1 + |\Gamma_V|)^2 - 4|\Gamma_V|\left[\sin\left(\beta(l-x) - \frac{\theta}{2}\right)\right]^2\right\}^{1/2} \tag{6.1.30}$$

The result shows that the magnitude of the voltage $|V|$ oscillates back and forth between the maximum values of $|V_A|(1+|\Gamma_V|)$ when $\beta(l-x)-\theta/2 = n\pi$, and the minimum values $|V_A|(1-|\Gamma_V|)$ when $\beta(l-x)-\theta/2 = n\pi + \pi/2$, where n is an integer. Intuitively, these results states that voltage maxima occur when the incident and reflected waves add in phase, and that voltage minima occur when they add 180° out of phase. Successive maxima and minima are spaced a distance $\pi/\beta = \lambda/2$ apart, where λ is the wavelength for the TEM wave in the medium surrounding the conductors. The distance between a maximum and the nearest minimum is $\lambda/4$.

The ratio of the maximum line voltage to the minimum line voltage is called the *voltage standing-wave ratio* (*VSWR*), defined by

$$VSWR = \frac{1 + |\Gamma_V|}{1 - |\Gamma_V|} \tag{6.1.31}$$

The *VSWR* is a useful concept when dealing with single frequencies and is often used as a figure of merit in specifying commercial high-frequency equipments.

It has been shown that a lossless line terminated in its characteristic impedance will always have an input impedance which is a pure resistance of value $(L/C)^{1/2}$. If a transmission line is not terminated in its characteristic impedance, a reflection will occur, resulting in a standing wave. When such standing waves are present on a line, the input impedance is no longer a pure resistance but rather can take on various values, including reactive components, depending on the termination and electrical length of the line. To find the input impedance for any line which is not properly terminated, we may start from the determination of the impedance of a line at any point x, looking toward the load with coordinates as in Fig. 6.2, by

$$Z_x = Z_o \frac{V_A e^{-\gamma x} + V_B e^{\gamma x}}{V_A e^{-\gamma x} - V_B e^{\gamma x}} \tag{6.1.32}$$

which can also be written in terms of the reflection coefficient Γ_V as

$$Z_x = Z_o \frac{1 + \Gamma_V e^{-2\gamma(l-x)}}{1 - \Gamma_V e^{-2\gamma(l-x)}} \tag{6.1.33}$$

If the input impedance of a line of length l is desired, then Z_x is the impedance at the input terminals; thus, $x = 0$ and Eq.(6.1.33) becomes

$$Z_{in} = Z_o \frac{1 + \Gamma_V e^{-2\gamma l}}{1 - \Gamma_V e^{-2\gamma l}} \tag{6.1.34}$$

In general, Eq.(6.1.34) can also be expressed in terms of the load and characteristic impedances by using Eq.(6.1.22):

$$Z_{in} = Z_o \frac{Z_l + Z_o \tanh(\gamma l)}{Z_o + Z_l \tanh(\gamma l)} \tag{6.1.35}$$

which shows that the input impedance of the line with given length l depends not only on the characteristic and termination impedance, but also on frequencies since the complex propagation constant γ is generally a function of frequency.

When the load impedance $Z_l = 0$, that is, the line is shorted at $x=l$, the input impedance becomes

$$Z_{in} = Z_o \tanh(\gamma l) \tag{6.1.36}$$

For a lossless line, we have

$$Z_{in} = Z_o \tanh(i\beta l) = iZ_o \tan\left(\frac{2\pi l}{\lambda}\right) \tag{6.1.37}$$

which shows that the input impedance of a quarter-wavelength, short-circuited at its load, looks like an infinite impedance to the source generator.

Similarly, if the line is open-circuited so that Z_l is infinite, the input impedance for an lossless line becomes

$$Z_{in} = Z_o \coth(i\beta l) = -iZ_o \cot\left(\frac{2\pi l}{\lambda}\right) \tag{6.1.38}$$

which shows that a quarter-wavelength, open line looks like a short circuit (zero input impedance). For the general case when the load impedance has a finite non-zero value, the input impedance will contain real and imaginary parts. Evaluation of these parts was a tedious task. Some methods, such as the well-known Smith chart, had been developed in the past. Today, with the aid of personal computers, such an evaluation becomes a relatively easy task. Nevertheless, the Smith chart is still used widely among electrical engineers. Readers who are interested in the method of Smith chart may refer to standard text books, such as Collin (1966).

6.1.3 Pulse Propagation in a Transmission Line

We have studied the characteristics of transmission lines with time-harmonic (sinusoidal) excitation. For a uniform, lossless and dispersionless line, sine waves may propagate along the line without change and will propagate at the same velocity for all frequencies. Thus, for an ideal line, a pulse will propagate with no distortion, arriving at the load with a waveform identical to that at the input end but delayed in time by an amount equal to the length of the line divided by the velocity of propagation, that is, the delay time of the line. This is understandable since a single pulse is composed of an infinite Fourier frequency spectrum with the amplitude of the spectrum decreasing for higher frequencies. These frequencies all add together at the input end to form the proper waveform. If all these frequency components travel along the line at the same speed and experience no change, then it is obvious that they must all arrive at the same time at the load and will add together to give the same waveform as at the input end. If, however, the line is non-ideal, having losses and dispersions, the phenomenon of pulse distortion will arise because the various frequency components travel at different speeds due to dispersion, and/or suffer different amounts of attenuation due to frequency-dependent losses.

In this section, we shall formulate the propagation of Gaussian pulses over a transmission line of length l, using the circuit shown in Fig. 6.2. Other types of pulses may be studied similarly. We consider the Gaussian voltage pulse that has the following form:

$$V_s(t) = V_o e^{-t^2/\tau^2} \tag{6.1.39}$$

where V_o is the amplitude and τ the half width of the pulse at the $1/e$ points. For such a pulse, we take its Fourier transformation:

$$V_s(\nu) = \int_{-\infty}^{+\infty} V_s(t) e^{-i2\pi\nu t} dt = V_o \tau \sqrt{\pi} e^{-(\pi\tau\nu)^2} \tag{6.1.40}$$

with ν the frequency (Hz), related to the radian frequency ω by $\omega = 2\pi\nu$ (rad/s).

In the frequency domain, we may solve the circuit shown in Fig. 6.2 and find the voltage response at the location l of the transmission line by

$$\frac{V_l(\nu)}{V_s(\nu)} = \left[\left(1 + \frac{Z_s}{Z_l}\right)\cosh(\gamma l) + \left(\frac{Z_s}{Z_o} + \frac{Z_o}{Z_l}\right)\sinh(\gamma l)\right]^{-1} \tag{6.1.41}$$

where Z_o is the characteristic impedance of the line. γ is the complex propagation constant. Thus the pulse obtained at the load takes the form:

$$V_l(t) = \int_{-\infty}^{+\infty} V_o\tau\sqrt{\pi}e^{-(\pi\tau\nu)^2}\left[\left(1 + \frac{Z_s}{Z_l}\right)\cosh(\gamma l) + \left(\frac{Z_s}{Z_o} + \frac{Z_o}{Z_l}\right)\sinh(\gamma l)\right]^{-1} e^{i2\pi\nu t}d\nu \tag{6.1.42}$$

The formulas derived here may be used to simulate the dispersion and attenuation behaviors of Gaussian pulses propagating along a transmission line once the line parameters of the transmission line are specified. In the following sections, we shall introduce some approaches to determine the line parameters of a transmission line from its geometric configuration and its material properties.

6.1.4 Scattering Matrix Representation of Transmission Line Parameters

The concept of scattering (matrix) parameters was introduced in the 1960s, a detailed discussion of which may be found in the work of, for instance, Kurokawa (1965). This concept has become very useful in characterizing properties of transmission lines among others at microwave frequencies. Standard measurement systems now exist to examine directly the scattering (S) parameters of, for instance, a transmission line. It is therefore of interest to introduce here the concept of the S–parameters.

The S–parameters describe the inter-relationship of a set of variables called power waves (a_k, b_k). Consider a N-port network, as shown in Fig. 6.3. The incident and reflected power waves a_k and b_k at the kth port of the network are defined by

$$a_k = \frac{V_k + Z_k I_k}{2\sqrt{|Re(Z_k)|}}, \qquad b_k = \frac{V_k - Z_k^* I_k}{2\sqrt{|Re(Z_k)|}} \tag{6.1.43}$$

where V_k and I_k are the voltage and the current flowing into the kth port, and Z_k is the impedance (reference impedance) looking out from the kth port. The positive real value of Z_k is chosen for the square root in the denominators. The asterisk denotes the complex conjugate.

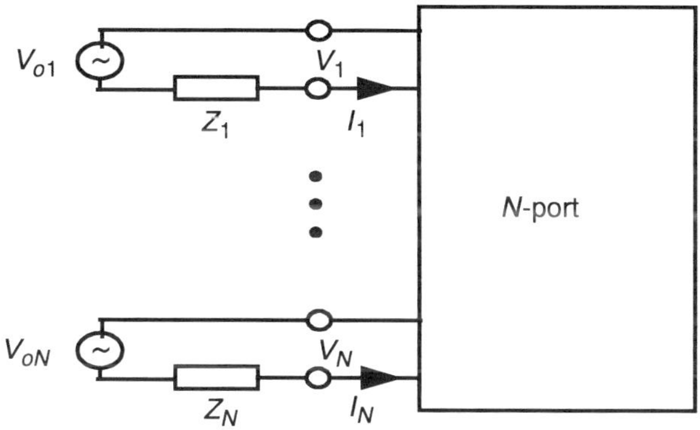

Figure 6.3 Scheme of a multiport network with N ports.

Since the waves defined by Eq.(6.1.43) are closely related with the exchangeable power of a generator, we consider now the equivalent circuit of a linear generator, as shown in Fig. 6.4, in which Z_k is the internal impedance and V_o is the open circuit voltage of the generator. The load impedance is denoted by Z_L. Thus we may find the following relation:

$$V_k = V_o - Z_k I_k \tag{6.1.44}$$

Inserting this into Eq.(6.1.43), we obtain

$$|a_k|^2 = \frac{|V_o|^2}{4|Re(Z_k)|} \tag{6.1.45}$$

which represents the maximum power that the generator can supply.

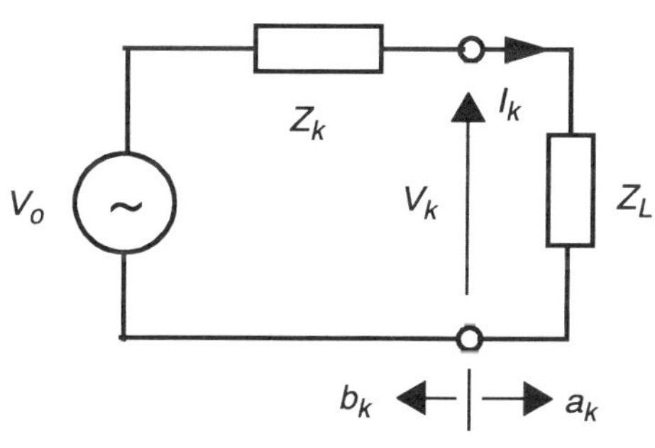

Figure 6.4 Equivalent circuit of a linear generator.

By defining a power wave reflection coefficient Γ as

$$\Gamma = \frac{b_k}{a_k} = \frac{V_k - Z_k{}^* I_k}{V_k + Z_k I_k} = \frac{Z_L - Z_k{}^*}{Z_L + Z_k} \tag{6.1.46}$$

noting $V_k/I_k = Z_L$, we may have

$$|a_k|^2 - |b_k|^2 = |a_k|^2 (1 - |\Gamma|^2) \tag{6.1.47}$$

It can be seen that, at the matching condition: $Z_L = Z_k{}^*$, we have $b_k = 0$ and Γ=0, the total power sent by the generator dissipates in the load. When the load is not matched ($\Gamma \neq 0$), a part of the incident power is reflected back to the generator.

In general, the power wave scattering matrix of a N-port network can be defined by

$$[b_k] = [S_{kl}][a_l] \tag{6.1.48}$$

where $\boldsymbol{S}$ is a square matrix of order N, and $\boldsymbol{a}$ and $\boldsymbol{b}$ are vectors, respectively, of input and output power wave variables at ports of the multiport network. In what follows, we shall not discuss the S-parameters for general multiport networks, which may be found elsewhere. Here we shall concentrate ourselves in discussing the S-parameters for transmission lines.

The transmission line may be treated as a two-port network, as shown in Fig. 6.5. For a two-port network, we have

$$\begin{bmatrix} b_1 \\ b_2 \end{bmatrix} = \begin{bmatrix} S_{11} & S_{12} \\ S_{21} & S_{22} \end{bmatrix} \begin{bmatrix} a_1 \\ a_2 \end{bmatrix} \tag{6.1.49}$$

Using this equation, we may find

$$S_{11} = \left. \frac{b_1}{a_1} \right|_{a_2 = 0} \tag{6.1.50}$$

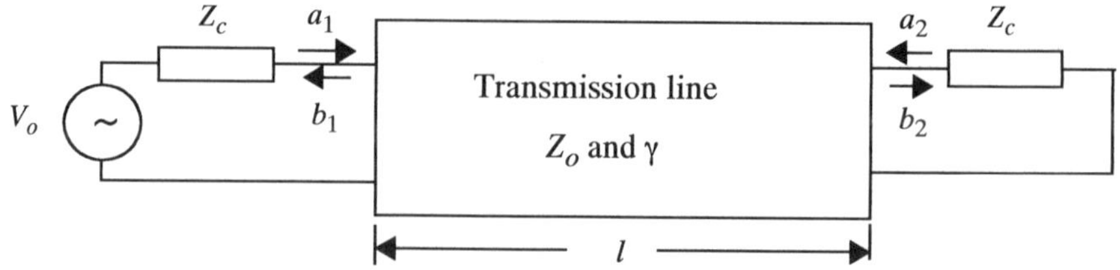

Figure 6.5 Network representation of S-parameter measurement for a transmission line.

which implies that S_{11} is the input reflection coefficient when the output is matched (so that there is no reflected power from the output port), and

$$S_{22} = \left.\frac{b_2}{a_2}\right|_{a_1 = 0} \tag{6.1.51}$$

which implies that S_{22} is the output reflection coefficient when the input is matched, and

$$S_{12} = \left.\frac{b_1}{a_2}\right|_{a_1 = 0} \tag{6.1.52}$$

which implies that S_{12} is the reverse transfer (or transmission) coefficient with the input matched, and

$$S_{21} = \left.\frac{b_2}{a_1}\right|_{a_2 = 0} \tag{6.1.53}$$

which implies that S_{21} is the forward transfer (or transmission) coefficient with the output matched. It may be noticed that since the transmission line is represented by a symmetrical two-port network, we have $S_{11} = S_{22}$, and $S_{12} = S_{21}$.

The transmission lines, such as microstrip lines, are usually tested in a controlled impedance ($Z_c = 50\ \Omega$) microwave measurement system. The S-parameters responses measured from a lossy unmatched transmission line with its line parameters γ and Z_o in the Z_c impedance system, as shown by Fig. 6.5, can be given by [Gupta et al. (1981) and Dobrowolski (1991)]

$$[S_{kl}] = \frac{1}{D_o}\begin{bmatrix} Z_o^2 - Z_c^2 & 2Z_oZ_c\cosh(\gamma l) \\ 2Z_oZ_c\operatorname{csch}(\gamma l) & Z_o^2 - Z_c^2 \end{bmatrix} \tag{6.1.54}$$

with

$$D_o = Z_o^2 + Z_c^2 + 2Z_oZ_c\coth(\gamma l)$$

in which l is the length of the transmission line.

Often, for transmission lines where unique voltage and current waves can be defined, the concept of so-called *ABCD* matrix is used. The *ABCD* matrix for a transmission line, as shown in Fig. 6.6, is defined by

$$\begin{bmatrix} V_1 \\ I_1 \end{bmatrix} = \begin{bmatrix} A & B \\ C & D \end{bmatrix}\begin{bmatrix} V_2 \\ -I_2 \end{bmatrix} \tag{6.1.55}$$

which relates directly the output terminal voltage and current (V_2, I_2) to the input voltage and current (V_1, I_1). For the transmission line shown in Fig. 6.6, the *ABCD* matrix is given by

$$\begin{bmatrix} A & B \\ C & D \end{bmatrix} = \begin{bmatrix} \cosh(\gamma l) & Z_o \sinh(\gamma l) \\ \frac{1}{Z_o}\sinh(\gamma l) & \cosh(\gamma l) \end{bmatrix} \tag{6.1.56}$$

where Z_o is the characteristic impedance and γ the complex propagation constant of the transmission line.

There is a relationship between the *ABCD* parameters and the *S*-parameters, which can be expressed by [Gupta et al. (1981)]

$$A = \frac{1}{2S_{21}}(1 + S_{11} - S_{22} - \Delta S) \tag{6.1.57a}$$

$$B = \frac{Z_o}{2S_{21}}(1 + S_{11} + S_{22} + \Delta S) \tag{6.1.57b}$$

$$C = \frac{1}{2Z_o S_{21}}(1 - S_{11} - S_{22} + \Delta S) \tag{6.1.57c}$$

$$D = \frac{1}{2S_{21}}(1 - S_{11} + S_{22} - \Delta S) \tag{6.1.57d}$$

where

$$\Delta S = S_{11}S_{22} - S_{21}S_{12} \tag{6.1.57e}$$

To extract the transmission line parameters γ and Z_o from the *S*-parameters, Eisenstadt and Eo (1992) have found the following relations from Eqs.(6.1.54), (6.1.56), and (6.1.57):

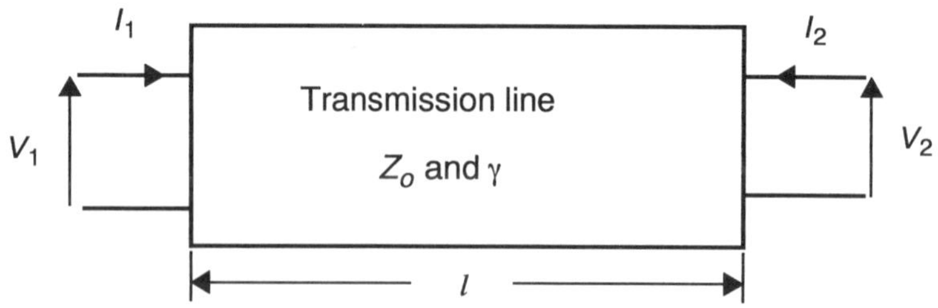

Figure 6.6 Voltage–current diagram for a transmission line.

$$e^{-\gamma l} = \left(\frac{1 - S_{11}^2 + S_{21}^2}{2S_{21}} \pm K\right)^{-1} \tag{6.1.58}$$

and

$$Z_o^2 = Z_c^2 \frac{(1 + S_{11})^2 - S_{21}^2}{(1 - S_{11})^2 - S_{21}^2} \tag{6.1.59}$$

where

$$K = \sqrt{\frac{(S_{11}^2 - S_{21}^2 + 1)^2 - 4S_{11}^2}{4S_{21}^2}} \tag{6.1.60}$$

During the extraction of complex parameters γ and Z_o from $e^{-\gamma l}$ and Z_o^2, extracted parameters with values that are not physically real, such as negative attenuation constants, should be corrected properly to be real solutions. Once γ and Z_o are determined, we may obtain the line resistance R, the line inductance L, the line conductance G, and the line capacitance C, respectively, by:

$$R = Re\{\gamma Z_o\} \tag{6.1.61}$$

$$L = Im\{\gamma Z_o\}/\omega \tag{6.1.62}$$

$$G = Re\{\gamma / Z_o\} \tag{6.1.63}$$

$$C = Im\{\gamma / Z_o\}/\omega \tag{6.1.64}$$

6.2 ELECTROMAGNETIC FIELD ANALYSIS OF TRANSMISSION LINE PARAMETERS

6.2.1 TEM Wave at Quasistatic Approximation

It has been shown in above sections that propagation behaviors of signals (electromagnetic waves) along a transmission line can be determined from the set of transmission line equations (6.1.2) and (6.1.3) with proper boundary conditions, provided that the transmission line parameters R, L, C, and G are specified. As we may expect, the transmission line parameters depend not only on material properties but also on geometric construction of the transmission line. It is known that normal conduction currents flowing in an electrical transmission line may cause noticeable losses, especially in high-frequency ac fields, due to the finite conductivity of the conductors in the transmission line. In a superconductor in ac field and at finite temperature, there exists also normal conduction current component, as we have already known from the classical two-fluid model discussed in Chapter 2. When power is absorbed in a conductor, there must be a component of the Poynting vector directed into the conductor. Therefore, not only transverse electromagnetic field components, but also longitudinal components of both electric and magnetic fields may generally exist inside the conductors of transmission lines.

According to the Maxwell theory, a longitudinal component of the magnetic field will have associated with the transverse currents on the conductor. Since these transverse currents arise only because of the perturbation of the TEM mode into a mode with longitudinal field components in a transmission line, they are small in comparison with the longitudinal currents. Thus the losses associated with the transverse currents are also small in comparison with the power loss arising from the longitudinal current. At the first approximation we may, however, assume that the transverse currents are negligible, and the wave mode is quasi-TEM. Thus it is often practice of using a perturbation method to calculate the line parameters of the transmission lines at the quasistatic approximation. The basic assumption in this perturbation method is that the introduction of a small (dielectric and/or conduction) loss does not substantially perturb the electromagnetic (EM) fields from their loss-free values, determined by a potential function Φ satisfying the Laplace equation:

$$\nabla_t^2 \Phi(x, y) = 0 \tag{6.2.1}$$

with proper boundary conditions. Here, ∇_t^2 is the transverse Laplace operator: $\nabla_t^2 = \partial^2/\partial x^2 + \partial^2/\partial y^2$, if we suppose that the transmission line is along the z-axis direction. It is assumed that the material properties as well as the cross-sectional shape of the transmission line do not depend on z-axis coordinate. The electric and magnetic fields for the lossless solution may be expressed by

$$\boldsymbol{E} = -\nabla_t \Phi e^{\mp ikz}, \qquad \boldsymbol{H} = \pm\sqrt{\varepsilon_r}\sqrt{\frac{\varepsilon_o}{\mu_o}} z^o \times \boldsymbol{E} \tag{6.2.2}$$

with $k = \omega(\mu_0 \varepsilon_0 \varepsilon_r)^{1/2}$. Here, the minus represents a wave propagating in the $+z$ direction, and the plus representing a wave propagating in the $-z$ direction. ε_r is the relative permittivity of the homogeneous dielectric medium of the transmission line. z^0 denotes the unit directional vector along the z-axis direction.

For an isotropic lossy dielectric with the complex permittivity $\varepsilon = \varepsilon_0 \varepsilon_r(1 - i\tan\delta)$, we may find the power loss in the dielectric per unit length of the transmission line by [Collin (1966)]

$$P_d = \frac{\omega \varepsilon_o \varepsilon_r \tan\delta}{2} \int_S \boldsymbol{E} \cdot \boldsymbol{E}^* dS \tag{6.2.3}$$

where the integration is over the cross-section area S of the dielectric of the line, and the loss-free solution for $\boldsymbol{E}$ is used in the perturbation method.

Similarly, for the lossy conductors, we may find the power loss P_c by

$$P_c = \frac{R_s}{2} \oint_{L_1+L_2} \boldsymbol{J}_s \cdot \boldsymbol{J}_s^* dL = \frac{R_s}{2} \oint_{L_1+L_2} \boldsymbol{H} \cdot \boldsymbol{H}^* dL \tag{6.2.4}$$

where the integration is taken around the periphery L_1+L_2 of the two conductors of the line. R_s denotes the high-frequency surface resistance, and $\boldsymbol{H}$ is the loss-free magnetic field. Here, we notice that the surface current $\boldsymbol{J}_s$ is taken equal to $\boldsymbol{n}\times\boldsymbol{H}$, where $\boldsymbol{n}$ is the unit outward normal to the conductor surface.

Furthermore, by noting that the energy stored in the magnetic field is accounted for by the series inductance L per unit length, and the energy stored in the electric field is accounted for by the distributed shunt capacitance C per unit length, we may find that the line parameters, that is, the line inductance L, the line capacitance C, the line resistance R, and the line conductance G can be determined, respectively, by

$$L = \frac{\mu}{II^*} \int_S \boldsymbol{H} \cdot \boldsymbol{H}^* dS \qquad (\mathrm{H/m}) \tag{6.2.5}$$

$$C = \frac{\varepsilon_o \varepsilon_r}{VV^*} \int_S \boldsymbol{E} \cdot \boldsymbol{E}^* dS \qquad (\mathrm{F/m}) \tag{6.2.6}$$

$$R = \frac{R_s}{II^*} \oint_{L_1+L_2} \boldsymbol{H} \cdot \boldsymbol{H}^* dL \qquad (\Omega/\mathrm{m}) \tag{6.2.7}$$

$$G = \frac{\omega \varepsilon_o \varepsilon_r \tan\delta}{VV^*} \int_S \boldsymbol{E} \cdot \boldsymbol{E}^* dS \qquad (\mathrm{S/m}) \tag{6.2.8}$$

in which I is the total current on the line (on one conductor), V the potential difference, and μ the permeability. For a superconducting transmission line, the surface resistance R_s may be given by, for instance, Eq.(3.3.49) for superconductors in weak fields. For normal conducting line, we have $R_s = 1/(\sigma_n \delta_n)$ with σ_n being the normal conductivity and δ_n the normal skin depth by Eq.(3.3.46).

Illustratively, let us consider a coaxial line, as shown in Fig. 3.8. For this problem, the potential Φ, defined in the dielectric medium ($a < r < b$), can be found by

$$\Phi = \frac{V_o \ln(r/b)}{\ln(a/b)} \tag{6.2.9}$$

where V_o is the specified potential value at the inner boundary $r = a$, besides $\Phi = 0$ on the outer boundary $r = b$. The electric and magnetic fields of a TEM wave propagating in the $+z$ direction can be given by Eq.(6.2.2) and are

$$\boldsymbol{E} = \frac{V_o}{\ln(b/a)} \frac{\boldsymbol{r}^o}{r} e^{-ikz} \tag{6.2.10}$$

$$\boldsymbol{H} = \frac{I_o}{2\pi r} \boldsymbol{r}_\phi^o e^{-ikz} \tag{6.2.11}$$

where $\boldsymbol{r}^{\text{o}}$ and $\boldsymbol{r}_\phi^{\text{o}}$ are the unit directional vectors in a cylindrical coordinate system (r, ϕ, z). Here, I_o is the total current, apart from the factor e^{-ikz}, which is obtained by noting that the current density on the inner conductor is given by $\boldsymbol{J}_s = \boldsymbol{n}\times\boldsymbol{H}$.

By Eqs.(6.2.5)–(6.2.8), we may find the following line parameters for the coaxial line:

$$L = \frac{\mu_o}{2\pi} \ln\left(\frac{b}{a}\right) \tag{6.2.12}$$

$$C = \frac{2\pi\varepsilon_o\varepsilon_r}{\ln(b/a)} \tag{6.2.13}$$

$$R = \frac{R_s(b+a)}{2\pi ab} \tag{6.2.14}$$

$$G = \frac{2\pi\omega\varepsilon_o\varepsilon_r \tan\delta}{\ln(b/a)} \tag{6.2.15}$$

where ε_r and $\tan\delta$ are, respectively, the relative permittivity and the tangent of the loss angle of the dielectric, which is assumed here to be non-magnetic with the permeability μ_{o}. We may notice that the line inductance L given by Eq.(6.2.12) is only the external inductance since the inductance inside the conductors is not included within the approximation. To account for the internal inductance L_{in}, a simple method may be used, which gives $L_{in} = R/\omega$ [Collin (1966)].

6.2.2 Formulation of Skin Effect at Quasi-TEM Approximation

It is known that electromagnetic fields may penetrate into normal conductors as well as superconductors in ac fields with the characteristic penetration depths δ_n and λ_L, according to the normal skin effect in normal conductors and the field penetration of London´s theory for superconductors, respectively. Today, with the rapid development of monolithic integrated microwave circuits, the metallization thickness of the planar microwave integrated circuits has become often of the order of the skin depth even at very high frequencies. Besides, the trends of miniaturization of integrated circuits may soon lead us to a characteristic size, even on the order of the London penetration depth λ_L (~ 0.1 μm). When the geometric sizes of conductors (or superconductors) in transmission lines are in the order of the characteristic penetration depth δ_n (or λ_L), we can imagine that the simple perturbation method discussed in the above section may no longer be applied since the surface resistance R_s cannot represent accurately the effect of actual field distribution within the conductors. Furthermore, the internal inductance of the conductors has to be calculated properly, which cannot be treated simply by the above perturbation method.

In order to analyze the effect of the skin effect in conductors (normal or superconductors), we introduce here a method based on the quasi-TEM approximation, in which the axial component of electric field may exist inside conductors of a transmission line. We assume that both the physical properties and the geometric structure of the transmission line are homogeneous along the z-axis in the longitudinal (axial) direction of the line. To study the problem, let us first decompose all fields into transverse and axial components. In particular, suppressing the time dependence $e^{i\omega t}$, we may write

$$\boldsymbol{H}(x, y, z) = \boldsymbol{H}_t(x, y)e^{-i\beta z} + z^o H_z(x, y)e^{-i\beta z} \tag{6.2.16}$$

where β is the propagation phase constant, and z^o is the unit vector along the direction of z-axis. H_z is the axial component, and $\boldsymbol{H}_t$ is the transverse component, $\boldsymbol{H}_t(x, y) = (H_x(x, y), H_y(x, y), 0)$, of the magnetic intensity field. Introducing the gradient operator ∇ by

$$\nabla = \nabla_t + z^o \frac{\partial}{\partial z} \tag{6.2.17}$$

where ∇_t is the transverse part of the gradient operator ∇ and can be written as $\nabla_t = (\partial/\partial x, \partial/\partial y, 0)$ in rectangular coordinates, we may write

$$\begin{aligned}\nabla \times \boldsymbol{H} &= \left(\nabla_t + z^o \frac{\partial}{\partial z}\right) \times [\boldsymbol{H}_t(x, y)e^{-i\beta z} + z^o H_z(x, y)e^{-i\beta z}] \\ &= [\nabla_t \times \boldsymbol{H}_t - i\beta z^o \times \boldsymbol{H}_t - z^o \times \nabla_t H_z]e^{-i\beta z}\end{aligned}$$

Thus we may express the Maxwell equation: $\nabla \times \boldsymbol{H} = \boldsymbol{J}$ (ignoring displacement current) in conductors as the following set of equations:

$$\nabla_t \times \boldsymbol{H}_t = z^o J_z \tag{6.2.18}$$

$$-i\beta z^o \times \boldsymbol{H}_t - z^o \times \nabla_t H_z = \boldsymbol{J}_t \tag{6.2.19}$$

where $\boldsymbol{J}_t$ denotes the transverse component, and J_z the axial component of the current density vector $\boldsymbol{J}$ ($\boldsymbol{J} = \boldsymbol{J}_t + z^o J_z$).

At the quasi-TEM approximation, we assume that the magnetic field is transverse ($H_z = 0$). Thus, from Eq.(6.2.19), we may find the transverse current density component by

$$\boldsymbol{J}_t = -i\beta z^o \times \boldsymbol{H}_t \tag{6.2.20}$$

which shows that the transverse current component is generally not zero in the conductors of the transmission lines. However, in many cases, the magnitude of the transverse current component is much smaller than the axial current component, and therefore, often ignored in many cases of practical interest.

Introducing the magnetic vector potential $\boldsymbol{A}$, we may write $\boldsymbol{B} = \nabla \times \boldsymbol{A}$. For the case of transverse magnetic field, we may introduce $\boldsymbol{A} = (0, 0, A_z)$, and obtain $\boldsymbol{B} = (\partial A_z/\partial y, -\partial A_z/\partial x, 0)$. Thus, for isotropic material with $\boldsymbol{B} = \mu \boldsymbol{H}$, we may derive from Eq.(6.2.18) the following two-dimensional field equation:

$$\frac{\partial}{\partial x}\left(\frac{1}{\mu}\frac{\partial A_z}{\partial x}\right) + \frac{\partial}{\partial y}\left(\frac{1}{\mu}\frac{\partial A_z}{\partial y}\right) = -J_z \tag{6.2.21}$$

In particular, for non-magnetic conductors, Eq.(6.2.21) becomes

$$-\nabla_t^2 A_z(x, y) = \mu_o J_z(x, y) \tag{6.2.22}$$

where μ_o is the permeability in vacuum, and ∇_t^2 is the transverse Laplace operator.

Similarly, by decomposing the electric field into transverse and axial components,

$$\boldsymbol{E}(x, y, z) = \boldsymbol{E}_t(x, y)e^{-i\beta z} + z^o E_z(x, y)e^{-i\beta z} \tag{6.2.23}$$

we may find from the Maxwell equation: $\nabla \times \boldsymbol{E} = -i\omega \boldsymbol{B}$ the following set of equations:

$$\nabla_t \times \boldsymbol{E}_t = -i\omega z^o B_z \tag{6.2.24}$$

$$-i\beta z^o \times \boldsymbol{E}_t - z^o \times \nabla_t E_z = -i\omega \boldsymbol{B}_t \tag{6.2.25}$$

At the quasi-TEM approximation, we have $B_z = 0$, and therefore, $\nabla_t \times \boldsymbol{E}_t = 0$. In particular, for isotropic conductors, the transverse component of the electric field can be obtained by using Ohm's law, $\boldsymbol{E}_t = \boldsymbol{J}_t/\sigma$. Thus we may rewrite Eq.(6.2.25) in the form:

$$z^o \times \left[\nabla_t(E_z + i\omega A_z) + \frac{\beta^2}{\mu\sigma}\nabla_t A_z\right] = 0 \tag{6.2.26}$$

where we have noticed that the transverse current density $\boldsymbol{J}_t$ can be expressed by

$$\boldsymbol{J}_t = -\frac{i\beta}{\mu}\nabla_t A_z \tag{6.2.27}$$

and the transverse magnetic field $\boldsymbol{B}_t$ by

$$\boldsymbol{B}_t = -z^o \times \nabla_t A_z \tag{6.2.28}$$

Noting $\beta^2 \sim \omega^2\varepsilon_0\varepsilon_r\mu$, we can make an estimation of the following numerical factor:

$$\frac{\beta^2}{\omega\mu\sigma} \approx \frac{\omega\varepsilon_o\varepsilon_r}{\sigma}$$

which is much less than 1 for good conductors ($\sigma \sim 10^7$ S/m), operating even at 100 GHz frequencies. Thus we may see that the correction of the transverse current effect on the axial electric field E_z is usually negligible at microwave frequencies.

Ignoring the effect of transverse current, we may find from Eq.(6.2.26) that the axial component of electric field E_z can be given by

$$E_z = -i\omega A_z - \frac{\partial V}{\partial z} \tag{6.2.29}$$

where $\partial V/\partial z$ is a constant, independent of space coordinates. For a uniform straight conductor, we may consider

$$\frac{\partial V}{\partial z} = \frac{V(L) - V(0)}{L} \tag{6.2.30}$$

where $V(0)$ and $V(L)$ are the electrical potential values at the end surfaces of the conductor at $z = 0$ and $z = L$, respectively. In general, the value of $\partial V/\partial z$ is, however, not known since $V(0)$ and $V(L)$ are not given. The presence of the axial component of the electric field E_z indicates that the electric field $\boldsymbol{E}$ is generally not transverse in the conductor.

In the dielectric medium surrounding the straight conductors, we have the following Laplace equation for the magnetic vector potential component A_z:

$$\nabla_t^2 A_z = 0 \tag{6.2.31}$$

at the quasi-TEM approximation, where displacement current is ignored. The set of equations (6.2.22) and (6.2.31) with proper boundary conditions can be used to determine the magnetic vector potential component A_z and, therefore, the magnetic field $\boldsymbol{B}$ and the axial current density J_z inside conductors. They may then be used to calculate the line inductance L and line resistance R of the transmission line, as we shall show later.

To determine the electric field in dielectric media of the transmission line, we may note that at the quasi-TEM approximation, we have $B_z = 0$, and therefore, $\nabla_t \times \boldsymbol{E}_t = 0$. Thus we may introduce an electric potential function $\Phi(x, y)$ by $\boldsymbol{E}_t = -\nabla_t\Phi$. By the Maxwell equation: $\nabla \cdot \boldsymbol{D} = 0$ (assuming no free charges in the dielectric), we may find that the electric potential Φ satisfies the following equation:

$$\nabla_t \cdot (\varepsilon \nabla_t \Phi) = 0 \tag{6.2.32}$$

where ε is the complex permittivity of an isotropic dielectric in ac fields. This equation can be solved with proper boundary conditions. The obtained electric potential function Φ can then be used to find the electric field in the dielectric, with which the line capacitance C and line conductance G of the transmission line can be obtained from Eq.(6.2.6) and Eq.(6.2.8), respectively.

It is shown that, at the quasi-TEM approximation, the calculation of the line parameters of a transmission line can be separated into two parts. One part involves the calculation of the line capacitance C and the line conductance G, which is relatively simple since we are solving essentially an electrostatic problem at the quasi-TEM approximation. In what follows we shall concentrate ourselves in another part on the calculation of the line inductance L and the line resistance R of the transmission line by solving the set of field equations (6.2.22) and (6.2.31).

Let us now consider some cases. If the conductors in a transmission line are normal conductors, we may write the axial current density by $J_z = \sigma E_z$, according to Ohm´s law, where σ is the normal conductivity. If, however, the conductor is a superconductor, we have the total axial current density $J_z = J_z^{(s)} + J_z^{(n)}$, with the axial normal current density $J_z^{(n)} = \sigma_n E_z$, following Ohm´s law, and the axial superconducting current density $J_z^{(s)} = E_z/(i\mu_0\omega\lambda_L^2)$, according to the London theory for superconductors. In such a case, we may introduce a complex conductivity σ:

$$\sigma = \sigma_n - \frac{i}{\mu_o \omega \lambda_L^2} \tag{6.2.33}$$

where σ_n and λ_L are, respectively, the normal conductivity and the London penetration depth of the superconductor, which is supposed here to be non-magnetic and isotropic.

Thus, by noting Eq.(6.2.29), we may write Eq.(6.2.22) in the following form:

$$-\nabla_t^2 A_z + i\omega\mu_o\sigma A_z = -\mu_o\sigma\frac{\partial V}{\partial z} \tag{6.2.34}$$

Furthermore, we may introduce the total (axial) current I flowing in the conductor by

$$I = \int_S J_z dS = -S\sigma\frac{\partial V}{\partial z} - \int_S i\sigma\omega A_z dS \tag{6.2.35}$$

where S is the cross-sectional area of the conductor, assuming σ is constant in S. Here, we may consider the total current I as the sum of the source current, defined by $I_s = -S\sigma\partial V/\partial z$, and the total induced or eddy current flowing inside the conductor, defined by the second term on the right-hand side of Eq.(6.2.35). By Eq.(6.2.35), we may write Eq.(6.2.34) in the following form:

$$-\nabla_t^2 A_z + i\omega\mu_o\sigma A_z - \frac{i\omega\mu_o}{S}\int_S \sigma A_z dS = \frac{\mu_o}{S}I \tag{6.2.36}$$

where the integration is over the cross-sectional area S of the conductor. Equation (6.2.36) is an intergrodifferential equation for the magnetic vector potential component A_z. Once A_z is determined for given total current I and proper boundary conditions, by noting Eqs.(6.2.29) and (6.2.35), the total current density J_z inside the conductor can then be calculated by

$$J_z = \sigma E_z = \frac{I}{S} - i\omega\sigma A_z + \frac{i\omega}{S}\int_S \sigma A_z dS \tag{6.2.37}$$

The obtained current distribution is generally non-uniform inside the conductor due to the normal skin effect for normal conductors and the field penetration for superconductors. Once the current distribution is known, we may calculate the line resistance as well as the line inductance of the conductor. It can be seen that the solution of the integrodifferential equation (6.2.36) may only be obtained, in general, by using some numerical methods. In what follows, we shall introduce an integrodifferential finite element method, which was first proposed by Konrad (1981, 1982) for studying steady-state normal skin-effect problems, and later studied by Costache (1987) for normal strip transmission lines.

6.2.3 Finite Element Formulation of Multi-Conductor Systems

Generally, if we consider a system of multi-conductor strip lines, as shown illustratively in Fig. 6.7, we may find a set of integrodifferential equations:

$$-\nabla_t^2 A_z + i\omega\mu_o\sigma_k A_z - \frac{i\omega\mu_o\sigma_k}{S_k}\int_{S_k} A_z dS = \frac{\mu_o}{S_k} I_k \quad \text{in } S_k \tag{6.2.38}$$

with S_k being the cross-sectional area of the kth conductor (k = 1, 2,..., Q), and I_k the total current flowing in the kth conductor, given by

$$I_k = -S_k\sigma_k\frac{\partial V}{\partial z} - \int_{S_k} i\sigma_k\omega A_z dS \tag{6.2.39}$$

which may be considered as the sum of the source current (defined by $I_s = -S_k\sigma_k\partial V/\partial z$) and the total induced or eddy current flowing inside the kth conductor. Besides, we have

$$\nabla_t^2 A_z = 0 \quad \text{in } S_D \tag{6.2.40}$$

where S_D denotes the dielectric and/or free-space region.

It is seen that Eq.(6.2.38) is an integrodifferential equation for the magnetic vector potential component A_z for the system of multi-conductors. To introduce a finite element formulation, we may apply the Galerkin integral form:

$$f_b = \sum_{k=1}^{Q}\int_{S_k}\left[(\nabla_t w)\cdot\nabla_t A_z + i\omega\mu_o\sigma_k\left(A_z - \frac{\eta_k}{S_k}\int_{S_k} A_z dS\right)w - \frac{\mu_o}{S_k} I_k w\right]dS$$
$$+\int_{S_D}(\nabla_t w)\cdot\nabla_t A_z dS \tag{6.2.41}$$

in which we have introduced the parameter η_k for the convenience of numerical

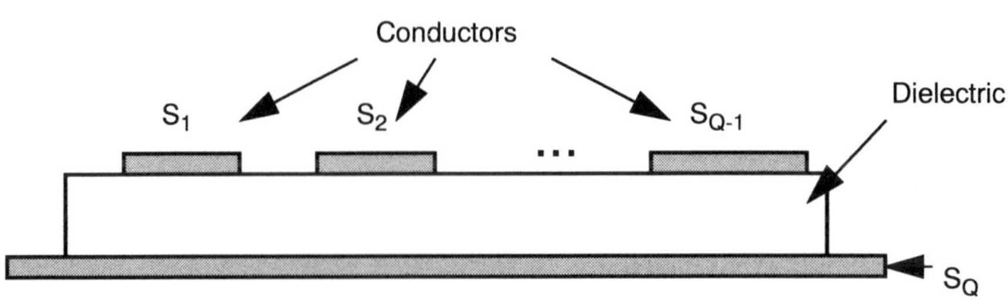

Figure 6.7 Scheme of a multi-conductor system.

calculation. The parameter η_k equals one ($\eta_k = 1$) for the kth conductor carrying non-zero source current (I_k is specified). If the kth conductor carrying no source current, we set $\eta_k = 0$ and $I_k = 0$ in the equation. In such a case, the total current flowing in the conductor is equal to the induced (eddy) current in the absence of the source current. In Eq.(6.2.41), f_b is given by

$$f_b = \int_\Gamma w(\nabla_t A_z) \cdot \boldsymbol{n} d\Gamma = \int_\Gamma w \frac{\partial A_z}{\partial n} d\Gamma \tag{6.2.42}$$

where w is a testing function, and $\boldsymbol{n}$ is the outward unit vector normal to the outer boundary Γ of the domain of the transmission line. For a homogeneous boundary condition (i.e., if the stripline is shielded by a magnetic material of high permeability (noting: $\boldsymbol{H}\times\boldsymbol{n} = \partial A_z/\partial n$)), we have $f_b = 0$. For a perfect conducting boundary, the magnetic vector potential component A_z is constant, which can be chosen equal to zero. Thus the perfect conducting shield surface corresponds a homogeneous Dirichlet boundary condition. For open boundary value problems, such as microstrip lines, in many cases of practical interest, we may often take a closed (i.e., perfect conducting) boundary to enclose the open boundary problems, provided that the closed outer boundary is taken to be sufficiently far from the region (i.e., near the signal) line), where fields are mostly concentrated at the magneto-quasistatic approximation when radiation losses are negligible.

By the well-known finite element method, we may discretise the domain of interest, and approximate the unknown function A_z within each element. If, for instance, linear triangular elements are used, the unknown function A_z within each element can be approximated by

$$A_z^e(x, y) = \sum_{j=1}^{3} N_j^e(x, y) A_{zj}^e \tag{6.2.43}$$

where A_{zj}^e ($j = 1, 2, 3$) denote the node potential, and N_j^e are the interpolation or expansion functions. Taking $w = N_p^e$ ($p = 1, 2, 3$) as the testing function in each element, we may then obtain a set of linear algebraic equations for the determination of discrete values of the magnetic potential component A_z at all nodes of the domain of interest. Details about the finite element methods may be found in many standard text books [e.g., Zienkiewicz and Taylor (1989), Silvester and Ferrari (1990), and Jin (1993)].

For a system of Q superconductors with M signal lines ($M < Q$), the ac resistance matrix $\boldsymbol{R}$ and the inductance matrix $\boldsymbol{L}$ may be obtained by computing, in general, the magnetic potential distributions for $M(M+1)/2$ independent excitation states. However, for a system of having certain symmetry, the number of computations can be reduced. In general, the matrices $\boldsymbol{R}$ and $\boldsymbol{L}$ can be determined, respectively, by the following relations:

$$\sum_{p,j=1}^{M} R_{pj} I_p I_j^* = \sum_{k=1}^{Q} \int_{S_k} \frac{1}{\sigma_k^{(n)}} |J_z^{(n)}|^2 dS \tag{6.2.44}$$

and

$$\sum_{p,j=1}^{M} L_{pj} I_p I_j^* = \sum_{k=1}^{Q} \left[Re\left\{ \int_{S_k} A_z J_z^* dS \right\} + \int_{S_k} \mu_o \lambda_k^2 |J_z^{(s)}|^2 dS \right] \tag{6.2.45}$$

where $\sigma_k^{(n)}$ and λ_k are, respectively, the normal conductivity and the London penetration depth of the kth superconductor. It can be seen that only the normal conduction current components $J_z^{(n)}$ causes the ac loss of the superconductors, and the second term on the right-hand side of Eq.(6.2.45) expresses the kinetic inductance per unit length of the superconductor.

6.2.4 Finite Element Analysis of Coaxial Line at High Frequencies

With the aid of the finite element method (FEM) and the formulation given in the above section, a computational code can be developed to analyze numerically line properties of superconducting transmission lines. In this section, we shall make use of the FEM code, developed according to the above method and can be run on a personal computer, to analyze numerically the transmission line properties of a coaxial line, and to compare the numerical results obtained with some well-known analytical results for the coaxial line both for the line resistance and the line inductance at different frequencies. Adaptive mesh generation is used to study effectively the skin effect. Some results on both normal and superconducting coaxial lines are discussed.

We have shown in Section 3.3.3 that the line inductance of a normal coaxial line may be obtained from Eqs.(3.3.20) and (3.3.24), provided that the fields inside both the inner and outer conductors of the coaxial line are uniform or approximately uniform, which corresponds to low-frequency cases where the skin depth is much larger than the geometrical sizes in the cross-section of the coaxial line. The geometric parameters a, b, and c are the radii of the coaxial line defined in Fig. 3.8. The conductors are assumed to be non-magnetic. In Fig. 6.8, the numerical results for the line inductance of the normal coaxial line at different frequencies are shown in comparison with the line inductance L_o obtained from the low-frequency approximation, and the external line inductance L_{ext} from Eq.(6.2.12). In the numerical example, we take the following parameters for the coaxial transmission line: $a = 0.1$ mm, $b = 0.4$ mm, $c = 0.45$ mm, and $\sigma_n = 5.8\times10^7$ S/m.

It is shown that the numerical results from the FEM code not only can give reasonable values of the line inductance for the coaxial line at both the low-frequency limit and the high-frequency limit, but also can predict the smooth transition of the inductance value in a wide frequency range between the low and

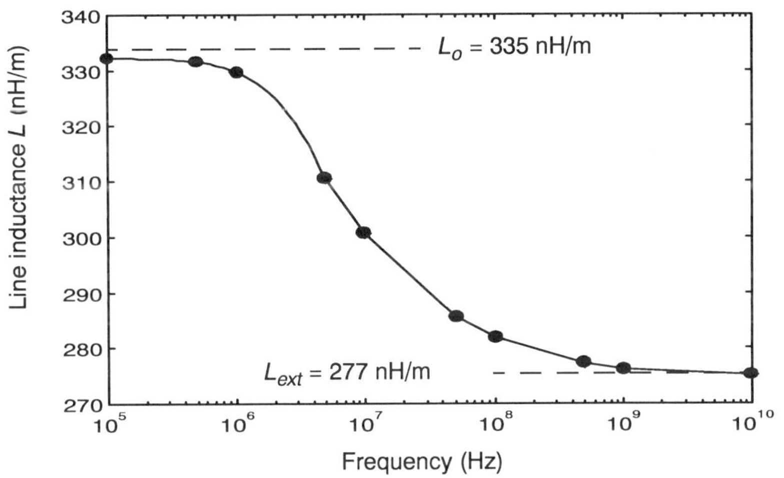

Figure 6.8 Line inductance of a normal coaxial line versus operating frequency.

high frequency limits. In particular, it is shown that the inductance value decreases with increasing frequency. This is physically reasonable since the magnetic field lines is expelled from both the inner and outer conductors of the coaxial line due to the normal skin effect at high frequencies.

Shown in Fig. 6.9 is the numerical results from both the FEM code and formula (6.2.14) for the line resistance R. It is shown that the numerical result of the FEM code predicts a value of $R = 0.695\ \Omega/\text{m}$ for the coaxial line at frequency of 500 kHz, which is quite close to the dc value of $R_{dc} = 0.68\ \Omega/\text{m}$. The dc line resistance of the coaxial line is given by the following formula:

$$R_{dc} = \frac{1}{\pi\sigma_n}\left[\frac{1}{a^2} + \frac{1}{c^2 - b^2}\right] \qquad (\Omega/\text{m}) \qquad (6.2.46)$$

where σ_n is the normal conductivity of both the inner and outer conductors of the coaxial line. This result is reasonable since the skin depth $\delta = (\pi f \mu_0 \sigma_n)^{-1/2}$ at the frequency of 500 kHz is about 93 μm, which is about the size of the radius a (= 100 μm) of the inner conductor, and is larger than the thickness (= 50 μm) of the outer conductor of the coaxial line. At high frequencies, Eq.(6.2.14) is used to characterize the rf line resistance of the coaxial line according to the perturbation method presented in Section 6.2.1, in which R_s is the surface resistance of the normal conductor, defined by $R_s = (\pi f \mu_0/\sigma_n)^{1/2}$, where f denotes the operating frequency (Hz).

It can be seen from Fig. 6.9 that the numerical value of the line resistance R predicted by the FEM code approaches to the value of R_{rf} from (6.2.14) at high frequencies. However, at relatively lower frequencies, the value of the line

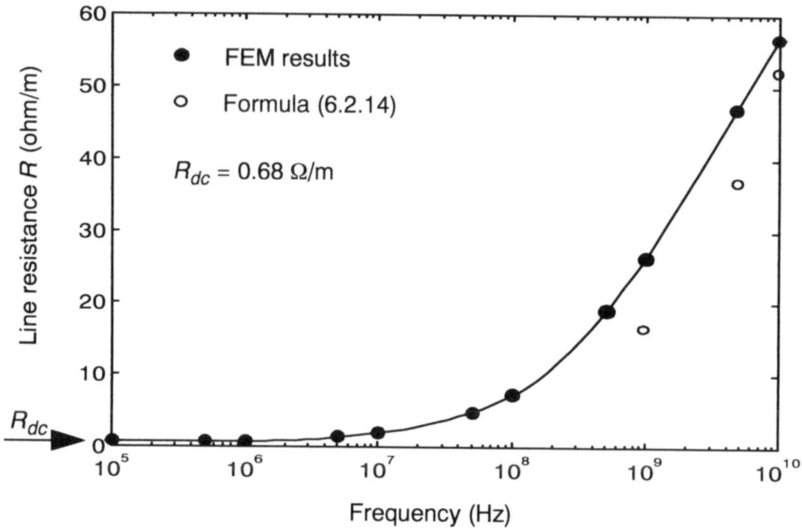

Figure 6.9 Line resistance of a normal coaxial line versus operating frequency.

resistance predicted by the FEM code is higher than the R_{rf} value from (6.2.14). This is reasonable since formula (6.2.14) is only good approximation for cases where the normal skin depth is much smaller than the geometric sizes of the conductors involved. For cases at relatively low frequencies, in which the normal skin depth involved is comparable to the size of the conductors, formula (6.2.14) underestimates the value of the line resistance. Again, we may observe that the FEM code is capable of predicting reasonably the frequency-dependence behavior of the line resistance in a wide frequency range from very low frequencies to microwave frequencies.

Some numerical calculations are also made for a superconducting coaxial line having the same geometric size as those in the above example. Shown in Fig. 6.10 is the obtained FEM results as compared with those from formula (6.2.14) with, however, the surface resistance given by the following well-known formula:

$$R_s = 2\pi^2 \sigma_n f^2 \mu_o^2 \lambda_L^3 \tag{6.2.47}$$

for superconductors, where σ_n and λ_L denote, respectively, the normal conductivity and London penetration depth of the superconductor at given temperature. In the numerical calculation, we have assumed, according to the classical two-fluid model, $\sigma_n(\Theta)=\sigma_{no}\Theta^4$, and $\lambda_L(\Theta)=\lambda_0/(1-\Theta^4)^{1/2}$ with $\sigma_{no} = 5\times10^5$ S/m, $\lambda_0 = 150$ nm, and the reduced temperature $\Theta = 0.85$. It is shown that the numerical results from the FEM code are consistent with the results predicted

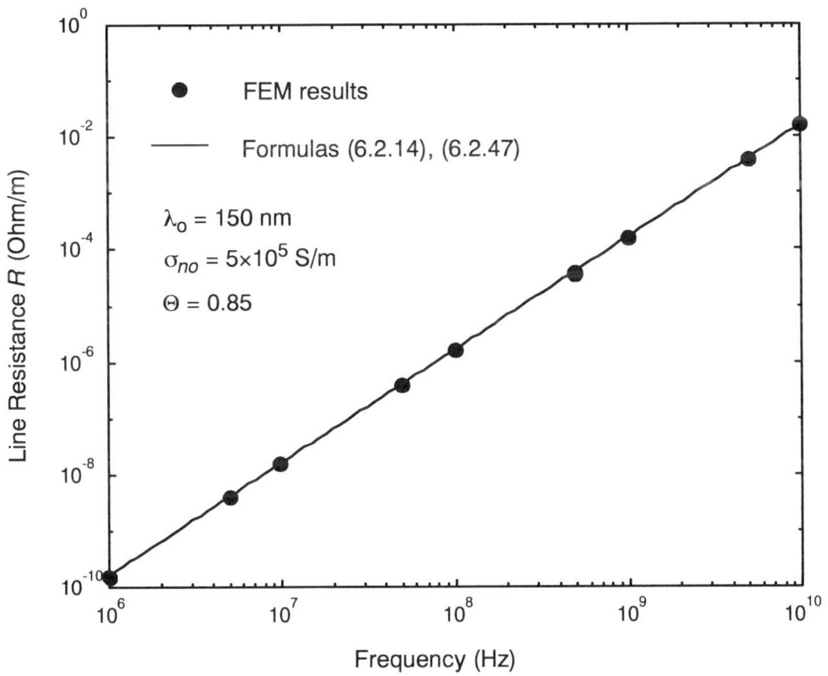

Figure 6.10 Line resistance of a superconducting coaxial line versus operating frequency.

from Eqs.(6.2.14) and (6.2.47) within the validity of the London model. It is noticed that, in order to obtain good accuracy of the numerical results, the conducting area has to be discretized sufficiently to model the "skin-effect" at high frequencies. This implies that the mesh size in the skin-effect region has to be smaller (roughly a few times smaller) than the field penetration depth at the given temperature and frequency so that the actual field distribution within the skin-effect region can be properly simulated.

6.2.5 Analysis of Composite Superconducting Striplines

In this section, we shall formulate and analyze the electromagnetic field problem of composite superconducting striplines. We consider a composite stripline in which the signal path (line) may comprise several different types of superconducting or normal conducting, or even dielectric materials. All material components are supposed to be uniform along the longitudinal direction of the stripline: Their geometric domain sizes, defined by the cross-sectional area of the stripline as shown in Fig. 6.11, are constant along the longitudinal direction of the stripline.

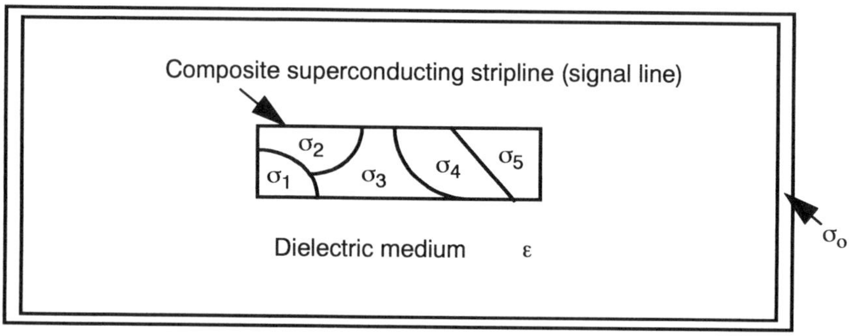

Figure 6.11 Scheme of a composite superconducting stripline.

At the quasi-TEM approximation, the basic formulation given in Section 6.2.2 may also be applied here. It is however noticed that the complex conductivity σ is now not a constant in the cross-sectional area of the composite superconducting stripline, as shown in Fig. 6.11. If we now introduce the total current I flowing in the strip (signal) line by

$$I = -\int_S \sigma \frac{\partial V}{\partial z} dS - \int_S i\omega\sigma A_z dS \qquad (6.2.48)$$

where S denotes the cross-sectional area of the strip (signal) line, we may write Eq.(6.2.34) in the following form:

$$-\nabla_t^2 A_z + i\omega\mu_o\sigma\left(A_z - \frac{1}{\Pi}\int_S \sigma A_z dS\right) = \frac{\sigma\mu_o}{\Pi} I \qquad (6.2.49)$$

in which the parameter Π is defined by

$$\Pi = \int_S \sigma dS \qquad (6.2.50)$$

where the integration is taken over the cross-sectional area S of the strip (signal) line.

For a superconducting composite with the cross-sectional area S having M subdomains, in which σ is constant in each subdomain S_k (k = 1, 2,..., M), as shown illustratively in Fig. 6.11, Eq.(6.2.50) can be written

$$\Pi = \sum_{k=1}^{M} \sigma_k S_k \qquad (6.2.51)$$

where σ_k is generally the complex conductivity of the kth subdomain with the area S_k. If the kth subdomian is a superconductor, we have

$$\sigma_k = \sigma_n^{(k)} - \frac{i}{\mu_o \omega (\lambda_L^{(k)})^2} \tag{6.2.52}$$

where $\sigma_n{}^{(k)}$ and $\lambda_L{}^{(k)}$ are, respectively, the normal conductivity and the London penetration depth of the superconductor. If the kth subdomain is a normal conductor, σ_k is real, denoting the conductivity of the normal conductor. If the kth subdomain is a dielectric, we set $\sigma_k = 0$.

Equation (6.2.49) is, in general, an integrodifferential equation for the magnetic vector potential component A_z, which can further be written

$$-\nabla_t^2 A_z + i\omega\mu_o\sigma\left(A_z - \frac{1}{\Pi}\sum_{k=1}^{M}\int_{S_k}\sigma_k A_z dS\right) = \frac{\mu_o\sigma}{\Pi}I \tag{6.2.53}$$

in the composite superconducting strip (signal) line.

Similarly, in the homogeneous outer (superconducting or normal) conductor wall, we have

$$-\nabla_t^2 A_z + i\omega\mu_o\sigma_o\left(A_z - \frac{1}{S_o}\int_{S_o} A_z dS\right) = -\frac{\mu_o}{S_o}I \tag{6.2.54}$$

where σ_0 is the conductivity of the outer conductor (complex for superconducting wall and real for normal conducting wall). S_0 denotes the cross sectional area of the outer conductor wall. Here, we have considered that the return current in the outer conductor has the same value as the input current I in the signal line of the composite stripline, but it flows in an opposite direction.

Once the magnetic vector potential component A_z is determined for given total current I and proper boundary conditions by the set of integrodifferential equations (6.2.53) and (6.2.54) together with the Laplace equation (6.2.31) for A_z in dielectrics, the total current density J_z can be calculated by

$$J_z = \sigma\left[\frac{I}{\Pi} - i\omega A_z + \frac{i\omega}{\Pi}\left(\sum_{k=1}^{M}\int_{S_k}\sigma_k A_z dS\right)\right] \tag{6.2.55}$$

in the composite superconducting strip (signal) line, and

$$J_z = -\frac{I}{S_o} - i\omega\sigma_o A_z + \frac{i\omega\sigma_o}{S_o}\left(\int_{S_o} A_z dS\right) \tag{6.2.56}$$

in the homogeneous outer conductor wall.

The superconducting current density component $J_z^{(s)}$ can be obtained by

$$J_z^{(s)} = Im(\sigma)\left[\frac{iI}{\Pi} + \omega A_z - \frac{\omega}{\Pi}\left(\sum_{k=1}^{M}\int_{S_k} \sigma_k A_z dS\right)\right] \tag{6.2.57}$$

and the normal conduction current density component $J_z^{(n)}$ by

$$J_z^{(n)} = Re(\sigma)\left[\frac{I}{\Pi} - i\omega A_z + \frac{i\omega}{\Pi}\left(\sum_{k=1}^{M}\int_{S_k} \sigma_k A_z dS\right)\right] \tag{6.2.58}$$

where $Im(\sigma)$ and $Re(\sigma)$ denote, respectively, the imaginary part and the real part of the complex conductivity σ, which is generally not uniform in the composite superconducting stripline. Similarly, we can find the superconducting and normal current density components inside the homogeneous outer conductor wall if it is at the superconducting state.

From the solutions of current components, we may then find the ac resistance R per unit length of the composite superconducting stripline by

$$R = \frac{1}{|I|^2}\left(\int_{S_o}\frac{1}{\sigma_n^{(o)}}|J_z^{(n)}|^2 dS + \sum_{k=1}^{M}\int_{S_k}\frac{1}{\sigma_n^{(k)}}|J_z^{(n)}|^2 dS\right) \tag{6.2.59}$$

where $\sigma_n^{(o)}$ denotes the normal conductivity of the outer conductor wall, and $\sigma_n^{(k)}$ denotes the normal conductivity for the kth subdomain S_k ($k = 1, 2,..., M$) in the composite superconducting strip (signal) line. Equation (6.2.59) implies that only the normal conduction current components $J_z^{(n)}$ causes the ac loss of the composite superconducting stripline. Theoretically, the ac losses in both the outer wall and the signal line contribute to the line resistance R of the composite superconducting stripline. In practice, since the outer wall conductor (or the ground plane) has usually a cross-sectional area much larger than that of the signal line, the line resistance R of the stripline is mainly determined by the ac loss in the signal line.

Furthermore, we may find the inductance L per unit length of the composite superconducting stripline by

$$L = \frac{1}{|I|^2}Re\left\{\int_{S+S_o} A_z J_z^* dS\right\} + \frac{\mu_o}{|I|^2}\left(\int_{S_o}(\lambda_L^{(o)})^2|J_z^{(s)}|^2 dS + \sum_{k=1}^{M}\int_{S_k}(\lambda_L^{(k)})^2|J_z^{(s)}|^2 dS\right) \tag{6.2.60}$$

where the asterisk denotes complex conjugate. It is shown that the second term on the right-hand side of Eq.(6.2.60) denotes the kinetic inductance per unit length of the composite superconducting stripline. Here, $\lambda_L^{(o)}$ is the London penetration depth of the outer wall, and $\lambda_L^{(k)}$ is the London penetration depth of the kth subdomain in the composite strip (signal) line if they are at the superconducting state. In the case of the kth subdomain being non-superconducting medium, the kth term of the kinetic inductance vanishes since the superconducting current component $J_z^{(s)}$ becomes zero.

Illustratively, let us use the FEM computational code developed based on above formulation to simulate numerically transmission line properties of some composite superconducting striplines. The first example is shown in Fig. 6.12, in which the width W and thickness h of the strip (signal) line is fixed to be $W = 16$ μm and $h = 2$ μm, centered in a straight superconducting tube of rectangular cross-section with the internal size of $W_o \times h_o$ ($W_o = 60$ μm and $h_o = 40$ μm), and the thickness $t_o = 3$ μm for the outer conductor shell. The outer conductor is supposed to be grounded. Other geometric parameters a_1, a_2, and a_3 may, however, vary in order to investigate the size effect of the composite superconducting stripline.

Some numerical results are shown in Table 6.1, in which σ_s is the complex conductivity of the superconductor, $\sigma_s = \sigma_n - i/(\mu_o \omega \lambda_L^2)$ with σ_n and λ_L being, respectively, the normal conductivity and the London penetration depth at temperature T ($< T_c$). In all the calculations, we assume, according to the classical two-fluid model, $\sigma_n = \sigma_{no}\Theta^4$, and $\lambda_L = \lambda_o/(1-\Theta^4)^{1/2}$ with $\sigma_{no} = 5\times10^5$ S/m and $\lambda_o = 150$ nm. The reduced temperature $\Theta = T/T_c$ is taken to be 0.85, and the ac frequency f (= $\omega/2\pi$) is taken to be $f = 2$ GHz. In the table, case 1 is for the stripline with its signal line being a homogeneous superconductor in the superconducting state, while case 5 is for the same stripline at the normal state. It is shown that the line resistance R of the stripline at the normal state has a value of about 65 kΩ/m, which is quite large since the normal conductivity of high-temperature cuprate superconductors is quite low.

The numerical result for case 3 indicates that if the homogeneous superconductor stripline is separated into two parts by a small good conductor (e.g., copper with $\sigma = 5\times10^8$ S/m at 77 K and $\sigma = 5.8\times10^7$ S/m at 300 K), the line resistance R has a value of about 5 Ω/m, which is larger than that of the

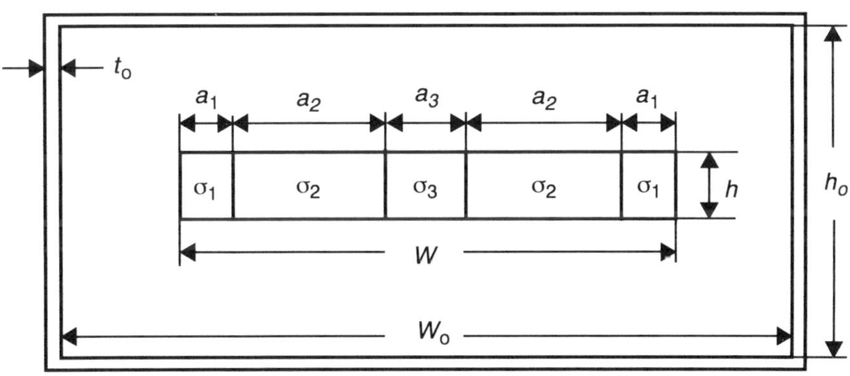

Figure 6.12 Scheme of an edge-side composite superconducting stripline.

Table 6.1 Numerical Results of *R* and *L* for Edgeside Composite Stripline

Case	a_1 (μm)	a_2 (μm)	a_3 (μm)	σ_1 (S/m)	σ_2 (S/m)	σ_3 (S/m)	R (Ω/m)	L (nH/m)
1	0	8	0	–	σ_s	–	0.022	334
2	2	6	0	5×10^8	σ_s	–	137	336
3	0	7	2	–	σ_s	5×10^8	4.9	334
4	0	7	2	–	σ_s	0	0.023	335
5	0	8	0	–	5×10^5	0	65662	380
6	0	7	2	–	5×10^5	6×10^7	7315	636
7	0	8	0	–	5×10^8	–	239	336

stripline with a homogeneous superconducting signal line (case 1), but is quite small as compared with the line resistance R of a corresponding homogeneous copper stripline at 77 K, as shown in case 7. One result that could be of interest is that if such a composite superconductor–normal conductor stripline is turned into the normal state (at room temperature), the line resistance R of the stripline becomes about 7 kΩ/m, as shown in case 6, which is about 10 times smaller than the value of a homogenous superconductor stripline at the normal state, as shown in case 5. This is because the most of current in the stripline is now flowing in the region of the good conductor (copper) even though it occupies only a very small part of the cross-sectional area of the stripline.

Furthermore, we may notice that the line inductance L of such a composite stripline has much larger value when the stripline becomes normal than that of the stripline in the superconducting state, as shown in case 6 and case 3. This is physically understandable since the effective area of the conducting stripline is reduced due to the fact that the current is concentrated in the good conductor (copper) area as shown from the simulation result.

According to the simulation result, we may think a possible way of making electrically and/or thermally controllable switching of inductors, made from superconductor-normal conductor composite microstrip lines (or planar striplines). The following is an example of making a switchable inductor [Zhou and Wikborg (1998)], in which a microstrip (signal) line is made from two high-temperature oxide superconductors (HTS) sandwiched by a normal good conductor (copper, silver, or gold), as shown in Fig. 6.13. The width W_c of the normal conductor is smaller than the total width W of the microstrip signal line, the exact value of which depends on the specified switching value of the

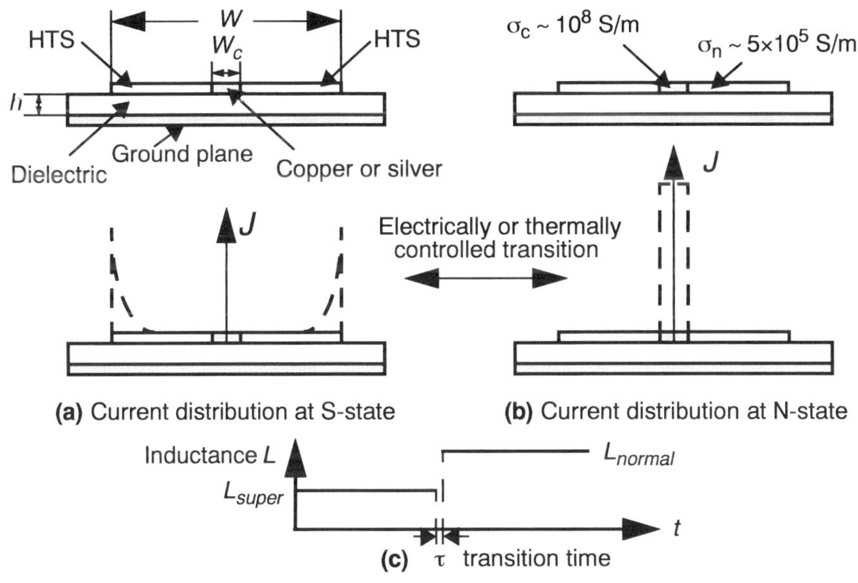

Figure 6.13 Scheme of a switchable inductor.

inductance required. When the microstrip line is in the superconducting state, the current distribution in the microstrip line is shown approximately in Fig. 6.13(*a*) due to the Meissner effect of the HTS, and the inductance of the microstrip line is thus determined by the total width W among others. However, when the microstrip line is at the normal state, the current distribution is shown in Fig. 6.13(*b*) due to the very low normal-state conductivity of HTS (~5×10^5 S/m) so that the most part of the current is now flowing in the normal (good) conductor which has a much high conductivity on the order of 10^8 S/m, especially at low temperature if such a superconducting-normal transition is made at, for instance, 77 K. This could be done by, for instance, passing a current higher than the critical current of the HTS. In practice, HTS materials with low critical currents could be chosen for this purpose if the device is not required to work at high power. By assigning a bias current slight lower than the critical current, the switching of the transition could be made extremely fast by using a high-speed current pulse. The switching of inductance is also reversible.

As we may know, the inductance of a microstrip line is approximately proportional to $1/W$ for a fixed height h to its ground plane. At the normal state, since almost all the current is flowing in the good normal conductor region, the effective width of the microstrip signal line is reduced to about W_c, shown in the Fig. 6.13(*b*), which increases thus the line inductance. It has been shown by the numerical simulation that the inductance can be easily doubled with such a method, which can be used at microwave frequencies, depending on its geometric sizes of the composite microstrip line.

Let us now continue our discussion on the simulation results. In case 4 of Table 6.1, we consider a small non-conducting medium in the middle of the superconducting stripline. It is shown that both the line resistance R and the line inductance L have negligible changes, as compared with the homogeneous superconducting stripline in case 1. Furthermore, in case 2, we consider a superconductor stripline with its two ends covered by two good conductors (copper here). It is shown that the line resistance R has a relatively large value even when the most part of the stripline is superconducting. This is because a large part of the current is now flowing inside the normal conductor regions at the two edges of the stripline due to the skin effect at high frequencies.

In the next example, we study the composite superconducting stripline having a structure shown in Fig. 6.14, in which the width W of the stripline is fixed to be $W = 16$ μm, centered in a free-space box of size $W_o \times h_o$ ($W_o = 60$ μm and $h_o = 40$ μm), bounded by a superconducting grounded tube with the shell thickness of $t_o = 3$ μm. Some numerical results are listed in Table 6.2.

It is shown from case 1 and case 3 that it is possible to insert a good (electric and thermal) conductor (e.g., copper) between two superconductor films with a little increase in the line resistance R of the composite superconducting stripline in the superconducting state. Such a composite stripline may, however, reduce significantly the line resistance R of the stripline when it becomes at the normal state, even at room temperature.

It is shown in case 4 and case 5 in Table 6.2 that there is about 10 times reduction of the line resistance if the composite strip line is used. Such a property of the composite superconducting stripline could be of useful for high-power superconducting stripline devices for various reasons, such as the protection of destruction of the superconducting device due to some unexpected failure in the cooling system, or overloads.

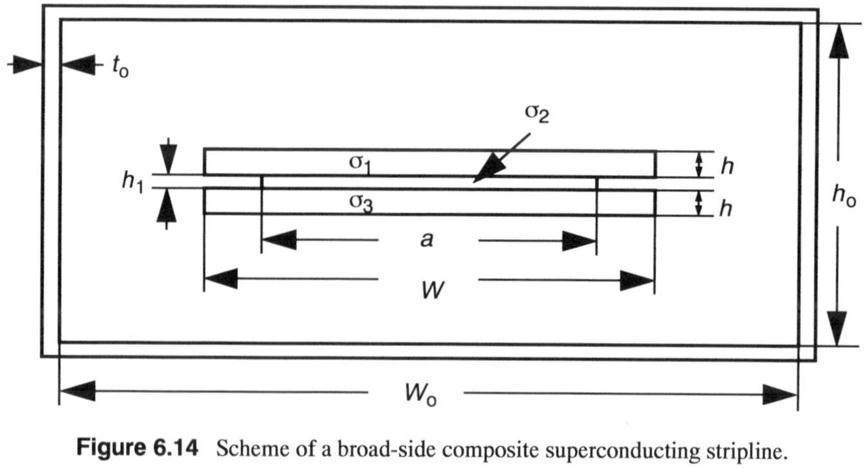

Figure 6.14 Scheme of a broad-side composite superconducting stripline.

Table 6.2 Numerical Results of *R* and *L* for Broadside Composite Stripline

Case	h (μm)	h_1 (μm)	a (μm)	σ_1 (S/m)	σ_2 (S/m)	σ_3 (S/m)	R (Ω/m)	L (nH/m)
1	1	1	16	σ_s	σ_s	σ_s	0.019	320
2	1	1	16	σ_s	5×10^8	σ_s	9.0	321
3	1	1	12	σ_s	5×10^8	σ_s	0.039	321
4	1	1	16	5×10^5	5×10^5	5×10^5	44830	372
5	1	1	16	5×10^5	6×10^7	5×10^5	4308	385
6	1	1	16	5×10^8	σ_s	5×10^8	155	323

Besides, the numerical result shown in case 6 indicates that the superconducting stripline covered by two layers of normal highly conducting films has quite high line resistance R. This is because most of current in this type of composite superconducting stripline is carried by normal electrons in the normal conducting film due to the skin effect at microwave frequencies. Obviously, this type of composite structure is not proper from the point view of effectively utilizing the superconducting properties (very low losses) of the superconductors.

6.2.6 Analysis of Nonlinear Superconducting Microstrip Lines

Since the discovery of high-temperature oxide (cuprate) superconductors (HTS), very compact high-Q resonators and filters using planar HTS stripline and patch structures have been widely studied. They have been deployed recently in microwave subsystems for high-performance radio communication systems, where filters with very sharp skirts and low insertion loss, as well as size and weight reductions are of great importance. At present, the applications are limited to microwave receive filters for radio base stations while transmitter filter applications are still too difficult to be realized because of the very high power levels that these HTS structures are required to handle. The nonlinear material properties of HTS at high power levels are currently being extensively investigated. In this section, we introduce a method for calculating the line resistance and inductance of nonlinear (normal or superconducting) microstrip lines from electromagnetic field analysis with the aid of the finite element method. Both nonlinear physical properties and geometric effects of the microstrip lines are taken into account in the basic formulation, which generally results in a nonlinear integrodifferential equation for vector magnetic potential. A finite element numerical code can then be developed to solve the integro-differential equation. Illustratively, a numerical example is given to show how the nonlinear line resistance and inductance of a superconducting microstrip line can be extracted from the electromagnetic field solution.

We assume that the electrical conduction property of a conductor can be described by the form of Ohm's law: $J_z = \sigma E_z$, where the conductivity σ may be complex if the conductor is a superconductor in time-harmonic fields, as shown by Eq.(6.2.33). To model nonlinear property of the conductor, we may adopt a simple approach by assuming that the conductivity σ is a phenomenological function of the local current density inside the conductor. In particular, for superconductors, as one of possible models, we may write

$$\lambda_L^2(\Theta, J) = \lambda_o^2\left[1 - \Theta^4\left(1 + \sum_{k=1}^{M} a_k(\Theta)J^{2k}\right)\right]^{-1} \tag{6.2.61}$$

and

$$\sigma_n(\Theta, J) = \sigma_N\Theta^4\left(1 + \sum_{k=1}^{M} a_k(\Theta)J^{2k}\right) \tag{6.2.62}$$

which is consistent with the classical two-fluid model. This model implies that the number of normal electrons may increase with increasing current because the probability of breaking-up of the Cooper-pair superelectrons may increase with increasing velocities of electrons in high fields. Besides, we have the following restrictive condition due to the conservation of total number of electrons:

$$\Theta^4\left(1 + \sum_{k=1}^{M} a_k(\Theta)J^{2k}\right) \le 1 \tag{6.2.63}$$

where J may be defined phenomenologically in one way as the normalized current density $J = J_z/J_c$ with J_c being the critical current density of the superconductor. In the expansion of (6.2.61) and (6.2.62), only even powers are used by considering the response of a superconductor to the current does not depend on the direction of the current. The parameters σ_N and λ_o are, respectively, the normal state conductivity and the penetration depth of the superconductor at absolute zero. $a_k(\Theta)$ (k = 1, 2,..., M) are phenomenological parameters characterizing the nonlinear properties of the superconductor. Θ is the reduced temperature ($\Theta = T/T_c$) with T_c being the critical temperature of the superconductor. Here we shall not discuss possible microscopic (or mesoscopic) mechanisms related to the nonlinear behavior of the superconductor. Thus the parameters a_k are treated here as experimentally fitting parameters at a given temperature.

By noting the inhomogeneity of the conductivity σ induced generally by non-uniform distribution of current density, we may find the following integro-differential equation for the magnetic vector potential component A_z:

$$-\nabla_t^2 A_z + i\omega\mu_o\sigma(J)\left(A_z - \frac{1}{\Pi}\int_S \sigma(J)A_z dS\right) = \frac{\sigma(J)\mu_o}{\Pi}I \tag{6.2.64}$$

in which the parameter Π is defined by

$$\Pi = \int_S \sigma(J)dS \tag{6.2.65}$$

where the integration is taken over the cross-sectional area S of the conductor. In general, Eq.(6.2.64) is nonlinear and should be solved together with the heat conduction equation since the local temperature T (or Θ) may depend on the local electrical power dissipation. This is a coupled thermomagnetic problem.

Once A_z is determined for the total given current I and with proper boundary conditions for the nonlinear integro-differential equation (6.2.64) together with the Laplace equation (6.2.31) for A_z in dielectrics, the total current density J_z inside the conductor can be calculated by

$$J_z = \frac{\sigma I}{\Pi} - i\omega\sigma A_z + \frac{i\omega\sigma}{\Pi}\int_S \sigma A_z dS \tag{6.2.66}$$

In particular, for superconductors, the superconducting current density $J_z^{(s)}$ can be given by

$$J_z^{(s)} = -\frac{iI}{\mu_o\omega\Pi\lambda_L^2} - \frac{1}{\mu_o\lambda_L^2}\left(A_z - \frac{1}{\Pi}\int_S \sigma A_z dS\right) \tag{6.2.67}$$

and the normal conduction current component $J_z^{(n)}$ by

$$J_z^{(n)} = \frac{\sigma_n I}{\Pi} - i\omega\sigma_n\left(A_z - \frac{1}{\Pi}\int_S \sigma A_z dS\right) \tag{6.2.68}$$

Thus the ac resistance R per unit length of the superconducting microstrip line is given by

$$R = \frac{1}{|I|^2}\int_{S_o+S_1} \frac{1}{\sigma_n}\left|J_z^{(n)}\right|^2 dS \tag{6.2.69}$$

and the inductance L per unit length of the superconducting microstrip line by

$$L = \frac{1}{|I|^2}Re\left\{\int_{S_o+S_1} A_z J_z^* dS\right\} + \frac{\mu_o}{|I|^2}\int_{S_o+S_1} \lambda_L^2\left|J_z^{(s)}\right|^2 dS \tag{6.2.70}$$

where S_o and S_1 denote the cross-sectional areas of the ground plane and of the microstrip (signal) line, respectively. It is shown that the second term on the right-hand side of Eq.(6.2.70) denotes the line kinetic inductance, which vanishes for normal conductors. It is noticed that both the ac resistance R and the inductance L are generally function of the current in the nonlinear case, which implies the nonlinearity of the corresponding transmission line.

To introduce a finite element formulation, we apply the Galerkin integral form to Eqs.(6.2.64) and (6.2.31):

$$\int_{S_1}\left[(\nabla_t w)\cdot(\nabla_t A_z) + i\omega\sigma\mu_o\left(A_z - \frac{1}{\Pi}\int_{S_1}\sigma A_z dS\right)w - \frac{\sigma\mu_o I}{\Pi}w\right]dS$$
$$+\int_{S_o}\left[(\nabla_t w)\cdot(\nabla_t A_z) + i\omega\sigma_o\mu_o\left(A_z - \frac{1}{S_o}\int_{S_o}A_z dS\right)w + \frac{\mu_o I}{S_o}w\right]dS$$
$$+\int_{S_D}(\nabla_t w)\cdot(\nabla_t A_z)dS = \int_{\Gamma} w\frac{\partial A_z}{\partial n}d\Gamma \qquad (6.2.71)$$

where w is the testing function, S_D denotes dielectric (or air) domains, and $\boldsymbol{n}$ is the outward unit vector normal to the outer boundary Γ of S_D enclosing the microstrip line. For some simplicity, in Eq.(6.2.71), we have assumed that the conductivity σ_o of the ground plane is constant at given temperature since the current density is usually small so that the nonlinear effect may be negligible there. A finite element numerical code can be developed based on the formulation to simulate nonlinear properties of nonlinear microstrip lines.

Illustratively, let us consider a superconducting microstrip line, as shown in Fig. 6.15(a). We assume that the conductivity σ is effectively a function of the reduced current $J = |I|/I_c$, where $|I|$ and I_c are, respectively, the magnitude of the total current and the critical current of the superconducting microstrip (signal) line. We take the following data: $\sigma_N = 5\times10^5$ S/m, $\lambda_o = 150$ nm, $I_c = 1.6$ A, $\Theta = 0.85$, $t = 4$ μm, w = 50 μm, and $h = 200$ μm, which are chosen to be illustrative rather than to describe any particular material sample. The numerical results on the current dependence of the nonlinear line resistance R and nonlinear line inductance L are shown in Fig. 6.15(b) and Fig. 6.15(c), respectively. It is shown that the nonlinearity of both the line resistance R and line inductance L of the microstrip line can be studied consistently by this method.

It can be seen that the finite element method presented here based on electromagnetic field analysis is flexible, and may be used to analyze nonlinear transmission lines of complex geometry as well as multi-conductor structures at microwave frequencies.

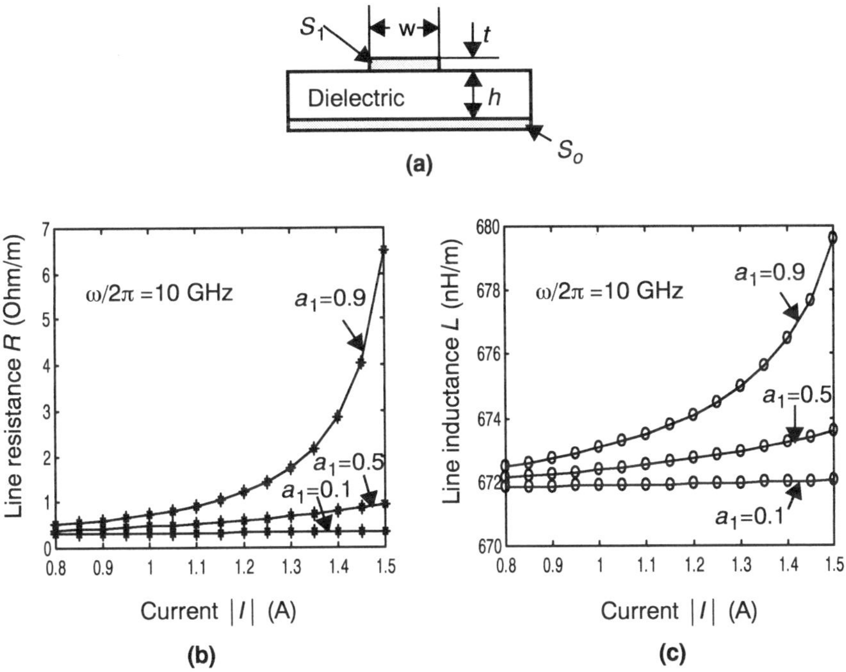

Figure 6.15 EM simulation of a nonlinear microstrip line.

6.2.7 Harmonic Generation and Two-Tone Intermodulation

The nonlinearity of materials may lead to nonlinear effects of microwave components, such as harmonic generation and intermodulation distortion. To illustrate these effects, let us consider a simple nonlinear resistor in a microwave circuit. We assume that the nonlinear resistor can be characterized by the following I–V relation:

$$I(V) = \frac{1}{R_L}V + G_N V^3 \tag{6.2.72}$$

where R_L is the linear resistance, and the parameter G_N characterizes effectively the nonlinearity of the resistor. Here, we have make use of the assumption that the I–V relation should be symmetrical with respect to the polarity, that is, $I(-V) = -I(V)$. Equation (6.2.72) implies that only the lowest harmonic, being the third order one, is taken into account here.

Let us first consider a sinusoidal signal

$$V = V_o \sin(\omega t) \tag{6.2.73}$$

as the single-tone test signal. Substitution of Eq.(6.2.73) into (6.2.72) gives

$$I = \frac{1}{R_L} V_o \sin(\omega t) + G_N V_o^3 [\sin(\omega t)]^3 \tag{6.2.74}$$

By noting $\sin^3(\omega t) = (3\sin(\omega t) - \sin(3\omega t))/4$, we may write Eq.(6.2.74) in the form:

$$I = \frac{V_o}{R_L} \sin(\omega t) + \frac{G_N V_o^3}{4} [3\sin(\omega t) - \sin(3\omega t)] \tag{6.2.75}$$

which shows that there is the third harmonic term being generated due to the nonlinearity of the resistor. The logarithm of the amplitude of the third harmonic term is given by

$$y = \log\left(\frac{G_N V_o^3}{4}\right) = \log\left(\frac{G_N}{4}\right) + 3x \tag{6.2.76}$$

with $x = \log(V_0)$. Equation (6.2.76) represents a straight line with a slope equal to 3 in the x-y plane.

We may also take the logarithm of the amplitude of the first order term, that is, the dominant fundamental frequency component:

$$y = \log\left(\frac{V_o}{R_L}\right) = -\log(R_L) + x \tag{6.2.77}$$

which also represents a straight line with however a slope equal to 1 in the x-y plane. These two lines (6.2.76) and (6.2.77) intercept at

$$x_c = \frac{1}{2}\log\left(\frac{4}{R_L G_N}\right) \tag{6.2.78}$$

$$y_c = \frac{1}{2}\log\left(\frac{4}{R_L^3 G_N}\right) \tag{6.2.79}$$

Often, the third order intercept in terms of decibels:

$$I_3 = 20 y_c = 10 \times \log\left(\frac{4}{R_L^3 G_N}\right) \quad \text{(dB)} \tag{6.2.80}$$

is used to describe quantitatively the nonlinearity of a microwave circuit [Shen (1994)]. Shown in Fig. 6.16 is a typical plot for the third order intercept.

Let us now consider two sinusoidal signals of different frequencies, but with the same amplitude and initial phase for some simplicity:

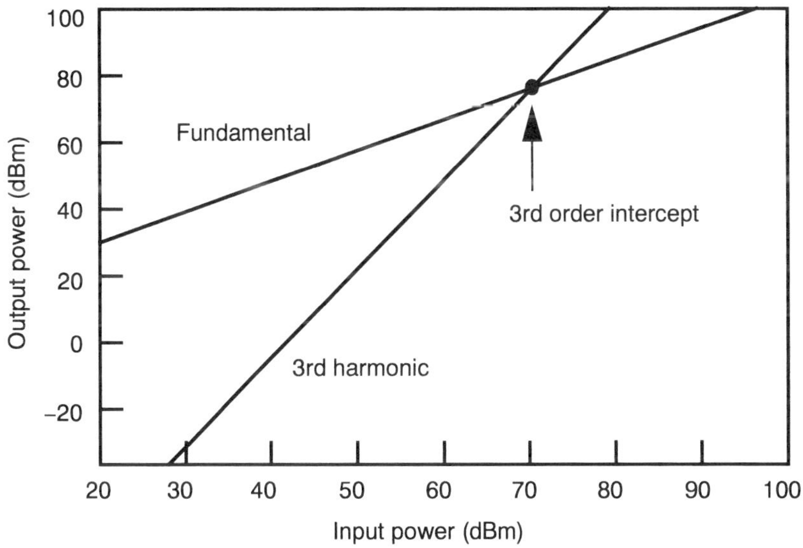

Figure 6.16 The third-order intercept plot at a certain fundamental frequency.

$$V = V_o[\sin(\omega_1 t) + \sin(\omega_2 t)] \tag{6.2.81}$$

as the input of the two-tone test. Substitution of Eq.(6.2.81) into Eq.(6.2.72) gives

$$I = \frac{1}{R_L}V_o[\sin(\omega_1 t) + \sin(\omega_2 t)] + G_N V_o^3\{[\sin(\omega_1 t)]^3 + [\sin(\omega_2 t)]^3\}$$
$$+ G_N V_o^3\{3[\sin(\omega_1 t)]^2\sin(\omega_2 t) + 3[\sin(\omega_2 t)]^2\sin(\omega_1 t)\} \tag{6.2.82}$$

By noting the mathematical relation:

$$[\sin(\omega_1 t)]^2\sin(\omega_2 t) = \frac{1}{2}\sin(\omega_2 t) + \frac{1}{4}\sin[(2\omega_1 - \omega_2)t] - \frac{1}{4}\sin[(2\omega_1 + \omega_2)t] \tag{6.2.83}$$

we may find that the intermodulation frequency $(2\omega_1 - \omega_2)$ term is given by

$$\frac{3}{4}G_N V_o^3 \sin[(2\omega_1 - \omega_2)t] \tag{6.2.84}$$

We take the logarithm of its amplitude:

$$y = \log\left(\frac{3G_N V_o^3}{4}\right) = \log\left(\frac{3G_N}{4}\right) + 3x \tag{6.2.85}$$

with $x = \log(V_0)$. Equation (6.2.85) represents a straight line with a slope of 3.

If we take the logarithm of the amplitude of the first-order term with frequency ω_1,

$$y = \log\left(\frac{V_o}{R_L}\right) = -\log(R_L) + x \tag{6.2.86}$$

we may find the third order intercept $I_{1,2}$ in terms of decibels for the intermodulation as

$$I_{1,2} = 10 \times \log\left(\frac{4}{3R_L^3 G_N}\right) \quad \text{(dB)} \tag{6.2.87}$$

from the set of Eqs.(6.2.85) and (6.2.86).

The presence of intermodulation indicates that the effects of nonlinearity not only generate harmonics, but also cause interaction among different frequency components. Physically, this implies that modulated signals in different frequency channels may get mixed up. The two-tone intermodulation is often used as a measure of the nonlinear effects in an a microwave circuit. Studies of two-tone intermodulation are therefore of practical interest. Some measurements of the two-tone intermediation on YBCO microstrip filters may be found in the work of, for instance, Lyons and Withers (1990), and Oates et al. (1991).

6.2.8 Method of Equivalence between Distributed-Circuit and Full-Wave Analyses

It is known that, to model wave propagation properties of electrical waveguides consisting of two or more parallel conductors (or superconductors), the well-known transmission line model has been used successfully in microwave device and circuit designs for a long time. Well-developed measurement instruments also exist in examining the transmission line parameters of various types of waveguides. Thus, for use with device models in circuit simulation, we are used to adopt the distributed-circuit analysis in which various types of electrical waveguides, such as the microstrip lines, are modeled as equivalent transmission lines. To apply the transmission line model, an important problem is to determine the line parameters, such as the distributed (line) resistance, inductance, conductance, and capacitance for given geometric configuration and material properties of the waveguide. Analyses have been undertaken with various approximations, as we have shown in above sections. These approximations, such as the quasi-TEM approximation, may not be valid for analyzing properties of some transmission lines operating at extremely high frequencies. Although a full-wave (field) analysis, which solves rigorously the electromagnetic field boundary-value problem, may be conducted to study behaviors of these transmission lines, the full-wave analysis is however carried out to determine propagation constants as well as electromagnetic field components rather than the transmission line parameters used in distributed-circuit analyses. Recently some efforts have been made to model lossy electrical waveguides as equivalent transmission lines [Brews (1987), Collin (1991), and Zhou (1996)].

In this section, we shall introduce a method of equivalence to determine the distributed-circuit parameters of transmission lines from full-wave (field) analyses. We shall assume that the electrical transmission line studied is composed of two or more parallel conductors of arbitrary geometric cross sections. In planar microwave or digital integrated circuits, the used transmission line (waveguide) structures are typically microstrip lines and/or coplanar lines of rectangular cross-section. It is known that the wave mode of propagation on a microstrip line is generally not a TEM mode and, hence, the line integral of the electric field between the ground plane and the strip of the line may depend on the path. Thus a unique value of voltage may not exist and, consequently, the usual expression for the characteristic impedance Z_o (= V/I) may also be not unique from theoretical point of view. Therefore, an alternative definition based on power consideration has been proposed [see, e.g., Collin (1991)]:

$$Z_o = \frac{2P_x}{|I|^2} \tag{6.2.88}$$

where P_x denotes the power flow along the line in the x-direction, and I is the total x-directed current flowing on the strip, that is,

$$|I|^2 = \int_{A_c} |J_x|^2 dA \tag{6.2.89}$$

where A_c is the cross-section of the strip and J_x is the current density flowing in the strip along x-axis direction. There exists also another definition based on the power consideration, which is $Z_o=|V|^2/(2P_x)$ with V being a defined voltage. In general, the three definitions discussed above do not give the same values for Z_o except in the dc limit. However, it can be shown that all three definitions given above for the characteristic impedance Z_o are equivalent provided that the complex power P_x satisfies the following relation:

$$P_x = \frac{1}{2}\int_A (\boldsymbol{E} \times \boldsymbol{H}^*)_x dA = \frac{1}{2} VI^* \tag{6.2.90}$$

where A is the area of the whole cross-section of the waveguide on the y-z plane. $\boldsymbol{E}$ and $\boldsymbol{H}$ are, respectively, the electric field and the magnetic field. Indeed, it has been shown by Brews (1987) that the relation (6.2.90) can be satisfied, at least, for the case where only one mode propagates.

Thus, to correlate the transmission line model with the full-wave analysis, we may use Eq.(6.2.88) to find the characteristic impedance Z_o of the equivalent transmission line for any one particular waveguide, noting that both the power flow P_x and the current I are directly related to the field components, $\boldsymbol{E}$, $\boldsymbol{H}$ and the current density $\boldsymbol{J}$ by Eqs.(6.2.89) and (6.2.90). If we now write the complex characteristic impedance by $Z_o = Z_{or} + iZ_{oi}$ and identify the propagation constant γ from the full-wave solution with the corresponding propagation constant for the equivalent transmission line by Eq.(6.1.7) for any one particular mode, we may obtain the following set of equations:

$$R = \alpha Z_{or} - \beta Z_{oi} \tag{6.2.91}$$

$$L = \frac{(\alpha Z_{oi} + \beta Z_{or})}{\omega} \tag{6.2.92}$$

$$G = \frac{(\alpha Z_{or} + \beta Z_{oi})}{|Z_o|^2} \tag{6.2.93}$$

$$C = \frac{(\beta Z_{or} - \alpha Z_{oi})}{\omega |Z_o|^2} \quad . \tag{6.2.94}$$

This set of equations (6.2.91)–(6.2.94) may be used to determine the complete set of equivalent transmission line parameters: R, L, G, and C by knowing the real propagation constant β, the attenuation constant α, and the characteristic impedance Z_o, which can generally be obtained from a full-wave analysis for the waveguide to be studied. Since the complex propagation constant γ and Z_o are different for each mode of propagation of wave in the waveguide, a different set of transmission line parameters, R, L, G, and C results for each mode of propagation. As we may notice, this method is general and can be used to analyze waveguides consisting of two or more parallel conductors (or superconductors) of arbitrary geometric cross-sections, although numerical field solutions may be required. As a concrete example, in the following section, we shall study quantitatively the equivalent transmission line parameters for modeling an anisotropic superconducting planar waveguide by using this method.

6.2.9 Full-Wave Analysis of Anisotropic Superconducting Planar Waveguide

In this section, we shall study an anisotropic superconducting planar waveguide of finite thickness, as shown in Fig. 6.17, in which the waveguide is made of a dielectric layer of $2d$ thickness, covered by two identical anisotropic superconducting strips of thickness h and width W. We shall look for a field solution for the anisotropic superconducting planar waveguide. To start with, let us assume that the surfaces of two superconductors are parallel to the a-b plane of the anisotropic superconductor. We choose a reference frame aligned with the principal axes of the superconductors such that the z-axis is along the c-axis, which is perpendicular to the surface of the superconducting strip. The x-axis and y-axis are assumed to be along, respectively, the a-axis and b-axis directions of the superconductor. The origin of the z-axis is located at the middle of the dielectric layer. Furthermore, we assume that the thickness $2d$ of the dielectric layer is much less than the width W of the waveguide so that the fringing field effect at the edges of the waveguide may be ignored here for some simplicity.

Since we are studying problems of involving extremely high frequencies at which a quasistatic (quasi-TEM) approximation may be not appropriate, we use the full-wave analysis. We shall look for a possible field solution having the magnetic field components, $\boldsymbol{B} = (0, B_y, 0)$, and the electric field components, $\boldsymbol{E} = (E_x, 0, E_z)$, which describes a type of electromagnetic (TM) wave propagating in the superconducting planar waveguide. We assume that no external magnetic fields are applied to the waveguide and the self-generated magnetic field by currents flowing in the waveguide is small so that the superconductor is in the Meissner state. Thus no magnetic flux-flow effects will be considered here. We also assume that the wave is propagating and attenuated in the positive x-direction ($\sim \exp(i\omega t - \gamma x)$) so that we may write $\gamma = \alpha + i\beta$ with β (> 0) being the propagation constant and α (≥ 0) the attenuation coefficient.

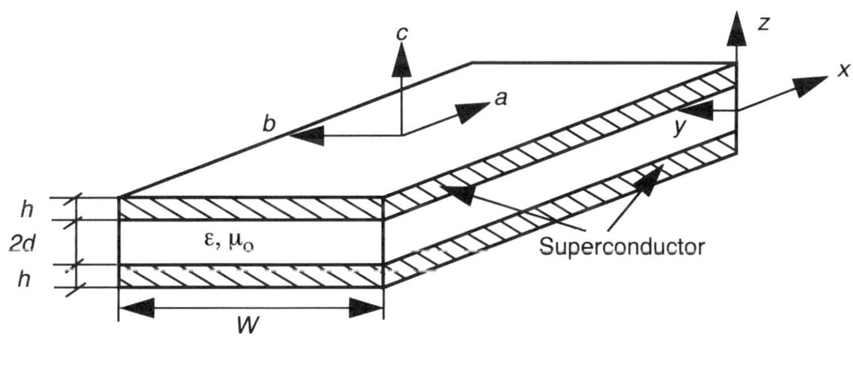

Figure 6.17 Scheme of an anisotropic superconducting planar waveguide.

Thus, from Maxwell's equations and London's equations for the anisotropic superconductor (see Section 3.5), we may find the following field equations for the wave problem:

$$\left(\lambda_c^2\frac{\partial^2}{\partial x^2}+\lambda_a^2\frac{\partial^2}{\partial z^2}\right)B_y-\mu_o\left(\lambda_c^2\sigma_c\frac{\partial E_z}{\partial x}-\lambda_a^2\sigma_a\frac{\partial E_x}{\partial z}\right) = B_y \qquad (6.2.95)$$

in the superconductor, and

$$\frac{\partial^2 B_y}{\partial z^2}+\kappa^2 B_y = 0 \qquad (6.2.96)$$

with $\kappa^2 = \mu_0\varepsilon\omega^2+\gamma^2$ in the dielectric layer. The parameter κ is in general complex and can be expressed as $\kappa = \kappa_R+i\kappa_I$. Here, $\varepsilon = \varepsilon_0\varepsilon_r(1 - i\tan\delta)$ denotes the complex permittivity of the dielectric layer with $\tan\delta$ being the loss tangent characterizing the dielectric loss of the dielectric layer. ε_0 and μ_0 are, respectively, the permittivity and the permeability in vacuum. ε_r denotes the relative dielectric constant of the dielectric layer, which is assumed to be isotropic for simplicity. In Eq.(6.2.95), λ_a and λ_c are the London penetration depths with their corresponding currents in the a-axis and the c-axis, respectively. σ_a and σ_c are the normal conductivities in the a-axis and the c-axis, respectively. Here, a common set of the principal axes has been assumed to exist for those material coefficient tensors involved for the anisotropic superconductor.

By noting the geometric symmetry of the waveguide studied, after some manipulations, we can find the following magnetic field solution of the form:

$$B_y = \frac{\mu_o I}{2W\sinh(\Gamma h)}\left[e^{-\Gamma(|z|-d-h)}-e^{\Gamma(|z|-d-h)}\right] \qquad (d<|z|<d+h) \quad (6.2.97)$$

$$B_y = \frac{\mu_o I}{W\cos(\kappa d)}\cos(\kappa z) \qquad (|z|<d) \qquad (6.2.98)$$

where I is the total current flowing in the superconducting strip of thickness h and width W along the x-axis direction. Γ is a parameter to be determined, which characterizes effectively the penetration of the magnetic field into the superconductor. Here, we consider only the case with $W >> (2d)$, $h \geq \lambda_a$, and $\omega \leq 10^{12}$ rad/s so that the magnetic field outside the waveguide (that is, at $|z| > d+h$) is negligible.

By Ampere´s circuital law and the first London equation (which, in our case, may be written as $J_x^{(s)} = E_x/(i\omega\mu_o\lambda_a)$ and $J_z^{(s)} = E_z/(i\omega\mu_o\lambda_c)$) together with the two-fluid model of $\boldsymbol{J} = \boldsymbol{J}^{(s)} + \boldsymbol{J}^{(n)}$ with $\boldsymbol{J}^{(s)}$ being the superconducting current density and $\boldsymbol{J}^{(n)}$ the normal conduction current density, we may find that the electric field components, E_x and E_z in the superconductors can be related to the magnetic field component B_y by

$$E_x = \frac{-i\omega\lambda_a^2}{1 + i\omega\mu_o\lambda_a^2\sigma_a}\frac{\partial B_y}{\partial z} \tag{6.2.99}$$

and

$$E_z = \frac{i\omega\lambda_c^2}{1 + i\omega\mu_o\lambda_c^2\sigma_c}\frac{\partial B_y}{\partial x} \tag{6.2.100}$$

The presence of the electric field component E_z indicates that there is a current component flowing along the c-axis direction, which is the case even for isotropic superconducting strip waveguides, as we know. Substitution of Eq.(6.2.97) and Eqs.(6.2.99)–(6.2.100) into Eq.(6.2.95), we may find that the parameter Γ, which is generally a complex quantity ($\Gamma = \Gamma_R + i\Gamma_I$), can be determined by

$$\Gamma^2 = \frac{1}{\lambda_a^2} + i\omega\sigma_a\mu_o - \frac{\gamma^2\lambda_c^2}{\lambda_a^2}\left(\frac{1 + i\omega\mu_o\lambda_a^2\sigma_a}{1 + i\omega\mu_o\lambda_c^2\sigma_c}\right) \tag{6.2.101}$$

in which γ is the complex propagation constant. With the use of the interface conditions, namely, the continuity of tangential components of the fields B_y and E_x at $|z| = d$, we may obtain the following eigenvalue equation:

$$\tan(\kappa d) = -\frac{\varepsilon\mu_o\omega^2\Gamma\lambda_a^2}{\kappa(1 + i\omega\mu_o\sigma_a\lambda_a^2)}\coth(\Gamma h) \tag{6.2.102}$$

for the determination of the complex propagation constant γ. For very thick superconductor layer ($h \to \infty$), we may find that Eq.(6.2.102) may recover the result given in the work of Zhou (1994c). Equation (6.2.102) is a transcendental equation, the analytical solution of which can only be found approximately. However, if we notice the fact of $|\gamma|^2 \sim \omega^2\mu_o\varepsilon$ for guided wave mode, we may find the following complex propagation parameter γ at the first-order approximation by

$$\gamma^2 = \frac{-\varepsilon_r(\omega/c)^2(1 - i\tan\delta)[1 + \coth(\Gamma h)/(\Gamma d)]}{1 - \varepsilon_r(\omega/c)^2(1 - i\tan\delta)\lambda_c^2\coth(\Gamma h)/\zeta} \tag{6.2.103}$$

with $\zeta = \Gamma d[1 + 2i(\lambda_c/\delta_c)^2]$. Here, the complex parameter Γ is determined approximately by

$$\Gamma = \sqrt{\frac{1}{\lambda_a^2} + i\frac{2}{\delta_a^2} - \frac{\gamma_o^2\lambda_c^2}{\lambda_a^2}\left(\frac{1 + 2i(\lambda_a/\delta_a)^2}{1 + 2i(\lambda_c/\delta_c)^2}\right)} \tag{6.2.104}$$

with γ_o being given by

$$\gamma_o^2 = -\varepsilon_r\left(\frac{\omega}{c}\right)^2(1 - i\tan\delta)\left[1 + \frac{\lambda_a \coth(h\sqrt{1/\lambda_a^2 + 2i/\delta_a^2})}{d\sqrt{1 + 2i(\lambda_a/\delta_a)^2}}\right] \qquad (6.2.105)$$

Here, $\delta_a = [2/(\omega\mu_o\sigma_a)]^{1/2}$ and $\delta_c = [2/(\omega\mu_o\sigma_c)]^{1/2}$ are the normal skin depths in a-axis and c-axis, respectively. $c = 1/(\mu_o\varepsilon_o)^{1/2}$ denotes the speed of light in vacuum. It may be noticed that γ_o from Eq.(6.2.105) has actually the explicit form of the complex propagation parameter for an isotropic superconducting planar waveguide with the London penetration depth λ_a and the normal conductivity σ_a [Matick (1969)]. It can be shown that for isotropic superconducting waveguides, the term with γ_o^2 in Eq.(6.2.106) is negligible even at millimeter-wave frequencies. This gives

$$\Gamma_R = \frac{1}{\sqrt{2}\lambda_a}\sqrt{1 + \sqrt{1 + 4(\lambda_a/\delta_a)^4}} \qquad (6.2.106)$$

$$\Gamma_I = \frac{1}{\sqrt{2}\lambda_a}\sqrt{\sqrt{1 + 4(\lambda_a/\delta_a)^4} - 1} \qquad (6.2.107)$$

for isotropic superconducting waveguides, which is the well-known result. However, for anisotropic superconducting waveguides, this term is generally not negligible at high frequencies.

To show quantitatively how the parameter Γ, which characterizes effectively the field penetration, is affected by the anisotropy of the superconductor at high frequencies, in Fig. 6.18, we plot the normalized magnetic field (magnitude) $|B_y(z)/B_y(d)|$ versus the normalized distance z/d from the superconductor surface in the range of $d \le z \le h+d$. Here, $B_y(d)$ denotes the magnetic field on the superconductor surface at $z = d$. The numerical results shown in Fig. 6.18 are obtained by using Eq.(6.2.97) and Eqs.(6.2.104)–(6.2.105), where we have set $\varepsilon_r = 25$ and $\tan\delta = 10^{-5}$ and other parameters given in the figure. The result indicates clearly the effect of anisotropy of the superconductor on the magnetic field penetration into the superconductor at high frequency ($\omega = 10^{12}$ rad/s in this numerical example). It is shown in the example that the larger ratio of the London penetration depths λ_c/λ_a in the anisotropic superconductor results in less field penetration since Γ_R increases with increasing the anisotropy ratio λ_c/λ_a. Furthermore, it can be shown that Γ_R may increase with increasing frequency ω due both to the normal skin-effect and to the effect of anisotropy of the superconductor. At low frequencies, both the normal skin-effect and the effect of anisotropy of superconductor becomes negligible.

Using the method of equivalence, we may model the anisotropic superconducting planar waveguide as an equivalent transmission line. By Eq.(6.2.88) and Eq.(6.2.90), after some manipulation with the aid of Eqs.(6.2.97)–(6.2.100), we can derive the following complex characteristic impedance Z_o for the equivalent transmission line:

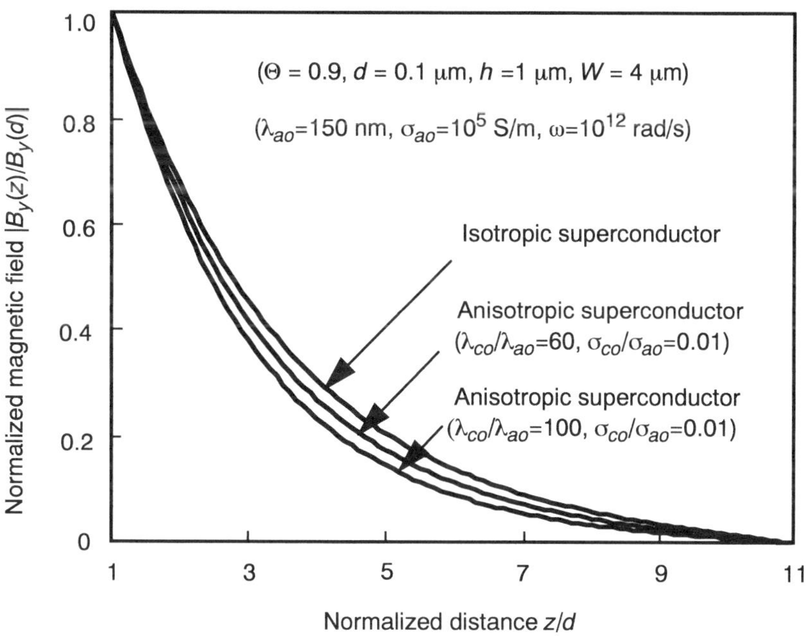

Figure 6.18 Effect of anisotropy on magnetic field penetration in a superconductor.

$$Z_{or} = \frac{[2\alpha(\lambda_c/\delta_c)^2 - \beta](\lambda_c/\delta_c)^2}{\sigma_c W|\sinh(\Gamma h)|^2[1+4(\lambda_c/\delta_c)^4]}\left[\frac{\sinh(2\Gamma_R h)}{\Gamma_R} + \frac{\sin(2\Gamma_I h)}{\Gamma_I}\right]$$
$$+ \frac{\beta + \alpha\tan\delta}{2\omega\varepsilon_o\varepsilon_r W|\cos(\kappa d)|^2[1+(\tan\delta)^2]}\left[\frac{\sinh(2\kappa_I d)}{\kappa_I} + \frac{\sin(2\kappa_R d)}{\kappa_R}\right] \quad (6.2.108)$$

$$Z_{oi} = \frac{[\alpha + 2\beta(\lambda_c/\delta_c)^2](\lambda_c/\delta_c)^2}{\sigma_c W|\sinh(\Gamma h)|^2[1+4(\lambda_c/\delta_c)^4]}\left[\frac{\sinh(2\Gamma_R h)}{\Gamma_R} + \frac{\sin(2\Gamma_I h)}{\Gamma_I}\right]$$
$$+ \frac{\beta\tan\delta - \alpha}{2\omega\varepsilon_o\varepsilon_r W|\cos(\kappa d)|^2[1+(\tan\delta)^2]}\left[\frac{\sinh(2\kappa_I d)}{\kappa_I} + \frac{\sin(2\kappa_R d)}{\kappa_R}\right] \quad (6.2.109)$$

By the derived characteristic impedance Z_o and the complex propagation constant γ from Eq.(6.2.103), we may then find the full set of equivalent transmission line parameters, R, L, G, and C from Eqs.(6.2.91)–(6.2.94).

In what follows, we shall show graphically some numerical results for the full set of equivalent transmission line parameters for the anisotropic superconducting planar waveguide. In the following numerical calculations, we choose $\varepsilon_r = 25$ and $\tan\delta = 10^{-5}$, which are supposed to be independent of both temperature and frequency in the numerical range of interest for some simplicity. Other material properties and geometric parameters for the anisotropic superconducting waveguide are shown in relevant figures. To account for the effect of temperature, the classical two-fluid model is used in the calculation, in which the London penetration depths, λ_a and λ_c, and the normal conductivities, σ_a and σ_c of the superconductor at temperature T are given, respectively, by $\lambda_a = \lambda_{ao}/(1-\Theta^4)^{1/2}$, $\lambda_c = \lambda_{co}/(1-\Theta^4)^{1/2}$, $\sigma_a = \sigma_{ao}\Theta^4$, and $\sigma_c = \sigma_{co}\Theta^4$, with $\Theta = T/T_c$ being the reduced temperature and T_c the critical temperature of the superconductor. Here, λ_{ao}, λ_{co}, σ_{ao}, and σ_{co} denote values of corresponding parameters at $T = 0$. It should be noted that the numerical data taken here are merely illustrative rather than representative for any specific material.

Shown in Fig. 6.19 is the frequency-dependent behaviour of the real part of characteristic impedance Z_o of the anisotropic superconducting planar waveguide, calculated by using Eq.(6.2.108) with relevant geometric and material parameters given in the figure.

Shown in Fig. 6.20(*a*)–(*d*) is the frequency-dependent behaviors of the distributed (line) resistance, inductance, conductance, and capacitance of the anisotropic superconducting planar waveguide from Eqs.(6.2.91)–(6.2.94), respectively. Comparison with results of a corresponding isotropic superconducting planar waveguide is also shown in Fig. 6.19 and Fig. 6.20, in which some well-known results from a quasi-static approximation are included. Here, we denote the characteristic impedance from the quasistatic (lossless) approximation by $Z_{oq} = (L_q/C_q)^{1/2}$, where L_q and C_q are the distributed inductance and capacitance given, respectively, in Fig. 6.20(*b*) and Fig. 6.20(*d*).

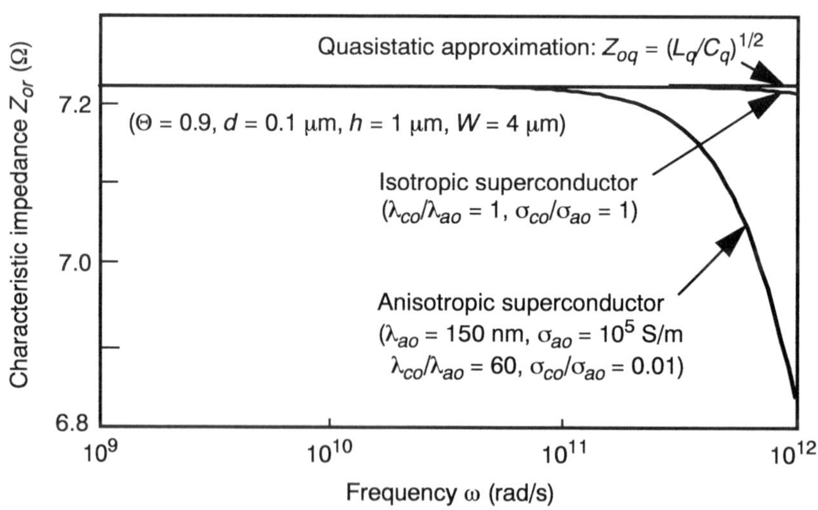

Figure 6.19 Real part of characteristic impedance Z_{or} vs. frequency ω.

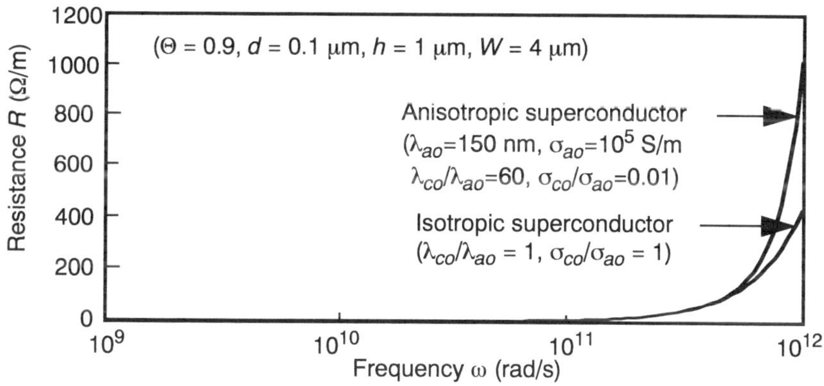

Figure 6.20(a) The equivalent line resistance R vs. frequency ω.

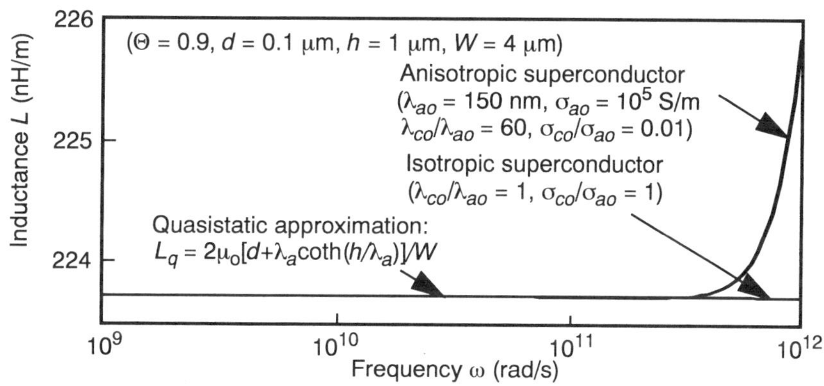

Figure 6.20(b) The equivalent line inductance L vs. frequency ω.

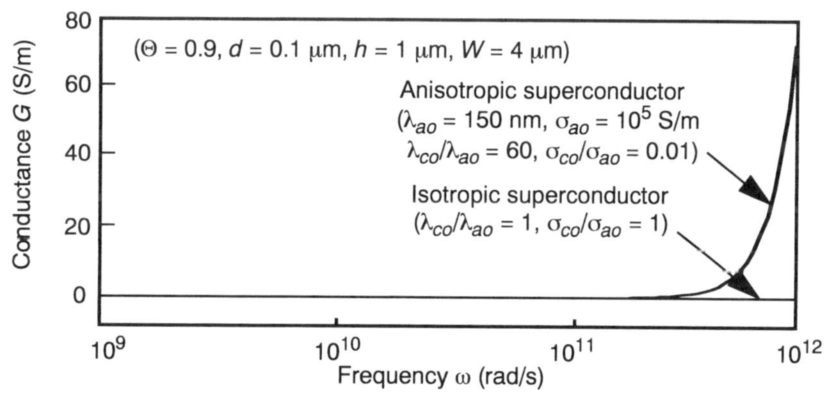

Figure 6.20(c) The equivalent line conductance G vs. frequency ω.

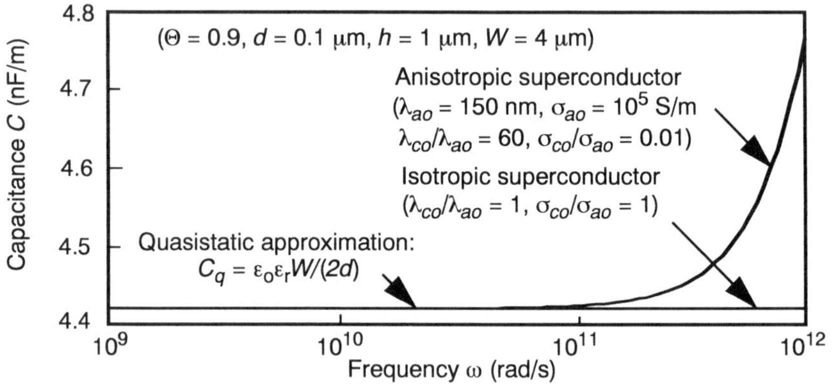

Figure 6.20(d) The equivalent line capacitance C vs. frequency ω.

It is shown in the numerical examples that, at frequencies less than about 20 GHz, the effect of anisotropy is negligible, and the equivalent transmission line properties of the anisotropic superconducting waveguide are essentially the same as those of the isotropic superconducting waveguide of the same geometry. The well-known quasistatic (TEM) approximation is found to be reasonable for describing the transmission properties of the isotropic superconducting waveguide in a wide frequency range, while it deviates noticeably from the full-wave solution for the anisotropic superconducting waveguide at frequencies over 20 GHz.

It is also shown that the real part of the characteristic impedance, Z_{or} decreases with increasing frequency ω for the anisotropic superconducting planar waveguide though its inductance L may increase as shown in Fig. 6.20(*b*). This is because both the capacitance C and conductance G increase here with increasing frequency, as shown in Fig. 6.20(*d*) and Fig. 6.19(*c*). The slight increase of inductance L with increasing frequency differs from our usual expectation for normal electrical waveguide, where the inductance is normally decreased with increasing frequency due to the normal skin-effect. The reason is that, although the normal skin-effect also exists in the superconductor at microwave frequencies, the total inductance of the anisotropic superconducting waveguide may increase with increasing frequency since the part of the kinetic inductance in the anisotropic superconductor may increase with increasing frequency, depending on the material properties and waveguide geometry (see also Section 3.5.3).

In Fig. 6.20(*a*), it is shown that the strong anisotropy of the superconductor may increase the resistance of the waveguide at very high frequencies, as we may expect. At low frequencies, results for the equivalent transmission line parameters derived by the proposed method are shown to be all consistent with those from the well-known quasistatic approximation as expected.

In the following Fig. 6.21 and Fig. 6.22, we have shown numerically how the real part of the characteristic impedance, Z_{or} depends on the geometric ratio W/d of the anisotropic superconducting waveguide and on the reduced temperature Θ, respectively. It may be seen that Z_{or} can be increased by reducing the geometric ratio W/d, and/or by increasing temperature.

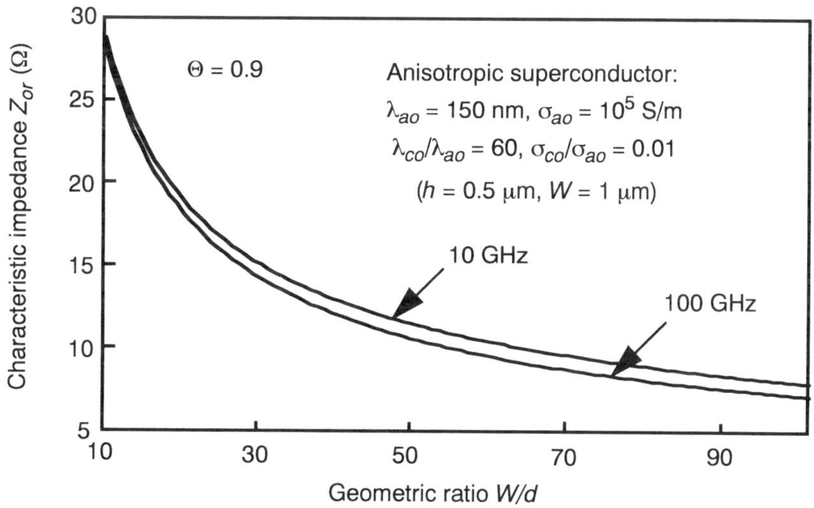

Figure 6.21 Real part of characteristic impedance Z_{ro} vs. waveguide geometric ratio W/d.

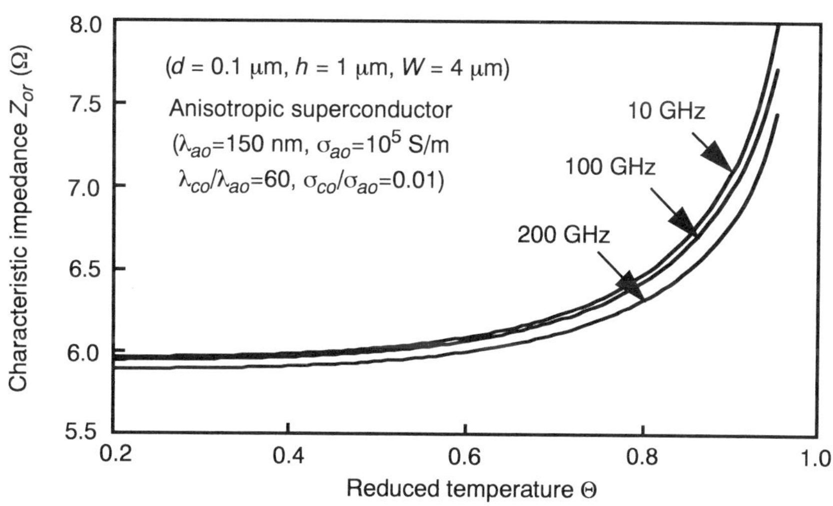

Figure 6.22 Real part of characteristic impedance Z_{ro} vs. the reduced temperature Θ.

6.3 ANALYSIS OF COUPLED TRANSMISSION LINES AND DIRECTIONAL COUPLERS

6.3.1 Formulation of Coupled Transmission Lines

In this section, we shall introduce a method for analyzing coupled transmission lines. In this method, the behavior of, for instance, two lossy coupled transmission lines is described generally by the following set of linear differential equations:

$$\frac{d\boldsymbol{V}}{dz} = -(\boldsymbol{R} + i\omega\boldsymbol{L}) \cdot \boldsymbol{I} \tag{6.3.1}$$

$$\frac{d\boldsymbol{I}}{dz} = -(\boldsymbol{G} + i\omega\boldsymbol{C}) \cdot \boldsymbol{V} \tag{6.3.2}$$

where $\boldsymbol{V} = (V_1, V_2)$ and $\boldsymbol{I} = (I_1, I_2)$ are, respectively, the voltage vector and the current vector, in which V_1, I_1 and V_2, I_2 are the voltage and current on the first and second transmission lines, respectively. $\boldsymbol{R}$, $\boldsymbol{L}$, $\boldsymbol{G}$, and $\boldsymbol{C}$ denote, respectively, the matrices of the distributed resistance (Ω/m), inductance (H/m), conductance (S/m), and capacitance (F/m) of the coupled transmission lines. ω is the radian frequency of the propagating wave.

Introducing the complex wave propagation constant $\gamma = \alpha + i\beta$ by

$$\boldsymbol{V} = \boldsymbol{V}_o e^{-\gamma z} \quad \text{and} \quad \boldsymbol{I} = \boldsymbol{I}_o e^{-\gamma z} \tag{6.3.3}$$

we may arrive at the following eigenvalue equation:

$$det[(\boldsymbol{R} + i\omega\boldsymbol{L}) \cdot (\boldsymbol{G} + i\omega\boldsymbol{C}) - \gamma^2\delta] = 0 \tag{6.3.4}$$

in which δ_{ij} is the Kronecker delta matrix. This equation gives, in general, four possible values of the propagation constant γ. Each of the values of the propagation constant corresponds to a distinct mode of propagation. These are the incident and reflected waves for an even mode, where the voltages as well as the currents on the two lines are in phase, and the incident and reflected waves of an odd mode, where the voltages as well as the currents on the two lines are out of phase with each other. Here, the even mode and odd mode are defined in generalized sense. When the two coupled lines are identical (symmetric), then the two lines have voltages of equal magnitude and currents of equal magnitude.

It can be seen from Eq.(6.3.3) that, in the complex propagation constant $\gamma = \alpha + i\beta$, the parameter α characterizes the attenuation of the wave and β is the real propagation (phase) constant relating to the propagation velocity of the wave. In general, the real propagation constants for the even mode β_e and the odd mode β_o are different, which implies that the phase velocities of the even and odd modes are different. Such a difference may come from the inhomogeneous dielectric media used as well as the unsymmetrical geometry of the coupler.

For each eigenvalue of Eq.(6.3.4), we may obtain its corresponding eigenvector $(\boldsymbol{V}, \boldsymbol{I})$, from which the corresponding mode characteristic impedance can be determined. In general, for asymmetric lossy coupled lines, the even (and odd) mode characteristic impedances for the two lines are different and can be expressed by $\boldsymbol{Z}_{\text{even}} = (Z_{\text{even1}}, Z_{\text{even2}})$ (and $\boldsymbol{Z}_{\text{odd}} = (Z_{\text{odd1}}, Z_{\text{odd2}})$). They are defined by the ratio of voltage to current of its corresponding eigenvector components. For symmetrical coupled lines, we have then simply $Z_{oe} = Z_{\text{even1}} = Z_{\text{even2}}$, and $Z_{oo} = Z_{\text{odd1}} = Z_{\text{odd2}}$. Furthermore, for symmetrical lossless coupled lines, both the even and odd mode characteristic impedances are real-valued quantities. In such a case, it can be said that Z_{oe} is defined as the characteristic impedance between one line and ground when equal and in phase voltages are impressed on the two lines, and Z_{oo} is defined as the characteristic impedance between one line and ground when equal but opposite phase voltages are impressed on the lines.

Using eigenvalues and eigenvectors obtained, we may write the general solution of Eqs.(6.3.1) and (6.3.2) in the following form:

$$\begin{bmatrix} \boldsymbol{V}(z) \\ \boldsymbol{I}(z) \end{bmatrix} = \sum_{k=1}^{4} B_k \boldsymbol{F}_k e^{-\gamma_k z} \tag{6.3.5}$$

where B_k ($k = 1, 2, 3, 4$) are unknown constants, which can be determined by proper boundary conditions. Here, $\boldsymbol{F}_k$ is the eigenvector (voltage and current) corresponding to the kth eigenvalue γ_k ($k = 1, 2, 3, 4$). Here, the four components of the eigenvector $\boldsymbol{F}_k$ are, respectively, $F_k(1) = V_{1k}$, $F_k(2) = V_{2k}$, $F_k(3) = I_{1k}$, and $F_k(4) = I_{2k}$ for $k = 1, 2, 3, 4$.

As an example, consider two coupled transmission lines shown schematically in Fig. 6.23. Here, the load is assumed to be at $z = L$ and the generator at $z = 0$. Thus the set of coupled transmission lines has the following

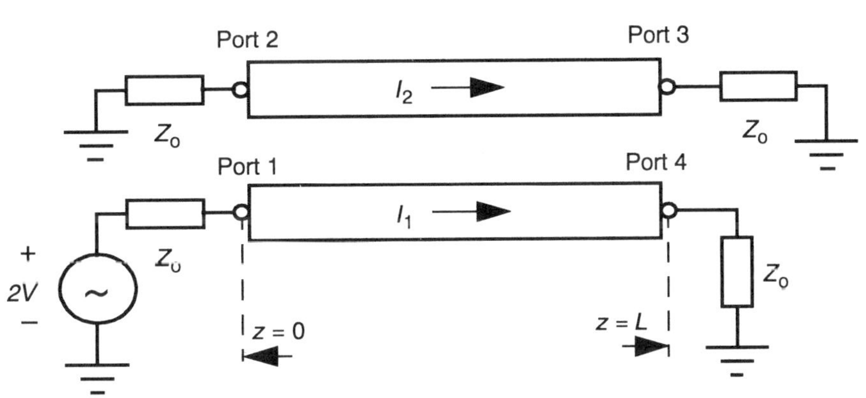

Figure 6.23 Scheme of two coupled transmission lines.

boundary conditions:

$$I_1(0)Z_o + V_1(0) = 2V \quad \text{at port 1} \tag{6.3.6}$$

$$I_2(0)Z_o + V_2(0) = 0 \quad \text{at port 2} \tag{6.3.7}$$

$$I_2(L)Z_o = V_2(L) \quad \text{at port 3} \tag{6.3.8}$$

$$I_1(L)Z_o = V_1(L) \quad \text{at port 4} \tag{6.3.9}$$

where Z_o denotes the matching impedance, usually taken to be 50 Ω. Applying the boundary conditions, we may obtain the following set of four linear algebraic equations:

$$\sum_{k=1}^{4} [Z_o F_k(3) + F_k(1)]B_k = 2V \tag{6.3.10}$$

$$\sum_{k=1}^{4} [Z_o F_k(4) + F_k(2)]B_k = 0 \tag{6.3.11}$$

$$\sum_{k=1}^{4} [Z_o F_k(3) - F_k(1)]e^{-\gamma_k L} B_k = 0 \tag{6.3.12}$$

$$\sum_{k=1}^{4} [Z_o F_k(4) - F_k(2)]e^{-\gamma_k L} B_k = 0 \tag{6.3.13}$$

from which the four unknown constants B_k (k = 1, 2, 3, 4) can be determined. Thus the behaviors of the two coupled transmission lines can be fully described by Eq.(6.3.5), from which the phase information at ports, the coupling efficiency (or crosstalk), as well as the wave attenuation and dispersion properties can be evaluated.

6.3.2 Characterization of Directional Couplers

With the use of the wave solution for coupled transmission lines discussed above, we may study properties of directional couplers, which have found many uses in practical applications. Very often directional couplers are made from two symmetrical coupled transmission lines that are one-quarter wavelength long. Also, it is often useful in practice to consider ideal cases for getting quantitatively feeling about the behavior of a directional coupler.

Consider a symmetrical directional coupler, as shown illustratively in Fig. 6.23. Ignoring losses, in the ideal case of symmetric lines with $\beta_e = \beta_o = \beta$, and using the matching relation $Z_o=(Z_{oe}Z_{oo})^{1/2}$, we may obtain the port voltages: $V_1 = V$, $V_3 = 0$,

$$V_2 = \frac{iVC_v \sin(\beta L)}{\sqrt{1-C_v^2}\cos(\beta L) + i\sin(\beta L)} \tag{6.3.14}$$

$$V_4 = \frac{V\sqrt{1-C_v^2}}{\sqrt{1-C_v^2}\cos(\beta L) + i\sin(\beta L)} \tag{6.3.15}$$

where the coupling coefficient C_v for the lossless symmetrical coupled line reads simply

$$C_v = \frac{Z_{oe} - Z_{oo}}{Z_{oe} + Z_{oo}} \tag{6.3.16}$$

For a $3dB$ directional coupler, we require that $-20\log_{10}(C_v) = 3$ (dB), resulting $C_v = 0.707$. Thus, for the 3dB directional coupler with matching impedance $Z_o = 50\ \Omega$, we may get the even and odd mode impedances required for the design: $Z_{oe} = 120.7\ \Omega$ and $Z_{oo} = 20.7\ \Omega$.

For a quadrature coupler (the line is a quarter wavelength long), we have $\beta L = \pi/2$, and $V_1=V$, $V_2 = C_vV$, $V_3 = 0$, and $V_4 = -\ i(1-C_v^2)^{1/2}$. Thus, in the ideal case, we find that there is no output from port 3, only a small output from port 2 depending on the magnitude of the coupling coefficient C_v, and the principle output from port 4. We may notice that the output at port 4 is 90° out of phase with the input. Besides, energizing the lines at port 4 results in outputs at ports 1 and 3, but no output at port 2. Thus the pair of coupled lines acts as an ideal directional coupler.

As we may know, the directivity of a coupler expresses the ratio of the unwanted signal in the isolated port to the coupled signal. The result for the ideal case implies that the coupler would have infinite directivity. In practice, however, properties of directional couplers deviates usually from their ideal cases due to losses and dielectric inhomogeneities involved in the couplers, which may cause the difference between the propagation velocities of the two (even and odd) modes. Besides, there are also, inevitably, reflections at the ends of the coupler where the modes are mismatched to connecting lines. These factors implies that practical couplers of this kind inherently possess a low directivity. To improve the properties of directional couplers, proper design is required, where the CAD tool can play an important role.

In general, we may define the coupling (in dB) of a directional coupler having losses by

$$C_{dB} = 10 \times \log_{10}\left(\frac{P_1}{P_2}\right) \tag{6.3.17}$$

and the directivity (in dB) of the coupler by

$$D_{dB} = 10 \times \log_{10}\left(\frac{P_2}{P_3}\right) \tag{6.3.18}$$

and the isolation (in dB) of the coupler by

$$O_{dB} = 10 \times \log_{10}\left(\frac{P_1}{P_3}\right) \tag{6.3.19}$$

and the insertion loss (in dB) of the coupler by

$$\Pi_{dB} = 10 \times \log_{10}\left(\frac{P_1}{P_4}\right) \tag{6.3.20}$$

where P_1, P_2, P_3, and P_4 denote, respectively, the power at port 1, port 2, and port 3 of the coupler, as shown in Fig. 6.23.

In most of practical cases, directional couplers are designed to have certain symmetry, and experimental measurements are usually made on scattering parameters at ports of the coupler. Thus it is of interest to introduce a scattering matrix approach for the determination of the characteristic parameters of directional couplers. In this approach, we introduce the following scattering matrix $\boldsymbol{S}$:

$$\boldsymbol{S} = \begin{bmatrix} S_{11} & S_{12} & S_{13} & S_{14} \\ S_{12} & S_{11} & S_{14} & S_{13} \\ S_{13} & S_{14} & S_{11} & S_{12} \\ S_{14} & S_{13} & S_{12} & S_{11} \end{bmatrix} \tag{6.3.21}$$

with its components given, respectively, by [e.g., Dobrowolski (1991)]

$$S_{11} = \frac{S_{11e} + S_{11o}}{2}, \qquad S_{12} = \frac{S_{11e} - S_{11o}}{2} \tag{6.3.22}$$

$$S_{13} = \frac{S_{12e} - S_{12o}}{2}, \qquad S_{14} = \frac{S_{12e} + S_{12o}}{2} \tag{6.3.23}$$

$$S_{11e} = \frac{Z_{oe}^2 - Z_o^2}{Z_{oe}^2 + Z_o^2 + 2Z_{oe}Z_o\coth(\gamma_e L)} \tag{6.3.24}$$

$$S_{12e} = \frac{2Z_{oe}Z_o\text{csch}(\gamma_e L)}{Z_{oe}^2 + Z_o^2 + 2Z_{oe}Z_o\coth(\gamma_e L)} \tag{6.3.25}$$

$$S_{11o} = \frac{Z_{oo}^2 - Z_o^2}{Z_{oo}^2 + Z_o^2 + 2Z_{oo}Z_o\coth(\gamma_o L)} \tag{6.3.26}$$

$$S_{12o} = \frac{2Z_{oe}Z_o\text{csch}(\gamma_o L)}{Z_{oo}^2 + Z_o^2 + 2Z_{oo}Z_o\coth(\gamma_o L)} \tag{6.3.27}$$

in which $\gamma_e = \alpha_e + i\beta_e$ and $\gamma_o = \alpha_o + i\beta_o$ are the complex propagation constants for the even mode and odd mode, respectively. Z_{oe} and Z_{oo} are the characteristic impedances associated with the even and odd modes, respectively. L is the length of the directional coupler. Once the scattering matrix associated with the coupler is known, we may then obtain the characteristic parameters (in terms of logarithmic attenuation levels (dB)) of the coupler by:

$$C_{dB} = -20 \times \log_{10}(|S_{12}|) \tag{6.3.28}$$

$$D_{dB} = -20 \times \log_{10}\left(\frac{|S_{13}|}{|S_{12}|}\right) \tag{6.3.29}$$

$$O_{dB} = -20 \times \log_{10}(|S_{13}|) \tag{6.3.30}$$

$$\Pi_{dB} = -20 \times \log_{10}(|S_{14}|) \tag{6.3.31}$$

It can be seen that the isolation O_{dB} of the coupler can be expressed as the sum of the coupling C_{dB} and directivity D_{dB} of the coupler, that is, $O_{dB} = C_{dB} + D_{dB}$.

6.3.3 Finite Element Analysis of 3dB Directional Couplers

This section presents some numerical examples of analyzing directional couplers with the aid of the finite element method introduced in Section 6.2 for the determination of transmission line parameters of multi-conductor systems. Let us first study a 3dB directional coupler with the broadside coupled suspended substrate stripline structure, as shown in Fig. 6.24. We suppose that the geometrical and material parameters of the directional coupler operating at the frequency of 1.8 GHz have the following values. The relative dielectric constant: $\varepsilon_1 = \varepsilon_2 = 9.5$ (Al_2O_3) and $\varepsilon_s = 3$ (polymide); and the geometric parameters: t = 12 μm, W = 140 μm, W_o = 3 mm, h = 1 mm, and s=64 μm.

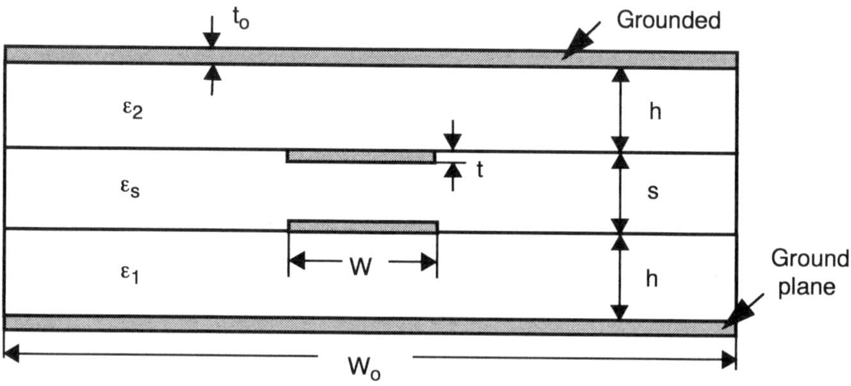

Figure 6.24 A broadside coupled suspended substrate stripline.

By using the finite element code developed, we may obtain the following numerical results for the coupler with a length of L_d = 17 mm and the matching impedance Z_o = 50 Ω: (a) Propagation velocity of even and odd modes: V_{even} = 1.021×10^8 m/s and V_{odd} = 1.534×10^8 m/s, (b) Characteristic impedances of even and odd modes: Z_{oe} = 126.0 Ω and Z_{oo} = 18.3 Ω, (c) Coupling coefficient: C_{dB} = 3.0 dB, Isolation: O_{dB} = 16.8 dB, Directivity: D_{dB} = 13.8 dB, (d) Phases at outer port 2 and port 4: θ_2=0.05 degree and θ_4= 90.6 degree, if losses are ignored.

When the conduction loss and the dielectric loss (assuming conductivity $\sigma = 5.8\times10^7$ S/m (copper) and tanδ = 0.008) are taken into account, we obtain the attenuation constants: α_{even} = 0.098 dB/cm for the even mode, and α_{odd} = 0.31 dB/cm for the odd mode at 1.8 GHz. The insertion loss is about 3.7 dB. Only small changes (within 10%) in other parameters of the coupler are found in the case considered.

It is shown that the propagation velocities as well as the attenuation constants for the even and odd modes are different due to the different dielectrics used in the coupler. Besides, due to some impedance mismatching at ports, all these factors have resulted in the very low directivity of the coupler. To improve the directivity, we could simply replace the dielectric material between the coupled lines by using the same dielectric material with ε_s = 9.5 (Al_2O_3) as the substrate. In such a case, we may get, by our numerical code, the following results for the quadrature coupler with s = 100 μm and W = 115 μm: (a) Propagation velocity of even and odd modes: $V_{even} = V_{odd} = 9.75\times10^7$ m/s, (b) Characteristic impedances of even and odd modes: Z_{oe} = 119.8 Ω and Z_{oo} = 20.6 Ω, (c) Coupling coefficient: C_{dB} = 3.0 dB, Isolation: O_{dB} = 47.8 dB, Directivity: D_{dB} = 44.8 dB, if losses are ignored.

The result indicates that the directivity of the coupler can be much improved provided that a homogeneous dielectric material could be used in the construction of the coupler. Note that calculations given here are not optimized.

It is only illustrative.

In some cases, to make use of available processing facilities for low-cost production, both the material and geometric sizes of the couplers are limited in some respect. For instance, the width W of the coupled lines (signal lines) might have a minimum value achievable. In what follows, an example is shown for the 3dB coupler with a homogeneous dielectric materials having the relative dielectric constant $\varepsilon_r = 4.5$ (Pyrex glass) and the loss angle about 0.008. In such a case, to achieve the 3 dB coupling, we find the following geometric parameters for the 3dB quadrature coupler: t = 12 μm, W = 420 μm, W_o = 3 mm, h = 1 mm, and s = 176 μm. The length of this quadrature coupler is L_d = 19.6 mm.

This coupler has the following designed properties: the coupling coefficient: C_{dB} = 3.0 dB; the isolation: O_{dB} = 49.4 dB; and the directivity: D_{dB} = 46.4 dB if losses are ignored. When losses are taken into account, we have the attenuation constants: α_{even} = 0.056 dB/cm for the even mode, and α_{odd} = 0.119 dB/cm for the odd mode at 1.8 GHz. The insertion loss is 3.1 dB. The attenuation constants for even and odd modes are shown to be different even when the propagation velocities of the even and odd modes are the same, which is about 1.42×10^8 m/s for this coupler. The directivity of the coupler is reduced to about 40 dB in the case, which is still a quite good value.

Next, we consider a 3dB directional coupler with the edge coupled microstrip line structure, as shown in Fig. 6.25. We assume that the geometrical and material parameters of the coupler operating at the frequency of 1.8 GHz have the following values. The relative dielectric constant: $\varepsilon_1 = \varepsilon_2 = 9.5$ (Al_2O_3) and $\varepsilon_3 = 1$ (in free-space); and the geometric parameters: t = 12 μm, W = 95 μm, W_o = 4 mm, $h_1 = h_2$ = 1 mm, and s = 20 μm. By using the finite element code, we can obtain the following numerical results for the coupler with a length of L_d = 14.3 mm and the matching impedance Z_o= 50 Ω: (a) Propagation velocity of even and odd modes: $V_{even} = 9.876\times10^7$ m/s and $V_{odd} = 1.074\times10^8$ m/s, (b)

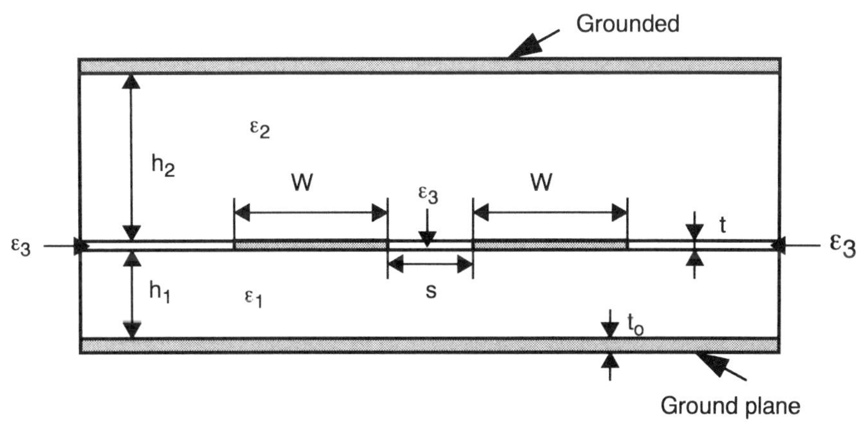

Figure 6.25 An edge-coupled microstrip line.

Characteristic impedances of even and odd modes: Z_{oe} = 120.8 Ω and Z_{oo} = 20.8 Ω, (c) Coupling coefficient: C_{dB} = 3.0 dB, Isolation: O_{dB} = 29.5 dB, Directivity: D_{dB} = 26.5 dB, (d) Phases at outer port 2 and port 4: θ_2 = 0.05 degree and θ_4 = 90.0 degree, if losses are ignored.

When the conduction loss and the dielectric loss (assuming conductivity $\sigma = 5.8\times10^7$ S/m (copper) and $\tan\delta$ = 0.008) are taken into account, we obtain the attenuation constants: α_{even}=0.1 dB/cm for the even mode, and α_{odd}=0.44 dB/cm for the odd mode at 1.8 GHz. The insertion loss is about 3.2 dB.

Similarly, superconducting directional couplers may also be analyzed with the use of above approach [Zhou and Hesselbom (1998)]. Some numerical results have indicated that superconducting directional couplers may have higher directivity and are less frequency-sensitive than normal conducting ones due to their lower resistance and nearly frequency-independent inductance up to a few GHz frequencies.

6.4 FULL-WAVE ANALYSIS OF WAVEGUIDE WITH CONDUCTING BOUNDARY

In the above sections, we have studied transmission lines made of two or more conductors, such as coaxial lines or microstrip lines, which are capable of supporting electromagnetic waves for the TEM mode or quasi-TEM mode. These transmission lines have the advantage of having relatively small geometrical sizes. They have, however, a disadvantage that, at high power, the current density on the signal line (e.g., the center conductor of a coaxial line) is very high and losses may approach intolerable levels, especially as the frequency increases. Besides, there are electromagnetic radiation losses and interference problems for transmission lines of open structures, such as microstrip lines. The solution to these problems is to use guided wave in some closed waveguide structures, such as a hollow conductor of constant cross-sectional dimension. The following three subsections are, therefore, devoted to the analysis of a typical one-conductor guiding structure, the rectangular waveguide.

6.4.1 Rectangular Waveguide

The rectangular waveguide with a cross-section, shown in Fig. 6.26, is one of the most important types of hollow-pipe waveguides with closed conducting boundaries. Here we shall only discuss the rectangular waveguide. Other types of waveguides of various geometrical structures can be analyzed similarly [see, e.g., Ramo et al. (1994)]. As in Fig. 6.26, a dielectric region of width a and height b extends indefinitely in the axial (z) direction and is closed by conducting boundaries on the four sides.

In this type of one-conductor guiding structures, there can be no transverse

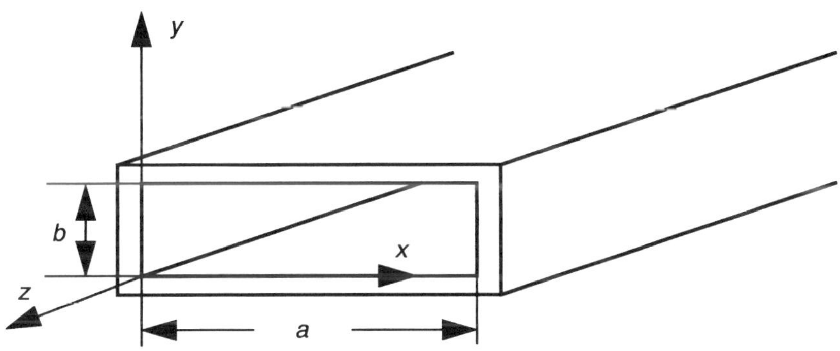

Figure 6.26 Scheme of a rectangular waveguide.

electromagnetic (TEM) wave since a TEM wave does not have any axial field components which are necessary for the existence of transverse electric or magnetic field inside the hollow waveguide. The types of waves that can be propagated in a hollow waveguide, such as the rectangular waveguide, are the TE and TM modes.

The TM waves have zero H_z but nonzero E_z. The wave equation for E_z (~ $\exp(i\omega t-\gamma z)$) can be expressed in rectangular coordinates [Collin (1966)]:

$$\left(\frac{\partial^2}{\partial x^2}+\frac{\partial^2}{\partial y^2}+k_c^2\right)E_z = 0 \tag{6.4.1}$$

in which

$$k_c^2 = \gamma^2+\mu_r\varepsilon_r\left(\frac{\omega}{c}\right)^2 \tag{6.4.2}$$

where μ_r and ε_r denote, respectively, the relative permeability and the relative permittivity of the dielectric medium inside the hollow waveguide ($\mu_r = 1$ and $\varepsilon_r = 1$ in vacuum). c is the speed of light in vacuum. For waveguides with good conducting boundaries, it is often convenient in the initial analysis to ignore both the conduction loss and the dielectric loss. The effect of losses may be computed later by using a perturbation method.

Equation (6.4.1) can be solved by the procedure of separation of variables:

$$E_z(x, y) = X(x)Y(y) \tag{6.4.3}$$

Substitution of Eq.(6.4.3) into (6.4.1) gives the following two separated equations:

$$\left(\frac{\partial^2}{\partial x^2}+k_x^2\right)X(x) = 0 \tag{6.4.4}$$

$$\left(\frac{\partial^2}{\partial y^2} + k_y^2\right) Y(y) = 0 \tag{6.4.5}$$

with the condition

$$k_x^2 + k_y^2 = k_c^2 \tag{6.4.6}$$

Thus we may find solutions of the general form:

$$E_z = (C_1 \cos(k_x x) + D_1 \sin(k_x x))(C_2 \cos(k_y y) + D_2 \sin(k_y y)) \tag{6.4.7}$$

which have to also satisfy proper boundary conditions. In the case of assuming perfect conduction boundary conditions, we may find that the axial electrical field E_z has the following non-zero solutions:

$$E_z = A_{mn} \sin\left(\frac{m\pi x}{a}\right) \sin\left(\frac{n\pi y}{b}\right), \qquad m, n = 1, 2, 3, \ldots \tag{6.4.8}$$

where A_{mn} is an arbitrary amplitude constant associated with the mnth mode.

The propagation constant for the mnth mode can be found from Eq.(6.4.2)

$$\gamma_{mn} = i\beta_{mn} = i\left[\mu_r\varepsilon_r\left(\frac{\omega}{c}\right)^2 - \left(\frac{m\pi}{a}\right)^2 - \left(\frac{n\pi}{b}\right)^2\right]^{1/2} \tag{6.4.9}$$

The cutoff wave number for the mnth mode is given by

$$k_{cmn} = \left[\left(\frac{m\pi}{a}\right)^2 + \left(\frac{n\pi}{b}\right)^2\right]^{1/2} \tag{6.4.10}$$

which is a function of the waveguide dimensions only.

It can be seen that, when k_o (defined by $k_o = \omega(\mu_r\varepsilon_r)^{1/2}/c$) > k_{cmn}, β_{mn} is pure real and the mode propagates; when $k_o < k_{cmn}$, then β_{mn} is imaginary and the mode decays rapidly with distance $|z|$ from the point at which it is excited. This decay is not associated with energy loss, but is a characteristic feature of the solution. The frequency separating the propagation and non-propagation bands is designated the cutoff frequency f_{cmn}, which is given by (the solution of $k_o = k_{cmn}$):

$$f_{cmn} = \frac{c}{2\pi\sqrt{\mu_r\varepsilon_r}}\left[\left(\frac{m\pi}{a}\right)^2 + \left(\frac{n\pi}{b}\right)^2\right]^{1/2} \tag{6.4.11}$$

The corresponding cutoff wavelength λ_{cmn} can be found by $\lambda_{cmn} = c/f_{cmn}$. It is shown that the lowest cutoff frequency for the TM modes is

$$f_{cmn} = \frac{c}{2\sqrt{\mu_r\varepsilon_r}}\left[\left(\frac{1}{a}\right)^2 + \left(\frac{1}{b}\right)^2\right]^{1/2} \tag{6.4.12}$$

The other field components for the TM modes can be found from Maxwell's equations with the use of the solution for E_z. They are

$$E_x = -\frac{\gamma}{k_c^2}\frac{\partial E_z}{\partial x}, \quad \text{and} \quad E_y = -\frac{\gamma}{k_c^2}\frac{\partial E_z}{\partial y} \tag{6.4.13}$$

$$H_x = \frac{i\omega\varepsilon}{k_c^2}\frac{\partial E_z}{\partial y}, \quad \text{and} \quad H_y = -\frac{i\omega\varepsilon}{k_c^2}\frac{\partial E_z}{\partial x} \tag{6.4.14}$$

Similarly, we may find the solution for transverse electric (TE) waves, for which E_z is zero, but H_z is non-zero. In this case, we have the following equation in rectangular coordinates:

$$\left(\frac{\partial^2}{\partial x^2} + \frac{\partial^2}{\partial y^2} + k_c^2\right)H_z = 0 \tag{6.4.15}$$

After some manipulates by imposing the perfect conduction boundary conditions, we may find that the axial magnetic field H_z have the following non-zero solutions:

$$H_z = B_{mn}\cos\left(\frac{m\pi x}{a}\right)\cos\left(\frac{n\pi y}{b}\right), \qquad m, n = 0, 1, 2, \ldots \tag{6.4.16}$$

where B_{mn} is an arbitrary amplitude constant associated with the mnth mode. It is shown that in contrast to the TM waves, one but not both of m and n may be zero without the wave's vanishing. The cutoff frequency f_{cmn} for the mnth (TE_{mn}) mode are found to be given by the same formulae as Eq.(6.4.11) for the corresponding same order TM_{mn} mode. Modes that have different field distributions but the same cutoff frequencies are said to be degenerate modes.

The lowest cutoff frequency for the TE modes is

$$f_{c10} = \frac{c}{2a\sqrt{\mu_r\varepsilon_r}} \tag{6.4.17}$$

which corresponds to the TE_{10} mode if we assume $a > b$. The corresponding cutoff wavelength (defined here by $\lambda_c = c/f_c$) is then

$$\lambda_{c10} = 2a\sqrt{\mu_r\varepsilon_r} \tag{6.4.18}$$

It can be seen that TE_{10} wave has the lowest cutoff frequency among all other TE or TM modes. Thus it is possible to construct the rectangular waveguide supporting only the TE_{10} wave mode by choosing proper geometric sizes and the dielectric material in the waveguide. Equation (6.4.18) indicates the maximum wavelength which can be supported by the waveguide. Since the maximum wavelength depends generally on the geometric size of the waveguide, due to practical limits, these types of waveguides (with one-conductor guiding structures) are usually not used to propagate radio waves of long wavelengths, but often used in the *cm*- or *mm*-wave bands.

6.4.2 Energy Transmission in a Waveguide

Let us consider the energy transmission in the rectangular waveguide. For a propagating TE_{mn} mode, the rate of energy flow (power) in the positive z direction is given by

$$P_{mn} = \frac{1}{2}Re\left[\int_0^a\int_0^b(\boldsymbol{E}\times\boldsymbol{H}^*)_z dxdy\right] = \frac{1}{2}Re\left[\int_0^a\int_0^b(E_xH_y^* - E_yH_x^*)dxdy\right] \quad (6.4.19)$$

For TE mode, the electromagnetic field components can be found from H_z by

$$H_x = -\frac{\gamma}{k_c^2}\frac{\partial H_z}{\partial x} = \frac{\gamma m\pi}{ak_{cmn}^2}B_{mn}\sin\left(\frac{m\pi x}{a}\right)\cos\left(\frac{n\pi y}{b}\right), \qquad m, n = 0, 1, 2, \ldots \quad (6.4.20)$$

$$H_y = -\frac{\gamma}{k_c^2}\frac{\partial H_z}{\partial y} = \frac{\gamma n\pi}{bk_{cmn}^2}B_{mn}\cos\left(\frac{m\pi x}{a}\right)\sin\left(\frac{n\pi y}{b}\right), \qquad m, n = 0, 1, 2, \ldots \quad (6.4.21)$$

and

$$E_x = \frac{i\omega\mu}{\gamma}H_y = \frac{i\omega\mu n\pi}{bk_{cmn}^2}B_{mn}\cos\left(\frac{m\pi x}{a}\right)\sin\left(\frac{n\pi y}{b}\right), \qquad m, n = 0, 1, 2, \ldots \quad (6.4.22)$$

$$E_y = -\frac{i\omega\mu}{\gamma}H_x = -\frac{i\omega\mu m\pi}{ak_{cmn}^2}B_{mn}\sin\left(\frac{m\pi x}{a}\right)\cos\left(\frac{n\pi y}{b}\right), \qquad m, n = 0, 1, 2, \ldots \quad (6.4.23)$$

In particular, for the TE_{10} mode, we have

$$H_x = \frac{\gamma\pi}{ak_{c10}^2}B_{10}\sin\left(\frac{\pi x}{a}\right) \quad (6.4.24)$$

$$H_y = E_x = 0 \quad (6.4.25)$$

$$E_y = -\frac{i\omega\mu\pi}{ak_{c10}^2}B_{10}\sin\left(\frac{\pi x}{a}\right) \tag{6.4.26}$$

and

$$P_{10} = \frac{1}{2}Re\left[\int_0^a\int_0^b(-E_yH_x^*)dxdy\right] = \frac{ab|E_{10}|^2}{4c\mu_o}\sqrt{\frac{\varepsilon_r}{\mu_r}}\left[1-\left(\frac{f_{c10}}{f}\right)^2\right]^{1/2} \quad (f > f_{c10}) \tag{6.4.27}$$

in which we have set

$$E_{10} = -\frac{i\omega\mu\pi}{ak_{c10}^2}B_{10} \tag{6.4.28}$$

and f_{c10} is the cutoff frequency for the TE_{10} mode given by Eq.(6.4.17). It is clearly shown by Eq.(6.4.27) that, when the frequency f (= $\omega/2\pi$) of the propagating wave is approaching to the cutoff frequency, the averaged transmitted power becomes zero, as we may expect.

Since power is a flow of energy, there exists a velocity of energy flow such that the averaged electromagnetic energy per unit length in the waveguide multiplied by this velocity is equal to the power. In a waveguide, it turns out that this velocity of energy flow is equal to the group velocity. To see this result, let us first find the averaged electromagnetic energy per unit length of the waveguide. For a propagating mode in a lossless waveguide, it can be readily shown from the complex Poynting vector theorem that the averaged electric energy W_e per unit length and the averaged magnetic energy W_m per unit length are equal. They are given here for the TE_{10} mode by

$$W_{m10} = W_{e10} = \int_0^a\int_0^b \frac{1}{4}\varepsilon|E_{10}|^2\left(\sin\left(\frac{\pi x}{a}\right)\right)^2 dxdy = \frac{1}{8}\varepsilon ab|E_{10}|^2 \tag{6.4.29}$$

Thus the total averaged electromagnetic energy per unit length of the waveguide is $\varepsilon ab|E_{10}|^2/4$ for the TE_{10} mode. The velocity of the energy flow is then given by

$$\mathrm{v} = \frac{P_{10}}{W_{e10}+W_{m10}} = \frac{c}{\sqrt{\varepsilon_r\mu_r}}\left[1-\left(\frac{f_{c10}}{f}\right)^2\right]^{1/2} \tag{6.4.30}$$

which is shown to be equal to the group velocity v_g, defined by

$$\mathrm{v}_g = \left(\frac{\partial\beta_{10}}{\partial\omega}\right)^{-1} \tag{6.4.31}$$

for the TE_{10} mode. Here, β_{10} is given by Eq.(6.4.9) for $m = 1$ and $n = 0$, i.e.,

$$\beta_{10} = \frac{\omega\sqrt{\varepsilon_r \mu_r}}{c}\sqrt{1-\left(\frac{f_{c10}}{f}\right)^2} \qquad (6.4.32)$$

It is shown that the average transmitted power in the waveguide is equal to the product of the averaged energy per unit length times the group (signal) velocity, as we have expected intuitively.

In the case of electromagnetic wave propagating in the waveguide, it is relatively easy to understand the fact that energy is transmitted through the electromagnetic field. However, in cases of steady-state or low-frequency ac circuits, one often ignores this fact since one usually solves only electronic circuit equations without the need of studying electromagnetic fields. Actually, for these cases, electromagnetic energy is also transmitted through the field.

To see more clearly this fact, we may consider an electronic circuit, in which the transmitted energy includes the kinetic energy of conduction electrons moving inside conductor wires and the electromagnetic field energy inside and outside of the wires. Let us first look at the kinetic energy of conduction electrons. Consider that the current density $\boldsymbol{J}$ of the conductor wire is given by $\boldsymbol{J} = n_e e\mathrm{v}$, where v is the mean drift velocity of conduction electrons, and n_e is the density of the conduction electrons. Assuming $n_e \sim 10^{29}/\mathrm{m}^3$ for a normal metal conductor, and $\boldsymbol{J} = 10^6$ A/m^3, we may find that v $\sim 6\times10^{-5}$ m/s, which shows that the mean drift velocity v of conduction electrons is very small, and consequently, the kinetic energy of conduction electrons is also small. Besides, in steady state, the electric current I has the same value in the circuit, including the load resistor. The kinetic energy of conduction electrons is, therefore, not the energy supplied to the load. The power (energy) dissipated at the load resistor and in conductor wires is transmitted through the electromagnetic field. During the energy transmission, a part of the energy enters into the conductor wires, which becomes the Joule heat dissipated. At the load, the electromagnetic energy from the field flows into the load resistor, and provides it with the energy required.

6.4.3 Wave Attenuation in Lossy Waveguide

We study now the effect of lossy waveguide. For lossy non-magnetic ($\mu_r = 1$) dielectric filling the rectangular waveguide, we may replace the permittivity ε by its complex form: $\varepsilon = \varepsilon_0\varepsilon_r(1 - i\tan\delta)$. In such a case, we have the complex propagation constant for the *mn*th mode

$$\gamma_{mn} = \alpha_{dmn} + i\beta_{mn} = \left[k_{cmn}^2 - \varepsilon_r(1 - i\tan\delta)\left(\frac{\omega}{c}\right)^2\right]^{1/2} \qquad (6.4.33)$$

where α_{dmn} is the attenuation constant due to the dielectric loss, and β_{mn} the propagation constant for the *mn*th mode of the TM (or TE) wave. Equation (6.4.33) can further be written

$$\gamma_{mn} = \left[k_{cmn}^2 - \varepsilon_r\left(\frac{\omega}{c}\right)^2\right]^{1/2} \times \left[1 + \frac{i\varepsilon_r \tan\delta(\omega/c)^2}{k_{cmn}^2 - \varepsilon_r(\omega/c)^2}\right]^{1/2} \tag{6.4.34}$$

When the dielectric loss is small and when wave frequency f (= $\omega/2\pi$) is sufficiently larger than the cutoff frequency f_{cmn}, defined by Eq.(6.4.11), we may find the attenuation constant approximately by

$$\alpha_{dmn} = \frac{\pi f \times \sqrt{\varepsilon_r}\tan\delta}{c\sqrt{1-(f_{cmn}/f)^2}} \tag{6.4.35}$$

where we have made use of a binomial expansion of the first. Higher-order corrections to the formulae can be obtained by analyzing Eq.(6.4.34). It may be noticed that the form of the attenuation caused by an imperfect dielectric is the same for all modes and all shapes of the guides, though the amount of attenuation is a function of the cutoff frequency, which does depend on the guide and the mode.

The calculation of the effect of the conduction loss in the conductor on the wave attenuation constant α_c is somewhat more involved. Illustratively, let us consider again in more detail about the TE_{10} wave mode due to its importance in engineering applications. At the first-order approximation, the currents on the lossy walls are assumed to be the same as the loss-free currents, and hence are given by

$$\boldsymbol{J}_s = \boldsymbol{n} \times \boldsymbol{H} \tag{6.4.36}$$

where $\boldsymbol{n}$ is a unit inward directed normal at the guide wall. Thus the power loss per unit length of the waveguide is

$$P_L = \frac{R_s}{2}\int_{l_w} |J_s|^2\, dl = R_s|B_{10}|^2\left[b + \frac{a}{2}\left(1 + \left(\frac{\beta_{10}a}{\pi}\right)^2\right)\right] \tag{6.4.37}$$

where R_s is the surface resistance of the conductor, and l_w denotes the periphery of the waveguide. The power, or rate of energy flow for the TE_{10} mode in the positive z direction, is given by

$$P_{10} = \frac{1}{2}Re\left\{\int_0^a\int_0^b (\boldsymbol{E} \times \boldsymbol{H}^*)_z\, dxdy\right\} = \frac{ba^3\omega\mu_o\beta_{10}|B_{10}|^2}{4\pi^2} \tag{6.4.38}$$

To calculate the attenuation due to the conduction loss, we may consider that, if P_0 is the power at $z = 0$, then $P_{10} = P_0\exp(-2\alpha_c z)$ is the power in the guide at any z. The rate of decrease of power propagated is then

$$-\frac{dP_{10}}{dz} = 2\alpha_c P_{10} = P_L \tag{6.4.39}$$

and equals the power loss. Thus the attenuation constant α_{c10} for the TE_{10} mode is found by

$$\alpha_{c10} = \frac{P_L}{2P_{10}} = \frac{R_s[2b(\pi/a)^2 + a(\beta_{10}^2 + \pi^2/a^2)]}{ab\omega\mu_o\beta_{10}} \quad \text{(nepers/m)} \tag{6.4.40}$$

which may also be written in the following form:

$$\alpha_{c10} = \frac{R_s}{b\eta\sqrt{1-(f_{c10}/f)^2}}\left[1 + \frac{2b}{a}\left(\frac{f_{c10}}{f}\right)^2\right] \tag{6.4.41}$$

where the parameter η, defined by $\eta = (\mu_o/(\varepsilon_o\varepsilon_r))^{1/2}$, has the dimension of ohm and is called the characteristic or intrinsic impedance of the medium. f_{c10} is the cutoff frequency of the TE_{10} mode given by Eq.(6.4.17). For superconductors, the surface resistance R_s may be given by, for instance, Eq.(3.3.49) for the superconductors in weak fields. For normal conductors, we have $R_s = 1/(\sigma_n\delta_n)$ with σ_n being the normal conductivity and δ_n the normal skin depth by Eq.(3.3.46).

Shown in Fig. 6.27 is the attenuation of the TE_{10} wave mode in the

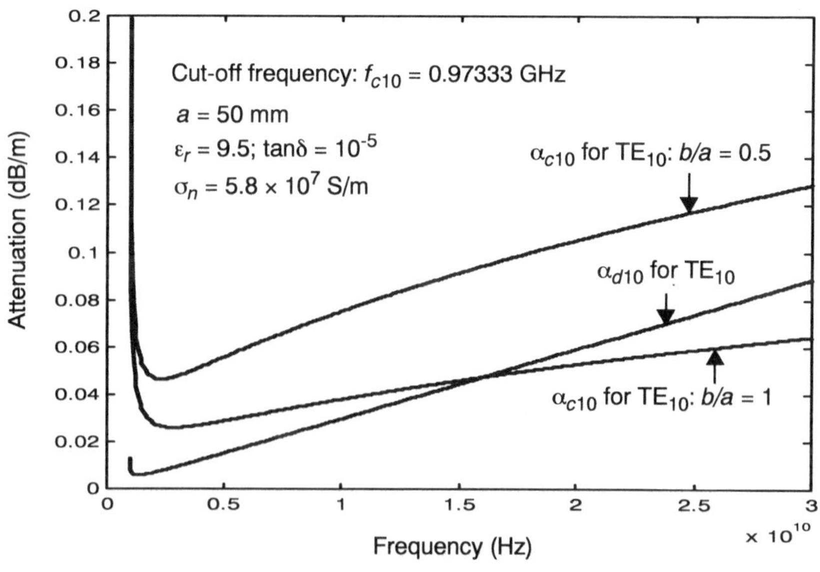

Figure 6.27 Attenuation due to conduction and dielectric losses.

rectangular waveguide due to the conduction loss from (6.4.41) and the dielectric loss from (6.4.35) for two different values of b/a. It is shown that the attenuation from the conduction loss depends strongly on the geometric ratio b/a of the waveguide. Near the cutoff frequency, the wave attenuation increases rapidly, as we expect.

6.5 MICROWAVE RESONATORS

Microwave resonators are electromagnetic components which exhibit resonant behaviors with frequency, and are used in filters, oscillators, and tuned amplifies. Microwave resonators are often made from sections of waveguides or transmission lines. Today, with the advance of integrated circuit technology, it has become possible to construct resonators with elements that behave like lumped circuit elements at microwave frequencies. In the following three subsections, we shall introduce some methods for the analyses of electromagnetic behaviors of these resonators.

6.5.1 Lumped Element Resonant Circuits

At radio frequencies, resonators are usually designed using discrete lumped elements, such as a series or a parallel combination of inductors and capcitors in order to have a relatively compact circuit. Two typical one-port resonators with discrete elements may be shown in Fig. 6.28.

First, let us consider the series RLC discrete element resonant circuit, shown in Fig. 6.28(a). According to Kirchhoff's voltage law, we have the voltage equation:

$$L\frac{dI}{dt} + RI + \frac{1}{C}\int I dt = V_s \tag{6.5.1}$$

where L, R, and C are the inductor, resistor, and capacitor, respectively. V_S is the source voltage, and I is the current flowing through each element in the resonant

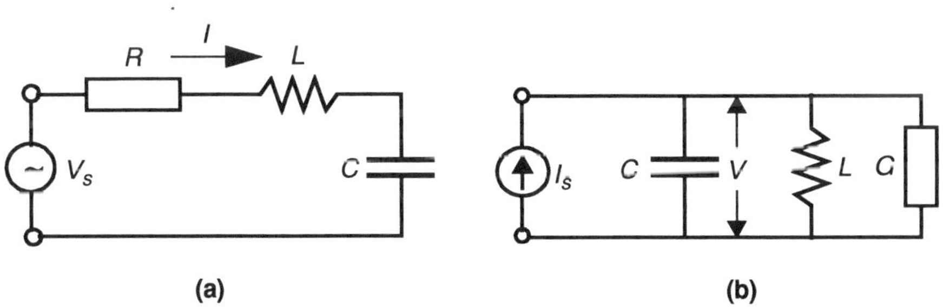

Figure 6.28 Series and parallel discrete element resonant circuits.

circuit. Since we are interested mainly in the behavior of the resonant circuit near resonance, we may focus on the frequency domain, assuming that all of the signals in the circuit have time dependence e^{st} with s being generally complex. Thus Eq.(6.5.1) can be expressed in the frequency domain by

$$V_s(s) = \left(sL + R + \frac{1}{sC}\right) I(s) \tag{6.5.2}$$

Here, we assume that R, L, and C are frequency-independent for some simplicity. Equation (6.5.2) can also be written in the following form:

$$V_s(s) = \frac{L(s - s_1)(s - s_2)}{s} I(s) \tag{6.5.3}$$

in which s_1 and s_2 are given by

$$s_1 = -\alpha_o + i\sqrt{\omega_o^2 - \alpha_o^2}\,, \qquad s_2 = -\alpha_o - i\sqrt{\omega_o^2 - \alpha_o^2} \tag{6.5.4}$$

with

$$\alpha_o = \frac{R}{2L} \tag{6.5.5}$$

$$\omega_o = \frac{1}{\sqrt{LC}} \tag{6.5.6}$$

Here, s_1 and s_2 are called the complex natural frequencies of the *RLC* circuit.

From Eq.(6.5.3), we may also find that the circuit impedance Z is given by

$$Z = \frac{L(s - s_1)(s - s_2)}{s} = Y^{-1} \tag{6.5.7}$$

where Y denotes the admittance of the circuit.

It can be seen from Eq.(6.5.5) that the parameter α_0 characterizes the loss of the resonant circuit, caused by the lossy resistance element R. Since we are usually interested in resonators with small losses, the condition $\alpha_0 << \omega_0$ is usually satisfied. In such a case, the circuit is said to be highly oscillatory. It has the resonant behavior near ω_0 and the rate at which power is dissipated is characterized by α_0 [Guillemin (1953)]. Explicitly, the power dissipation in the resonant circuit can be given by

$$P_L(s) = \frac{1}{2} R |I(s)|^2 = \frac{1}{2} R |V_s(s)|^2 |Y(s)|^2 \tag{6.5.8}$$

A typical resonance curve of the simple *RLC* circuit can be shown in Fig. 6.29, which shows illustratively the power dissipation in the circuit with respect to the operating frequency ω. In practice (ignoring possible small shifts caused

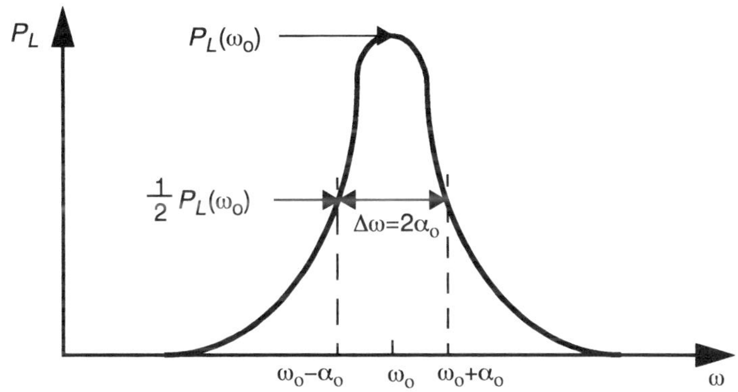

Figure 6.29 Resonance curve of a simple *RLC* circuit.

by losses), we call ω_0, given by Eq.(6.5.6), the resonance frequency of the series *RLC* circuit. The frequencies $\omega = \omega_0+\alpha_0$ and $\omega = \omega_0-\alpha_0$ are called the half-power frequencies of the resonance, at which the dissipated power drops to one-half its peak power, as shown in Fig. 6.29. $\Delta\omega = 2\alpha_0$ is called the half-power bandwidth or full width at half maximum (FWHM).

To show the behavior of the impedance Z at the resonance, we may let $s = i\omega$. We can find from Eq.(6.5.2)

$$Z = R + i\frac{L}{\omega}(\omega^2 - \omega_o^2) \tag{6.5.9}$$

which shows that at the resonance $\omega = \omega_0$, the impedance $Z = R$ is real. Thus a simple way of stating the resonance condition is that the resonance occurs when the reactive part of Z is zero. This implies that, at resonance, the reactive power of the inductor and capacitor are equal and the source has to supply only the active power of a resonant circuit.

Besides, by noting that the time-average stored magnetic energy in the inductor L is

$$W_m = \frac{1}{4}L|I|^2 \tag{6.5.10}$$

and the time-average stored electric energy in the capacitor C is

$$W_e = \frac{1}{4}C|V|^2 \tag{6.5.11}$$

we may find that, at the resonance, the stored magnetic energy W_m is equal to the stored electric energy W_e by noting also the capacitance voltage $|V| = |I|/(C\omega)$.

Since energy in a resonator oscillates back and forth between its electric and magnetic forms, some of this energy may heat the resonator or escape to its outside environment. Thus the total electromagnetic energy W_T may decay over time in a lossy resonant circuit. Approximately, we may write

$$W_T(t) = W_T(0)\exp\left(-\frac{\omega_o t}{Q}\right) \tag{6.5.12}$$

where ω_o is the resonant frequency, and Q is the quality factor, which measures how well the resonator stores energy. From Eq.(6.5.12), we have

$$-\frac{dW_T}{dt} = \frac{\omega_o W_T}{Q} \tag{6.5.13}$$

which is the power loss. Thus the Q factor for a resonant circuit may be obtained by

$$Q = \frac{\omega_o W_T}{P_L} \tag{6.5.14}$$

where P_L is the power loss, averaged over one cycle in the time-harmonic resonant circuit.

For the series RLC resonant circuit, its Q factor can be obtained by

$$Q = \frac{\omega_o 2W_m}{P_L} = \frac{\omega_o L|I|^2/2}{R|I|^2/2} = \frac{\omega_o L}{R} = \frac{1}{R}\sqrt{\frac{L}{C}} \tag{6.5.15}$$

which shows that Q becomes infinite as the lossy element R vanishes. Furthermore, by Eq.(6.5.5), we may find

$$Q = \frac{\omega_o}{2\alpha_o} = \frac{\omega_o}{\Delta\omega} = \frac{f_o}{\Delta f} \tag{6.5.16}$$

which indicates that a resonant circuit can select a certain band of frequencies. An increase in the Q of the resonant circuit increases the selectivity by decreasing the bandwidth.

Similarly, for a parallel GLC discrete element resonant circuit, shown in Fig. 6.28(b), we have the following current equation:

$$C\frac{dV}{dt} + GV + \frac{1}{L}\int V dt = I_s \tag{6.5.17}$$

according to Kirchhoff's current law. Here, G is the conductance, I_s is the source current, and V is the voltage on each element in the resonant circuit. In the frequency domain, Eq.(6.5.17) can be expressed by

$$I_s(s) = \left(sC + G + \frac{1}{sL}\right) V(s) \tag{6.5.18}$$

from which we may find the impedance Z of the circuit by

$$Z = \left(sC + G + \frac{1}{sL}\right)^{-1} \tag{6.5.19}$$

In particular, for $s = i\omega$, we have

$$Z = \left[G + i\frac{C}{\omega}(\omega^2 - \omega_o^2)\right]^{-1} \tag{6.5.20}$$

in which ω_0 is the resonant frequency given by Eq.(6.5.6).

At the resonance, the impedance is real, $Z = 1/G$. The Q factor of the parallel GLC resonant circuit can be found by

$$Q = \frac{\omega_o 2W_e}{P_L} = \frac{\omega_o C|V|^2/2}{G|V|^2/2} = \frac{\omega_o C}{G} = \frac{1}{G}\sqrt{\frac{C}{L}} \tag{6.5.21}$$

where the power is dissipated in the conductance G in the parallel resonant circuit. It should be noticed that the Q factors calculated here and in all the following sections are unloaded Q, which account for the internal losses in the resonator only.

The concepts introduced here for discrete lumped-element resonant circuits are often useful in studying distributed electromagnetic resonators, such as transmission line resonators or cavity resonators. As we shall show later, these distributed electromagnetic resonators used at microwave frequencies exhibit an infinite number of resonances, making it difficult to model their global behavior in a simple way. However, in many cases, we are interested in the frequency response of such resonators near some specific isolated resonances, for which equivalent discrete element circuits may be found. These equivalent discrete element circuits enable us to study resonators with the concepts from the circuit theory, which is relatively simple to be used, especially for discussing the coupling of resonators to external world [see, e.g., Staelin et al. (1994)].

6.5.2 Transmission Line Resonators

To study transmission line resonators, let us first consider a lossless TEM-line resonator formed from a piece of uniform TEM transmission line of length l, as shown in Fig. 6.30. According to the general solution of the transmission line given in Eq.(6.1.6) for the voltage $V(x)$ and Eq.(6.1.10) for the current $I(x)$, we have

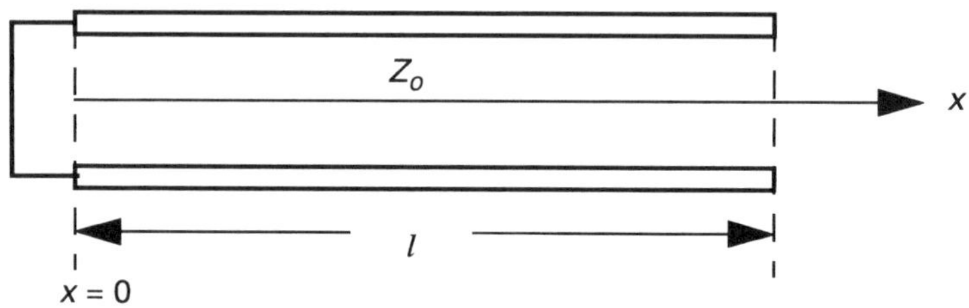

Figure 6.30 A short-circuited transmission line of length l.

$$V(x) = -2iV_A \sin(\beta x) \tag{6.5.22}$$

$$I(x) = \frac{2\beta}{\omega L} V_A \cos(\beta x) = 2\sqrt{\frac{C}{L}} V_A \cos(\beta x) \tag{6.5.23}$$

for the short-circuit boundary condition $V(x=0) = 0$. Here, the propagation constant β is given by

$$\beta = \omega\sqrt{LC} \tag{6.5.24}$$

where L and C are, respectively, the inductance and capacitance per unit length of the TEM line.

The impedance Z of the TEM line at $x = l$ can be found by

$$Z = \frac{V(l)}{I(l)} = -i\sqrt{\frac{L}{C}} \tan(\beta l) \tag{6.5.25}$$

As a function of the length l of the TEM line, this impedance can take on any value of reactance from minus infinity through zero to infinity. Resonance may occur when the reactance is zero or infinity. This corresponds to the resonance behaving as a series resonant circuit, or the resonance behaving as a parallel resonant circuit, as discussed in the above section. The first resonance occurs when $\beta l = \pi/2$ for $l = \lambda/4$, where λ is the wavelength ($\lambda = 2\pi/\beta$). Thus a TEM transmission line of length $l/4$ gives a circuit at microwave frequencies that is essentially equivalent to the lumped element parallel resonant circuit at lower frequencies. If the length l is increased to $\lambda/2$, at which the reactance is zero, the TEM transmission line behaves as a series resonant circuit. This behavior repeats every $\lambda/2$ in the ideal case.

To calculate the Q factor of the transmission line resonator, we must take into account the distributed nature of electromagnetic energy storage and dissipation. The power dissipation of the line comes from the series resistance R per unit length of the line and from the shunt conductance G per unit length of the line. R and G account for the conduction loss and the dielectric loss in the transmission line, respectively. It can be found that the Q factor of the transmission line resonator is given by its definition

$$Q = \frac{\omega_o W_T}{P_L} = \frac{\omega_o L|I|^2/2}{R|I|^2/2 + G|Z_o|^2|I|^2/2} \approx \frac{\omega_o LC}{RC + GL} \qquad (6.5.26)$$

where we have assumed that the losses are small so that the characteristic impedance of the line Z_0 may be taken approximately as $Z_0 = (L/C)^{1/2}$. By noting that the propagation constant γ for the transmission line is given by $\gamma = \alpha + i\beta$, we may find that the Q factor for low-loss transmission line resonator may also be expressed by

$$Q \approx \frac{\beta}{2\alpha} \qquad (6.5.27)$$

with the aid of Eqs.(6.1.13) and (6.1.14), where α is the attenuation constant.

Since the conduction loss in superconductors is much smaller than normal conductors at microwave frequencies (~ 2 GHz) used in recent mobile radio communications, the one-dimensional transmission line resonators using high-temperature superconductors have been developed to achieve high-Q resonators with small sizes. These high-Q resonators can then be used as building blocks for constructing, for instance, narrowband filters, which can have very sharp skirts, very high rejection in the stop band, and very low insertion loss in the passing band [Talisa et al. (1991), Shen (1994), Liang et al. (1995), and Zhang et al. (1995)].

6.5.3 Cavity Resonators

A cavity is considered here as a volume enclosed by a conducting surface and within which an electromagnetic field can be excited by means of, for instance, small coaxial line probes or loops. Figure 6.31 illustrates a rectangular cavity of height b, width a, and length d. Obviously, this rectangular cavity can be considered as a section of a rectangular waveguide terminated in short circuits at both $z = 0$ and $z = d$ (see Fig. 6.26).

If we assume that the two ends of the cavity at $z = 0$ and $z = d$ are perfect conducting boundaries, we may write

$$E_z(x, y, z) = E_z(x, y)\cos\left(\frac{l\pi z}{d}\right), \qquad l = 0, 1, 2, \ldots \qquad (6.5.28)$$

for TM modes in order to satisfy the boundary conditions that the tangential component of the electric field $\boldsymbol{E}$ has to be zero at $z = 0$ and $z = d$. Similarly, for TE modes, we may write

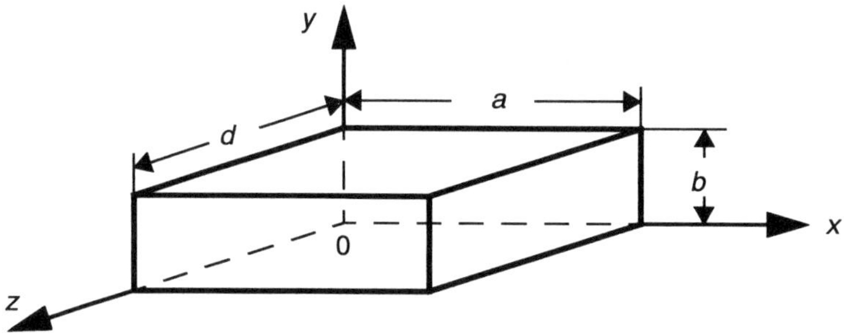

Figure 6.31 A rectangular cavity.

$$H_z(x, y, z) = H_z(x, y)\sin\left(\frac{l\pi z}{d}\right), \qquad l = 1, 2, 3, \ldots \tag{6.5.29}$$

in order to satisfy the boundary conditions that the normal component of the magnetic field $\boldsymbol{H}$ has to be zero at $z = 0$ and $z = d$.

We may now notice that the field solution for the cavity can be obtained directly from the corresponding waveguide solutions obtained in Section 6.4.1. In particular, for the TE or TM mode, we may find the following equation:

$$f_{mnl} = \frac{c}{2\pi\sqrt{\mu_r\varepsilon_r}}\sqrt{\left(\frac{m\pi}{a}\right)^2 + \left(\frac{n\pi}{b}\right)^2 + \left(\frac{l\pi}{d}\right)^2} \tag{6.5.30}$$

for the determination of the resonant frequencies f_{mnl} of the lossless rectangular cavity. It is shown that there is a triply infinite number of resonant frequencies corresponding to different field distributions.

Illustratively, for TE_{101} mode, which is the mode corresponding to the lowest resonant frequency if $b < a < d$, we may find the field solution:

$$H_z = B_{101}\cos\left(\frac{\pi x}{a}\right)\sin\left(\frac{\pi z}{d}\right) \tag{6.5.31}$$

$$H_x = -\frac{aB_{101}}{d}\sin\left(\frac{\pi x}{a}\right)\cos\left(\frac{\pi z}{d}\right) \tag{6.5.32}$$

$$E_y = -i2a\mu f_{101}B_{101}\sin\left(\frac{\pi x}{a}\right)\sin\left(\frac{\pi z}{d}\right) \tag{6.5.33}$$

and

$$H_y = E_x = E_z = 0 \tag{6.5.34}$$

In Eq.(6.5.33), f_{101} is the resonant frequency for the TE_{101} mode, given by

$$f_{101} = \frac{c}{2\pi\sqrt{\mu_r\varepsilon_r}}\sqrt{\left(\frac{\pi}{a}\right)^2 + \left(\frac{\pi}{d}\right)^2} \tag{6.5.35}$$

where μ_r and ε_r are the relative permeability and relative permittivity of the dielectric inside the cavity, respectively.

At resonance, the time-averaged electric and magnetic energy stored in the cavity are found to be equal, and the total averaged electromagnetic field in the cavity with volume V can be obtained by

$$W_{101} = 2\int_V \left(\frac{\varepsilon}{4}E_y E_y{}^*\right) dV = \frac{\varepsilon\mu^2}{2} a^3 b d f_{101}^2 |B_{101}|^2 \tag{6.5.36}$$

for the TE_{101} mode.

So far, we have considered only the lossless cavity. In practice, there exist always losses in the cavity due to the dielectric loss and conduction loss, even for superconducors. For small losses, a simple perturbation method can be used to determine the power loss in the cavity. To illustrate the method, we first consider the case in which the cavity is filled with a lossy dielectric with the complex permittivity: $\varepsilon = \varepsilon_0\varepsilon_r(1 - i\tan\delta)$, but the walls of the cavity are assumed to be perfect conducting. In such a case, the averaged power dissipated in one cycle for TE_{101} mode can be obtained by

$$P_{Ld} = \frac{1}{2}\int_V \boldsymbol{J}\cdot\boldsymbol{E}^* dV \approx \frac{\omega\varepsilon_o\varepsilon_r\tan\delta}{2}\int_V \left|E_y^{(0)}\right|^2 dV \tag{6.5.37}$$

since the lossy dielectric has an effective conductivity $\omega\varepsilon_0\varepsilon_r\tan\delta$. In Eq.(6.5.37), $E_y^{(0)}$ denotes the electric field in the loss-free case, given by Eq.(6.5.33), which approximates the actual field in the lossy case. Thus we may find

$$P_{Ld} = 2\pi W_{101} f_{101}\tan\delta \tag{6.5.38}$$

at the resonant frequency f_{101}.

Now, let us study the power dissipation in the lossy walls of the cavity due to the finite conductivity. We shall assume that the walls are good conductors. Since the resonator walls are good conductors, most of the current flows within a skin depth δ_n of the wall surfaces. Thus the power loss in the imperfectly conducting walls may be found by

$$P_{Lc} = \frac{R_s}{2}\int_S |J_s|^2\, dS = \frac{R_s}{2}\int_S \left|\boldsymbol{n}\times\boldsymbol{H}^{(0)}\right|^2 dS \tag{6.5.39}$$

where $\boldsymbol{J}_s$ denotes the surface current, and $\boldsymbol{H}^{(0)}$ is the loss-free field solution (6.5.31) and (6.5.32) for the TE_{101} mode, which approximates the actual field for the lossy case. R_s is the surface resistance, $R_s = 1/(\sigma_n\delta_n)$, where σ_n is the normal conductivity of the walls. Substituting from Eqs.(6.5.31) and (6.5.32) into Eq.(6.5.39), we may find the power loss in the walls by

$$P_{Lc} = \frac{R_s|B_{101}|^2(ad^3 + da^3 + 2a^3b + 2d^3b)}{4d^2} \tag{6.5.40}$$

for the TE_{101} mode.

Introducing the total electromagnetic energy W_{Tp} stored in the cavity for a particular mode p, and the power loss P_{Lp} for the pth mode, averaged over one cycle, we may find the quality factor Q_p for the pth mode by

$$Q_p = \frac{2\pi f_p W_{Tp}}{P_{Lp}} \tag{6.5.41}$$

where f_p is the resonant frequency (Hz) of the pth mode. The quality factor Q_p typically differs for each mode, with larger Q-values associated with high-quality resonances and smaller losses.

In particular, for the TE_{101} mode of the rectangular cavity with small losses, we may find the quality factor:

$$Q_{101} = \frac{2\pi f_{101} W_{101}}{P_{Ld} + P_{Lc}} = \left(\frac{1}{Q_{d101}} + \frac{1}{Q_{c101}}\right)^{-1} \tag{6.5.42}$$

where Q_{d101} denotes the quality factor for the TE_{101} mode of the cavity with lossy dielectric, but perfect conducting walls, and Q_{c101} denotes the quality factor for the TE_{101} mode of the cavity with imperfect conducting walls but lossless dielectric. By using Eqs.(6.5.36), (6.5.38), and (6.5.40), we may find

$$Q_{d101} = \frac{2\pi f_{101} W_{101}}{P_{Ld}} = \frac{1}{\tan\delta} \tag{6.5.43}$$

and

$$Q_{c101} = \frac{2\pi f_{101} W_{101}}{P_{Lc}} = \frac{4\pi a^3 d^3 b \varepsilon_o \varepsilon_r \mu^2 f_{101}^3}{R_s(ad^3 + da^3 + 2a^3b + 2d^3b)} \tag{6.5.44}$$

As a typical numerical example, consider a cavity with copper walls ($\sigma_n = 5.8\times10^7$ S/m), the filling dielectric ($\varepsilon_r = 9.6$ and $\tan\delta = 5.2\times10^{-5}$), and the size: $b = 1$ cm, $a = 2$ cm, and $d = 3$ cm, we may obtain for the TE_{101} mode, the resonant frequency: $f_{101} = 2.9$ GHz from Eq.(6.5.35); the quality factor $Q_{d101} = 19\,230$; and the quality factor $Q_{c101} = 4\,273$. The total quality factor Q_{101} can then be found from Eq.(6.5.42) as Q_{101} = 3 495.

If the walls of the cavity is made from superconductors with some typical values of λ_L = 200 nm, and $\sigma_n = 10^6$ S/m, we may find the surface resistance $R_s = 2.1\times10^{-6}\ \Omega$ from Eq.(3.3.49) for superconductors in weak fields and at the resonant frequency f_{101}. This value can be compared with the surface resistance $R_s = 0.014\ \Omega$ for the copper walls. In the case of superconducting walls, we may find the quality factor $Q_{c101} = 2.8\times10^7$, which indicates the limiting factor for having a high Q-value of the superconducting cavity is the dielectric loss in the case considered.

Recently, ceramic dielectric resonators are often utilized in making filters or oscillators. At room temperature and 10 GHz, the typical dielectric quality factor is about 2×10^4 for ceramics and 2.6×10^5 for sapphire. At liquid-nitrogen temperature (77 K), the Q_d value of sapphire at 10 GHz is close to 6×10^7, and at 4.2 K, it is 8×10^9 [Hartemann (1998)]. Thus, with normal-metal electrodes, it is impossible to benefit from this extremely high dielectric quality factor using conventional resonating modes. Shown in Fig. 6.32 is a dielectric resonator with high-T_c superconducting electrodes and a sapphire cylinder diameter of about 9 mm. An unloaded Q value of about 10^6 is obtained at 77 K and 10 GHz for the resonator. A sapphire-loaded cavity in superconducting niobium which shows the same general configuration and exhibits a Q value greater than 10^9 at 4.2 K and 10 GHz has been reported [Tobar and Blair (1991)], where the so-called whispering-gallery (WG) mode is used to obtain the highest confinement of fields in the sapphire resonator.

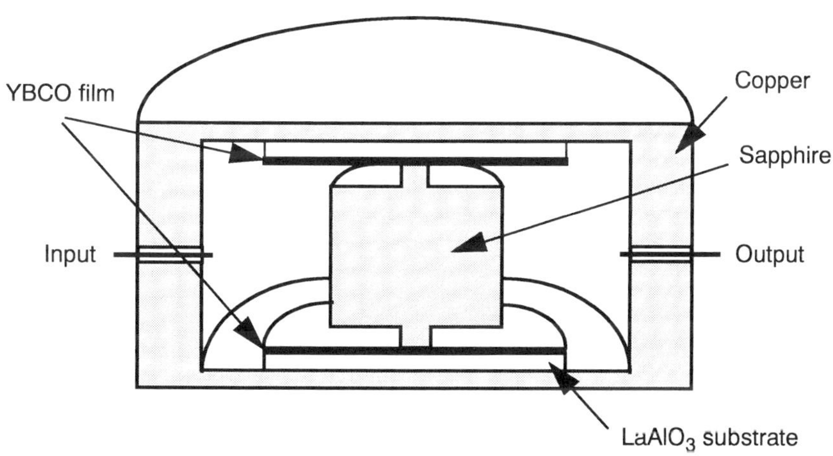

Figure 6.32 A cross-section of a dielectric resonator with high-T_c superconducting electrodes.

7

Electrodynamics of Deformable Superconductors

In many engineering applications of superconductivity, superconductors may not only conduct electric currents but they may also be subject to thermal and mechanical loadings. Since the discovery of hard superconductors, the mechanical behaviors of superconducting materials have been of interest and concern to researchers for many years because of large mechanical forces expected in foreseen applications of superconducting devices, such as magnets for plasma confinement and energy storage, superconducting generator, electrical transmission lines, magnetic levitating trains, electromagnetic-propulsion ship, or still unimagined new devices. In particular, the latest generation of superconductors uses ceramic materials which are brittle and much harder to work with than metals, where considerable amount of work is needed to improve the mechanical behaviors of the new superconductors. Furthermore, dynamic effects in nonequilibrium superconductors can introduce interesting acoustic phenomena in superconductors, which are of importance for potential technological applications. Obviously, the possibility for wider applications of superconductivity hinges on success in improving our understanding on the superconductivity and its related electrodynamic and mechanical problems. This chapter is, therefore, devoted to the introduction of some theoretical models developed recently for the study of interaction between electromagnetics and mechanics of superconductors.

7.1 MODELING OF MOVING DEFORMABLE SUPERCONDUCTORS

7.1.1 On Limitation of Minkowski's Theory

The correct formulation of electrodynamics of moving media has always been a challenge to scientists in the pre-relativistic era as well as in modern times. As we have learned from Section 1.6, the electrodynamics of moving media was founded in 1908 by Minkowski on the basis of Einstein's special theory of relativity. Although Minkowski's theory seems unsatisfactory in that it can be applied rigorously only to electromagnetic phenomena in material media moving with uniform velocity, attempts [Dunkin and Eringen (1963), Penfield and Hermann (1967), Bladel (1976, 1984)] have been made to extend Minkowski´s theory to cases where material media are performing accelerated motion with respect to inertial reference frames. As a working hypothesis, it has often been assumed that the macroscopic electromagnetic properties of a moving medium in the instantaneous rest-frame (an instantaneous co-moving inertial frame in which the observer performs his measurements) attached to the medium are unaffected by the acceleration. Thus Minkowski´s theory has been applied without reservation in many cases to deformable bodies in motion, despite its original assumption that the material body is rigid and moves with constant velocity. Naturally, a question may arise: is Minkowski´s theory applicable in modeling moving deformable superconductors? In order to answer the question, let us first start with a brief review of some basic concepts (terminologies) on reference frames with coordinate systems, which will be used in this chapter.

In the study of electrodynamics of moving bodies, it is common to introduce a reference frame, called the laboratory frame, which is an inertial frame with a coordinate system K. When a rigid body moves with a constant velocity $\boldsymbol{V}$, relative to the laboratory frame, we may attach a reference frame with the coordinate system K' to the moving body, as shown in Fig. 1.12. This reference frame is called the rest frame because to an observer situated at the rest frame, the rigid body appears to be at rest. According to Einstein´s special theory of relativity, Maxwell´s equations in these two coordinate systems K and K' should be invariant. Being aware of the consideration of only proper relativistic formulation of fundamental physical laws, such as Newton´s laws and Maxwell´s equations, is not sufficient for the complete description of a physical system involving material media, for which proper formulation of constitutive laws is also of vital importance, Minkowski proposed a theoretical model to formulate constitutive equations of moving media based on Einstein´s special theory of relativity. His logic is simple. Firstly, the Maxwell equations formulated originally for stationary media are also valid for a medium moving with a constant (uniform) velocity $\boldsymbol{V}$ with respect to a "fixed" reference frame, the laboratory frame K. Secondly, constitutive equations for a material body moving in the laboratory frame K can be obtained from those valid in the rest frame K' of the body by effecting a field transformation from the rest frame K' to the laboratory frame K, as discussed in Section 1.6.

To check the possibility of applying Minkowski´s theory to moving deformable superconductors, let us first consider a case, in which a superconductor is assumed to be rigid and move with a constant velocity $\boldsymbol{V}$ with respect to a laboratory frame K. By introducing a rest-frame K', co-moving with the superconductor, we may write the second London equation (3.2.9) at the rest-frame K' as

$$\mu_o \lambda_L^2 \nabla' \times \boldsymbol{J}_s' = -\boldsymbol{B}' \tag{7.1.1}$$

where J_S' denotes the supercurrent density vector in the rest-frame K'. According to Minkowski´s theory, we may find that the second London equation for the moving superconductor in the laboratory frame K should become

$$\mu_o \lambda_L^2 \nabla \times \boldsymbol{J}_S = -\boldsymbol{B} \tag{7.1.2}$$

by using the well-known field transformation relations: $\boldsymbol{J}_S = \boldsymbol{J}_S'$, $\boldsymbol{B} = \boldsymbol{B}'$, and $\nabla = \nabla'$ at the magneto-quasistatic approximation. This result indicates that the London equation for a rigid superconductor moving with a constant (uniform) velocity $\boldsymbol{V}$ is invariant under the Galilean transformation at the magneto-quasistatic approximation, as we may expect.

If we now generalize the Minkowski theory to model a superconductor moving with a non-uniform velocity according to the commonly used working hypothesis that the electrical properties in the instantaneous rest-frame K' are unaffected by the accelerations [Bladel (1984)], we would conclude that Eq.(7.1.2) applies also to the superconductor moving with the non-uniform velocity with respect to the laboratory frame K. Such a conclusion is however proven to be incorrect according to the well-known experiments on the small magnetic field generated upon rotation of a superconductor, as shown illustratively in Fig. 7.1 [Hildebrandt (1964), Brickman (1969), and Verheijen et al. (1990)]. This fact indicates, therefore, that it is not appropriate to simply generalize Minkowski´s theory to model superconductors moving with non-uniform velocities, even at the low velocity and low acceleration approximation.

The reason for the failure of the commonly used working hypothesis stated above may be explained intuitively by noting the fact that the force of inertia of superelectrons is not negligible due to the disappearance of the dissipative

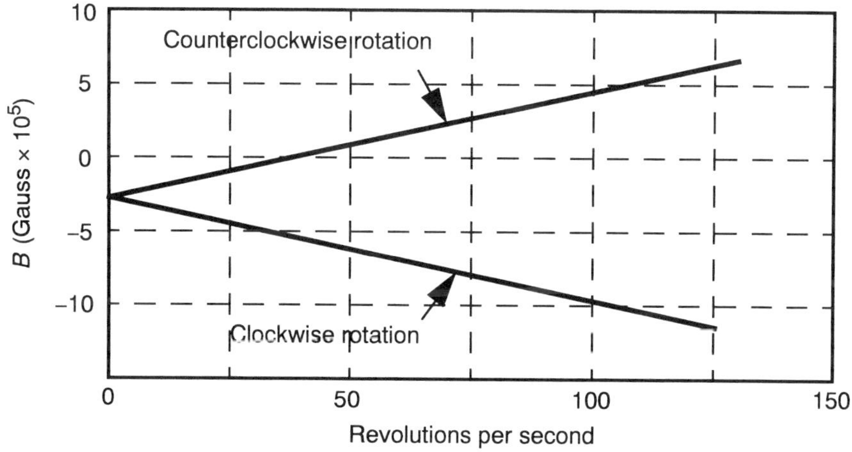

Figure 7.1 Recording of magnetic field strength vs. rotation rate [Hildebrandt (1964)].

resistance (viscous) force on superelectrons in superconductors. This is true even for the moving superconductor with low accelerations. While in normal conductors, the force of inertia of normal electrons is much smaller than the resistance force on the electrons in the moving conductors under the exertion of electric fields in usual engineering applications for sufficiently low accelerations and, therefore, is negligible in accordance with the commonly used working hypothesis. Quantitatively, a simple comparison of the resistance force with the force of inertia of normal electrons in a rotating normal conductor shows that the force of inertia is negligible as long as the angular velocity Ω of the rotating normal conductor is less than 10^5 rev/s even if we assume that the momentum relaxation time of the normal electrons is around 1 ps, the mean velocity of conduction electrons is 1 m/s, and the largest size of the conductor is less than 1 m. On the other hand, the London moment has been observed even at the angular velocity Ω less than 10 rev/s for small superconducting samples with rotation [Hildebrandt (1964) and Brickman (1969)]. Indeed, as we already know, the effect of inertia of electrons is practically negligible in the description of conductivity of normal conductors in a wide frequency range from dc up to GHz frequencies according to the classical Drude´s electron model. Some studies on the validity of Minkowski´s theory for rotating dielectrics by the instantaneous rest frame method have also been performed by Shiozawa (1973), who showed that the influence of the force of inertia is negligible as long as the operating frequency of a rotating dielectric is much less than the resonant frequency lying typically in the infrared or the ultraviolet.

7.1.2 London´s Formulation for Rotating Superconductors

Different from Minkowski´s approach, London (1950) proposed first a theoretical model for rotating (rigid) superconductors by generalizing his macroscopic theory for stationary superconductors. Based on Einstein´s special theory of relativity, London showed that equations of supercurrent fields contain no reference to the state of motion of the superconductor since the superelectrons were treated as entirely free in the free-electron model. In particular, at the linear approximation, London argued that the following field equations:

$$\frac{\partial \mathbf{v}_s}{\partial t} = \frac{e^*}{m^*}\boldsymbol{E} \tag{7.1.3}$$

$$\nabla \times \mathbf{v}_s = -\frac{e^*}{m^*}\boldsymbol{B} \tag{7.1.4}$$

apply for the stationary as well as for moving (rotating) superconductors. Here, $\mathbf{v}_s$ is the velocity field of the superelectrons. m^* and e^* denote, respectively, the mass and charge of the superelectron.

To account for the state of motion of a superconductor, London argued that the total supercurrent density $\boldsymbol{J}_s$ in the rotating superconductor should be given by

$$\boldsymbol{J}_s = e^* n_s(\mathbf{v}_s - \boldsymbol{V}) \tag{7.1.5}$$

where n_s denotes the number of superelectrons per unit volume, and $\boldsymbol{V}$ is the local velocity of the rotating superconductor with respect to the laboratory frame K. London's argument was originated from the consideration of Becker et al. (1933) that, if a superconducting body of axial symmetry is brought into a state of rotation around its symmetry axis, the superelectrons should not lag behind by virtue of their inertia, as might be expected, but almost rigidly follow the movement of the body. This was understood, according to London (1950), as follows: Initially the positive rest charges consisting of the non-superconducting electrons and nuclei advance a little without the superelectrons and thus generate a small current. This current generates a variable magnetic field which, in turn, generates an electric field by induction. This electric field is just strong enough and of such duration to drag the superelectrons in such a way that they move almost everywhere exactly in phase with the body. Only quite near the surface of the body do the superelectrons lag a little, thus generating a very weak surface current. But the bulk of the superelectrons follows the movement of the body exactly in phase. Based on these arguments, London derived the following set of equations:

$$\mu_o \lambda_L^2 \frac{\partial \boldsymbol{J}_s}{\partial t} = \boldsymbol{E} + \frac{m^*}{|e^*|}\frac{\partial \boldsymbol{V}}{\partial t} \tag{7.1.6}$$

$$\mu_o \lambda_L^2 \nabla \times \boldsymbol{J}_s - \frac{m^*}{|e^*|}\nabla \times \boldsymbol{V} = -\boldsymbol{B} \tag{7.1.7}$$

by substitution of (7.1.5) into (7.1.3) and (7.1.4). This set of equations, according to London, characterizes electrodynamic behaviors of the rotating superconductor at the linear approximation. Here, λ_L is the London penetration depth, defined by $\lambda_L = (m^*/\mu_0 n_s e^{*2})^{1/2}$. Furthermore, by Eq.(7.1.7) and Maxwell's equations, London analyzed the magnetic field distribution generated by a rotating superconductive sphere. His result indicates that, except for a layer of the depth λ_L near the surface of the superconductor, there exists a uniform magnetic field inside the rotating superconductor, reading

$$\boldsymbol{B} = \frac{2m^*}{|e^*|}\Omega \tag{7.1.8}$$

where $\Omega = \nabla \times \boldsymbol{V}/2$ is the angular velocity of rotation of the superconductor. Thus, by his model, London showed theoretically that a small magnetic field may be generated upon rotation of a superconductor. London´s prediction was later confirmed experimentally by Hildebrandt (1964). Recently, the small magnetic field generated upon rotation of a superconductor (often called the London moment) has also been observed for rotating high-T_c oxide superconductors [Verheijen et al. (1990)]. Besides, London also showed analytically that the supercurrent density $\boldsymbol{J}_s$ flows only within the thin layer of the depth λ_L near the surface of the rotating superconductor. Thus, by Eq.(7.1.5), London showed quantitatively that, in the rotating superconducting sphere, the superelectrons are rotating in synchronism with the lattice ions except for those within the penetration depth; where they rotate somewhat slower, and it is this lag which gives rise to a net surface current which, in turn, produces the London moment field.

Although, London´s theoretical prediction on the generation of a small magnetic field upon rotation of a superconductor has been confirmed experimentally, London´s theoretical treatment seems to be not conceptually consistent with Einstein´s theory of relativity. This is because, on one hand, London´s argument on the equation of supercurrent fields containing no reference to the state of motion of the superconductor is based on the special theory of relativity, which is only valid for inertial systems. On the other hand, the rest frame K' attached to the rotating superconductor is non-inertial. In London´s treatment, it is not clear if and how the acceleration of the rotating superconductor may influence macroscopic electrodynamic behaviors of the moving superconductor. Besides, the introduction of expression (7.1.5) on supercurrent density relating to the motion of superelectrons and the moving superconducting medium was based on a plausible argument, the general validity of which for moving deformable superconductors is not obvious since deformable superconductors which we are considering may have arbitrary geometry and have arbitrary motions of no axial symmetry. It is therefore not clear whether or not London´s model for rotating rigid superconductors could be generalized to study moving deformable superconductor, where the localized non-uniform motion of material elements may play an important role.

7.1.3 Formulation of Moving Deformable Superconductors

In this section, different from Minkowski´s theory and London´s approach, we shall formulate a set of field equations for moving deformable superconductors at the magneto-quasistatic approximations and within the framework of the generalized Galilean relativity [Zhou (1998)]. In the generalized Galilean relativity, we introduce the following Galilean transformation, generalized to include the accelerated case:

$$\boldsymbol{x}' = \boldsymbol{x} - \int_0^t \boldsymbol{V} d\tau \tag{7.1.9}$$

with time t being universal, that is, $t' = t$. Here, V is the velocity of an origin O′ in the frame of reference K', as measured with respect to the origin O in the frame of reference K, as shown in Fig. 1.12. According to the principle of relativity, by asserting that Maxwell´s equations at the magneto-quasistatic approximation:

$$\nabla \times \boldsymbol{H} = \boldsymbol{J}, \qquad \nabla \times \boldsymbol{E} = -\frac{\partial \boldsymbol{B}}{\partial t}, \qquad \nabla \cdot \boldsymbol{B} = 0 \tag{7.1.10}$$

are covariant or form-invariant with respect to the generalized Galilean transformation, we may find the field transformation relations: $\boldsymbol{B}' = \boldsymbol{B}, \boldsymbol{J}' = \boldsymbol{J}$, and $\boldsymbol{E}' = \boldsymbol{E} + \boldsymbol{V}\times\boldsymbol{B}$. Obviously, since the Galilean relativity is adopted, we are dealing with relatively simple laws of Euclidean geometry. There is no length contraction nor clock slow-down. However, some complications may be involved in studying moving deformable media. This is because, in a moving deformable medium, the velocity of material elements may not only change in time, but also vary from point to point (macroscopic continuum hypothesis) in space. It has been customary to introduce a reference frame attached to each and every material element (point) at each instant, called the pointwise instantaneous rest-frame K' to study the deformable medium moving with a velocity $\boldsymbol{V}$ with respect to a laboratory frame of reference K. This instantaneous rest-frame is usually supposed to be an inertial frame. Each point P has its own rest frame having a velocity $\boldsymbol{V}(P)$ with respect to the laboratory frame.

The theory of the "instantaneous rest-frame" has found wide acceptance in the literature for studying moving deformable media with sufficiently low velocities and low accelerations. However, to account for the effect of superelectron inertia in moving deformable superconductors, we shall generalize here the usual assumptions on the instantaneous rest-frame by stating that the rest frame is a frame of reference which may generally be non-inertial. Thus, at the linear approximation, we may write the following dynamic equation for superelectrons in an electric field $\boldsymbol{E}'$ in the generalized instantaneous rest-frame K':

$$m^* \frac{\partial \mathbf{v}_s'}{\partial t'} = e^* \boldsymbol{E}' - m^* \frac{\partial \boldsymbol{V}}{\partial t'} \tag{7.1.11}$$

where $\mathbf{v}_s'$ denotes the local (mean) velocity of superelectrons measured in the generalized instantaneous rest-frame K'. According to the classical Newton´s idea, the second term on the right-hand side of Eq.(7.1.11) may be considered to be a "fictitious force" acting on the superelectron due to the effect of its inertia, which measures the inertial resistance to the acceleration of the moving superconductor. We may notice that the "fictitious force" has no connection whatever with the physical properties of the mechanical system itself. It depends exclusively on the acceleration relative to the systems of inertia of the system of reference introduced. It was for just this reason that Newton introduced the concept of absolute space, which should represent the system of reference where the laws of nature assume the simplest and most natural form. However, as we may know, Newton´s notion of absolute space lost its physical meaning as soon as Einstein´s special principle of relativity was generally accepted. To generalize his special theory of relativity, Einstein has advocated a new interpretation of the fictitious forces in accelerated systems of reference: instead of regarding them as an expression of a difference in principle between the fundamental equations in uniformly moving and in accelerated systems, he considered both kinds of systems of reference to be completely equivalent as regards the form of the fundamental equations; and the "fictitious" forces were treated as real forces on the same footing as any other force of nature [see Möller (1970)]. Thus we could treat the second term on the right-hand side of Eq.(7.1.11) as a type of "gravitational" force if Einstein´s principle of equivalence were applied.

By defining the superconducting current density vector $\boldsymbol{J}_S' = e^* n_S \mathbf{v}_S'$ in the generalized instantaneous rest-frame K', and by taking the curl of both sides of Eq.(7.1.11), and using Maxwell´s equation of Faraday´s law of induction, we may obtain

$$\mu_o \lambda_L^2 \nabla' \times \boldsymbol{J}_s' - \frac{m^*}{|e^*|} \nabla' \times \boldsymbol{V} + \boldsymbol{B}' = \text{Constant } (in \ \ time) \qquad (7.1.12)$$

after integration with respect to time. Here, we have noticed that at the London approximation, the density of superelectrons (n_S) is supposed to be constant for given temperature, which is valid for superconductors in weak fields, as we have discussed in Chapter 3. Due to the well-known Meissner effect in stationary superconductors ($\boldsymbol{V} = 0$), and the fact that the velocity $\boldsymbol{V}$ is generally a function of time and space coordinates, the constant on the right-hand side of Eq.(7.1.12) can be set to be zero for all time. Thus we may derive the modified second London equation, having the same form as Eq.(7.1.7), by using the field transformation relations at the magneto-quasistatic approximation within the framework of the generalized Galilean relativity.

Furthermore, by $\boldsymbol{J}_S = \boldsymbol{J}_S' = e^* n_S \mathbf{v}_S' = e^* n_S(\mathbf{v}_S - \boldsymbol{V})$ at the magneto-quasistatic approximation, we thus arrive naturally at London´s expression (7.1.5) for the superconducting current density vector relating to the local velocity of the superelectrons and the local velocity of the moving superconductor. The modified first London equation (7.1.6) may then be obtained straightforward by using the acceleration equation (7.1.3) and noting the local (mean) velocity of superelectrons expressed by $\mathbf{v}_S = \boldsymbol{V} + \boldsymbol{J}_S/(e^* n_S)$ in the laboratory frame of reference K.

It should be point out that the weak magnetization effect due to the rotational motion of a superconductor was early calculated by Becker, Heller, and Sauter in 1933, based on their acceleration theory. However, the Becker, Heller, and Sauter result depends on the initial state of the superconductor. According to the Becker, Heller, and Sauter acceleration theory, no effect would occur if the body became superconducting when already rotating, while in the modified London model, the body would suddenly acquire the same weak magnetization (London moment) as if it had been set rotating when already superconducting. Here, we have shown that, by taking into account of the effect of superelectron inertia, we may derive a set of constitutive equations for characterizing electrodynamic behaviors of moving deformable superconductors in weak fields within the framework of generalized Galilean relativity. Although our formulation at the magneto-quasistatic approximation arrives at the same mathematical expressions as those of London (7.1.6) and (7.1.7) for rotating (rigid) superconductors, the same set of equations has now wider applicability. It may be used not only for analyzing rotating (rigid) superconductors of axial symmetry, but also for studying moving deformable superconductors of arbitrary geometry according to the new formulation. In Eqs.(7.1.6) and (7.1.7), $\boldsymbol{V}$ denotes the local velocity of the superconducting medium, $\boldsymbol{J}_s$ and $\boldsymbol{B}$ are, respectively, the superconducting current density vector and the magnetic induction field in the superconducting medium, measured in the laboratory frame K.

To see the consequence of this new model, let us study the problem on phenomena of wave propagation in an elastic superconductor. Specifically, if a plane elastic wave with a displacement vector $\boldsymbol{U} = \boldsymbol{U}_0 \exp[i(\omega t - \boldsymbol{k}\cdot\mathbf{x})]$ propagates inside an infinitely large elastic superconductor (so that boundary effects are ignored), we may find according to our model that there may appear an induced magnetic field inside the superconductor, given by

$$\boldsymbol{B} = \frac{\omega m^*}{|e^*|(1+\lambda_L^2|\boldsymbol{k}|^2)}\boldsymbol{k}\times\boldsymbol{U} \tag{7.1.13}$$

where $\boldsymbol{U}_0$ denotes the amplitude of the elastic displacement. $\boldsymbol{k}$ is the wave number vector, and ω is the radian frequency of the elastic wave. Besides, inside the superconductor, there may appear an induced superconducting current given by

$$\boldsymbol{J}_s = \frac{i\omega m^*}{\mu_o|e^*|(1+\lambda_L^2|\boldsymbol{k}|^2)}[|\boldsymbol{k}|^2\boldsymbol{U} - (\boldsymbol{k}\cdot\boldsymbol{U})\boldsymbol{k}] \tag{7.1.14}$$

Both the induced magnetic field and the induced superconducting current are found to be time- and space-varying. In particular, we may find from Eqs.(7.1.13) and (7.1.14) that only the transverse part of the elastic plane wave contributes to the magnetic field and to the superconducting current density induced inside the elastic superconductor. This means that neither the magnetic field nor the superconducting current will be generated when a longitudinal elastic plane waves propagates inside an elastic superconductor according to this model. Such a special response of the superconductor for acoustic waves may be used to construct some kind of novel acoustomagnetic devices, such as the acoustic "filter", which we might imagine. Quantitatively, for a transverse elastic plane wave with the amplitude $|\boldsymbol{U}| = 10$ nm, the phase velocity $\omega/|\boldsymbol{k}| = 2$ km/s, and the frequency $\omega/2\pi = 100$ MHz propagating in the superconductor with $\lambda_L =$ 0.2 μm, the magnetic field induced has the magnitude of $|\boldsymbol{B}|$ about 1.12×10^{-5} Tesla, which is presumably measurable.

Furthermore, it is shown by Eq.(7.1.14) that, different from rotating superconductors, there exists a non-zero induced superconducting current inside a deformable superconductor in which a transverse elastic plane wave is propagating. This implies that the bulk of the superelectrons inside a moving deformable superconductor does not follow, in general, the movement of the body exactly in phase. This statement is clearly different from the argument by Becker et al. and by London for rotating superconductors. Intuitively, we may understand that, in a moving deformable superconductor, not only the direction but also the magnitude of the local velocity $\boldsymbol{V}$ may vary with time and space coordinates. Due to their inertia, superelectrons can not follow the local movement of the body exactly in phase. For a transverse elastic plane wave propagating inside an elastic superconductor, the superconducting current induced is found to be 90° out of phase, while the magnetic field induced is in phase with the elastic displacement $\boldsymbol{U}$ of the propagating elastic wave, as shown by Eqs.(7.1.13) and (7.1.14).

From the modified first London equation (7.1.6), we may notice that a voltage across a moving superconductor may be developed when the superconductor is driven by an ac current and/or by its non-uniform (accelerated) motion. This voltage drop occurs due to the fact that the motion of superelectrons is out of phase with the time-varying driving source because of the inertia of superelectrons. Such a voltage drop may drive normal (unpaired) electrons in the superconductor to flow at finite temperature. Thus, in the presence of normal conduction currents, we may introduce the following modified Ohm´s law for the normal conduction current density $\boldsymbol{J}_n$ in the moving medium by

$$\boldsymbol{J}_n = \sigma_n(\boldsymbol{E} + \boldsymbol{V}\times\boldsymbol{B}) \tag{7.1.15}$$

where σ_n denotes the conductivity of the superconductor at the normal state. Thus, by writing the total current density vector $\boldsymbol{J} = \boldsymbol{J}_s + \boldsymbol{J}_n$ in Maxwell´s equations, and by using the modified second London equation (7.1.7), and the mechanical motion equation for an elastic (non-magnetic) superconductor, we may derive the following set of coupled wave equations:

$$\frac{\partial \boldsymbol{B}}{\partial t} = \nabla\times\left(\frac{\partial \boldsymbol{U}}{\partial t}\times\boldsymbol{B}\right) + \frac{1}{\mu_o\sigma_n}\nabla^2\boldsymbol{B} - \frac{1}{\mu_o\sigma_n\lambda_L^2}\left(\boldsymbol{B} - \frac{m^*}{|e^*|}\nabla\times\frac{\partial \boldsymbol{U}}{\partial t}\right) \tag{7.1.16}$$

$$\rho\frac{\partial^2 \boldsymbol{U}}{\partial t^2} = (\lambda^{(s)} + G^{(s)})\nabla(\nabla\cdot\boldsymbol{U}) + G^{(s)}\nabla^2\boldsymbol{U} + \frac{1}{\mu_o}\left[\boldsymbol{B}\cdot\nabla\boldsymbol{B} - \nabla\left(\frac{B^2}{2}\right)\right] \tag{7.1.17}$$

for the description of possible magnetoelastic waves in elastic superconductors with proper boundary conditions in the Meissner state [Zhou (1993)]. Here, $\boldsymbol{U}$ is the elastic displacement vector, and ρ is the mass density of the superconductor. $\lambda^{(s)}$ and $G^{(s)}$ denote the effective Lamé constants of the superconductor at the superconducting state. It is noted that a simple Hooke's law is used here to describe effectively the elastic behavior of the superconductor.

7.2 SOME WAVE PHENOMENA IN ELASTIC SUPERCONDUCTORS

7.2.1 Magnetic Wave Induced by Elastic Wave in Superconductors

In the above section, we have derived a set of coupled electromagnetic and mechanical motion equations (7.1.16) and (7.1.17) for elastic superconductors at the magneto-quasistatic approximation. In general, this set of equations are nonlinear differential equations. In some cases of practical interest, linearization of this set of equations is possible. For instance, we may consider a case in which an elastic superconductor is located in a static magnetic field $\boldsymbol{B}_a$. The superconductor is then subjected to a small time-varying disturbance, which can be of either mechanical or electromagnetic origin. Electromagnetic-elastic waves may thus be generated. To study properties of wave propagation in such an elastic superconductor, we may use the following linearization approach. Suppose that the total magnetic induction field $\boldsymbol{B}$ in the superconductor can be expressed by

$$\boldsymbol{B}(\boldsymbol{x}, t) = \boldsymbol{B}_o(\boldsymbol{x}) + \boldsymbol{b}(\boldsymbol{x}, t) \tag{7.2.1}$$

where $\boldsymbol{b}$ denotes the disturbed time-varying magnetic induction field, and $\boldsymbol{B}_o$ is the local static magnetic field, which satisfies the following equation

$$\lambda_L^2 \nabla^2 \boldsymbol{B}_o - \boldsymbol{B}_o = 0 \tag{7.2.2}$$

in the superconductor. This equation shows the characteristic of the Meissner effect in the superconductor. The static force induced by the static magnetic field $\boldsymbol{B}_o$ may cause static elastic deformation of the superconductor, for which the static elastic displacement vector, $\boldsymbol{U}_o(\boldsymbol{x})$, can be determined by the following equilibrium equation:

$$(\lambda^{(s)} + G^{(s)})\nabla(\nabla \cdot \boldsymbol{U}_o) + G^{(s)}\nabla^2 \boldsymbol{U}_o + \frac{1}{\mu_o}\left[\boldsymbol{B}_o \cdot \nabla \boldsymbol{B}_o - \nabla\left(\frac{B_o^2}{2}\right)\right] = 0 \tag{7.2.3}$$

For superconductors of finite dimension, some boundary conditions have to be also satisfied, as discussed in Chapter 2. In the presence of the time-varying perturbation magnetic field $\boldsymbol{b}$ ($|\boldsymbol{b}| << |\boldsymbol{B}_o|$), the total displacement vector $\boldsymbol{U}$ may be expressed by

$$U(x, t) = U_o(x) + u(x, t) \tag{7.2.4}$$

where u is the time-varying elastic displacement vector field.

Substitution of Eqs.(7.2.1) and (7.2.4) into Eqs.(7.1.16) and (7.1.17), and using Eqs.(7.2.2) and (7.2.3), we may find that u and b have to satisfy the following set of coupled linearized differential (wave) equations:

$$\rho\frac{\partial^2 u}{\partial t^2} = (\lambda^{(s)} + G^{(s)})\nabla(\nabla \cdot u) + G^{(s)}\nabla^2 u + \frac{1}{\mu_o}[B_o \cdot \nabla b + b \cdot \nabla B_o - \nabla(B_o \cdot b)] \tag{7.2.5}$$

$$\frac{\partial b}{\partial t} = \nabla\times\left(\frac{\partial u}{\partial t} \times B_o\right) + \frac{1}{\mu_o\sigma}\nabla^2 b - \frac{1}{\mu_o\sigma\lambda_L^2}\left(b - \frac{m^*}{|e^*|}\nabla\times\frac{\partial u}{\partial t}\right) \tag{7.2.6}$$

This set of coupled linearized wave equations may be used to study some wave phenomena in elastic superconductors.

Let us first consider the case in which the normal conduction current and the nonlinear magnetic stress effect are supposed to be negligible at the first order approximation. In such a case, we are studying an uncoupled wave problem, and Eq.(7.1.16) may then be reduced to be

$$\lambda_L^2\nabla^2 B = B - \frac{m^*}{|e^*|}\nabla\times\frac{\partial U}{\partial t} \tag{7.2.7}$$

which is a linear differential equation.

In the case of the absence of the applied static magnetic field and the neglect of nonlinear magnetic stress effect, we may find that a transverse elastic wave, $U = (0, U_0 exp(ik(x-c_T t)), 0)$, with the phase velocity c_T $(=(G^{(s)}/\rho)^{1/2})$ in the elastic superconductor may generate a transverse magnetic wave of the following particular form, $B = (0, 0, B_z)$ with

$$B_z = \frac{m^* c_T\omega^2}{|e^*|(c_T^2 + \lambda_L^2\omega^2)}U_o exp(ik(x - c_T t)) \tag{7.2.8}$$

where ω is the radian frequency ($\omega = kc_T$). This magnetic wave propagates with the same phase velocity as the elastic wave, but its magnetic field vector is perpendicular to the displacement vector of the elastic wave. Figure 7.2 shows how the normalized magnetic field amplitude of the magnetic wave may depend on the excitation frequency ω of the elastic wave in the elastic superconductor for two different values of the London penetration depth λ_L.

It is shown that the magnetic field amplitude is very small at low frequencies. For excitation frequency larger than a certain frequency, which depends on the value of the London penetration depth λ_L, $|B_z|$ increases quickly with increasing

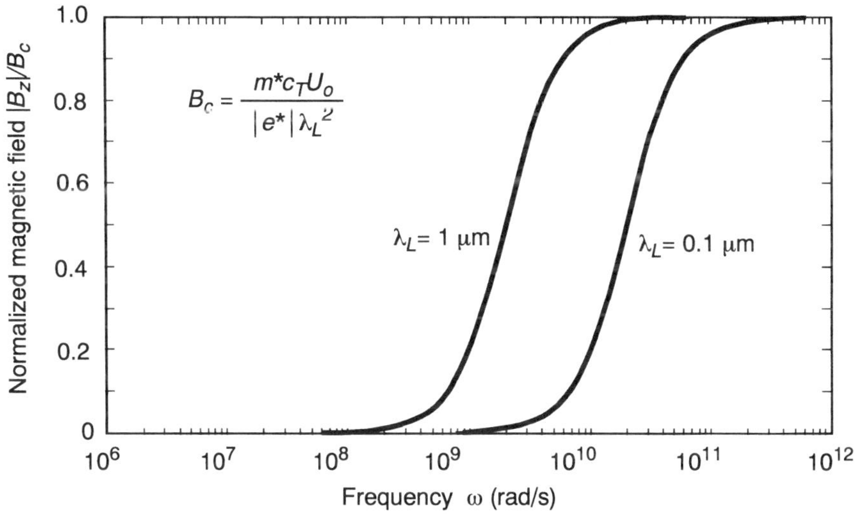

Figure 7.2 Dependence of the magnetic field amplitude on excitation frequency ω.

frequency and reaches a limiting value, B_c, as shown in Fig. 7.2. Quantitatively, this limiting magnetic field, B_c could be as high as 0.5 tesla for $c_T = 2$ km/s, $U_o = 1$ μm, and $\lambda_L = 0.15$ μm in theory. This probably imply that superconductivity may also be destroyed by elastic waves of very high frequencies. We shall, however, notice that the theoretical model used here is only valid at the magneto-quasistatic approximation, which means that the frequency range of interest in this model is limited (to about 100 GHz, depending on materials). Besides, we notice that B_c decreases quickly as the London penetration depth increases.

7.2.2 Magnetic Surface Wave Induced by the Rayleigh Elastic Surface Wave

In this section, we shall further consider the wave problem in a semi-infinite non-magnetic superconductor as shown in Fig. 7.3. We consider the case in which the normal conduction current is negligible in the semi-infinite elastic superconductor. No static magnetic fields are supposed to be applied to the superconductor. Thus, by ignoring the magnetic stress effect, we may look for a solution of magnetic surface waves, which may be generated by the Rayleigh elastic surface wave. Mathematically, we shall consider the Rayleigh elastic surface wave of the following form:

$$U_x = A_1 e^{-\alpha_1 x} e^{ik(z - vt)} + A_2 e^{-\alpha_2 x} e^{ik(z - vt)} \tag{7.2.9}$$

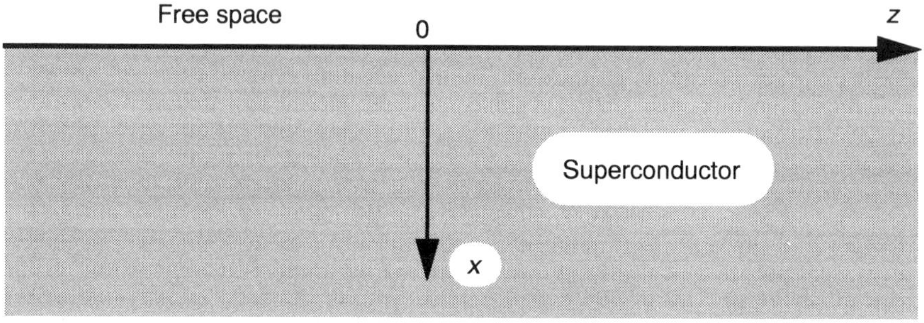

Figure 7.3 Wave propagation in a semi-infinite superconductor.

$$U_z = C_1 e^{-\alpha_1 x} e^{ik(z - vt)} + C_2 e^{-\alpha_2 x} e^{ik(z - vt)} \tag{7.2.10}$$

where the coefficients α_1 and α_2 read

$$\alpha_1 = k\sqrt{1 - \frac{v^2}{c_T^2}} \quad \text{and} \quad \alpha_2 = k\sqrt{1 - \frac{v^2}{c_L^2}} \tag{7.2.11}$$

with c_L and c_T denote the characteristic velocity of longitudinal elastic wave and the velocity of transverse elastic wave defined, respectively, by

$$c_L = \sqrt{\frac{\lambda + 2G}{\rho}} \quad \text{and} \quad c_T = \sqrt{\frac{G}{\rho}} \tag{7.2.12}$$

Here, ρ, λ and G denote, respectively, the mass density, and the Lamé constants of the elastic superconductor. At the superconducting state, we have $\lambda = \lambda^{(s)}$ and $G = G^{(s)}$ while, at the normal state, $\lambda = \lambda^{(n)}$ and $G = G^{(n)}$. The coefficients A_1, A_2, C_1, and C_2 satisfy the following relations:

$$\frac{A_1}{C_1} = \frac{ik}{\alpha_1} \quad \text{and} \quad \frac{A_2}{C_2} = \frac{i\alpha_2}{k} \tag{7.2.13}$$

The phase velocity of the Rayleigh elastic surface wave v is determined by the Rayleigh equation:

$$\frac{v^2}{c_T^2}\left[\frac{v^6}{c_T^6} - 8\frac{v^4}{c_T^4} + v^2\left(\frac{24}{c_T^2} - \frac{16}{c_L^2}\right) - 16\left(1 - \frac{c_T^2}{c_L^2}\right)\right] = 0 \tag{7.2.14}$$

which may be solved numerically.

By Eq.(7.2.7), we find that the Rayleigh elastic surface wave may generate a transverse magnetic surface wave in the superconductor, which has the particular form of $\boldsymbol{B} = (0, B_y, 0)$ with

$$B_y = B_1 e^{-\alpha_1 x} e^{ik(z - vt)} \qquad (0 \le x) \tag{7.2.15}$$

where the coefficient B_1 is found to be

$$B_1 = \frac{m^* v\omega^2 A_1}{|e^*|[c_T^2 + \lambda_L^2\omega^2]} \tag{7.2.16}$$

Now, let us look for the part of the solution in free-space. The electromagnetic fields in free-space can be found by

$$\boldsymbol{E} = (E_x, 0, E_z)e^{qx}e^{ik(z - vt)} \qquad (-\infty < x < 0) \tag{7.2.17}$$

$$\boldsymbol{B} = (0, B_1, 0)e^{qx}e^{ik(z - vt)} \qquad (-\infty < x < 0) \tag{7.2.18}$$

where q can be obtained from Maxwell's equations in free-space

$$q = k\sqrt{1 - \frac{v^2}{c_o^2}} \tag{7.2.19}$$

where c_o denotes the velocity of light in free-space. A relation between $\boldsymbol{E}$ and $\boldsymbol{B}$ in the free-space can be expressed by

$$E_x = \frac{c_o^2}{v} B_1 \qquad \text{and} \qquad E_z = -\frac{iqc_o^2}{kv} B_1 \tag{7.2.20}$$

where the continuity condition of the magnetic field at the boundary is used.

It is shown that a transverse magnetic surface wave can be generated by the Rayleigh elastic surface wave in a semi-infinite elastic superconductor. This magnetic surface wave propagates with the same phase velocity as the Rayleigh surface wave, but its magnetic field vector is perpendicular to the elastic displacement vector of the Rayleigh surface wave. The normalized amplitude of the magnetic field associated with the electromagnetic wave in free space may be calculated by

$$\frac{|B_y|}{B_c} = \frac{\lambda_L^2 \omega^2 e^{\frac{\omega}{v}x}}{c_T^2 + \lambda_L^2 \omega^2} \qquad (-\infty < x < 0) \tag{7.2.21}$$

where we have introduced the characteristic magnetic field B_c, defined by

$$B_c = \frac{m^* \mathrm{v} |A_1|}{|e^*| \lambda_L^2} \tag{7.2.22}$$

It is noted that the approximation $q \approx \omega/\mathrm{v}$ has been used in Eq.(7.2.21) since the phase velocity of the Rayleigh elastic surface wave is much smaller than the velocity of light.

It is seen that the magnitude of the magnetic field induced by the Rayleigh elastic surface wave decreases exponentially with the increase of $|x|$. In Fig. 7.4, we have shown how the normalized magnetic field amplitude of the induced magnetic wave may depend on the excitation frequency ω of the Rayleigh elastic surface wave in the elastic superconductor for two different values of $|x|$. Here, we have quantitatively chosen some typical values of $\lambda_L = 1\ \mu\text{m}$, $c_T = 2$ km/s, and $\mathrm{v} = 0.935 c_T$. It is shown that there exists a maximum value of the amplitude of the magnetic field found at a certain frequency. This characteristic frequency, denoted here by ω_m, may be determined by

$$\omega_m = \left[\sqrt{\left(\frac{\mathrm{v} c_T^2}{x\lambda_L^2}\right)^2 + \left(\frac{c_T^2}{3\lambda_L^2}\right)^3} - \frac{\mathrm{v} c_T^2}{x\lambda_L^2}\right]^{1/3} - \left[\sqrt{\left(\frac{\mathrm{v} c_T^2}{x\lambda_L^2}\right)^2 + \left(\frac{c_T^2}{3\lambda_L^2}\right)^3} + \frac{\mathrm{v} c_T^2}{x\lambda_L^2}\right]^{1/3} \quad (x < 0) \tag{7.2.23}$$

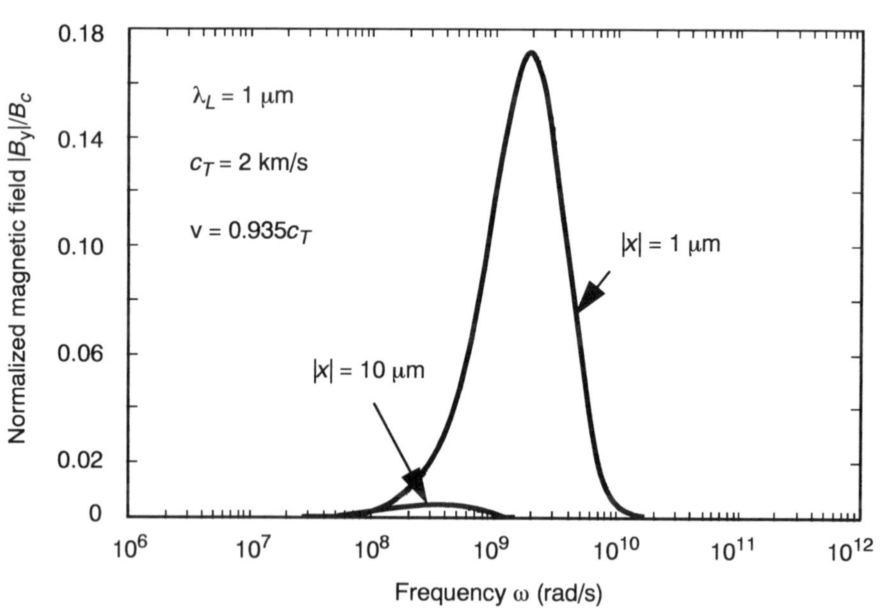

Figure 7.4 Dependence of the magnetic field amplitude on excitation frequency ω.

In the numerical example shown in Fig. 7.4, we may find $\omega_m = 1.93\times10^9$/s from Eq.(7.2.23) for $x = -1$ μm. It can be seen that the frequency ω_m is a function of x, which may be shown quantitatively in Fig. 7.5.

At either low frequencies or very high frequencies, the amplitude of the magnetic field in free-space is shown to be quite small in Fig. 7.4. Corresponding to the frequency ω_m, we find the maximum amplitude of the magnetic field, $|B_y|_{max}$ in free-space. Numerically, shown in Fig. 7.6 is the maximum

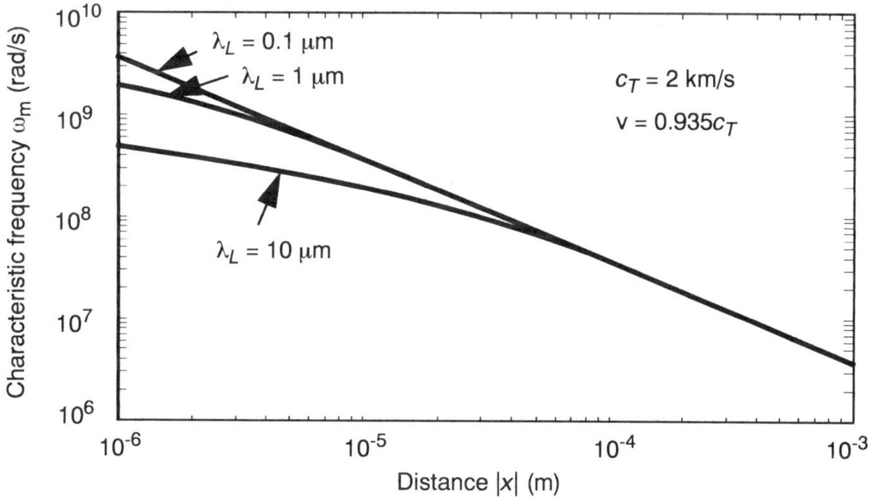

Figure 7.5 Dependence of ω_m on the distance $|x|$ from the superconductor surface.

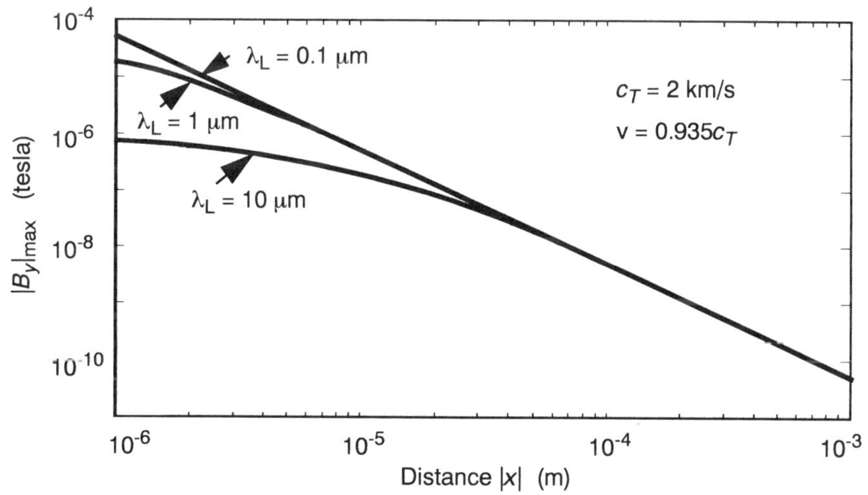

Figure 7.6 Maximum magnetic field amplitude $|B_y|_{max}$ as a function of distance $|x|$.

amplitude of the magnetic field in free-space as a function of the distance $|x|$ from the superconductor surface. The maximum amplitude of the magnetic field at the surface of the superconductor can be calculated for the superconductor with the London penetration depth λ_L, and for the Rayleigh elastic surface wave of v = $0.935c_T$, $c_T = 2$ km/s, and $|A_1| = 10$ nm.

It is shown that the electromagnetic fields in free-space generated by the Rayleigh elastic surface wave in an elastic superconductor are presumably measurable if the experiment is properly designed. So far, it seems that such an experiment has not yet been made. Considering the possible device application of this effect and its basic theoretical interest, experiments to detect such an effect is very desirable.

7.2.3 Electromagnetoelastic Surface Wave

In this section, we shall consider the general case in which both the effect of normal conduction current and the effect of magnetic stresses are taken into account. We shall look for some types of coupled electromagnetoelastic surface wave solutions. Let us consider the wave problem in a semi-infinite non-magnetic superconductor as shown in Fig. 7.3. A static magnetic field $\boldsymbol{B}_a$ is supposed to be applied along the y-direction. A simple solution for the static magnetic field $\boldsymbol{B}_o$ in the superconductor can be found as $\boldsymbol{B}_o = (0, B_{oy}, 0)$ with

$$B_{oy}(x) = B_a exp(-\gamma x) \tag{7.2.24}$$

where γ equals $1/\lambda_L$ for the superconductor at the superconducting state, and is zero at the normal state. Now, we look for a solution of electromagnetoelastic surface wave of the form, $\boldsymbol{b} = (0, b_y, 0)$ and $\boldsymbol{u} = (u_x, 0, u_z)$, with

$$b_y = Be^{-\beta x}e^{ik(z - vt)} \tag{7.2.25}$$

$$u_x = Ae^{-\alpha x}e^{ik(z - vt)} \tag{7.2.26}$$

$$u_z = Ce^{-\alpha x}e^{ik(z - vt)} \tag{7.2.27}$$

where i denotes the imaginary unit and A, B, and C are complex constants. The coefficients α and β are supposed to be real and positive so that the amplitude of the wave decreases exponentially with increasing x, and tend to zero as $x \to \infty$. Substituting Eqs.(7.2.25)–(7.2.27) into Eqs.(7.2.5) and (7.2.6), we obtain the following equations:

$$ik\mathrm{v}\left[(\alpha+\gamma)B_{oy}+\frac{ikm^*}{|e^*|\mu_o\sigma\lambda_L^2}\right]u_x+k\mathrm{v}\left(kB_{oy}+\frac{i\alpha m^*}{|e^*|\mu_o\sigma\lambda_L^2}\right)u_z$$

$$-\left(ik\mathrm{v}+\frac{\beta^2-k^2}{\mu_o\sigma}-\frac{1}{\mu_o\sigma\lambda_L^2}\right)b_y = 0 \tag{7.2.28}$$

$$[c_L^2\alpha^2+(\mathrm{v}^2-c_T^2)k^2]u_x-i\alpha k(c_L^2-c_T^2)u_z+\frac{\beta+\gamma}{\rho\mu_o}B_{oy}b_y = 0 \tag{7.2.29}$$

$$-i\alpha k(c_L^2-c_T^2)u_x+[c_T^2\alpha^2+(\mathrm{v}^2-c_L^2)k^2]u_z-\frac{ik}{\rho\mu_o}B_{oy}b_y = 0 \tag{7.2.30}$$

From Eqs.(7.2.28)–(7.2.30), we may find that, to have a coupled solution of this set of equations, we have $\gamma = 0$ and $\alpha = \beta$, which implies that this type of electromagnetoelastic surface wave exists only for the superconductor at a normal state. In such a case, we have $B_{oy} = B_a$, and the set of equations (7.2.28)–(7.2.30) becomes

$$-ik\mathrm{v}\alpha B_aA-k^2\mathrm{v}B_aC+\left[ik\mathrm{v}+\frac{\alpha^2-k^2}{\mu_o\sigma}\right]B = 0 \tag{7.2.31}$$

$$[c_L^2\alpha^2+(\mathrm{v}^2-c_T^2)k^2]A-i\alpha k(c_L^2-c_T^2)C+\frac{\alpha}{\rho\mu_o}B_aB = 0 \tag{7.2.32}$$

$$-i\alpha k(c_L^2-c_T^2)A+[c_T^2\alpha^2+(\mathrm{v}^2-c_L^2)k^2]C-\frac{ik}{\rho\mu_o}B_aB = 0 \tag{7.2.33}$$

with the use of Eqs.(7.2.25)–(7.2.27). The condition for the existence of a non-trivial solution is the vanishing of the determinant of the coefficients of the set of equations (7.2.31)–(7.2.33) for the unknown A, B, and C, which may be written in the following form:

$$\left\{i\frac{k\mathrm{v}B_a^2}{\rho\mu_o}(\alpha^2-k^2)+\left(ik\mathrm{v}+\frac{\alpha^2-k^2}{\sigma\mu_o}\right)\times[c_L^2\alpha^2-(c_L^2-\mathrm{v}^2)k^2]\right\}$$

$$\times[c_T^2\alpha^2-(c_T^2-\mathrm{v}^2)k^2] = 0 \tag{7.2.34}$$

From this equation, we may get the following roots for α

$$\alpha_1 = k\sqrt{1-\frac{\mathrm{v}^2}{c_T^2}} \quad \text{and} \quad \alpha_2 = k\sqrt{1-\frac{\mathrm{v}^2}{c_L^2+c_A^2}} \tag{7.2.35}$$

in a perfect conductor ($\sigma \rightarrow \infty$). Here, $c_A = B_a/(\rho\mu_o)^{1/2}$ denotes the Alfvén wave velocity. The assumption that α is real requires that $\mathrm{v} < c_T$ ($< c_L$).

By considering the wave solution in free-space and the boundary (interface) conditions, we may derive the following a characteristic equation for the determination of the velocity v of the coupled electromagnetoelastic surface wave in an elastic perfect conductor

$$\left[\frac{(2c_T^2/c_L^2)\sqrt{1-\mathrm{v}'^2}}{2-\mathrm{v}'^2} - \sqrt{1-\frac{\mathrm{v}'^2 c_T^2}{c_L^2+c_A^2}}\right] \times \frac{(c_L^2+c_A^2)\sqrt{1-\mathrm{v}'^2 c_T^2/(c_L^2+c_A^2)}}{c_T^2 \mathrm{v}'^2}$$

$$+\left[1-\frac{2c_T^2}{c_L^2}+\frac{(2c_T^2/c_L^2)\sqrt{1-\mathrm{v}'^2}\sqrt{1-\mathrm{v}'^2 c_T^2/(c_L^2+c_A^2)}}{2-\mathrm{v}'^2}\right] \times \frac{c_L^2+c_A^2}{c_T^2\mathrm{v}'^2}+\frac{c_A^2}{c_L^2} = 0 \tag{7.2.36}$$

where the normalized velocity v′ is defined by $\mathrm{v}' = \mathrm{v}/c_T$. This equation can in general be solved numerically. A calculation shows that the wave velocity increases slightly with the increasing magnetic field B_a, and it recovers its classical value for a pure elastic body in the absence of the magnetic field. Thus we have shown that the Rayleigh electromagnetoelastic surface wave may exist in an elastic perfect conductor [Kaliski and Rogula (1960)]. However, this type of the Rayleigh electromagnetoelastic surface wave solution is not admitted in a semi-infinite superconductor at the superconducting state due to the onset of the Meissner effect.

7.3 MACROSCOPIC QUANTUM EFFECTS IN MOVING DEFORMABLE SUPERCONDUCTORS

7.3.1 Effect of Non-Uniform Motion of Superconductors on Flux-Quantization

In the 1940s, London has already realized that superconductivity is an inherently quantum mechanical phenomenon that manifests itself on macroscopic scales. In the macroscopic quantum wave model developed later based on the BCS microscopic theory for superconductivity (see Section 3.2.2), it is postulated that there exists one macroscopic quantum wave-function, $\psi = |\psi| e^{i\theta}$, which describes the behavior of the entire ensemble of superelectrons in the superconductor. Here, $|\psi|$ denotes the amplitude of the wave-function, which characterizes the local density of superelectrons, and θ is a real function representing the phase of the wave-function. To formulate the macroscopic quantum phenomena in moving deformable superconductors, let us start the investigation on the effect of inertia of superelectrons on the flux-quantization condition in a moving multiply connected superconductor. We begin with the modified second London equation (7.1.7). Introducing a magnetic vector potential $\boldsymbol{A}$ by $\boldsymbol{B} = \nabla\times\boldsymbol{A}$, we may obtain from Eq.(7.1.7) the following relation:

$$\frac{\hbar}{e*}\nabla\theta = \mu_o\lambda_L^2 \boldsymbol{J}_s + \boldsymbol{A} + \frac{m*}{e*}\boldsymbol{V} \tag{7.3.1}$$

where $\hbar$ is the reduced Plank constant, and θ is identified as the phase of the macroscopic quantum wave-function, $\psi=|\psi|e^{i\theta}$ in the macroscopic quantum wave model. Taking the time-derivative of Eq.(7.3.1) and using the modified first London equation (7.1.6), we may get

$$\boldsymbol{E} = \frac{\hbar}{e*}\nabla\left(\frac{\partial\theta}{\partial t}\right) - \frac{\partial \boldsymbol{A}}{\partial t} \tag{7.3.2}$$

which indicates that the phase θ of the macroscopic quantum wave-function ψ may be related to the electric potential function ϕ ($\boldsymbol{E} = -\partial\boldsymbol{A}/\partial t - \nabla\phi$) by the relation: $\phi = -(\hbar/e*)\partial\theta/\partial t$.

It can be shown that the gauge invariance of the superconducting current equation (7.3.1) may be fulfilled by the following gauge transformations:

$$\boldsymbol{A}' = \boldsymbol{A} + \nabla\chi\ , \qquad \phi' = \phi - \frac{\partial\chi}{\partial t}\ , \qquad \theta' = \theta + \frac{e*}{\hbar}\chi \tag{7.3.3}$$

Consider now a moving multiply connected superconductor in a weak magnetic field, as shown in Fig. 7.7. Integrating Eq.(7.3.1) along a closed contour Γ

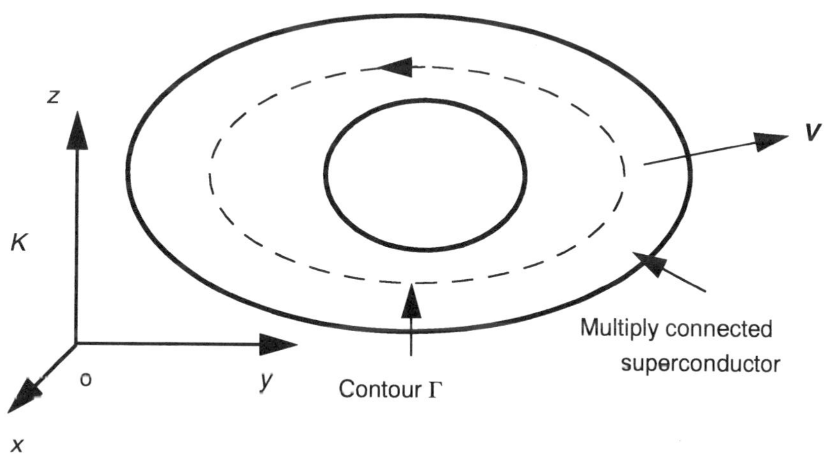

Figure 7.7 A multiply connected superconductor in motion with respect to a laboratory frame K.

within the superconductor, we may obtain

$$\Phi_F = \int_S \left(\boldsymbol{B} - \frac{2m^*}{|e^*|}\Omega\right) \cdot d\boldsymbol{S} + \oint_\Gamma \mu_o \lambda_L^2 \boldsymbol{J} \cdot d\Gamma = N\Phi_o \qquad (7.3.4)$$

since the wave-function $\psi = |\psi|e^{i\theta}$ is physically a single-valued function. Here, S is the surface defined by the contour Γ. Φ_0 is the magnetic flux quantum. N is an integer ($N = 0, \pm 1, \pm 2, \ldots$). In a simply connected superconducting region, we have $N = 0$. Equation (7.3.4) indicates the quantization condition for the fluxoid Φ_F through the loop surface S when the superconductor is moving. Here, we may notice that the fluxoid Φ_F has the same value for any loop around a given hole (or normal region).

In general, if we choose a closed contour Γ deep within, for instance, the thick wall of a hollow superconducting cylinder, where the superconducting current density is negligibly small, Eq.(7.3.4) may be reduced to the following form:

$$\Phi = \int_S \boldsymbol{B} \cdot d\boldsymbol{S} = N\Phi_o + \int_S \frac{2m^*}{|e^*|}\Omega \cdot d\boldsymbol{S} \qquad (7.3.5)$$

where Φ denotes the total magnetic flux through the loop surface S. For a stationary superconductor or a superconductor moving with uniform velocity (even with accelerated translational rigid-body motion), we have $\Omega = 0$, and therefore, recover the well-known result on the flux–quantization from Eq.(7.3.5). In general, we find that the total magnetic flux Φ through the loop surface S is quantized for any given loop Γ within a multiply connected superconductor moving with a given angular velocity Ω. The London moment due to the rotational motion of the superconductor contributes to the total magnetic flux Φ, as shown by the second term on the right-hand side of Eq.(7.3.5).

As we may see from Eq.(7.3.5), the total magnetic flux Φ through the loop surface S can be time-dependent since the London moment induced by the non-uniform motion of a deformable superconductor may vary with time. Besides, the total magnetic flux Φ is now not a single quantity associated with a hole, but depends on the loop chosen around the "hole". Illustratively, we may consider a case in which a superconducting ring is rotating with an angular velocity $\Omega = 10$ rev/s, and the radius of the loop Γ chosen is 1 mm. Thus the contribution of the London moment [the second term on the right-hand side of Eq.(7.3.5)] to the magnetic flux has a value of 2.2×10^{-15} Wb, which is about the same value of one single magnetic flux-quantum Φ_0. This result indicates that the mechanical motion, such as vibrations of a multiply connected superconductor, may affect the behavior of some superconducting electronic devices based on the single flux quantum, depending on both the size of the device and the (local) angular velocity of the mechanical motion of the device.

7.3.2 Formulation of Moving Deformable Josephson Junctions

It is known that, in a superconductor at a finite temperature ($T < T_c$), there is a dynamic equilibrium between the paired electrons and quasi-particles. Pairs are constantly breaking up via the absorption of thermal phonons of energy greater than or equal to $2\Delta(T)$, while quasi-particles are constantly recombining with one another to form pairs with the emission of phonons of energy $2\Delta(T)$ [Rothwarf and Cohen (1963), and Taylor (1968)]. Thus tunnel junctions may be used as generators and detectors of phonons of very high frequency (several hundred GHz) [Albeles and Goldstein (1965), Lax and Vernon (1965), and Eisenmenger and Dayem (1967)]. The classical way of detecting phonons by superconducting acoustic sensors is using the phenomenon of phonon–assisted tunneling, in which the incident phonon breaks up an electron pair into two quasiparticles, one becoming part of the normal fluid in the first superconductor and the other tunneling across the barrier and becoming part of the normal fluid in the second superconductor. By the bias voltage, the energy of the incident phonon can be less than $2\Delta(T)$ so that the minimum detectable frequency for a one-phonon process is given by $\hbar\omega = 2\Delta - |e|V$. However, practical considerations related to the sharpness of the I–V curve at $|e|V = 2\Delta(T)$ may limit the minimum detectable frequency. Obviously, the superconducting acoustic sensors based on the phenomenon of phonon-assisted tunneling are essentially normal electron tunneling devices.

In what follows, we shall study the possible phenomenon of the Josephson (electron) pair tunneling induced by some elastic wave propagating across a deformable Josephson junction, which may be utilized for novel superconducting acoustic sensors. We shall show how the electrodynamic equation of the Josephson junction is modified by the effect of local dynamic deformation of the elastic superconductors that form the junction. We shall then show how the Josephson pair tunneling current could be induced by transverse elastic wave with phonon energy $< 2\Delta(T)$ (or $< 2\Delta - |e|V$ in the presence of the bias voltage V).

For the Josephson junction that lies in the x-y plane, as shown in Fig. 7.8, we may find, by integrating Eq.(7.3.1) along the contour Γ (= $\Gamma_1+\Gamma_2+P_2P_1+Q_1Q_2$), the gauge-invariant relative pair phase φ as

$$\varphi(P) - \varphi(Q) = \frac{2\pi\Phi_\Gamma}{\Phi_o} - \frac{m^*}{\hbar}\oint_\Gamma \left(\frac{\partial \boldsymbol{u}}{\partial t} \cdot d\Gamma\right) \tag{7.3.6}$$

where Φ_Γ is the magnetic flux enclosed by the contour Γ, Φ_0 is the flux quantum, and $\boldsymbol{u}$ is the local elastic displacement vector of the elastic superconductor ($\boldsymbol{V} = \partial\boldsymbol{u}/\partial t$ at the linear approximation).

If we now let P and Q be close to each other, we may then arrive at

$$\frac{\partial\varphi}{\partial x} = \frac{|e^*|\delta}{\hbar}B_y - \frac{2m^*\delta}{\hbar}\Omega_y \tag{7.3.7}$$

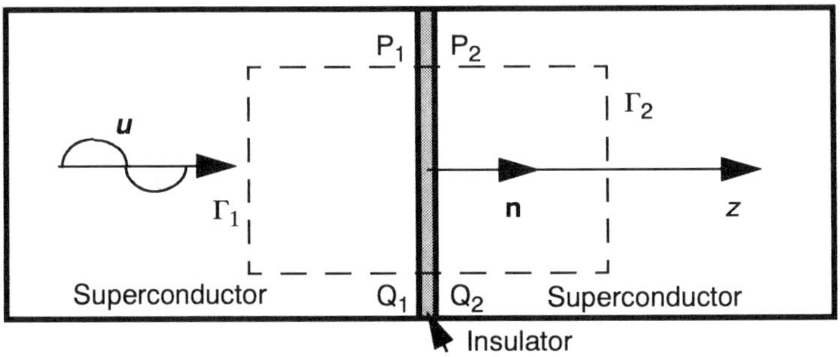

Figure 7.8 An elastic Josephson junction which is constructed by placing a thin insulator at the junction between two elastic superconductors.

$$\frac{\partial \varphi}{\partial y} = -\frac{|e^*|\delta}{\hbar}B_x + \frac{2m^*\delta}{\hbar}\Omega_x \tag{7.3.8}$$

where φ is the gauge-invariant relative pair phase. Here, the parameter δ is defined by $\delta = d+2\lambda_L$ with d being the thickness of the insulator and λ_L being the effective London penetration depth of the two superconductors forming the junction. Here, we have assumed that these superconductors are identical. $\boldsymbol{B}$ is the actual magnetic field in the plane of the junction including both externally applied magnetic field and the field induced by the current flowing in the junction. The vorticity vector Ω may be taken to be its value at the plane of the junction ($z = 0$) if the vorticity vector Ω is approximately uniform along z-axis for $|z| \leq \delta/2$ in the cases where the wavelength of elastic wave considered is much larger than δ. Otherwise, we may have

$$\Omega_x = \frac{1}{2\delta}\int_{-\delta/2}^{\delta/2}\left(\frac{\partial V_z}{\partial y} - \frac{\partial V_x}{\partial z}\right)dz \tag{7.3.9}$$

$$\Omega_y = \frac{1}{2\delta}\int_{-\delta/2}^{\delta/2}\left(\frac{\partial V_x}{\partial z} - \frac{\partial V_z}{\partial x}\right)dz \tag{7.3.10}$$

At the magneto-quasistatic approximation, the Josephson current-phase relation (5.1.7), and the Josephson voltage-phase relation (5.1.11) are supposed to be applicable to the deformable Josephson junction with the use of the gauge-invariant relative pair phase φ. If we now use the Maxwell equation of Ampere's circuital law and Eqs.(7.3.7) and (7.3.8), we may derive the following electrodynamic equation:

$$\frac{1}{\lambda_J^2}\sin\varphi = \frac{\partial^2\varphi}{\partial x^2} + \frac{\partial^2\varphi}{\partial y^2} - \frac{\beta}{\bar{c}^2}\frac{\partial\varphi}{\partial t} - \frac{1}{\bar{c}^2}\frac{\partial^2\varphi}{\partial t^2} - \frac{m^*\delta}{\hbar}\left(\frac{\partial^2 V_z}{\partial x^2} + \frac{\partial^2 V_z}{\partial y^2} - \frac{\partial^2 V_x}{\partial x\partial z} - \frac{\partial^2 V_y}{\partial y\partial z}\right) \tag{7.3.11}$$

with $\lambda_J = (\hbar/|e^*|\delta\mu_o J_o)^{1/2}$ for the deformable Josephson junction at the magneto-quasistatic approximation. The Josephson penetration depth λ_J gives a measure of the distance in which dc Josephson currents are confined at the edges of the junction. Here, we have taken into account of the effect of the normal electron tunneling current and the displacement current. The parameters introduced in Eq.(7.3.11) are, respectively, $\bar{c} = c(d/\varepsilon_r\delta)^{1/2}$ with c being the velocity of light in vacuum, ε_r being the effective relative dielectric constant of the insulator, $\beta = \sigma_o/\varepsilon$ with σ_o being the effective normal conductivity of the junction, and $\varepsilon = \varepsilon_r\varepsilon_o$ with ε_o being the permittivity of vacuum.

It is shown that the last term on the right-hand side of Eq.(7.3.11) characterizes the effect of local dynamic deformation of the Josephson junction. We may notice that this term will disappear if the vorticity vector Ω is a constant vector, as in the case of a rigid-body rotating Josephson junction. Equation (7.3.11) may be considered as a generalized form of the classical electrodynamic equation for stationary Josephson junctions shown by Eq.(5.1.27). In general, Eq.(7.3.11) has to be solved together with mechanical motion equations for the deformable Josephson junction. However, in most of Josephson´s junction applications, where the magnetic field is not large, we may ignore the effect of magnetic force on the deformation of the junction so that we could simplify the problem into an uncoupled electrodynamic-mechanical problem.

7.3.3 Rotating Superconducting Interferometer

To study the effect of mechanical motion of the Josephson junctions, let us first consider an example of a superconducting loop interrupted by two Josephson junctions in a magnetic field which is a constant field along z-axis direction, as shown in Fig. 7.9. The loop is supposed to lie at the x-y plane, and it is rotating rigidly in the plane with the vorticity vector $\Omega = (0, 0, \Omega)$. Such a device shown is known to be the superconducting interferometer. To study electrodynamic behavior of such an rotating superconducting interferometer, it is convenient to choose a coordinate system moving with the device. In this case, we may find that the gauge invariant phase difference across the junctions, φ_a and φ_b satisfy the following equation:

$$\varphi_a - \varphi_b = \frac{2\pi\Phi_\Gamma}{\Phi_o} - \frac{2m^*}{\hbar}\Omega A_\Gamma \tag{7.3.12}$$

where Φ_Γ is the magnetic flux enclosed by the loop Γ, and A_Γ is the area enclosed

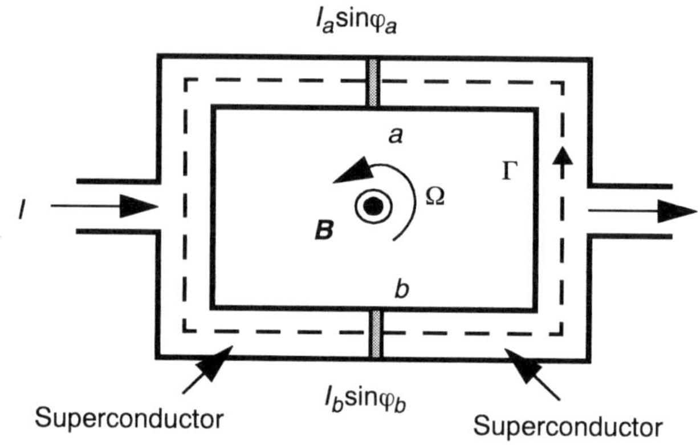

Figure 7.9 A rotating superconducting interferometer with double junctions.

by the loop Γ. The total supercurrent I flowing into the device is then given by

$$I = I_a \sin\varphi_a + I_b \sin\varphi_b \tag{7.3.13}$$

Assuming that these two junctions are identical ($I_a = I_b = I_o$), and the loop inductance is negligible, we may derive the following expression:

$$I = 2I_o \cos\left(\frac{\pi\Phi_\Gamma}{\Phi_o} - \frac{m^*}{\hbar}\Omega A_\Gamma\right) \sin\left(\varphi_b + \frac{\pi\Phi_\Gamma}{\Phi_o} - \frac{m^*}{\hbar}\Omega A_\Gamma\right) \tag{7.3.14}$$

from which we may find that the maximum supercurrent flowing through the device is given by

$$I_{max} = 2I_o \left|\cos\left(\frac{\pi\Phi_\Gamma}{\Phi_o} - \frac{m^*}{\hbar}\Omega A_\Gamma\right)\right| \tag{7.3.15}$$

This result may be of practical interest. For instance, a proposal was made to measure the Compton wavelength of superconducting electrons by measuring the maximum supercurrent across the junctions as a function of rotation rate [Zimmerman and Mercereau (1965)].

7.3.4 Effect of Acoustic Wave on the Josephson Junction

Next, let us study the effect of an elastic (acoustic) wave incident upon a deformable Josephson junction. For simplicity, we shall suppose that no external magnetic field is applied to the junction. We shall also assume that the effect of magnetic force caused by the self-induced magnetic field on possible deformation of the Josephson junction is negligible. Thus we are dealing with an uncoupled problem in which an elastic wave is prescribed. We assume now that the incident elastic plane wave can be described by the elastic displacement vector:

$$\boldsymbol{u}(z, t) = \boldsymbol{u}_o \sin\left[\frac{2\pi}{\lambda_e}(z - c_e t)\right] \tag{7.3.16}$$

where $\boldsymbol{u}_o$ denotes the amplitude vector of the elastic displacement, which is a constant vector for the plane wave. λ_e is the wavelength and c_e is the phase velocity of the elastic wave propagating along z-axis direction, which is perpendicular to the junction plane lying in the x-y plane (see Fig. 7.8). By Eq.(7.3.16), we may find from Eqs.(7.3.7) and (7.3.8) the relative pair phase φ at z = 0 by

$$\varphi = \frac{m^* \delta \omega^2}{\hbar c_e} \sin(\omega t) \times (u_{ox} x + u_{oy} y) + \alpha \tag{7.3.17}$$

where ω denotes the radian frequency of the elastic wave, and α is an integration constant. Here, we have ignored the possible effect of the thickness of the insulator on the behavior of the elastic wave by assuming perfect mechanical bonding between two superconductors that form the junction. This simplification may be justified for cases where the elastic wave frequency is not too high. For instance, the wavelength λ_e of an elastic wave of 1 GHz frequency is typically on the order of 1 μm, which is much larger than the thickness of the insulator (on the order of 1 nm) in the Josephson junction. Otherwise, for extremely high-frequency problems, details about mechanical interface conditions as well as the elastic properties of the insulator have to be taken into account. Here, for simplicity, we shall only consider the simplified case where these details may be ignored.

We may see, from Eqs.(7.3.16) and (7.3.17), that only the transverse elastic wave ($u_{oz} = 0$) contributes to the relative pair phase φ. The longitudinal elastic wave does not contribute to the relative pair phase φ and, therefore, has no effect on the Josephson pair tunneling current, though it may cause phonon-assisted normal electron tunneling if its phonon energy is sufficiently high, as we know.

By Eq.(7.3.17), we may find the Josephson pair tunneling current induced by the transverse elastic wave by

$$I = \int J_o \sin\varphi dS \tag{7.3.18}$$

where the integration is taken over the area of the junction. If we choose a special coordinate system in which we have $u_{ox} = 0$, we may find the maximum electron pair tunneling current by

$$I_{max}(t) = I_o \left| \frac{\sin[\pi\Phi_{eff}(t)/\Phi_o]}{\pi\Phi_{eff}(t)/\Phi_o} \right| \tag{7.3.19}$$

where the effective magnetic flux is given by

$$\Phi_{eff}(t) = \frac{2\delta W m^* \omega^2 u_{oy}}{|e^*| c_e} \sin(\omega t) \tag{7.3.20}$$

Here, Φ_0 is the flux quantum, and I_o is given by $I_o = J_o L W$ with L and W being the linear dimension of the junction area along the x-axis direction and the y-axis direction, respectively.

It is shown that a transverse elastic wave propagating across a deformable Josephson junction may induce an ac Josephson pair tunneling current due to the effect of inertia of superconducting electron pairs in the dynamically deformed Josephson junction with local non-uniform motion. Theoretically, such a tunneling pair current induced should exist for any acoustic frequency. Numerically, if we take some values of $c_e = 2$ km/s, $\delta = 0.5$ μm, W = 50 μm, and $u_{oy} = 1$ μm, we may find that the maximum value of the effective flux Φ_{eff} is about $0.68\times\Phi_0$ at frequency $\omega = 10^8$ rad/s, and $68\times\Phi_0$ at frequency $\omega=10^9$ rad/s. We may notice that the predicted phenomenon is essentially due to the effect of non-uniform (local or global) motion of the Josephson junction. The phenomenon will simply disappear if the Josephson junction is either in a static state, or in a uniform (translational) motion. It can be seen that, for transverse elastic wave incident upon the Josephson junction, the induced Josephson pair tunneling current may appear theoretically at any non-zero frequency, which is clearly different from the well-known phonon-assisted tunneling phenomenon, where a minimum phonon frequency exists.

By the Josephson voltage–phase relation, we may find that there is an ac voltage across the junction when the transverse elastic wave propagating across the Josephson junction. If we now consider a situation in which there is a dc (bias) voltage, V_o, applied to the junction, we have then

$$\varphi = \omega_o t + \frac{m^* \delta \omega^2 u_{oy} y}{\hbar c_e} \sin(\omega t) + \varphi_o \tag{7.3.21}$$

with $\omega_o = |e^*| V_o/\hbar$. We may notice that a similar result can be found in the well-known effect of external microwave (photon) radiation on the Josephson junction biased by a dc voltage [Shapiro (1963)], as discussed in Section 5.1.3, though there is a difference in the details of the frequency dependent behavior of the relative pair phase φ between these two situations.

By using the Josephson current–phase relation, we may find the Josephson pair tunneling current density induced by the transverse elastic wave by

$$J(t) = J_o \sin\left[\omega_o t + \frac{m^* \delta \omega^2 u_{oy} y}{\hbar c_e} \sin(\omega t) + \varphi_o\right] \tag{7.3.22}$$

This expression can also be written

$$J(t) = J_o[\sin(\omega_o t + \varphi_o)\cos(q\sin(\omega t)) + \cos(\omega_o t + \varphi_o)\sin(q\sin(\omega t))] \tag{7.3.23}$$

in which

$$q = \frac{m^* \delta \omega^2 u_{oy} y}{\hbar c_e} \tag{7.3.24}$$

Similar to the case discussed in Section 5.1.3, we may find that whenever the relation: $\omega_0 = 2k\omega$ ($k = \pm 1, \pm 2, \ldots$) is satisfied, a dc current component is present whose maximum amplitude of the kth step is given by

$$I_{dc}^{(2k)} = I_o \left| \int_0^1 J_{2k}(\Gamma\zeta) d\zeta \right| \tag{7.3.25}$$

where $J_{2k}(x)$ are Bessel functions of the first kind of integer order $2k$. The parameter Γ is given by $\Gamma = m^* \delta W u_{oy} \omega^2/(2\hbar C_e)$, which is a frequency dependent quantity. Obviously, the dc current component $I_{dc}^{(2k)}$ is frequency-dependent. Numerically, if we let $c_e = 2$ km/s, $\delta = 0.5$ μm, $W = 50$ μm, and $u_{oy} = 1$ μm, we have $\Gamma = 1.083$ for $\omega = 10^8$ rad/s, and $\Gamma = 108.3$ for $\omega = 10^9$ rad/s. In particular, for $k = 1$, we may find that the ratio of $I_{dc}^{(2)}/I_o$ is about 0.048 at the frequency of $\omega = 10^8$ rad/s. The dc current steps predicted here are quite similar to those from the Josephson junction irradiated with microwaves (photons), which were first observed by Shapiro (1963), as shown illustratively in Fig. 5.4. So far, it seems that experimental observation of the dc current steps or ac Josephson pair tunneling current induced by the transverse elastic wave incident upon the Josephson junction has not yet been made. Further effort seems to be desirable to verify experimentally the phenomenon predicted here since it may be of not only theoretical interest in understanding the electrodynamic behavior of deformable Josephson junctions, but also practical interest for possible applications of superconducting acoustic sensor technology [Kuttruff (1991)].

7.4 MAGNETOELASTIC THEORY OF TYPE II SUPERCONDUCTORS IN THE MIXED STATE

In this section, we shall present a macroscopic theory for analyzing magnetoelastic response of elastic type II superconductors in the mixed state. The theory includes not only effects of vortex dynamics and the effect of normal current, but also the effect of the London moment induced by the local motion of the deformable superconductor in the mixed state. The theory is considered to be valid within the framework of the generalized Galilean relativity at the magneto-quasistatic approximation. By this theory, a set of linearized coupled wave equations are formulated to study some problems concerning magnetoelastic wave propagation in type II superconductors in the mixed state. Attenuation and dispersion behaviors of the magnetoelastic wave are then analyzed. It is shown that the effect of the London moment induced by dynamic deformation can be of significance as compared with the effect of the normal Lorentz current on the magnetoelastic coupling behavior of elastic type II superconductors in the mixed state. Furthermore, a phase change between a transverse elastic wave and the induced magnetic wave propagating in the elastic type II superconductor in the mixed state is predicted to exist according to the theory, which is found to be closely related to the vortex dynamic properties of the superconductor.

7.4.1 Formulation of Elastic Type II Superconductors in the Mixed State

During the past, much effort has been made both theoretically and experimentally to study the dynamic behavior of vortex structures in type II superconductors in the mixed state due to its importance in understanding macroscopic electrodynamic properties of the type II superconductors. Recently, Coffey and Clem (1991) have developed a theoretical model, using a self-consistent approach to vortex dynamics, to describe electrodynamic responses of the type II superconductors in the mixed state. In their model, a continuum approximation of the London equation with a vortex term is used and the effect of quasi-particle excitations is included by a normal current density contribution. Their model is thought to be applicable to wide range of electrodynamic phenomena involving vortex dynamics and over a wide range of frequencies. However, the Coffey-Clem model can not be simply used to analyze electromagneto-mechanical behavior of deformable type II superconductors in the mixed state, where the effect of the motion of the superconductor has to be properly taken into account. In this section, we shall formulate the magnetoelastic response of elastic type II superconductors in the mixed state, in which both the effect of the London moment induced by the local motion of the elastic type II superconductor in the mixed state and the effect of the Lorentz body force on the magnetoelastic behavior of the elastic type II superconductor are taken into account.

It is known that, at the microscopic level, a flux line consists of superelectrons moving with a certain density and velocity distribution around the center of the flux line. This means that forces on a flux line will actually be experienced by the electrons, and can therefore be caused by electric and magnetic fields. At the macroscopic level, we are dealing with phenomena that can be described in terms of macroscopic (average) fields and currents. In the continuum theory, we may introduce phenomenologically the following equation of motion for the flux-line lattice continuum:

$$m^{(v)}\frac{\partial^2 \boldsymbol{w}}{\partial t'^2} = -\eta\frac{\partial \boldsymbol{w}}{\partial t'} - \kappa \boldsymbol{w} + \boldsymbol{J}' \times \boldsymbol{B}'^{(v)} \tag{7.4.1}$$

in a reference frame $K'(\mathbf{x}', t')$ attached to the material medium. Here, $m^{(v)}$ denotes the effective mass density associated with the flux-line lattice. The vector $\boldsymbol{w}$ denotes the vortex displacement vector, measured from an equilibrium pinning site in the medium in the reference frame K'. η is the flux-flow viscosity for the isotropic superconductor. The current density vector $\boldsymbol{J}'$, and the vortex magnetic field $\boldsymbol{B}'^{(v)}$ are averaged quantities over microscopic fields and current distributions in the superconductor. Here, we have adopted a simplified model in which, for the case of local vibration of flux lines without global flux-flow motion, the effect of the pinning force and the deformation of the flux-line lattice is modelled by a simple restoring force of the form $\boldsymbol{f}^{rest} = -\kappa \boldsymbol{w}$ for the isotropic type II superconductor, where κ denotes the spring coefficient. Besides, we have ignored here possible thermomagnetoelectric effects (see Section 4.6) by assuming that the superconductor is at a constant temperature for some simplicity.

When the flux-line lattice moves in the reference frame K', the vortex magnetic field $\boldsymbol{B}'^{(v)}$ and the vortex electric field $\boldsymbol{E}'^{(v)}$ have to obey a continuity equation [Goodman (1966)]:

$$\nabla' \times \boldsymbol{E}'^{(v)} = -\frac{\partial \boldsymbol{B}'^{(v)}}{\partial t'} \tag{7.4.2}$$

In many cases, electrodynamic phenomena of superconductors can be analyzed with the aid of knowledge gained from studying the behavior of the superconductor in time-harmonic fields ($\sim e^{i\omega t}$). In the time-harmonic fields, we may get from Eq.(7.4.1) the following relation:

$$w_k = \frac{1}{i\omega\sigma_v (B'^{(v)})^2} e_{kpq} J_p' B'^{(v)}_q \tag{7.4.3}$$

in which ω is the radian frequency, and e_{ijk} is the permutation symbol. $\sigma_V = \sigma_{V1} - i\sigma_{V2}$ is the complex flux-flow conductivity, defined by

$$\sigma_{V1} = \frac{\eta}{(B'^{(v)})^2} \quad \text{and} \quad \sigma_{V2} = \frac{\kappa - m^{(v)}\omega^2}{\omega (B'^{(v)})^2} \tag{7.4.4}$$

If the direction of the vortex magnetic field $\boldsymbol{B}'^{(v)}$ is supposed to be along the z'-axis, we may find the following expressions for the electric current density components:

$$J'_{x'} = \sigma_V E'^{(v)}_{x'} \tag{7.4.5}$$

$$J'_{y'} = \sigma_V E'^{(v)}_{y'} \tag{7.4.6}$$

The current density component along z'-axis is zero, implying that the local current density vector $\boldsymbol{J}'$ is perpendicular to the direction of the local vortex magnetic field $\boldsymbol{B}'^{(v)}$ according to the model.

Let us now discuss the electrodynamic and mechanical field equations for the study of elastic type II superconductors in the mixed state in the laboratory frame of reference, $K(\mathbf{x}, t)$. To model the electrodynamics of the type II superconductor in the mixed state, we use the following Maxwell equations at the magneto-quasistatic approximation:

$$\nabla \times \boldsymbol{H} = \boldsymbol{J} \quad \text{and} \quad \nabla \times \boldsymbol{E} = -\frac{\partial \boldsymbol{B}}{\partial t} \tag{7.4.7}$$

and the modified second London equation

$$\mu_o \lambda_L^2 \nabla \times \boldsymbol{J}_s - \frac{m^*}{|e^*|} \nabla \times \boldsymbol{V} = -(\boldsymbol{B} - \boldsymbol{B}^{(v)}) \tag{7.4.8}$$

where $\boldsymbol{J}_s$ is the supercurrent density, $\boldsymbol{V}$ is the local velocity of the superconductor ($\boldsymbol{V} = \partial \boldsymbol{U}/\partial t$ with $\boldsymbol{U}$ being the elastic displacement vector of the superconducting medium), $\boldsymbol{B}$ is the magnetic induction field, and $\boldsymbol{B}^{(v)}$ is the local vortex magnetic field which is, in general, not equal to the magnetic induction field $\boldsymbol{B}$ due to the nonlocal effect on the length scale of λ_L and to the presence of the normal conducting fluid. It can be seen from Eq.(7.4.8) that the Coffey-Clem formulation (1991) has been modified here by the introduction of a new term, the second term on the left-hand side of Eq.(7.4.8). The inclusion of this new term is due to the effect of the London moment induced by the local motion (dynamic deformation) of the elastic superconductor, as discussed in above sections. Here, the London approximation has been used to model electrodynamic phenomena in type II superconductors in the mixed state. Thus we may expect that the model is appropriate for the field region $B_{c1} << B << B_{c2}$. Such an intermediate field region is of interest, especially for recently discovered high-temperature oxide superconductors due to their small lower critical field and large upper critical field.

In addition, to account for the effect of normal conduction currents, we may use the classical two-fluid model, which can be expressed by

$$\boldsymbol{J} = \boldsymbol{J}_s + \boldsymbol{J}_n \tag{7.4.9}$$

where $\boldsymbol{J}_n$ denotes the normal conduction current density, given by

$$\boldsymbol{J}_n = \sigma_n(\boldsymbol{E} + \boldsymbol{V} \times \boldsymbol{B}) \tag{7.4.10}$$

with σ_n being the local electric conductivity of the normal fluid.

We now consider the equation of motion for the elastic type II superconductor in the presence of magnetic field, which may be expressed by

$$\rho\frac{\partial^2 \boldsymbol{U}}{\partial t^2} = \nabla \cdot \boldsymbol{t} + \frac{1}{\mu_o}\left[(\boldsymbol{B} \cdot \nabla)\boldsymbol{B} - \frac{1}{2}\nabla(\boldsymbol{B} \cdot \boldsymbol{B})\right] \tag{7.4.11}$$

where ρ is the mass density of the superconductor, $\boldsymbol{U}$ the elastic displacement vector, and $\boldsymbol{t}$ the Cauchy stress tensor, which, at the linear approximation, may be expressed by Hooke's law $t_{ij} = C_{ijkl}U_{k,l}$ with C_{ijkl} being the elastic modulus tensor (see Chapter 2). For isotropic elastic superconductors, the elastic modulus tensor can be expressed by

$$C_{ijkl} = \left(K - \frac{2}{3}G\right)\delta_{ij}\delta_{kl} + G(\delta_{ik}\delta_{jl} + \delta_{il}\delta_{jk}) \tag{7.4.12}$$

where K is the elastic bulk modulus and G is the elastic shear modulus of the superconductor. We may note that for the type II superconductor in the mixed state, the elastic moduli C_{ijkl} may differ from their values in the normal state due to possible anomalies arising from the superconducting phase transition and from the effect of magnetic flux structures in the superconductor in the mixed state [see Zhou (1991a)]. The second term on the right-hand side of Eq.(7.4.11) comes from the Lorentz body force.

It is shown that the given set of self-consistent field equations for analyzing magnetoelastic behavior of the elastic type II superconductors in the mixed state are, in general, nonlinear and coupled. These field equations should be supplemented with appropriate boundary conditions and initial conditions in order to solve some mixed boundary-value, evolution problems. In what follows, two illustrative examples will be given to study some problems of wave propagation in elastic type II superconductors in the mixed state.

7.4.2 Magnetoelastic Plane Wave in Type II Superconductors

As the first example, let us study the problem of magnetoelastic wave propagating in an elastic type II superconductor in the mixed state. Here, we are particularly interested in the wave problem in which the superconductor is initially in a static magnetic field $\boldsymbol{B}_0$ and it may have the static elastic displacement $\boldsymbol{U}_0$, which can be caused by some static mechanical force and/or the static magnetic force. When the type II superconductor is in the mixed state, magnetoelastic waves may be induced by a small perturbed time-varying loading, which can be of electromagnetic and/or mechanical origin. In such a case, we may write the total magnetic induction field by $\boldsymbol{B} = \boldsymbol{B}_o + \boldsymbol{b}$, and the total elastic displacement by $\boldsymbol{U} = \boldsymbol{U}_o + \boldsymbol{u}$ in the superconductor. Here, $\boldsymbol{b}$ and $\boldsymbol{u}$ are, respectively, the small perturbed time-varying magnetic field and elastic displacement field to be determined. At the first-order approximation, we assume that all material properties concerned are independent of the small perturbed fields $\boldsymbol{b}$ and $\boldsymbol{u}$, and are determined only in those static fields at a constant and uniform temperature.

By using Eqs.(7.4.7)–(7.4.10), we may derive the following equation:

$$\nabla^2 \boldsymbol{B} = \mu_o \sigma_n \frac{\partial \boldsymbol{B}}{\partial t} + \frac{1}{\lambda_L^2}(\boldsymbol{B} - \boldsymbol{B}^{(v)}) - \frac{m^*}{|e^*|\lambda_L^2} \nabla \times \boldsymbol{V} - \mu_o \sigma_n \nabla \times (\boldsymbol{V} \times \boldsymbol{B}) \quad (7.4.13)$$

from which we may see that the third term and the fourth term on the right-hand side of this equation are caused by the local motion of the elastic superconductor. They represent, respectively, the effect of the London moment induced by dynamic deformation and the effect of the normal Lorentz current. A rough estimation on the relative importance of these two terms can be made by choosing some typical values of $\sigma_n = 5 \times 10^6/\Omega\cdot\text{m}$, $\lambda_L = 1\ \mu\text{m}$, and $B = 1$ T, which shows that the effect of the London moment is, at least, of the same order of importance as compared with the effect of the normal Lorentz current. If the London penetration depth λ_L is smaller, for instance, $\lambda_L = 0.1\ \mu\text{m}$ for some high-T_c oxide superconductors (and/or σ_n is smaller) at low temperature $T << T_c$, the effect of the London moment may become even more important.

Now, let us study further Eq.(7.4.13). By integrating the vortex continuity equation (7.4.2) with respect to time, we may find

$$\boldsymbol{B}^{(v)} = \boldsymbol{B}_o + \boldsymbol{b}^{(v)} = \boldsymbol{B}_o - \nabla \times [\boldsymbol{B}_o \times (\boldsymbol{w} + \boldsymbol{u})] \quad (7.4.14)$$

at the linear and magneto-quasistatic approximation. Thus, by Eq.(7.4.3), where $\boldsymbol{B}'^{(v)}$ is replaced by $\boldsymbol{B}_0$ and $\boldsymbol{J}'$ is replaced by $\boldsymbol{J}$ at the linear and magneto-quasistatic approximation, we may derive, after some manipulations, at the following equation in time-harmonic fields:

$$\tilde{\lambda}^2 \nabla^2 \boldsymbol{b} = \boldsymbol{b} - \nabla \times (\boldsymbol{u} \times \boldsymbol{B}_o) + \frac{\frac{i}{2}\delta_{VC}^2 \nabla \times [(\nabla \times \boldsymbol{b}) \cdot \boldsymbol{n}]\boldsymbol{n} - \frac{i\omega m^*}{|e^*|} \nabla \times \boldsymbol{u}}{1 + 2i\lambda_L^2/\delta_N^2} \quad (7.4.15)$$

where $\tilde{\lambda}$ is the complex penetration depth defined by [Coffey and Clem (1991)]

$$\tilde{\lambda}^2(\omega, B_o, T) \equiv \frac{\lambda_L^2 - i\delta_{VC}^2/2}{1 + 2i\lambda_L^2/\delta_N^2} \quad (7.4.16)$$

and $\delta_N = (2/\mu_o\omega\sigma_n)^{1/2}$ is the normal skin depth, and $\delta_{VC} = (2/\mu_o\omega\sigma_V)^{1/2}$ is the complex skin depth. In Eq.(7.4.15), $\boldsymbol{n}$ is the unit direction vector of $\boldsymbol{B}_o$, that is, $\boldsymbol{n} = \boldsymbol{B}_o/B_o$. Here, $\boldsymbol{B}_o$ is assumed to be a constant and uniform magnetic field. It is shown that the appearance of the second and third terms on the right-hand side of Eq.(7.4.15) is due partly to the effect of current flow parallel to the magnetic field $\boldsymbol{B}_o$ and partly to the effect of the local mechanical motion (dynamic deformation) of the elastic type II superconductor in the mixed state.

Furthermore, from Eq.(7.4.11), we may get

$$-\rho\omega^2\boldsymbol{u} = \left(K+\frac{G}{3}\right)\nabla(\nabla\cdot\boldsymbol{u}) + G\nabla^2\boldsymbol{u} + \frac{1}{\mu_o}[\boldsymbol{B}_o\cdot\nabla\boldsymbol{b} - \nabla(\boldsymbol{B}_o\cdot\boldsymbol{b})] \qquad (7.4.17)$$

Equations (7.4.15) and (7.4.17) constitute a set of coupled linearized field equations for the determination of the perturbed magnetic field $\boldsymbol{b}(\mathbf{x}, t)$ and the perturbed elastic displacement field $\boldsymbol{u}(\mathbf{x}, t)$ in the elastic type II superconductor in the mixed state.

We now look at the following type of magnetoelastic plane wave by

$$u_x = u_o exp[i(\omega t - kx)] \qquad (7.4.18)$$

$$b_z = b_o exp[i(\omega t - kx)] \qquad (7.4.19)$$

where u_o and b_o are constants relevant to the wave amplitudes of the elastic displacement and of the magnetic field, respectively. Other field components are $u_y = u_z = 0$ and $b_x = b_y = 0$. ω is the radian frequency of the wave, and k is the propagation constant which is, in general, a complex quantity, $k = k_o - i\alpha$ with α being the attenuation coefficient. Here, we have chosen z-axis being along the direction of the magnetic field $\boldsymbol{B}_o$. Equations (7.4.18) and (7.4.19) describe a type of magnetoelastic wave being longitudinal in the elastic displacement $\boldsymbol{u}$ but transverse in the magnetic field $\boldsymbol{b}$. This wave propagates along the x-axis direction, being perpendicular to the direction of the magnetic field $\boldsymbol{B}_o$.

Substitution of Eqs.(7.4.18) and (7.4.19) into Eqs.(7.4.15) and (7.4.17) gives the following set of equations:

$$(\omega^2 - k^2c_L^2)u_o + i\frac{kB_o}{\mu_o\rho}b_o = 0 \qquad (7.4.20)$$

$$ikB_ou_o - (1 + k^2\tilde{\lambda}^2)b_o = 0 \qquad (7.4.21)$$

for the determination of u_o and b_o. Here, $c_L = [(K+4G/3)/\rho]^{1/2}$ is the phase velocity of the longitudinal elastic wave. The condition for the existence of a non-trivial solution of Eqs.(7.4.20)–(7.4.21) gives the following dispersion equation:

$$(\omega^2 - k^2 c_L^2)(1 + k^2\tilde{\lambda}^2) = k^2 c_A^2 \tag{7.4.22}$$

where $c_A = B_o/(\rho\mu_o)^{1/2}$ is the Alfvén wave velocity, which is a measure of the relative importance of the magnetic effect in comparison with mechanical ones insofar as wave propagation is concerned. It is noted that the wavelength of the magnetoelastic wave considered is usually much longer than 1 μm in most of practical cases where the wave frequency is less than 1 GHz. The continuum model presented here may work well. For higher frequency applications, the model may work in some cases if, however, the effect of microstructures of the material could be taken into account effectively.

Two special cases may be of interest. First, if we let $\lambda_L \to \infty$ mathematically, by noting Eq.(7.4.16), we may find that Eq.(7.4.22) recovers its classical form for a normal elastic conductor. Secondly, if we let $\lambda_L \to 0$, we may find that Eq.(7.4.22) recovers the result obtained by Shapira and Neuringer (1967) as long as the longitudinal elastic wave is concerned. In the general case, we may introduce an effective complex conductivity, $\sigma_{eff} = \sigma_{e1} - i\sigma_{e2}$, defined by

$$\sigma_{eff} \equiv \frac{1}{i\mu_o\omega\tilde{\lambda}^2} \tag{7.4.23}$$

from which, by noting Eq.(7.4.16), we may find after some manipulations:

$$\sigma_{e1} = \frac{(\sigma_{V1} + \mu_o\omega\sigma_n\sigma_{V2}\lambda_L^2)(1 + \mu_o\omega\sigma_{V2}\lambda_L^2)}{(1 + \mu_o\omega\sigma_{V2}\lambda_L^2)^2 + (\mu_o\omega\sigma_{V1}\lambda_L^2)^2} - \frac{\mu_o\omega\sigma_{V1}\lambda_L^2(\sigma_{V2} - \mu_o\omega\sigma_n\sigma_{V1}\lambda_L^2)}{(1 + \mu_o\omega\sigma_{V2}\lambda_L^2)^2 + (\mu_o\omega\sigma_{V1}\lambda_L^2)^2} \tag{7.4.24}$$

and

$$\sigma_{e2} = \frac{(\sigma_{V2} - \mu_o\omega\sigma_n\sigma_{V1}\lambda_L^2)(1 + \mu_o\omega\sigma_{V2}\lambda_L^2)}{(1 + \mu_o\omega\sigma_{V2}\lambda_L^2)^2 + (\mu_o\omega\sigma_{V1}\lambda_L^2)^2} + \frac{\mu_o\omega\sigma_{V1}\lambda_L^2(\sigma_{V1} + \mu_o\omega\sigma_n\sigma_{V2}\lambda_L^2)}{(1 + \mu_o\omega\sigma_{V2}\lambda_L^2)^2 + (\mu_o\omega\sigma_{V1}\lambda_L^2)^2} \tag{7.4.25}$$

Mathematically, for $\lambda_L \to 0$, we have $\sigma_{e1} = \sigma_{V1}$ and $\sigma_{e2} = \sigma_{V2}$, and for $\lambda_L \to \infty$, we have $\sigma_{e1} = \sigma_n$ and $\sigma_{e2} = 0$ as expected. If the effect of the normal conduction fluid is negligible in some cases, we may set $\sigma_n = 0$ mathematically, which results in the following expressions:

$$\sigma_{e1} = \frac{\sigma_{V1}}{(1 + \mu_o\omega\sigma_{V2}\lambda_L^2)^2 + (\mu_o\omega\sigma_{V1}\lambda_L^2)^2} \tag{7.4.26}$$

$$\sigma_{e2} = \frac{\sigma_{V2} + \mu_o\omega(\sigma_{V1}^2 + \sigma_{V2}^2)\lambda_L^2}{(1 + \mu_o\omega\sigma_{V2}\lambda_L^2)^2 + (\mu_o\omega\sigma_{V1}\lambda_L^2)^2} \tag{7.4.27}$$

It is shown that the nonlocal effect on the length scale of λ_L leads to effectively the reduction of the flux-flow conductivity of the superconductor in the mixed state.

To solve the dispersion equation (7.4.22), for some simplicity, we make the approximation by replacing the term k^2 which appears in the second parenthesizes on the left-hand side of Eq.(7.4.22) by its zero-field value $(\omega/c_L)^2$, which may be justified by noting the fact that the effect of the magnetic field modifies only slightly the phase velocity of the elastic wave. With this approximation, we may obtain the attenuation coefficient α by

$$\alpha = \frac{\omega}{c_L}\sqrt{\frac{a}{2}[\sqrt{1 + (b/a)^2} - 1]} \tag{7.4.28}$$

and the phase velocity c_{me} of the magnetoelastic wave by

$$c_{me} = \frac{\omega}{k_o} = c_L\left\{\frac{a}{2}[\sqrt{1 + (b/a)^2} + 1]\right\}^{-1/2} \tag{7.4.29}$$

where the parameters a and b are given by

$$a = \frac{1 + (1 + c_A^2/c_L^2)(\sigma_{e1}^2/\sigma_o^2 + \sigma_{e2}^2/\sigma_o^2) + (\sigma_{e2}/\sigma_o)(2 + c_A^2/c_L^2)}{1 + (1 + c_A^2/c_L^2)^2(\sigma_{e1}^2/\sigma_o^2 + \sigma_{e2}^2/\sigma_o^2) + (2\sigma_{e2}/\sigma_o)(1 + c_A^2/c_L^2)} \tag{7.4.30}$$

and

$$b = \frac{(c_A^2/c_L^2)(\sigma_{e1}/\sigma_o)}{1 + (1 + c_A^2/c_L^2)^2(\sigma_{e1}^2/\sigma_o^2 + \sigma_{e2}^2/\sigma_o^2) + (2\sigma_{e2}/\sigma_o)(1 + c_A^2/c_L^2)} \tag{7.4.31}$$

where σ_0 is defined by $\sigma_0 = \omega/(\mu_0 c_L{}^2)$.

It may be noticed that Shapira and Neuringer's result (1967) has been modified here by accounting for the nonlocal effect on the length scale of λ_L and the effect of normal conduction current, which may be seen from the complex conductivity given by Eqs.(7.4.24) and (7.4.25). In this example, we may also notice that the effect of the London moment is not shown up since the wave is longitudinal in the elastic displacement. Numerically, some results are shown in Figs. 7.10 and 7.11, where material parameters are chosen to be illustrative rather than to describe any particular material sample.

Figure 7.10 shows the variation of attenuation coefficient α with respect to the normalized frequency ω/ω_0, calculated from Eq.(7.4.28). Here, ω_0 is the

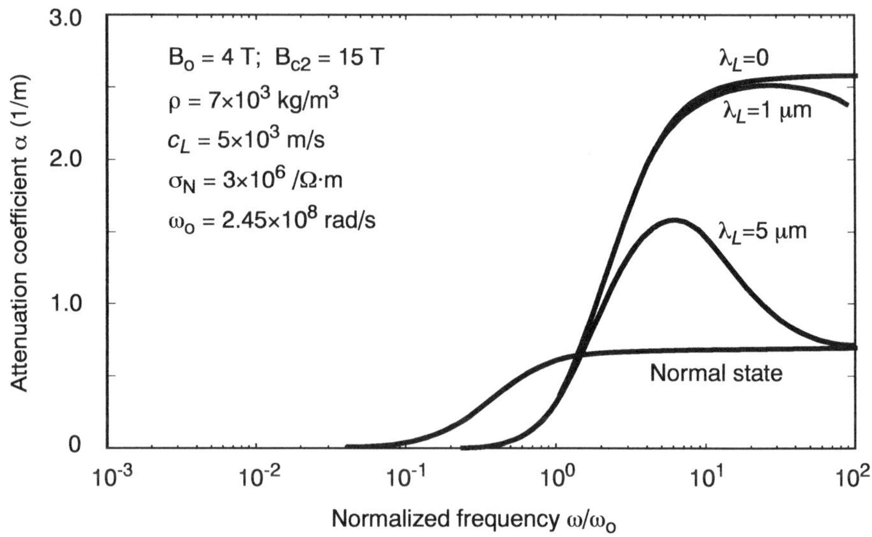

Figure 7.10 Variation of the attenuation coefficient α with normalized frequency ω/ω_o.

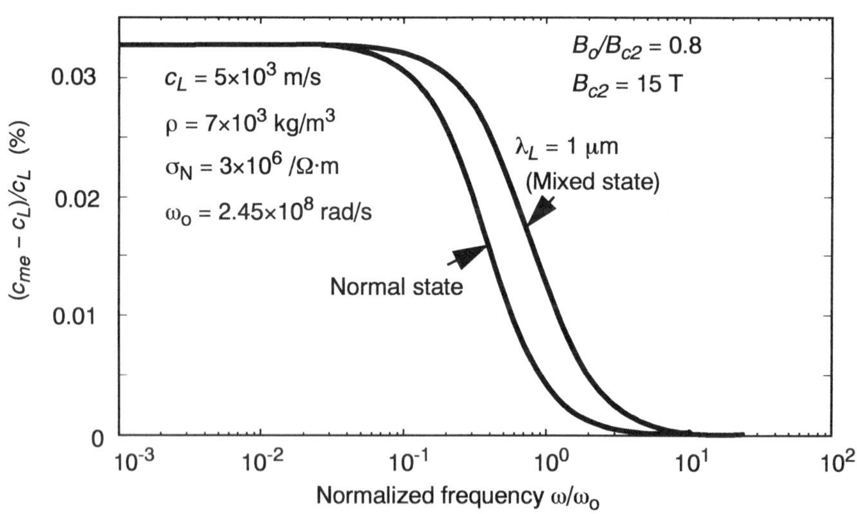

Figure 7.11 Dispersion behavior of the magnetoelastic wave.

depinning frequency, defined by the relation $\sigma_{V2}/\sigma_{V1} = \omega_o/\omega$. If the Bardeen-Stephen model (1965) and the Gittleman-Rosenblum model (1966) are used, ω_o is expressed by Eq.(4.5.16). Intuitively, we might argue that ω_o should not increase with increasing the magnetic field B because the surface resistance should not decrease for increasing magnetic field B. If this argument, which has yet to be verified experimentally, is correct, we may expect that the field dependence of the dc critical current density J_c should be of the form $J_c \propto B^{-\gamma}$ with $\gamma \geq 1/2$ if the model is proper. Indeed, Kim et al. (1962) has proposed a model in which one has $J_c \propto (B+B_1)^{-1}$ with B_1 being a small constant. The Bean model (1962) in which J_c is supposed to be independent of the field B (i.e., $\gamma = 0$) seems to not obey the criterion of $\gamma \geq 1/2$. However, if the operating frequencies of concrete problems are far from (e.g., much lower than) the depinning frequency, the Bean model may still work because the result will not be sensitive to the exact value of the depinning frequency. Other forms of the field dependence of the dc critical current density were also proposed in the past, which are, for instance, $J_c \propto B^{-1/2}$ [Yasukochi et al. (1964)], and $J_c \propto B^{-1/2}(\mu_o H_{c2} - B)$ [Campbell et al. (1968)].

It can be seen from Fig. 7.10 that the attenuation of the magnetoelastic wave in the type II superconductor in the mixed state is smaller than that in the normal state when the wave frequency ω is less than the depinning frequency ω_o. However, for wave frequencies larger than the depinning frequency, the attenuation of the magnetoelastic wave in the type II superconductor in the mixed state can be larger than that in the normal state. In particular, Fig. 7.10 shows the effect of the London penetration depth λ_L. The Shapira-Neuringer model corresponds to the case of $\lambda_L = 0$. It can be seen that, when the effect of the London penetration depth is taken into account, the result predicted by Eq.(7.4.28) coincides with the result from the Shapira-Neuringer model at frequencies near or less than the depinning frequency ω_o, but the results shows their difference at higher frequencies. At very high frequencies much larger than the depinning frequency, Eq.(7.4.28) shows that the attenuation of the magnetoelastic wave approaches to its normal-state value for non-zero London penetration depth, which is physically reasonable. It would be valuable that one could verify experimentally the high-frequency behavior of attenuation of the magnetoelastic wave in the type II superconductor in the mixed state. So far, it seems that no such experimental results have been reported due probably to the fact that high-frequency (GHz) ultrasonic devices are not easily obtainable. Recent progress in researches on hypersound (or quantum acoustics) may help to improve the situation [Kuttruff (1991)].

In Fig. 7.11, the dispersion behavior of the magnetoelastic wave is shown. It can be seen that the strong dispersion of the magnetoelastic wave in the elastic superconductor in the mixed state appears at frequencies near the depinning frequency ω_o. When the superconductor is in the normal state, the wave becomes strongly dispersive at frequencies lower than that in the mixed state. We may note that, in the above example, the effect of the London moment is not shown up. This is because the wave considered is longitudinal in the elastic displacement. In the following section, we shall study the effect of the London moment by considering transverse waves in the type II superconductor in the mixed state.

7.4.3 Phase Shift Between Propagating Elastic Wave and Induced Magnetic Wave

As the second example, we consider the problem of a magnetic plane wave induced by a transverse elastic plane wave propagating in an elastic type II superconductor in the mixed state. We assume that the elastic type II superconductor is located in a static and uniform magnetic field $\boldsymbol{B}_0$ and at a uniform temperature $T << T_c$ so that the effect of the normal conduction current due to thermal excitation is negligible. The static magnetic field $\boldsymbol{B}_0$ is supposed to be along the z-axis direction, and the elastic type II superconductor is in the mixed state. The superconductor is subject to an elastic disturbance, described by a transverse elastic plane wave of the following form:

$$u_y = u_o exp\left[i\omega\left(t - \frac{x}{c_T}\right)\right] \tag{7.4.32}$$

with $u_x = u_z = 0$. Here, u_0 (> 0) denotes the amplitude of the elastic plane wave, and $c_T = (G/\rho)^{1/2}$ is the speed of the transverse elastic plane wave. We assume that the effect of the time-varying magnetic force on the behavior of the elastic wave in the case considered is small and negligible.

To study the electrodynamic response of the superconductor by such a given mechanical disturbance, we look for the following type of magnetic wave:

$$b_z = b_o exp\left[i\omega\left(t - \frac{x}{c_T}\right)\right] \tag{7.4.33}$$

with $b_x = b_y = 0$. By substituting Eqs.(7.4.32) and (7.4.33) into Eq.(7.4.15) in the absence of normal conduction currents, we may obtain, after some manipulations, the following result:

$$b_o = |b_o| e^{i\gamma} \tag{7.4.34}$$

where the amplitude of the magnetic wave, $|b_0|$ is expressed by

$$|b_o| = \frac{m^* \omega^2 u_o}{|e^*| c_T} \sqrt{a_1^2 + a_2^2} \tag{7.4.35}$$

and the phase factor γ by

$$\gamma = tan^{-1}\left(\frac{a_2}{a_1}\right) \tag{7.4.36}$$

Here, the two parameters a_1 and a_2 are given, respectively, by

$$a_1 = \frac{1 + (\omega/c_T)^2[\lambda_L^2 + \sigma_{V2}/(\mu_o\omega(\sigma_{V1}^2 + \sigma_{V2}^2))]}{\left\{1 + \left(\frac{\omega}{c_T}\right)^2\left[\lambda_L^2 + \frac{\sigma_{V2}}{\mu_o\omega(\sigma_{V1}^2 + \sigma_{V2}^2)}\right]\right\}^2 + \frac{\omega^2\sigma_{V1}^2}{c_T^4\mu_o^2(\sigma_{V1}^2 + \sigma_{V2}^2)^2}} \tag{7.4.37}$$

$$a_2 = \frac{\omega\sigma_{V1}/[c_T^2\mu_o(\sigma_{V1}^2 + \sigma_{V2}^2)]}{\left\{1 + \left(\frac{\omega}{c_T}\right)^2\left[\lambda_L^2 + \frac{\sigma_{V2}}{\mu_o\omega(\sigma_{V1}^2 + \sigma_{V2}^2)}\right]\right\}^2 + \frac{\omega^2\sigma_{V1}^2}{c_T^4\mu_o^2(\sigma_{V1}^2 + \sigma_{V2}^2)^2}} \tag{7.4.38}$$

The result shows that the transverse elastic plane wave propagating in the elastic type II superconductor in the mixed state may induce a transverse magnetic plane wave with the same phase velocity propagating in the superconductor due to the effect of the London moment induced by the local motion of the elastic type II superconductor. Furthermore, the result shows that there is a phase change between the elastic plane wave and the magnetic plane wave induced, characterized by the phase factor γ. This phase factor γ is found to be dependent of the material properties of the elastic type II superconductor in the mixed state. It is also frequency dependent. Mathematically, if we let σ_{V1}, $\sigma_{V2} \rightarrow \infty$ ($B_o \rightarrow 0$), which corresponds to the Meissner state, we may find that γ becomes zero, and Eq.(7.4.35) recovers the result [see Eq.(7.2.8)] derived earlier in Section 7.2.1 for elastic superconductors in the Meissner state, where there is no such a phase change between the transverse elastic plane wave and the transverse magnetic plane wave induced. This indicates that the effect of the phase change is closely related to the vortex dynamic properties of the elastic type II superconductor in the mixed state.

7.5 MAGNETOTHERMOELASTIC EFFECTS IN TYPE II SUPERCONDUCTORS IN THE MIXED STATE

In the above section, we have formulated magnetoelastic behaviors of type II superconductors in the mixed state, where a constant and uniform temperature field is assumed. In the following two subsections, we shall consider cases where temperature distribution in the superconductor can be non-uniform and may change with time. In particular, we shall take into account of the effect of mechanical motion on the electrodynamic behaviors of anisotropic deformable type II superconductors in the mixed state.

7.5.1 Generalized London Equations for Anisotropic Type II Superconductors

We first introduce a set of generalized London equations for characterizing electrodynamic behaviors of anisotropic deformable type II superconductors in the mixed state at the London approximation. We start with the following generalized second London equation:

$$\mu_o e_{kpq}\frac{\partial}{\partial x_p}(\Lambda_{qn}J_n^{(s)}) - \frac{m^*}{|e^*|}e_{kpq}\frac{\partial}{\partial x_p}(M_{qn}V_n) = B_k^{(v)} - B_k \tag{7.5.1}$$

which is generalized from Eq.(7.4.8) for isotropic type II superconductors in the mixed state. Here, μ_o is the permeability of free space, $\boldsymbol{J}^{(s)}$ is the supercurrent density vector, $\boldsymbol{V}$ is the local velocity vector of the moving superconductor, $\boldsymbol{B}$ is the magnetic induction field, which is, in general, not equal to the local vortex magnetic field $\boldsymbol{B}^{(v)}$. Λ_{kl} is a coefficient tensor characterizing effectively the penetration depths of the anisotropic superconductor. For isotropic superconductors, we have simply $\Lambda_{kl} = \lambda_L{}^2\delta_{kl}$ with λ_L being the London penetration depth. For anisotropic superconductors, Λ_{kl} is diagonal if the reference frame chosen is aligned with the principal axes. The coefficient m^* in Eq.(7.5.1) denotes the effective mass of the superconducting electron in the phenomenological model (the Cooper pair of electrons in the BCS model). M_{kl} is the normalized effective mass tensor. When M_{kl} is diagonal in the reference frame aligned with the principal axes of the superconductor, we have $M_1M_2M_3 = 1$. It can be seen that the second term on the left-hand side of Eq.(7.5.1) characterizes the effect of the London moment induced by the motion of the superconductor. Here, we have considered that the direction of the London moment induced by local motion of the superconductor is, in general, not necessarily parallel to the vorticity vector $\Omega = (\nabla\times\boldsymbol{V})/2$ due to the anisotropy of the superconductor. We may notice that the generalization of the classical second London equation (3.2.9) has been made by including both the effect of anisotropy and the effect of local dynamic deformation of the superconductor in the mixed state.

Furthermore, in the phenomenological model, we assume that the total electric field in the superconductor can be expressed by

$$\boldsymbol{E} = \boldsymbol{E}^{(v)} + \mu_o\Lambda\cdot\frac{\partial\boldsymbol{J}^{(s)}}{\partial t} - \frac{m^*}{|e^*|}\boldsymbol{M}\cdot\left(\frac{\partial\boldsymbol{V}}{\partial t}\right) \tag{7.5.2}$$

with $\boldsymbol{E}^{(v)}$ being the vortex electric field. Equation (7.5.2) may be considered as the generalized first London equation for the anisotropic deformable type II superconductor in the mixed state, which is independent of the modified second London equation (7.5.1). Indeed, if we combine the generalized second London equation (7.5.1) with Faraday's law and the continuity equation (4.4.27), we may obtain the following equation:

$$\boldsymbol{E} = \boldsymbol{E}^{(v)} + \mu_o\Lambda\cdot\frac{\partial\boldsymbol{J}^{(s)}}{\partial t} - \frac{m^*}{|e^*|}\boldsymbol{M}\cdot\left(\frac{\partial\boldsymbol{V}}{\partial t}\right) + \nabla\phi \tag{7.5.3}$$

with an arbitrary scalar function ϕ. By noting the arguments discussed in Section 3.2.1 on the disappearance of the $\nabla\phi$ term in the classical first London equation for superconductors in the Meissner state, we may argue that the disappearance of the $\nabla\phi$ term in Eq.(7.5.2) is an independent assumption valid also for the superconductor in the mixed state. Of course, the results of such an argument can only be validated by experimental evidence, just like those from the classical first London equation. Thus, at present, we consider the two generalized London equations as two independent equations.

Obviously, Eq.(7.5.2) may recover the classical first London equation (3.2.3) for isotropic (rigid-body) superconductors in the Meissner state, as it should be. In particular, if the superconductor is in a stationary state (with steady electric current and/or thermal current), we may find from Eq.(7.5.2) that the electric field in the superconductor in the mixed state is just equal to the vortex electric field, as we may expect. If the flux-lines are not moving due to some strong pinning forces, there exists, then, no electric field in the superconductor in the dc case and at constant temperature $T < T_c$. This is indeed the experimental fact that some type II superconductors with strong pinning forces in the mixed state may also carry large dc superconducting current with essentially no Joule losses.

7.5.2 Formulation of Magnetothermoelastic Effects in Type II Superconductors

Let us now consider a homogeneous type II superconductor, which is initially in an applied dc magnetic field $\boldsymbol{B}_0$ and a constant temperature T_0 such that the superconductor is in the mixed state. The magnitude of the applied magnetic field B_o is supposed to be not close to the upper critical field B_{c2} so that the London approximation can be used. In the superconductor, there may also exist a static displacement field $\boldsymbol{U}_0$, which can be caused by static mechanical, thermal, and/or static magnetic forces. We shall study a situation in which the superconductor is subject to some small time-varying loading, which may be of electromagnetic, mechanical, and/or thermal origin. We would like to model the magnetothermoelastic response of the superconductor in the case considered.

To start with, we may write the total magnetic induction field $\boldsymbol{B} = \boldsymbol{B}_o + \boldsymbol{b}$, the total elastic displacement field $\boldsymbol{U} = \boldsymbol{U}_o + \boldsymbol{u}$, and the total temperature field $T = T_o + \theta$ in the superconductor. Here, $\boldsymbol{b}$, $\boldsymbol{u}$, and θ are, respectively, the small perturbed time-varying magnetic induction field ($|\boldsymbol{b}| << |\boldsymbol{B}_0|$), the small perturbed time-varying elastic displacement field, and the small perturbed time-varying temperature field ($\theta << T_o$) to be determined. In many cases, physical phenomena of superconductors can be analyzed with the aid of knowledge gained from studying the behavior of the superconductor in time-harmonic fields ($\sim e^{i\omega t}$). In what follows, we shall deal with time-harmonic field problems.

With above considerations in mind, we may obtain from the vortex continuity equation (4.4.27) the following relation at the linear approximation:

$$\boldsymbol{B}^{(\mathrm{v})} = \boldsymbol{B}_o - \nabla \times [\boldsymbol{B}_o \times (\boldsymbol{w} + \boldsymbol{u})] \tag{7.5.4}$$

where $\boldsymbol{w}$ is the local vortex displacement vector of the flux–lattice continuum, which may be related to the local vortex velocity $\mathbf{v}$ by $\mathbf{v} = i\omega \boldsymbol{w}$ in the time-harmonic case. Here, the effect of local dynamic deformation of the superconductor has been taken into account by the presence of the elastic displacement vector $\boldsymbol{u}$ as shown in Eq.(7.5.4). It is shown that the vortex magnetic field $\boldsymbol{B}^{(\mathrm{v})}$ in the superconductor deviates from the applied magnetic field $\boldsymbol{B}_0$. Such a deviation is usually small if B_o is not too close to B_{c1}. Thus, as the first-order approximation, we may ignore the effect of the small perturbed fields on the material properties, and assume that all material properties of the superconductor under consideration are functions of the applied dc magnetic field $\boldsymbol{B}_0$ and the constant temperature T_o, which are supposed to be uniform in the superconductor.

To model the flux-flow galvanomagnetic and thermomagnetic effects in the anisotropic deformable type II superconductor in the mixed state in the time-harmonic field (see also Section 4.6), we shall ignore, in this section, the possible effect of normal conduction fluid in the superconductor for some simplicity. In practice, this approximation can be reasonable if one is limited to cases where the temperature of a superconductor is not close to the critical temperature and the frequency involved is not too close to the gap frequency of the superconductor so that the number of quasi-particles is not large and their effect is negligible. Thus, if the effect of normal conduction fluid is negligible, we may obtain, from Eqs.(4.6.5) and (7.5.2), the following relation at the linear approximation:

$$E_k = \rho_{kl} J_l + \Pi_{kl}^{(\mathrm{v})} \frac{\partial \theta}{\partial x_l} + \left(\frac{\omega^2 m^*}{|e^*|} M_{kl} + i\omega e_{kml} B_{om} \right) u_l \tag{7.5.5}$$

where ρ_{kl} may be called the effective complex resistivity tensor, defined here by

$$\rho_{kl} \equiv i\omega\mu_o \Lambda_{kl} + \rho_{kl}^{(\mathrm{v})} \tag{7.5.6}$$

The complex flux-flow thermoelectric coefficient tensor $\Pi_{kl}^{(\mathrm{v})}$ in Eq.(7.5.5) is given by Eq.(4.6.7) with $\boldsymbol{B}^{(\mathrm{v})}$ being now replaced by $\boldsymbol{B}_0$ at the linear approximation. It is shown in Eq.(7.5.5) that both effects of superelectron inertia and of the Lorentz field $\boldsymbol{V} \times \boldsymbol{B}_0$ contribute to the total electric field $\boldsymbol{E}$ inside the deformable type II superconductor in the mixed state due to the dynamic deformation of the superconductor.

To describe the heat conduction phenomenon, we have the heat conduction equation:

$$\rho_M C_{\mathrm{v}} \frac{\partial T}{\partial t} = -\nabla \cdot \boldsymbol{q} - T_o \frac{\partial}{\partial t}(\beta_{ij} U_{i,j}) + Q_h \tag{7.5.7}$$

where ρ_M is the mass density of the superconductor, C_v is the specific heat per unit mass, β_{ij} is the effective thermal modulus tensor, $\boldsymbol{q}$ is the heat current density vector, and Q_h is the intensity of heat source per unit volume, which may be due to Joule's heat or other heat sources. For the heat current density vector $\boldsymbol{q}$, we may notice that since the entropy density within the normal core of a flux line is larger than that in the surrounding superconducting phase, the motion of the flux-lines will be associated with the transport of entropy [Solomon and Otter (1967)]. By Eq.(4.6.16), the total heat current density $\boldsymbol{q}$ may be expressed by

$$q_k = T_o \Gamma_{kl}^{(v)} J_l - \kappa_{kl} \frac{\partial \theta}{\partial x_l} \tag{7.5.8}$$

where κ_{kl} is the effective thermal conductivity tensor, defined by Eq.(4.6.17), and $\Gamma_{kl}^{(v)}$ is defined by Eq.(4.6.18).

Furthermore, to describe the mechanical motion of the superconductor, we have the following equation of motion:

$$\rho_M \frac{\partial^2 \boldsymbol{U}}{\partial t^2} = \nabla \cdot \boldsymbol{t} + \frac{1}{\mu_o}\left[(\boldsymbol{B} \cdot \nabla)\boldsymbol{B} - \frac{1}{2}\nabla(\boldsymbol{B} \cdot \boldsymbol{B})\right] \tag{7.5.9}$$

where $\boldsymbol{U}$ is the elastic displacement vector, and $\boldsymbol{t}$ is the Cauchy stress tensor, which may be related to the elastic strain and temperature difference by the following linearized constitutive relation:

$$t_{ij} = C_{ijkl} u_{k,l} - \beta_{ij} \theta \tag{7.5.10}$$

where t_{ij} is the Cauchy stress tensor and C_{ijkl} is the effective elastic modulus tensor. It is noticed that both the effective elastic moduli and the effective thermal moduli of the anisotropic thermoelastic type II superconductor in the mixed state may differ from their corresponding values in the normal state due to possible anomalies arising from the superconducting phase transition [Zhou (1991a)].

Equations (7.5.5), (7.5.8), and (7.5.10) may be considered as the constitutive equations for describing phenomenologically the flux-flow galvanomagnetic, the flux-flow thermomagnetic, and thermoelastic properties of the anisotropic deformable type II superconductor in the mixed state in time-harmonic fields at the linear approximation.

The complete set of field equations for analyzing electromagnetic, temperature, and elastic deformation fields in thermoelastic type II superconductors in the mixed state at the linear approximation can be given as follows: In the case of that the effect of normal conduction fluid is negligible, we may derive, with the aid of the generalized second London equation, the heat conduction equation, and the mechanical motion equation for the superconductor, the following set of linearized field equations:

$$e_{kpq}e_{mst}\Lambda_{qm}\frac{\partial^2 b_t}{\partial x_p \partial x_s} + \frac{1}{i\omega\mu_o}e_{kmn}e_{pst}\rho^{(\mathrm{v})}_{np}\frac{\partial^2 b_t}{\partial x_m \partial x_s} + b_k$$

$$= -\frac{1}{i\omega}e_{kmn}\Pi^{(\mathrm{v})}_{ns}\frac{\partial^2\theta}{\partial x_m \partial x_s} + e_{kmn}\left(\frac{i\omega m^*}{|e^*|}M_{np} + e_{npq}B_{oq}\right)\frac{\partial u_p}{\partial x_m} \tag{7.5.11}$$

and

$$\kappa_{kl}\frac{\partial^2\theta}{\partial x_k \partial x_l} - i\omega\rho_M C_{\mathrm{v}}\theta = \mu_o^{-1}T_o\Gamma^{(\mathrm{v})}_{kl}e_{lmn}\frac{\partial^2 b_n}{\partial x_m \partial x_k} + i\omega T_o\beta_{pq}\frac{\partial u_p}{\partial x_q} \tag{7.5.12}$$

and

$$-\rho_M\omega^2 u_k = C_{klmn}\frac{\partial^2 u_m}{\partial x_n \partial x_l} - \beta_{kl}\frac{\partial\theta}{\partial x_l} + \frac{1}{\mu_o}\left(B_{om}\frac{\partial b_k}{\partial x_m} - B_{om}\frac{\partial b_m}{\partial x_k}\right) \tag{7.5.13}$$

for the determination of the magnetic induction field $\boldsymbol{b}$, the elastic displacement $\boldsymbol{u}$ and the temperature field θ. It is shown that the derived complete set of linearized field equations (7.5.11)–(7.5.13) is, in general, coupled for the determination of the magnetic induction field $\boldsymbol{b}$, the elastic displacement field $\boldsymbol{u}$, and the temperature field θ. This set of linearized field equations may be used to analyze some magnetothermoelastic response of the type II superconductors in the mixed state in time-harmonic fields.

As a special case, for an isotropic thermoelastic type II superconductor in the mixed state, we may find that Eqs.(7.5.11)–(7.5.13) are reduced to have the following form:

$$\left(\lambda^2 - \frac{i}{\mu_o\omega\sigma_{\mathrm{v}}}\right)\nabla^2\boldsymbol{b} = \boldsymbol{b} + \frac{i}{\mu_o\omega\sigma_{\mathrm{v}}}\nabla\times[(\nabla\times\boldsymbol{b})\cdot\boldsymbol{n}]\boldsymbol{n} + \frac{iS_f}{\omega B_o\sigma_{\mathrm{v}}}\nabla\times(\boldsymbol{n}\times\nabla\theta)$$

$$-\nabla\times(\boldsymbol{u}\times\boldsymbol{B}_o) - \frac{i\omega m^*}{|e^*|}\nabla\times\boldsymbol{u} \tag{7.5.14}$$

and

$$\left(\kappa^{(o)} + \frac{T_o S_f^2}{B_o^2\sigma_{\mathrm{v}}}\right)\nabla^2\theta - i\omega\rho_M C_{\mathrm{v}}\theta = -\frac{T_o S_f}{\mu_o B_o\sigma_{\mathrm{v}}}[\boldsymbol{n}\cdot(\nabla^2\boldsymbol{b})] + i\omega T_o\beta_T(\nabla\cdot\boldsymbol{u}) \tag{7.5.15}$$

and

$$-\omega^2\boldsymbol{u} = (c_L^2 - c_T^2)\nabla(\nabla\cdot\boldsymbol{u}) + c_T^2\nabla^2\boldsymbol{u} - \frac{\beta_T}{\rho_M}\nabla\theta + \frac{1}{\rho_M\mu_o}[\boldsymbol{B}_o\cdot\nabla\boldsymbol{b} - \nabla(\boldsymbol{B}_o\cdot\boldsymbol{b})] \tag{7.5.16}$$

in which $\boldsymbol{n}$ is the unit direction vector of the applied dc magnetic field $\boldsymbol{B}_o$. $c_T =$

$(G/\rho_M)^{1/2}$ and $c_L = [(K+4G/3)/\rho_M]^{1/2}$ are, respectively, the phase velocities of the transverse elastic wave and of the longitudinal elastic wave, where K and G are, respectively, the elastic bulk modulus and the elastic shear modulus. Here, we have ignored the effect of normal conduction fluid, the flux-flow Hall effect, and the flux-flow Righi-Leduc effect.

In the case where the effect of the normal conduction fluid in the superconductor is not negligible, we may derive the following set of linearized field equations:

$$e_{kpq}e_{mst}\Lambda_{qm}\frac{\partial^2 b_t}{\partial x_p \partial x_s} + \frac{1}{i\omega\mu_o}e_{kmn}e_{pst}\rho_{np}^{(v)}\frac{\partial^2 b_t}{\partial x_m \partial x_s} - \mu_o e_{kpq}\Lambda_{qm}\sigma_{mn}\frac{\partial E_n}{\partial x_p} + b_k$$

$$= e_{kmn}\left(\frac{i\omega m^*}{|e^*|}M_{np} + e_{npq}B_{oq} + i\mu_o\omega e_{qpt}\Lambda_{ns}\sigma_{sq}B_{ot}\right)\frac{\partial u_p}{\partial x_m} \tag{7.5.17}$$

and

$$-\rho_M\omega^2 u_k = C_{klmn}\frac{\partial^2 u_m}{\partial x_n \partial x_l} + \frac{1}{\mu_o}\left(B_{om}\frac{\partial b_k}{\partial x_m} - B_{om}\frac{\partial b_m}{\partial x_k}\right) \tag{7.5.18}$$

where we have ignored the thermal effect. This set of Eqs.(7.5.17) and (7.5.18) for anisotropic elastic type II superconductors in the mixed state is the generalization of the magnetoelastic field equations derived for isotropic elastic type II superconductors in the mixed state, as shown in Section 7.4. Here, the effect of anisotropy has been formulated. In general, we have to solve the complete set of coupled partial differential equations, including the generalized first London equation (7.5.2) and relevant Maxwell's equations for the electric field, the magnetic field, and the elastic displacement field simultaneously for the anisotropic superconductor in either the mixed state or the Meissner state. An illustrative example on an anisotropic superconducting planar waveguide has been discussed in Section 3.5. Another example on a superconducting planar waveguide in a strong dc magnetic field has also been analyzed in Section 4.5.

It is worth notice that, in the study of moving type II superconductors, there exist some experimentally observed phenomena, requiring special attention. Since the flux pinning centers may move together with the moving superconductor, we may expect that flux lines pinned at these pinning centers may also move, and therefore, cause not only ac loss, but also an effective resistance force on the moving superconductor in the mixed state. Indeed, the phenomenon of a continuous range of stable equilibrium positions of a magnetic disk levitated above a high-T_c (type II) superconducting plate has been observed experimentally, and has been attributed to flux pinning by some researchers [Brandt (1989) and Davis (1990)]. Interestingly, there is an another experimentally observed phenomenon of that almost no resistance torque is found on a rotating circularly symmetric magnetic disk levitated above a high-T_c superconducting plate, which seems to contradict the intuitive expectation from a pictorial study of flux-line pinning. However, if we notice that ignoring possible effect of inertia, the macroscopic magnetic field "pattern" is actually not changed during the rotation of the circularly symmetric magnetic disk on the axis of rotation, levitated above the superconducting plate, we may understand why there is no resistant torque on the magnetic disk from the flux-pinning mechanism in the superconducting plate.

7.6 MAGNETOELASTIC PROPERTIES OF SUPERCONDUCTORS

In above sections, we have investigated some electrodynamic models and mathematical formulations of basic equations for analyzing electrodynamic behaviors of deformable superconductors, assuming given material properties which are supposed to be able to be determined experimentally. In this section, we shall study theoretically in some details how the material properties of superconductors may change due to material deformation and/or superconducting-normal phase transitions of the superconductors.

7.6.1 Effects of Elastic Deformation on Superconductive Properties

To study effects of mechanical deformation on superconductive properties, let us first introduce a continuum model for an elastic superconductor. We assume that the superconductor is in thermodynamic equilibrium and its deformation is elastic. No dissipative processes, such as plasticity, normal electric and heat conductions are being considered here. Thus we may start with the introduction of the following virtual variation equation for the elastic superconductor in the thermodynamic equilibrium at a constant temperature [Zhou (1991a)]

$$\int_V \delta F dV = \int_{\partial V} \boldsymbol{t}^{(n)} \cdot \delta \boldsymbol{x} dS + \int_V (\boldsymbol{f}^{me} + \boldsymbol{f}^{em}) - \int_V \boldsymbol{J} \cdot \delta \boldsymbol{A} dV \tag{7.6.1}$$

where V and ∂V denote, respectively, the volume and boundary surface of the superconductor. $\boldsymbol{J}$ is the supercurrent density, $\boldsymbol{t}^{(n)}$ the surface traction, $\boldsymbol{f}^{me}$ the mechanical body force and $\boldsymbol{f}^{em}$ the Lorentz body force given by

$$\boldsymbol{f}^{em} = \boldsymbol{J} \times \boldsymbol{B} \tag{7.6.2}$$

The thermodynamic function of free energy density F may be expressed by

$$F = F_n + F_{sn} \tag{7.6.3}$$

where F_n is the free energy density of the superconductor at the normal state, being only a functional of the Lagrangian strain E_{KL} defined by Eq.(2.2.7) at a given temperature T. F_{sn} denotes the free energy density difference between the superconducting state and the normal state, which, near superconducting transition temperature, may be expressed by

$$F_{sn} = \alpha(E_{KL})|\psi|^2 + \frac{1}{2}\beta(E_{KL})|\psi|^4$$

$$+ \frac{1}{2m^*_{pq}}\left[\left(-i\hbar\frac{\partial}{\partial x_p} - e^*A_p\right)\psi\right]\left[\left(i\hbar\frac{\partial}{\partial x_q} - e^*A_q^*\right)\psi^*\right] \quad (7.6.4)$$

where A is the magnetic vector potential. The last term on the right-hand side of Eq.(7.6.4) represents presumably the kinetic energy of superelectrons with its macroscopic quantum generalization. The necessity of such a macroscopic quantum generalization was confirmed by Josephson-junction experiments. The anisotropy of the superconductor is taken into account here by introducing the effective mass tensor $\boldsymbol{m}^*$, which could, in general, be function of the elastic strain tensor E_{KL}. However, to simplify the formulation for practical applications, we may assume that the effective mass is a material constant with negligible effects from the elastic lattice deformation by noting the fact that $\boldsymbol{m}^*$ plays essentially the role of the mass of one Cooper pair of electrons rather than the mass density of the Cooper pair superelectrons [Gor´kov and Melik-Barkhudarov (1964) and Zimmerman and Mercereau (1965)]. Nevertheless, there would be, in principle, no difficult to consider mathematically more general cases where such dependence could be taken into account if it were necessary.

After carrying out the variation procedure for Eq.(7.6.1), we can derive the following set of field equations and boundary conditions for the elastic superconductor at the equilibrium state.

The mechanical equilibrium equation is

$$t_{mn,m} + f_n^{me} + f_n^{em} = 0 \quad \text{in } V \quad (7.6.5)$$

where t_{mn} is the Cauchy stress tensor given by

$$t_{mn} = \frac{\partial F}{\partial E_{KL}} x_{n,K} x_{m,L} \quad \text{in } V \quad (7.6.6)$$

with $t_{mn} = t_{nm}$, that is, the Cauchy stress tensor t_{mn} is a symmetric tensor.

The generalized anisotropic Ginzburg-Landau equation is

$$\alpha\psi + \beta|\psi|^2\psi + \frac{1}{2m^*_{pq}}\left(-i\hbar\frac{\partial}{\partial x_p} - e^*A_p\right)\left(-i\hbar\frac{\partial}{\partial x_q} - e^*A_q\right)\psi = 0 \quad (7.6.7)$$

and the equation for the superconducting current density is

$$J_p = \frac{e^*\hbar}{i2m^*_{pq}}\left(\psi^*\frac{\partial\psi}{\partial x_q} - \psi\frac{\partial\psi^*}{\partial x_q}\right) - \frac{e^{*2}}{m^*_{pq}}|\psi|^2 A_q \tag{7.6.8}$$

in the superconductor. The model presented here is general and may ensure a consistent treatment of different-order effects. The resulting field equations are, however, highly non-linear and coupled. In order to obtain solutions of any practical interest for applications, simplifications are often required.

Let us now study the effects of mechanical deformation on superconducting properties of an elastic superconductors by using the above model. We suppose that the superconductor is in the thermodynamic equilibrium and is in a uniform temperature field, being constant in time. Possible applied electromagnetic fields are static. Furthermore, the superconductor is assumed to be homogeneous, and its elastic deformation is small. In such a case, the mechanical constitutive equation (7.6.6) of the elastic superconductor at a constant temperature T may be written at the first-order approximation

$$t_{mn} = C^{(n)}_{mnkl}\varepsilon_{kl} + \alpha'(a_{mn} + a'_{mnkl}\varepsilon_{kl})|\psi|^2 + \frac{\beta'}{2}(b_{mn} + b'_{mnkl}\varepsilon_{kl})|\psi|^4 \tag{7.6.9}$$

where ε_{kl} is the infinitesimal strain tensor defined by Eq.(2.2.10), and $C^{(n)}_{mnkl}$ the elastic moduli of the superconductor in the normal state. The phenomenological coefficients α', a_{mn}, a'_{mnkl} and β', b_{mn}, b'_{mnkl} are defined from the expansion coefficients of α and β by

$$\alpha(T, \varepsilon) = \alpha'(T)\left[1 + a_{mn}(T)\varepsilon_{mn} + \frac{1}{2}a''_{mnkl}(T)\varepsilon_{mn}\varepsilon_{kl}\right] \tag{7.6.10}$$

$$\beta(T, \varepsilon) = \beta'(T)\left[1 + b_{mn}(T)\varepsilon_{mn} + \frac{1}{2}b''_{mnkl}(T)\varepsilon_{mn}\varepsilon_{kl}\right] \tag{7.6.11}$$

It is noticed that the coefficients a'_{mnkl} and b'_{mnkl} in Eq.(7.6.9) are generally different from the expansion constants a''_{mnkl} and b''_{mnkl}, which is due to the influence of the deformation gradient $\boldsymbol{F}$ [defined by Eq.(2.2.2)] in the expansion calculation.

For isotropic superconductors, we have

$$a_{mn} = a_o\delta_{mn} \tag{7.6.12}$$

$$a''_{mnkl} = a''_o\delta_{mn}\delta_{kl} + a''_1(\delta_{mk}\delta_{nl} + \delta_{ml}\delta_{nk}) \tag{7.6.13}$$

and

$$b_{mn} = b_o\delta_{mn} \tag{7.6.14}$$

$$b''_{mnkl} = b''_o\delta_{mn}\delta_{kl} + b''_1(\delta_{mk}\delta_{nl} + \delta_{ml}\delta_{nk}) \tag{7.6.15}$$

and

$$a'_{mnkl} = a''_o\delta_{mn}\delta_{kl} + (a_o + a''_1)(\delta_{mk}\delta_{nl} + \delta_{ml}\delta_{nk}) \tag{7.6.16}$$

$$b'_{mnkl} = b''_o\delta_{mn}\delta_{kl} + (b_o + b''_1)(\delta_{mk}\delta_{nl} + \delta_{ml}\delta_{nk}) \tag{7.6.17}$$

and

$$C^{(n)}_{mnkl} = \frac{2G^{(n)}\nu^{(n)}}{1-2\nu^{(n)}}\delta_{mn}\delta_{kl} + G^{(n)}(\delta_{mk}\delta_{nl} + \delta_{ml}\delta_{nk}) \tag{7.6.18}$$

where $G^{(n)}$ and $\nu^{(n)}$ are, respectively, the elastic shear modulus and Poisson´s ratio of the superconductor at the normal state.

By noting the fact that the phenomenological coefficient $\alpha(T, \varepsilon)$ must change its sign at the transition temperature T_c, we may also make an expansion of α in the following form

$$\alpha(T, \varepsilon) = \alpha_o(\varepsilon)(T - T_c) \tag{7.6.19}$$

with α_0 being a real and positive quantity. Here, the expansion with respect to temperature is made near the superconducting transition temperature T_c defined at the actual configuration in the absence of applied magnetic field. In general, the transition temperature T_c of the superconductor is also strain-dependent. However, such a dependence of T_c on elastic strain of the superconductor cannot be derived simply from the macroscopic model. Empirically, it may be assumed that the strain-dependent transition temperature T_c may be described up to the second-order approximation of the elastic strain by

$$T_c(\varepsilon) = T_c^o\left(1 + \Delta_{mn}\varepsilon_{mn} + \frac{1}{2}\Xi_{mnkl}\varepsilon_{mn}\varepsilon_{kl}\right) \tag{7.6.20}$$

where $T_c^{\,o}$ may be defined as the "ideal" superconducting transition temperature for the superconductor in a natural reference state where the material is strain-free and temperature field is uniform. Here, Δ_{mn} and Ξ_{mnkl} are dimensionless material constants.

Also, we may make an expansion of $\alpha_0(\varepsilon)$ by

$$\alpha_o(\varepsilon) = \alpha_{\varepsilon o} + \alpha_{\varepsilon mn}\varepsilon_{mn} + \frac{1}{2}\alpha_{\varepsilon\varepsilon mnkl}\varepsilon_{mn}\varepsilon_{kl} \tag{7.6.21}$$

Up to the second-order approximation with respect to the elastic strain, we may find the following relations among the coefficients defined in Eqs.(7.6.10), (7.6.20), and (7.6.21):

$$\alpha'(T) = \alpha_{\varepsilon o}(T - T_c^o) \tag{7.6.22}$$

$$a_{mn}(T) = \frac{1}{\alpha_{\varepsilon o}}\alpha_{\varepsilon mn} - \frac{T_c^o}{T - T_c^o}\Delta_{mn} \tag{7.6.23}$$

$$a''_{mnkl}(T) = \frac{1}{\alpha_{\varepsilon o}}\alpha_{\varepsilon\varepsilon mnkl} - \frac{T_c^o}{T - T_c^o}\Xi_{mnkl} - \frac{2T_c^o}{\alpha_{\varepsilon o}(T - T_c^o)}\alpha_{\varepsilon mn}\Delta_{kl} \tag{7.6.24}$$

For an isotropic superconductor, we have $\Delta_{kl} = \Delta_o\delta_{kl}$, where the constant Δ_o may be either positive or negative depending on the materials. For instance, Smith and Chu (1967) observed a linear hydrostatic pressure dependence of T_c for Al, Cd, Zn, Sn, In, and Pb with $\Delta_o > 0$, which means that the transition temperature decreases with increasing pressure. The experimental studies of the pressure effects on superconductivity has been made since 1926 when Sizoo and Kamerlingh Onnes (1926) first reported a small decrease in T_c of In and Sn when subjected to pressure generated by liquid helium. Quite many experimental data for some conventional superconductors have been summarized in the work of Narlikar and Ekbote (1983). Recently, Borges et al. (1987), however, found that the transition temperatures of some high temperature oxide superconductors, such as $YBa_2Cu_3O_x$ and $RBa_2Cu_3O_x$ (R = Gd, Er, and Yb) are enhanced by pressure, which implies a negative value of Δ_o [see also Phillips (1989)]. For relatively large deformation, it may be necessary to include the term of ε^2 by considering the second-order approximation [Welch (1980)]. Using the elastic stress–strain relations, we may also expressed the strain-dependence of the transition temperature T_c in terms of the stress-dependence of T_c.

With the use of Eqs.(7.6.10), (7.6.11), and (7.6.22), the thermodynamic critical field B_c defined in Eq.(4.2.9) can be expressed, at the first-order approximation with respect to the elastic strain, by

$$B_c = B_{co}\left[1 + \frac{1}{2}(2a_{mn} - b_{mn})\varepsilon_{mn}\right] \tag{7.6.25}$$

with B_{co} being defined as the thermodynamic critical field of the superconductor at zero strain and near transition temperature by

$$B_{co} = \sqrt{\frac{\mu_o\alpha_{\varepsilon o}^2}{\beta_o}\left|T - T_c^o\right|} \tag{7.6.26}$$

with $\beta_o = \beta'(T_c^o) > 0$.

In a weak magnetic field ($B < B_{c1}$) for a homogeneous isotropic superconducting material, to the first order in $\boldsymbol{B}$, $|\psi|^2$ may be replaced by $|\psi_o|^2$ ($= |\alpha|/\beta$). Thus, by taking into account elastic deformation, we may find that the penetration depth λ also depends on the elastic strain. As the first-order approximation with respect to the elastic strain, the penetration depth λ may be expressed, near the transition temperature, by

$$\lambda = \sqrt{\frac{m^*}{\mu_o e^{*2} |\psi_o|^2}} \approx \lambda_{\varepsilon o}\left[1 + \frac{1}{2}(b_o - a_o)\varepsilon_{kk}\right] \tag{7.6.27}$$

with $\lambda_{\varepsilon o}$ being defined as the penetration depth of the superconductor at zero strain, that is,

$$\lambda_{\varepsilon o} = \sqrt{\frac{m^* \beta_o}{\mu_o e^{*2} \alpha_{\varepsilon o} |T - T_c^o|}} \tag{7.6.28}$$

Similarly, at the first-order approximation with respect to the elastic strain, we may find the strain-dependent coherence length ξ, near the transition temperature, by

$$\xi = \xi_{\varepsilon o}\left(1 - \frac{1}{2} a_o \varepsilon_{kk}\right) \tag{7.6.29}$$

with $\xi_{\varepsilon o}$ being defined as the G-L coherence length of the superconductor at zero strain

$$\xi_{\varepsilon o} = \sqrt{\frac{\hbar^2}{2m^* \alpha_{\varepsilon o} |T - T_c^o|}} \tag{7.6.30}$$

The strain-dependent G-L parameter κ may also be found, at the first-order approximation with respect to the elastic strain, by

$$\kappa = \kappa_{\varepsilon o}\left(1 + \frac{1}{2} b_o \varepsilon_{kk}\right) \tag{7.6.31}$$

with $\kappa_{\varepsilon o}$ being defined as the G-L parameter of the superconductor at zero strain

$$\kappa_{\varepsilon o} = \frac{m^*}{\hbar |e^*|} \sqrt{\frac{2\beta_o}{\mu_o}} \tag{7.6.32}$$

The upper critical field B_{c2} can be found, at the first-order approximation with respect to the elastic strain, as

$$B_{c2} = B_{co2}(1 + a_o \varepsilon_{kk}) \tag{7.6.33}$$

with B_{co2} being defined as the upper critical field of the superconductor at zero strain

$$B_{co2} = \sqrt{2}\kappa_{\varepsilon o} B_{co} \tag{7.6.34}$$

with $\kappa_{\varepsilon o}$ and B_{co} defined by Eq.(7.6.32) and Eq.(7.6.26), respectively.

It is shown that, at the first-order approximation, the superconducting properties of an isotropic superconducting solid, such as the transition temperature T_c, the penetration depth λ, the coherence length ξ, the G-L parameter κ as well as the upper critical field B_{c2} are only influenced by the dilatation ($\varepsilon_{kk} = \Delta V/V$) of the superconductive body. Thus the effect of shearing strain on the superconducting properties of isotropic superconductors can only be taken into account by considering the second-order approximation with respect to the elastic strain.

7.6.2 Elastic Properties of Superconductors

In the above section, some studies have been made on the problem of how superconductive properties of the material can be influenced by mechanical (elastic) deformation. We are now going to consider the problem of how the elastic properties of the superconductor is influenced by superconducting-normal phase transition. Here, we assume that the superconductor considered is a homogeneous medium.

We first consider the case of a superconductor in a weak magnetic field ($B < B_{c1}$) and in a constant temperature field. Using a perturbation method, we may write, at the first-order approximation with respective to the elastic strain,

$$|\psi|^2 = -\frac{\alpha'}{\beta'}[1 + (a_{mn} - b_{mn})\varepsilon_{mn}], \qquad |\psi|^4 = \frac{\alpha'^2}{\beta'^2}[1 + 2(a_{mn} - b_{mn})\varepsilon_{mn}] \tag{7.6.35}$$

where α', β', a_{mn}, and b_{mn} are material coefficients defined in Eqs.(7.6.10) and (7.6.11). From Eq.(7.6.35), we may write

$$\alpha' a_{mn}\frac{\partial|\psi|^2}{\partial x_n} + \frac{1}{2}\beta' b_{mn}\frac{\partial|\psi|^4}{\partial x_n} = -\frac{\alpha'^2}{\beta'}(a_{mn} - b_{mn})(a_{kl} - b_{kl})\frac{\partial \varepsilon_{kl}}{\partial x_n} \tag{7.6.36}$$

where we have assumed that the deformation of the superconductor may be non-uniform, but the temperature field is uniform.

Furthermore, from Eq.(7.6.9), we have

$$t_{mn,n} = C^{(n)}_{mnkl}\varepsilon_{kl,n} + \alpha' a'_{mnkl}\varepsilon_{kl,n}|\psi|^2 + \frac{\beta'}{2}b'_{mnkl}\varepsilon_{kl,n}|\psi|^4$$
$$+ \alpha' a_{mn}\frac{\partial|\psi|^2}{\partial x_n} + \frac{1}{2}\beta' b_{mn}\frac{\partial|\psi|^4}{\partial x_n} \tag{7.6.37}$$

Thus, by Eq.(7.6.36), we may find

$$t_{mn,n} = \left\{C^{(n)}_{mnkl} - \frac{\alpha'^2}{\beta'}\left[a'_{mnkl} - \frac{1}{2}b'_{mnkl} + (a_{mn} - b_{mn})(a_{kl} - b_{kl})\right]\right\}\varepsilon_{kl,n} \tag{7.6.38}$$

from which, we may write

$$t_{mn} = C^{(s)}_{mnkl}\varepsilon_{kl} \tag{7.6.39}$$

where $C^{(s)}_{mnkl}$ may be called the effective elastic modulus tensor of the superconductor in the superconducting state in weak fields ($B < B_{c1}$), defined by

$$C^{(s)}_{mnkl} = C^{(n)}_{mnkl} - \frac{1}{\mu_o}B^2_{co}\left[a'_{mnkl} - \frac{1}{2}b'_{mnkl} + (a_{mn} - b_{mn})(a_{kl} - b_{kl})\right] \tag{7.6.40}$$

Here, B_{co} is the thermodynamic critical magnetic field of the superconductor at the zero strain, defined by Eq.(7.6.26). The relation (7.6.40) is in agreement with the result derived by Labusch (1968) in a somewhat different way. However, there exists a slight modification here among those coefficients defined by Eqs.(7.6.12)–(7.6.18).

We now consider the case of type II superconductors in a strong applied magnetic field B^e which is close to the upper critical magnetic field B_{c2}. In such a case, a mixed state presents in the superconductor which splits into some fine-scale mixture of superconducting and normal regions whose boundaries lie parallel to the applied field, and the arrangement being such as to give the maximum boundary area relative to volume of normal material since the surface energy associated with the boundary between the normal and superconducting region is negative for type II superconductors. The structure of such a mixed state is, in general, on a very fine scale with a periodicity less than 1 μm. Thus, for macroscopic strain fields slowly varying compared with $|\psi|^2$ in the mixed state, we may introduce the local averages $<t_{kl}>$, $<\varepsilon_{kl}>$, $<|\psi|^2>$ and $<|\psi|^4>$ over a small macroscopic volume. We then have, at the zero-order approximation, the classical results [see Eqs.(4.2.64) and (4.2.65)]

$$\langle|\psi|^2\rangle = -\frac{\alpha}{\beta}\frac{2\kappa^2}{\beta_A(2\kappa^2 - 1)}\left(1 - \frac{B^e}{B_{c2}}\right) \tag{7.6.41}$$

$$\langle|\psi|^4\rangle = \beta_A\langle|\psi|^2\rangle^2 \tag{7.6.42}$$

with $\beta_A = 1.16$ for the triangular flux line lattice, which is known as the most thermodynamically stable lattice among all other possible periodic lattices [Essmann and Träuble (1967)].

Thus, with the use of Eqs.(7.6.10), (7.6.11), (7.6.31), and (7.6.33), we can derive, at the first-order approximation with respective to the elastic strain,

$$\langle|\psi|^2\rangle = -\frac{2\alpha'\kappa_{\varepsilon o}^2}{\beta'\beta_A(2\kappa_{\varepsilon o}^2-1)}\left(1-\frac{B^e}{B_{co2}}\right)$$

$$\times\left[1+\left(\frac{B_{co2}}{B_{co2}-B^e}a_{mn}-\frac{2\kappa_{\varepsilon o}^2}{2\kappa_{\varepsilon o}^2-1}b_{mn}\right)\langle\varepsilon_{mn}\rangle\right] \tag{7.6.43}$$

and

$$\alpha' a_{mn}\frac{\partial\langle|\psi|^2\rangle}{\partial x_n}+\frac{1}{2}\beta' b_{mn}\frac{\partial\langle|\psi|^4\rangle}{\partial x_n} = -\left[a_{mn}-\frac{2\kappa_{\varepsilon o}^2}{2\kappa_{\varepsilon o}^2-1}\left(1-\frac{B^e}{B_{co2}}\right)b_{mn}\right]$$

$$\times\frac{2\alpha'^2\kappa_{\varepsilon o}^2}{\beta'\beta_A(2\kappa_{\varepsilon o}^2-1)}\left[a_{kl}-\frac{2\kappa_{\varepsilon o}^2}{2\kappa_{\varepsilon o}^2-1}\left(1-\frac{B^e}{B_{co2}}\right)b_{kl}\right]\frac{\partial\langle\varepsilon_{kl}\rangle}{\partial x_n} \tag{7.6.44}$$

From Eq.(7.6.9), we can then obtain

$$\frac{\partial\langle t_{mn}\rangle}{\partial x_n} = C_{mnkl}^{(n)}\frac{\partial\langle\varepsilon_{kl}\rangle}{\partial x_n}-\frac{2\alpha'^2\kappa_{\varepsilon o}^2}{\beta'\beta_A(2\kappa_{\varepsilon o}^2-1)}\left\{\left(1-\frac{B^e}{B_{co2}}\right)\left(a'_{mnkl}-\frac{\kappa_{\varepsilon o}^2}{2\kappa_{\varepsilon o}^2-1}b'_{mnkl}\right)\right.$$

$$\left.+\left(a_{mn}-\frac{2\kappa_{\varepsilon o}^2}{2\kappa_{\varepsilon o}^2-1}\left(1-\frac{B^e}{B_{co2}}\right)b_{mn}\right)\left(a_{kl}-\frac{2\kappa_{\varepsilon o}^2}{2\kappa_{\varepsilon o}^2-1}\left(1-\frac{B^e}{B_{co2}}\right)b_{kl}\right)\right\}\frac{\partial\langle\varepsilon_{kl}\rangle}{\partial x_n} \tag{7.6.45}$$

from which we may define an effective elastic modulus tensor of the superconductor in the mixed state ($B_{c1} << B < B_{c2}$) by

$$C_{mnkl}^{(s)} = C_{mnkl}^{(n)}-\frac{2B_{co}^2\kappa_{\varepsilon o}^2}{\mu_o\beta_A(2\kappa_{\varepsilon o}^2-1)}\left\{\left(1-\frac{B^e}{B_{co2}}\right)\left(a'_{mnkl}-\frac{\kappa_{\varepsilon o}^2}{2\kappa_{\varepsilon o}^2-1}b'_{mnkl}\right)\right.$$

$$\left.+\left(a_{mn}-\frac{2\kappa_{\varepsilon o}^2}{2\kappa_{\varepsilon o}^2-1}\left(1-\frac{B^e}{B_{co2}}\right)b_{mn}\right)\left(a_{kl}-\frac{2\kappa_{\varepsilon o}^2}{2\kappa_{\varepsilon o}^2-1}\left(1-\frac{B^e}{B_{co2}}\right)b_{kl}\right)\right\} \tag{7.6.46}$$

Equation (7.6.46) indicates a jump of the elastic moduli at $B^e = B_{co2}$, which is given by

$$C_{mnkl}^{(n)}-C_{mnkl}^{(s)} = \frac{2B_{co}^2\kappa_{\varepsilon o}^2}{\mu_o\beta_A(2\kappa_{\varepsilon o}^2-1)}a_{mn}a_{kl} \tag{7.6.47}$$

The relation (7.6.47) is also in agreement with the result obtained by Labusch (1968), with some slight modification. It is seen that most of calculations given above have been restricted to the first-order approximation: however, it is, in principle, not difficult mathematically to generalize such calculations to higher order approximations if it were required.

Some experiments of studying elastic properties of superconductors have been made by a number of researchers. Bourne et al. (1987) performed Young´s modulus measurements of the $La_{2-x}Sr_xCuO_4$ superconducting compound by using a resonant vibration technique, where changes in Young´s modulus were determined directly from the changes in vibration frequency. It was observed that, at the superconducting transition temperature (T_c = 35 K), there was an anomaly in Young´s modulus. The anomalous singularities in the elastic shear modulus and Young´s modulus of polycrystalline $La_{1.85}Sr_{0.15}CuO_4$ were also observed in the vicinity of their superconducting transitions by Xiang et al. (1988). Other observations on various anomalies in elastic behavior of, for instance, Y-Ba-Cu-O superconducting materials at varying low temperatures were also made by Ledbetter et al. (1987) and Bhattachary et al. (1988a, b). A relatively recent review on ultrasonic and related experiments in high-T_c superconductors may be found in the work of Dominec (1993). Shown in Table 7.1 are some experimental data on the elastic properties reported for some high-T_c oxide superconductors. Further efforts are, however, needed not only to improve experiments for collecting reliable data, but also to correlate experimental observations to theoretical models when sufficient experimental data are available.

Table 7.1 Some Data on Elastic Properties of High-T_c Superconductors

Materials	T_c (K)	Mass Density (kg/m^3)	Bulk Modulus (GPa)	Shear Modulus (GPa)
YBCO	95	6370	52 ~ 120	36 ~ 64
BiSCCO	85	6450	11 ~ 32	16 ~ 36
TlBCCO	125	6960	46 ~ 69	40 ~ 44
LaSCO	37	6990	67	66

For polycrystalline materials around room temperature.

Appendix

A1 PHYSICAL CONSTANTS

ε_o	permittivity of free space	8.854×10^{-12} F/m
μ_o	permeability of free space	1.257×10^{-6} H/m
$\lvert e \rvert$	elementary charge	1.6022×10^{-19} C
m_e	rest mass of an electron	9.1×10^{-31} kg
m_p	rest mass of a proton	1.6726×10^{-27} kg
c	speed of light in free space	2.998×10^{8} m/s
k_B	Boltzmann constant	1.3805×10^{-23} J/K
h	Plank constant	6.62617×10^{-34} J·s
$\hbar$	reduced Plank constant ($\hbar = h/2\pi$)	1.05458×10^{-34} J·s
N_A	Avogadro´s constant	6.022×10^{26} kmole^{-1}
Φ_o	magnetic flux quantum	2.07×10^{-15} T·m^2 (Wb)
G	gravitational constant	6.673×10^{-11} N·m^2/kg^2
μ_B	Bohr magneton	9.274×10^{-24} J/T

A2 USEFUL CONVERSIONS

1 Å = 10 nm

1 inch = 2.540 cm

1 eV = 1.602×10^{-19} J

273 K = 0 °C

1 tesla = 10^4 gauss

A3 SI UNITS

Quantity	Name of Unit	Symbol for Unit	In Terms of Base Units
Length	meter	m	m
Mass	kilogram	kg	kg
Time	second	s	s
Electric current	ampere	A	A
Temperature	kelvin	K	K
Frequency	hertz	Hz	s^{-1}
Force	newton	N	$m \cdot kg \cdot s^{-2}$
Energy	joule	J (= N·m)	$m^2 \cdot kg \cdot s^{-2}$
Power	watt	W (= J/s)	$m^2 \cdot kg \cdot s^{-3}$
Pressure, stress	pascal	Pa (= N/m^2)	$m^{-1} \cdot kg \cdot s^{-2}$
Charge	coulomb	C (= A·s)	$s \cdot A$
Resistance	ohm	Ω (= V/A)	$m^2 \cdot kg \cdot s^{-3} \cdot A^{-2}$
Conductance	siemens	S (= A/V)	$m^{-2} \cdot kg^{-1} \cdot s^3 \cdot A^2$
Capacitance	farad	F (= C/V)	$m^{-2} \cdot kg^{-1} \cdot s^4 \cdot A^2$
Inductance	henry	H (= Wb/A)	$m^2 \cdot kg \cdot s^{-2} \cdot A^{-2}$
Magnetic flux	weber	Wb (= V·s)	$m^2 \cdot kg \cdot s^{-2} \cdot A^{-1}$
Magnetic field (***B***)	tesla	T (= Wb/m^2)	$kg \cdot s^{-2} \cdot A^{-1}$
Electric potential	volt	V (= J/C)	$m^2 \cdot kg \cdot s^{-3} \cdot A^{-1}$
Electric field (***E***)		(V/m)	$m \cdot kg \cdot s^{-3} \cdot A^{-1}$
Magnetic intensity (***H***)		(A/m)	$m^{-1} \cdot A$
Electric displacement (***D***)		(C/m^2)	$m^{-2} \cdot s \cdot A$
Vector potential (***A***)		(Wb/m)	$m \cdot kg \cdot s^{-2} \cdot A^{-1}$
Velocity		(m/s)	$m \cdot s^{-1}$

A4 VECTOR FORMULAS

$$\boldsymbol{A}\cdot(\boldsymbol{B}\times\boldsymbol{C}) = \boldsymbol{B}\cdot(\boldsymbol{C}\times\boldsymbol{A}) = \boldsymbol{C}\cdot(\boldsymbol{A}\times\boldsymbol{B}) \tag{A.1}$$

$$\boldsymbol{A}\times(\boldsymbol{B}\times\boldsymbol{C}) = (\boldsymbol{A}\cdot\boldsymbol{C})\boldsymbol{B}-(\boldsymbol{A}\cdot\boldsymbol{B})\boldsymbol{C} \tag{A.2}$$

$$\nabla(\phi+\psi) = \nabla\phi+\nabla\psi \tag{A.3}$$

$$\nabla\cdot(\boldsymbol{A}+\boldsymbol{B}) = \nabla\cdot\boldsymbol{A}+\nabla\cdot\boldsymbol{B} \tag{A.4}$$

$$\nabla\times(\boldsymbol{A}+\boldsymbol{B}) = \nabla\times\boldsymbol{A}+\nabla\times\boldsymbol{B} \tag{A.5}$$

$$\nabla(\phi\psi) = \phi\nabla\psi+\psi\nabla\phi \tag{A.6}$$

$$\nabla\cdot(\psi\boldsymbol{A}) = \boldsymbol{A}\cdot\nabla\psi+\psi\nabla\cdot\boldsymbol{A} \tag{A.7}$$

$$\nabla\cdot(\boldsymbol{A}\times\boldsymbol{B}) = \boldsymbol{B}\cdot(\nabla\times\boldsymbol{A})-\boldsymbol{A}\cdot(\nabla\times\boldsymbol{B}) \tag{A.8}$$

$$\nabla\times(\boldsymbol{A}\times\boldsymbol{B}) = \boldsymbol{A}(\nabla\cdot\boldsymbol{B})-\boldsymbol{B}(\nabla\cdot\boldsymbol{A})+(\boldsymbol{B}\cdot\nabla)\boldsymbol{A}-(\boldsymbol{A}\cdot\nabla)\boldsymbol{B} \tag{A.9}$$

$$\nabla\cdot(\nabla\psi) = \nabla^2\psi \tag{A.10}$$

$$\nabla\cdot(\nabla\times\boldsymbol{A}) = 0 \tag{A.11}$$

$$\nabla\times(\nabla\psi) = 0 \tag{A.12}$$

$$\nabla\times(\nabla\times\boldsymbol{A}) = \nabla(\nabla\cdot\boldsymbol{A})-\nabla^2\boldsymbol{A} \tag{A.13}$$

$$\nabla(\boldsymbol{A}\cdot\boldsymbol{B}) = (\boldsymbol{A}\cdot\nabla)\boldsymbol{B}+(\boldsymbol{B}\cdot\nabla)\boldsymbol{A}+\boldsymbol{A}\times(\nabla\times\boldsymbol{B})+\boldsymbol{B}\times(\nabla\times\boldsymbol{A}) \tag{A.14}$$

A5 THEOREMS FROM VECTOR CALCULUS

Divergence theorem:

$$\int_V \nabla \cdot \boldsymbol{A} dV = \int_S \boldsymbol{A} \cdot \boldsymbol{n} dS \tag{A.15}$$

$$\int_V \nabla \psi dV = \int_S \psi \boldsymbol{n} dS \tag{A.16}$$

$$\int_V \nabla \times \boldsymbol{A} dV = \int_S \boldsymbol{n} \times \boldsymbol{A} dS \tag{A.17}$$

Green's first identity:

$$\int_V (\phi \nabla^2 \psi + \nabla \phi \cdot \nabla \psi) dV = \int_S \phi \boldsymbol{n} \cdot \nabla \psi dS \tag{A.18}$$

Green's theorem:

$$\int_V (\phi \nabla^2 \psi - \psi \nabla^2 \phi) dV = \int_S (\phi \nabla \psi - \psi \nabla \phi) \cdot \boldsymbol{n} dS \tag{A.19}$$

where ϕ, ψ, and $\boldsymbol{A}$ are well-behaved functions. V is the volume and S is a closed surface bounding V, with the unit outward normal vector $\boldsymbol{n}$.

Stokes's theorem:

$$\int_S (\nabla \times \boldsymbol{A}) \cdot \boldsymbol{n} dS = \oint_L \boldsymbol{A} \cdot \boldsymbol{dl} \tag{A.20}$$

$$\int_S \boldsymbol{n} \times \nabla \psi dS = \oint_L \psi \boldsymbol{dl} \tag{A.21}$$

where S is an open surface and L is the contour bounding it, with line element $\boldsymbol{dl}$. The normal $\boldsymbol{n}$ to S is defined by the right-hand side rule in relation to the sense of the line integral around L.

A6 FORMS OF VECTOR OPERATORS IN CURVILINEAR COORDINATE SYSTEMS

In cylindrical coordinate system (r, θ, z) with $\boldsymbol{e}_r$, $\boldsymbol{e}_\theta$, and $\boldsymbol{e}_z$ being the corresponding unit directional vectors:

$$\nabla\psi = \frac{\partial\psi}{\partial r}\boldsymbol{e}_r + \frac{1}{r}\frac{\partial\psi}{\partial\theta}\boldsymbol{e}_\theta + \frac{\partial\psi}{\partial z}\boldsymbol{e}_z \tag{A.22}$$

$$\nabla\cdot\boldsymbol{A} = \frac{1}{r}\frac{\partial}{\partial r}(rA_r) + \frac{1}{r}\frac{\partial A_\theta}{\partial\theta} + \frac{\partial A_z}{\partial z} \tag{A.23}$$

$$\nabla\times\boldsymbol{A} = \left(\frac{1}{r}\frac{\partial A_z}{\partial\theta} - \frac{\partial A_\theta}{\partial z}\right)\boldsymbol{e}_r + \left(\frac{\partial A_r}{\partial z} - \frac{\partial A_z}{\partial r}\right)\boldsymbol{e}_\theta + \frac{1}{r}\left[\frac{\partial}{\partial r}(rA_\theta) - \frac{\partial A_r}{\partial\theta}\right]\boldsymbol{e}_z \tag{A.24}$$

$$\nabla^2\psi = \frac{1}{r}\frac{\partial}{\partial r}\left(r\frac{\partial\psi}{\partial r}\right) + \frac{1}{r^2}\frac{\partial^2\psi}{\partial\theta^2} + \frac{\partial^2\psi}{\partial z^2} \tag{A.25}$$

In spherical coordinate system (r, θ, φ) with $\boldsymbol{e}_r$, $\boldsymbol{e}_\theta$, and $\boldsymbol{e}_\varphi$ being the corresponding unit directional vectors:

$$\nabla\psi = \frac{\partial\psi}{\partial r}\boldsymbol{e}_r + \frac{1}{r}\frac{\partial\psi}{\partial\theta}\boldsymbol{e}_\theta + \frac{1}{r\sin\theta}\frac{\partial\psi}{\partial\varphi}\boldsymbol{e}_\varphi \tag{A.26}$$

$$\nabla\cdot\boldsymbol{A} = \frac{1}{r^2}\frac{\partial}{\partial r}(r^2A_r) + \frac{1}{r\sin\theta}\frac{\partial}{\partial\theta}(A_\theta\sin\theta) + \frac{1}{r\sin\theta}\frac{\partial A_\varphi}{\partial\varphi} \tag{A.27}$$

$$\nabla\times\boldsymbol{A} = \frac{1}{r\sin\theta}\left[\frac{\partial}{\partial\theta}(A_\varphi\sin\theta) - \frac{\partial A_\theta}{\partial\varphi}\right]\boldsymbol{e}_r + \frac{1}{r}\left[\frac{1}{\sin\theta}\frac{\partial A_r}{\partial\varphi} - \frac{\partial}{\partial r}(rA_\varphi)\right]\boldsymbol{e}_\theta$$
$$+ \frac{1}{r}\left[\frac{\partial}{\partial r}(rA_\theta) - \frac{\partial A_r}{\partial\theta}\right]\boldsymbol{e}_\varphi \tag{A.28}$$

$$\nabla^2\psi = \frac{1}{r^2}\frac{\partial}{\partial r}\left(r^2\frac{\partial\psi}{\partial r}\right) + \frac{1}{r^2\sin\theta}\frac{\partial}{\partial\theta}\left(\sin\theta\frac{\partial\psi}{\partial\theta}\right) + \frac{1}{r^2(\sin\theta)^2}\frac{\partial^2\psi}{\partial\varphi^2} \tag{A.29}$$

A7 TENSOR NOTATION AND SUMMATION CONVENTION

In tensor analysis, one makes extensive use of indices. A set of n variables, x_1, x_2, ..., x_n is usually denoted as x_i, $i = 1, 2, ..., n$. A set of $n \times n$ variables, t_{11}, t_{12}, t_{21}, ..., t_{nn} is usually denoted as t_{ij}, $i,j = 1, 2, ..., n$.

Summation convention: The repetition of an index (whether superscript or subscript) in a term will denote a summation with respect to that index over its range. The range of an index i is generally the set of n integer values 1 to n. An index that is summed over is called a *dummy index*; one that is not summed out is called a *free index*.

Examples are

$$a_i x_i \equiv \sum_{i=1}^{n} a_i x_i \tag{A.30}$$

$$a_i b^i \equiv \sum_{i=1}^{n} a_i b^i \tag{A.31}$$

$$t_{ii} \equiv \sum_{i=1}^{n} t_{ii} \tag{A.32}$$

$$df(x_1, x_2, \ldots, x_n) = \frac{\partial f}{\partial x_i} dx_i \equiv \sum_{i=1}^{n} \frac{\partial f}{\partial x_i} dx_i \tag{A.33}$$

$$a_i t_{ij} \equiv \sum_{i=1}^{n} a_i t_{ij} \qquad (j = 1, 2, \ldots, n) \tag{A.34}$$

where i is the dummy index, and j is the free index with its range from 1 to n. Here, we emphasize that b^1, b^2, ..., b^n are n independent variables and not the first n powers of the variable b.

Since a dummy index just indicates summation, it is immaterial which symbol is used. Thus $a_i x_i$ may be replaced by $a_k x_k$. For instance,

$$t_{ij} a_i c_j \equiv t_{kl} a_k c_l \equiv \sum_{k=1}^{n} \sum_{l=1}^{n} t_{kl} a_k c_l \tag{A.35}$$

if the range of the dummy index i (or k) is from 1 to n, and the range of the dummy index j (or l) is also from 1 to n.

Kronecker delta δ_{ij} is defined by the equations:

$$\delta_{11} = \delta_{22} = \delta_{33} = 1 \tag{A.36}$$

$$\delta_{12} = \delta_{21} = \delta_{13} = \delta_{31} = \delta_{23} = \delta_{32} = 0 \tag{A.37}$$

with the understanding that the range of the indices i and j is 1 to 3 here.

Permutation symbol e_{kmn} is defined by the equations:

$$e_{111} = e_{222} = e_{333} = e_{112} = e_{113} = e_{221} = e_{223} = e_{331} = e_{332} = 0 \tag{A.38}$$

$$e_{123} = e_{231} = e_{312} = 1 \tag{A.39}$$

$$e_{213} = e_{321} = e_{132} = -1 \tag{A.40}$$

In other words, e_{kmn} vanishes whenever the values of any two indices coincide; $e_{kmn} = 1$ when the subscripts permute like 1, 2, 3; and $e_{kmn} = -1$ otherwise.

There is a relation between the Kronecker delta δ_{ij} and the permutation symbol e_{kmn}, which is given by

$$e_{ijk}e_{ist} = \delta_{js}\delta_{kt} - \delta_{jt}\delta_{ks} \tag{A.41}$$

Bibliography

Abrikosov, A.A., *Sov. Phys. JETP* **5**, 1174 (1957).
Abrikosov, A.A., *Fundamentals of the Theory of Metals*, North-Holland, Amsterdam (1988).
Adams, R. et al., in: *SQUID'80* (eds. H.D. Hahlbohm and H. Lubbig), pp. 509–517, Walter de Gruyter, Berlin (1980).
Aharonov, Y. and Bohm, D., *Phys. Rev.* **115**, 485 (1959).
Aifantis, E.C. and Gittus, J., *Phase Transformations*, Elsevier Appl. Sci. Publ., London (1986).
Ainslie, B.J. and Day, C.R., *J. Lightwave Tech.* **LT-4**, 967 (1989).
Albeles, B. and Goldstein, Y., *Phys. Rev. Lett.* **14**, 595 (1965).
Alblas, J.B., in: *Electromagnetic Interactions in Elastic Solids, CISM Course* (ed. H. Parkus), pp. 1–104, Springer-Verlag, Wien (1979).
Amaldi, E., in: *Old and New Problems in Elementary Particles* (ed. G. Puppi), Academic Press, New York (1968).
Ambegaokar, V. and Baratoff, A., *Phys. Rev. Lett.* **10**, 486 (1963); **11**, 104 (E) (1963).
Anderson, P.W. and Suhl, H., *Phys. Rev.* **116**, 898 (1959).
Anderson, P.W., *Phys. Rev. Lett.* **9**, 309 (1962).
Andersson, P.W. and Kim, Y.B., *Rev. Mod. Phys.* **36**, 39 (1964).
Anderson, P.W. and Rowell, J.M., *Phys. Rev. Lett.* **10**, 230 (1963).
Arkadiev, V., *J. Phys. U.S.S.R.* **9**, 148 (1945).
Arkadiev, V., *Nature* **160**, 330 (1947).
Aronov, A.G., Gal'perin, Yu. M., Gurevich, V.L. and Kozub, V.I., in: *Nonequilibrium Superconductivity* (eds. D.N. Langenberg and A.I. Larkin), pp. 325–376, North-Holland, Amsterdam (1986).
Aslamazov, L.G. and Larkin, A.I., *Phys. Lett.* **A 26**, 238 (1968a).
Aslamazov, L.G. and Larkin, A.I., *Zh. Eksp. Teor. Fiz.* **55**, 1477 (1968b).
Atkins, P.W., *Molecular Quantum Mechanics* (2nd ed.), Oxford University Press, Oxford (1983).
Atsumi, K., Kotani, M., Ueno, S., Katila, T. and Williamson, S.J. (eds.), *Biomagnetism' 87*, Tokyo Denki University Press, Tokyo (1988).
Auld, B.A., *Acoustic Fields and Waves in Solids*, Vol. 1, Wiley-Interscience, New York (1973).
Averin, D.V. and Likharev, K.K., in: *Mesoscopic Phenomena in Solids* (eds. B. Al'tshuler et al.), Ch.6, Elsevier, Amsterdam (1991).
Bardeen, J., *Phys. Rev.* **79**, 167 (1950).
Bardeen, J., *Phys. Rev.* **81**, 829, 1070 (1951).
Bardeen, J. *Rev. Mod. Phys.* **34**, 667 (1962).
Bardeen, J., *IEEE Trans. Mag.* **Mag-13**, 13 (1977).
Bardeen, J., Cooper, L.N. and Schrieffer, J.R., *Phys. Rev.* **108**, 1175 (1957).

Bardeen, J. and Stephen, M.J., *Phys. Rev.* **A 140**, A1197 (1965).
Barnett, S.J., *Phys. Rev.* **6**, 239 (1915).
Barone, A., Esposito, F., Likharev, K., Semenov, V.K., Todorov, B.N. and Vaglio, R., *J. Appl. Phys.* **53**, 5802 (1982).
Barone, A. and Paterno, G., *Physics and Applications of the Josephson Effect,* Wiley, New York (1982).
Bartolo, B.D., *Classical Theory of Electromagnetism*, Prentice-Hall, Englewood Cliffs, N.J. (1991).
Bate, R.T., in: *VLSI Handbook*, pp. 615-628, Academic Press (1985).
Baule, G.M. and McFee, R., *Am. Heart J.* **66**, 95 (1963).
Bean, C.P., *Phys. Rev. Lett.* **8**, 250 (1962).
Becker, R., Sauter, F., and Heller, G., *Z. Physik* **85**, 772 (1933).
Beckmann, P., *Einstein Plus Two,* Golem Press, Boulder, CO (1987).
Bednorz, J.G. and Mÿller, K.A., *Zeitschrift fur Physik* **B 64**, 189 (1986).
Bensoussan, A., Lions, J.L. and Papanicolaou, G., *Asymptotic Analysis for Periodic Structures*, North-Holland, Amsterdam (1978).
Beran, M.J., *Statistical Continuum Theories*, Wiley Interscience, New York (1968).
Bhattacharya, S., Higgins, M.J., Johnston, D.C., Jacobson, A.J., Stokes, J.P., Goshorn,D.P., and Lewandowski, J.T., *Phys. Rev. Lett.* **60**, 1181 (1988a).
Bhattacharya, S., Higgins, M.J., Johnston, D.C., Jacobson, A.J., Stokes, J.P., Lewandowski, J.T. and Goshorn, D.P., *Phys. Rev.* **B 37**, 5901 (1988b).
Biondi, M.A., Forrester, A.T. and Garfunkel, M.P., *Phys. Rev.* **108**, 497 (1957).
Biondi, M.A. and Garfunkel, M.P., *Phys. Rev. Lett.* **2** (1959) 143–145.
Bishop, D.J., Ramirez, A.P., Gammel, P.L., Batlogg, B., Rietman, E.A., Cava, R.J. and Millis, A. J., *Phys. Rev.* **B 36**, 2408 (1987).
Bladel, J.V., *Proc. IEEE* **64**, 301 (1976).
Bladel, J.V., *Relativity and Engineering*, Springer-Verlag, Berlin (1984).
Blatt, J.M., *Theory of Superconductivity*, Academic Press, New York (1964).
Blazey, K.W., in: *Proc. of ICTPS'90 on Transport Properties of Superconductors* (ed. R. Nicolsky), pp. 313–318, World Scientific, Singapore (1990).
Bloch, F., *Z. Physik* **61**, 206 (1930).
Blount, E.I. and Varma, C.M., *Phys. Rev. Lett.* **42**, 1079 (1979).
Bodmer, A., Essmann, U. and Trauble, H., *Phys. Status Solidi (a)* **13**, 471 (1972).
Bon Mardion, G., Goodman, B.B. and Lacaze, A., *Phys. Lett.* **8**, 15 (1964).
Borges, H.A., Kwok, R., Thompson, J.D., Well, G.L., Smith, J.L, Fisk, Z. and Peterson, D.E., *Phys. Rev.* **B 36**, 2404 (1987).
Born, M. and Huang, K., *Dynamical Theory of Crystal Lattices*, Clarendon Press, Oxford (1954).
Born, M. and Oppenheimer, R., *Ann. d. Phys.* **84**, 457 (1927).
Bourne, L.C., Zettl, A., Chang, K.J., Cohen, M.L., Stacy, A.M. and Ham, W.K., *Phys. Rev.* **B 35**, 8785 (1987).
Brailsford, F., *Physical Principles of Magnetism*, D. Van Norstrand Ltd, London (1966).
Bradley, F.N., *Materials for Magnetic Functions*, Hayden Book Co., New York (1971).
Brandt, E.H., *Science* **243**, 349 (1989).
Brandt, E.H., *J. Supercond.* **6**, 201 (1993).
Brickman, N.F., *Phys. Rev.* **184**, 460 (1969).
Brown Jr., W. F., *Micromagnetism*, Wiley, New York (1963).
Brown Jr., W.F., *Magnetoelastic Interactions*, Springer-Verlag, Berlin (1966).
Brulin, O. and Hjalmars, S., *Int. J. Eng. Sci.* **19**, 1731 (1981).

Buckingham, M.J., *Phys. Rev.* **101**, 1431 (1956).
Budiansky, B., *J. Mech. Phys. Solids* **13**, 223 (1965).
Bulaevskii, L.N., Ginzburg, V.L. and Sobyanin, A.A., *Physica* **C 152**, 378 (1988).
Byers and Chang, *Phys. Rev. Lett.* **7**, 46 (1961).
Böttcher, C.J.F., *Rec. Trav. Chim. Pays-Bas* **64**, 47 (1945).
Böttcher, C.J.F., *Theory of Electric Polarization*, Elsevier, Amsterdam (1973).
Cagniard, L.G., *Geophys.* **18**, 605 (1953).
Campbell, A.M. and Evetts, J.E., *Adv. in Phys.* **21**, 199 (1972).
Campbell, A.M., Evetts, J.E. and Deew-Hughes, D., *Phil. Mag.* **18**, 313 (1968).
Cardona, M. and Rosenblum, M.B., *Phys. Lett.* **8**, 308 (1964).
Caroli, C. and Maki, K., *Phys. Rev.* **164**, 591 (1967).
Carr, Jr., W.J., *AC Loss and Macroscopic Theory of Superconductors*, Gordon & Breach, London (1983).
Carr, Jr. W.J., *J. Appl. Phys.* **54**, 5911 (1983).
Chambers, R.G., *Proc. Roy. Soc.* (London) **A65**, 458 (1952).
Chang, W.H., *J. Appl. Phys.* **50**, 8129 (1979).
Chang, W.H., *IEEE Trans. Magn.* **17**, 764 (1981).
Chen, C.W., *Magnetism and Metallurgy of Soft Magnetic Materials*, North-Holland, Amsterdam (1977).
Chengren, L. and Larbalestier, D.C., *Cryogenics* **27**, 171 (1987).
Chikazumi, S., *Physics of Magnetism,* Wiley, New York (1964).
Chin, C.C., Oates, D.E., Dresselhaus, G. and Dresselhaus, M.S., *Phys. Rev.* **B 45**, 4788 (1992).
Christensen, R.M., *Mechanics of Composite Materials*, Wiley, New York (1979).
Ciccarello, I., Fazio, C., Guccione, M. and Li Vigni, M., *Physica* **C 159**, 769 (1989).
Claeson, T., in: *Advances in Superconductivity* (eds. B. Deaver and J. Ruvalds), pp. 341–277, Plenum Press, New York (1983).
Clark, T.D., Prance, R.J. and Grassie, A.D.C., *J. Appl. Phys.* **51**, 2736 (1980).
Clarke, D.R., *Advanced Ceramic Materials* **2**, 273 (1987).
Clarke, J., *Phil. Mag.* **13**, 115 (1966).
Clarke, J., *Science,* **184**, 1235 (1974).
Clarke, J., in: *Superconductor Applications: SQUIDs and Machines* (eds. B.B. Schwartz and S. Foner), pp. 67–124, Plenum Press, New York (1977).
Clarke, J., in: *Advances in Superconductivity* (eds. B. Deaver and J. Ruvalds), pp. 13–50, Plenum Press, New York (1983).
Clarke, J., in: *Superconducting Electronics* (eds. H. Weinstock and M. Nisenoff), pp. 87–148, Springer-Verlag, Berlin (1989).
Clayhold, C., Ong, N.P., Hor, P.H. and Chu, C.W., *Phys. Rev.* **B 38**, 7016 (1988).
Cleland, A.N. and Roukes, M.L., *Nature* **392**, 160 (1998).
Clem, J.R. and Coffey, M.W., *J. Supercond.* **5**, 313 (1992).
Cody, G.D. and Miller, R.E., *Phys. Rev.* **173**, 481 (1968).
Coffey, M.W., *Phys. Rev.* **B 46**, 567 (1992).
Coffey, M.W. and Clem, J.R., *Phys. Rev.* **B 44**, 6903 (1991).
Coffey, M.W. and Clem, J.R., *Phys. Rev.* **B 45**, 10527 (1992a).
Coffey, M.W. and Clem, J.R., *Phys. Rev.* **B 46**, 11757 (1992b).
Coffey, M.W. and Clem, J.R., *Phys. Rev.* **B 48**, 342 (1993).
Coffey, M.W. and Clem, J.R., *Phys. Rev. Lett.* **67**, 386 (1991).
Cohen, D., *Science* **161**, 784 (1968).
Cohen, D., *IEEE Trans. Magn.* **MAG-11**, 694 (1975).

Cohen, D. and Givler, E., *Appl. Phys. Lett.* **21**, 114 (1972).
Collin,R.E., *Foundations for Microwave Engineering*, McGraw-Hill, New York (1966).
Collin, R.E., *Field Theory of Guided Waves*, IEEE Press, New York (1991).
Cooper, L.N., *Phys. Rev.* **104**, 1189 (1956).
Cosserat, E. and Cosserat, F., *Theorie des Corps Deformable,* Hermann, Paris (1909).
Costache, G.I., *IEEE Trans. Microwave Theory Tech.* **MTT-35**, 280 (1987).
Crabtree, G.W., Behroozi, F., Campbell, S.A. and Hinks,. D.G., *Phys. Rev. Lett.* **49**, 1342 (1982).
Crabtree, G.W., Liu, J.Z., Umezawa, A., Kwok, W.K., Sowers, C.H., Malik, S.K., Veal, B.W., Lam, D.J., Brodsky, M.B. and Downey, J.W., *Phys. Rev.* **B 36**, 4021 (1987).
Curie, P., *Oeuvres*, p. 129, Paris (1908).
Cyrot, M., *Rep. Progr. Phys.* **36**, 103 (1973).
Davis, J.R. and Nisenoff, M., in: *SQUID and Applications* (eds. H.D. Hahlbohm and H. Lubbig), pp. 439–484, Walter de Gruyter, Berlin (1977).
Davis, L.C., *J. Appl. Phys.* **67**, 2631 (1990).
Dayem, A.H. and Wiegrand, J.J., *Phys. Rev.* **155**, 418 (1967).
Deaver, B.S. and Fairbank, W.M., *Phys. Rev. Lett.* **7**, 43 (1961).
De Bruyn Ouboter, R., in: *Superconductor Applications: SQUIDs and Machines* (eds. B.B. Schwartz and S. Foner), pp. 21–65, Plenum Press, New York (1977).
De Gennes, P.G., *Rev. Mod. Phys.* **36**, 225 (1964).
De Gennes, P.G., *Superconductivity of Metals and Alloys*, Benjamin, New York (1966).
De Gennes, P.G. and Matricon, J., *Rev. Mod. Phys.* **36**, 45 (1964).
De Groot, S.R. and Mazur, P., *Non-Equilibrium Thermodynamics*, North-Holland, Amsterdam (1962).
De Groot, S.R. and Vlieger, J., *Physica* **31**, 254 (1965).
De Haas, W.J. and Voogd, J., *Commun. Phys. Lab. Univ. Leiden*, no. 214c (1931).
Deutscher, G. and de Gennes, P.G., in: *Superconductivity* (ed. R.D. Parks), pp. 1005–1034, Marcel Dekker, New York (1969).
Dirac, P.A.M., *Proc. Roy. Soc.* **A 133**, 60 (1931).
Dmitriev, V.M., Gubankov, V.N. and Nad, F. Ya., in: *Nonequilibrium Superconductivity* (eds. D.N. Langenberg and A.I. Larkin), pp. 163–209, North-Holland, Amsterdam (1986).
Dobrowolski, J.A., *Introduction to Computer Methods for Microwave Circuit Analysis and Design*, Artech House, Boston (1991).
Dolan, G.J., *J. Low Temp. Phys.* **15**, 133 (1974).
Dolgov, O.V., Kirzhnits, D.A. and Maksimov, E.G., in: *Superconductivity, Superdiamagnetism, Superfluidity* (ed. V.L. Ginzburg), pp. 18–68, MIR Publ., Moscow (1987).
Doll, R. and Nabauer, M., *Phys. Rev. Lett.* **7**, 51 (1961).
Dominec, J. *Supercond. Sci. Technol.* **6**, 153 (1993).
Douglass, D.H., Jr., *IBM J. Res. Dev.* **6**, 44 (1962).
Dunkin, J.W. and Eringen, A.C., *Int. J. Eng. Sci.* **1**, 461 (1963).
Duzer, T.V. and Turner, C.W., *Principles of Superconductive Devices and Circuits*, Elsevier, North-Holland, New York (1981).
Einstein, A., *Ann. Phys. (Leipzig)* **17**, 891 (1905).
Eisenmenger, W. and Dayem, A.H., *Phys. Rev. Lett.* **18**, 125 (1967).
Eisenstadt, W.R. and Yungseon, E., *IEEE Trans. Comp. Hybrids & Manufact. Tech.* **15**, 483 (1992).
Eliashberg, G.M. and Ivlev, B.I., in: *Nonequilibrium Superconductivity* (eds. D.N. Langenberg and A.I. Larkin), pp. 211–251, North-Holland, Amsterdam (1986).

Elliott, R.S., *An Introduction to Guided Waves and Microwave Circuits*, Prentice-Hall, Englewood Cliffs, NJ (1993).
Eringen, A.C., *Int. J. Eng. Sci.* **2**, 205 (1964).
Eringen, A.C., in: *Developments in Mechanics* **3** (eds. Huang T.C. and Johnson, Jr. M.W.), Wiley, New York (1965).
Eringen, A.C., *J. Math. Mech.* **15**, 909 (1966).
Eringen, A.C., *Mechanics of Continua*, Wiley, New York (1967).
Eringen, A.C., *Tensor Analysis, in: Continuum Physics* Vol.I (ed. A.C. Eringen), pp. 1–155, Academic Press, New York (1971).
Eringen, A.C., in: *Nonlinear Equations in Physics and Mathematics* (ed. A.O. Barut), pp.271-318, D. Reidel Publ., Dordrecht, Holland (1978).
Eringen, A. and Maugin, G.A., *Electrodynamics of Continua*, Vols. I–II, Springer, New York (1990).
Eringen, A.C. and Kim, B.S., *Mech. Res. Commun.* **1**, 233 (1974).
Eshelby, J.D., *Prog. Solid Mech.* **2**, 89–139 (1961).
Essmann, U. and TrSuble, H., *Phys. Lett.* **24A**, 526 (1967).
Faber, T.E. and Pippard, A.B., *Proc. Roy. Soc.* **A 231**, 336 (1955).
Feldman, M.R., *Appl. Optics* **27**, 1742 (1988).
Fertiq, W.A., Johnston, D.C., Delong, L.E., McCallum, R.W., Maple, M.B. and Matthias, B.T., *Phys. Rev. Lett.* **38**, 387 (1977).
Fertiq, W.A., Johnston, D.C., Delong, L.E., McCallum, R.W., Maple, M.B. andMatthias, B.T., *Phys. Rev. Lett.* **38**, 987 (1977).
Feynman, R.P., *Lectures on Physics III*, Addison-Wesley, Reading, MA (1965).
Finlayson, T.R. and Milne, I., *Solid State Comm.* **9**, 1339 (1971).
Fischer, O. and Peter, M., in: *Magnetism,* Vol.V (ed. H. Suhl), pp. 327–352, Academic Press, New York (1973).
Forgacs, R.L. and Warnick, A., *Rev. Sci. Instrum.* **38**, 214 (1967).
Frank, D.J., Brady, M.J. and Davidson, A., *IEEE Trans. Magn.* **21**, 721 (1985).
Freimuth, A., Hohn, C. and Galffy, M., *Phys. Rev.* **B 44**, 10396 (1991).
Fröhlich, H., *Phys. Rev.* **79**, 845 (1950).
Fulton, T.A. and Dolan, D.J., *Phys. Rev. Lett.* **59**, 109 (1987).
Fung, Y.C., *Foundations of Solid Mechanics*, Prentice-Hall, Englewood Cliffs, NJ (1965).
Galffy, M., Freimuth, A. and Murek, U., *Phys. Rev.* **B 41**, 11029 (1990).
Gambling, W.A. and Matsumura, H., *Electron. Lett.* **13**, 691 (1977).
Geckeler, S., *Optical Fiber Transmission Systems*, Artech House, Inc., Norwood (1987).
Gibbs, H.M., *Optical Bistability: Controlling Light with Light*, Academic, Orlando, FL (1985).
Ginsberg, D.M., *Physical Properties of High Temperature Superconductors I*, World Scientific, Singapore (1989).
Ginzburg, V.L. and Landau, L.D., *Zh. Eksp. Teor. Fiz.* **20**, 1064 (1950).
Ginzburg, V.L., *Zh. Eksp. Teor. Fiz.* **23**, 236 (1952).
Ginzburg, V.L., *Soviet Phys. JETP* **4**, 153 (1957).
Ginzburg, V.L., *Soviet Phys. Solid State* **2**, 1824 (1961).
Gittleman, J.I. and Rosenblum, B., *Phys. Rev. Lett.* **16**, 734 (1966).
Gittleman, J.I. and Rosenblum, B., *J. Appl. Phys.* **39**, 2617 (1968).
Gloge, D., *Appl. Opt.* **10**, 2442 (1971).
Gloge, D., *IEEE Trans. Microwave Theory Tech.* **MTT-23**, 106 (1975).
Glover, R.E. and Tinkham, M., *Phys. Rev.* **108**, 243 (1957).

Glover, R.E., in: *Superconductivity* (ed. F. Chilton), pp. 3–23, North-Holland, Amsterdam (1971).
Golosovsky, M.A., Snortland, H.J. and Beasley, M.R., *Phys. Rev.* **B 51**, 6462 (1995).
Golubov,A.A., in: *Handbook of Applied Superconductivity* (ed. B. Seeber), pp. 3–36, IOP, Bristol (1998).
Goodman, B.B., *IBM J. Res. Dev.* **6**, 63 (1962).
Goodman, B.B., *Reports on Progress in Physics* **29**, 445 (1966).
Gor'kov, L.P., *Soviet Phys. JETP* **7**, 505 (1958).
Gor'kov, L.P., *Zh. Eksp. Teor. Fiz.* **36**, 1918 (1959a). (*Soviet Phys. JETP* **9**, 1364).
Gor'kov, L.P., *Zh. Eksp. Teor. Fiz.* **37**, 1407 (1959b). (Soviet Phys. JETP **10**, 998).
Gor'kov, L.P., *Soviet Phys. JETP* **9**, 1364; 10, 593 (1960).
Gor'kov, L.P. and Melik-Barkhudarov, T.K., *Soviet Phys. JETP* **18**, 1031 (1964).
Gorter, C.J. and Casimir, H.B.G., *Phys. Z.* **35**, 963 (1934).
Gorter, C.J. and Casimir, H.B.G., *Physica* **1**, 306 (1934).
Goubau, W.M., in: *SQUID'80* (eds. H.D. Hahlbohm and H. Lubbig), pp. 603–613, Walter de Gruyter, Berlin (1980).
Gray, K.E., *J. Low Temp. Phys.* **15**, 335 (1974).
Green, A.E. and Rivlin, R.S., *Arch. Rat. Mech. Anal.* **17**, 113 (1964).
Grynszpan, F. and Geselowitz, D.B., *Biophys. J.* **13**, 911 (1973).
Guillemin, E.A., *Introductory Circuit Theory*, Wiley, New York (1953).
Gupta, K.C., Grag, R. and Chada, R., *Computer Aided Design of Microwave Circuits*, Artech House, Dedham, MA (1981).
Guyon, E., Caroli, C. and Martinet, A., *J. Phys. Radium* **25**, 683 (1964).
Hagen, S.J., Lobb, C.J., Greene, R.L., Forrester, M.G. and Kang, J.H., *Phys. Rev.* **B 41**, 11630 (1990).
Hake, R.R., *Phys. Rev.* **168**, 442 (1968).
Halbritter, J., *Z. Physik* **266**, 209 (1974).
Hale, D.K., *J. Materials Sci.* **11**, 2105 (1976).
Hammond, P., *Energy Methods in Electromagnetism*, Clarendon Press, Oxford (1981).
Hao, Z. and Clem, J.R., *IEEE Trans. Mag.* **27**, 1086 (1991).
Hao, Z., Clem, J.R., McElfresh, M.W., Civale, L., Malozemoff, A.P. and Holtzberg, F., *Phys. Rev.* **B 43**, 2844 (1991).
Hari, R. and Ilmoniemi, R.J., *Biomed. Eng.* **14**, 93 (1986).
Hartemann, P., in: *Handbook of Applied Superconductivity* (ed. B. Seeber), pp. 1835–1859, IOP, Bristol (1998).
Hartwig, W.H. and Passow, C., in: *Applied Superconductivity*, Vol.II (ed. V.L. Newhouse), pp. 541–639, Academic Press, New York (1975).
Hashin, Z., *J. Comp. Mater.* **2**, 284 (1968).
Hashin, Z. and Shtrikman, S., *J. Appl. Phys.* **33**, 3125 (1962).
Hasnain, G., Dienes, A. and Whinnery, J.R., *IEEE Trans. Microwave Theory Tech.* **MTT-34**, 738 (1986).
Hay, H.J., Schiffer, J.P., Cranshaw, T.E., and Egelstaff, P.A., *Phys. Rev. Lett.* **4** (1960) 165.
Heidrich, H. and Mataew, P., in: *SQUID'80* (eds. H.D. Hahlbohm and H. Lubbig), Walter de Gruyter, Berlin (1980).
Heisenberg, W., *Z. Physik* **49**, 619 (1928).
Hershey, A.V., *J. Appl. Mech.* **21**, 236 (1954).
Hettinger, J.D. and Steel, D.G., in: *High-Temperature Superconducting Materials Science and Engineering* (ed. D. Shi), pp. 21–79, Pergmon, Oxford (1995).
Hildebrandt, A.F., *Phys. Rev. Lett.* **12**, 190 (1964).

Hill, R. *J. Mech. Phys. Solids* **13**, 213 (1965).
Hippel, Von A.R., *Dielectrics and Waves*, Wiley, New York (1954).
Horacek, B.M., *IEEE Trans. Magn.* **MAG-9**, 440 (1973).
Hori, M., *J. Math. Phys.* **14**, 514 (1973).
Hosoya, M., Goto, E., Shimizu, N. and Harada, Y., *IEEE Trans. Magn.* **25**, 1111 (1989).
Hosoya, M., Hioe, W., Casas, J., Kamikawai, R., Harada, Y., Wada, Y., Nakane, H., Suda, R. and Goto, E., *IEEE Trans. Appl. Supercond.* **1**, 77 (1991).
How, H., Widom, A. and Vittoria, C., *J. Supercond.* **2**, 479 (1989).
Huebener, R.P., *Magnetic Flux Structures in Superconductors*, Springer-Verlag, Berlin (1979).
Huebener, R.P. and Seher, S., *Phys. Rev.* **181**, 701 (1969).
Huebener, R.P., Ustinov, A.V. and Kaplunenko, V.K., *Phys. Rev.* **B 42**, 4831 (1990).
Hughes, R. and Priest, R., *Appl. Opt.* **19**, 1477 (1980).
Hutter, K. and van de Ven, A.A.F., *Field Matter Interactions in Thermoelastic Solids*, Springer-Verlag, Berlin (1978).
Inderhees, S.E., Salamon, M.B., Goldenfeld, N., Rice, J.P., Pazol, B.G., Ginsberg, D.M., Liu, J.Z. and Crabtree, G.W., *Phys. Rev. Lett.* **60**, 1178 (1988).
Ishikawa, M. and Fischer, O., *Solid State Commun.* **23**, 37 (1977).
Ising, E., *Z. Physik* **31**, 253 (1925).
Ismail, K.E., Bagwell, P.F., Orlando, T.P., Antoniadis, D.A. and Smith, H.I., *Proc. IEEE* **79**, 1106 (1991).
Iye, Y., in: *Physical Properties of High Temperature Superconductors III*, (ed. D.M. Ginsberg), p. 285, World Scientific, Singapore (1992).
Izyumov, Yu. A. and Skryabin, Yu. N., *Phys. Met. Metall.* **49**, 1 (1980).
Jackson, J.D., *Classical Electrodynamics* (2nd ed.), Wiley, New York (1975).
Jeans, J., *The Mathematical Theory of Electricity and Magnetism*, 5th ed., Cambridge University Press, Cambridge (1925).
Jeffrey, D.J., *Proc. Roy. Soc. London* **A 335**, 355 (1973).
Jeffrey, D.J., *Proc. Roy. Soc. London* **A 338**, 503 (1974).
Jeunhomme, L.B., *Single-Mode Fiber Optics*, Marcel Dekker, New York (1983).
Jin, J., *The Finite Element Method in Electromagnetics*, Wiley, New York (1993).
Johnk, C.T.A., *Engineering Electromagnetic Fields and Waves*, Wiley, New York (1988).
Josephson, B.D., *Phys. Lett.* **1**, 251 (1962).
Josephson, B.D., in: *Superconductivity* (ed. R.D. Parks), pp. 423–447, Marcel Dekker, New York (1969).
Kaiser, A.B. and Uher, C., in: *Handbook of Applied Superconductivity* (ed. B. Seeber), pp. 1140–1164, IOP, Bristol (1998).
Kakani, S.L. and Upadhyaya, U.N., *J. Low Temp. Phys.* **70**, 5 (1988).
Kaliski, K. and Rogula, D., *Proc. Vibr. Probl.* **1**, 63 (1960).
Kanamori, J., in: *Magnetism*, Vol. I (ed. G.T. Rado and H. Suhl), pp. 127–203, Academic Press, New York (1963).
Kaplunenko, V.K., Khabipov, M.I., Koshelets, V.P., Likharev, K.K., Mukhanov, O.A., Semenov V.K., Serpuchenko, I.L. and Vystavkin, A.N., *IEEE Trans. Magn.* **25**, 861 (1989).
Kautz, R.L., *J. Appl. Phys.* **49**, 308 (1978).
Kautz, R.L., *J. Res. NBS* **84**, 247 (1979).
Keesom, W.H., *Rapp et Disc. 4^e Congr. Phys. Solvay*, p. 288 (1924).
Kessler, J., Dill, R. and Russer, P., *IEEE Trans. Microwave Theory & Tech.* **MTT-39**, 1566 (1991).
Keyes, R.W., *Science* **168**, 796 (May, 1970).

Keyes, R.W., *Proc. IEEE* **63**, 740 (1975).
Keyes, R.W., in: *Molecular Electronics and Molecular Electronic Devices* (ed. K. Sienicki), pp. 1–30, CRC Press, Florida (1993).
Keyes, R.W., Harris, E.P. and Konnerth, K.L., *Proc. IEEE* **58**, 1914 (1970).
Khazan, A.D., *Transducers and Their Elements*, PTR Prentice-Hall, Englewood Cliffs, NJ (1994).
Khurana, A., *Physics Today*, pp. 17–21, July (1987).
Kikoin, I.K. and Goobar, S.V., *C.R. Acad. Sci. U.R.S.S.* **19**, 249 (1938).
Kikoin, I.K. and Goobar, S.V., *J. Phys. U.S.S.R.* **3**, 333 (1940).
Kim, Y. B., Hempstead, C.F. and Strnad, A.R., *Phys. Rev. Lett.* **9**, 306 (1962).
Kim, Y. B., Hempstead, C.F. and Strnad, A.R., *Phys. Rev.* **139**, A1163 (1965).
Kim, Y.B. and Stephen, M.J., in: *Superconductivity* (ed. R.D. Parks), p. 1107, Marcel Dekker, New York (1969).
King, W.P. and Prasad, S., *Fundamental Electromagnetic Theory and Applications*, Prentice-Hall, Englewood Cliffs, NJ (1986).
Kleinsasser, A.W., in: *Heterostructures on Silicon: One Step Further with Silicon* (eds. Y.I. Nissim and E. Rosencher), pp. 187–201, Kluwer Academic Publ., Dordrecht (1989).
Kobayashi, S., Shibata, N., Shibata, S. and Izawa, T., *Rev. ECL* **26**, 453 (1978).
Koch, C.C. and Easton, D.S., *Cryogenics* **17**, 391 (1977).
Kogan, V.G., *Phys. Rev.* **B 24**, 1572 (1981).
Kok, J.A., *Physica* **1**, 1103 (1934).
Konrad, A., *IEEE Trans. Magn.* MAG-17, 1148 (1981).
Konrad, A., *IEEE Trans. Magn.* **MAG-18**, 284 (1982).
Kortum, G., *Treatise on Electrochemistry*, Elsevier, Amsterdam, London (1965).
Kostorz, G., *Phys. Status Solidi* **B 58**, 9 (1973).
Krey, U., *Int. J. Magn.* **3**, 65 (1972).
Krivoruchko, V.N., *J. Low Temp. Phys.* **93**, 217 (1993).
Kroeger, D.M., Easton, D.S. and Moazed, A., *IEEE Trans. on Magn.* **Mag-13**, 120 (1977).
Kröner, E., *Z. Phys.* **151**, 504 (1958).
Kröner, E., *Arch. Rational Mech. Anal.* **4**, 275 (1959).
Kröner, E., *J. Phys.: (F) Metal Phys.* **8**, 2261 (1978).
Kröner, E., in: *Mechanics Today* 5 (ed. S. Nemat-Nasser), pp. 155–159, Pergamon, Oxford (1980).
Kröner, E., in: *Modelling Small Deformations of Polycrystals* (eds. J. Gittus and J. Zarka), pp. 229–292, Elsevier Appl. Sci. Publ., London (1986).
Kröner, E. and Datta, B.K., *Z. Phys.* **196**, 203 (1966).
Kunin, I.A., *Elastic Media with Microstructure I, II*, Springer-Verlag, Berlin (1982, 1983).
Kunze, U., Lischke, B. and Rodewald, W., *Phys. Status Solidi* (*b*) **62**, 377 (1974).
Kuper, C.G., *An Introduction to the Theory of Superconductivity*, Clarendon, Oxford (1968).
Kurokawa, K., *IEEE Trans. Microwave Theory Tech.* **MTT-13**, 194 (1965).
Kuttruff, H., *Ultrasonics Fundmentals and Applications*, Elsevier, London (1991).
Kwon, O.K., Langley, B.W., Pease, R.F.W. and Beasley, M.R., *IEEE Electron Device Lett.* **EDL-8**, 582 (1987).
Labusch, R., *Phys. Rev.* **170**, 470 (1968).
Lagakos, N., Bucaro, J.A. and Jarzynski, *J., Appl. Opt.* **20**, 2305 (1981).
Lam, C.-W., Sheen, D.M., Ali, S.M. and Oates, D.E., *IEEE Trans. Appl. Supercond.* **2**, 58 (1992).
Landau, L.D., *Zh. Eksp. Teor. Fiz.* **7**, 371 (1937).
Landau, L.D. and Lifshitz, E.M., *Statistical Physics*, Pergamon, London (1958).

Landau, L.D. and Lifshitz, E.M., *Quantum Mechanics* (3rd ed.), Pergamon, Oxford (1977).

Landau, L.D., Lifshitz, E.M. and Pitaevskii, L.P., *Electrodynamics of Continuous Media* (2nd ed.), Pergamon, Oxford (1984).

Landauer, R., *IBM J. Res. Develop.* **5**, 183 (1961).

Langenburg, D.N., Scalapino, D.J. and Taylor, B.N., *Proc. IEEE* **54**, 560 (1966).

Langenburg, D.N. and Larkin, A.I. (eds), *Nonequilibrium Superconductivity*, North-Holland, Amsterdam (1986).

Lax, E. and F.L. Vernon, Jr. F.L., *Phys. Rev. Lett.* **14**, 256 (1965).

Ledbetter, H.M., Austin, M.W., Kim, S.A., Datta, T. and Violet, C.E., *J. Mat. Res.* **2**, 790 (1987).

Levy, T. and Sanchez-Palencia, E., *J. Non-Newtonian Fluid Mech.* **13**, 63 (1983).

Liang, G.-C., Zhang, D., Shih, C.-F., Johansson, M.E., Withers, R.S., Andersson, A.C. and Oates, D.E., *IEEE Trans. Appl. Supercond.* **5**, 2652 (1995).

Libby, H.L., *Introduction to Electromagnetic Nondestructive Test Methods*, Wiley-Interscience, New York (1971).

Lifshitz, E.M. and Pitaevskii, L.P., *Statistical Physics* (3rd ed.), Part 1, Pergamon, Oxford (1980).

Likharev, K.K., *Rev. Mod. Phys.* **51**, 101 (1979).

Likharev, K.K., *Dynamics of Josephson Junctions and Circuits*, Gordon & Breach, New York (1986).

Likharev, K.K., *IBM J. Res. Devl.* **32**, 144 (1988).

Likharev, K.K. and Claeson, T., *Scientific American*, June Issue, **50** (1992).

Likharev, K.K., Mukhanov, O.A. and Semenov, V.K., in: *SQUID'85*, pp. 1103–1108, Walter de Gruyter & Co., Berlin (1985).

Likharev, K.K. and Semenov, V.K., *IEEE Trans. Appl. Supercond.* **1**, 3 (1991).

Lions, J.L., *Some Methods in the Mathematical Analysis of Systems and Their Control*, Gordon & Breach, New York (1981).

Livingston, J.D., *Phys. Rev.* **129**, 1943 (1963).

Lock, J.M., *Proc. Roy. Soc.* **A 208**, 391 (1951).

London, F. and London, H., *Proc. Roy. Soc.* **A 149**, 71 (1935).

London, F. and London, H., *Physica* **2**, 341 (1935).

London, F., *Phys. Rev.* **74**, 562 (1948).

London, F., *Superfluids* Vol.1, Wiley, London (1950).

London, H., *Proc. Roy. Soc. London* **A 155**, 102 (1936).

London, H., *Proc. Roy. Soc.* **A 176**, 522 (1940).

London, H., *Phys. Lett.* **6**, 162 (1962).

Lorentz, H.A., *The Theory of Electrons and Its Application to the Phenomena of Light and Radiant Heat* (2nd ed.), Leipzig (1916).

Lorrain, P. and Corson, D.R., *Electromagnetic Fields and Waves* (2nd ed.), W.H. Freeman & Company, New York (1970).

Lundqvist, S., Tosatti, E., Tosi, M.P. and Lu, Y., *Progress in High Temperature Superconductivity*, Vol.1, World Scientific, Singapore (1987).

Lynn, J.W., Shirane, G., Thomlinson, W., Shelton, R.N., and Mocton, D.E., *Phys. Rev.* **B 24**, 3817 (1981).

Lyons,W.G. and Withers, R.S., *Microwave J.*, November Issue, **85** (1990).

Maki, K., *Physics* **1**, 21 (1964).

Malozemoff, A.P., et al., in: *Strong Correlation and Superconductivity* (ed. H. Fukuyama, S. Maekawa, and A.P. Malozemoff), Springer Series in Solid-State Sciences, Vol. 89, pp. 349–360, Springer, Berlin (1989).

Maple, M.B., in: *Advances in Superconductivity* (eds. B. Deaver and J. Ruvalds), pp. 279–346, Plenum Press, New York (1983).

Marcuse, D., *Bell Syst. Tech. J.* **56**, 703 (1977).

Marcuse, D., *J. Opt. Soc. Am.* **68**, 103 (1978).

Marcuse, D., *Light Tranmission Optics* (2nd ed.), Van Nostrand Reinhold Co., New York (1982).

Mason, W.P., *Piezoelectric Crystals and Their Applications to Ultrasonics*, Van Nostrand, New York (1950).

Mason, W.P., *Crystal Physics and Interaction Processes*, Academic Press, New York (1966).

Matick, R.E., *Transmission Lines for Digital and Communication Networks*, McGraw-Hill, New York (1969).

Matsubara, T. and Kotani, A. (eds.), *Superconductivity in Magnetic and Exotic Materials*, Springer-Verlag, Berlin (1984).

Mattis, D.C. and Bardeen, J., *Phys. Rev.* **111**, 412 (1958).

Matveev, A.N., *Electricity and Magnetism*, Mir Publishers, Moscow (1986).

Maugin, G.A., (ed.) *Mechanical Behavior of Electromagnetic Solid Continua*, IUTAM Symp., North-Holland, Amsterdam (1984).

Maugin, G.A., *Nonlinear Electromechanical Effects and Applications*, World Scientific, Singapore (1985).

Maugin, G.A., *Continuum Mechanics of Electromagnetic Solids*, North-Holland, Amsterdam (1988).

Maxwell, E., *Phys. Rev.* **78**, 477 (1950).

Mazur, P. and Nijboer, B.R.A., *Physica* **19**, 971 (1953).

McCoy, J.J. and Beran, M.J., *Int. Eng. Sci.* **14**, 7 (1976).

McCumber, D.E., *J. Appl. Phys.* **39**, 3113 (1968).

Mei, K.K. and Liang, G.-C., *IEEE Trans. Microwave Theory Tech.* **39**, 1545 (1991).

Meilikhov, E.Z. and Farzetdinova, R.M., *Physica* **C 221**, 27 (1994).

Meissner, H., *Phys. Rev.* **109**, 686 (1958).

Meissner, H., *Phys. Rev. Lett.* **2**, 458 (1959).

Meissner, H., *Phys. Rev.* **117**, 672 (1960).

Meissner, H. and Ochsenfield, R., *Naturwiss* **21**, 787 (1933).

Mercereau, J.E., in: *Tunneling Phenomena in Solids*, (eds. E. Burstein and S. Lundqvist), Chap. 31, Plenum Press, New York (1969).

Mercereau, J.E., *Rev. Phys. Appl.* **5**, 13 (1970).

Meyers, N.H., *Proc. IRE* **49**, 1640 (1961).

Miller, D.A.B., *Optics Lett.* **14**, 146 (1989).

Miller, M.N., *J. Math. Phys.* **10**, 1988 (1969).

Miller, P.B., *Phys. Rev.* **118**, 928 (1960).

Miller, R.E. and Cody, G.D., *Phys. Rev.* **173**, 494 (1968).

Mindlin, R.D. and Tiersten, H.F., *Arch. Rat. Mech. Anal.* **11**, 415 (1962).

Minkowski, H., *Göttinger Nachrichten*, pp. 53–116 (1908).

Mints, R.G., in: *Handbook of Applied Superconductivity* (ed. B. Seeber), pp. 99–138, IOP, Bristol (1998).

Mints, R.G. and Rakhmanov, A.L., *J. Phys. D: Appl. Phys.* **15**, 2297 (1982).

Mints, R.G. and Rakhmanov, A.L., *J. Phys. D: Appl. Phys.* **20**, 826 (1988).

Miya, K. (ed.), *Applied Electromagnetics in Materials*, Pergamon, Oxford (1989).

Mo, T.C., *J. Math. Phys.* **11**, 2589 (1970).

Moon, F.C., *Magneto-Solid Mechanics*, Wiley, New York (1984).

Morisue, M., Furusawa, S., Asahina, J. and Kanasugi, A., *IEEE Trans. Magn.* **27,** 2805 (1991).
Morrish, A.H., *The Physical Principles of Magnetism*, Wiley, New York (1965).
Morse, P.M. and Feshbach, H., *Methods of Theoretical Physics*, McGraw-Hill, New York (1953).
Muller, G., Klein, N., Brust, A., Chaloupka, H., Hein, M., Orbach, S., Piel, H. and Reschke, D., *J. Supercond.* **3**, 235 (1990).
Mura, T., *Micromechanics of Defects in Solids*, Martinus Nijhoff Publishers, The Hague (1982).
Möller, C., *The Theory of Relativity* (2nd ed.), Clarendon, Oxford (1972).
Narasimhamurty, T.S., *Photoelastic and Electro-Optic Properties of Crystals*, Plenum, New York (1981).
Narlikar, A.V. and Ekbote, S.N., *Superconductivity and Superconducting Materials*, South Asian Publ., New Delhi (1983).
Nelson, D.F., *Electric Optic and Acoustic Interactions in Dielectrics*, Wiley, New York (1979).
Newhouse, V.L., *Applied Superconductivity*, Wiley, New York (1964).
Nghiem, D., Williams, J.T. and Jackson, D.R., *IEEE Trans. Microwave Theory and Tech.* **MTT-39**, 1553 (1991).
Niessen, A.K. and Staas, F.A., *Phys. Lett.* **15**, 26 (1965).
Niessen, A.K. and Van Vijfeijken, A.G., *Phys. Lett.* **16**, 23 (1966).
Nishino, T., Miyake, M., Harada, Y., and Kawabe, U., *IEEE Electron Device Lett.* **EDL-6**, 297 (1985).
Nowacki, W., *Theory of Micropolar Elasticity*, CISM Courses and Lectures No. 25, Springer-Verlag, Vienna and New York (1970).
Nowacki, W., in: *Electromagnetic Interactions in Elastic Solids* (ed. H. Parkus), CISM Course, pp. 158–183, Springer-Verlag, New York (1976).
Nowacki, W., *Thermoelasticity*, Pergamon, Oxford (1986).
Nuss, M.C. and Goossen, K.W., *IEEE J. Quantum Electron.* **25**, 2596 (1989).
Nuss, M.C., Mankiewich, P.M., O´Malley, M.L., Westerwick, E.H. and Littlewood, P.B., *Phys. Rev. Lett.* **66**, 3305 (1991).
Nye, J.F., *Physical Properties of Crystals*, Clarendon Press, Oxford (1960).
Oates, D.E., Anderson, A.C., Sheen, D.M. and Ali, S.M., *IEEE Trans. Microwave Theory Tech.* **39**, 1522 (1991).
Oates, D.E., Nguyen, P.P., Dresselhaus, G., Dresselhaus, M.S., Lam, C.W. and Ali, S.M., *J. Supercond.* **5**, 361 (1992).
Oates, D.E., Nguyen, P.P., Dresselhaus, G., Dresselhaus, M.S. and Chin, C.C., *IEEE Trans. Appl. Supercond.* **3**, 1114 (1993).
Oates, J.H., Shin, R.T., Oates, D.E., Tsuk, M.J. and Nguyen, P.P., *IEEE Trans. Appl. Supercond.* **3**, 17 (1993).
Ollendorff, F., *Arch. Elektrotechnik* **56**, 1 (1974).
Ollendorff, F., *Arch. Elektrotechnik* **59**, 133 (1977).
Olsen, J.L. and Rohrer, H., *Helv. Phys. Acta* **30**, 49 (1957).
Olsen, J.L. and Rohrer, H., *Helv. Phys. Acta* **33**, 872 (1960).
Olshansky, R. and Keck, D.B., *Appl. Opt.* **15**, 483 (1976).
Onnes, H.K., *Commun. Phys. Lab. Univ. Leiden*, no.119b (1911).
Onnes, H.K., *Commun. Phys. Lab. Univ. Leiden*, no.139f (1914).
Oohira, T., *Nonlinear Analysis of Magnetic Field and Stress for a Thin Ferromagnetic Material*, Master Thesis, University of Tokyo (1989).

Orlando, T.P. and Delin, K.A., *Foundations of Applied Superconductivity*, Addison-Wesley, Reading, MA (1991).

Osamura, K. (ed.), *Composite Superconductors*, Marcel Dekker, New York (1994).

Oshinowo, J., Nishioka, M., Ishida, S. and Arakawa, Y., *Appl. Phys. Lett.* **65**, 1421 (1994).

Ott, H.R., Fertig, W.A., Johnston, D.C., Maple, M.B. and Matthias, B.T., *J. Low Temp. Phys.* **33**, 159 (1978).

Otter, F.A. and Solomon, P.R., *Phys. Rev. Lett.* **16**, 681 (1966).

Oussena, M., Gagnon, R., Wang, Y. and Aubin, M., *Phys. Rev.* **B 46**, 528 (1992).

Pambianchi, M.S., Wu, D.H., Ganapathi, L. and Anlage, S.M., *IEEE Trans. Appl. Supercond.* **3**, 2774 (1993).

Pao, Y-H., in: *Mechanics Today* 4 (ed. S. Nemat Nasser), pp. 209–305, Pergamon Press, Oxford (1978).

Pedersen, N.F., in: *SQUID'80* (eds. H.D. Hahlbohm and H. Lubbig), pp. 739–762, Walter de Gruyter, Berlin (1980).

Penfield, P. and Haus, H.A., *Electrodynamics of Moving Media*, MIT Press, Cambridge, MA (1967).

Peshkin, M. and Tonomura, A., *The Aharonov-Bohm Effect*, Lecture Notes in Physics, Springer-Verlag, Berlin (1989).

Pillips, J.C., *Physics of High-Tc Superconductors*, Academic Press, Boston (1989).

Pippard, A.B., *Proc. Roy. Soc.* **A 203**, 210 (1950a).

Pippard, A.B., *Proc. Roy. Soc.* **A 203**, 98 (1950b).

Pippard, A.B., *Proc. Roy. Soc.* **A 216**, 547 (1953).

Pippard, A.B., *Adv. in Electronics & Electron Phys.* (ed. L. Marton) vol. **6**, 1–45 (1954).

Pires, P.S.M., Rogers, D.A., Bochove, E. and Souza, R.F., *IEEE Trans. Microwave Theory Tech.* **30**, 131 (1982).

Podney, W., *J. Geophys. Res.* **80**, 2977 (1975).

Poole, Jr., C.P., Farach, H.A., and Creswick, R.J., *Superconductivity*, Academic Press, San Diego (1995).

Popovic, R.S., *Hall Effect Devices*, Adam Hilger, Bristol (1991).

Post, E.J., *Formal Structure of Electromagnetics*, North-Holland, Amsterdam (1962).

Potter, R.I. and Schmulian, R.J., *IEEE Trans. Magn.* **MAG-7**, 873 (1971).

Primak, W. and Post, D., *J. Appl. Phys.* **30**, 779 (1959).

Pöpel, R. in: *Superconducting Quantum Electronics* (ed. V. Kose), pp. 44–78, Springer-Verlag, Berlin (1989).

Ramo, S., Whinnery, J.R. and Van Duzer, T., *Fields and Waves in Communication Electronics* (2nd ed.), Wiley, New York (1984) (3rd ed. 1994).

Rayleigh, J.W., *Phil. Mag.* **34**, 481 (1892).

Reed, W.A., Fawcett, E. and Kim, Y.B., *Phys. Rev. Lett.* **14**, 790 (1965).

Reuter, G.E.H. and Sondheimer, E.H., *Proc. Roy. Soc.* **A 195**, 336 (1948).

Reynolds, C.A., Serin, B., Wright, W.H. and Nesbitt, L.B., *Phys. Rev.* **78**, 487 (1950).

Ri, H.-C., Kober, F., Cross, R., Huebener, R.P. and Gupta, A., *Phys. Rev.* **B 43**, 13739 (1991).

Ri, H-C., Cross, R., Gollnik, F., Beck, A., Huebener, R.P., Wagner, P. and Adrian, H., *Phys. Rev.* **B 50**, 3312 (1994).

Richards, P.L., in: *Semiconductors and Semimetals* (eds. R.K. Willardson and A.C. Beer), Vol. 12, Chap. 6, Academic, New York (1977).

Roas, B., Schultz, L. and Saemann-Ischenko, G., *Phys. Rev. Lett.* **64**, 479 (1990).

Robinson, F.N.H., *Macroscopic Electromagnetism*, Pergamon Press, Oxford (1973).

Robinson, S.E., *IEEE Trans. Nuc. Sci.* **28**, 272 (1981).

Rogalla, H. in: *Handbook of Applied Superconductivity* (ed. B. Seeber), pp. 1759–1775, IOP, Bristol (1998).
Rose-Innes, A.C. and Rhoderick, E.H., *Introduction to Superconductivity*, Pergamon Press, Oxford (1969).
Rosenfeld, L., *Theory of Electrons*, North-Holland, Amsterdam (1951).
Rosser, W.G.V., *An Introduction to The Theory of Relativity*, Butterworths, London (1964).
Rothwarf, A. and Cohen, M., *Phys. Rev.* **130**, 1401 (1963).
Rowell, J.M., *Phys. Rev. Lett.* **11**, 200 (1963).
Rutgers, A.J., *Physica* **1**, 1055 (1934).
Russakoff, G., *Am. J. Physics* **38**, 1188 (1970).
Saint-James, D. and de Gennes, P.G., *Phys. Lett.* **7**, 306 (1963).
Saint-James, D., Thomas, E.J. and Sarma, G., *Type II Superconductivity*, Pergamon, Oxford (1969).
Sakurai, A., *Solid State Commun.* **25**, 867 (1978).
Sanchez-Palencia, E., *Nonhomogeneous Media and Vibration Theory*, Springer, Heidelberg (1980).
Scherer, G.W., *J. Non-Cryst. Solids* **34**, 223 (1979).
Schieber, D., *Electromagnetic Induction Phenomena*, Springer-Verlag, Berlin (1986).
Schiff, L.I., *Quantum Mechanics* (3rd. ed.), McGraw-Hill, New York (1968).
Schmid, A., *Phys. Kondens. Materie* **5**, 302 (1966).
Seeger, A. and Kronmuller, H., *Phys. Stat. Sol.* **23**, 371 (1968).
Seraphim, D.P. and Marcus, P.M., *IBM J. Res. Develop.* **6**, 94 (1962).
Shadowitz, A., *The Electromagnetic Field*, McGraw-Hill, New York (1975).
Shapira, Y. and Neuringer, L.J., *Phys. Rev.* **154**, 375 (1967).
Shapiro, S., *Phys. Rev. Lett.* **11**, 80 (1963).
Sheen, D.M., Ali, S.M., Oates, D.E., Withers, R.S. and Kong, J.A., *IEEE Trans. Appl. Supercond.* **1**, 108 (1991).
Shen, Z.-Y., *High-Temperature Superconducting Microwave Circuits*, Artech House, Boston (1994).
Shenoy, G.K., Dunlap, B.D., Fradin, F.Y., Sinha, S.K., Kimball, C. W., Potzel, W., Probst, F. and Kalvius, G.M., *Phys. Rev.* **B 21**, 3886 (1980).
Shiozawa, T., *Proc. IEEE* **61**, 1694 (1973).
Shoenberg, D., *Superconductivity* (2nd ed.), Cambridge Univ. Press, Cambridge (1952).
Shibata, N., *Optical Characteristics of High Silica Glass Fibers for Communication*, Ph.D. thesis (in Japanese), Nagoya University (1983).
Silvester, P.P. and Ferrari, R.L., *Finite Elements for Electrical Engineers* (2nd ed.), Cambridge Univ. Press, Cambridge (1990).
Simon, I., *J. Appl. Phys.* **24**, 19 (1953).
Sizoo, G.J. and Kamerlingh Onnes, H., *Commu. Phys. Lab. Leiden Univ.*, No.180 (1926).
Smith, P.W., *Bell Syst. Tech. J.* **61**, 1975 (1982).
Smith, T.F. and Chu, C.W., *Phys. Rev.* **159**, 353 (1967).
Smith, T.I., *Phys. Rev. Lett.* **15**, 460 (1965).
Smythc, W.R., *Static and Dynamic Electricity*, McGraw-IIill, New York (1950).
Snyder, A.W. and Love, J.D., *Optical Waveguide Theory*, Chapman & Hall, London (1983).
Solomon, P.R. and Otter, F.A., *Phys. Rev.* **164**, 608 (1967).
Solymar, L., *Superconductive Tunnelling and Applications*, Chapman & Hall, London (1972).
Sommerfeld, A., *Partial Differential Equations in Physics*, Academic Press, New York (1949).

Sommerfeld, A., *Electrodynamics*, Academic Press, New York (1952).
Stacey, F.D., *Pure and Appl. Geophys.* **58**, 5 (1964).
Staelin, D.H., Morgenthaler, A.W. and Kong, J.A., *Electromagnetic Waves*, Prentice-Hall, Englewood Cliffs, NJ (1994).
Stephen, M.J., *Phys. Rev.* **139**, A197 (1965).
Stewart, W.C., *Appl. Phys. Lett.* **12**, 277 (1968).
Stogryn, D.E. and Stogryn, A.P., *Mol. Phys.* **11**, 371 (1966).
Stoletov, A.G., *Pogg. Ann.* **144**, 439 (1873a).
Stoletov, A.G., *Phil. Mag.* (4) **45**, 40 (1873b).
Stratton, J.A., *Electromagnetic Theory*, McGraw-Hill, New York (1941).
Suhl, H., *Phys. Rev. Lett.* **14**, 226 (1965).
Suhl, H., *J. Less Common Metals* **62**, 225 (1978).
Swihart, J.C., *J. Appl. Phys.* **32**, 461 (1961).
Talisa, S.H., Janocko, M.A., Moskowitz, C., Talvacchio, J., Billing, J.F., Brown, R., Buck, D.C., Jones, C.K., McAvoy, B.R., Wagner, G.R. and Watt, D.H., *IEEE Trans. Microwave Theory Tech.* **39**, 1448 (1991).
Taylor, B.N., *J. Appl. Phys.* **39**, 2490 (1968).
Tazoh, Y., *IEEE Trans. Magn.* **27**, 3316 (1991).
Tedrow, P.M., Farazi, G. and Meservey, R., *Phys. Rev.* **B 4**, 74 (1971).
Testardi, L.R., *Phys. Rev.* **B 3**, 95 (1971).
Testardi, L.R., Kunzler, J.E., Levinstein, H.J., Maita, J.P. and Wernick, J.H., *Phys. Rev.* **B 3**, 107 (1971).
Thomas, H., in: *Earlier and Recent Aspects of Superconductivity* (eds. J.G. Bendnorz and K.A. Müller), p. 2, Springer, Berlin (1990).
Tiersten, H.F., *J. Math. and Phys.* **5**, 1 (1964).
Tiersten, H.F., *A Development of the Equations of Electromagnetism in Material Continua*, Springer-Verlag, New York (1990).
Tilley, D.R., *Proc. Phys. Soc.* **85**, 1177 (1965).
Timp, G.L. and Howard, R.E., *Proc. IEEE* **79**, 1188 (1991).
Tinkham, M., *Phys. Rev.* **129**, 2413 (1963).
Tinkham, M., *Phys. Rev. Lett.* **13**, 804 (1964).
Tinkham, M., *Introduction to Superconductivity*, McGraw-Hill, New York (1975).
Tinkham, M., *Phys. Rev. Lett.* **61**, 1658 (1988).
Tobar, M.E. and Blair, D.G., *IEEE Trans. Microwave Theory Tech.* **MTT-39**, 1582 (1991).
Tomasch, W.J. and Joseph, A.S., *Phys. Rev. Lett.* **12**, 148 (1964).
Truesdell, C., in: *Principles of Continuum Mechanics*, Socony Mobil Oil Co. (1961).
Turneaure, J.P., Ph.D. Dissertation, Stanford University (1967).
Turneaure, J.P., Halbritter, J. and Schwettman, H.A., *J. Supercond.* **4**, 341 (1991).
Tyler, F., *Phil. Mag.* **11**, 596 (1931).
Ullmaier, H., *Irreversible Properties of Type II Superconductors*, Springer-Verlag, Berlin (1975).
Ulrich, B.T. and Tutter, M., in: *SQUID'80* (eds. H.D. Hahlbohm and H. Lubbig), pp. 831–839, Walter de Gruyter, Berlin (1980).
Van Beek, L.K.H., *Progress in Dielectrics* **7**, 69 (1967).
Van Vlack, J.H., *The Theory of Electric and Magnetic Susceptibilities*, Oxford (1932).
Van Vlack, L.H., *Elements of Materials Science and Engineering* (3rd ed.), Addison-Wesley, Reading, MA (1975).
Vendik, O.G., Vendik, I.B. and Samoilova, T.B., *IEEE Trans. Microwave Theory Tech.* **45**, 173 (1997).

Verheijen, A.A., Van Ruitenbeek, J.M., De Bruyn Ouboter, R. and De Jongh, L.J., *Nature* **345**, 418 (1990).
Vonsovskii, S.V., *Magnetism*, Vol.2, Wiley, New York (1974).
Wang, Z.D. and Ting, C.S., *Phys. Rev.* **B 46**, 284 (1992).
Weinstock, H. and Ralston, R.W. (eds.), *The New Superconducting Electronics*, Kluwer Academic Press, Dordrecht (1993).
Werthamer, N.R., *Phys. Rev.* **132**, 2440 (1963).
Werthamer, N.R., in: *Superconductivity* (ed. R.D. Parks) Vol.1, pp. 321–370, Marcel Dekker, New York (1969).
Weiss, P., *J. Phys.* **6**, 667 (1907).
Welch, D.O., *Advances in Cryogenic Engineering Materials* **26**, 48 (1980).
Weller, W., *Phys. Stat. Sol.* **30**, 373 (1968).
Whitaker, J.F., Norris, T.B., Mourou, G.A. and Hsiang, T.Y., *IEEE Trans. Microwave Theory Tech.* **MTT-35**, 41 (1987).
Willis, J.R., *J. Mech. Phys. Solids* **25**, 185 (1977).
Willis, J.R., *J. Appl. Mech.* **50**, 1202 (1983).
Wilson, M.N., *Superconducting Magnets*, Clarendon Press, Oxford (1983).
Wolf, P., in: *SQUID and Applications* (eds. H.D. Hahlbohm and H. Lubbig), pp. 519–540, Walter de Gruyter, Berlin (1977).
Woolf, L.D., Torar, M., Hamaker, H.C. and Maple, M.B., *Phys. Lett.* **71A**, 137 (1979).
Wyatt, A.F.G., Dmitriev, V.M., Moore, W.S. and Sheard, F.W., *Phys. Rev. Lett.* **16**, 1166 (1966).
Wynn, W.M. Frahm, C.P., Carroll, P.J., Clarke, R.H., Wellhoner, J. and Wynn, M.J., *IEEE Trans. Magn.* **Mag-11**, 701 (1975).
Xiang, X.D., Brill, J.W., Delong, C.J., Bourne, L.C., Zettl, A., Jones, J.C. and Rice, L.A., *Solid State Comm.* **65**, 1073 (1988).
Yamomoto, Y. and Miya, K., (eds.) *Electromagnetomechanical Interactions in Deformable Solids and Structures*, IUTAM Symp., North-Holland, Amsterdam (1987).
Yasukochi, K., Ogasawara, T., Usui, N. and Ushio, S., *J. Phys. Soc. Japan* **19**, 1649 (1964).
Yeh, N.-C., *Phys. Rev.* **B 43**, 523 (1991).
Yeshurun, Y. and Malozemoff, A.P., *Phys. Rev. Lett.* **60**, 2202 (1988).
Zhang, D., Liang, C.-G., Shih, C.G., Withers, R.S., Johansson, E. and Cruz, A.D., *IEEE Trans. Appl. Supercond.* **5**, 2656 (1995).
Zhou, S.-A., *Material Multipole Mechanics of Electromagnetoelastic Solids with Defects*, Ph.D. thesis, Royal Institute of Technology, Stockholm (1987).
Zhou, S.-A., *Fiber Mechaoptics*, TRITA-TFY 90-06, KTH, Stockholm (1990).
Zhou, S.-A., *Electrodynamic Theory of Superconductors*, Peter Peregrinus Ltd. (IEE), London (1991a).
Zhou, S.-A., *Int. J. Solids Structures* **28**, 423 (1991b).
Zhou, S.-A., *J. Appl. Phys.* **74**, 1912 (1993).
Zhou, S.-A., *J. Tech. Phys.* **35**, 161 (1994a).
Zhou, S.-A., *Physica* **C 228**, 122 (1994b).
Zhou, S.-A., *Physica* **C 232**, 379 (1994c).
Zhou, S.-A., *Phys. Rev.* **B 50**, 354 (1994d).
Zhou, S.-A., *J. Supercond.* **8**, 113 (1995a).
Zhou, S.-A., *SPIE Proc.*, vol. **2559**, pp.47-58 (1995b).
Zhou, S.-A., *Int. J. Appl. Electromag. Mech.* **6**, 55 (1995c).
Zhou, S.-A., *Int. J. Eng. Sci.* **36**, 1511 (1998).

Zhou, S.-A. and Hesselbom, H., in: *Non-Linear Electromagnetic Systems* (eds. V. Kose and J. Sievert), pp. 629–632, IOS Press, Amsterdam (1998).
Zhou, S.-A. and Hsieh, R.K.T., in: *Nonlinear Mechanics* (ed. W. Chien), pp. 1367–1375, Science Press, Beijing (1985).
Zhou, S.-A. and Hsieh, R.K.T., *Int. J. Eng. Sci.* **24**, 1195 (1986).
Zhou, S.-A. and Hsieh, R.K.T., *Int. J. Eng. Sci.* **26**, 13 (1988).
Zhou, S.-A., Hsieh, R.K.T. and Maugin, G.A., *Int. J. Solids Structures* **22**, 1411 (1986).
Zhou, S.-A., Köszegi, L., Björkman, J. and Svensson, T., in: *Appled Electromagnetics in Materials* (ed. K. Miya), pp. 305–316, Pergamon, Oxford (1989).
Zhou, S.-A. and Miya, K., *Int. J. Appl. Electromag. Mater.* **1**, 1 (1990).
Zhou, S.-A. and Miya, K., *Int. J. Appl. Electromag. Mater.* **2**, 21 (1991).
Zhou, S.-A. and Wikborg, E., *A Swichable Inductor*, Ericsson Patent E10397, Stockholm (1998).
Zhou, S.-P., Jabbar, A., Bao, J.-S., Wu, K.-Q. and Jin, B.-J., *J. Appl. Phys.* **71**, 2789 (1992).
Zaitsev, R.O., *Soviet Phys. JETP* **21**, 1178 (1965).
Zaitsev, R.O., *Soviet Phys. JETP* **23**, 702 (1966).
Zienkiewicz, O.C. and Taylor, R.L., *The Finite Element Method* (4th ed.), McGraw-Hill, New York (1989).
Zimmerman, J.E. and Mercereau, J.E., *Phys. Rev. Lett.* **14**, 887 (1965).
Zimmerman, J.E. and Silver, A.H., *Phys. Rev.* **141**, 367 (1966).
Zimmerman, J.E., Thiene, P. and Harding, J.T., *J. Appl. Phys.* **41**, 1572 (1970).

Index